THE ROLE OF RNA IN DEVELOPMENT AND REPRODUCTION

Proceedings of the Second International Symposium,
April 25-30, 1980

Edited by

M. C. NIU

Department of Biology, Temple University, Philadelphia, U. S. A.

H. H. CHUANG

The Director of Shanghai Institute of Cell Biology and Institute of Developmental Biology, Beijing, China.

SCIENCE PRESS

Beijing, 1981

Distributed by

VAN NOSTRAND REINHOLD COMPANY

New York, Cincinnati, Toronto, London, Melbourne

Published by Science Press,Beijing
Distributed by Van Nostrand Reinhold Company
New York, Cincinnati, Toronto, London, Melbourne
Printed by C & C Joint Printing Co., (H.K.) Ltd.

First published 1981
ISBN 0-442-20090-0
Science Press Book No. 2281-10

PREFACE

It was eight years ago that the first International Symposium on RNA in Development and Reproduction was held in Washington, D. C. The second International Symposium took place in Beijing (Peking), China, April 25-30, 1980. During the intervening 8 years, striking progress has been made in various aspects of RNA biology and chemistry. Some striking examples are the relation between the structure of DNA and RNA, nucleic acid synthetase and restriction enzymes, RNA processing from which derived the concepts of introns-exons, slicing, leader RNA and flanker RNA, stored mRNAs in eggs and seeds, gene biosynthesis, molecular control of gene expression, RNA in differentiation, immune RNA and mRNA-mediated genetic changes. In this period, one of the editors (MCN) had the opportunity to collaborate with the late Professor Tung Tichow and his colleagues on the study of the role of mRNA in the development of goldfish. We felt the necessity to correlate our work with similar studies of others. One of the ways to accomplish this would be holding an international meeting.

The rationale for choosing Beijing as the site of the symposium is twofold: (1) Beijing is a unique and beautiful city with many interesting and historical landmarks, and (2) after the normalization of diplomatic relations between the People's Republic of China and the United States, many scientists from the U. S. and other Western countries have had an internal interest in coming to Beijing. Unfortunately, Prof. Tung could not fulfill his ardent wish to be the host for so many of his foreign colleagues because he died on March 31, 1979. However, his enthusiasm for the symposium had inspired us to hold it as scheduled. There were 8 sessions: (1) the relation between the structure of DNA and RNA, (2) RNA processing, (3) nucleic acid synthetase and restriction enzymes, (4) molecular control of gene expression, (5) stored mRNA in eggs and seeds, (6) RNA in differentiation, (7) immune RNA and (8) mRNA-mediated genetic changes. Six speakers were scheduled for each session. As the papers were presented, the materials covered were diverse and thus could hardly be included under the heading of the designated session. Therefore, we published the papers in this symposium volume in alphabetical order by their first authors.

The Symposium was co-sponsored by the Chinese Zoological Society and the American Society for Developmental Biology. Every presentation was followed by open discussion. All participants could not only learn about the experimental design, and the significance of the results, but could also hear the views expressed by colleagues. Due to the limitation of pages, this Symposium volume contains only papers presented and, regrettably, not the discussion. Our appreciation is extended to all speakers who contributed so freely to the rich discussion that helped to make the RNA Symposium in Development and Reproduction a success. Also, we are grateful to those who labored through day and night to make our daily programs enjoyable and to make the stay in Beijing comfortable. We are obliged to the colleagues at the Institute of Developmental Biology who spent much of their time working toward the preparation of the Symposium and the publication of this volume. We wish to thank, also, Drs. Sheldon J. Segal, John Chen and Paul Gross for their willingness to extend consultation and aid at times of need. Finally, our appreciation goes to the Academia Sinica, the United Nations Fund for Population Activity and the Rockefeller Foundation. These organizations provided the financial support that made this symposium possible.

M. C. Niu
S. H. Chuang
1980

Meeting with the President of the Chinese Academy of Sciences, Fang Yi.

Opening Session

Audience

Dr. S. Ochoa

At the Great Wall

LIST OF PARTICIPANTS

Allfrey, Vincent G.
(The Rockefeller University, New York, N. Y. 10021)

Anderson, W. French
(Laboratory of Molecular Hematology, National Heart, Lung, and Blood Institute, National Institutes of Health, Bethesda, Maryland 20205)

Baltimore, David
(Center for Cancer Research and Department of Biology, Massachusetts Institute of Technology, Cambridge, Massachusetts 02139)

Bauer, Georg
(Institut für Virologie, Zentrum für Hygiene, Universität Freiberg, Hermann-Herder-Str. 11, 78 Freiburg, West Germany)

Bei, Shizhang (Pei, Shih-Chang 贝时璋)
(Institute of Biophysics, Academia Sinica, Beijing)

Beljanski, Mirko
(Laboratoire de Pharmacodynamie, Faculté de Pharmacie, 92290 Châtenay-Malabry, France)

Beljanski, Monique
(Laboratoire de Pharmacodynamie, Faculté de Pharmacie, 92290 Châtenay-Malabry, France)

Chang, Chih-Ye (张致一)
(Institute of Zoology, Academia Sinica, Beijing)

Chen John H. (陈 享)
(Department of Biochemistry, New York University, Dental Center N. Y. 10010)

Chien, Yen-wen (钱燕文)
(Institute of Zoology, Academia Sinica, Beijing)

Chuang, Shao-hui (庄孝僡)
(Institute of Developmental Biology, Beijing; Shanghai Institute of Cell Biology, Academia Sinica, Shanghai)

Darnell, James, E. Jr.
(The Rockefeller University, New York, N. Y. 10021 U. S. A.)

Dong, Lin (董 霖)
(Shanghai Institute of Biochemistry, Academia Sinica, Shanghai)

Fishman, Marvin
(Division of Immunology, St. Jude Children's Research Hospital, 332 North Lauderdale, P. O. Box 318, Memphis, Tennessee, 38101, U. S. A.)

Fraenkel-Conrat, Heinz
(Department of Molecular Biology and Virus Lab., University of California, Berkeley)

Galston, Arthur W.
(Department of Biology, Yale University, New Haven, Connedicut, 06520 U. S. A.)

Gilbert, Walter
(Harvard University)

Goldberger, Robert
(National Institutes of Health, Bethesda, Maryland. 20205)

Gross, Paul R.
(Marine Biological Laboratory, Woods Hole, Mass., 02543, and Department of Biology, University of Rochester, Rochester, New York, U. S. A.)
Guo, Chan (郭 婵)
(Department of Biochemistry & Molecular Biology, Shanghai Cancer Institute)
Guo, Xing-xian (过兴先)
(No. 1 Bureau, Academia Sinica, Beijing)
Hsiao, Shu-hsi (萧淑熙)
(Institute of Developmental Biology, Academia Sinica, Beijing)
Hu, Jun (胡 钧)
(Kunming Institute of Zoology, Academia Sinica, Kunming)
Huang, Alice S. (黄诗厚)
(Department of Microbiology and Molecular Genetics, Harvard Medical School and Division of Infectious Diseases. Children's Hospital Medical Center, 300 Longwood Avenue, Boston, Massachusetts 02115)
Huang, Hua-Zhang (黄华漳)
(Laboratory of Experimental Tumor Research, Tianjin Medical College, Tianjin,)
Jiang, Xi-ming (江希明)
(Department of Biology, Hangzhou University, Hangzhou)
Jin, Jia-rui (靳加瑞)
(Shanghai Institute of Biochemistry, Academia Sinica, Shanghai)
Kedes, Laurence H.
(The Howard Hughes Medical Institute Laboratory and Department of Medicine, Stanford University School of Medicine and Veterans Administration Hospital, Palo Alto, California 94304)
Koide, Samuel S.
(Center for Biomedical Research, The Population Council, The Rockefeller University, New York, New York 10021)
Lengyel, Peter
(Department of Moleculer Biophysics and Biochemisty, Yale University, New Haven, Connecticut 06511 U. S. A.)
Li, Irene Yi (李 漪)
(Laboratory of Experimental Tumor Research, Tianjin Medical College, Tianjin, China)
Li, Su (李 苏)
(Academia Sinica, Beijing)
Lin, Zhong-ping (林忠平)
(Institute of Botany, Academia Sinica, Beijing)
Liu, Xiang-ling (刘祥麟)
(Zhejiang Medical University)
Ma, Cheng (马 诚)
(Institute of Botany, Academia Sinica, Beijing)
Marcus, Abraham
(The Institute for Cancer Research, 7701, Burholme Avenue, Philadelphia, Pennsylvania 19111)
Mishra N. C.
(Department of Biology, University of South Carolina, Columbia S. C. 29208, U. S. A.)
Niu, Lillian Chang (张葆英)
(Department of Biology, Temple University, Philadelphia, PA 19117 and Institute of

Developmental Biology, Academic Sinica, Beijing)

Niu, M. C. （牛满江）

(Department of Biology, Temple University, Philadelphia, PA 19122 and Institute of Developmental Biology, Academia Sinica, Beijing)

Ochoa, Severo

(Roche Institute of Molecular Biology, Nutley, New Jersey 07110 U. S. A.)

Paque, Ronald E.

(Department of Microbiology, The University of Texas Health Science Center, San Antonio, Texas)

Pederson, Thoru

(Cell Biology Group, Worcester Foundation for Experimental Biology, Shrewsbury, Massachusetts 01545 U. S. A.)

Pilch, Yosef H.

(Surgical Oncology Service, Department of Surgery, University of California, San Diego, School of Medicine, San Diego, California, 92103, U. S. A.)

Pottathil, Raveenbran

(Department of Pediatrics, Duke University Medical Center, Durham, North Carolina 27710)

Roeder, Robert

(Washington U.iversity)

Rosbach, Michael

(Department of Biology and Rosenstiel Basic Medical Sciences Research Center, Brandeis University, Waltham, Massachusetts 02254 U. S. A.)

Segal, Sheldon

(Rockefeller Foundation, 1133 Avenue of America, New York, N. Y. 10036)

Shen, Xiao-Zhou （沈孝宙）

(Institute of Zoology, Academia Sinica, Beijing)

Shih, Ying-hsien （史瀛仙）

(Institute of Developmental Biology, Academia Sinica, Beijing)

Siddiqui, M. A. Q.

(Department of Biochemistry, Roche Institute of Molecular Biology, Nutley, N. J. 07110 U. S. A.)

Slavkin, Harold C.

(Laboratory for Developmental Biology, Graduate Program in Craniofacial Biology, and Department of Biochemisty, School of Dentistry, University of Southern California, Los Angeles, California 90007 U. S. A.)

Song, De-Xiu （宋德秀）

(Institute od Developmental Biology, Academia Sinica, Beijing)

Sperelakis, Nick

(Department of Physiology, School of Medicine University of Virginia Charlottesvill, VA 22908, U. S. A.)

Tan, Li-ling （谭丽玲）

(Institute of Genetics, Academia Sinica, Beijing)

Tilghman, Shirley M.

(Institute for Cancer Research, The Fox Chase Cancer Center, Philadelphia, Pa. 19111)

Tu, Miao （杜　淼）

(Institute of Developmental Biology, Academia Sinica, Beijing)

Villee, Claud A.

(Department of Biological Chemistry, and Laboratory of Human Reproduction and Reproductive Biology, Harvard Medical School, Boston, Massachusetts)

Viza, Dimitri

(Laboratoire d'Immunobiologie, Faculté de Médecine 15, rue de l'Ecole de Médicine, 75006 Paris, France)

Wang, Bosco S. (王 翔)

(Department of Surgery, Peter Bent Brigham Hospital, Harvard Medical School, Boston, Massachusetts 02115 and Department of Surgery, Boston University. School of Medicine, Boston, Massachusetts 02118)

Weinberg, Eric S.

(Department of Biology, University of Pennylvania, Philadelphia, Pa. 19104 U. S. A.)

Woo, Savio L. C. (胡流清)

(Department of Cell Biology and Howard Hughes Medical Institute, Baylor College of Medicine, Houston, Texas, 77030 U. S. A.)

Xing, Rui (邢 芮)

(Department of Societies, Chinese Association of Science and Technology)

Xue, Guo-xiang (薛国雄)

(Institute of Developmental Biology, Academia Sinica, Beijing)

Xue, Yugu (薛禹谷)

(Institute of Microbiology, Academia Sinica, Beijing)

Yan, Shao-yi (严绍颐)

(Institute of Developmental Biology, Academia Sinica, Beijing)

Yu, Jian-kang (于建康)

(Institute of Developmental Biology, Academia Sinica, Beijing)

Zhao, Hui-Zhi (赵惠智)

(Institute of Biophysics, Academia Sinica, Beijing)

Zhao, Ji-ying (赵季英)

(Institute of Biophysics, Academia Sinica, Beijing)

Zhu, Zhi-ping (朱治平)

(Shanghai Institute of Plant Physiology, Academiâ Sinica, Shanghai)

CONTENTS

* Speaker.

---

* Speaker.

* Speaker.

STRUCTURAL ANALYSIS OF A EUKARYOTIC RIBOSOMAL GENE AND OBSERVATIONS ON DIFFERENCES IN CHROMATIN STRUCTURE IN ITS TRANSCRIBING AND NON-TRANSCRIBING REGIONS

Vincent G. Allfrey, Irene Yi-Chi Sun and Edward M. Johnson

The Rockefeller University

New York, New York

SUMMARY

The ribosomal genes of the simple eukaryotic organism, Physarum polycephalum, occur within extrachromosomal rDNA molecules of molecular weight 39.2 x 10^6 daltons. The rDNA can be isolated in high purity in milligram amounts, and its genetic organization has been studied by selective cleavage with restriction nucleases and hybridization of the resulting rDNA fragments to radioactively-labeled 19S, 5.8S and 26S ribosomal RNA's. Each rDNA molecule contains two ribosomal genes arranged in inverted order at the ends of a long, non-transcribed central 'spacer'.

Isolated nuclei of Physarum microplasmodia are very active in rRNA chain elongation and they also initiate the synthesis of preribosomal RNA molecules in vitro. The fidelity of transcription and the map positions of the coding sequences have been established by hybridization of the newly synthesized RNA molecules to rDNA restriction fragments. Using ATP-γ-S to selectively label the 5'-ends of newly-initiated ribosomal RNA chains, we have purified the thio-derivatized rRNA (HS-rRNA) by chromatography on organomercurial-Sepharose columns. Hybridization of the HS-rRNA to restriction fragments of the rDNA has identified two sites of rRNA chain initiation on the rDNA molecule, each placed symmetrically about the central 'spacer' at about 17 kilobase-pairs from each end of the rDNA palindrome.

The presence of a long non-transcribed 'spacer' between the two ribosomal genes on the rDNA molecule permits comparisons of chromatin structural organization in the template-active and template-inactive regions of the same DNA strand. The coding and non-coding regions of rDNA chromatin have been compared using DNase I and staphylococcal nuclease as probes of DNA accessibility and nucleosome structure. It was found that the non-coding regions in the central 'spacer' are organized into beaded nucleosomal arrays, each nucleosome 'core' containing 144 base-pairs of DNA in a compact coil. The coding regions of rDNA chromatin are cut at regular intervals by staphylococcal nuclease to generate subunits which also contain 144 base-pairs of DNA, but differ in sedimentation properties and composition. Electron microscopy of these subunits show that the DNA is nearly fully extended. When the synthesis of ribosomal RNA is suppressed in the microsclerotial form of the organism, the ribosomal genes revert to the beaded, compact configuration.

INTRODUCTION

The ribosomal genes of eukaryotic cells occur in multiple copies, most of which are expressed at some stage in development. Their transcription is usually blocked in mitosis and reactivated in the G_1 phase of the ensuing cell cycle. The control of so many ribosomal genes would be expected to require a corresponding amplification of proteins involved in the formation of the initiation complex on each of the discrete transcription units. The reiteration of the ribosomal genes and their compartmentation in the nucleolus thus offers an enhanced potential for the isolation of ribosomal chromatin and the identification of proteins involved in transcriptional control.

As a prelude to the identification of DNA-binding proteins which take part in the initiation and termination of pre-ribosomal RNA synthesis, we have analyzed the structure of the ribosomal genes in a simple eukaryotic organism which offers particular advantages for such studies.

RIBOSOMAL GENE STRUCTURE IN PHYSARUM

As is the case for other eukaryotes, the ribosomal genes of the myxomycete, Physarum polycephalum, are reiterated several hundred-fold (1,2) and localized in the nucleolus (3,4). Virtually all of the ribosomal genes of Physarum are extrachromosomal, and their replication proceeds independently of the bulk of the chromosomal DNA (5,6). Each of the diploid nuclei of the multinucleate syncytial forms of Physarum contains about 1.2 pg of DNA (7,8) of which 0.4% (calculated as double-stranded DNA) is hybridizable to the 19S and 26S ribosomal RNA sequences of the organism (9-11). Our recent estimates place this figure at 0.360 ± 0.002 % in the a x i strain of Physarum polycephalum in the plasmodial stage (12,13). This corresponds to about 400 copies per nucleus of the genes coding for 26S, 19S, and 5.8S ribosomal RNA's. The number of gene copies varies with the stage in the cell cycle. Except for a brief pause in the early S phase (1), ribosomal DNA (rDNA) synthesis occurs at all stages of the cell cycle and about 50% of the ribosomal RNA-coding sequences are synthesized during G_2 phase, after the bulk of chromosomal DNA synthesis has ceased (1,6,9,14). Because of the highly synchronous growth and division cycle of the plasmodial form of the organism (15-17), the late replication of the ribosomal genes makes it possible to selectively label the rDNA sequences using isotopic precursors or density labels in the G_2 phase of the naturally synchronized nuclear population. The rapid doubling time of *Physarum* nuclei and the ease of cultivation of the microplasmodia (17) permit the preparation of large quantities of nuclei (18, 19) and gram amounts of nucleoli (4, 18, 20, 21) for subsequent isolation of the ribosomal DNA (4, 13, 21-23).

The ribosomal genes occur in the nucleolus as sequences within a dense satellite DNA (rDNA) (1, 4, 9, 11, 12, 21-26) which may comprise as much as 95% of the total DNA in a pure nucleolar preparation. (This is far beyond the corresponding proportion of ribosomal DNA in nucleoli isolated from mammalian cells, which is usually about 0.5%). The rDNA is extracted by lysis of the isolated nucleoli in sodium dodecylsulfate and treatment with proteinase K and ribonuclease prior to purification by isopycnic centrifugation on a CsCl gradient. The rDNA of *Physarum* has a density of 1.713 g/ml as compared with 1.703 g/ml for the main-band DNA (21). We have recently modified the procedure for isolation of the nucleoli to avoid shearing of the long rDNA molecules by sonication, a hazard in the previous isolation procedures.

Intact rDNA molecules have a median length of 18-20 μm (25, 26, 28) and a molecular weight of 39×10^6 daltons (23, 24). The sequence organization of the rDNA has been analyzed by selective cleavage by a variety of restriction nucleases, followed by electrophoretic separation of the fragments (23-25) and hybridization to purified 26S and 19S ribosomal RNA's labeled in vivo with ^{32}P-phosphate (23, 24), using the Southern procedure (29).

Each rDNA molecule contains two copies each of the 19S and 26S genes (24,25); these comprise 21.4% of the total DNA sequence. Each set of ribosomal genes is separated by a central 'spacer' which contains multiple repeat sequences (25). Most of the 'spacer' is not transcribed (30). Hybridization analyses of the rDNA fragments produced by digestion with the restriction nucleases, Eco R1 and Hind III, have established the polarity of the 19S and 26S coding regions: the 19S genes are located toward the center of the molecule, and the 26S sequences towards the ends (24). Thus, the rDNA of *Physarum* is a giant palindrome, as judged by hybridization analysis (23,24), by R-loop mapping (23), and by the rapid reannealing of the denatured rDNA molecules to form hairpin duplexes about one-half the size of native rDNA (31). The palindromic arrangement of the ribosomal genes in *Physarum* has been confirmed by electron microscopic observations of nucleolar chromatin which clearly show the inverted polarity of the transcription units at each end of the rDNA (26). (A similar palindrome-like arrangement of ribosomal genes is seen in another simple eukaryote, *Tetrahymena pyriformis*, in which the 17S and 25S genes occur in inverted polarity in extrachromosomal DNA molecules of molecular weight 13×10^6 (32, 33).)

In the past year we have completed a more detailed map of the rDNA which shows two new aspects of its organization. First is the localization of the 5.8S ribosomal RNA coding sequence in a transcribed intervening sequence of 1.66 ± 0.12 kb between the 19S and 26S genes (23). Second is the observation that the 26S gene consists of 3 coding regions (α, β, γ) interrupted by two intervening sequences (23). (Although intervening sequences have been reported in the ribosomal genes of Tetrahymena pyriformis (34) and in two strains of Drosophila (35-40), this is the first example of two intervening sequences in a ribosomal gene.) The two intervening sequences are present in at least 88% of the ribosomal genes from the transcriptionally active

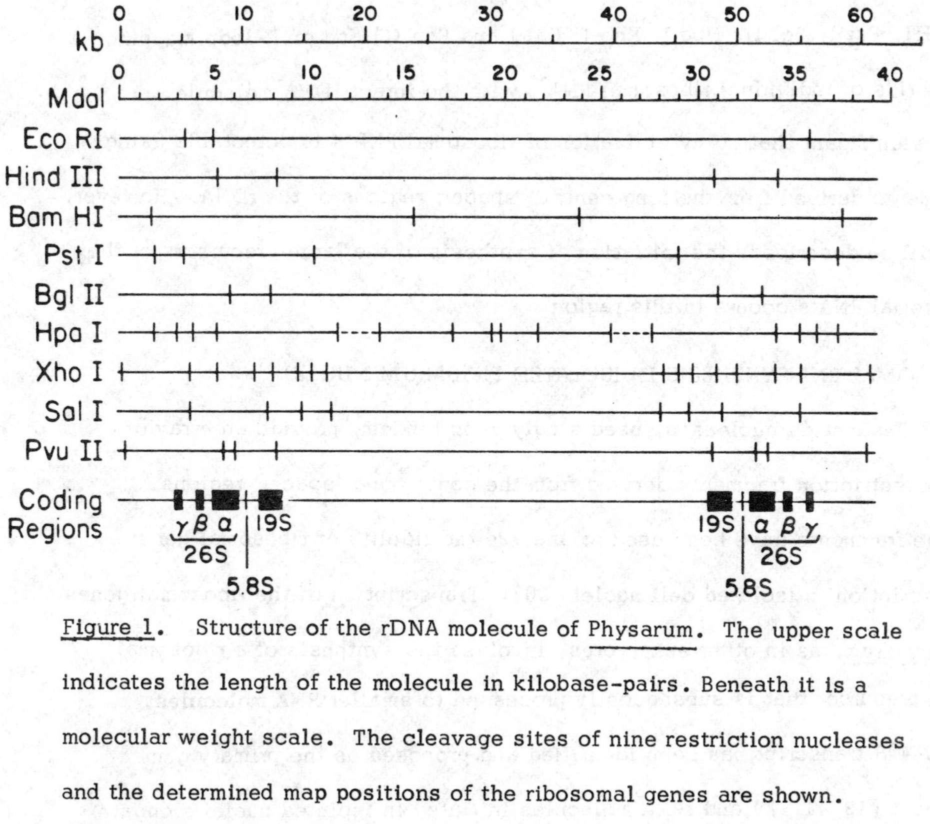

Figure 1. Structure of the rDNA molecule of Physarum. The upper scale indicates the length of the molecule in kilobase-pairs. Beneath it is a molecular weight scale. The cleavage sites of nine restriction nucleases and the determined map positions of the ribosomal genes are shown. Note the two intervening sequences in the 26S gene and the palindromic arrangement of the coding regions for 19S, 5.8S and 26S rRNA's.

plasmodial form of Physarum. This implies that they are actively transcribed during the synthesis of pre-ribosomal RNA, a conclusion which is now being tested directly by R-loop mapping of the precursor itself to purified rDNA.

The positions of the ribosomal genes and the intervening sequences between (and within) them are shown in FIGURE 1. The map is based on results obtained with a battery of restriction nucleases (Eco R1, Hind III, Bam H1, Pst I, Bgl II, Hpa I, Xho I, Sal I and Pvu II) and on R-loop mapping of hybrids of individual ribosomal RNA's with the intact rDNA molecule. It is significant that no hybridization of ribosomal RNA's is detectable using fragments derived from the long central 'spacer' regions of the rDNA. However, as will be described, the initiation of synthesis of the large precursors of the ribosomal RNA's occurs in this region.

RIBOSOMAL RNA SYNTHESIS IN ISOLATED PHYSARUM NUCLEI

Restriction nucleases, used singly or in tandem, provide an array of rDNA restriction fragments derived from the coding and 'spacer' regions. These fragments have been used to analyze the fidelity of ribosomal gene transcription in isolated cell nuclei (30). Transcription of the ribosomal genes in Physarum, as in other eukaryotes, involves the synthesis of a ribosomal RNA precursor that is subsequently processed to smaller RNA molecules. A 40-44S transcript has been identified and proposed as the primary gene product (19,41,42) and rRNA molecules initiated in isolated nuclei elongate to a size greater than 11 kb (30). This approximates the length of the Physarum rDNA transcription unit observed in the electron microscope; 4.0-4.2 μm, corresponding to 12.0-12.6 kb of extended DNA (26). A 30S molecule, possibly an intermediate in the processing of the original 44S transcript, has also been detected (43).

Ribosomal gene transcription can be studied in vitro in nuclei isolated from the Physarum microplasmodia (19,22,30,44). The nucleoli contain an α-amanitin-resistant RNA polymerase I analogous to RNA polymerase I of other eukaryotic cells (45-47). Isolated nuclei incorporate radioactively-labeled nucleoside triphosphates for periods in excess of one hour, and α-amanitin inhibits total incorporation by about 50% without effect on rRNA synthesis (30). The identification of the nascent rRNA chains is based on their hybridization to rDNA and its restriction fragments (30,44). It has been consistently observed that RNA synthesized in isolated nuclei hybridizes to restriction fragments known to contain rRNA coding sequences, but little or no hybridization is obtained to rDNA fragments derived from a very large segment, about 27 kb, in the non-transcribed central 'spacer' (30). This evidence for fidelity of transcription in isolated Physarum nuclei is in accord with observations on other eukaryotic systems. For example, nuclei from cultured Xenopus cells have been observed to transcribe the correct coding strand of rDNA (48), whereas fidelity of strand transcription was not observed in isolated nucleolar chromatin, despite the use of the homologous RNA polymerase (49). Isolated mouse myeloma nuclei correctly reinitiate the synthesis of 4.5S and 5S RNA's (50). Novikoff hepatoma nuclei synthesize ribosomal RNA with sequence fidelity (51) and HeLa nuclei can complete the synthesis of the 45S rRNA precursor (52). In isolated yeast nuclei, the addition of the homologous RNA polymerase I leads to selective transcription of the ribosomal genes, without effect on the synthesis of 4.5S pre-tRNA or 5S RNA's (53). Significantly, the transcription of rDNA by added yeast polymerase I was shown to be 8-fold more selective than that obtained with the RNA polymerase from E.coli.

The fidelity of transcription in isolated nuclei in the presence of the appropriate polymerase, and the loss of specificity in chromatin fractions suggest the presence of factors in the nucleolus which specifically direct the binding of RNA polymerase I to the correct sites of rRNA chain initiation on the rDNA template. In attempting to isolate and characterize such factors there is a clear need to identify the rDNA sequences involved. We have used incorporation of nucleoside (5'-γ-S) triphosphates (54, 55) (FIGURE 2) and

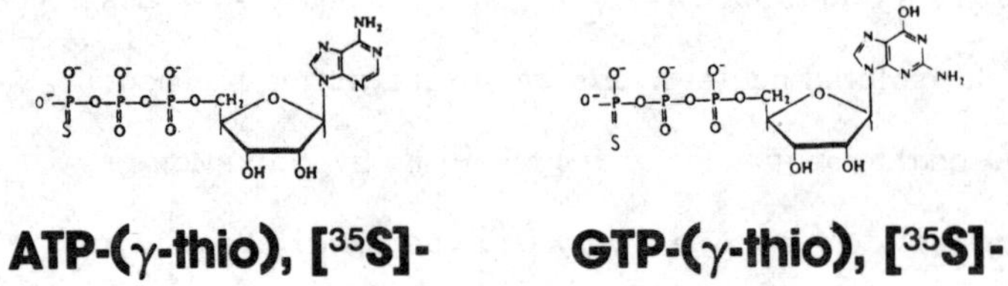

Figure 2. Chemical formulas of the ATP-γ-S and GTP-γ-S analogues of ATP and GTP. The retention of the thio-triphosphate at the 5'-ends of newly initiated RNA chains allows their recovery by chromatography on organomercurial-Sepharose columns, as shown in Figure 3.

nucleoside 5' (α-^{32}P) triphosphates to study initiation of ribosomal gene transcription in isolated Physarum nuclei. The nuclei were incubated with ATP-γ-S (or GTP-γ-S) and radiolabeled precursors in the presence of α-amanitin. The newly-initiated RNA molecules, now sulfur-derivatized at their 5'-ends, were isolated by organomercurial column chromatography (30). In a typical experiment, using a 25 minute pulse with ATP-γ-S and ^{3}H-UTP, $5.5 \pm 2.8\%$ of the incorporated radioactivity could be recovered as newly initiated chains.

These are retained on the organomercurial column and subsequently eluted with dithiothreitol, as shown in FIGURE 3. The initiation of transcription of transcription of RNA complementary to rDNA preferentially utilizes ATP-γ-S over GTP-γ-S by a factor of nearly 15-fold, indicating that pppA- is the primary initiating nucleotide in *Physarum* rRNA transcription. This is in accord with the finding that pppA- is also the 5'-terminal nucleotide in the *Xenopus* rRNA precursor (56). (We believe that some of the newly initiated chains labeled with GTP-γ-S represent 5S RNA sequences, since pppG- has been reported to be the initiating nucleotide in 5S RNA from *Dictyostelium* (57).)

By hybridizing nuclear transcripts labeled with ATP-γ-S to a variety of rDNA restriction fragments of known map positions, it was found that most rDNA transcripts initiate near a Xho I cleavage site at 17.0 kb from each end of the molecule. FIGURE 4 summarizes the hybridization studies, and maps the direction and length of the transcription unit. If initiation begins in this region near the Xho I *g* fragment and elongation is allowed to proceed, then short HS-rRNA chains should hybridize selectively to fragments near this site while longer chains should hybridize to additional fragments located more distally in the transcription unit. This was found to be the case. HS-rRNA was isolated following a 20-minute pulse with ATP-γ-S and subjected to sucrose gradient centrifugation. Fractions containing HS-rRNA chains of different lengths were annealed to Xho I rDNA fragments labeled to high specific activity with ^{32}P and purified electrophoretically.

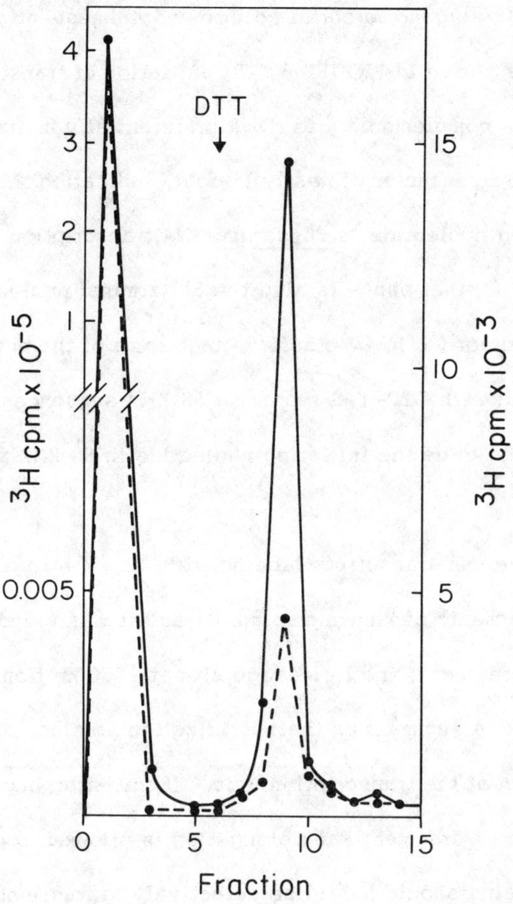

Figure 3. Organomercurial column chromatography of RNA synthesized in the presence of purine nucleoside 5'-(γ-S) triphosphates. Nuclei were pulse labeled for 10 minutes with ^{3}H-UTP in the presence of either ATP-γ-S (——) or GTP-γ-S (---). RNA was purified and subjected to Hg-Sepharose column chromatography. Fractions 1-6: 9 ml per fraction. Fractions 7-15: 1 ml per fraction, eluted in the presence of 10 mM dithiothreitol (DTT). Radioactivity of each fraction was assayed. The scale on the left of the figure refers to the initial peak of radioactivity, while the scale on the right refers to the radioactivity eluted after addition of DTT.

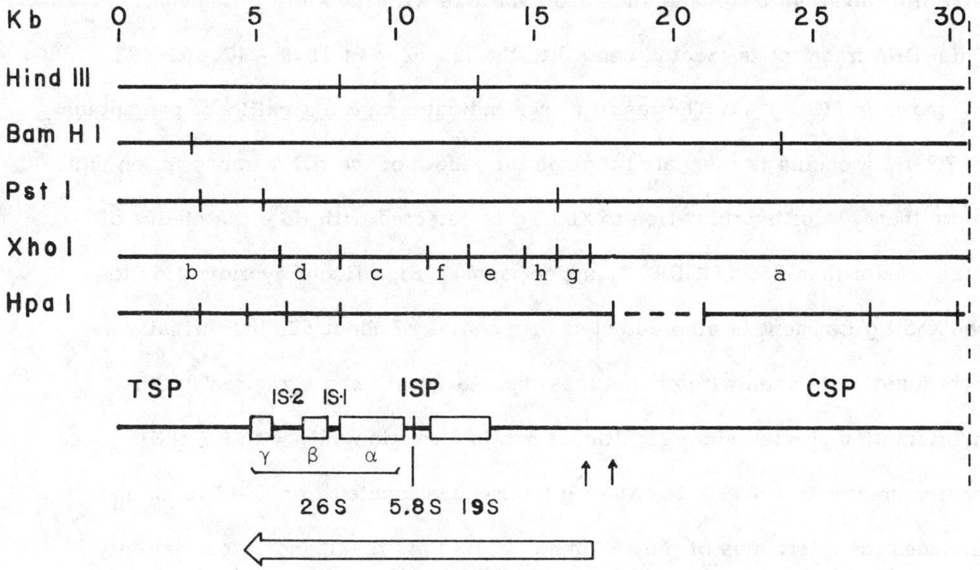

Figure 4. Restriction map showing *Physarum* rDNA coding regions and the approximate position of the ribosomal gene transcription unit, as judged by hybridization of 5'-end labeled rRNA to restriction fragments generated by Hind III, Bam H1, Pst I, Xho I and Hpa I. One half of the rDNA palindrome is depicted. The sequence delimiting the initiation site is shown by the small vertical arrows. The direction of transcription and the length of the transcription unit are indicated by the large horizontal arrow.

It can be seen in FIGURE 5 (upper panel) that HS-rRNA molecules ranging in size from about 4S to greater than 35S hybridize with the Xho I g fragment. This rDNA fragment is located central to the 19S gene at 15.8 - 17.0 kb (23) (as shown in FIGURE 4). The result shown indicates that a significant percentage of HS-rRNA chains is elongated through the length of the rDNA transcription unit. More than 40% of hybridization to Xho I g is detected with HS-rRNA chains of size greater than 35S (FIGURE 5, upper panel). Significant hybridization to the Xho I g fragment is also detected with chains of about 5S, but virtually no hybridization is seen with chains less than 4S. With Xho I fragment b, the hybridization studies show significant hybrid formation only with HS-rRNA chains greater than 35S. The Xho I b fragment is located from 0 - 5.9 kb and includes the 3'terminus of the 26S gene (23). This result indicates that only newly initiated chains which have elongated to great size can hybridize with fragments at the 3'-terminus of the transcription unit. Consistent with this view are the results obtained with Xho I fragment c which is located from 8.0 - 11.1 kb and hybridizes significantly only with HS-rRNA chains greater than 18S. These results are fully consistent with initiation of HS-rRNA near 17.0 kb followed by elongation through the rDNA transcription unit.

Our studies of rRNA synthesis in isolated nuclei show that a significant percentage of the newly initiated RNA is elongated throughout the length of the rDNA transcription unit. The rate of elongation was estimated at 5 to 6 nucleotides per second, but evidence was obtained that many of the newly initiated chains elongate to a size of about 2 kb and proceed no further.

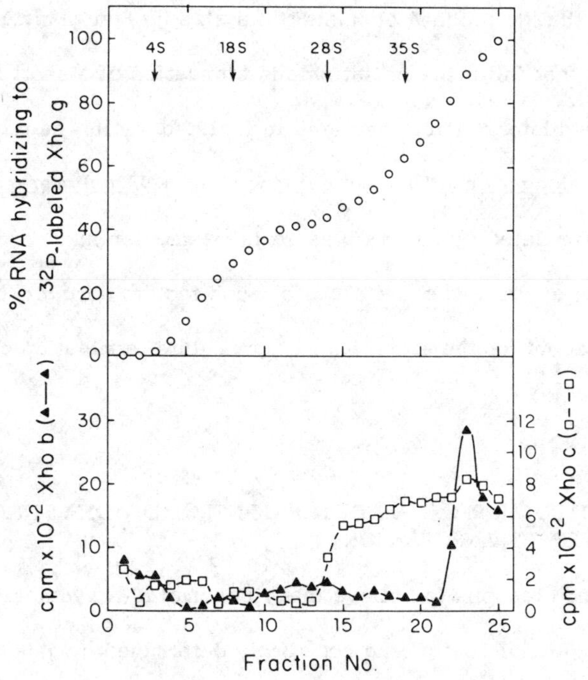

Figure 5. Hybridization of nick-translated rDNA Xho I fragments with HS-rRNA chains of different length. Upper panel: aliquots of the ^{32}P-labeled Xho I g fragment were hybridized to HS-rRNA chains separated according to size by sucrose density gradient centrifugation. Note that hybridization to the Xho I g fragment begins after the HS-rRNA chains reach a size of greater than 4S and that nearly 40% of detected hybridization is to Hs-rRNA chains greater than 35S. Lower panel: Hybridization of the ^{32}P-labeled Xho I b and c fragments to HS-rRNA chains separated by sucrose density gradient centrifugation. Note that the Xho I c fragment hybridizes to shorter chains than does the b fragment. The latter fragment contains the terminus of the rDNA transcription unit.

The bulk of those that do proceed elongates to a size greater than 11 kb (35S). It may be that following initiation and elongation of a short segment, some additional regulatory factor, depleted in isolated nuclei, is required for more efficient elongation. (There is evidence that RNA polymerase I activity can be stimulated by factors present in mammalian nuclei (58, 59).) It remains to be determined whether a loss of such factors during nuclear isolations may account for the attenuation or premature termination of rRNA chain growth.

THE ROLE OF NUCLEOLAR PROTEINS IN THE CONTROL OF rDNA STRUCTURE AND RIBOSOMAL GENE TRANSCRIPTION

There are good reasons to believe that ribosomal RNA synthesis is subject to transcriptional control and not simply determined by gene dosage. An extreme case is the shut-off of ribosomal RNA synthesis during nutritional deprivation without any change in the rDNA content of the *Physarum* nucleus (2, 60, 61). Reactivation of *Physarum* spores to form hatching amoeba is not accompanied by amplification of the ribosomal genes (61). The rate of ribosomal RNA synthesis in the plasmodial forms of *Physarum* varies throughout the cell cycle (16, 62); it stops at mitosis (63, 64) and resumes in the next cycle. The rate of rRNA synthesis is not tightly coupled to the number of rDNA molecules at different stages of the cell cycle. The rate of ribosomal RNA synthesis increases 5-fold as the number of templates is only doubled (62).

It is not known whether the rate of transcription of the ribosomal genes at different stages of Physarum growth and differentiation is dependent upon changes in RNA polymerase I levels in the nucleoli, but in another myxomycete, Dictyostelium discoideum, polymerase I activity does not change during differentiation from the exponential growth phase to the formation of fruiting bodies (65). Similarly, in the simple eukaryote, Acanthamoeba castellani, no significant difference in RNA polymerase I activity, extractability, or subunit composition could be detected during differentiation of actively growing trophozoites to metabolically inactive cysts (66, 67). This suggests that the rate of RNA synthesis is not primarily controlled by changing the number of RNA polymerase I molecules in Dictyostelium or Acanthamoeba. If this were the case for Physarum as well, control of rRNA synthesis could still be effected at other levels, such as polymerase binding to the template, chain initiation, chain elongation, attenuation and termination, and rate of processing of the pre-ribosomal RNA molecules.

In analyzing a system of this complexity, it is important that distinctions be made between mechanisms that control the accessibility of the rDNA template to RNA polymerase I and accessory factors, and mechanisms that modulate the activity of the polymerase itself. Among the former are enzymatic reactions, such as histone acetylation, which release constraints upon DNA in transcribing regions of chromatin (68-71); among the latter are factors which inhibit (72) or stimulate (58, 59) RNA polymerase I activity. The modification of polymerase I subunit structure by phosphorylation (73, 74) and the stimulation of polymerase activity by kinases (75) raise the potential for cascade control over rRNA synthesis by influencing the assembly of the holoenzyme or affecting its conformation.

Between these two levels of control - the first affecting the accessibility of the rDNA template, and the second influencing the intrinsic structure and activity of the polymerase - are factors which direct the binding of the polymerase to promoter or initiation sequences in the rDNA molecule. The assumption that such factors exist in eukaryotic cells is supported by recent evidence that a 37,000 dalton protein necessary for accurate transcription of cloned Xenopus 5S gene sequences in vitro interacts specifically with a region involved in transcription initiation (76). We have shown, by DNA-affinity chromatography and filter-binding assays, that Physarum nucleoli contain proteins which preferentially combine with rDNA sequences as compared to main-band DNA (44). A nucleolar phosphoprotein of MW 70,000 has been reported to stimulate the synthesis of rRNA in a deoxyribonucleoprotein complex prepared from Physarum nucleoli (77). The phosphoprotein was capable of combining with rDNA in vitro, but it did not discriminate in its binding to transcribed or non-transcribed sequences of the rDNA palindrome. This makes it unlikely to function as a specific factor in the formation of an initiation complex, but the 70,000 dalton phosphoprotein has other interesting properties. Dephosphorylation resulted in a loss of its DNA-binding properties and a failure to stimulate transcription (77). The positive correlation between the phosphorylated state and activity is in accord with observations on mammalian cells which relate high nucleolar kinase activities to increased nucleolar function in proliferating tissues (78).

A related question deals with the fidelity of rRNA chain termination. Nucleoli from Novikoff hepatoma cells (79) or *Tetrahymena pyriformis* (80) fail to terminate correctly after extraction with salt or detergents. A nucleolar protein fraction which restores normal termination to salt-extracted nucleoli from *Tetrahymena* has recently been described (80), but its DNA-binding properties have not been assessed.

It may be taken as axiomatic that proteins involved in transcriptional control will not be distributed at random along the rDNA strand. The presence of a large non-transcribed 'spacer' between the two ribosomal genes in the *Physarum* rDNA molecule permits comparisons of chromatin structural organization in the template-active and template-inactive regions of the same DNA molecule.

DIFFERENCES IN CHROMATIN SUBUNIT STRUCTURE IN TRANSCRIBING AND NON-TRANSCRIBING REGIONS OF rDNA CHROMATIN

What controls the accessibility of the rDNA template? In considering the structure of nucleolar chromatin, it is important to emphasize the differences already recognized between rapidly transcribing and non-transcribing genes. Electron microscopic evidence indicates that the template active rDNA sequences in *Physarum* are fully extended and not organized into compact 'beaded' nucleosome structures (13,22,26,44). In this respect the ribosomal transcription unit in *Physarum* resembles that in *Oncopeltus* (81), *Notophthalmus* (82), *Triturus* (83-85), *Xenopus* (83,84) and mammalian (CHO) cells (86).

In Physarum the long central 'spacer', most of which is not transcribed (30), appears to contain particles similar to nucleosomes (26) and it is degraded by Staphylococcal nuclease to yield the typical repeat pattern of nucleosomes and nucleosome oligomers (12, 13, 87). Beaded structures have also been observed in the 'spacer' regions between the tandem transcription units of amphibian nucleoli (83-85), but the beads are not evident in the template-active regions. When the ribosomal genes are inactivated, most of the nucleolar chromatin is beaded and indistinguishable from inactive, non-nucleolar chromatin (83). Failure to observe a beaded conformation in ribosomal transcription units is in accord with measurements of the length of the transcription unit relative to its DNA content; in Oncopeltus, for example, the compaction ratio is only 1.2 μm of B-structure DNA per μm of chromatin (88). It follows that the DNA cannot be coiled around nucleosomes in the usual two turns per nucleosome. The contrast between transcribing and non-transcribing chromatin is dramatically seen in plasmids containing cloned Xenopus ribosomal genes after injection into oocyte nuclei. The non-transcribed DNA of the vector (PMB9) is fully beaded, but the 8,200 base-pairs of the transcribed rDNA is smooth and has a compaction ratio of only 1.3 (89). Although the rDNA in active transcription units is not beaded in the typical nucleosomal conformation but seems relatively smooth, its diameter (70 A) greatly exceeds that of a free DNA strand (22 A), presumably due to the presence of associated proteins. We believe that this represents a reversible change in nucleosome organization which releases some of the constraints on

DNA in the transcription unit. When transcription is suppressed, the beaded structure reappears. The mechanisms underlying this reversible conversion of DNA topology from tightly-coiled to more accessible states probably involve post-synthetic modifications of proteins in the nucleosomes (68-71), as well as more specific associations with non-histone proteins.

Given the ultrastructural differences between the rDNA sequences in the transcriptionally active and inactive regions of nucleolar chromatin, one would expect differences in their susceptibilities to nuclease digestion. The coding and non-coding regions of Physarum rDNA chromatin have been compared, using DNase I and Staphylococcal nuclease as probes of DNA accessibility and nucleosome structure. In nucleoli from the transcriptionally active plasmodial form of Physarum, the rDNA coding sequences are more rapidly degraded by DNase I than is bulk DNA, suggesting a more accessible conformation of the ribosomal genes. This is in agreement with observations on the heightened DNase I sensitivity of DNA sequences committed to transcription in erythrocytes (90) and oviduct nuclei (91,92). The ribosomal coding sequences in Physarum are also more accessible to Staphylococcal nuclease digestion - which generates a series of rDNA chromatin subunits that can be separated on sucrose gradients (12,13). Two types of monomeric chromatin subunits - with sedimentation coefficients of 5S and 11S, respectively - were separated. Both forms contain 144 base-pairs of DNA, but the more slowly sedimenting peak (peak A) is greatly enriched in the sequences hybridizing to 19S and 26S ribosomal RNA's. Electron microscopy of the two types of chromatin subunits (shown in FIGURE 6) indicates that peak A monomers contain rDNA as an extended strand, while the 11S monomers are compact, beaded nucleosomes (13). Other studies have established that the non-transcribed central 'spacer' in rDNA chromatin is resistant to nuclease attack; its sequences are disproportionately represented in the higher oligomers of nucleosomes in the Staphylococcal nuclease digest (87).

Ribosomal RNA synthesis is suppressed in the microsclerotial forms of Physarum, and no peak A was observed in sucrose density gradient centrifugation of microsclerotial nucleosomes (13). The absence of peak A, and the slower kinetics of digestion of 19S and 26S rDNA coding sequences in microsclerotia, suggest that the change in nucleosome structure from the extended to the beaded form, and vice versa, reflects the template activity of the associated rDNA sequences. It seems plausible that changes in nucleosomal proteins underlie the transition from the active extended state of rDNA coding regions in the plasmodial forms to the beaded configuration of the suppressed ribosomal genes in microsclerotia. Similar changes in nucleosomal proteins are also likely to account for the differences in morphology and nuclease resistance of the coding regions and the non-transcribed 'spacer' of plasmodial rDNA chromatin.

Our results are consistent with the extended morphology of the rDNA transcription unit as seen in the electron microscope (26), but they also indicate that the rDNA is shielded from random nuclease attack by a periodic assembly of protecting proteins, which would account for the added thickness

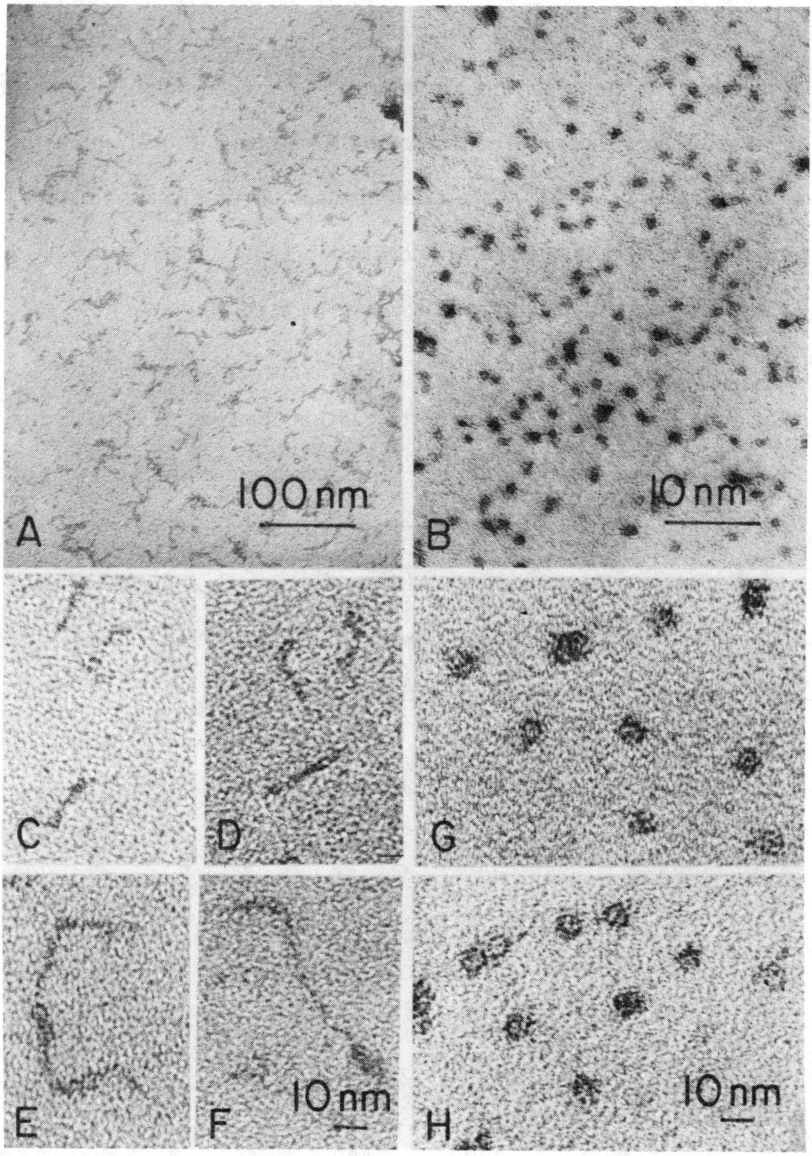

Figure 6. Electron micrographs from Peak A and the monomer nucleosome peak. All samples were tranferred to carbon-coated grids and negatively-stained with 1% aqueous uranyl acetate. A : peak A particles, separated on a sucrose gradient, fixed with formaldehyde, and further centrifuged through a CsCl gradient. B : monomeric nucleosome peak from the sucrose gradient. C and D : examples of the predominant smaller class of peak A particles with an average DNA content of 144 base-pairs at a compaction ratio of about 1.27. E. and F. examples of the minor, longer class of peak A particles, presumed to be dimers of the particles shown in C and D . G and H: particles from monomeric nucleosome peak.

of the DNA strand in the coding regions. A great advantage of the *Physarum* ribosomal chromatin is the opportunity it provides to compare nucleosome structure in the active and inactive regions of the same DNA strand. Higher order structures can also be compared. The sites of cleavage of *Physarum* rDNA by a variety of restriction nucleases are now known (23, 30) and it is possible to restrict nucleolar chromatin in much the same way. We have found that Eco R1 and Hind III cut rDNA in the chromatin at the same sites already mapped in the free rDNA. Chromatin fractions derived from the coding and 'spacer' regions are now being compared with respect to their content of histone and non-histone proteins (HMG proteins in particular) and the post-synthetic modification of histones and HMG proteins by the acetylation of lysine residues in their DNA-binding domains will be correlated with the transcriptional activity of the associated rDNA sequences.

REFERENCES

1. Zellweger,A.,Ryser,U. & Braun,R. (1972). J.Mol.Biol. 64 ,681.

2. Hall,L. & Braun,R. (1977). Eur.J.Biochem. 76 , 165.

3. Ryser,U.,Fakan,S. & Braun,R. (1973). Exptl.Cell Res. 78 , 89.

4. Bradbury,E.M.,Matthews,H.R.,McNaughton,J. & Molgaard,H.V. (1973). Biochim.Biophys.Acta 335 , 19.

5. Braun,R. & Evans,T.E. (1969). Biochim.Biophys.Acta 182, 511.

6. Vogt,V.M. & Braun,R. (1977). Eur.J.Biochem. 80 , 557.

7. Mohberg,J. & Rusch,H.P. (1969). Arch.Biochem.Biophys. 134 , 577.

8. Mohberg,J. (1977). J.Cell Sci. 24, 95.

9. Newlon,C.S.,Sonenschein,G.E. & Holt,C.E. (1973). Biochemistry 12 , 2338.

10. Ryser,U. & Braun,R. (1974). Biochim.Biophys.Acta 361 , 33.

11. Bohnert,H.S.,Schiller,B.,Bohme,R. & Sauer,W. (1975). Eur.J.Biochem. 57, 361.

12. Johnson,E.M.,Allfrey,V.G.,Bradbury,E.M. & Matthews,H.R. (1978). Proc. Natl.Acad.Sci.USA 75 , 1116.

13. Johnson,E.M., Matthews,H.R., Littau,V.C.,Lothstein,L.,Bradbury,E.M. & Allfrey,V.G. (1978). Arch.Biochem.Biophys. 191, 537.

14. Guttes,E. & Guttes,S. (1969). J.Cell Biol. 43 , 229.

15. Guttes,E., Guttes,S. & Rusch,H.P. (1961). Dev. Biol. 3 , 588.

16. Cummins,J.E. (1969). In "The Cell Cycle" (G.M.Padilla,G.L.Whitson & I. Cameron, eds.) pp. 141-158, Academic Press, New York.

17. Daniel,J.W. & Baldwin,H.H. (1964). In "Methods in Cell Physiology" (D.Prescott, ed.) pp. 9-41, Academic Press, New York.

18. Mohberg,J. & Rusch,H.P. (1971). Exptl.Cell Res. 66 , 305.

19. Davies,K.E. & Walker,I.O. (1977). J.Cell Sci. 26 , 267.

20. Seebeck,T.,Stalder,J. & Braun,R. (1979). Biochemistry 18 , 484.

21. Matthews,H.R.,Johnson,E.M.,Steer,W.M.,Bradbury,E.M. & Allfrey,V.G. (1978). Eur.J.Biochem. 82 , 569.

22. Allfrey,V.G.,Johnson,E.M.,Sun, I.Y-C.,Littau,V.C.,Matthews,H.R. & Bradbury, E.M. (1978). Cold Spring Harbor Symp.Quant.Biol. 42 , 505.

23. Campbell,G.R.,Littau,V.C.,Melera,P.W.,Allfrey,V.G. & Johnson,E.M. (1979). Nucleic Acids Res. 6 , 1433.

24. Molgaard,H.V.,Matthews,H.R. & Bradbury,E.M. (1976). Eur.J.Biochem. 68 , 541.

25. Vogt,V.M. & Braun,R. (1976). J.Mol.Biol. 106 , 567.

26. Grainger,R.M. & Ogle,R.C. (1978). Chromosoma 65 , 115.

27. Holt,C.E. & Gurney,E.C. (1969). J.Cell Biol. 40 , 484.

28. Grainger,R.M. (1976). J.Cell Biol. 70 , 327a.

29. Southern,E.M. (1975). J.Mol.Biol. 98 , 503.

30. Sun,I.Y-C.,Johnson,E.M. & Allfrey,V.G. (1979). Biochemistry 18 , 4572.

31. Hardman,N., Jack,P.L.,Brown,A.J.P. & MacLachlan,A. (1979). Biochim. Biophys.Acta 562 , 365.

32. Karrer,K.M. & Gall,J.G. (1976). J.Mol.Biol. 104 , 421.

33. Engberg,J.,Andersson,P.,Leick,V. & Collins,J. (1976). J.Mol.Biol. 104, 455.

34. Wild,M.A. & Gall,J.G. (1979). Cell 16 , 565.

35. Glover,D.M. & Hogness,D.S. (1977). Cell 10 , 167.

36. Wellauer,P.K. & Dawid,I. (1977). Cell 10 , 193.

37. Pellegrini,M.,Manning,J. & Davidson,N. (1977). Cell 10 , 213.

38. White,R.L. & Hogness,D.S. (1977). Cell 10 , 177.

39. Barnett,T. & Rae, P.M.M. (1979). Cell 16 , 763.

40. Wellauer,P.K., Dawid,I. & Tartoff,K.D. (1978). Cell 14 , 269.

41. Melera,P.W. & Rusch,H.P. (1973). Exptl.Cell Res. 82 , 197.

42. Jacobson,D.N. & Holt,C.E. (1973). Arch.Biochem.Biophys. 159 , 342.

43. Zellweger,A. & Braun,R. (1971). Exptl.Cell Res. 65 , 413.

44. Allfrey,V.G.,Johnson,E.M.,Sun,I.Y-C.,Littau,V.C.,Matthews,H.R. & Bradbury, E.M. (1977). In "Molecular Cytogenetics" (ICN-UCLA Symposia on Molecular and Cellular Biology)(R.S.Sparks, D.Comings and C.F.Fox, eds.) Vol. VII, pp. 159-175, Academic Press, New York.

45. Burgess,A.B. & Burgess,R.R. (1974). Proc.Natl.Acad.Sci.USA 71 , 1174.

46. Gornicki, S.Z., Vuturo, S.B., West, T.V. & Weaver, R.F. (1974). J.Biol.Chem. 249, 1792.

47. Weaver, R.F. (1976). Arch.Biochem.Biophys. 172, 470.

48. Reeder, R.H. & Roeder, R.G. (1972). J.Mol.Biol. 67, 433.

49. Honjo, T. & Reeder, R.H. (1974). Biochemistry 13, 1896.

50. Marzluff, W.F.Jr., Murphy, E.C. & Huang, R.-C.C. (1974). Biochemistry 13, 3689.

51. Ballal, N.R., Choi, Y.C., Mouche, R. & Busch, H. (1977). Proc.Natl.Acad.Sci. USA 74, 2446.

52. Zylber, E.A. & Penman, S. (1971). Proc.Natl.Acad.Sci.USA 68, 2861.

53. Tekamp, P.A., Valenzuela, P., Maynard, T., Bell, G.L. & Rutter, W.J. (1979). J.Biol.Chem. 254, 955.

54. Reeve, A.E., Smith, M.M., Pigiet, V. & Huang, R.-C.C. (1977). Biochemistry 16, 4464.

55. Smith, M.M., Reeve, A.E. & Huang, R.-C.C. (1978). Biochemistry 17. 493.

56. Reeder, R.H., Wahn, H.L., Botchan, P., Hipskind, R. & Sollner-Webb, B. (1978). Cold Spring Harbor Symp.Quant.Biol. 42, 1167.

57. Batts-Young, B. & Lodish, H.F. (1978). Proc.Natl.Acad.Sci.USA 75, 740.

58. Goldberg, M.I., Perriard, J.C. & Rutter, W.J. (1977). Biochemistry 16, 1648.

59. James, G.T., Yeoman, L.C., Matsui, S., Goldberg, A.H. & Busch, H. (1977). Biochemistry 16, 2384.

60. Hall, L. & Braun, R. (1977). Experientia 33, 820.

61. Affholter, H. & Braun, R. (1978). Biochim.Biophys.Acta 519, 118.

62. Hall, L. & Turnock, G. (1976). Eur.J.Biochem. 62, 471.

63. Birch, B. & Turnock, G. (1977). FEBS Letters 84, 317.

64. Fink, K. & Turnock, G. (1977). Eur.J.Biochem. 80, 93.

65. Pong, S.S. & Loomis, W.F. (1973). J.Biol.Chem. 248, 3933.

66. Detke, S. & Paule, M.R. (1975). Biochim.Biophys.Acta 83, 67.

67. Detke, S. & Paule, M.R. (1978). Arch.Biochem.Biophys. 185, 333.

68. Allfrey, V.G., Faulkner, R. & Mirsky, A.E. (1964). Proc.Natl.Acad.Sci.USA 51, 786.

69. Allfrey, V.G. (1977). In "Chromatin and Chromosome Structure" (H.J.Li & R.A.Eckhardt, eds.) pp. 167-191, Academic Press, New York.

70. Johnson, E.M. & Allfrey, V.G. (1978). In "Biochemical Actions of Hormones" Vol. 5 (G.Litwack, ed.) pp. 1-51, Academic Press, New York

71. Allfrey, V.G. (1980). In "Cell Biology: A Comprehensive Treatise", Vol. 3, (L.Goldstein & D.M.Prescott, eds.) pp. 347-437, Academic Press, New York.

72. Hildebrandt, A. & Sauer, H.W. (1977). Biochem.Biophys.Res.Commun. 74, 466.

73. Jungmann, R.A., Hiestand, P.C. & Schweppe, J.S. (1974). J.Biol.Chem. 249, 5444.

74. Bell, G.I., Valenzuela, P. & RuttermW.J. (1977). J.Biol.Chem. 252, 3082.

75. Dahmus, M. (1976). Biochemistry 15, 1821.

76. Engelke, D.R., Ng, S.Y., Shastry, B.S. & Roeder, R.G. (1980). Cell 19, 717.

77. Keuhn, G.D., Affholter, H.U., Atmar, V.J., Seebeck, T., Gubler, U. & Braun, R. (1979). Proc.Natl.Acad.Sci.USA 76, 2541.

78. Olson, M.O.J., Hatchett, S., Allan, R., Hawkins, T.C. & Busch, H. (1978). Cancer Res. 38, 3421.

79. Samal, B., Ballal, N.R., Choi, Y.C. & Busch, H. (1978). Biochem.Biophys.Res. Commun. 84, 328.

80. Leer, J.C., Tiryaki, D. & Westergaard, O. (1979). Proc.Natl.Acad.Sci.USA 76, 5563.

81. Foe, V.E., Wilkinson, L.E. & Laird, C.D. (1976). Cell 9, 131.

82. Woodcock, C.L.F., Frado, L.L.Y., Hatch, C.L. & Ricciardiello, L. (1976). Chromosoma 58, 33.

83. Scheer, U., Trendelenburg, M.F., Krohne, G. & Franke, W.W. (1977). Chromosoma 60, 147.

84. Franke, W.W., Scheer, U., Trendelenburg, M.F., Spring, H. & Zentgraf, H. (1976). Cytobiologie 13, 401.

85. Scheer, U. (1978) Cell 13, 535.

86. Puvion-Dutilleul, F., Bachellerie, J.P., Zalta, J.P. & Bernhard, W. (1977). Biol.Cellulaire 30, 183.

87. Johnson, E.M., Campbell, G.R. & Allfrey, V.G. (1979). Science 206, 1192.

88. Foe, V.E. (1978). Cold Spring Harbor Symp.Quant.Biol. 42, 732.

89. Trendelenburg, M.F. & Gurdon, J.B. (1978). Nature 276, 292.

90. Weintraub, H. & Groudine, M. (1976). Science 193, 848.

91. Garel, A. & Axel, R. (1976). Proc.Natl.Acad.Sci.USA 73, 3966.

92. Garel, A., Zolan, M. & Axel, R. (1977). Proc.Natl.Acad.Sci.USA 74, 4867.

Expression of Microinjected Eucaryotic Genes

W. French Anderson, M.D.

Laboratory of Molecular Hematology
National Heart, Lung, and Blood Institute
National Institutes of Health
Bethesda, Maryland 20205

ABSTRACT

A mixture of two recombinant plasmids was microinjected into mouse thymidine kinase (TK) L cells. One plasmid contained the herpes simplex TK gene and the other contained the human β globin gene. These microinjected cells were shown to produce functionally active herpes simplex TK enzyme, replicate the human β gene, and produce human β globin mRNA sequences at low levels. Thus, the genetic defect (lack of TK activity) was corrected by the microinjected TK gene, and a co-injected human β globin gene was replicated and weakly expressed.

Our laboratory has had a long-standing interest in the study of globin gene expression (1-3). Our work in recent years has centered on an attempt to transfer functional globin genes into tissue culture cells. Such an endeavor requires the ability to isolate a known gene and a procedure for transferring this specific gene into a living cell. Recombinant DNA technology has provided the ability to carry out the former. As for the latter, we have used various techniques (including extensive studies with somatic cell hybridization), but the most promising for injecting single genes appears to be the physical microinjection procedure developed by Dr. Elaine Diacumakos at Rockefeller University (4).

We wanted to transfer a thymidine kinase (TK) gene and a human beta-globin gene into a mouse TK^- L cell to determine if the cell could be genetically corrected for TK deficiency and if the globin gene would replicate and be expressed. Axel and his colleagues (5, 6) and others (7-9) have clearly shown that the TK gene can be transferred to TK^- L cells by DNA-mediated gene transfer. In this technique, picogram quantities of TK DNA are mixed with microgram quantities of salmon sperm carrier DNA and then layered onto mouse TK^- L cells in the presence of calcium phosphate. The TK gene is incorporated and expressed by approximately one cell in 10^6 (when 20 picograms TK DNA are used). These transformed cells can be recognized by their ability to grow in selective (HAT) medium. Other DNA fragments (e.g., containing a human or rabbit beta-globin gene) can be co-transformed along with the TK gene under appropriate conditions. In contrast, we wished to develop a procedure whereby a single copy of a specific gene could be inserted into the nucleus of a single cell under conditions whereby the injected cell could be grown in culture into a cloned population with or without selective pressure. This is the reason we chose to collaborate with Dr. Diacumakos and use her physical microinjection procedure. A minute volume (10^{-11} ml) of solution can be injected into any region of a cell with no apparent interference with cell function.

The two genes we microinjected are shown in Fig. 1. On the left is pRK1, which is a pBR322 plasmid (10) with a 4.4 kb Pst I fragment of human DNA [isolated from HβG1 (11)] containing the human beta-globin gene insert (12). In the middle is χ1, another pBR322 plasmid with a 3.5 kb Bam HI fragment of herpes simplex type 1 containing the TK gene. This latter plasmid was developed by Enquist and colleagues (13). Our experiment is outlined in Fig. 2. One copy (on the average) of pRK1 and one copy of χ1 were microinjected into the nucleus of individual mouse TK^- L cells attached to a coverslip. The coverslips were then incubated in HAT selective medium (14), where transformed cells grew as individual colonies. The individual colonies were isolated, grown to 10^8 - 10^9 cells and analyzed. The presence of the injected genes was detected by the technique of restriction endonuclease and Southern blot analysis while the expression of the genes was measured by analysis of herpes simplex TK enzyme activity and by liquid cDNA-mRNA molecular hybridization for the human beta-globin mRNA.

Lillian Killos, a Ph.D. graduate student in our laboratory, carried out a series of microinjections and obtained seven transformed colonies. A picture of one of the colonies is shown in Fig. 3. Each of the colonies has been grown in HAT medium over the past eight months. Growth of injected TK^- cells in HAT selective medium indicates that either the microinjected herpes simplex TK gene is functioning or the mouse TK gene itself has been activated. A

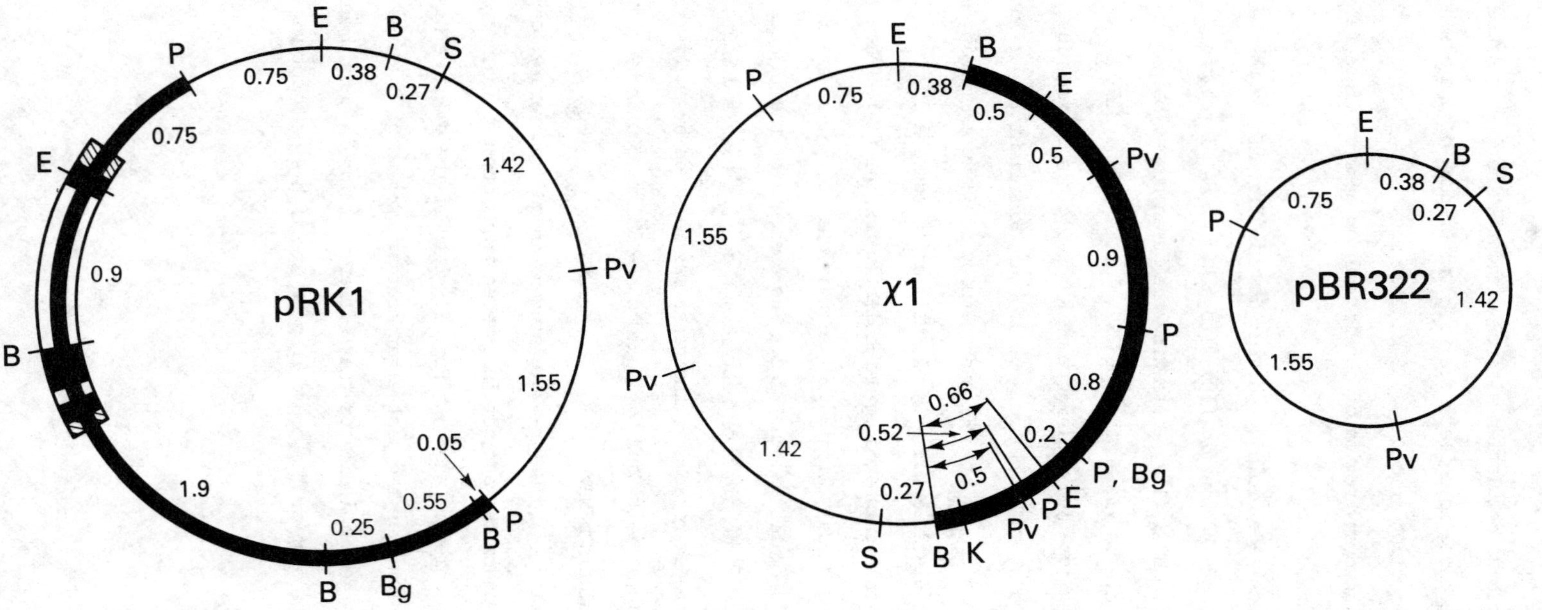

Fig. 1. Restriction endonuclease maps of pRK1, χ1, and pBR322.

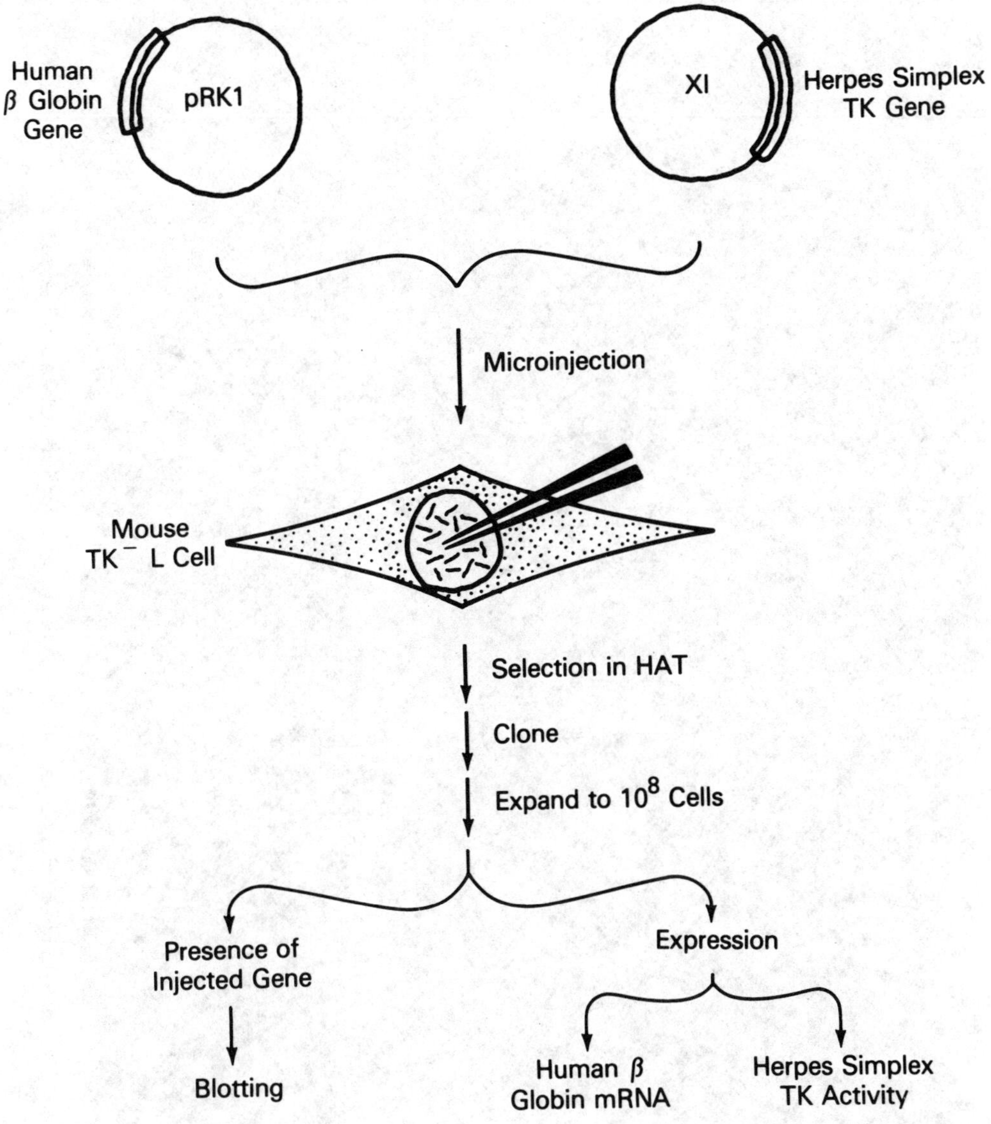

Fig. 2. Flow chart for microinjection experiment. See text.

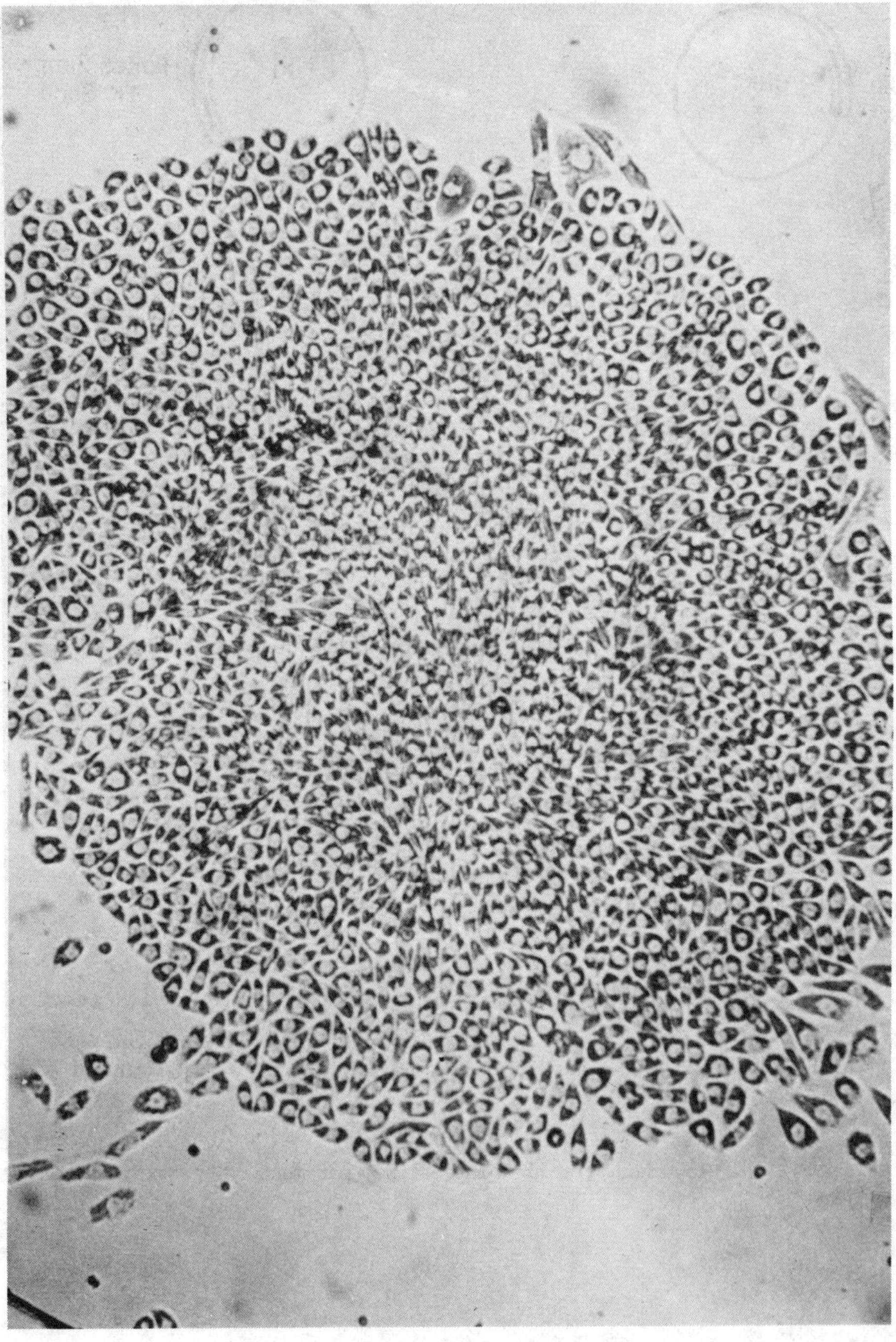

Fig. 3. Single colony of microinjected TK^- L cells growing in HAT selective medium.

definitive technique using ^{125}I-dC for detection of herpes simplex TK enzyme activity has been developed by Drs. William and Wilma Summers at Yale University (15). The herpes simplex TK enzyme can phosphorylate ^{125}I-dC, which is then incorporated into trichloroacetic acid insoluble DNA, while mouse TK enzyme cannot. However, mouse cells possess an enzyme, dC deaminase, which can convert ^{125}I-dC into a usable substrate. The inhibitor tetrahydrouridine (THU) blocks dC deaminase activity. Therefore, incorporation of ^{125}I-dC into DNA in the presence of THU is specific evidence for herpes simplex TK enzyme activity. Lillian Killos took colonies C1 and C2 to the laboratory of the Drs. Summers and they kindly analyzed the two colonies for the expression of herpes simplex TK enzyme with the results shown in Table 1. Their mouse TK^- L cell ($LMTK^-$) shows no incorporation of [^{125}I]-dC; wild type A9 cells do incorporate [^{125}I]-dC in the absence, but not in the presence, of THU; and TK^- mouse fibroblasts transformed with herpes simplex virus were positive for herpes simplex TK enzyme. Our parental TK^- L cells (probably identical to the Summers' $LMTK^-$) are negative for TK activity. Both colonies C1 and C2, however, are positive for herpes simplex TK enzyme activity. Colonies C2B, C2D, and C3 have also been shown to possess herpes simplex TK activity. Since the assays were done on clones of 5×10^4 cells expanded from, presumably, single cells, the injected herpes simplex TK gene is clearly replicating and functioning in these cloned cells.

Linda Sanders-Haigh, another Ph.D. graduate student in our laboratory, carried out Southern blot analysis (16) on the DNA extracted from the microinjected cells. The DNA was digested with the enzymes Eco RI, Bam HI and Pst I, electrophoresed on 1% aqueous gel slabs, transferred to nitrocellulose filters, and then annealed with the ^{32}P-labeled nick-translated Hha I fragment (17, 18) isolated from the human beta-globin cDNA plasmid clone JW-102 (19). Data for

Table 1. Measurement of Herpes Simplex Virus Type I Thymidine Kinase (HS TK) in Microinjected TK^- L Cells

Cell Line	Incorporation of ^{125}I-dC (cpm) -THU	
Controls		
$LMTK^-$	46	54
A9 (TK^+)	2,576	107
TFC3 (HS TK^+)	3,477	2,917
Clones		
Parental LTK^-	48	53
C1	1,541	1,227
C2	1,979	2,000

colony C1 are shown in Fig. 4 and Table 2. As can be seen, the human beta-globin gene is present in this colony. Linda Sanders-Haigh then analyzed for the presence of human beta-globin messenger RNA by hybridizing the RNA extracted from several of the microinjected colonies using ^{32}P-labeled anti-mRNA single stranded human beta-globin cDNA (isolated from JW-102) as probe. The data are shown in Fig. 5. XX-8 is a somatic cell hybrid (2S MEL x human fibroblast) containing a human chromosome 11 (18) and is known to contain roughly 500 molecules of human beta-globin mRNA per cell. The data for C2B, C3, and C2C indicate that approximately 7, 5, and 2 molecules of human beta-globin mRNA sequences are present per microinjected L cell. Thus, the human beta-globin gene is present and expressing, albeit at a very low level.

We then asked the question: In what form is the human beta-globin gene in the microinjected cell? In Table 2 are collected all of the data for the seven microinjected colonies with the three restriction endonucleases shown. Even though the patterns are different for different colonies, there are a surprisingly high number of fragments which are the same size as those obtained when the originally injected plasmid pRK1 is digested. The implication from these data is that perhaps covalently-closed-circular pRK1 itself is present in the microinjected cells. If this were the case, it should be possible to rescue the pRK1 plasmids by transforming E. coli with undigested microinjected mouse L cell DNA. Dr. Peter Kretschmer carried out these experiments using cells grown for sixty generations and was able to rescue pRK1 from the undigested DNA with moderate efficiency (approximately one transformant per µg of DNA). Surprisingly, however, not only was he able to extract pRK1, but in addition, a number of other unique plasmids which appear to be recombination events occurring within the mouse cell. For example, as shown in Fig. 6,

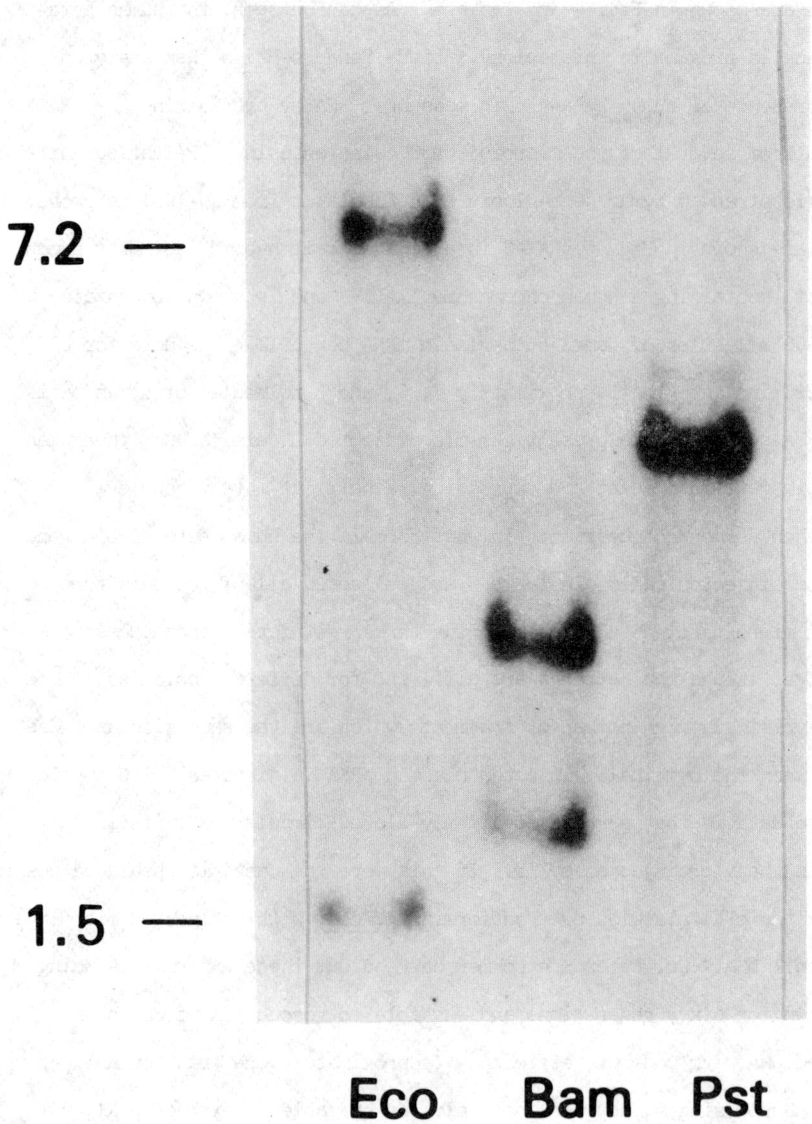

Fig. 4. Southern blot analysis of the DNA isolated from microinjected colony C1 using a ^{32}P-labeled human β globin cDNA as probe. Thirty to forty μg DNA were loaded into each lane. Restriction endonucleases used: Eco (Eco RI), Bam (Bam HI), and Pst (Pst I).

Table 2. Restriction Bands of DNA from Microinjected Colonies (in kb)

DNA Source	Eco RI	Bam HI	Pst I
C1	7.2	2.8	4.4
	1.5	1.9	
C2	7.2	2.8	7.0
	1.5	1.9	6.0
			4.4
			1.5
C2A	7.2	2.8	4.4
	1.5	1.9	3.2
C2B	9.0	2.8	
	7.2	1.9	4.4
	4.5		
	1.5		
C2C	14.0	3.2	4.4
	7.2	2.8	
	1.5	1.9	
C2D	7.2	3.2	4.4
	4.5	2.8	
	1.5	1.9	
C3	7.2	4.2	4.4
	4.5	2.8	4.2
	1.5	1.9	
pRK1	7.2	2.8	4.4
	1.5	1.9	

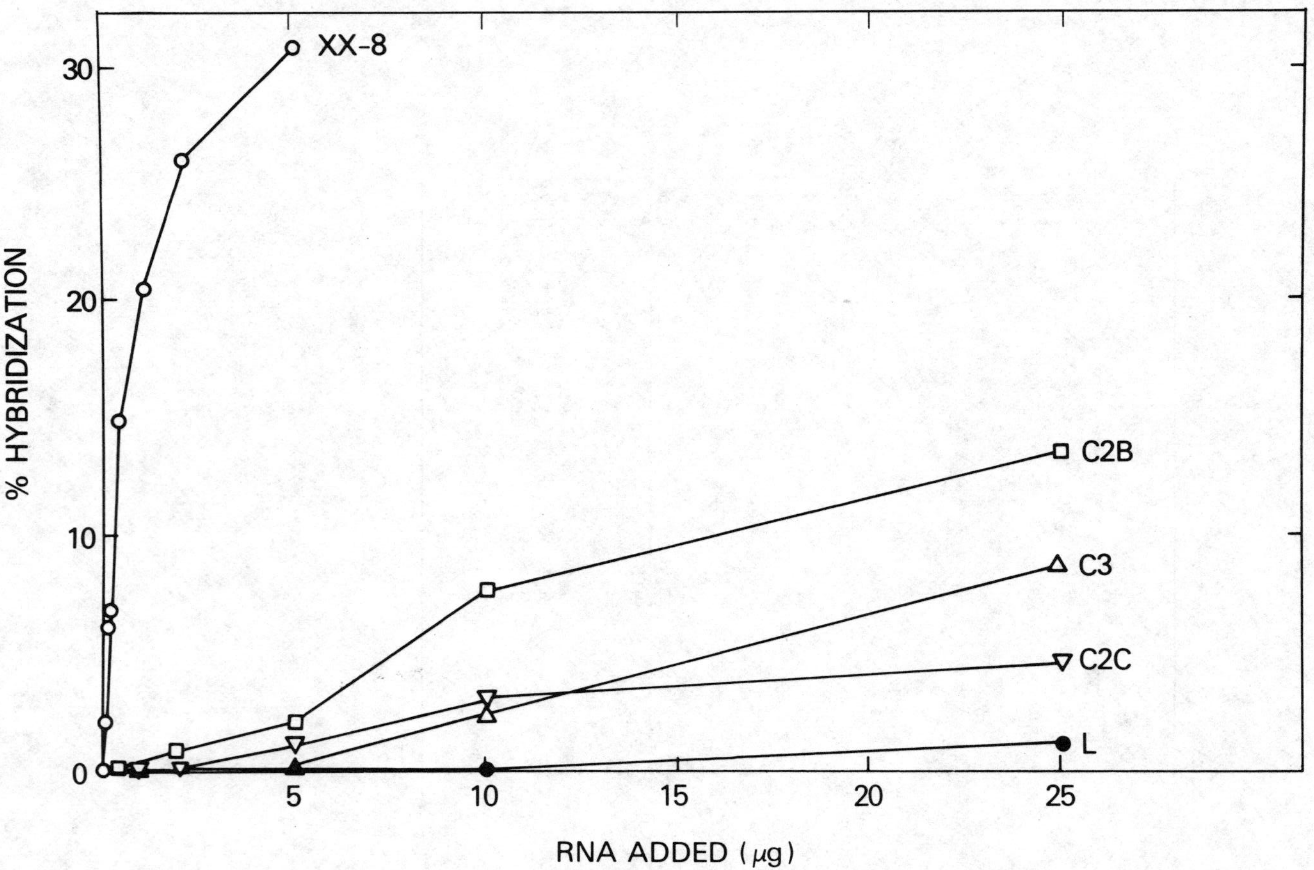

Fig. 5. Liquid hybridization of RNA extracted from microinjected L cells using a single-stranded ^{32}P-labeled human β globin cDNA as probe.

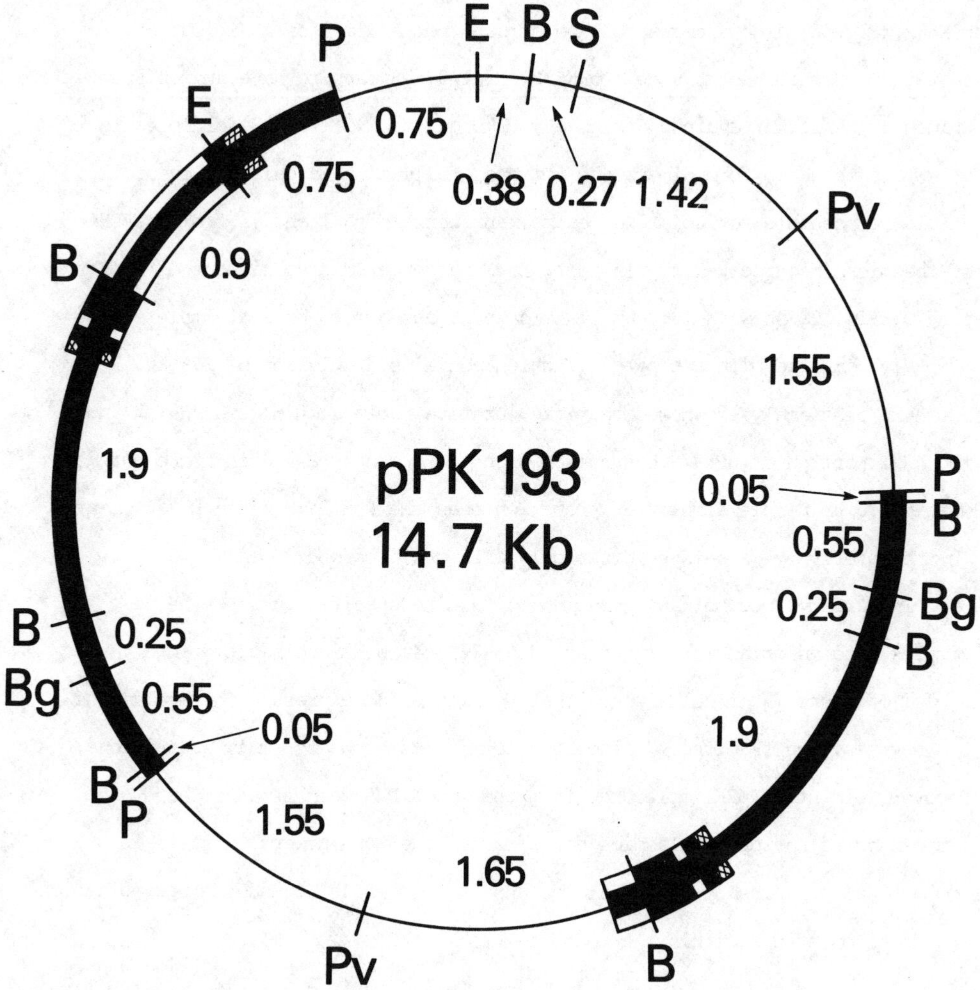

Fig. 6. Restriction endonuclease map of pPK193.

pPK193 appears to be a plasmid of 14.7 kb containing an intact copy of pRK1 and a second copy of pRK1 which contains a deletion of approximately 2.8 kb. pPK682, Fig. 7, also appears to have arisen from a dimer of pRK1, but contains a deletion of 5.6 kb. pPK578, Fig. 8, appears to have arisen from a trimer of pRK1, but again contains a 5.6 kb deletion. If the pRK1 molecule is able to combine to form dimers and trimers, followed by a deletion event, it should be possible to isolate a recombinant between pRK1 and $\chi 1$ from the DNA of the microinjected cells. In fact, Dr. Kretschmer did so without difficulty. pPK579, Fig. 9, shows a 13.7 kb plasmid which appears to have a complete copy of pRK1, combined with a copy of $\chi 1$ which contains a deletion of 2.9 kb. Southern blot analyses showed that the putative human globin gene region in the pRK1 portion of pPK579 hybridizes with the human beta-globin probe but not with a 3.5 kb herpes simplex TK probe, whereas the 0.5 Bam HI/Eco RI fragment in the putative $\chi 1$ area hybridizes only with the TK probe.

In summary, recombinant plasmids containing specific human or herpes simplex gene inserted into the plasmid pBR322 can be microinjected into the nuclei of mouse TK^- L cells where they replicate and express. These plasmids can be recovered from the DNA isolated from cells grown for many generations. Recovery of intact pRK1 molecules from the mouse DNA demonstrates that microinjected bacterial plasmids can exist in the non-integrated state in mouse L cells. Finally, the plasmids appear to be able to undergo recombination and deletion events within the mouse cell.

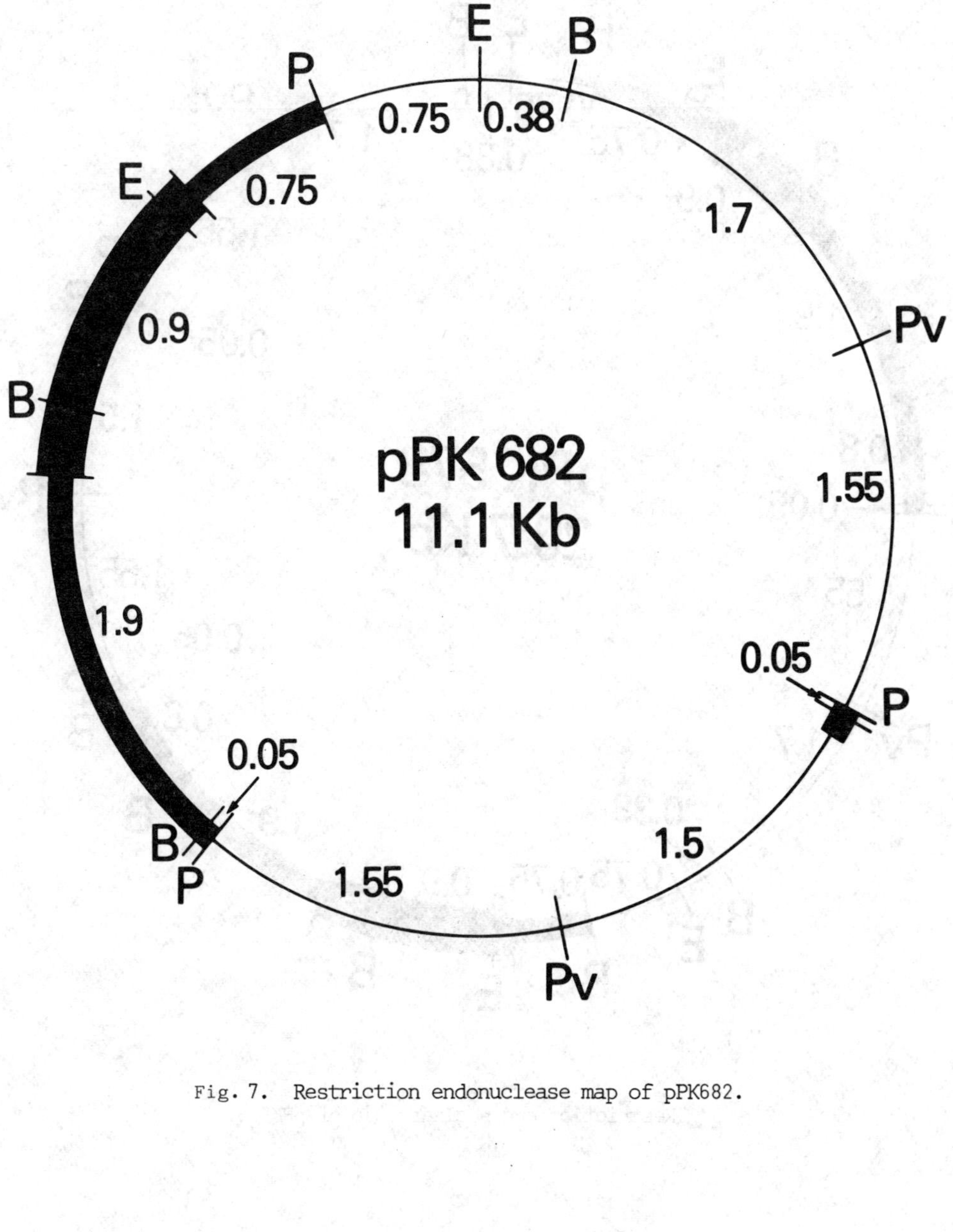

Fig. 7. Restriction endonuclease map of pPK682.

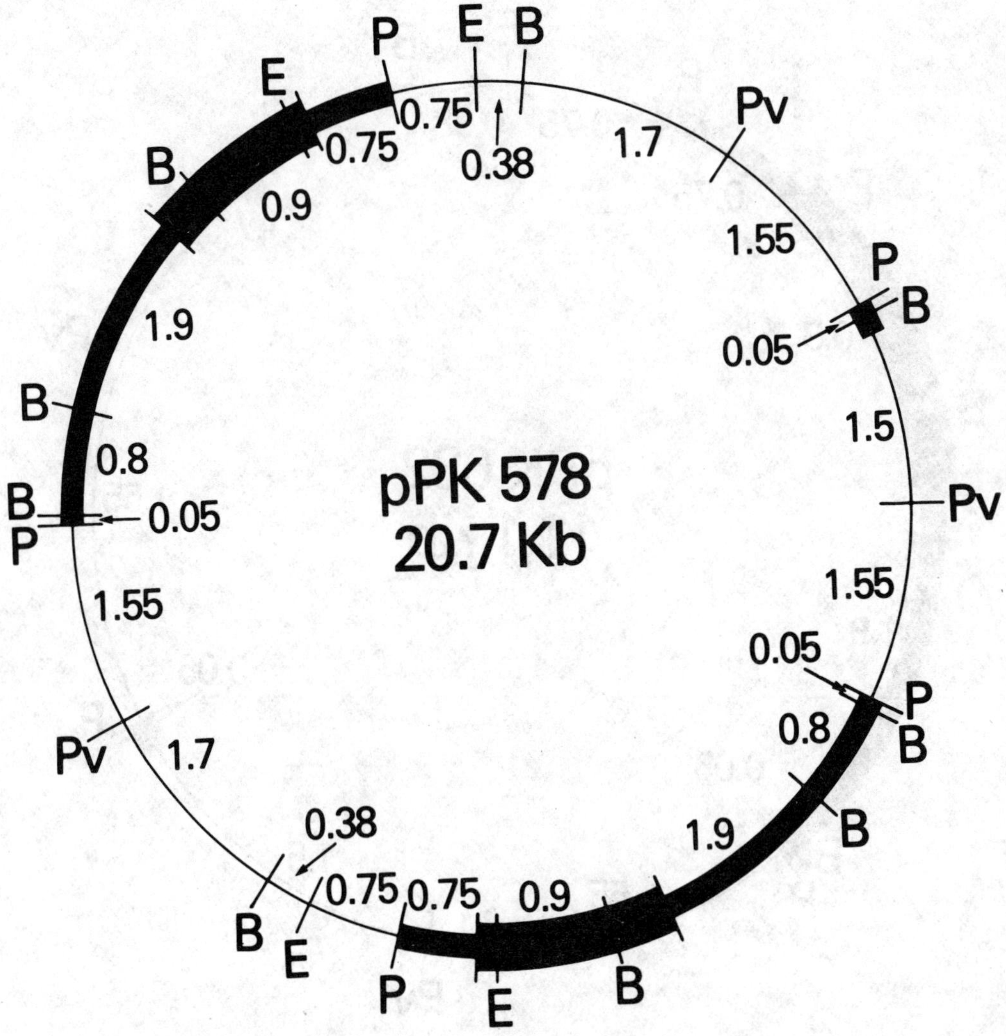

Fig. 8. Restriction endonuclease map of pPK578.

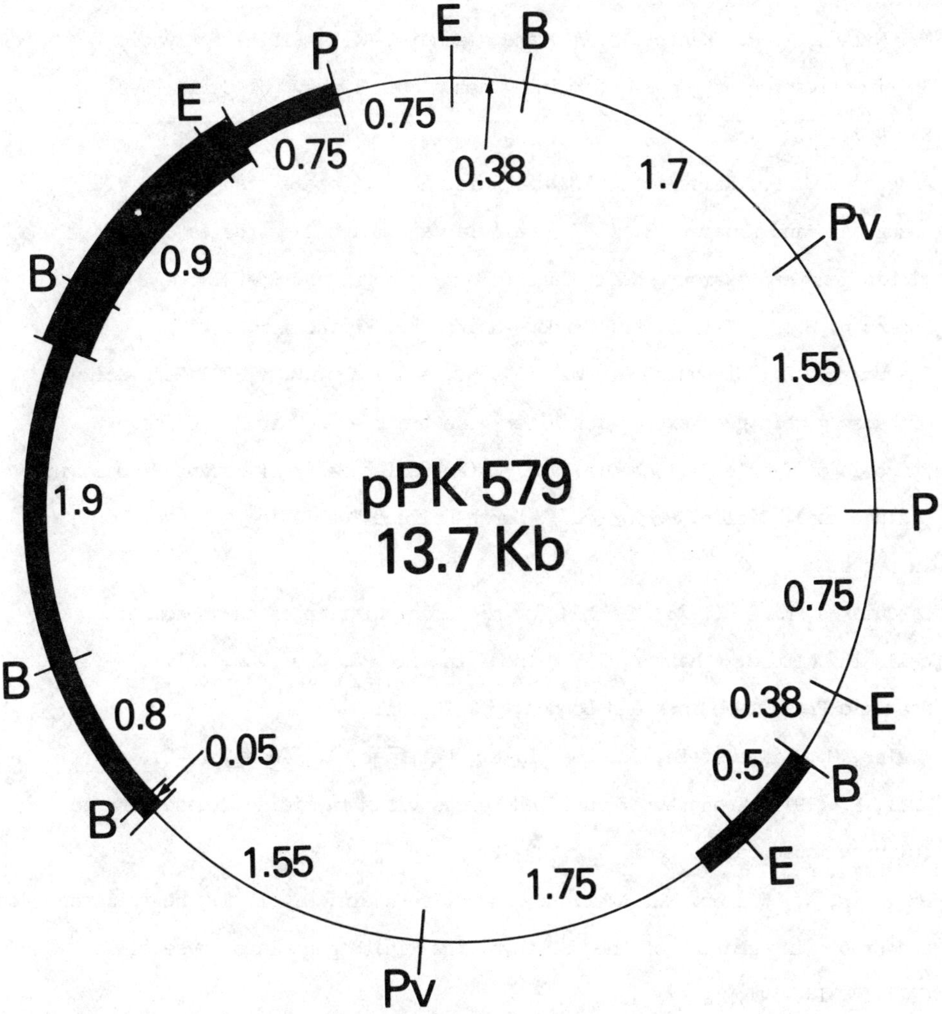

Fig. 9. Restriction endonuclease map of pPK579.

References

1. Anderson, W. F., Barker, J. E., Elson, N. A., Merrick, W. C., Steggles, A. W., Wilson, G. N., Kantor, J. A., and Nienhuis, A. W. (1975) Activation and inactivation of genes determining hemoglobin types. J. Cell. Physiol. 85, 477-494.

2. Anderson, W. F., Deisseroth, A., Nienhuis, A. W., Gopalakrishnan, T. V., Huang, A., and Krueger, L. (1978) Cellular and molecular studies on globin gene expression. Natl. Cancer Inst. Monogr. 48, 65-73.

3. Anderson, W. F., Killos, L., Sanders-Haigh, L., Kretschmer, P., Diacumakos, E., Nienhuis, A., Willing, M., and Vembu, D. (1980) Regulation of human globin gene expression after gene transfer. Univ. of Chicago Symposia on Sickle Cell Anemia, Vol. 1 (Molecular Basis of Mutant Hemoglobin Dysfunction), Sigler, P. B. (ed.), North-Holland Publishing Co., New York, in press.

4. Diacumakos, E. G. (1973) Methods for micromanipulation of human somatic cells in culture. Chapter, 15, Methods in Cell Biology, Vol. VII, Academic Press, New York and London, pp. 287-311.

5. Wigler, M., Silverstein, S., Lee, L.-S., Pellicer, A., Cheng, Y.-C., and Axel, R. (1977) Transfer of purified herpes virus thymidine kinase gene to cultured mouse cells. Cell 11, 223-232.

6. Pellicer, A., Wigler, M., Axel, R., and Silverstein, S. (1978) The transfer and stable integration of the HSV thymidine kinase gene into mouse cells. Cell 14, 133-141.

7. Bacchetti, S. and Graham, F. L. (1977) Transfer of the gene for thymidine kinase to thymidine kinase-deficient human cells by purified herpes simplex viral DNA. Proc. Natl. Acad. Sci. USA 74, 1590-1594.

8. Maitland, N. J. and McDougall, J. K. (1977) Biochemical transformation of mouse cells by fragments of herpes simplex virus DNA. Cell 11, 233-241.

9. Huttner, K. M., Scangos, G. A., and Ruddle, F. H. (1979) DNA-mediated gene transfer of a circular plasmid into murine cells. Proc. Natl. Acad. Sci. USA 76, 5820-5824.

10. Bolivar, F., Rodriguez, R. L., Greene, P. J., Betlach, M. C., Heyneker, H. L., and Boyer, H. W. (1977) Construction and characterization of new cloning vehicles. II. A multipurpose cloning system. Gene 2, 95-113.

11. Lawn, R. M., Fritsch, E. F., Parker, R. C., Blake, G., and Maniatis, T. (1978) The isolation and characterization of linked δ- and β-globin genes from a cloned library of human DNA. Cell 15, 1157-1174.

12. Kaufman, R. E., Kretschmer, P. J., Adams, J. W., Coon, H. C., Anderson, W. F., and Nienhuis, A. W. (1980) Cloning and characterization of DNA sequences surrounding the human γ, δ and β globin genes. Proc. Natl. Acad. Sci. USA, in press.

13. Enquist, L. W., Vande Woude, G. F., Wagner, M., Smiley, J. R., and Summers, W. C. (1979) Construction and characterization of a recombinant plasmid encoding the gene for the thymidine kinase of herpes simplex type 1 virus. Gene 7, 335-342.

14. Littlefield, J. W. (1964) Selection of hybrids from matings of fibroblasts in vitro and their presumed recombinants. Science 145, 709-710.

15. Summers, W. C., and Summers, W. P. (1977) [^{125}I]deoxycytidine used in a rapid, sensitive, and specific assay for herpes simplex virus type 1 thymidine kinase. J. Virol. 24, 314-318.

16. Southern, E. M. (1975) Detection of specific sequences among DNA fragments separated by gel electrophoresis. J. Mol. Biol. 98, 503-517.

17. Benz, E. J., Kretschmer, P. J., Geist, C. E., Kantor, J. A., Turner, P. A., and Nienhuis, A. W. (1979) Hemoglobin switching in sheep. Synthesis, cloning, and characterization of DNA sequences coding for the β^B-, β^C-, and γ-globin mRNAs. J. Biol. Chem. 254, 6880-6888.

18. Willing, M. C., Nienhuis, A. W., and Anderson, W. F. (1979) Selective activation of human β- but not γ-globin gene in human fibroblast x mouse erythroleukemia cell hybrids. Nature 277, 534-538.

19. Wilson, J. T., Wilson, L. B., deRiel, J. K., Villa-Komaroff, L., Efstratiadis, A., Forget, B. G., and Weissman, S. M. (1978) Nucleic Acid Res. 5, 563-581.

RETROVIRUS INTEGRATION

David Baltimore

Center for Cancer Research

and Department of Biology

Massachusetts Institute of Technology

Cambridge, Massachusetts 02139

We think of the mammalian genome as a relatively fixed entity but retroviruses provide us with a clear violation of that stability. When a retrovirus infects a cell, the virus is able to integrate its genetic information into that of a host cell. Furthermore the virus can carry from one cell to another genes that have been picked up from a cell by the virus and thus provides a potential mechanism for the movement of genes from one cell to another and from one organism to another.

The ability of retroviral DNA to integrate into cellular DNA is obviously central to the ability of the virus to move genes from one cell to another. We have recently been deriving molecular clones of circular forms of viral DNA found in infected cells (Shoemaker et al., 1980). In the process of analyzing the structure of these clones we have come across a class of clones that provide evidence for a highly active integration system localized at the ends of retroviral DNA. Detailed analysis of these clones has provided an explicit model for the integration of viral DNA into cellular DNA; this model is closely analogous to the process by which bacterial transposons and mu bacteria phage are thought to integrate into bacterial DNA (Bukhari et al., 1977; Shapiro, 1979).

The ends of the retroviral DNA are formed by the process of reverse transcription of the viral RNA (Gilboa et al., 1979b). Because the ends are so important we will first review how the ends come about and then describe the evidence for a specific integration mechanism.

This discussion will focus on the Moloney murine leukemia virus

(M-MuLV) and a derivative of this virus called the Abelson murine leukemia virus (A-MuLV).

Reverse transcription

The M-MuLV genome consists of a single molecule of RNA approximately 8300 bases in length. About 60 bases of this RNA are found in a direct repeat at either end of the RNA molecule (Coffin et al., 1978). The RNA has base paired to it a specific transfer RNA molecule that acts as the initiator for reverse transcription (the synthesis of a DNA copy of the viral RNA) (Taylor, 1977). This RNA is diagrammed as structure I in Figure 1. The rest of this Figure describes the various steps by which the RNA is transformed by reverse transcriptase into a DNA copy. The details of this mechanism have been described elsewhere (Gilboa et al., 1979b) and will only be very briefly described here.

The first step of reverse transcription is the synthesis of a DNA complementary in base sequence to the viral RNA (minus strand DNA). This DNA starts to be made on the tRNA initiator, is copied to the end of the RNA, and then synthesis starts again by the re-hybridization of the short piece of DNA to the other end of the viral genome (structures II and II). Once minus strand synthesis is in progress plus strand synthesis is initiated at a specific position on the minus strand (structure IV) (Mitra et al., 1979) and that DNA synthesis is continued by a jump to the other end of the minus strand. Ultimately, these jumping mechanisms lead to a DNA molecule that is longer at both ends than the original viral RNA molecule and has directly repeated sequences of approximately 600 bases at either end. These sequences

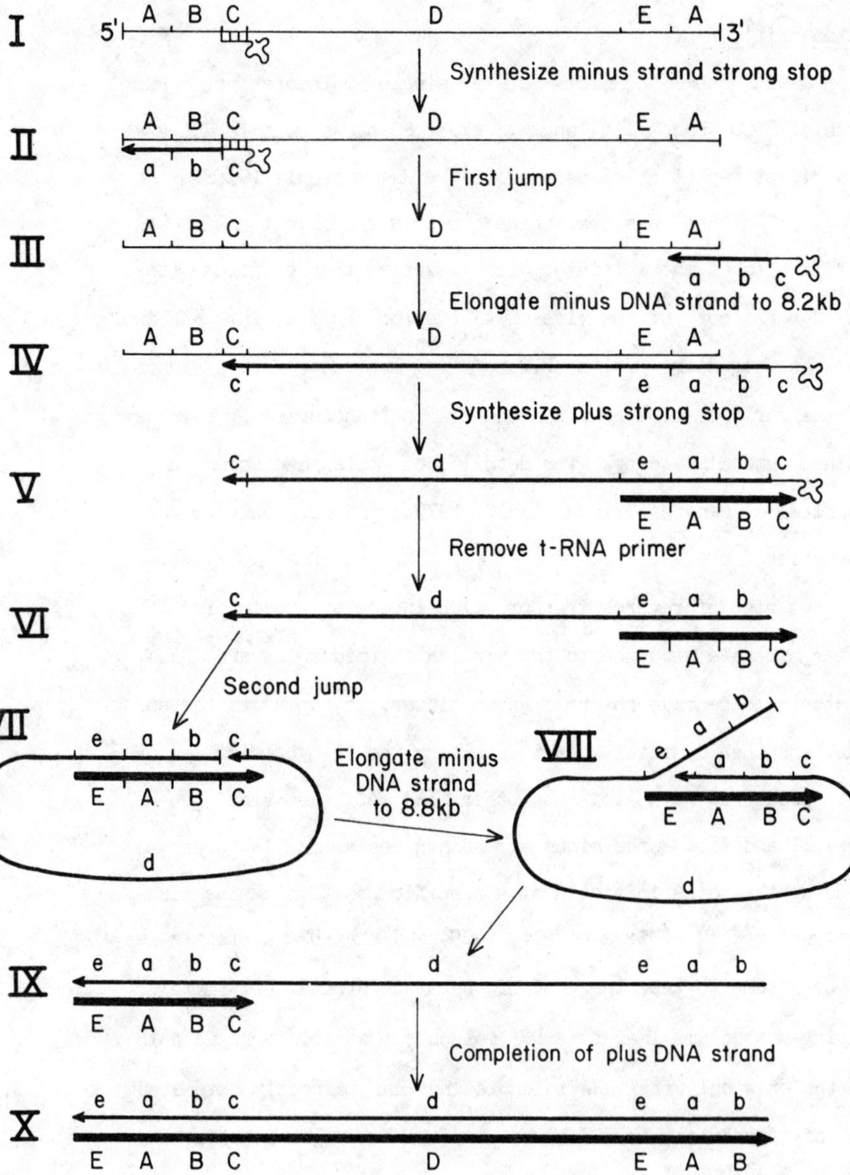
I
5'
A
B
C
D
E
A
3'
Synthesize minus strand strong stop
II
First jump
III
Elongate minus DNA strand to 8.2kb
IV
Synthesize plus strong stop
V
Remove t-RNA primer
VI
Second jump
VII
Elongate minus
DNA strand
to 8.8kb
VIII
IX
Completion of plus DNA strand
X

Fig. 1

are known as the long terminal repeat or LTR sequence. Formation of the LTR by joining sequences from both ends of the RNA is shown schematically in Figure 2.

Integration of viral DNA

When reverse transcription is complete the linear viral DNA that is formed can circularize in the cell nucleus to form two kinds of circles (Fig. 2). One kind of circle is 8.2 kb in length and has one copy of the LTR; the other circle is 8.8 kb in length and has two copies of the LTR. It is not known whether the linear or one another of the circular forms is the directly integrating form of viral DNA but the integrated proviral DNA is known to have LTR's at both ends (Fig. 2) (Sabran et al., 1979).

There is a single Hind III restriction endonuclease site in viral DNA (Gilboa et al., 1979a). We have derived molecular clones of the population of circles in infected cells by cutting the circles at the single Hind III site and cloning them into the lambda bacteriophage Charon 21A. As expected, 8.2 and 8.8 kb viral DNA inserts were found in the lambda bacteriophage population but we found two other types of DNA that we had not expected. One class of molecules had deletions and the other had an inversion of an area of DNA sequence.

We have characterized the inversion clone by restriction enzyme analysis as well as sequencing of specific regions of the DNA (Shoemaker et al., 1980). We found that the structure was consistent with an origin of the inversion clone by a form of integration in which a circular DNA integrates into itself (Shapiro, 1979). Rather than describing that process in detail here, its implication for viral

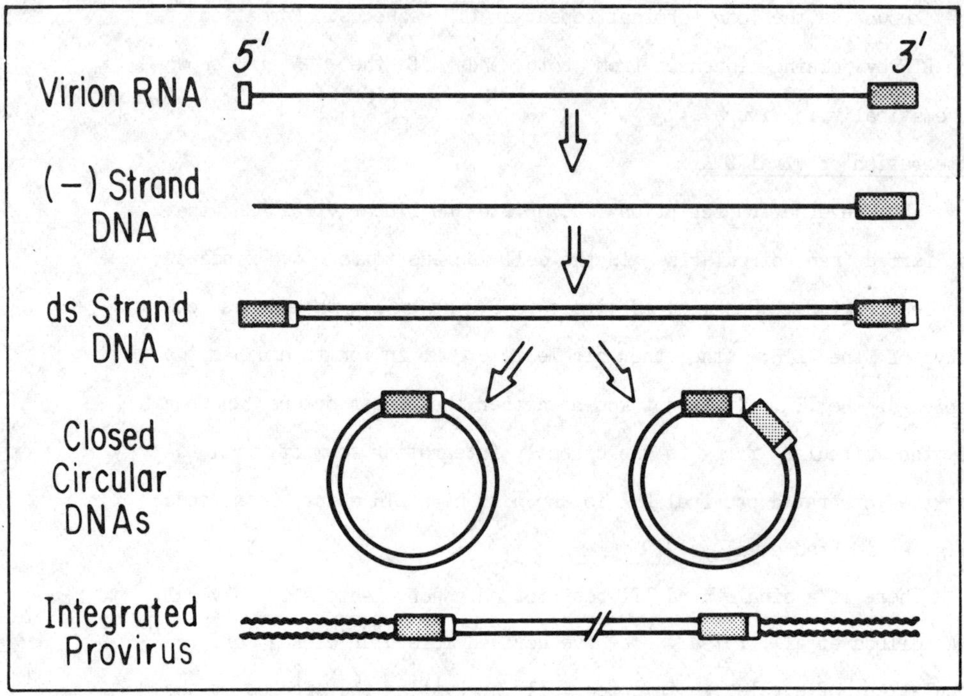
5′
3′
Virion RNA
(−) Strand DNA
ds Strand DNA
Closed Circular DNAs
Integrated Provirus

Fig. 2

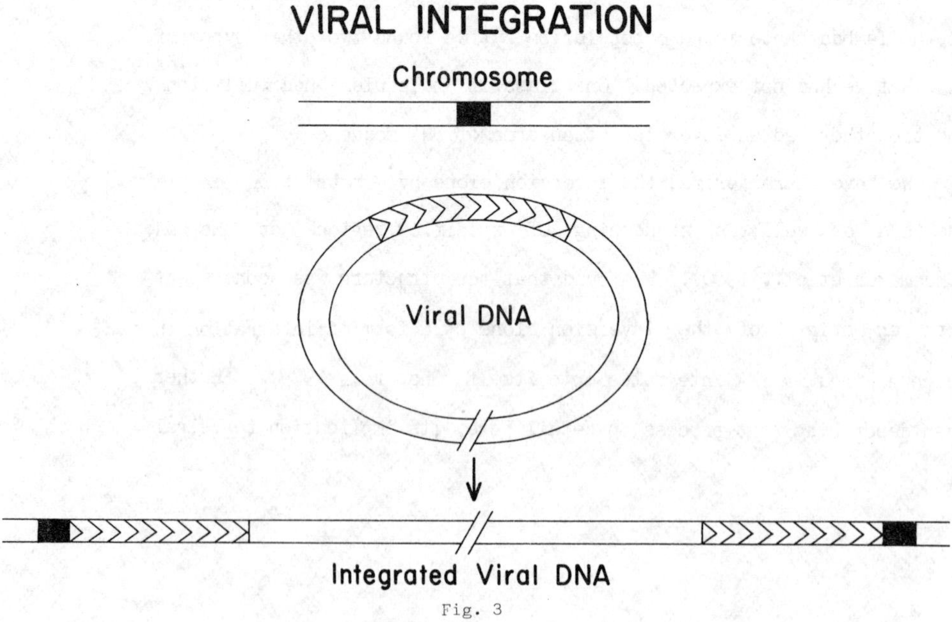
VIRAL INTEGRATION
Chromosome
Viral DNA
Integrated Viral DNA

Fig. 3

integration into a host cell chromosome is shown in Figure 3. In that Figure, the black box in the host cell DNA represents a sequence of four base pairs and the viral DNA is shown as a circular molecule with a single copy of the LTR. If one imagines that an enzyme were able to make single strand nicks on either side of the four bases in the host cell DNA and on either end of the LTR, then joining of those nicked DNAs would lead to the integrated structure shown at the bottom of the Figure. This, it will be recognized, is exactly the structure of the known integrated form of viral DNA with the very important proviso that four bases of host cell DNA will be repeated at either end of the integrated DNA form. Such a form has been found by Dhar et al. (1980) who have analyzed the integration of Moloney sarcoma virus into its host cell DNA.

A very important aspect of this mechanism of viral integration is that the ends of the LTR sequences must be highly active as sites of DNA joining reactions. In fact, the ends of the LTR sequences appear to be inverted repeats of one another (Shoemaker et al., 1980) implying that enzymes which recognize one end of the LTR will also recognize the other end of the LTR. In this sense the LTRs are very similar to IS sequences known to exist at either end of bacterial transposons and known to be intimately involved in the transposition of DNA elements from one place on the bacterial chromosome to another (Starlinger and Saedler, 1976). It seems likely that the LTR sequences have a similar type of role in mammalian cells and may represent the first molecular representation of the type of transposible DNA elements first recognized by McClintock (1965) in her studies on transposible elements in maize.

Transcription of proviral DNA

Although detailed studies of viral transcription are still in progress, it is evident from the sequence of viral RNA and from the structure of proviral DNA that a strong promoter must exist in the viral DNA. Because viral RNA is capped at its 5'-end we can expect that the position at which viral RNA synthesis is initiated is the cap proximal sequence. We have found this sequence in the LTR region and have found 20 bases upstream from this sequence an A·T-rich sequence that looks similar to the "Hogness box" (Gannon et al., 1979) that is found upstream from most RNA polymerase initiation sites in eukaryotic systems (Fig. 4). We interpret this to mean that the proviral DNA is a self-contained integration and transcription system designed to put a series of genes into the cell DNA and then to express those genes at a high level.

The ability of viral DNA to become integrated into the host cell DNA and express itself in RNA suggests that if the viral genes were replaced by other genes the virus might be able to carry those genes into cells and cause their expression.

The Abelson murine leukemia virus (A-MuLV)

A-MuLV is a virus that transforms bone marrow cells into continuous lines of leukemic pre-B-lymphocytes (Rosenberg and Baltimore, 1976; Siden et al., 1979). As opposed to M-MuLV, A-MuLV contains a genetic region that allows it to be a directly transforming virus acting with great rapidity to cause disease in infected animals (Rosenberg and Baltimore, 1980). The A-MuLV genome represents an

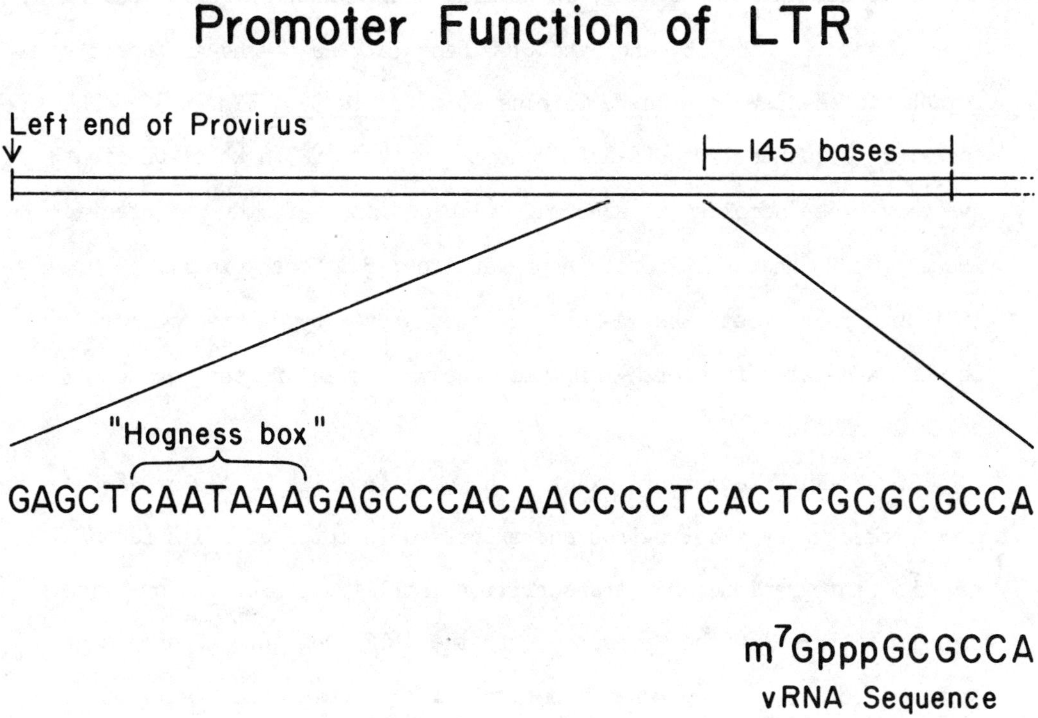
Promoter Function of LTR
Left end of Provirus
145 bases
"Hogness box"
GAGCTCAATAAAGAGCCCACAACCCCTCACTCGCGCGCCA
m7GpppGCGCCA
vRNA Sequence

Fig. 4

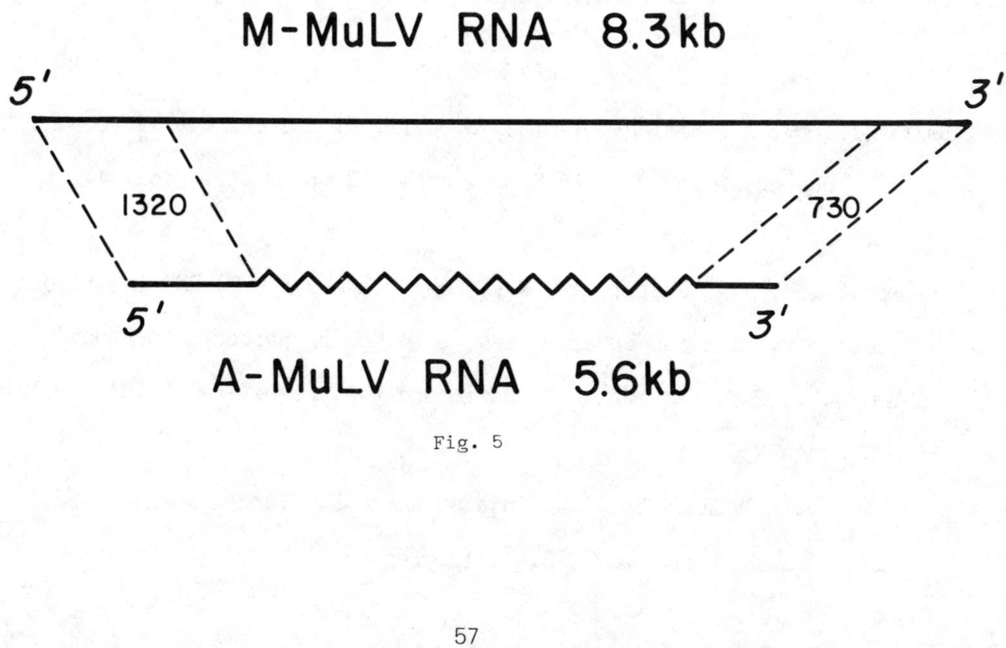
M-MuLV RNA 8.3kb
5'
3'
1320
730
5'
3'
A-MuLV RNA 5.6kb

Fig. 5

insertion and deletion mutant of M-MuLV. The central 6.3 kb of the M-MuLV genome has been deleted in forming the A-MuLV genome leaving 1300 bases at the 5'-end and 700 bases at the 3'-end as the only remnants of M-MuLV in A-MuLV (Shields et al., 1979) (Fig. 5). The inserted sequence in the A-MuLV genome consists of 3.6 kb of DNA that has very close homology to a normal cellular gene of the uninfected mouse (Goff and Baltimore, in preparation; Baltimore et al., 1979). It thus appears that the A-MuLV genome arose by a recombination between a deleted form of M-MuLV and a normal cellular gene (or normal cellular mRNA).

Thus A-MuLV represents exactly the case in which cellular genes are picked up by a retrovirus and expressed in infected cells because of the integration and transcription ability conferred on the sequences through the presence of the LTR region. There is no apparent reason why any genes might not find themselves part of a retrovirus in this fashion and thus retrovirus-catalyzed gene transfer between cells and between organisms may have an important role to play in evolution and/or differentiation.

REFERENCES

Baltimore, D., A. Shields, G. Otto, S. Goff, P. Besmer, O. Witte and N. Rosenberg (1979) Cold Spring Harbor Symp. Quant. Biol. 44, in press.

Bukhari, A.I., J.A. Shapiro and S.L. Adhya (eds.) (1977) DNA Insertion Elements, Plasmids and Episomes, Cold Spring Harbor Laboratory.

Coffin, J.M., T.C. Hageman, A.M. Maxam and W.A. Haseltine (1978) Cell 13, 761-773.

Dhar, R., W.L. McClements, L.W. Enquist and G.F. Vande Woude (1980) Proc. Nat. Acad. Sci. U.S.A., in press.

Gannon, F., K. O'Hare, R. Perrin, J.P. Le Pennec, C. Benoit, M. Cochet, R. Breathnach, A. Royal, A. Garapin, B. Carrie and P. Chambon (1979) Nature 278, 428-434.

Gilboa, E., S. Goff, A. Shields, F. Yoshimura, S. Mitra and D. Baltimore (1979a) Cell 16, 863-874.

Gilboa, E., S.W. Mitra, S. Goff and D. Baltimore (1979b) Cell 18, 93-100.

McClintock, B. (1965) In Brookhaven Symposia in Biology, no. 18, pp. 162-184.

Mitra, S.W., S. Goff and E. Gilboa and D. Baltimore (1979) Proc. Nat. Acad. Sci. U.S.A. 76, 4355-4359.

Rosenberg, N. and D. Baltimore (1976) J. Exp. Med. 143, 1953-1963.

Rosenberg, N. and D. Baltimore (1980) In Viral Oncology, G. Klein (ed.), Raven Press, New York, in press.

Sabran, T.L., T.E. Hsu, C. Yeater, A. Kaji, W.S. Mason and J.M. Taylor (1979) J. Virol. 29, 170-178.

Shapiro, J.A. (1979) Proc. Nat. Acad. Sci. U.S.A. 76, 1933-1937.

Shields, A., S. Goff, M. Paskind, G. Otto and D. Baltimore (1979) Cell 18, 955-962.

Shoemaker, C., S. Goff, E. Gilboa, M. Paskind, S.W. Mitra and D. Baltimore (1980) Proc. Nat. Acad. Sci. U.S.A., in press.

Siden, E., D. Baltimore, D. Clark and N. Rosenberg (1979) Cell 16, 389-396.

Starlinger, P. and H. Saedler (1976) In Current Topics in Microbiology and Immunology, vol. 75, 111-152.

Taylor, J.M. (1977) Biochim. Biophys. Acta 473, 57

RNA-DEPENDENT DNA POLYMERASE RELEASED BY NORMAL AVIAN CELLS*

Georg Bauer

Institut für Virologie, Zentrum für Hygiene, Universität Freiburg, Hermann-Herder-Str. 11, 78 Freiburg, West Germany

Abstract We have isolated particle - associated RNA-dependent DNA polymerase from the allantoic fluid of embryonated virus free chicken eggs and the supernatant of primary embryonic chicken fibroblast cultures, as well as from the supernatant of a goose embryo fibroblast culture.

The particles containing the DNA polymerase were not infectious for avian cells, did not show endogenous DNA synthesis and had a protein pattern different from those of known retroviruses.

The purified DNA polymerases were shown to be true reverse transcriptases by the criterium that they could synthesize DNA complementary to the heteropolymeric part of globin mRNA. Immunological studies, including radioimmunoassays, and enzymological studies of the DNA polymerases, showed that they were different from the DNA polymerases of known avian retroviruses.

The occurence of RNA-dependent DNA polymerases in systems which are free of partial or complete expression of endogenous retroviruses and which are not infected by exogenous retroviruses indicates that they might play a role in the normal physiology of cells.

---

*The experiments presented in this paper have been performed at the Max-Planck-Institut für Biochemie, Martinsried bei München, West Germany and the McArdle Laboratory for Cancer Research, Madison, Wisconsin, U. S. A. I am grateful to Dr. P. H. Hofschneider and Dr. H. M. Temin for having the chance to work in their laboratories and for their support. I would like to thank my collegues from both institutions for gifts of materials and for discussions. The financial support by the Max-Planck Gesellschaft and the Deutsche Forschungsgemeinschaft is acknowlegded.

RNA-dependent DNA polymerase has been shown to be a constituent of retrovirus virions. (Baltimore, D. 1970. Nature 226, 1209-1211; Temin H. M. and Mizutani, S. 1970. Nature 226, 1211 - 1213). The function of this DNA polymerase and its role for the replication of the retrovirus genome is clearly established. The enzyme catalyzes the synthesis of a (finally) double - stranded DNA copy of the viral RNA genome, which subsequently may become integrated into the genome of the host cell.

A role for RNA - dependent DNA synthesis in the physiology of normal cells could not be demonstrated so far. However, it was speculated that reverse transcription of cellular mRNA together with subsequent integration of the DNA product into the genome might lead to amplification, translocation and mutation of cellular genes (Temin, H. M. 1971. J. Natl. Canc. Inst. 46, III-VII).

In this paper I will review our attempts to isolate and characterize RNA - dependent DNA polymerases from normal avian cells.

We have been able to demonstrate RNA - dependent DNA polymerases in two different avian species: chick and goose. The allantoic fluid of embryonated, virus - free chicken eggs, as well as the supernatant of primary cultures of embryonic chicken cells regularily contains particle - associated RNA - dependent DNA polymerase. The enzyme activity was found in the allantoic fluid of chicken eggs from chickens which did not release the endogenous retrovirus. The appearance of the DNA polymerase is independent of the activity of endogenous virus genes, i.e. there is release of particle - associated RNA - dependent DNA polymerase by chicken cells which may be positive or be negative for chick helper factor. All flocks of chicken tested so far have been positive for the presence of RNA - dependent DNA polymerase. Within one flock, each one of a series of eggs tested individually has been shown to be positive.

In the case of goose cells, one embryo (out of eight) was releasing RNA - dependent DNA polymerase. Again, like in the chicken system, the enzyme was inside particles which were released into the medium. Unlike chicken cells, goose cells continued to release enzyme after being passaged. For convenience, in this paper the particles carrying DNA polymerases will be called

"Chicken particles" and "goose particles", respectively.

The particles were sedimentable by ultracentrifugation and had a defined density in sucrose density gradients. Chicken particles banded at a density of 1.15 g/cm^3, goose particles at 1.18 g/cm^3. The lack of detectable enzyme activity in the absence of NP 40 indicated that the DNA polymerase was located inside the particles.

The particles were not infectious for avian cells and lacked endogenous DNA synthesis (i.e. DNA synthesis in the absence of added template-primer complexes). Nucleic acids could not be demonstrated to be present in the particles. The protein patterns of purified particles were different from those of known avian retroviruses.

Properties of the DNA polymerases from chicken particles and goose particles

Both DNA polymerase activities were shown to be strictly dependent on the concomitant presence of template and primer. The DNA polymerases were characterized as RNA-dependent DNA polymerases (reverse transcriptases) by their ability to utilize various homopolymeric and heteropolymeric RNA templates. Product analysis of the reaction product of a globin mRNA-directed DNA synthesis by chicken particle polymerase showed that a DNA molecule complementary to the heteropolymeric RNA had been synthesized. The primary product was a DNA-RNA hybrid, which was insensitive to single-strand specific nuclease S1. After alkali-treatment, single-stranded DNA remained. This DNA was hybridized to newly added globin mRNA to form a specific hybrid as judged by combined nuclease S1 treatment and Cs_2SO_4 density gradient analysis. The Tm of the hybrid was shown to be typical for a duplex of heteropolymeric nuclei acids. The specific utilization of heteropolymeric RNA as template for DNA synthesis characterized the particle polymerases as genuine RNA-depended DNA polymerases and thus distinguished them for DNA polymerases α, β and γ of avian cells.

As only small quantities of particle-associated DNA polymerase activity can be obtained, extensive purification of the enzymes

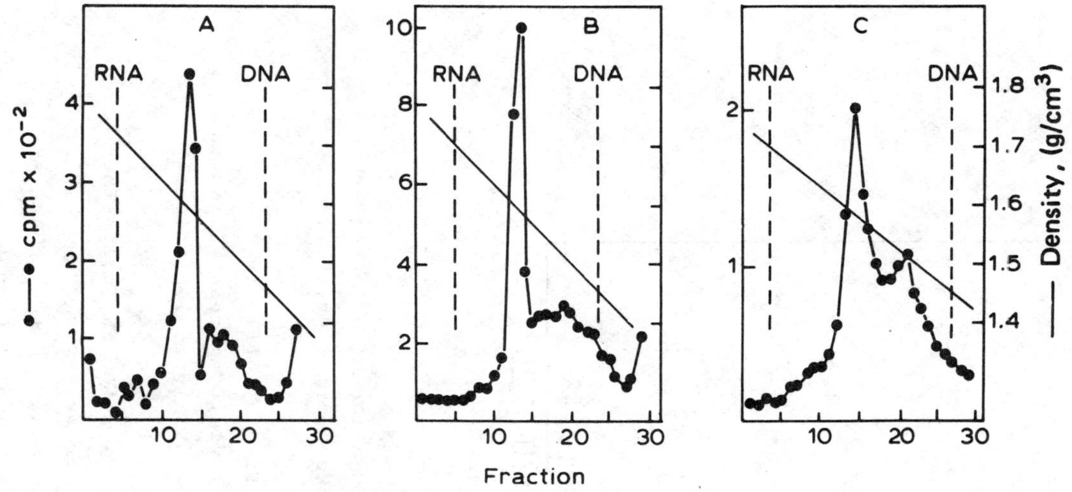

Figure 1 Cs_2SO_4 density gradient analysis of the reaction product of globin mRNA - directed DNA synthesis by chicken particle DNA polymerase ^{3}H-dGMP-labelled reaction product from globin mRNA-directed DNA synthesis (25 μg/ml globin mRNA, 10 μg/ml $(dT)_{12}$, 200 μg/ml purified, disrupted chicken particles, 50 μg/ml Actinomycin D, 0.6 mM dATP, dCTP, DTTP and 100 μCi/ml ^{3}H-dGTP (8 Ci/mMole)) was purified and analyzed on Cs_2SO_4 density gradients after different treatment: A) untreated; B) treated with nuclease S1; C) alkali-treated, backhybridized to newly added globin mRNA and subsequently treated with nuclease S1.

The T_m of the hybrid molecules shown in panel C was 86°C in standard saline citrate.

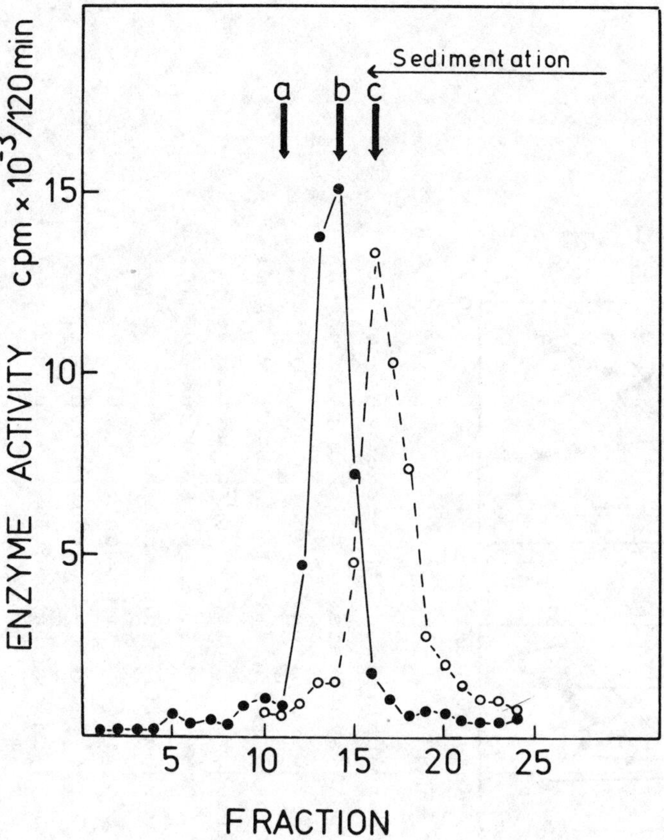

Figure 2 Glycerol gradient centrifugation of RNA dependent DNA polymerases from chicken particles, REV, and AMV.

DNA polymerases were purified by affinity chromatography on polycytidylate agarose. Aliquots were layered on top of 10-30% glycerol gradients (in the presence of 0.4 M KCl) and centrifuged for 17.5 hours at 45000 rpm in a rotor SW 50.1. The positions of the DNA polymerases were found by standard DNA polymerase assays with $(C)_n(dG)_{12}$ as template-primer. The positions of the protein markers in parallel gradients were determined by measuring the protein concentration.

(●) chicken particle DNA polymerase

(○) REV DNA polymerase

a: aldolase; b: AMV polymerase; c: bovine serum albumin.

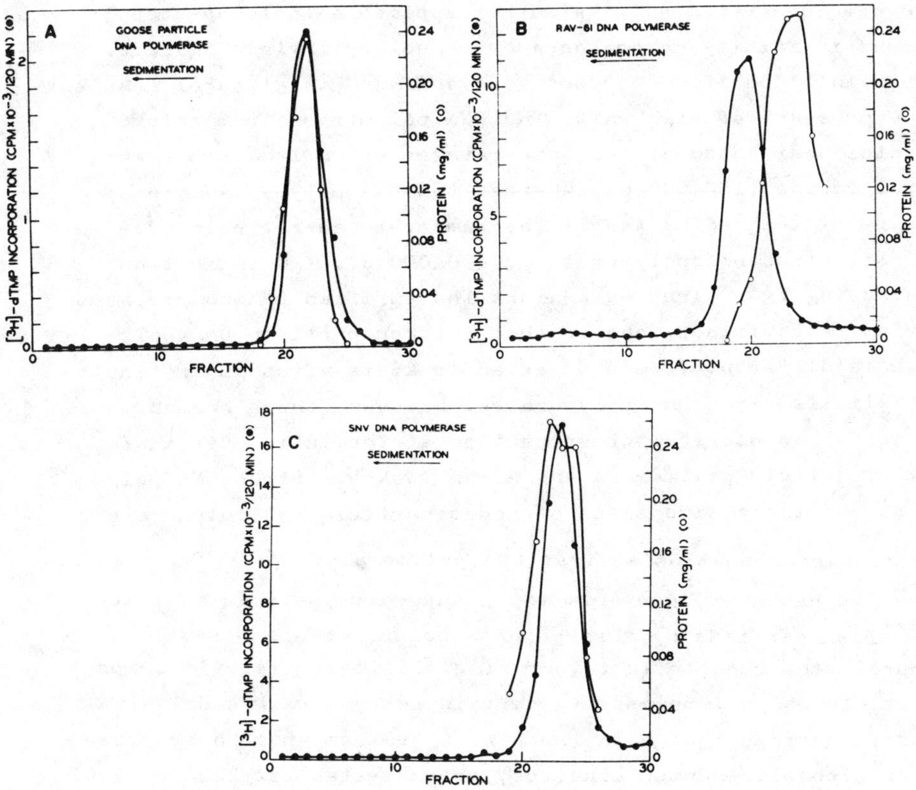

Figure 3 Glycerol gradient centrifugation of goose particle, RAV-61, and SNV DNA polymerases.

Goose particle DNA polymerase was partially purified by affinity chromatography on $(C)_n$-agarose. After purification, bovine serum albumin was added to a final concentration of 1 mg/ml, and a sample of 200μl was layered on a linear glycerol gradient (10 mM potassium phosphate pH 8.0, 10 to 30% glycerol, 0.5 M KCl, 1 mM dithiothreitol). A 200 μl amount of RAV-61 DNA polymerase and 200 μl of SNV DNA polymerase (each sample also containing 200 μg of bovine serum albumin) were layered on top of parallel gradients. Centrifugation was for 16.5 hours at 45,000 rpm at 4°C in an SW 50.1 rotor. Fractions were collected from the bottom. A 10 μl amount of each fraction was tested in a standard DNA polymerase assay for 120 minutes. A background of 200 cpm was subtracted from all values. The position of the marker bovine serum albumin was determined by the method of Lowry.

could not be carried out. We rather applied a fast one-step procedure: affinity chromatography on polycytidylate agarose. The partially purified enzymes were analyzed in glycerol gradients in the presence of high salt. The DNA polymerase from chicken particles sedimented at the same rate as avian leukosis virus DNA polymerase (150 000 d), whereas the DNA polymerase from goose particles sedimented at the same rate as reticuloendotheliosis virus DNA polymerase (70-80 000 d). IgG inhibition tests, using IgGs directed against the purified DNA polymerases of avian retroviruses, showed that chicken particle DNA polymerase is different from both avian leukosis virus and reticuloendotheliosis virus DNA polymerases. However, there seemed to be a weak immunological crossreaction at the active sites of chicken particle polymerase and avian leukosis virus DNA polymerase, and the active sites of goose particle DNA polymerase and reticuloendotheliosis virus DNA polymerase.

As we have not been able to demonstrate the specificities of these crossreactions by absorption studies, we are hesitating to interprete the results in favour of evolutionary relationships of particle DNA polymerases to certain retrovirus DNA polymerases. Enzymological data (such as K_M values and characteristics of Michaelis-Menten kinetics, thermal stability and optimal ion concentrations) gave additional support for the idea that the DNA polymerases from chicken particles and from avian leukosis viruses are different enzymes.

Radioimmunological studies

To clarify the relationship of particle DNA polymerases to the DNA polymerases of avian retroviruses and to study the relationship between the three avian retrovirus species, radioimmunoassays were developed. Purified DNA polymerases of spleen necrosis virus(a member of the REV species) and of avian myeloblastosis virus (a member of the avian leukosis virus species) were used in immunoprecipitation studies. ^{125}I - AMV DNA polymerase could be precipitated by antibody directed against DNA polymerases of avian leukosis viruses and pheasant virus, but not by antibody against reticuloendotheliosis virus or mammalian type C viruses. ^{125}I - SNV DNA polymerase was precipitated by antibody

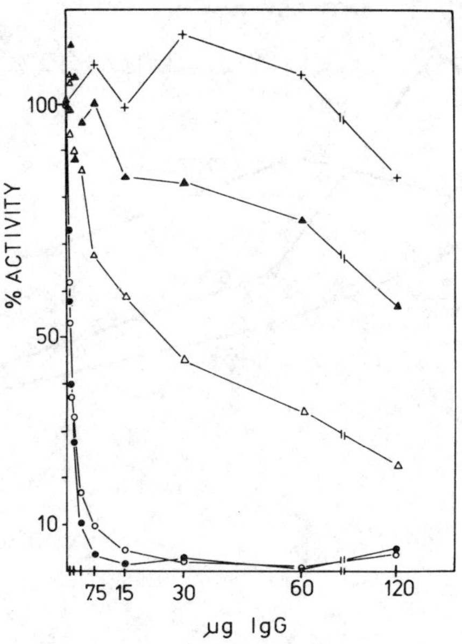

Figure 4 Comparative inhibition of RNA dependent DNA polymerases from chicken particles and AMV by IgG directed against AMV DNA polymerase. Two different IgG preparations (generous gifts of the laboratories of Dr. Gallo and Dr. Spiegelman) directed against AMV DNA polymerase were serially diluted. Constant amounts of either chicken particle DNA polymerase or AMV polymerase were preincubated with constant amounts of each dilution for 14 hours at 4°C. Then the residual activity was determined in 1 hour standard reactions. 100% activity (no IgG present was about 15000 cpm of ^{3}H-dGMP incorporated into TCA precipitable material per hour and 100 μl reaction volume for both enzymes. Parallel tests ensured that under the conditions of the test the activities were proportional to the amount of enzyme and were linear with respect to time.

(○) Anti-AMV IgG 1+AMV DNA polymerase

(△) Anti-AMV IgG 1+chicken particle DNA polymerase

(●) Anti-AMV IgG 2+AMV DNA polymerase

(▲) Anti-AMV IgG +ckicken particle DNA polymerase

(+) normal IgG + chicken particle DNA polymerase.

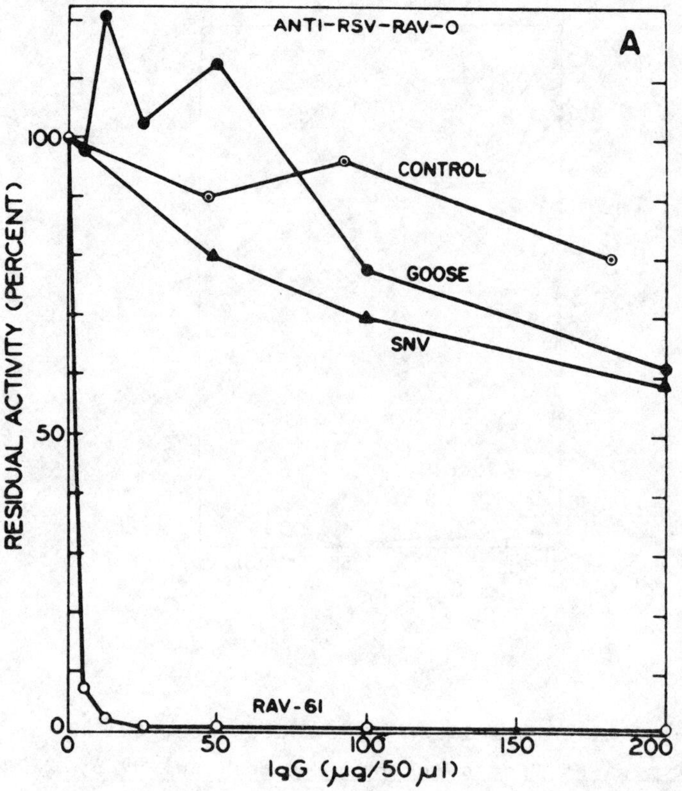
ANTI-RSV-RAV-O
A
CONTROL
GOOSE
SNV
RAV-61
RESIDUAL ACTIVITY (PERCENT)
100
50
0
0
50
100
150
200
IgG (μg/50 μl)

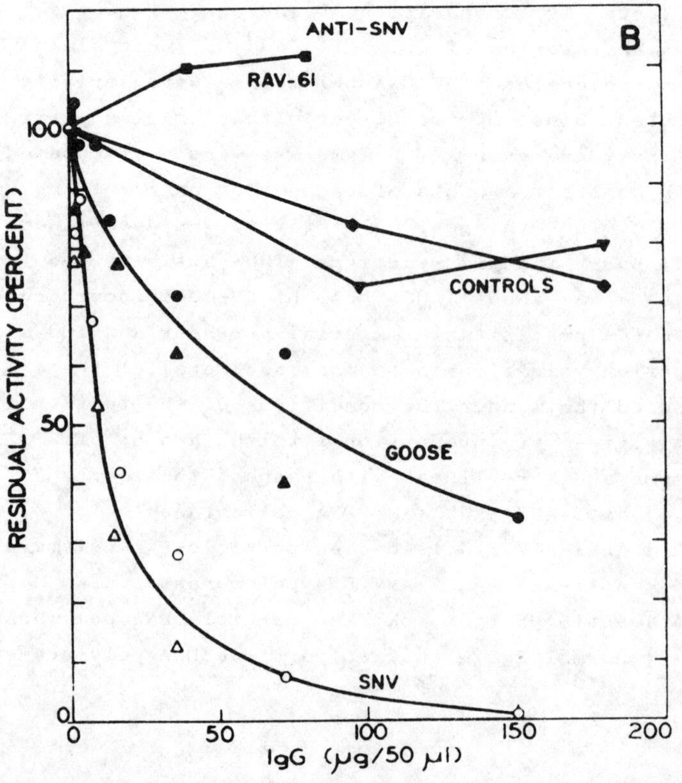
ANTI-SNV
B
RAV-61
CONTROLS
GOOSE
SNV
RESIDUAL ACTIVITY (PERCENT)
100
50
0
0
50
100
150
200
IgG (μg/50 μl)

Figure 5 Characterization of goose particle DNA polymerase by the IgG inhibition test.

A constant amount of either partially purified enzyme was preincubated with the indicated amounts of IgG for 30 min at room temperature and then the residual activity was determined. A background value of 400 cpm was subtracted from all values. 100% values (no IgG added) were obtained from duplicate experiments.

A:

(○) RAV-61 DNA polymerase/anti-RSV-RAV-C DNA polymerase IgG, 100% = 59,300 cpm incorporated /hour;

(●) goose particle DNA polymerase/anti-RSV-RAV-O DNA polymerase IgG; 100% = 10,000 cpm incorporated/hour;

(⊙) goose particle DNA polymerase/normal IgG, 100% = 10,000 cpm/hour;

(▲) SNV DNA polymerase/anti-RSV-RAV-O DNA polymerase IgG, 100% = 26,000 cpm/hour.

B: (○) SNV DNA polymerase/anti-SNV DNA polymerase IgG, experiment I, 100% - 26,000 cpm/80 min;

(△) SNV DNA polymerase/anti-SNV DNA polymerase IgG, Experiment II, 100% = 20,800 cpm/80 min;

(●) goose particle DNA polymerase/anti-SNV DNA polymerase IgG, Experiment I, 100% = 19,600 cpm/80 min,;

(▲) goose particle DNA polymerase/anti-SNV DNA polymerase IgG, Experiment II, 100% = 25,400 cpm/80 min;

(▼) SNV DNA polymerase/normal IgG, 100% = 25,000 cpm/80 min;

(◆) goose particle DNA polymerase/normal IgG, 100% = 12,600 cpm/80 min.

(■) RAV-61 DNA polymerase/anti SNV DNA polymerase IgG, 100% =26,000 cpm/80 min.

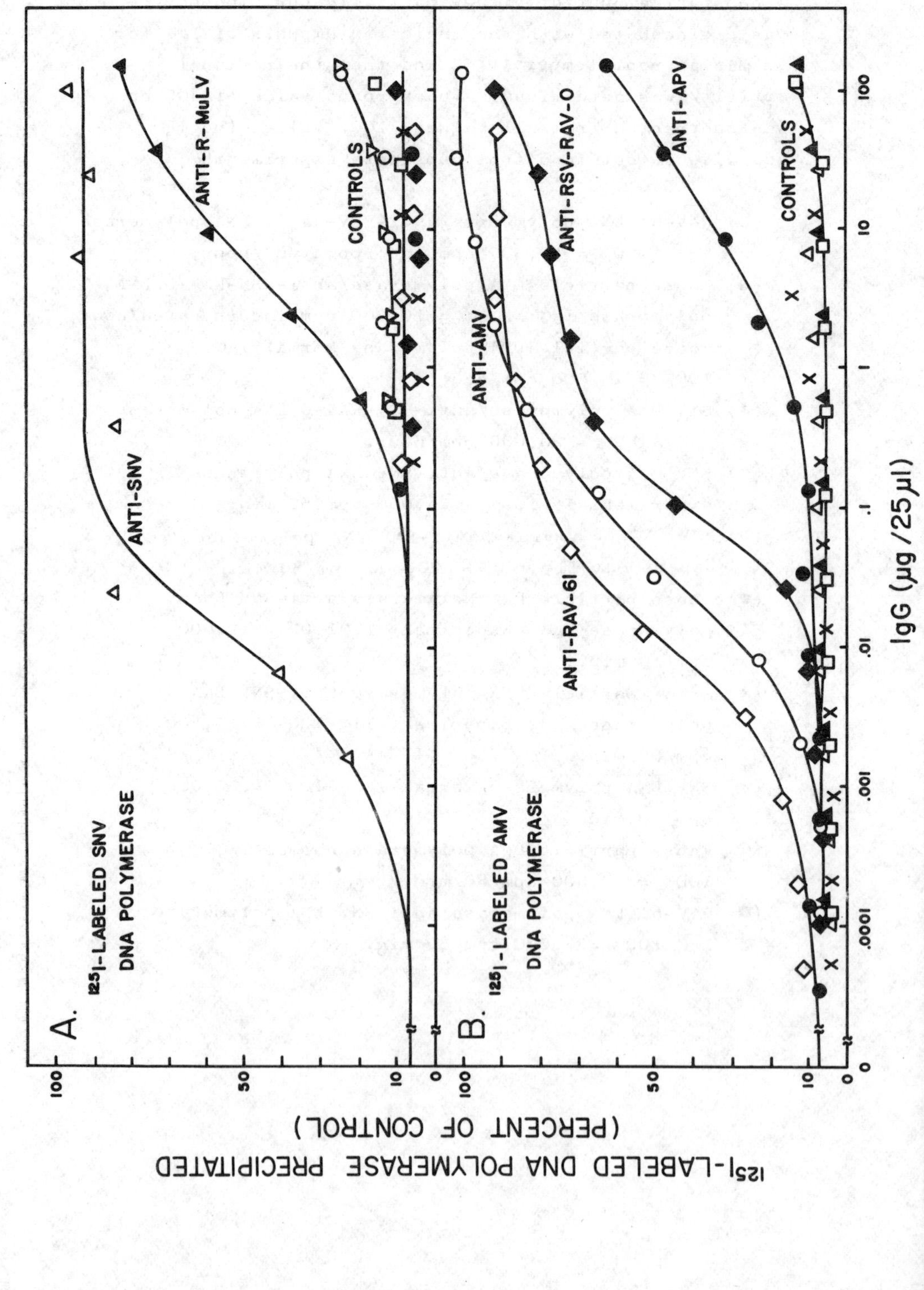
A.
125I-LABELED SNV
DNA POLYMERASE
ANTI-SNV
ANTI-R-MuLV
CONTROLS
B.
125I-LABELED AMV
DNA POLYMERASE
ANTI-RAV-61
ANTI-AMV
ANTI-RSV-RAV-O
ANTI-APV
CONTROLS
125I-LABELED DNA POLYMERASE PRECIPITATED
(PERCENT OF CONTROL)
100
50
10
0
.0001
.001
.01
.1
1
10
100
IgG (μg/25μl)

Figure 6 Radioimmunoprecipitation of SNV DNA polymerase and AMV DNA polymerase. Serial dilutions of purified IgGs were incubated with ^{125}I-labeled SNV DNA polymerase (panel A) or ^{125}I-labeled AMV DNA polymerase (panel B) in a volume of 25 μl for 18 h at 4°C in the presence of 150 mM KCl, 20 mM Tris-HCl, pH 8, 1 mg/ml BSA.

Normal serum and second antibody were added in amounts determined to be optimal for immunoprecipitation. In the case of assays containing rabbit IgG, the total amount of IgG present was brought to 50 μg. 150 μl of goat antibody to rabbit γ-globulin were added. Assays with higher IgG amounts received 300-400 μl of second antibody. In the case of goat IgG, assays were brought to 25 μg IgG, and 150 μl of rabbit antibody to goat γ-globulin were added. Assays with higher IgG amounts recieved 300 μl of second antibody. The assays were incubated for another 4-5 h at 4°C, and the immunoprecipitates were collected.

Rabbit IgG to SNV DNA polymerase (△)
Rabbit IgG to RAV-61 DNA polymerase (◇)
Rabbit IgG to RSV-RAV-0 DNA polymerase (◆)
Rabbit IgG to APV DNA polymerase (●)
Normal rabbit IgG (×)
goat IgG to R-MuLV DNA polymerae (▲)
goat IgG to BEV DNA polymerase (▽)
goat IgG to AMV DNA polymerase (○)
Normal goat IgG (□)

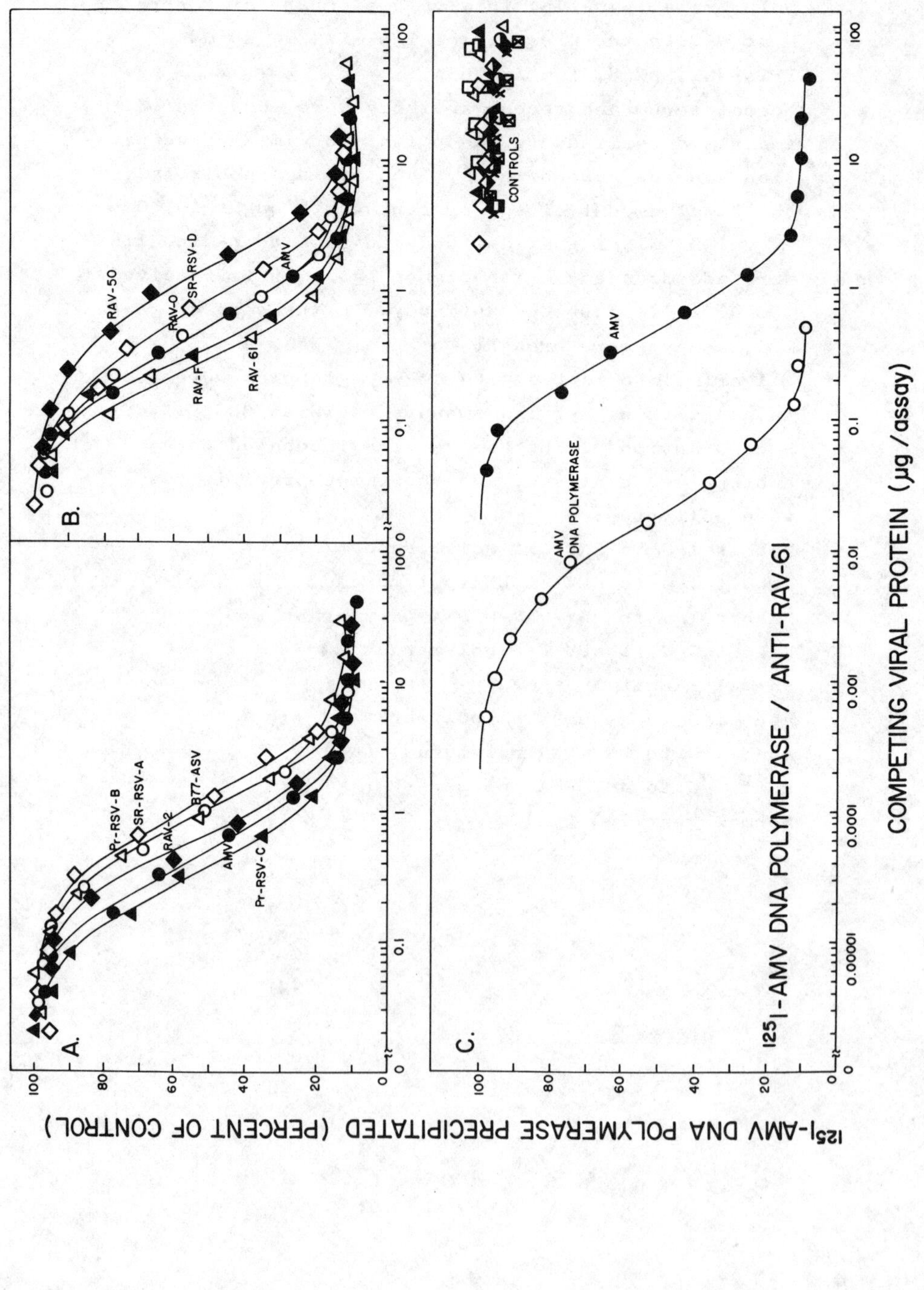

A.
B.
C.
Pr-RSV-B
SR-RSV-A
B77-ASV
RAV-2
AMV
Pr-RSV-C
RAV-50
RAV-O
SR-RSV-D
AMV
RAV-F
RAV-61
CONTROLS
AMV
AMV DNA POLYMERASE
125I-AMV DNA POLYMERASE / ANTI-RAV-61
125I-AMV DNA POLYMERASE PRECIPITATED (PERCENT OF CONTROL)
COMPETING VIRAL PROTEIN (μg / assay)

Figure 7 Species-specific antigenic determinants of the ALV species. Serial two-fold dilutions of competing material (in radioimmunoassay buffer) were preincubated with a 2.5 x 10^{-4} dilution of rabbit antiserum to RAV-61 DNA polymerase in 190 μl for 1 h at 37°C. The assays were cooled to 4°C, 20,000 cpm of ^{125}I-AMV DNA polymerase (in 20 μl radioimmunoassay buffer) were added, and incubation at 4°C was continued for 18 hours. After addition of 50 μl of a 1:10 dilution of normal rabbit serum and 150 μl goat antiserum to rabbit gamma globulin, incubation was continued for another 4-5 h at 4°C The immunoprecipitates were collected by low-speed centrifugation. Samples were counted for 5 min in a γ-counter.The 100% value (no competing material present) was calculated from 12 independent samples and was 48,500 counts per 5 minutes.

Symbols: ALV are indicated in the Figure.

controls: APV (⬓), SNV (△), DIAV (▲), CSV (◇), REV-T (◆), chicken particles with RNA-dependent DNA polymerase (✕), goose particles with RNA-dependent DNA polymerase (⊠), R-MuLV, BEV (⬓), MMTV (□), MPMV (⬘).

The data from one experiment are shown in three panels to allow clearer presentation.

In parallel experiments type-specific determinants of the ALV DNA polymerases were demonstrated by using labelled AMV DNA polymerase and antibody against AMV DNA polymerase. ALV DNA polymerases other than AMV DNA polymerase competed not completely and with a different slope compared to AMV DNA polymerase. None of the controls (including particle polymerases from chicken or goose) showed any competition in this assay.

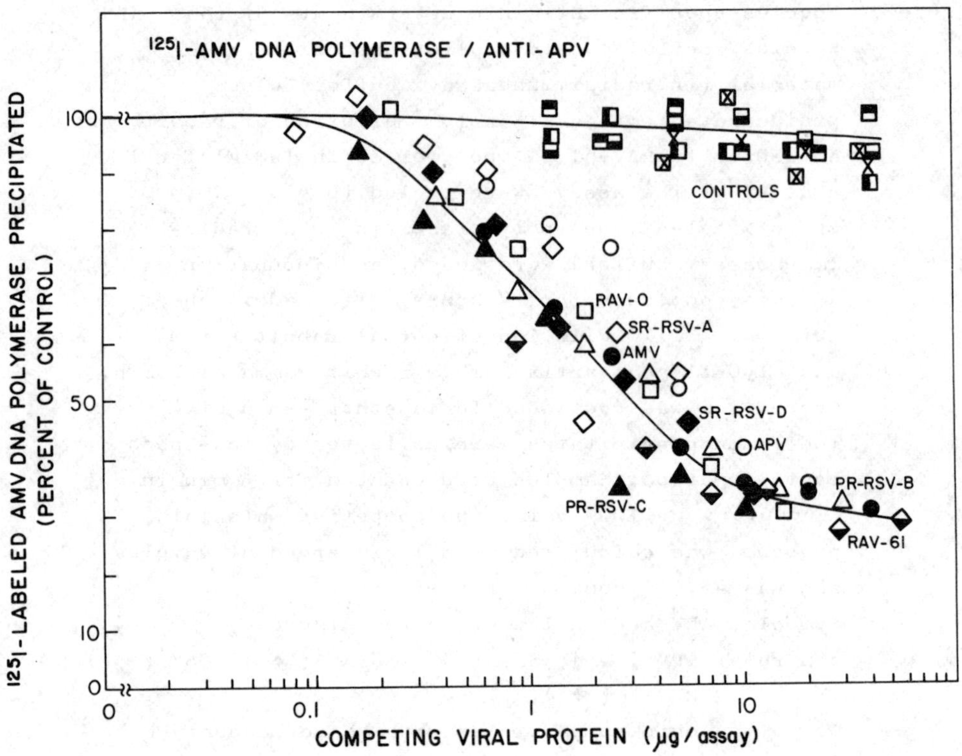

Figure 8 Demonstration of genus-specific antigenic determinants of avian retroviruses (ALV and PV species). The competition experiment was performed as described in Figure 3, except that rabbit antiserum to APV DNA polymerase (4×10^{-3} final dilution) was used. In this experiment 50 μl of a 1:50 dilution of normal rabbit serum and 150 μl goat antiserum to rabbit γ-globulin were used for immunoprecipitation. 100% precipitation was 30,000 counts per 5 min.

Symbols: ALV and APV as indicated in the Figure.

Controls: SNV (⬓)

DIAN (⬒)

REV (◧)

chicken particles with RNA-dependent DNA polymerase (×) goose particles with RNA-dependent DNA polymerase (⊠).

Not indicated in the figure are CSV, R-MuLV, BEV, MMTV and MPMV, that competed less than 5% at up to 40 μg viral protein per assay.

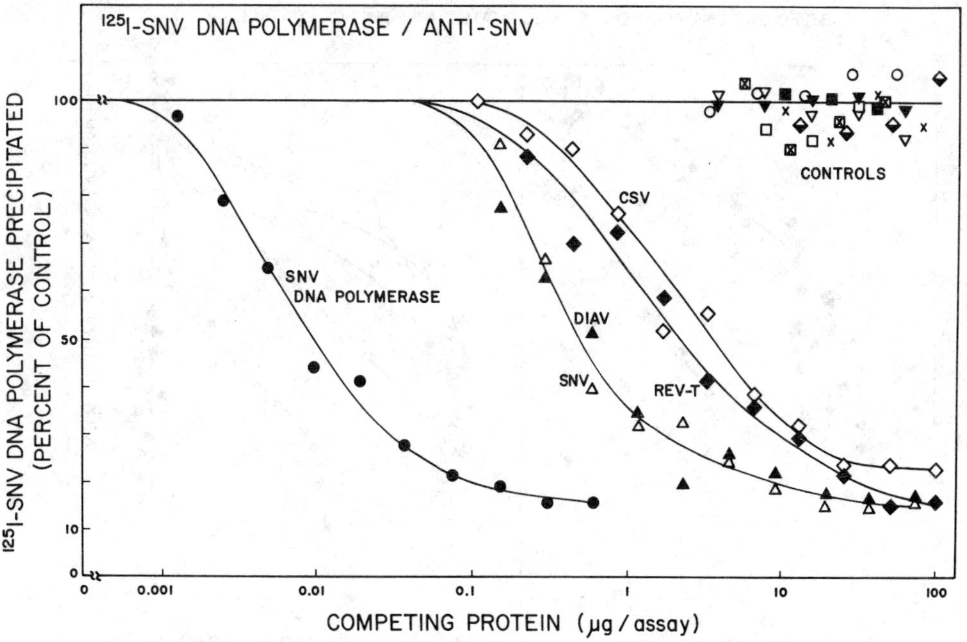

Figure 9 Homologous radioimmunoassay for SNV DNA polymerase. The experiment was performed as described in Figure 3, except that rabbit antiserum to SNV DNA polymerase (1×10^{-4} final dilution of antiserum obtained after 4th immunization of a rabbit with SNV DNA polymerase;) and ^{125}I-labeled SNV DNA polymerase (16,000 cpm per assay) were. used. The 100% value (no competing material present) was 38,000 counts per 5 min.

Symbols:

Reticuloendotheliosis viruses as indicated.

Controls: goose particles with RNA-dependent DNA polymerase (×)

AMV (▽)

SR-RSV-D (▼)

APV (◈)

MMTV (■)

R-MuLV (○)

chicken particles with RNA-dependent DNA polymerase (⊠).

Not indicated in the Figure are M-MuLV and Gross-MuLV, which competed less than 5% at up to 50 μg viral protein/assay.

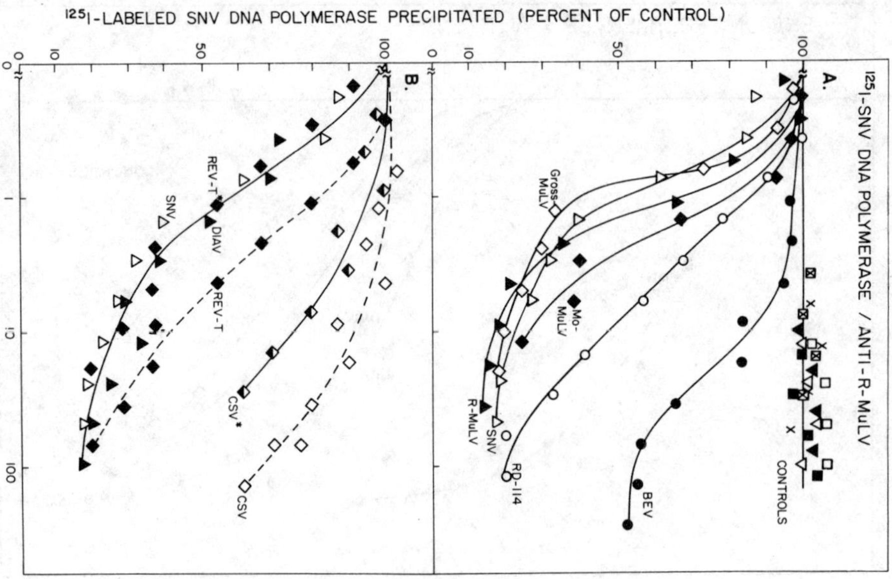

Figure 10 Demonstration of antigenic determinanats common to the DNA polymerase of avian reticuloendotheliosis viruses and mammalian type C retroviruses in a competition radioimmunoassay.

Serial two-fold dilutions of viruses or particles were incubated with 37.5 µg of goat antibody to R-MuLV DNA polymerase in a total volume of 40 µl in the presence of 190 mM KCl, 0.2% Triton X-100, 3 ng/ml BSA, 25 mM Tris-HCl, pH 7.8, for 1 h at 37°C. After cooling, 10 µl of rabbit antibody to goat γ-globulin and 150 µl radioimmunoassay buffer were added, and the immunoprecipitates were collected. after 4 h.

Panel A: competition by SNV, mammalian type C viruses and controls (AMV (▽), APV (▼), chicken particles with RNA-dependent DNA polymerase (⊠), goose particles with RNA-dependent DNA polymerase (×), MMTV (■), MPMV (□).

Panel B: competition by reticuloendotheliosis viruses Since the CSV and REV-T used in this study have been previously shown to contain less DNA polymerase per viral protein than SNV or DIAV (G. Bauer and H. M. Temin, manuscript submitted), curves normalized for an equivalent DNA polymerase content relative to SNV are included (CSV*, REV-T*).

against homologous enzyme and by antibody against R-MuLV DNA polymerase. These data show that ALV and PV DNA polymerases are related to each other but are unrelated to REV DNA polymerases. RNV DNA polymerase shares antigenic determinants with mammalian type C retrovirus DNA polymerases. Based on the results of the immunoprecipitation studies various sets of antigenic determinants were defined in specific radioimmunoassays: species- and type-specific antigenic determinants of ALV DNA polymerases, genus-specific antigenic determinants (those shared by ALV and PV DNA polymerases) species-specific antigenic determinants of REV DNA polymerases and the antigenic determinants shared by REV-DNA polymerases and mammalian type C retrovirus DNA polymerases. None of those determinants were shared by either chicken particle DNA polymerase or goose particle polymerase. These DNA polymerases therefore seem to be different from known retrovirus DNA polymerases. Their origin, in terms of evolution therefore remains unclear. The proof of a physiological function of these RNA - dependent DNA polymerases awaits the demonstration of their physiological template RNA.

References

Bauer, G., and P. H. Hofschneider. 1976. An RNA-dependent DNA polymerase, different from the known viral reverse transcriptases, in the chicken system. Proc. Natl. Acad. Sci. USA 73 , 3025 - 3029.

Bauer. G., G. Jilek, and P. H. Hofschneider. 1977. Purification and further characterization of an RNA-dependent DNA polymerase from the allantoic fluid of leukosis-virus-free chicken eggs. Eur. J. Biochem. 79, 345 -354.

Bauer, G., R. R. Friis, G. Jilek and P. H. Hofschneider. 1978. Purification and characterization of particles containing RNA-dependent DNA polymerase, in the allantoic fluid of uninfected leukosis virus-free chicken eggs. Biochimica et Biophysica Acta, 518, 125 - 137.

Bauer, G., R. R. Friis, H. Mattersberger, and P. H. Hofschneider. 1978. Controlled release of particle-associated RNA-dependent DNA polymerase by primary chick embryo cell cultures. Exp. Cell Res. 117, 383 - 392.

Bauer, G. and H. M. Temin. 1979. RNA-directed DNA polymerase from particles released by normal goose cells. J. Virol. 29, 1006 - 1013.

Bauer, G. and H. M. Temin. 1980. Radioimmunological comparison of the DNA polymerases of avian retroviruses. J. Vir. 33, 1046-1057.

Bauer, G. and H. M. Temin 1980. Specific antigenic relationships between the RNA-dependent DNA polymerases of avian reticuloendotheliosis viruses and mammalian type C retroviruses. J. Virol., April issue.

SHORT CHAIN RNA FRAGMENTS AS PROMOTERS OF LEUKOCYTE AND PLATELET GENESIS IN ANIMALS DEPLETED BY ANTI-CANCER DRUGS

Mirko Beljanski, Michel Plawecki, Pierre Bourgarel and Monique Beljanski
Laboratoire de Pharmacodynamie, Faculté de Pharmacie, 92290 Châtenay-Malabry, France

SUMMARY

We describe the biological activity of single-stranded short-chain RNA fragments obtained in large amounts by mild degradation of purified Escherichia coli ribosomal RNA with pancreatic RNase A . In vitro , these RNA fragments , termed $BLR_{(s)}$, can be used by DNA dependent DNA polymerase I as primers to initiate the replication of DNA isolated from rabbit bone marrow and spleen . They concentrate in these tissues when given i.v. or per os to rabbits and can restore normal leukocyte and platelet counts after these have been dangerously decreased by various antimitotic drugs. These actions are manifest within 24 h for leukocytes and 72 h for platelets . The RNA fragments are devoid of toxicity for animals and do not induce tolerance phenomenon . When polynuclear/lymphocyte ratio is upset by cyclophosphamide , the RNA fragments induce white blood cell differentiation and a differential increase of polynuclear and lymphocyte counts until their normal ratio and levels are fully restored . The RNA fragments do not act as primers for the in vitro replication of cancer DNA and do not stimulate tumor cell multiplication in mice . They appear to act on physiologically normal cells involved in leukopoiesis and platelet formation .

ABBREVIATIONS

RNA , ribonucleic acid ; DNA , deoxyribonucleic acid : Leuco-4 (adenine + phosphate) ; RNase , ribonuclease ; $BLR_{(s)}$, Beljanski leukocyte restorers ; DEDTC , diethyldithiocarbamate.

INTRODUCTION

It is well known that the primary step of cell division requires the replication of DNA by DNA dependent DNA polymerase which , in different biological systems , uses special naturally produced short chain RNA as primers to initiate a new DNA chain and thus make cell multiplication possible (1)(2)(3)(4)(5)(6)(7).

The transfer of genetic informations mediated by purified RNA in different biological systems has been reported by several authors (8) (9) (10)(11)(12)(13)(14). In higher organisms the development of specific tissue in response to exogenous RNA during organ culture has been described (15) (16) (17) (18). Various examples show that RNA plays a part in cell regulation . For instance , according to their nature and their affinity for a given DNA , RNA may , depending on the prevailing physiological conditions , force a cell to lose or to recover the ability to follow the normal regulation process . Thus , purine rich small size RNA (6 S) and RNA fragments (25 - 50 nucleotides) prepared by degradation of r- RNA with ribonuclease A may act on gene expression in plants (19) (5) and animals (20) (21) . When inoculated into young axenically cultured _Datura stramonium_ (19) or inverted stem sections (5) , these short-chain RNA initiate the appearance of transplantable tumors , but only if the solid culture medium contains _large amounts of auxine_ . Other RNA fragments , prepared using a different nuclease , can arrest tumor cell development in plants (22) . It has also been reported that special small molecular weight RNA can achieve specific tumor regression in animals (23) (24). Purine-rich RNA fragments can prevent DNA virus proliferation in rabbits (20).

In previous articles , we showed that specific short chain RNA fragments obtained by mild degradation of bacterial r-RNA with ribonuclease A can act as primers for *in vitro* replication of DNA. These primers are single stranded RNA fragments and their nucleotide compositions determine their specificity of action : some act as primers for the replication of viral DNA (20) and others for the replication of plant (5) , bacterial or mammalian DNA (7) (2) . Primer RNA are not transcribed into DNA but simply initiate the DNA replication process ; they are rapidly eliminated once the replication is in progress. We expected their priming action to involve an effect on gene regulation. It seemed possible to gain specific harmless control over the multiplication of traget cells by acting solely on the replication and expression of the genetic material without altering genes and cell structure. This is a very important point , because , in most cases , agents which inhibit cell multiplication have harmful side effects : for instance , antimitotic drugs used for cancer treatment have an inhibiting effect on leukocyte and blood platelet stem cells or their derivatives . We expected that , under appropriate conditions , those RNA fragments which selectively prime spleen and bone marrow DNA *in vitro* replication might be able to stimulate stem cell division (through DNA replication) , with the consequent restoration of normal amounts of circulating leukocytes and platelets in depleted animals.

In the present report , we describe the properties of these RNA fragments (named $BLR_{(s)}$) which exert a remarkable effect on leukocyte and blood platelet formation in laboratory animals.

BLR - RNA fragments may be obtained in large amounts by pancreatic ribonuclease digestion of purified *Escherichia coli* ribosomal RNA. These fragments have been fully characterized both physically and chemically and their *in vitro* and *in vivo* activities are described below.

MATERIALS and METHODS.=

Chemicals and biological products : Phenol (grade A) and chloroform (reagent grade) were obtained from Prolabo, Paris , France and sodium lauryl sulfate from Serlabo, Paris,France. Pancreatic RNase (4 x crystallized) and deoxyribonuclease (grade A) were supplied by Worthington , Freehold, N.J. USA ; poly AG , Poly I-Poly C and (^{3}H)-Poly A (s.act.94 µci/mole P , 3 S) by Miles Laboratories , Elkhart, Ind. USA ; Endoxan (cyclophosphamide) by Laboratories Lucien , Colombes, France; methotrexate (amethopterin) by Specia,Paris, France , and Leuco-4 (adenine + phosphate) by Leboratoire H.Villette,Paris France ; Daunorubicin (anthracyclin) was gift of Dr. R.Maral, Rhône-Poulenc, Ivry/Seine,France ; sodium-diethyldithiocarbamate and Endotoxin *E.coli* 026 B6 Difco were gifts of Dr. Roumiantzeff, Institut Mérieux,Lyon,France. (^{3}H)-uracil (44,5 ci/mmol) and (^{3}H)-guanine (7,5 ci/mmol) were obtained from the Commissariat à l'Energie Atomique, Saclay,France. Rabbits (3-4 kg) were white "Bouska", New Zealand. Mice were Swiss.

Bacteria and growth conditions. *Escherichia coli* T 3000 (K 12) a non pathogenic strain was used . Cells were grown aerobically in synthetic medium at 36°C (25) and harvested during the exponential phase of growth. When radioactive RNA were required cells were labelled for 6 h during exponential growth by addition of 500 µci (^{3}H)-guanine and 500 µci (^{3}H)-uracil per liter of synthetic medium .

Isolation of ribosomes and ribosomal RNA . Ribosomes were prepared from cells as previously described (26) and deproteinized by treatment with phenol and chloroform . Ribosomal RNA were dissolved in 0.01 M Na acetate, 0.1 M NaCl pH 5.0 and the solution was centrifuged for 30 min. at 40,000 g to remove most polysaccharide contaminants (27). The supernatant solution was dialysed against 0.1 M KCl for 18 h at 4°C and then against distilled water for 1 h at 4°. The yield of RNA was determined by measurement of A 260 nm and integrity of r-RNA was checked by polyacrylamide gel electrophoresis (28) (23 S and 16 S = 1.6 x 10^6 and 0.5 x 10^6 daltons)

Degradation of ribosomal RNA by pancreatic RNase A. Pancreatic RNase A (15 mg) was added to a solution of r-RNA (1 g) in distilled water(50 ml) and the mixture was incubated at 36°C for 20 min. Reaction was stopped by adding 1 volume of phenol containing 10% (v/v) distilled water, the mixture was stirred at room temperature for 10 min. and centrifuged at 12,000 g for 10 min. to separate phenolic and aqueous phases . The aqueous phase was removed and treated again as before with 1 volume of 10 % aqueous phenol . The second aqueous phase was then shaken for 10 min. with 1 volume of chloroform, and after centrifugation the aqueous phase was recovered . This step was repeated three times and the final aqueous phase was dialysed against 2.5 liter of distilled water for 16 h at 4°C. under axenic conditions. The amount of non dialysable RNA was determined by measurement of A 260 nm (yield 50-60 %); the dialysed solution was lyophilized and the residue was stored in a dry environment . The product which stimulates leukopoiesis and platelet formation are termed $BLR_{(s)}$ (see abbreviations). $BLR_{(s)}$ were characterized by determination of electrophoretic mobility in polyacrylamide gels and U.V. absorption spectra .

Analytical tehniques. Proteins were measured by Lowry method (29) and DNA by the diphenylamine reaction (30). RNA base ratios were determined as follows : 150 µg of $BLR_{(s)}$ were hydrolysed at 100°C in N HCl for 1 h (boiling water bath), the hydrolysate was evaporated to dryness in a dessicator and the residue was dissolved in 0.05 ml of distilled water. Samples of the solutions were analysed by chromatography on Ecteola cellulose plates according to Björk and Svensson (31). UV absorbing spots were located (purine bases and pyrimidine nucleotides) cut out and eluted with 0.1 N HCl . The concentrations of eluted compounds were determined using the following coefficients absorption maxima : A, 13 ; G , 12,8 ; C = 11 ; U, 10 (32).

Determination of leukocyte and platelet count in circulating blood of rabbits.

Blood taken from rabbit marginal veins was mixed immediately with EDTA and leukocyte and platelet counts were determined using a Coulter counter model F. Differential leukocyte counts were made using the May-Grünwald Giemsa staining method and 200 - 300 leukocytes were counted under the microscope by the same observer . Leukocytes were classified as polynuclears (including neutrophil, eosinophil and basophil cells) and lymphocytes (including monocytes).

Isolation of DNA and DNA polymerase. DNA was isolated from gently broken tissue by phenol/chloroform extraction in the presence of 2 M $NaClO_4$ and 0.01 M EDTA (33).

E.coli DNA dependent DNA polymerase I (EC 2.7.7.7.) was partially purified from bacterial extracts as previously described (34). This preparation contains some phosphatase activity for removing phosphate from 3'-ribonucleotides.

RESULTS

Biochemical properties of $BLR_{(s)}$. Chemical and physical characterization of

$BLR_{(s)}$ shows that they are single stranded RNA fragments (no hyperchromicity) almost devoid of DNA and protein contaminants(0.3 and 0.5 % respectively). Base ratio analysis shows that they contain an excess of purine over pyrimidine bases (G+A/C/U = 2.4 while in intact r-RNA this ratio is close to 1.12, Table 1). This is not surpirsing since RNase A cleaves RNA chains solely at C and U residues. The average electrophoretic mobility of $BLR_{(s)}$ is the same as that of poly A preparation which corresponds to m.w. of 1.7 x 10^4 daltons (Fig.1). This suggests that these RNA fragments contain on an average about 50 nucleotides which is within the size range of DNA replication primers . Their priming action on DNA replication in vitro was tested using bone marrow and spleen $DNA_{(s)}$. The choice of these $DNA_{(s)}$ was guided first by the observation that injected radioactive $BLR_{(s)}$ are found in relatively high amount in the bone marrow and spleen , and second by the need to ascertain wether $BLR_{(s)}$ act differently on DNA of normal and cancerous cells . When DNA polymerase I is incubated in a complete system containing a radioactive deoxyribonucleoside-5'-triphosphate but no added RNA primer it is able to synthesize a very limited amount of DNA in a reaction which quickly reaches a plateau. This synthesis is probably due to the fact that the enzyme used here is only partly purified (when highly purified it functions less well) and contains some material of primer RNA type . When $BLR_{(s)}$ are added to such a system using bone marrow (Fig. 2) or spleen DNA (results not shown) as template DNA replication is highly stimulated whereas when brain (Fig. 3) or kidney DNA (results not shown) is used as template no stimulation is seen . These results indicate that $BLR_{(s)}$ may be potential in vivo DNA replication primers in the bone marrow and spleen where are retained. We next asked whether $BLR_{(s)}$ stimulate DNA replication in the same way in healthy and cancerous tissue. When DNA from healthy breast and lung tissue was used as template in a complete system a slight stimulation of replication was observed. On the contrary when template DNA was prepared from cancerous tissue , replication was slightly slowed down by the presence of $BLR_{(s)}$ (Fig. 4). These results indicate, but do not constitute proof, that $BLR_{(s)}$ will not stimulate DNA replication in vivo and multiplication of malignant cells .

In vivo localization of radioactive $BLR_{(s)}$ in the rabbit. Preliminary experiments in mice had shown that , after i.v. injection , (^3H)-$BLR_{(s)}$ were mostly found in the spleen and in the bone marrow ; radioactivity slowly decreased and could not be detected for more than 2 to 3 weeks. When the distribution of radioactive $BLR_{(s)}$ in the rabbit was determined the results shown in Fig. 5 were obtained . The fact that high specific activity is

TABLE 1

Base ratio (G+A/C+U) of r-RNA and $BLR_{(s)}$ obtained by degradation of the same r RNA

Bases	r-RNA (23 S + 16 S)	Moles per 100 moles of analysed nucleotides $BLR_{(s)}$ obtained with pancreatic RNase A Sample 1	 Sample 2
G	28.0	42.0	40.5
A	24.9	28.8	29.8
C	24.1	15.7	16.5
U	23.0	13.5	14.2
G+A/C+U	1.12	2.42	2.36

$BLR_{(s)}$ and r-RNA were hydrolysed and analysed as described in Material and Methods.

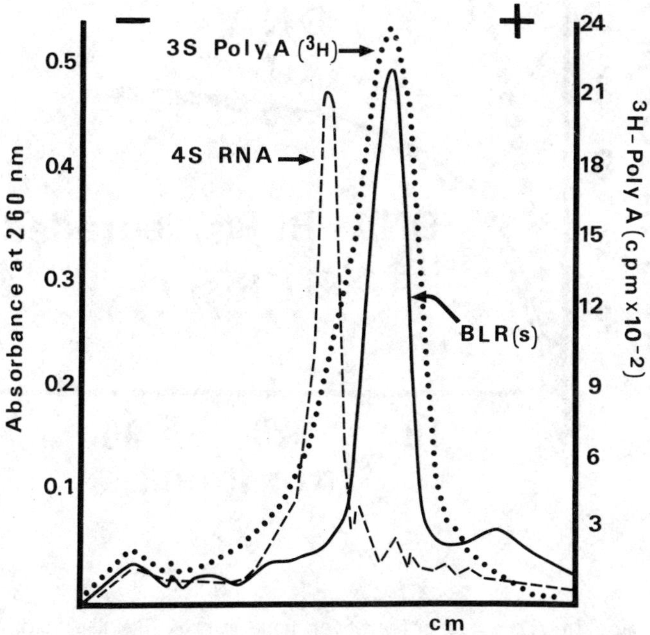

Fig.1 Polyacrylamide gel electrophoresis of $BLR_{(s)}$. $BLR_{(s)}$ (4 μg) were mixed with (^{3}H)-Poly A (3 S) (containing 14,000 CPM) used as a marker and submitted to electrophoresis on polyacrylamide gel (5 % (W/v) for 90 min. at 5 mA per tube at 4°C and the position of U.V. absorbing material for $BLR_{(s)}$ was detected as previously described (9). The position of (^{3}H)-Poly A was determined by measuring the radioactivity in very thin slices of polyacrylamide gel using a Packard liquid spectrometer (Prias). 4 S RNA was separately run as a marker and densitometer tracings are superposed on this Fig.

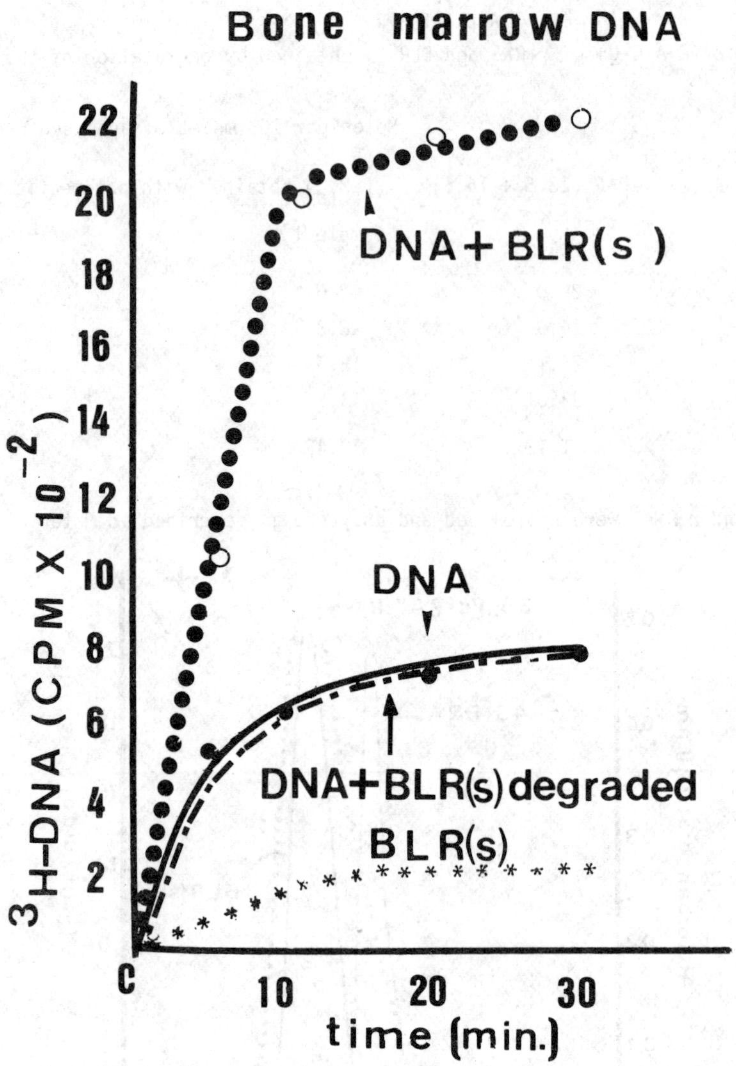

Fig.2 $BLR_{(s)}$ act in vitro as primers for bone marrow DNA replication. Incubation mixture contained per 0.15 ml : Tris-HCl buffer pH 7.65 , 25 μMoles; $MgCl_2$: 2 μMoles ; four d-XTP : each 5 nanomoles (+ (^3H)-TTP , 50,000 CPM) ; DNA : 0.2 μg ; $BLR_{(s)}$: 4 μg ; DNA dependent DNA polymerase I, 80 μg . Incubation : 10 , 20 and 30 min. at 36°C. TCA (trichloroacetic acid)-precipitable material was filtered on GF/C glass filter, washed, dried and radioactivity measured with a Packard liquid spectrometer (Prias). $BLR_{(s)}$ were degraded by incubation in 0.3 N KOH at 36°C for 16 h , then neutralized. Analysis were carried out in triplicate for each incubation time .

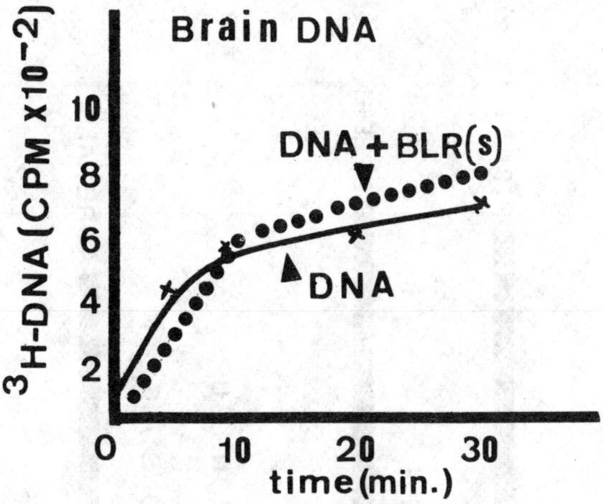

Fig.3 BLR$_{(s)}$ do not prime for monkey brain DNA replication ; see conditions described in the legend to Fig. 2

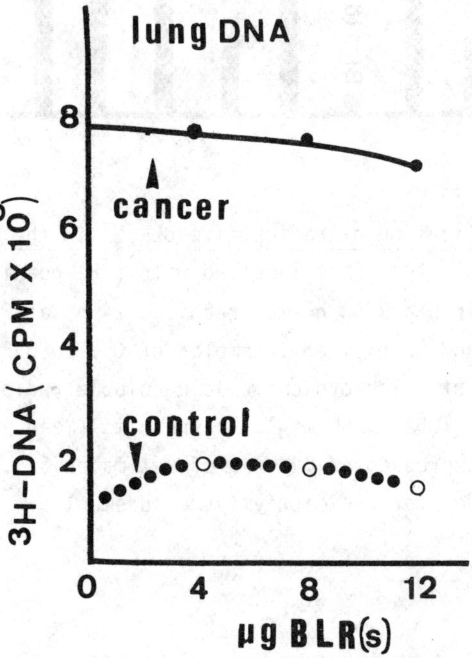

Fig.4 BLR$_{(s)}$ and *in vitro* replication of DNA from human lung , healthy and cancerous tissues . For incubation conditions , see legend to Fig.2. BLR concentrations : 4 , 8 and 12 µg (in duplicate). Incubation time 10 min. at 36°C .

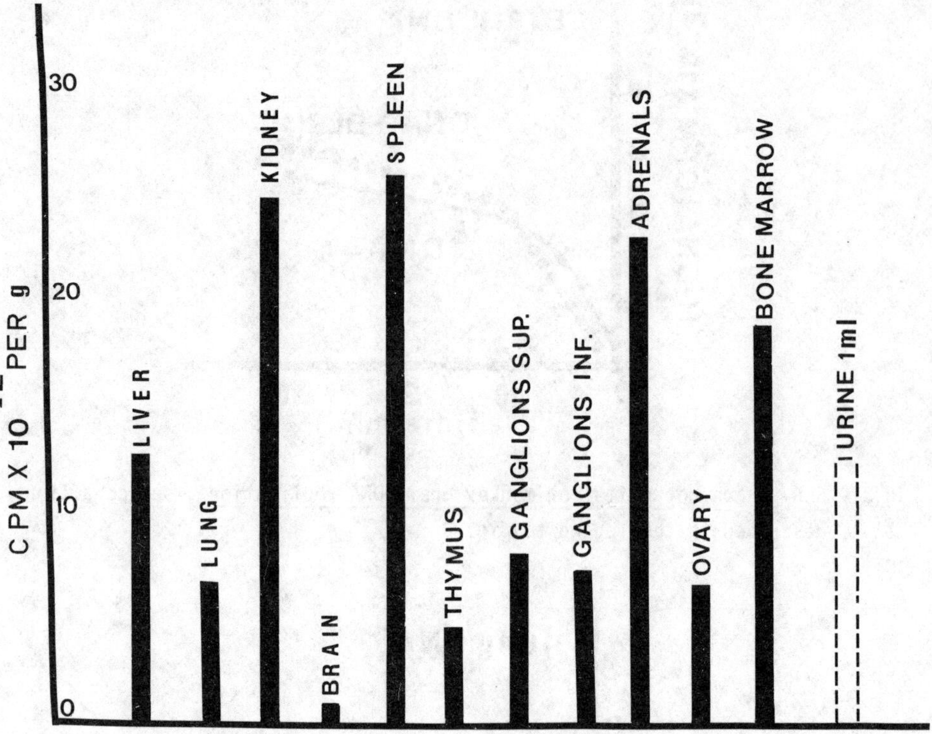

Fig.5 *In vivo* localization of radioactive $BLR_{(s)}$ in the rabbit . 5 mg of (^3H)-$BLR_{(s)}$ (3×10^6 CPM) (labelled with (^3H)-guanine and (^3H)-uracil) are injected into a 3 kg normal rabbit . 24 h later , the animal is anesthesized and sacrificed . Samples of 0.2 to 0.5 g of organs are analysed for hot trichloroacetic acid soluble radioactivity (5 % final conc. of TCA , 100°C , 30 min). The results (mean values for three rabbits) are expressed as CPM / g of wet material . Urine was collected from the bladder for radioactivity measurement .

found in the femur bone marrow and spleen suggested that $BLR_{(s)}$ might act on DNA replication in the cells which give rise to leukocytes and platelets. The presence of radioactive material in adrenal glands, kidney and liver may reflect blood flow or possibly reutilization of radioactive nucleotides by organs which synthesize $RNA_{(s)}$. The excretion of degraded radioactive material found in urine 24 h after (^{3}H)-BLR injection show that $BLR_{(s)}$ are biodegradable (Fig.5). However , in the adrenal glands, which contain stress stimulated haemetopoïetic cells (35), $BLR_{(s)}$ could exert a stimulating action.

Degradation of $BLR_{(s)}$ by rabbit plasma. In order to get some information about the extent of BLR degradation , plasma of circulating blood from healthy rabbits was prepared and incubated with (^{3}H)-$BLR_{(s)}$ at 36°C in function of time. Parallel experiments were performed with (^{3}H)-Poly C or Poly A. Alcohol precipitable material was filtered on millipore, dried and its radioactivity measured. The half life of (^{3}H)-$BLR_{(s)}$ was 8 min. and that of Poly C , 1 min., while Poly A was poorly degraded (Fig.6). $BLR_{(s)}$ which contain an excess of purine bases show a relatively important resistance to plasma RNase action. The activity of plasma RNase was also tested after rabbits had received $BLR_{(s)}$ i.v. Blood plasma was prepared just before injection and every h thereafter for 6-7 h. Fig.7 shows that after BLR injection, plasma RNase exhibit higher degradation activity toward (^{3}H)-BLR than that found in the plasma before BLR administration. Increased RNase activity is manifest during 5 - 6 h and then returns to normal value. These observations do not tell us if injected $BLR_{(s)}$ induced the increase of the amount of plasma RNase, or if some RNase activating factor (s) appeared in the presence of $BLR_{(s)}$. The period of increased plasma RNase activity corresponds to fever response induced by BLR injection to rabbits (Fig. 15).

$BLR_{(s)}$ activity and integrity. Hydrolysis of $BLR_{(s)}$ to nucleotides by incubation of 16 h in 0.3 N KOH at 36°C completely abolishes their stimulating activity *in vitro* (Fig.2) and *in vivo* (Fig.9). Treatment with RNase A for 2 h at 36° followed by deproteinisation and dialysis of the incubation mixture also reduces the activity of $BLR_{(s)}$ but does not completely abolish it.

Excellent priming activity of $BLR_{(s)}$ in DNA replication might be explained by the presence , in DNA dependent DNA polymerase as well as in RNase A preparations even crystallized several times, of phosphatase activity(verified for both types of enzymes). Phosphatase removes 3' phosphate giving rise to 3'OH group necessary for priming activity. The endogenous enzyme does it in animals. We have controlled that $BLR_{(s)}$ retreated with *E.coli* alkaline phosphatase exhibit normal leukopoietic activity.

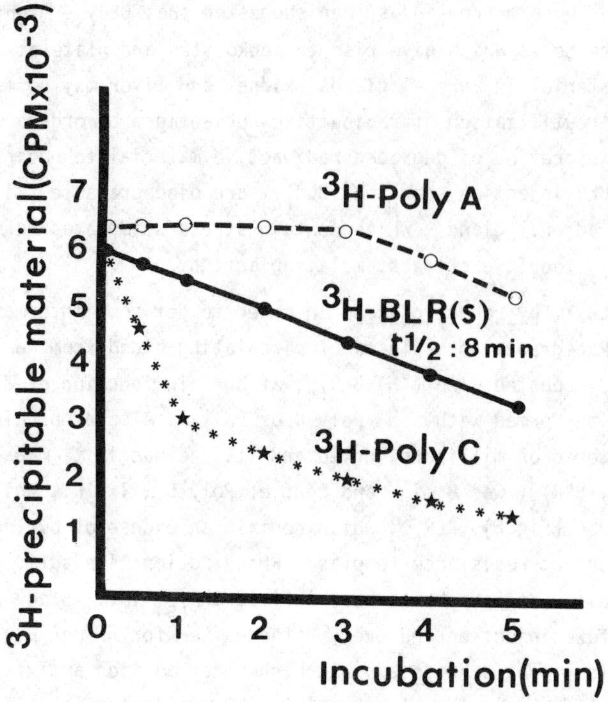

Fig. 6 In vitro degradation of (^{3}H)-BLR by blood plasma of healthy rabbits. For each time indicated on the Fig , incubation mixture (final volume 0.15 ml) contains 25 μMoles of Tris-HCl buffer pH 7.65 ; 200 μg of (^{3}H)-BLR$_{(S)}$(7,000 CPM) ; 0.01 ml of plasma (0.7 mg of protein). A parallel experiment was performed with (^{3}H)-Poly A (200 μg ,7.200 CPM) and (^{3}H)-Poly C (200 μg, 7,160 CPM) . Incubation at 36°C. ; time is indicated on the figure . To stop the reaction , alcohol (95°) was added in excess and 5 μMoles of KCl to precipitate the (^{3}H) labelled insoluble material . The precipitate was filtered on glass GF/C millipore filter , washed with alcohol , dried and its radioactivity was measured with Packard liquid scintillation counter (Prias). Results are expressed as CPM . Half life (t 1/2 = 0.693/K) was calculated using the values in the linear portion of the curves.

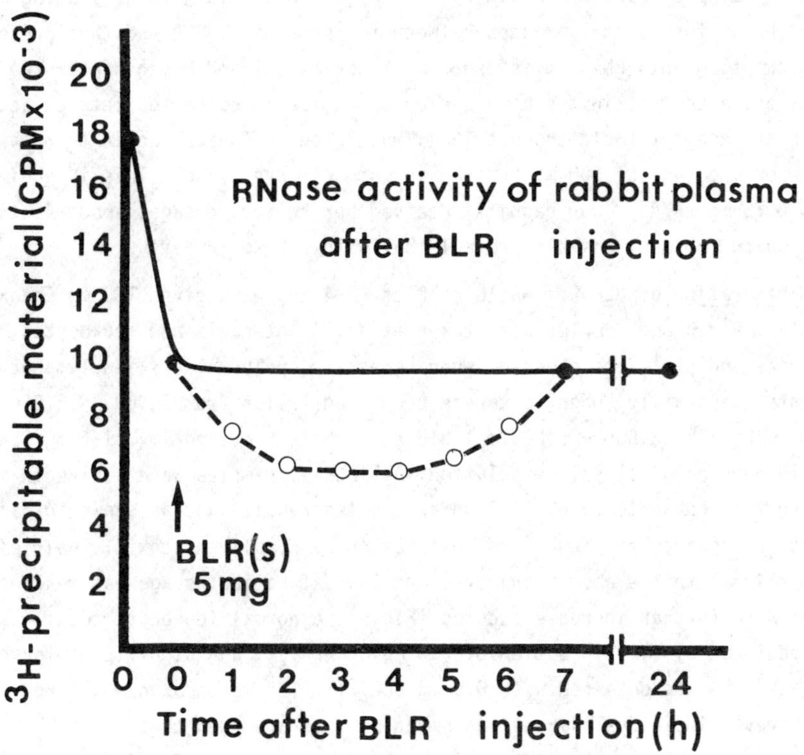

Fig.7 Increase of RNase activity in plasma of rabbits injected with $BLR_{(s)}$. For incubation conditions , see legend to figure 6 . 200 ug of (^{3}H)-$BLR_{(s)}$(18,000 CPM) were used for degradation with rabbit plasma (RNase) which was prepared from blood taken before BLR i.v. injection and every h thereafter for 6 - 7 h .

Leukopoietic activity of $BLR_{(s)}$ in rabbits permanently treated with Endoxan. Endoxan is a commercial preparation containing 75 % cyclophosphamide, an alkylating agent which can form alcoyl bridges between the strands of DNA and thus prevent its transcription into RNA. It is used for experimental and human cancer therapy and it is believed that in the body cyclophosphamide is converted into an active substance and non cytotoxic product (36). When used for human cancer therapy a total weekly dosage of 300 mg Endoxan is commonly prescribed. This is equivalent to about 0.75 mg/kg/day. Our rabbits received per kg i.v. dosages about 40 times higher , which for many animals prove lethal within 8 to 12 days.

I.V. administration of $BLR_{(s)}$. White rabbits (3-4 kg) were given 100 mg Endoxan i.v. daily , and blood samples were taken at daily intervals and leukocytes, erythrocytes and platelets counted. When necessary, polynuclears and lymphocytes were counted separately. When leukocyte count had fallen from 7,000 - 12,000 (healthy rabbit) to 3,000 - 5,000/mm^3 blood, rabbits received i.v. 1-6 mg $BLR_{(s)}$ in sterile physiological saline solution. All $BLR_{(s)}$ samples were active in increasing the leukocyte count in Endoxan treated rabbits and as shown in Fig.8 could not be replaced for leukocyte restoration by E.coli endotoxin. Their effect was proportional to the amount injected until a saturation dosage was reached after which no further increase occured (Fig.9). A normal leukocyte count was usually restored by i.v. injection of 2.5 mg of $BLR_{(s)}$, and significant leukopoïetic activity was obtained with 0.5 mg $BLR_{(s)}$ / 100 mg Endoxan / 3.5 kg rabbit. However individual variations among rabbits are such that the $BLR_{(s)}$ dosage may have to be doubled to obtain the same response. The increased leukocyte count obtained with $BLR_{(s)}$ reaches a normal value 24-48 h after their injection , remains at a high level for a few days and then returns to the value found before injection of $BLR_{(s)}$. The duration of cycle of $BLR_{(s)}$ induced changes in leukocyte count is about 7 days (Fig.10). Increased leukocytes obtained with $BLR_{(s)}$ never exceeded 250 % of the lowered count produced by Endoxan treated nor 100 % of the normal value in untreated rabbits. These results suggest that a very powerful cellular regulation process is involved. This process is not perturbed by excess dosages of $BLR_{(s)}$ since in Endoxan treated rabbits given 1-6 mg $BLR_{(s)}$ i.v. evry second day there is no additional increase in leukocyte counts as the dose is increased. In addition there is no decrease in the response such as might have been caused by a possible mass action of repeated doses, a possible toxic effect or the development of resistance to $BLR_{(s)}$. A fresh dose of $BLR_{(s)}$ is observed to stimulate leukopoiesis only when the preceeding dose has ceased to act (Fig.10). This numerous repeated doses of $BLR_{(s)}$ do not have a cumulative effect of any kind and do not lead to a loss of leukopoietic stimulating effect.

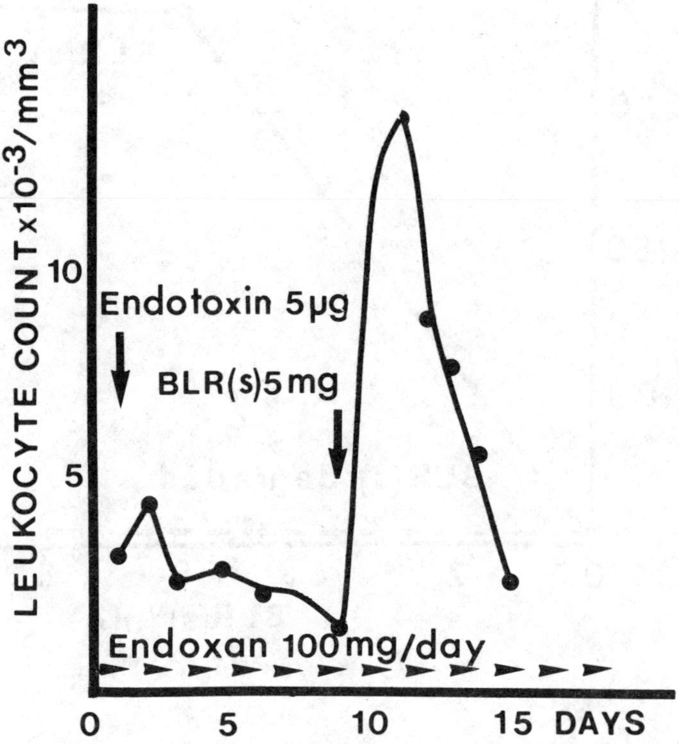

Fig.8 Effect of E.coli Endotoxin and $BLR_{(s)}$ on the leukocyte count in the Endoxan-treated rabbit . E. coli Endotoxin (5 µg) was given i.v. to 3 kg rabbit which have been treated for 7 days with 100 mg Endoxan/ day . No appreciable effect on leukopoïesis was noted. As shown by the arrow when these same rabbits received 5 mg $BLR_{(s)}$ i.v. the leukocyte count increased . The mean increase of leukocyte count (6 rabbits) after endotoxin i.v. injection was 14 % $\pm$ 3 % , $p < 0.02$.

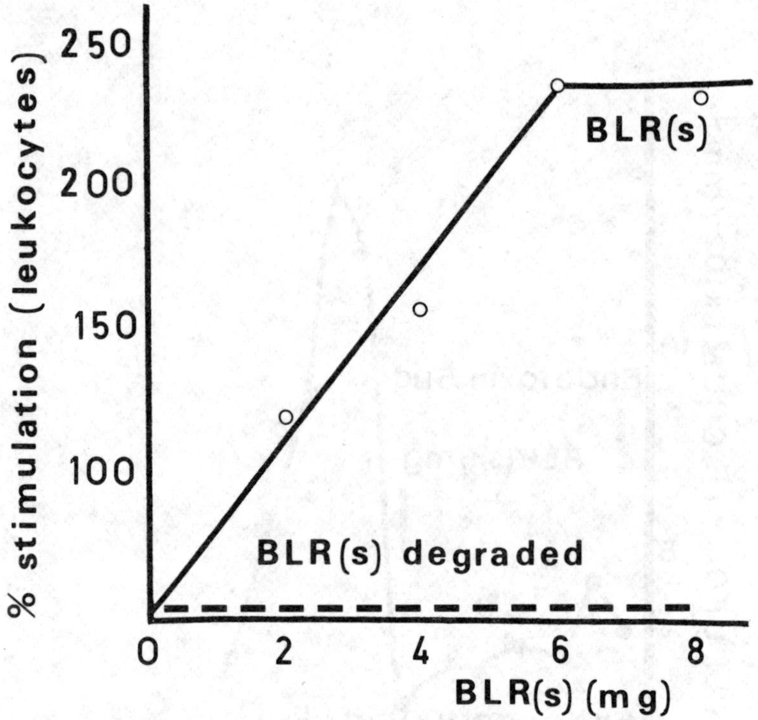

Fig.9 Increase in leukocyte count in Endoxan treated rabbits given various amounts of $BLR_{(s)}$. 3.5 kg rabbits were treated with Endoxan (100 mg injected i.v.) daily for four weeks . The first dose of $BLR_{(s)}$ (0.5 mg) was injected i.v. on day 1 , 10 min after Endoxan injection ; when the stimulating effect of the first dose ended , a second 1 mg dose was given i.v. etc... Circulating leukocytes were counted with a "Coulter counter ". The same results were obtained with three other rabbits. $BLR_{(s)}$ were degraded by incubation in 0.3 N KOH at 36°C for 16 h , then neutralized.

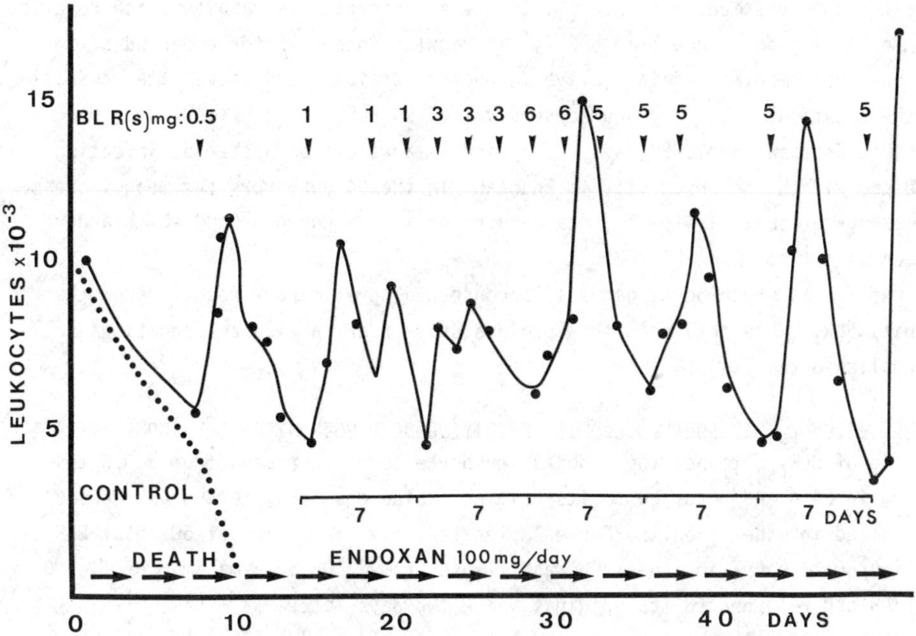

Fig. 10 Leukocyte count in the Endoxan treated rabbits receiving varying i.v. doses of $BLR_{(s)}$ every second day. After leukocyte count had been strongly decreased, a 3.5 kg Endoxan treated rabbit (100 mg/day) received varying doses of $BLR_{(s)}$ ranging from 1 to 6 mg every second day as shown by the arrows. Circulating leukocytes were counted daily with a Coulter Counter. The results given in this Fig. are an average obtained with 10 rabbits. The mean increase in leukocyte count was 172 % $\pm$ 17 % as standard error. The confidence interval calculated using paired sample student's t test : $p < 0.001$.

Although Endoxan dosage at the level used in these experiments proves lethal in many sensitive animals after 8-12 days this effect is suppressed by simultaneous treatment with $BLR_{(s)}$. In one experiment five rabbits which received 100 mg Endoxan daily and 5 mg $BLR_{(s)}$ at regular intervals for a period of two and one half months were all alive in good condition 2 - 3 years after cessation of the treatment. $BLR_{(s)}$ procured survival of 94 rabbits (over 100 used) which received Endoxan for 30-90 days . Six died as a result of bacterial infection combined with high sensitivity to Endoxan. In the 94 survivors the mean increase in leukocyte count was 154 % , standard error $\pm$ 11.% and $p < 0.001$ by standard Student's t test.

As far as erythrocyte genesis is concerned, preliminary results show that in rabbits $BLR_{(s)}$ are active in this process only under particular conditions (unpublished results).

Effect of $BLR_{(s)}$ in control rabbits receiving no drugs. After i.v. administration of 5 mg of $BLR_{(s)}$ to healthy rabbits leukocyte count increases from 8,000 to 14,000 in 24 h and returned to its original value one day later. Higher doses are controlled in other rabbits. For example after i.v. injection of 50 mg of $BLR_{(s)}$ the leukocyte count in a healthy rabbit rose from 7,300 to a maximum of 17,500 in 20 h and returned to its original value two days later.

After injection of $BLR_{(s)}$ a short transcient fall in cell count is observed before stimulation occurs. This is seen after a four h delay for leukocytes, and a 24 h delay for platelets. For example in the experiment mentioned above in which a rabbit received a single 50 mg dose the leukocyte count had fallen from 7,300 to 3,300 four h after injection but was 17,500 twenty h later. This reaction is apparently independent of the amount of $BLR_{(s)}$ administered (range tested : 3 - 50 mg per rabbit weighing 3-4 kg). Furthermore the same initial fall in leukocyte and platelet counts was observed when rabbits receive 3 mg of poly AG i.v. although this polynucleotide which was used because $BLR_{(s)}$ are relatively rich in A and G, has no leukopoietic activity. Finally it may be noted that oral administration of $BLR_{(s)}$ to rabbits (see below) does not cause an initial transcient fall in cell count.

Effect of $BLR_{(s)}$ administered orally. As shown in Fig.11 the effect of orally administered $BLR_{(s)}$ on leukocyte and platelet counts is manifest more slowly and maintained longer that that produced by i.v. injection (compare Fig.10 and 11). Four rabbits were prepared by treatment for 30 days with 100 mg of Endoxan daily and sufficient $BLR_{(s)}$ i.v. at the necessary intervals to ensure survival. Administration of first $BLR_{(s)}$ and then Endoxan were discontinued. Leukocyte counts were then 5,500-5,000/mm^3 . After this treatment rabbits cannot spontaneously restore

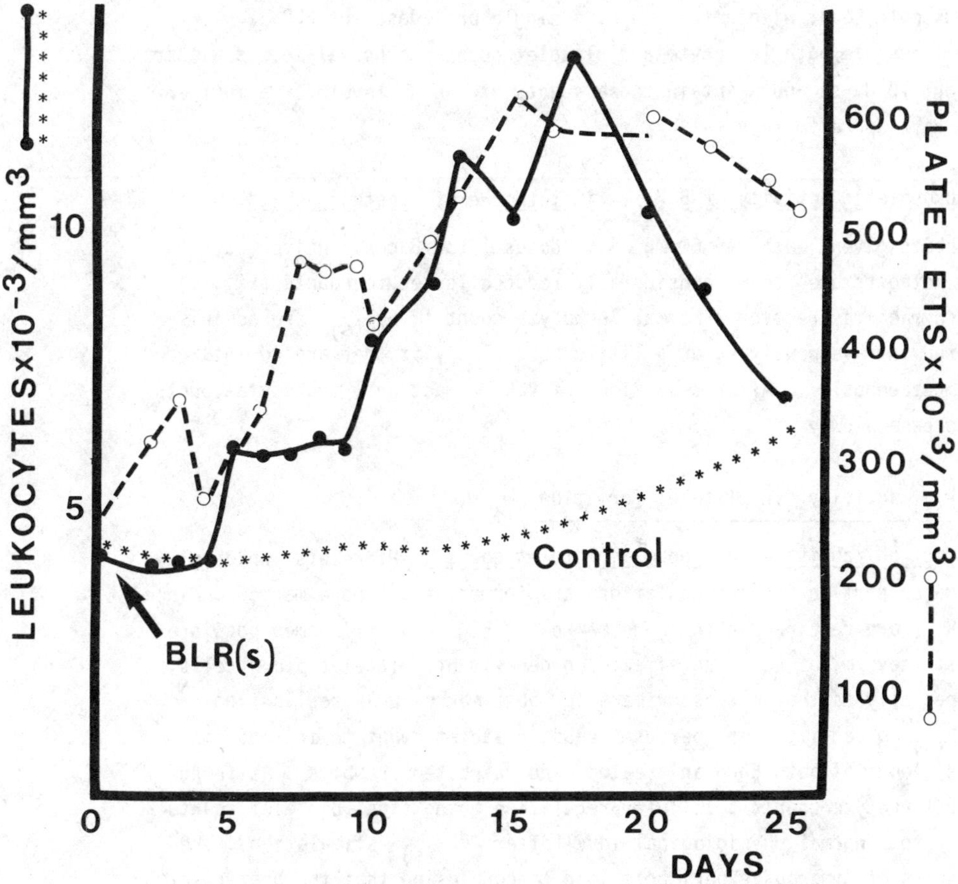

Fig.11 Leukocyte and platelet counts in the rabbit receiving $BLR_{(s)}$ per os.
This Fig. shows leukocyte and platelet counts increase in rabbits receiving 20 mg $BLR_{(s)}$ per os . Values represent an average of six rabbits . Initial leukocyte maximum count 3,7000 $\pm$ 100 / mm^3 (a) increased up to 9,250 $\pm$ 2,200 ($p < 0.02$) after BLR administration. Initial platelet maximum count 250,000 $\pm$ 50,000 increased up to 575,000 $\pm$ 25,000 / mm^3 ($p < 0.03$) after BLR administration.
(a) values shown are arithmetic mean $\pm$ 1 SD.

leukocyte count within two weeks; a single oral dose of $BLR_{(s)}$ restores both leukocyte and platelet counts to normal levels within about 10 days and maintain these counts at a high level for a further 15 days approximately .

Leukopoietic activity of $BLR_{(s)}$ in methotrexate tested rabbits. Rabbits given each day 60 mg i.v. doses (for 2 consecutive days) of methotrexate show considerably reduced leukocyte counts (Fig.12). Such rabbits recover a normal leukocyte count if $BLR_{(s)}$ are administered intraveinously , or a little more slowly if they are administered subcutaneously . After cessation of $BLR_{(s)}$ action , leukocyte counts decrease slowly .

$BLR_{(s)}$ activity in platelet formation

Effect on rabbits with a normal platelet count . Platelets which play a major part in blood coagulation are formed in the bone marrow . Since $BLR_{(s)}$ are retained in the bone marrow (Fig. 5) it seemed possible that they might have an effect on genesis of platelet blood cells, especially as they act as primers for bone marrow DNA replication . $BLR_{(s)}$ given i.v. or *per os* cause platelet count to increase in the blood of both Endoxan treated and untreated rabbits . As found for leukocyte counts , cellular regulation bring final platelet counts back to a normal physiological level after $BLR_{(s)}$ stimulation . The results of numerous experiments lead to conclusion that the best results are obtained when $BLR_{(s)}$ are given *per os* . Increased platelet counts then remain stable for about two weeks .

Effect on rabbits with a decreased platelet count . Daunorubicin (anthracyclin) causes a rapid and spectacular fall of leukocyte and platelet counts in 4 kg rabbits receiving 5 mg of the drug i.v. for four consecutive days only . Toxicity is so high that death occurs on the 6th - 8 th day after the start of the treatment . However , if $BLR_{(s)}$ are given on the 5 th and 7 th day the animal can be saved . Fig. 13 shows the evolution of platelet and leukocyte counts during such an experiment. Daunorubicin treatment reduces platelet

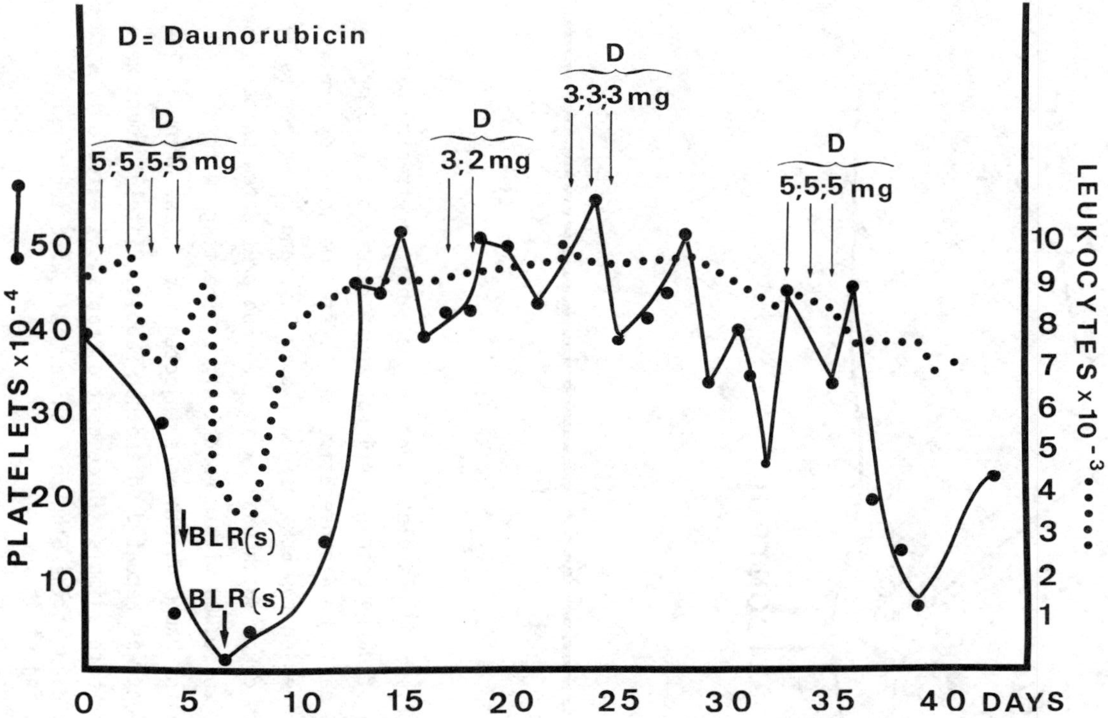

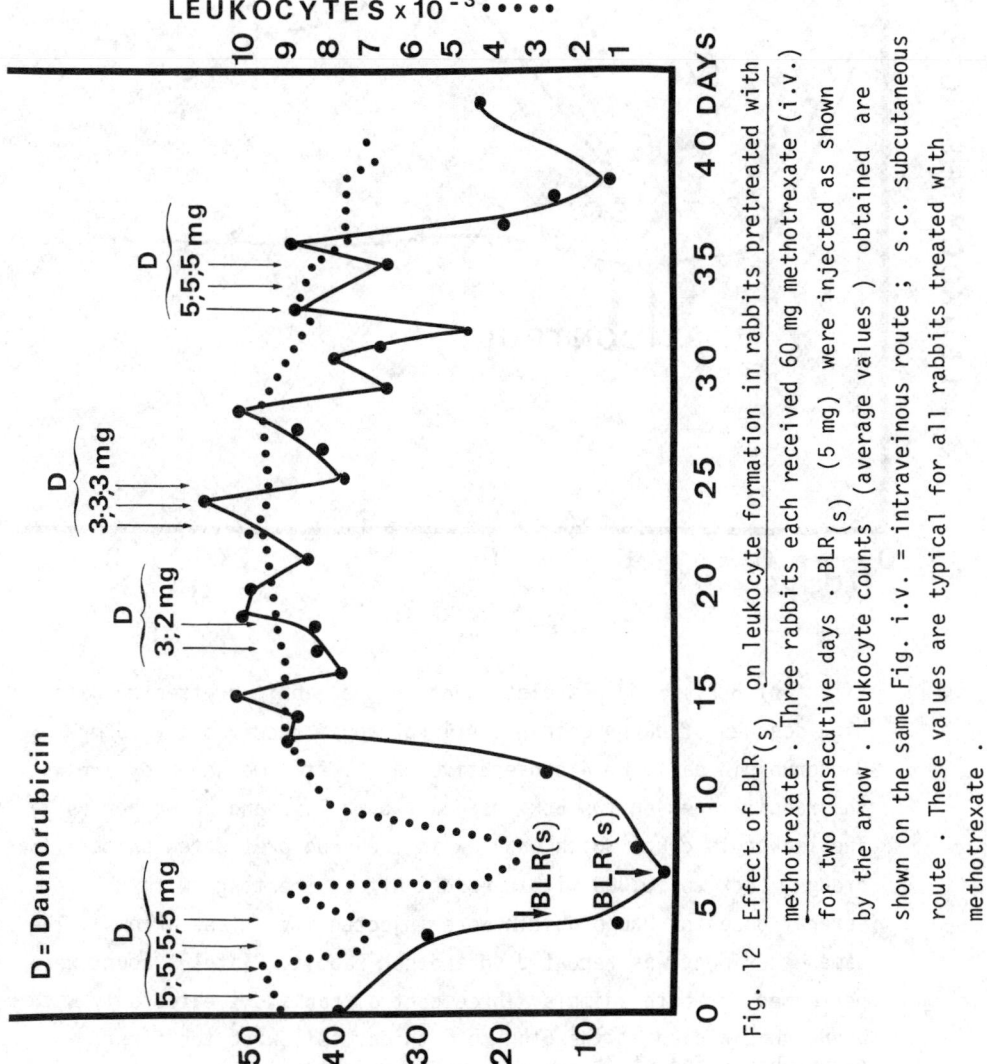

Fig. 12 Effect of $BLR_{(s)}$ on leukocyte formation in rabbits pretreated with methotrexate . Three rabbits each received 60 mg methotrexate (i.v.) for two consecutive days . $BLR_{(s)}$ (5 mg) were injected as shown by the arrow . Leukocyte counts (average values) obtained are shown on the same Fig. i.v. = intraveinous route ; s.c. subcutaneous route . These values are typical for all rabbits treated with methotrexate .

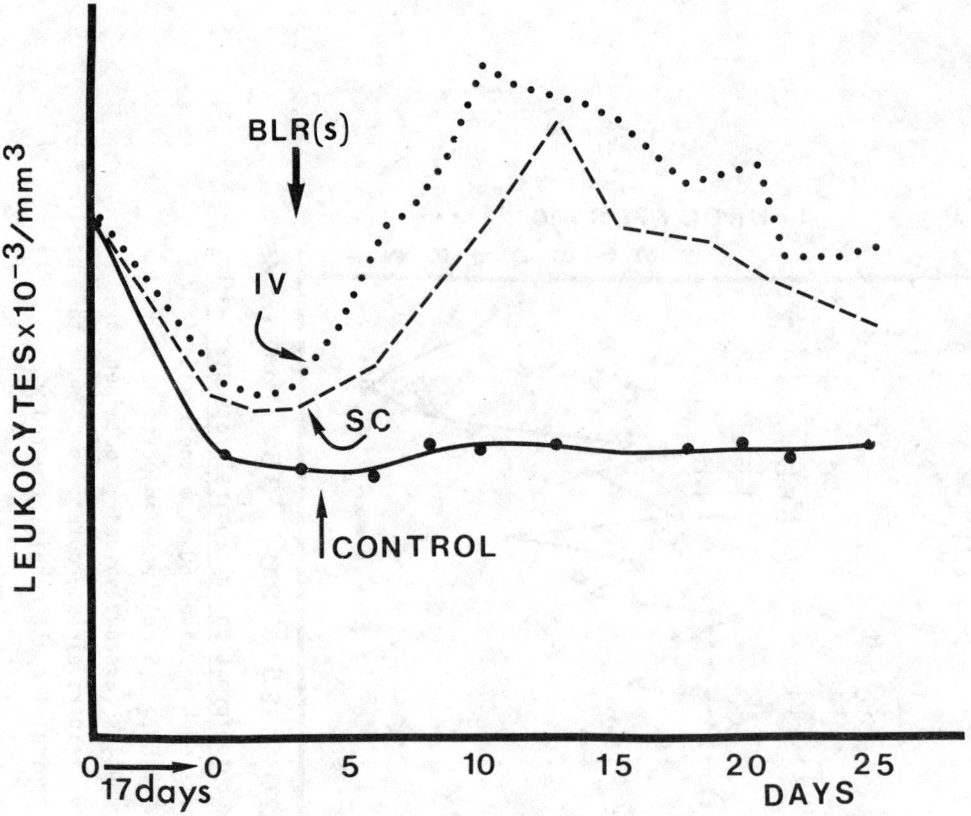

Fig.13 Effect of $BLR_{(s)}$ on platelet count in the rabbit pretreated with high dosages of daunorubicin . A 4 Kg rabbit received i.v. 5 mg Daunorubicin daily for 4 consecutive days . At time shown by arrows, the rabbit received two BLR dosages (5 mg i.v. and 20 mg per os). The leukocyte count which was low in the drug-pretreated animal , was brought back to 10,000 within 48 h . After treating with $BLR_{(s)}$, several doses of Daunorubicin were injected i.v. (see arrows). The same experiment was repeated on another rabbit. Platelet count were performed for both animals. Three control rabbits treated only with Daunorubicin died at the 6th to 9 th day following the first daunorubicin injection .

and leukocyte count from 400,000 and 10,000 to 15,000 and 3,600 respectively. After two doses of $BLR_{(s)}$, (arrows , Fig. 13) leukocyte count returned to normal (10,000/mm^3) in 48 h whereas platelet counts did not reach the initial value of 400,000 until 5 - 6 days later . Subsequent renewed doses of daunorubicin , totalizing 14 mg in 15 days did not cause any significant decrease in the platelet and leukocyte counts . Protection by $BLR_{(s)}$ therefore subsisted and even when 15 mg of daunorubicin was given three weeks after the last $BLR_{(s)}$ dose the fall in platelet counts was much less striking than that observed before administration of $BLR_{(s)}$.

$BLR_{(s)}$ / drug ratio Many chemotherapeutic agents decrease leukocyte and platelet counts and as we show here it is possible to slow down or arrest this effect by appropriate $BLR_{(s)}$ dosage . Thus $BLR_{(s)}$ may make a positive contribution to human chemotherapy and particularly to cancer therapy. The amount of $BLR_{(s)}$ necessary to achieve optimum results can be calculated by giving variable amounts to an animal receiving a constant drug dosage but it must be noted that the value obtained will vary somewhat from animal to animal . Thus for a rabbit receiving 100 mg of Endoxan i.v. daily a $BLR_{(s)}$ - drug ratio of 1 : 20 is normally sufficient to maintain a high leukocyte count for 3 - 4 days . However smaller amounts e.g. a 2 mg dose (instead of 5 mg of $BLR_{(s)}$) ($BLR_{(s)}$ - drug ratio 1 : 50) are sufficient in some cases . Because various drugs e.g. Endoxan , Methotrexate, Daunorubicin damage bone marrow and spleen cells to different extents , optimum $BLR_{(s)}$ - drug ratios must be determined in each case.

Comparison of the effect of $BLR_{(s)}$ and other polynucleotides . Administration of 10 - 20 mg *E.coli* ribosomal $RNA_{(s)}$ *per os* to Endoxan treated rabbits causes no noticeable increase in either platelet or leukocyte count. This shows that the high molecular weight r-RNA chains must be degraded to small products ($BLR_{(s)}$) in order to obtained material which can act on platelet and leukocyte formation . Administration of poly AG or poly A by i.v. injection to Endoxan treated rabbits also fails to stimulate leukocyte formation (Fig. 14). The products of deoxyribonuclease digestion of salmon DNA were also found to be inactive with respect to platelet and leukocyte formation in Endoxan treated rabbits.

$BLR_{(s)}$ restore the balance between lymphocytes and polynuclears . In the healthy human , leukocytes contain 60 - 70 % polynuclears and 25 - 40 % lymphocytes whereas in the normal rabbit the amounts of these two types

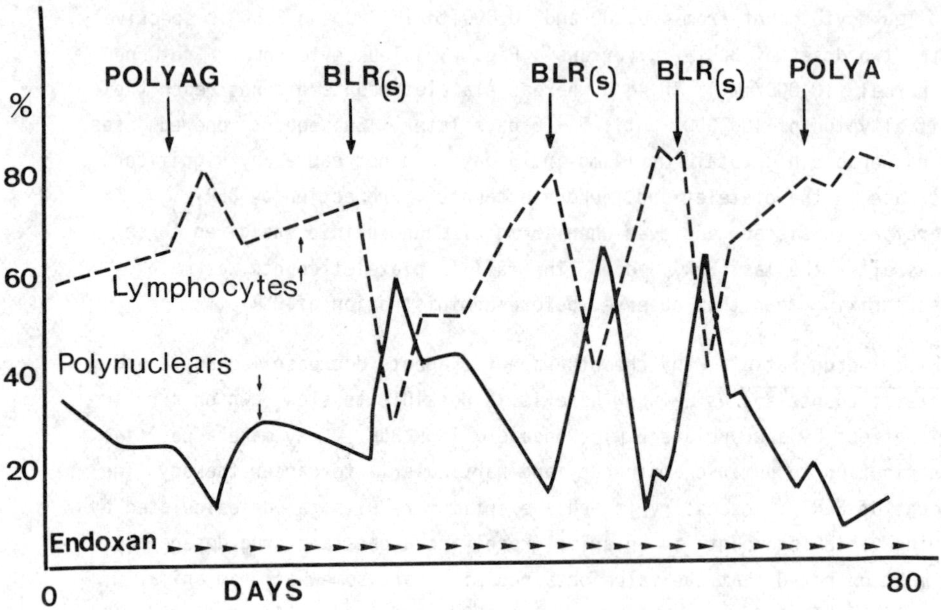

Fig. 14 <u>BLR$_{(s)}$ restore the balance between lymphocyte and polynuclear percentages</u>. A 3 kg rabbit was treated daily with 100 mg Endoxan i.v. for several weeks and periodically with BLR$_{(s)}$. When leukocyte count had fallen and polynuclear and lymphocyte imbalance had appeared , Poly AG (3 mg), BLR$_{(s)}$, (5 mg), Leuco-4 (5 mg), diethyldithiocarbamate (DEDTC)(5 mg), Poly A (5 mg) were given intravenously at the intervals shown on this Fig. The same results were obtained with two other rabbits . The balance-restoring effect of BLR$_{(s)}$ was observed in 10 other rabbits . Initial lymphocyte maximum count 3,530 $\pm$ 322 / mm^3 [a] increased up to 6,100 $\pm$ 520 / mm^3 ($p < 0.02$) after BLR injection . Initial polynuclear maximum count 1,080 $\pm$ 250 / mm^3 increased up to 6,990 $\pm$ 680 ($p < 0.001$) after BLR injection . Cyclophosphamide destroys polynuclears more efficiently than lymphocytes . BLR$_{(s)}$ reestablish first the poly-nuclear / lymphocyte balance and then normal amounts of both types

(a) values shown are arithmetic mean $\pm$ 1 SD

of cells are usually practically equal . In rabbits , daily Endoxan treatment (100 mg/day/3 kg rabbit), the considerable decrease in leukocyte count seen after 6 - 7 days is accompanied by a large differential decrease in the number of polynuclears which falls from 55 % to 18 % of the total white cell count (Fig.14). Endoxan and other drugs cause a similar effect in man.

When Endoxan treated rabbits are given $BLR_{(s)}$ i.v. polynuclear and lymphocyte counts increase differentially and the number of these cells become practically restored within 4 - 5 h and a normal ratio is fully re-established after 24 h. Six days later the effect of $BLR_{(s)}$ has ended and the Endoxan induced imbalance reappears (Fig.14) but can be again corrected by a further $BLR_{(s)}$ injection . This sequence of events can be repeated again and again . In one Endoxan treated rabbit the effect of $BLR_{(s)}$ on leukocyte formation was compared with that of sodium diethyldithiocarbamate (DEDTC) , poly A , Leuco-4 and poly AG (Fig.14). None of the latter four agents were able to stimulate leukocyte formation or restore a normal lymphocyte / polynuclear ratio .

<u>$BLR_{(s)}$ do not induce tolerance phenomenon in rabbits as Endotoxin does</u>.Since $BLR_{(s)}$ originate from <u>Escherichia coli</u> ribosomal RNA , it appeared essential to distinguish the physiological activity of BLR in leukopoiesis from the effect exhibited by <u>E.coli</u> endotoxin . The Shwartzman reaction (37) , a sensitive test for detection of small amounts of bacterial endotoxin in biological preparations gives in rabbit a positive reaction with 0.2 - µg of bacterial endotoxin or by a few µg of Poly I-Poly C (38). 250 µg of $BLR_{(s)}$ were injected at each of 5 different sites in shaved rabbits epiderm and 16 h later the animals received 5 mg $BLR_{(s)}$ i.v. No trace of inflamation was detected at the inoculation sites for three days. Several $BLR_{(s)}$ preparations were tested on different rabbits and none gave a positive Shwartzman reaction.

It has been well established that daily endotoxin injections to healthy rabbits result in a progressive decrease in the amount of fever and in the transient polynuclear liberation into circulating blood (39)(40). This tolerance state might be induced in rabbits and humans with small doses of endotoxin (1.0 - 5.0 ng/kg) (41)(42) . Rabbits received every 48 h , 1-2 mg BLR/kg i.v.,altogether eleven injections. Total white blood cell count was determined at intervals indicated in the legend to Fig15 and rectal temperature taken just before injection and every h thereafter for 6 - 8 h. Firstly, each BLR administration resulted in full leukocyte count increase . Secondly, the amount of fever produced was the same after the first or the last of the eleven $BLR_{(s)}$ injection (Fig. 15). These data clearly demonstrate failure of tolerance to develop after repeated i.v. administration of BLR , as already suggested by results illustrated in Fig. 10 . To these same rabbits , E.coli endotoxin (1.5 μg / kg) was injected i.v. every 48 h totalizing 7 injections) . The mean peak of polynuclear count declined on successive days in parallel with fever decrease (Fig. 15). Endotoxin tolerant rabbits respond normally to $BLR_{(s)}$: full expression of their leukopoietic activity and undiminished fever response . In rabbits treated with high dosages of Endoxan (100 mg/ 3 kg rabbit), endotoxin (1.0 μg/kg) does not induce transcient polynuclear liberation while under the same conditions 5 mg of $BLR_{(s)}$ induce the restoration of normal leukocyte level with peak response between 24-48 h without inducing tolerance phenomenon. The demonstrations that $BLR_{(s)}$ actively induce not only leukocyte but also platelet formation in rabbits treated with daunorubicin for instance , show that their physiological effects areclearly distinguishable from that exhibited by endotoxin .

Pyrogenicity of various polyribonucleotides. The pyrogenic effect of $BLR_{(s)}$ (1 mg / kg), of Poly AG (1 mg/kg) and of Poly I - Poly C (5 μg/kg) was determined in rabbits injected i.v. in duplicate for each substance. Anal temperature was measured . The mean values (ΔT maximum) were respectively 1.4°C , 1.9°C and 2.0°C and so called temperature period lasted for 4-5 h. The pyrogenic effect of $BLR_{(s)}$ is no higher for 20 mg BLR/kg than for 1 mg BLR/kg, which suggests that the pyrogenic effect of $BLR_{(s)}$ is due to the RNA fragments themselves and not to any possible contaminant. When given per os $BLR_{(s)}$ are not pyrogenic .

$BLR_{(s)}$ activity and animal behaviour . No modification of behaviour was observed in Endoxan-treated or control untreated rabbits which were given i.v. BLR dosages of 0.5 to 20 mg/kg. In 20-22 gr mice injected i.p. with 2 to 20 mg $BLR_{(s)}$, no adverse reactions were noted and the animals survived in normal condition.

$BLR_{(s)}$ and malignant cell development in mice under chemotherapy. The fact that $BLR_{(s)}$ do not stimulate the in vitro replication of DNA from cancerous tissues suggested that in vivo , they could not stimulate malignant

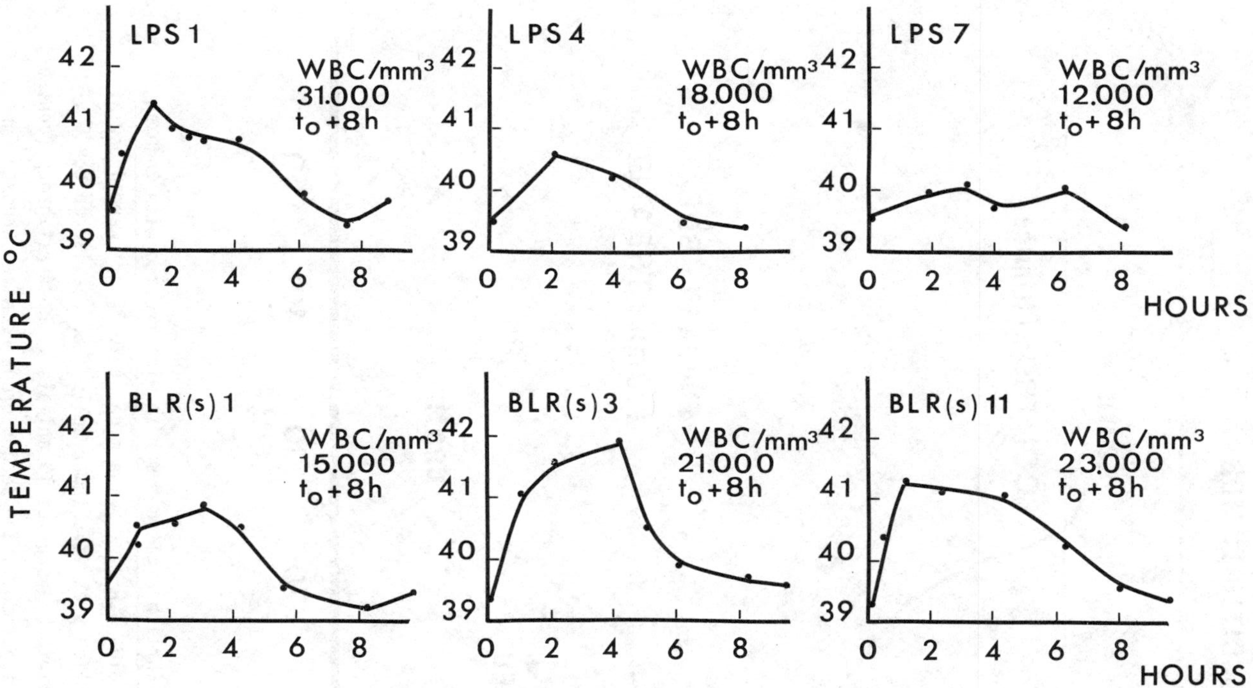

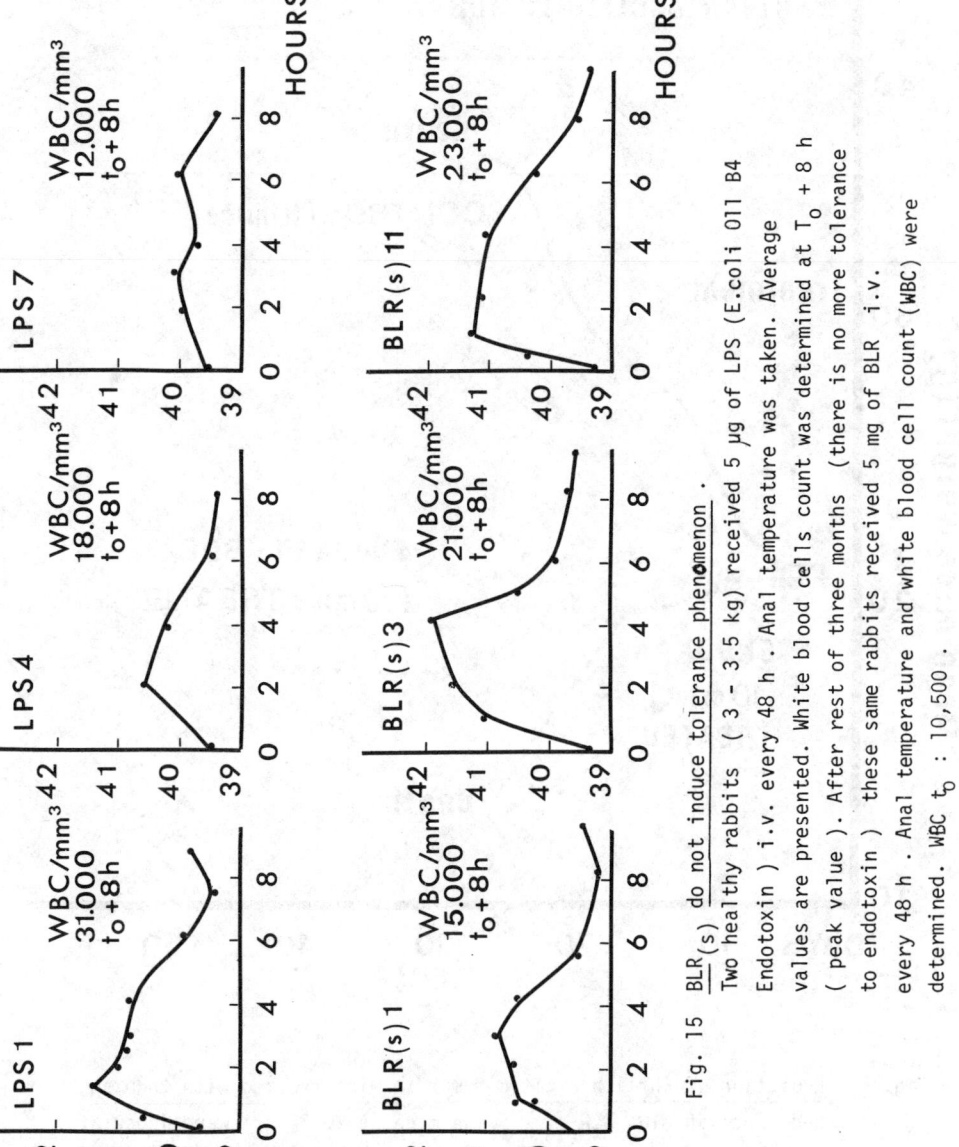

Fig. 15 BLR$_{(s)}$ do not induce tolerance phenomenon .

Two healthy rabbits (3 - 3.5 kg) received 5 µg of LPS (E.coli 011 B4 Endotoxin) i.v. every 48 h . Anal temperature was taken. Average values are presented. White blood cells count was determined at T_o + 8 h (peak value). After rest of three months (there is no more tolerance to endotoxin) these same rabbits received 5 mg of BLR i.v. every 48 h . Anal temperature and white blood cell count (WBC) were determined. WBC t_o : 10,500 .

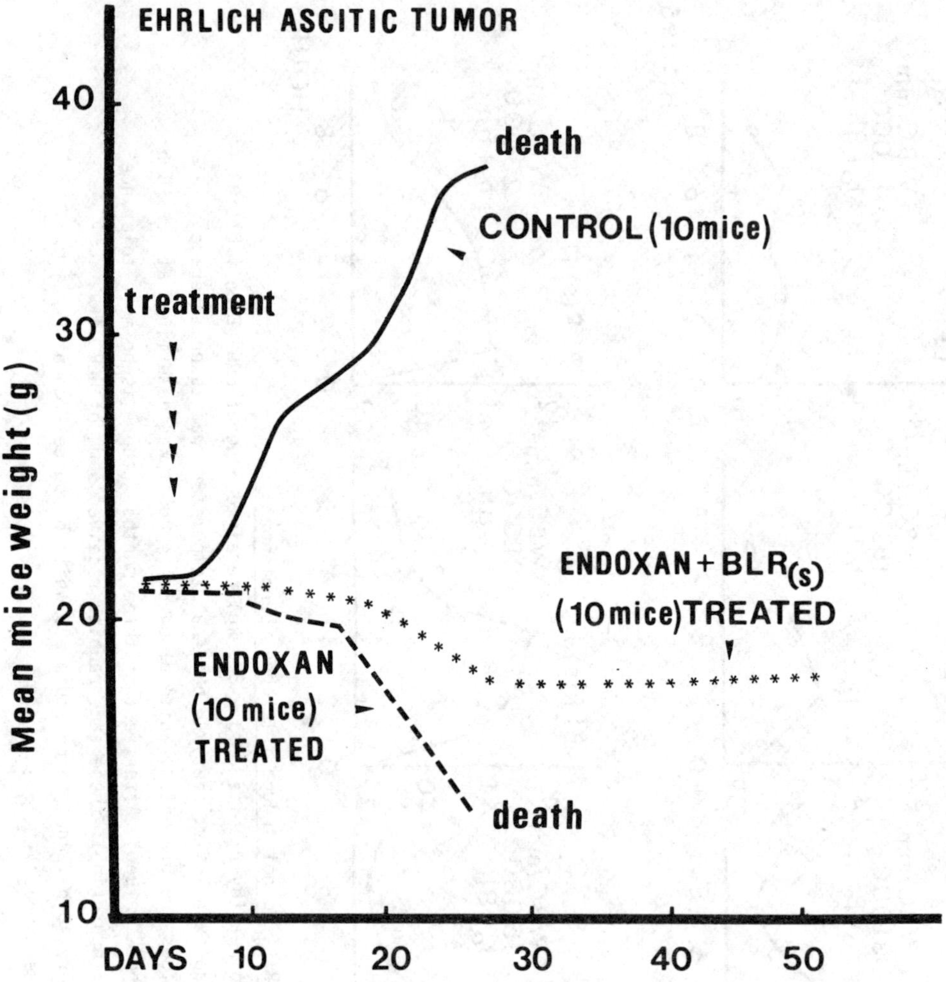

Fig. 16 Evolution of Ehrlich ascitic tumor in mice treated with Endoxan and Endoxan plus $BLR_{(s)}$. Swiss mice (20-22 gr) were inoculated i.p. with Ehrlich's ascitic tumor cells (50^5 cells) on day 0 , then divided into 3 groups of 10 animals . The first group (control) received no treatment . The second group received Endoxan for 15 consecutive days starting from day 3 : 2 mg i.p. twice a day (4 mg daily dosage). The third group was given Endoxan under identical conditions and received 1 mg $BLR_{(s)}$ i.p. every fourth day until Endoxan was discontinued . Weight and survival duration of the animals were noted during and after treatment .

cell multiplication and would not interfere with antimitotic drug activity. In order to demonstrate this, Swiss mice inoculated i.p. with Ehrlich ascitic tumor cells were treated with Endoxan dosages alone or complemented with BLR injections. Weight and survival of the animals were noted during and after treatment (Fig.16). All mice from the control group receiving neither Endoxan nor BLR died within 3 weeks after malignant cell inoculation. Mice treated with high dosages of Endoxan alone experienced an important weight decrease and died from leukopenia and toxic effects of the drug within the same time range as control mice (Fig.16). In contrast, mice which were given high dosages of Endoxan plus $BLR_{(s)}$ lost very little weight and generally survived for over three months. The same results were obtained with BALB C mice bearing lymphoma YC 8 ascitic cells. These data show that $BLR_{(s)}$ can protect the animals against the toxic effect of Endoxan without impeding the anti-cancer activity of the drug.

CONCLUSION and DISCUSSION

The biological effect of $BLR_{(s)}$ in leukopoiesis and platelet formation can be considered as a model illustrating how a given short chain RNA molecule can selectively exert its action on the expression of gene activity and cell reproduction without disorganizing physiological regulation in animals.
An <u>in vivo</u> activity of $BLR_{(s)}$ could be expected from their <u>in vitro</u> priming effect on the replication of DNA isolated from rabbit bone marrow and spleen. Moreover, <u>in vivo</u>, injected $^3H\text{-}BLR_{(s)}$ were found localized in stem cell containing tissues. The <u>in vivo</u> activity of $BLR_{(s)}$ copies the basic biochemical mechanism by which a cell can maintain its activity, ensure its multiplication and undergo differentiation. In rabbits treated with high dosages of antimitotic drugs such as cyclophosphamide (endoxan), methotrexate (amethopterin) and daunorubicin (anthracyclin), which inflict severe injury to the multiplying cell population, as they injure stem cells (43) are strongly leukopenic and/ or thrombocytopenic, $BLR_{(s)}$ promptly restore a normal level of circulating leukocytes and platelets. The high physiological activity

of $BLR_{(s)}$ is evidenced by various observations : 1) the increase of leukocyte and platelet counts produced by $BLR_{(s)}$ never exceeds normal physiological limits and there is no cumulative effect following repeated dosages ; 2) an excess BLR dosage never produces overhigh counts of circulating leukocytes and platelets (this is certainly connected to the finite life span of these blood cells) ; 3) a normal balance between polynuclears and lymphocytes is restored within a few hours following BLR injection ; 4) $BLR_{(s)}$ are not toxic ; 5) $BLR_{(s)}$ do not induce tolerance phenomenon ; 6) $BLR_{(s)}$ have no priming effect on in vitro replication of DNA from various cancerous tissues and stimulate neither cancer cells in mice nor pathological leukocytes .

All these results clearly show that these RNA-fragments ($BLR_{(s)}$) do not act by liberating leukocytes and platelets from a reserve pool , but instead act on stem cell DNA replication (stem cell compartment) , stem cell division , and also on differentiation in the multiplying compartment , thus giving rise to polynuclears and platelets . Preliminary experiments (Plawecki and al. , unpublished) show an increase of myelocyte count in the bone marrow and also in the circulating blood of BLR treated rabbits . The fact that in animals pretreated with $BLR_{(s)}$, daunorubicin has a far smaller decreasing effect on platelet and leukocyte counts confirms clearly that a competition exists between $BLR_{(s)}$ and daunorubicin for stem cells and their derivatives . This competition can also be observed during in vitro DNA replication .

It should be emphasized that rabbits , permanently treated with high cyclophosphamide doses which are lethal for sensitive animals within 8 to 12 days (leukocyte depletion) but receiving a weekly 2 - 6 mg BLR i.v. injection have been kept alive and healthy for 2 - 3 years and showed no ill effect after simultaneous cessation of both the antimitotic drug and $BLR_{(s)}$. It should be recalled that once $BLR_{(s)}$ have achieved their effect on leukocyte and platelet genesis they are degraded by endogeneous nucleases present in animal tissues . $BLR_{(s)}$, which are also active per os , do not impede the activity of anti-cancer drugs.

From our four year observation of many rabbits and mice under various conditions, $BLR_{(s)}$, which act on humans as they do on animals , may contribute very positively to the cure of leukocyte and platelet deficiences whether they are induced by chemotherapy (particularly cancer therapy), by various ailments or by genetic defects .

R E F E R E N C E S

1. Schekman,R., W.Wickner, O.Westergaard, D.Brutlag , K.Geider, L.L.Bertsch and A.Kornberg. Initiation of DNA Synthesis : synthesis of ϕ X174 Replicative form Requires RNA synthesis Resistant to Rifampicin . Proc.Nat.Acad.Sci. USA 69 : 2691 - 2695 (1972).

2. Fox,R.M., J.Mendelsohn , E.Barbosa and M.Goulian . RNA in Nascent DNA from cultured Human Lymphocytes . Nature , New Biology 245 : 234-237 (1973)

3. Geider, E.B. and H.Schaller . An RNA transcribed from DNA at the Origin of Phage fd single strand to Replicative Form conversion . Proc.Nat.Acad.Sci. USA 75 , 645 - 649 (1978)

4. Sugino,A. , S.Hirose and R.Okazaki . RNA-linked Nascent DNA fragments in Escherichia coli . Proc.Nat.Acad.Sci. USA 69 : 1863 - 1867 (1972)

5. Beljanski,M. and M.I.Aaron Da Cunha . Particular Small size RNA and RNA fragments from different Origins as Tumor Inducing Agents in Datura stramonium . Molecular Biology reports 2 : 497 - 506 (1976).

6. Plawecki,M. and M.Beljanski . Synthese in vitro d'un ARN utilisé comme Amorceur pour la Réplication de l'ADN . C.R.Acad.Sc.Paris Série D 278 : 1413-1416 (1974)

7. Beljanski,M., M.S.Beljanski , M.Plawecki and P.Bourgarel . ARN fragments , Amorceurs nécessaires à la Réplication in vitro des ADN . C.R.Acad.Sc.Paris Série D , 280 : 363 - 366 (1975).

8. Evans,A.H. Introduction of Specific drug resistance properties by purified RNA - containing fractions from Pneumococcus . Proc.Natl.Acad.Sci.USA 52 , 1442 - 1449 (1964)

9. Beljanski,M., M.S.Beljanski , P.Manigault and P.Bourgarel . Transformation of Agrobacterium tumefaciens into a non-oncogenic species by an Escherichia coli RNA. Proc.Natl.Acad.Sci. USA 69 : 191-195 (1972)

10. Beljanski,M and M.Plawecki . Transforming RNA as a template directing RNA and DNA synthesis in bacteria. In the Role of RNA in Reproduction and Development Niu M.C. and Segal, S.J. Eds. North Holland Publ.Co. 199-224 (1973).

11. Mishra,N.C., M.C.Niu and E.L. Tatum . Induction by RNA of Inositol Independence in Neurospora crassa . Proc.Natl.Acad.Sci. USA 72 , 642-645 (1975)

12. Tung, T.C. and Niu, M.C. Nucleic Acid-Induced transformation in Goldfish. Scientia Sinica XVI : 366-384 (1973)

13. Tung, T.C. and M.C.Niu . Transmission of the nucleic acid induced character , caudal fin , to the offspring in Goldfish . Scientia Sinica XVIII : 223-228 (1975).

14. Fishman,M. The Role of macrophage RNA in the immune response. In The Role of RNA in Reproduction and Development. Niu,M.C. and Segal,S.J. eds. North Holland Publ.Co. 127-136 (1973)

15. Niu,M.C. Cellular Mechanism in differentiation and Growth. Ed.Rudnick . Prince-ton Univ. Press. 155-171 (1956)

16. Niu,M.C. Thymus Ribonucleic acid and Embryonic differentiation. Proc.Natl.Acad. Sci. USA 44 : 1264-1274 (1958)

17. Lee,H and M.C.Niu . Studies on Biological potentiality of testis-RNA. I.Induction of axial structures in whole and excised chick blastoderms. In The Role of RNA in Reproduction and Development . Niu,M.C. and Segal,S.J. Eds. North Holland Publ. Co. 137-154 (1973)

18. Deshpande,A.K., S.B.Jakowlew , Arnold,H, P.A.Crawford, M.A.Q.Siddiqui. A novel RNA affecting embryonic gene functions in early chick blastoderm . J.Biol. Chem. 252 : 6521-6527 (1977)

19. Beljanski,M, M.I.Aaron Da Cunha, M.S.Beljanski, P.Manigault, and P.Bourgarel. Isolation of the Tumor Inducing RNA from oncogenic and non-oncogenic Agrobacterium tumefaicens . Proc. Natl. Acad.Sci. USA 71 : 1585-1589 (1974)

20. Beljanski,M. , L.Chaumont , C.Bonissol and M.S.Beljanski. ARN fragments , Inhibiteurs de la multiplication des virus du Fibrome de Shope et de la Vaccine. C.R.Acad.Sc. Paris Série D , 280 : 783-789 (1975)

21. Beljanski,M., M.Plawecki, P.Bourgarel and M.S.Beljanski. Nouvelles substances (RLB) actives dans la leucopoiese et la formation des plaquettes. Bull.Acad. Nat. Med. 162 : 475-481 (1978)

22. Le Goff,L. and M.Beljanski. Stimulation de l'induction ou inhibition du developpement des tumeurs de Crown gall par des ARN fragments U_2 . Interférence de l'auxine . C.R.Acad.Sci. Paris Série D 288 , 147-150 (1978)

23. Schlager,S.I., R.E.Paque and S.Dray. Complete and Apparently specific local tumor Regression using syngeneic or xenogeneic Tumor-Immune RNA extracts. Cancer Res. 35 : 1907-1914 (1975)

24. Pottathil,R. and H.Meier. Antitumor Effects of RNA isolated from murine Tumors and Embryos. Cancer Res. 37 : 3280 - 3286 (1977)

25. Pardee, A.B., F.Jacob and J.Monod. The Genetic Control and Cytoplasmic Expression of "Inductibility" in the Synthesis of β-Galactosidase by *Escherichia coli* J.Mol.Biol. 1 : 165-178 (1959)

26. Beljanski,M. , P.Bourgarel and M.S.Beljanski. Drastic Alteration of Ribosomal RNA and Ribosomal Proteins in Showdomycin-resistant *Escherichia coli* . Proc. Natl. Acad. Sci. USA 68 : 491-495 (1971)

27. Shine,J. and L.Dalgarno. Terminal-Sequence Analysis of Bacterial Ribosomal RNA. Correlation between the 3'-terminal Polypyrimidine. Sequence of 16 S RNA and Translational Specificity of the Ribosome . Eur.J.Bioch. 57 : 221-230 (1975).

28. Beljanski,M., P.Bourgarel and M.S.Beljanski. Showdomycine et Biosynthèse d'un ARN non complémentaire de l'ADN. Inst.Pasteur 118 : 253-276 (1970)

29. Lowry,H. , N.J.Rosebrough, A.L.Faar and R.J.Randall. Protein Measurement with Phenol Reagent . J.Biol.Chem. 193 : 265-269 (1951)

30. Burton,K. A Study of the Conditions and Mechanism of the Diphenylamine Reaction for the Colorimetric estimation of Deoxyribonucleic Acid. Biochem. J . 62 : 315-323 (1956)

31. Björk,G.R. and I.Svensson . Analysis of Methylated Constituents from RNA by Thin Layer Chromatography . Biochem. Biophys. acta 138 : 430-433 (1967)

32. Hori,M. Chromatography of Nucleic Acid Components on Dowex 1 Column by the use of an Automatic Recording System. In Methods in Enzymology. Nucleic Acids Part A. Ed.Grossman,L. and Moldave,K. Academic Press XII 381-390 (New York) (1967).

33. Schildkraut,C.L. and J.J.Maio. Fractions of HeLa DNA depending on their Content of Guanine + cytosine . J. Mol.Biol. 46 : 305-312 (1969)

34. Beljanski,M. and M.S.Beljanski . RNA-Bound Reverse Transcriptase in Escherichia coli and in vitro Synthesis of a Complementary DNA. Biochemical Genetics 12 : 163-180 (1974)

35. Bessis,M. Le Sang et la Lymphe . Encyclopédie Française (sous la Direction du Professeur Grassé). Librairie Larousse IV : 229-241 (1960)

36. Hill,D.L., W.R.Jr.Laster and R.F.Struck. Enzymatic Metabolism of Cyclophosphamide and Nicotine and Production of a Toxic cyclophosphamide metabolite. Cancer Res. 32 : 658 - 665 (1972)

37. Shwartzman,G. Phenomenon of Local Tissue Reactivity and its Immunological, Pathological and clinical Significance . New York : P.B. Hoeber Incorp. 1-37 (1937)

38. Lindsay,H.L., P.W.Trown. J.Brandt and M.Forbes. Pyrogenicity of Poly I-Poly C in rabbits . Nature 223 : 717-718 (1969)

39. Smith,W.W., I.M.Alderman and J.Cornfield . Granulocyte release by endotoxin in normal and irradiated mice. Am.J.Physiol. 201 : 396-402 (1961)

40. Fukuda,T. and O.Matsumato. Endogenous Factors concerning the febrile and the leukocytic response to bacterial endotoxin in relation to the adrenal cortex. Jap. J. Physiol. 9 : 274-281 (1959)

41. Mechanic,R.C., E.Frei III , M.Landy and W.W.Smith. Quantitative studies of human leukocytic and febrile response to single and repeated doses of purified bacterial endotoxin. J.Clin.Invest. 41 : 162-172 (1962)

42. Dale,C.D., A.S.Fauci , D.P.IV Guerry and S.M.Wolff. Comparison of Agents Producing a Neutrophilic Leukocytosis in Man. J.Clin.Invest. 56 : 808-813 (1975)

43. Henderson,E.S. The Granulocytopenic Effects of cancer chemotherapeutic agents in "Drugs and Hematologic Reactions ". Eds N.V.Dimitrov and J.H.Nodine. The Twenty ninth Hahnemann Symposium . 18 : 207-220 , 1974. Grune and Stratton New York , London

Adenosine Diphosphate Ribosylation and Regulation of Nucleic Acid Synthesis

(histones/adenosine diphosphate ribose/rat liver)

LUIS O. BURZIO*, PERE PUIGDOMENECH**, ADOLFO RUIZ-CARRILLO** and S.S. KOIDE***

*Institute of Biochemistry, Faculty of Science, Universidad Austral, Valdivia, Chile; **Max-Planck Institut fur Molekulare Genetik, 1 Berlin 33, Ihnestrasse 63-73; ***Center for Biomedical Research, The Population Council, The Rockefeller University, New York, New York 10021.

ABSTRACT This novel reaction is mediated by a synthetase located in the nucleus and chromatin of eukaryotic cells. Histones H1 and H2B are modified. The ADP-ribose moiety is attached to glutamic acid via an ester bond: two sites in H1, position 2 and 116, and one site in H2B, position 2.

The effect of ADP-ribosylation on the properties of H1 and H2B was investigated by measuring the rate of hydrolysis by chymotrypsin, sucrose gradient centrifugation and nuclear magnetic resonance (NMR) spectroscopy. ADP-ribosylation had a protective influence to chymotrypsin hydrolysis. At low salt concentration (0.15M NaCl) the sedimentation coefficients of untreated and ADP-ribosylated H1 were 1.4S and 3S, respectively. No difference in sedimentation velocity was observed when measurement was performed at high salt concentration (1.0M or 2M NaCl).

These findings suggest that ADP-ribosylation promotes aggregation of histone H1. High resolution NMR measurements indicate that histone H1 occurs as random coils and ADP-ribosylation confers structural order. Low field NMR spectrum of ADP-ribosylated H1 showed two modifying groups corresponding to ADP-ribose moieties. Signal for tyrosine appeared, indicating a higher order of structured state of ADP-ribosylated H1. Properties of H2B remained unchanged after ADP-ribosylation.

ADP-ribosylation of H1 markedly influenced its binding to DNA. Chromatography of ADP-ribosylated H1 on DNA-Sephadex G-25 revealed that highly ADP-ribosylated H1 (ratio of moles of ADP-Rib per mole of H1 $\geq$ 10) was eluted earlier than poorly modified H1 (ratio of about 1). ADP-ribosylated H1 was less effective in promoting the binding of DNA to millipore filter than unmodified H1. This reduced capacity disappeared upon hydrolysis of the ADP-Rib moieties from H1, i.e., alkali-treated ADP-ribosylated H1 recovered its original potency. NMR studies showed that ADP-ribosylation reduced the capacity of H1 to bind to DNA at the same salt concentration. The present results suggest that modification of H1 by ADP-ribosylation might influence structural transformation of chromatin.

Proteins can be modified as a post-translational process by several biochemical reactions: methylation, acetylation, phosphorylation, peptidylation and nucleotidylation. An interesting and novel modifying reaction is adenosine diphosphate (ADP)-ribosylation, catalyzed in eukaryotic cells by poly(ADP-ribosyl) synthetase (polymerase or synthase) and ADP-ribosyl transferase. The poly (ADP-ribosyl) synthetase is located in the cell nucleus and transfers ADP-ribose moiety of NAD^+ to specific proteins as monomers and polymers. The ADP-ribosyl transferase is situated in cytosol of avian erythrocytes and this enzymatic activity is also mediated by bacterial toxins. The acceptor proteins, e.g., cAMP cyclase and elongation factor 2, (EF-2) are modified by the attachment of monomers of ADP-ribose. Modification of acceptor proteins by ADP-ribosylation has been implicated in the regulation of DNA synthesis, repair, transcription, and also participates in the events associated with cellular differentiation and proliferation (1-3).

Possible involvement of ADP-ribosylation in cellular differentiation is based on observations that, during the process of differentiation and development, there is an alteration in the NAD level and in ADP-ribosylating activities. Studies have been conducted with the following systems showing this relationship: embryonic chick limb bud (4); progesterone-induced maturation of Xenopus oocyte (5,6); mouse ovum and embryo (7,8); mouse myeloid leukemic cells (9); erythroleukemic mouse spleen cells (Friend cells) (10, 11); and in rat cardiac muscle (12).

A recent study on the role of ADP-ribosylation in transcription and cellular differentiation was performed by determination of poly(ADP-ribosyl) synthetase activity in nuclei isolated from testis of mice of various ages (13). The enzyme activity increases during spermatogenesis. Total RNA polymerase activity in isolated testis nuclei is relatively constant during testicular development. On the other hand, RNA polymerase I activity is high in testis nuclei of the neonatal mouse, and decreases to a low level in those of the post-pubertal mouse. In contrast, polymerase II activity is highest in testicular nuclei of the post-pubertal mouse. The finding that ADP-ribosylating activity is inversely related to RNA polymerase I activity suggests its participation in ribosomal RNA synthesis.

NAD^+ metabolism and ADP-ribosylation of nuclear protein appear to vary inversely with DNA synthesis during the cell cycle. In general, poly(ADP-ribosyl) synthetase activity was lowest in S phase and highest in either G_2 or G_1 (1, 14-20). These results suggest that ADP-ribosylation is involved in the termination of DNA synthesis.

ADP-ribosylation of nuclear proteins has been implicated in cell replication and DNA synthesis (21-30). Involvement of ADP-ribosylating activity in DNA synthesis varies with the tissues studied. The thesis that ADP-ribosylation of nuclear proteins participates in DNA synthesis is based on the finding that pre-incubation of isolated rat liver nuclei with NAD^+ blocked ^3H-dTTP incorporation into DNA (21). It was concluded that inhibition of template activity for DNA synthesis of isolated nuclei and chromatin is due to a direct ADP-ribosylation of a Ca^{2+}, Mg^{2+}-dependent endonuclease that initiates nicks in the DNA (22). On the other hand, incorporation of ^3H-dTTP into DNA of isolated nuclei from Novikoff hepatoma, leukemic cells and lymphocytes are not influenced by ADP-ribosylation (31-34).

A recent finding revealed that poly(ADP-ribosyl) synthetase is located in the linker DNA region of HeLa cell chromatin and not with the nucleosomes (35, 36) and that a higher synthetase activity was found in the extended form of chromatin undergoing DNA replication and repair than in the inactive condensed form (38). The high ADP-ribosylating activity in extended chromatin and in nucleoli suggests that this enzymatic reaction may participate in the regulation of DNA and ribosomal RNA synthesis.

The major nuclear protein acceptors of ADP-ribosylation are histones H1, H2B, and, to a lesser extent, H2A, H3, HMG, and M_1-M_4 (35-37, 39-44). ADP-ribosylation of H1 will increase electrostatic attraction and promote cross-linking of histones. Cross-linking can also take place by covalent dimer formation of H1. The net effect is to sustain chromatin condensation, thereby modulating template activity of chromatin for nucleic acid synthesis (43, 45-50).

Information about the effect of ADP-ribosylation on the properties of the acceptor protein is limited. The best characterized systems are ADP-ribosylation of the elongation factor 2 catalyzed by diphtheria toxin (54) and the modification of the E. coli RNA polymerase (55). In the present study the effect of ADP-ribosylation on the properties of H1 and H2B were examined.

MATERIALS AND METHODS

Rat liver nuclei were purified and incubated with $[^{14}C]NAD^+$ as described recently (42). ADP-ribosylated H1 and H2B were purified by standard procedures (42, 43). Chromatography of ADP-ribosylated H1 on DNA-Sephadex G-25 was carried out as follows: a column of DNA-Sephadex (0.9cm X 20cm) was equilibrated with 0.1M NaCl, 5mM sodium phosphate buffer (pH 710) and 1mM EDTA at 4°C. The modified H1 was dissolved in the same buffer and applied to the column at very low flow rate (2ml/h). After the sample was added, the column was washed with the equilibration buffer (2 column volumes), H1 was eluted with a linear gradient between that buffer and 1M NaCl in 5mM sodium phosphate (pH 7.0) and 1mM EDTA (100 ml each). The amount of protein was measured by turbidimetry (56) and the radioactivity by scintillation counting. DNA binding to millipore filters was carried out according to the procedure described by Renz and Day (57). H1 and ADP-ribosylated H1 were hydrolyzed with 1/500th (w/w) of chymotrypsin at 25°C (42).

The rate of hydrolysis of H1 by chymotrypsin was determined by subjecting the reaction mixture to electrophoresis on acid/urea polyacrylamide gels (42).

The amount of H1 and the fragment were determined by densitometry of the stained gel in a Gilford 2400 spectrophotometer. Measurement of the sedimentation velocity of H1 in a sucrose gradient at low (0.15M NaCl) and high (2M) salt concentrations was performed according to a previously described procedure (58).

NMR measurements were carried out as described in a previous report (46, 59). Spectra a5 270 MHz were recorded in a Bruker WH-270 spectrometer (Biophysics Laboratory, Portsmouth Polytechnic, Portsmouth, England). The number of scans for protein and for complexes with DNA using the Fourier Transform Mode were between 15,000 and 35,000; and 2,000 respectively. Convolution difference methods (60) were used for resolution enhancement in some experiments. Calibrated tubes of 5mm were used in all studies.

For these experiments ADP-ribosylated H1 was prepared from rat liver nuclei incubated with 1mM cold NAD. The modulated protein was purified as described above. Protein samples were dissolved directly in 99.96% ^2H_2O. The pH was adjusted by the addition of 1M NaOD. Urea-d_4 (MSD isotopes) was added as a solid. DNA-histone complexes were prepared by stepwise dialysis as previously described (46). To obtain the spectrum of ADP-ribosylated H1 in 1M NaCl (Fig 8C), solid NaCl was added to the 0.35M NaCl DNA-histone complex. Calf thymus DNA (Type 1, Sigma Chemical Co.) was purified by treating with RNase and deproteinized in chloroform-isoamyl alcohol (24/1, v/v). The DNA was sheared in a Virsonic sonicator for 30 seconds at maximum power output.

RESULTS

Summary of our recent determination of the sites of ADP-ribosylation of H1 and H2B is depicted in Fig 1 (42, 43). H1 contains two sites of modifications: glutamic acid at position 2 and 116, and H2B containing one single site of modification of glutamic acid at position 2 (42). The average chain of ADP-ribose on H1 was composed of trimers while that of H2B consisted of only monomers (42, 43).

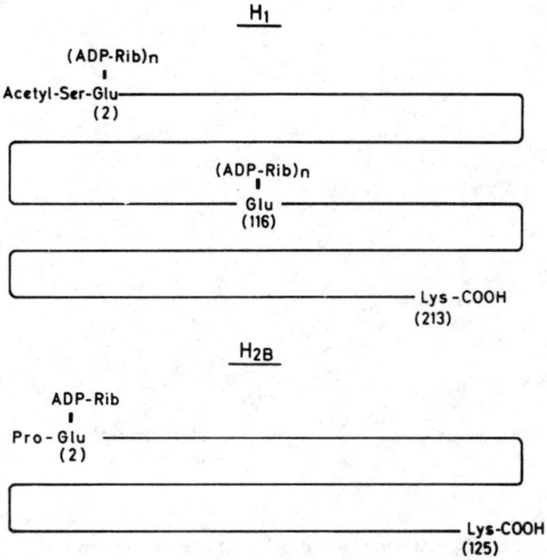

Fig 1. Sites of ADP-ribosylation on H1 and H2B.

Conformational Changes of ADP-ribosylated H1. The rate of hydrolysis was determined by electrophoretic analysis of each sample in acid/urea gels and by densitometric readings of the stained gel. Unmodified H1 was hydrolyzed completely into fragments upon six minutes of incubation. On the other hand, even after incubation for six minutes or longer about 12% of ADP-ribosylated H1 remained intact. This induced resistance to chymotryptic hydrolysis, however, disappeared when oligomers of ADP-ribosylated H1 were removed by alkali treatment (0.1N NaOH Fig 2).

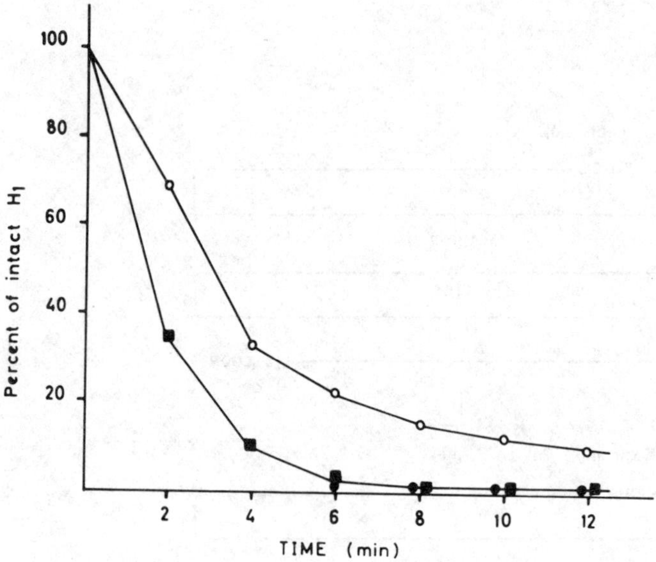

Fig 2. Rate of hydrolysis of ADP-ribosylated H1 with chymotrypsin. Unmodified and ADP-ribosylated H1 were incubated with 1/500th (w/w) of chymotrypsin at 25°C (42). Aliquots were removed at the indicated time and subjected to acid/urea gel electrophoresis. The amount of intact H1 was determined by densitometry of the stained gel. ADP-ribosylated H1 (O—O); unmodified H1 (■—■); alkali-treated ADP-ribosylated H1 (●—●).

Sedimentation of ADP-ribosylated H1 was determined by centrifugation in a sucrose gradient at low (0.15M NaCl) and high (2M NaCl) ionic strength. The front of the modified H1 sedimented faster than that of control H1. The average sedimentation coefficient of unmodified H1 in 0.15M NaCl was 1.4S while that of ADP-ribosylated H1 was 3S (Fig 3A). When the experiment was performed in a sucrose gradient with 2M NaCl, no significant difference was observed between the modified and the control H1 (Fig 3B). The faster sedimenting fractions of ADP-ribosylated H1 are probably aggregates formed by electrostatic attraction between the negative charge possessed by oligomers of ADP-Rib and the positive groups in the protein.

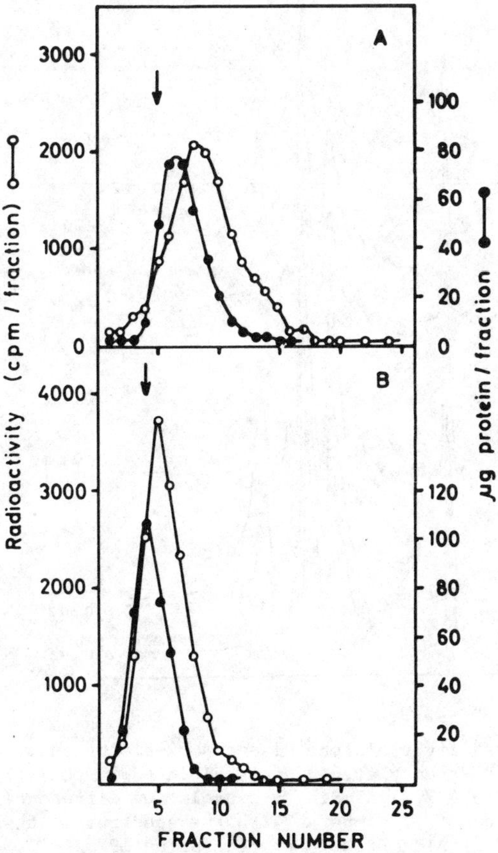

Fig 3. Sedimentation of ADP-ribosylated H1 in a linear sucrose gradient. A; sedimentation in the presence of 0.15M NaCl. B; in 2M NaCl. The position of unmodified H1 is indicated by arrow. Sedimentation is from left to right.

The conformation of histone in solution was studied by high resolution nuclear magnetic resonance spectroscopy. The spectrum of non-modified H1, $^{2}H_2O$ (Fig 4) indicates that its conformation is a random coil. This finding is in agreement with the results obtained with H1 purified from other sources under the same solvent conditions (46, 59). On the other hand, ADP-ribosylated H1 in $^{2}H_2O$ (Fig 4B) appeared to be somewhat structured since the spectrum showed signals at high field at the position of the ring-current shift peaks (Fig 4E, con diff) compared to non-modified H1. The spectrum of ADP-ribosylated

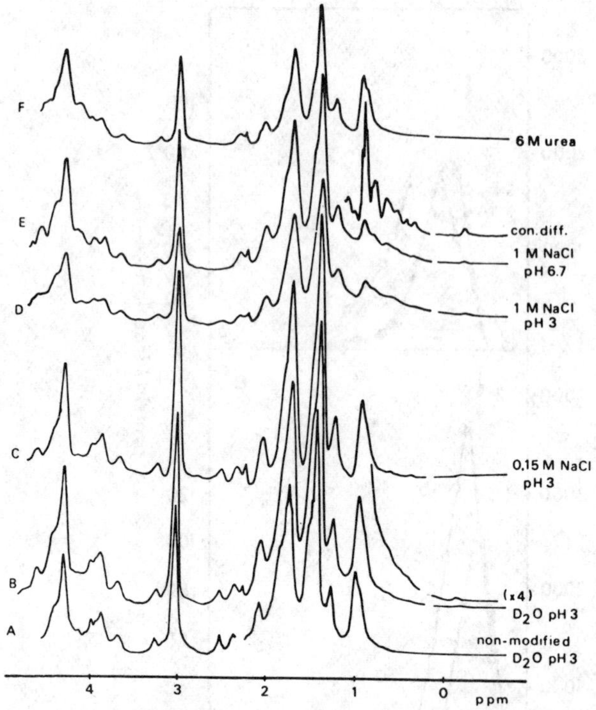

Fig 4. Conformation of rat liver histone H1 and ADPR-histone H1. High field 270 MHz spectrum of (A) ADP-ribosylated histone H1 in ^2H_2O, pH 3 (B); 0.15M NaCl, pH 3 (D); 1M NaCl, pH 6.7, including a convolution difference treated spectrum (E); and in 6M urea (F), compared with the spectrum of the non-modified protein in ^2H_2O, pH 3 (A). Protein concentration 19 mg/ml.

H1 is influenced by increasing the salt concentration to 0.15M (Fig 4C). The pattern is similar to that observed with non-modified H1 (result not shown). When the spectrum of ADPR-H1 is observed in the presence of urea, the ring-current shift signals disappeared (Fig 4F). Increasing the ionic strength to 0.1M NaCl (Fig 4D, 4E) produced a broadening of the apolar groups as typically observed with H1 obtained from other sources.

The low field spectrum of ADPR-H1 (Fig 5B) showed two signals. One at 6.2 and the other at 8.5 ppm due to the modifying groups (c.f. Fig 5A and 5B). Both signals consist of complex line shape that may relate to the presence of

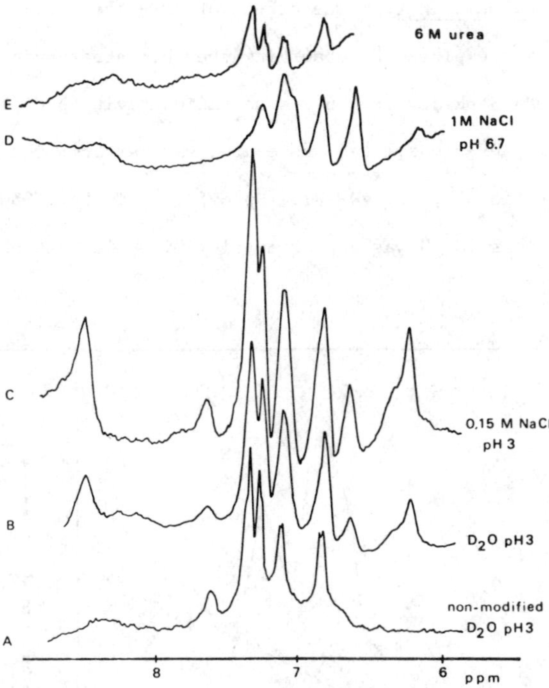

Fig 5. Conformation of rat liver histone H1 and ADPR-histone H1. Low field 270 MHz NMR spectrum of ADP-ribosylated and non-modified histone H1. Conditions are the same as those described in Fig 4.

ADP-Rib groups in two positions along the polypeptide chain. These signals are, however, difficult to study since the peaks broaden with time due to deuterium exchange. Structural effects due to ADP-Rib groups are also observed at low field (Fig 5B, 5C). A signal at 6.6 ppm probably arising from a tyrosine is easily discernible with ADP-ribosylated H1 in $^{2}H_{2}O$ and absent in unmodified H1 (c.f. Fig 5A and 5B). Based on the pattern of the signal which is prominent and well resolved, the degree of structured state was calculated to be about 25% for ADP-ribosylated H1. This estimate agrees reasonably well with the 30% ADP-Rib content of the sample analyzed.

Interaction of H1 and DNA. The effect of ADP-ribosylation on the affinity of H1 for DNA was explored by observing the chromatographic behavior of the modified H1 on DNA-Sephadex (Fig 6). The radioactivity profile showed that H1 most extensively ADP-ribosylated was eluted earlier from the column. For example, the mole ratio of [^{14}C] ADP-Rib (Specific activity: 850/cpm/nmole) to moles of H1 in fractions 48-50 was approximately 10, while that of fraction 70 was 1.3.

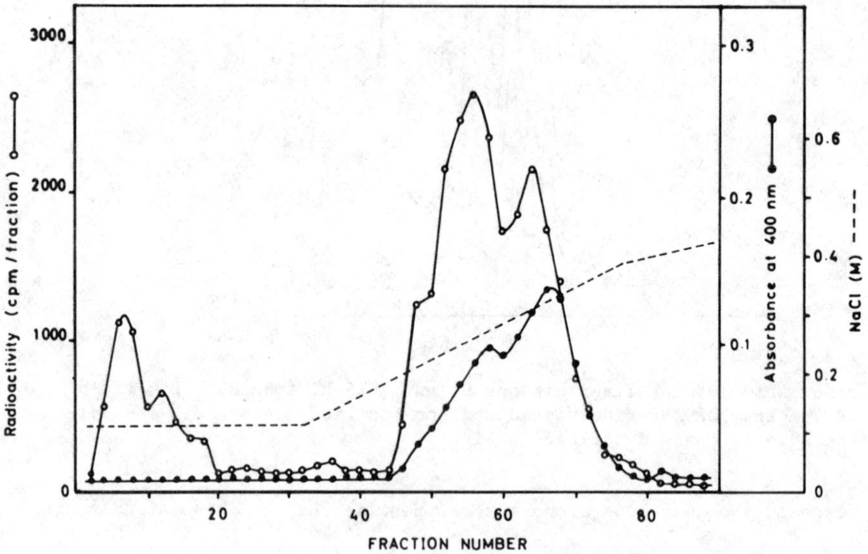

Fig 6. Chromatography of ADP-ribosylated H1 on DNA-Sephadex. Procedures are described under Methods.

Histone-dependent binding of DNA to millipore filters was determined. Although all histones are capable of promoting DNA binding, H1 is the most potent agent (57, 61). The capacity of ADP-ribosylated H1 to induce binding of DNA was determined (Fig 7). Both modified and control H1 achieved half-maximal binding of DNA, however, ADP-ribosylated H1 was less effective.

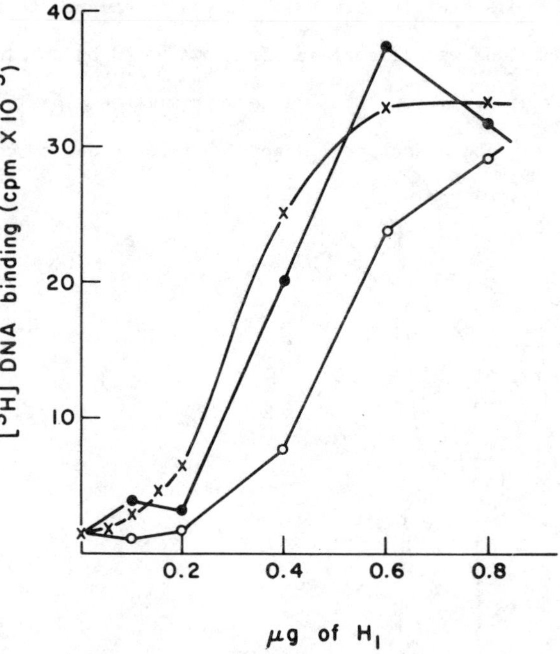

Fig 7. H1-mediated binding of [^{3}H]DNA to millipore filter. Unmodified H1 (X—X); ADP-ribosylated H1 (O—O); alkali-treated ADP-ribosylated H1 (●—●).

The ADP-ribosylated H1 used in these experiments was obtained after chromatography on DNA-Sephadex (fraction 47 to 52, Fig 6). These pooled fractions possessed a ratio of ADP-ribose to H1 of about 10. When similar binding experiment was performed with minimally modified H1 (fraction 65 to 74, Fig 6), the binding capacity was similar to that effected by control H1. The reduced capacity of modified H1 is due to the covalently attached ADP-ribose moieties because on incubation with 0.1M NaOH the alkali-treated H1 regained its potency in inducing DNA binding (Fig 7).

The capacity of H1 to bind to DNA was also studied by NMR spectroscopy. The results showed that ADP-Rib residue attached to H1 reduced the affinity of the protein for DNA (Fig 8). The spectrum of H1-DNA complex taken in 0.35M NaCl (Fig 8A) indicated little or no dissociation of unmodified H1 from DNA whereas at the same ionic strength, dissociation of the ADP-ribosylated H1-DNA complex was readily discernible (Fig 8B). It was calculated that approximately 40% of the histone was dissociated from DNA based on the intensities of the lysine peak at 3.02 ppm. This value approximates closely the degree of ADP-ribosylation. The signal at 2.2 ppm is probably due to contaminants such as acetate (Figs 8B and 8C).

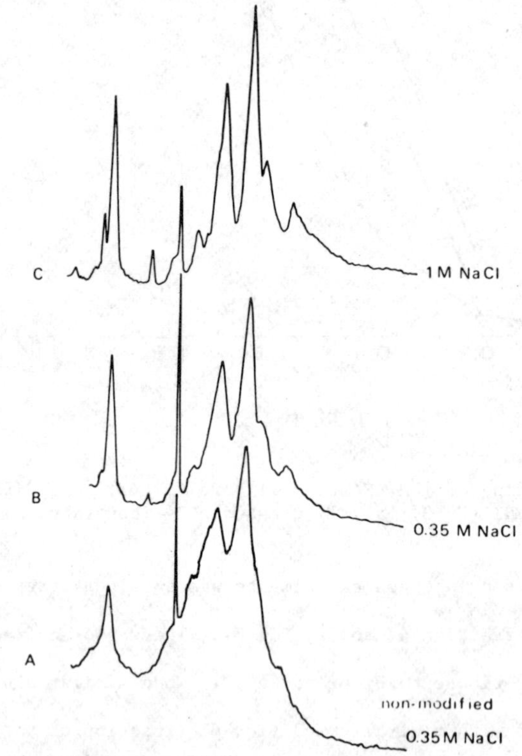

Fig 8. Rat liver histone H1 and ADPR-H1 affinity for DNA. 270 MHz spectrum of DNA: ADP-ribosylated histone H1 complex in 0.35M NaCl, pH 7 (B); and in 1M NaCl, pH 7 (C), compared with the spectrum of the complex of non-modified H1 in 0.35M NaCl, pH 7 (A). Protein concentration 1 mg/ml; histone/DNA ratio 0.2.

DISCUSSION

The present results showed that ADP-ribosylation of H1 confers some degree of structured organization to the protein and increases H1-H1 aggregation while decreasing the affinity of histone for DNA. It has been proposed that the primary role of H1 in chromatin structure is to induce packing of the 100 A° nucleofilament into a higher order of structure (46-48). Cross-linking of H1 histones may result in the folding of nucleosome chains to form the thick chromatin fiber as a solenoid structure (47) or promote clustered arrangement of nucleosomes (48, 62). ADP-ribosylation of H1 tends to diminish the affinity of H1 for DNA while increasing histone-histone interaction. In this manner the transformation of the thick 250 A° chromatin fiber to a nucleofilament of 100 A° can be modulated. The conformational changes of H1 induced by ADP-ribosylation might influence structural transformation of chromatin (Fig 9). It should be pointed

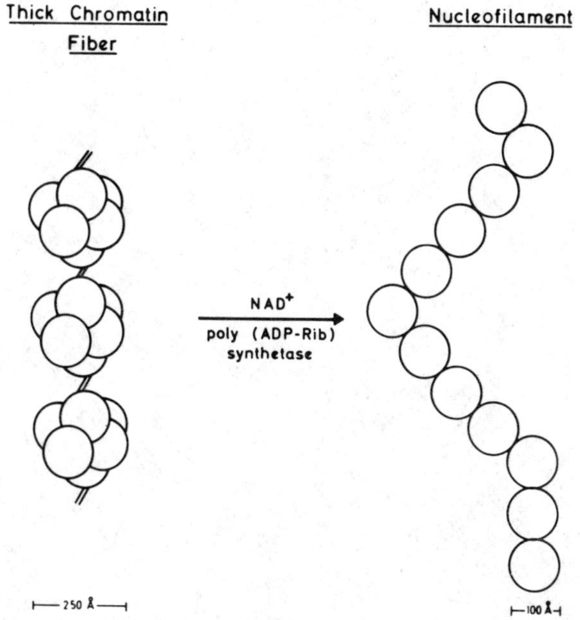

Fig 9. Summarizes our present hypothesis that extensive ADP-ribosylation of H1 will alter the continuation of chromatin to form a thick fiber to a nucleofilament.

out that in a similar series of study the conformational structure of H2B and its interaction with DNA appeared not to be altered by ADP-ribosylation.

ACKNOWLEDGEMENT

The authors are grateful to Dr. E.M. Bradbury (Biophysics Laboratory, Portsmouth Polytechnic, Portsmouth, England) for permitting the use of the NMR spectrometer. We thank Dr. Akira Inoue and Dr. Vincent Allfrey for the DNA-Sephadex G-25 and Dr. David Cowburn for reviewing the NMR spectra and paper. The study was supported in part by Grant 1 RO1 HD 13184-02 from NICHHD and from Grant S-79-20 from the Universidad Austral de Chile.

REFERENCES

1. Hilz, H., & Stone, P.R. (1976) Rev. Physiol. Biochem. Pharmacol. 76, 1-58.

2. Hayaishi, O., & Ueda, K. (1977) Ann. Rev. Biochem. 46, 95-116.

3. Purnell, M.R., Stone, P.R., & Whish, W.J.D. (1980) Biochem. Soc. Trans. 8, 215-227.

4. Caplan, A.I., & Rosenberg, M.J. (1975) Proc. Nat. Acad. Sci. USA, 72, 1852-1857.

5. Burzio, L.O., & Koide, S.S. (1977) Ann. N.Y. Acad. Sci. 286, 398-407.

6. Furneaux, H.M., & Pearson, C.K. (1978) Biochem. Soc. Trans. 6, 753-755.

7. Young, R.J., & Sweeney, K. (1978) Biochem. 17, 1901-1907.

8. Young, R.J., & Sweeney, K. (1979) J. Embryol. Exp. Morph. 49, 139-152.

9. Yamada, M., Shimada, T., Nakayasu, M., Okada, H., & Sugimura, T. (1978) Biochem. Biophys. Res. Commun. 83, 1325-1332.

10. Rastl, E., & Swetly, P. (1978) J. Biol. Chem. 253, 4333-4340.

11. Morioka, K., Tanaka, K., Nokuo, T., Ishizawa, M., & Ono, T. (1979) Gann 70, 37-46.

12. Claycomb, W.C. (1976) Biochem. J. 154, 387-393.

13. Momii, A., & Koide, S.S. (1980) Fed. Proc. 39 (Part II) 954.

14. Roberts, J.H., Stark, P., Gazzoli, M., & Smulson, M. (1973) Biochem. Biophys. Res. Commun. 52, 43-50.

15. Colyer, R.A., Burdette, K.E., & Kidwell, W.R. (1973) Biochem. Biophys. Res. Commun. 53, 960-966.

16. Kidwell, W.R., & Mage, M. (1976) Biochem. 15, 1213-1217.

17. Berger, N.A., Weber, G., Kaichi, A.S., & Petzold, S.J. (1978) Biochim. Biophys. Acta. 519, 105-117.

18. Berger, N.A., Kaichi, A.S., Steward, P.G., Klevecz, R.R., Forrest, G.L., & Gross, S.D. (1978) Exp. Cell Res. 117, 127-135.

19. Wielckens, K., Sachseomaier, W., & Hilz, H. (1979) Hoppe Seyler's Z. Physiol. Chem. 360, 39-43.

20. Caplan, A.I., Ord, M.G., & Stocken, L.A. (1978) Biochem. J. 174, 475-483.

21. Burzio, L.O., & Koide, S.S. (1970) Biochem. Biophys. Res. Commun. 40, 1013-1020.

22. Yoshihara, K., Tanigawa, Y., Burzio, L., & Koide, S.S. (1975) Proc. Natl. Acad. Sci. USA, 72, 289-293.

23. Yanigawa, Y., Kawamura, M., Kitamura, A., & Shimoyama, M. (1978) Biochem. Biophys. Res. Commun. 81, 1278-1285.

24. Tanigawa, Y., Kitamura, A., & Shimoyama, M. (1978) Eur. J. Biochem. 92, 261-269.

25. Janakidevi, K. (1978) Exp. Cell Res. 112, 345-351.

26. Ghani, Q.P., & Hollenberg, M. (1978) Biochem. J. 170, 387-394.

27. Suhadolnik, R.J., Baur, R., Lichtenwalner, D.M., Uematsu, T., Roberts, J.H., Sudhakar, S.K., & Smulson, M.E. (1977) J. Biol. Chem. 252, 4134-4144.

28. Berger, N.A., Weber, G., & Kaichi, A.S. (1978) Biochim. Biophys. Acta. 519, 87-104.

29. Berger, N.A., Adams, J.W., Sikorski, G.W., Petzold, S.J., & Shearer, W.T., (1978) J. Clin. Invest. 62, 111-118.

30. Bredehorst, R., Lengyel, H., & Hilz, H. (1979) Eur. J. Biochem. 99, 401-411.

31. Burzio, L., & Koide, S.S. (1972) FEBS Letters 20, 29-32.

32. Lehmann, A.R. & Shall, S. (1972) FEBS Letters 26, 181-184.

33. Roberts, J.H., Stark, P., Gazzoli, M., & Smulson, M. (1974) Fed. Proc. 33, 1491.

34. Burzio, L., Reich, L., & Koide, S.S. (1975) Proc. Soc. Exp. Biol. Med. 149, 933-938.

35. Mullins, D.W., Giri, C.P., & Smulson, M.E. (1977) Biochem. 16, 506-513.

36. Giri, C.P., West, M.H.P., Ramirez, M.L., & Smulson, M.E. (1978) Biochem. 17, 3501-3504.

37. Giri, C.P., West, M.H.P., & Smulson, M.E. (1978) Biochem. 17, 3495-3500.

38. Jump, D.B., Butt, T.R., & Smulson, M.E. (1979) Biochem. 18, 983-990.

39. Perrella, F.W., & Lea, M.A. (1978) Biochem. Biophys. Res. Commun. 82, 575-581.

40. Perrella, F.W., & Lea, M.A. (1979) Cancer Res. 39, 1382-1389.

41. Adamietz, P., Braeuer, H., & Hilz, H. (1979) In: Novel ADP-Ribosylation of Regulatory Enzymes and Proteins. (T. Sugimura, M. Smulson, eds.) (in press).

42. Riquelme, P., Burzio, L.O., & Koide, S.S. (1979) J. Biol. Chem. 254, 3018-3028.

43. Burzio, L.O., Riquelme, P.T., & Koide, S.S. (1979) J. Biol. Chem. 254, 3029-3037.

44. Ogata, N., Ueda, K., & Hayaishi, O. (1979) In: Novel ADP-Ribosylations of Regulatory Enzymes and Proteins. (T. Sugimura & M. Smulson, eds.) (in press).

45. Stone, P.R., Lorimer III, W.S., & Kidwell, W.R. (1977) Eur. J. Biochem. 81, 9-18.

46. Puigdomenech, P., Martinez, P., Cahre, O., Palau, J., Bradbury, E.M., & Crane-Robinson, C. (1976) Eur. J. Biochem. 65, 357-363.

47. Finch, J.T., & Klug, A. (1976) Proc. Nat. Acad. Sci. USA, 73, 1897-1901.

48. Hozier, J., Renz, M., & Nehis, P. (1977) Chromosoma 62, 301-317.

49. Lorimer III, W.S., Stone, P.R., & Kidwell, W.R. (1977) Exp. Cell Res. 106, 261-266.

50. Janakidevi, K., & States, J.C. (1980) In: Novel ADP-Ribosylations of Regulatory Enzymes and Proteins. (T. Sugimura & M. Smulson, eds.) (in press).

51. Nishimura, Y., Ueda, K., Honjo, T., & Hayaishi, O. (1968) J. Biol. Chem. 243, 3765-3767.

52. Otake, J., Miwa, M., Fujimura, S., & Sugimura, T. (1969) J. Biochem. (Tokyo) 65, 145-146.

53. Ueda, K., Omachi, A., Kawaichi, M., & Hayaishi, O. (1975) Proc. Natl. Acad. Sci. USA 72, 205-209.

54. Honjo, T., & Hayaishi, O. (1973) Curr. Topic Cell Regul. 7, 87-217.

55. Goff, C.G. (1974) J. Biol. Chem. 249, 6181-6190.

56. Burzio, L.O., & Koide, S.S. (1973) In: Poly(ADP-Rib), An Internat. Symposium, (M. Harris, ed.) Fogarty International Center, Proc. No. 26, Washington, D.C. pp. 117-140.

57. Renz, M., & Day, L.A. (1976) Biochemistry 15, 3220-3228.

58. Weintraub, J., Palter, K., & Van Lente, F. (1975) Cell 6, 85-110.

59. Puigdomenech, P., Martinez, P., Cahre, O., Palau, J., Bradbury, E.M., & Crane-Robinson, C. (1976) Eur. J. Biochem. 59, 237-243.

60. Campbell, J.D., Dodson, C.M., Williams, R.J.P., & Xavier, A.V. (1973) J. Magnet. Res. 11, 172-181.

61. Vogel, T., & Singer, M. (1975) J. Biol. Chem. 250, 796-798.

62. Olins, D.E. (1977) In: Molecular Human Cytogenetics, (R.S. Sparkes, D.E. Comings, & C.F. Fox, eds.) Academic Press, Inc., New York, pp. 1-16.

LENS MORPHOGENESIS INDUCED BY RETINAL FACTOR CONTAINING RIBONUCLEOPROTEIN COMPLEX

John H. Chen
Department of Biochemistry
New York University
Dental Center
New York, N.Y. 10010

ABSTRACT

We have observed that in cloned lens cell culture the elongated fiber-like lens cells can be morphologically transformed to rounded cell form (RF) by a set of specific conditions (Fig. 1). In an effort to reverse this process, we have isolated a membrane-associated ribonucleoglycoprotein factor (called lens morphogenic factor, lenmofin, LMF), from embryo neural retina, from vitreous fluid, as well as from retinal cell culture. When the factor was tested on various RF cell populations (e.g., normal, Nakano mutant and cAMP treated lens cells), three classes of cell-to-cell recognition specificity were observed: (1) lens cell aggregation without spreading; (2) lens cell spreading, elongation and lentoid body formation and (3) super lentoid body/miniature lens production. The molecular properties of the retinal factor have been determined, including molecular weight, RNA content, sugar binding specificity, heat stability and sensitivity to various enzymatic modifications. The physiological nature of the putative miniature lens have been studied by phase contrast, transmission-electron and scanning-electron microscopy as well as by chemical means. Results will be discussed in terms of lens morphogenesis and pathological development.

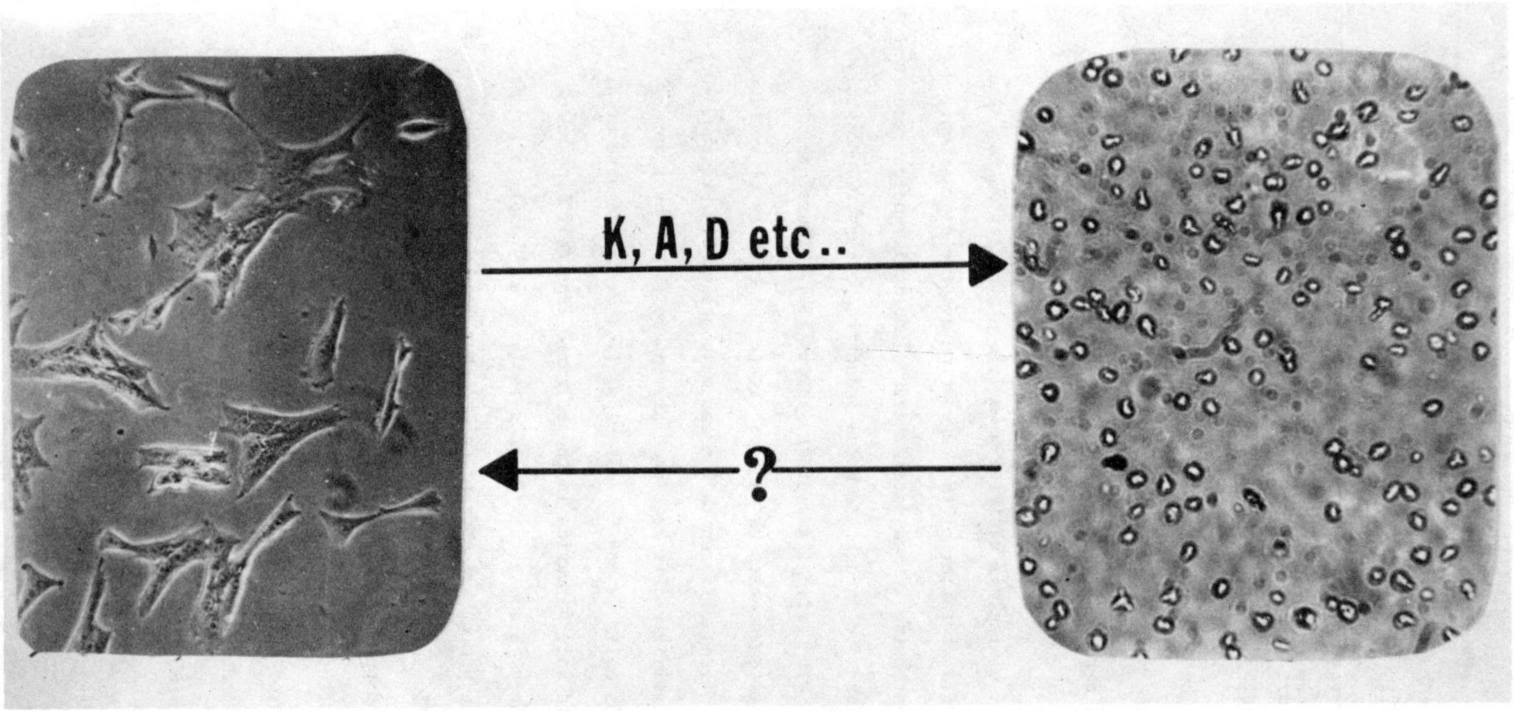

Figure 1. Phase micrograph of transformation of normal lens cells. Demonstration of elongated fiber-like lens cells can be transformed to round cell form by (1) K, High dose (500 μg/ml culture) of Kanomycin treatment. (2) A, aging more than six month in a deep freezer (-120°C) in the presence of D, demethylsulfoxide.

INTRODUCTION

Lens tissue has many unique physiological features and provides an ideal model system for investigating molecular differentiation. The morphological transformation process from epithelial cell to fiber cell constitutes a major challenge to investigators of lens morphogenesis. Lens cells are derived from a pure ectodermal cell line and are composed of its single cell type in various differentiated forms. The process of cellular differentiation results in the disappearance of nucleus and other functional organelles (1) and the majority of protein synthesis is directed toward the production of a single protein species, the crystallins. The lens clearly defined anatomical region and topological variation of biological activity make it a useful system for studying a) the control of cell division; b) the regulation of gene expression; c) organizational biosynthesis; and particularly d) molecular morphogenesis. However, lens morphogenesis is thought to be an induction process (2) possibly through the interaction with retinal and other systems (3). Therefore, the use of cloned culture system permits us to investigate the effects of various biological components on lens differentiation and morphogenesis. Previous work has established that terminally differentiated pigment cells can be transdifferentiated into lens cells (4). Russell et al (5) and Tsunematsu et al (6) have showed lentoid body formation in the cultured lens cell line monolayer. Most recently, Beebe et al (7) has isolated a cytosol factor from vitreous fluid which can promote lens fiber cell elongation. However, the interaction of a specific retinal factor with the growing monolayer of lens cells *in vitro* has not been studied. In this work, we present evidence that lens epithelial cells of normal and congenital cataractous mice can be induced by specific retinal factor into differentiated lens fibers in monolayer culture. The formation of super lentoid body or minilens in clones originating from single lens epithelial cells is also demonstrated.

MATERIALS AND METHODS

Materials and Methods are the same as described: mRNA and cDNA preparation (8,9); cell culture media and growth condition (5); RNA-cDNA hybridization (10). Isolation and purification of lenmofin: chick embryo neural retina tissue or cultured retinal cells were washed with buffer A, containing 100 mM NaCl, 3.5 mM KCl, 0.5 mM KH_2PO_4 and 0.1 mM Na_2HPO_4 10 mM $NaHCO_3$ pH 7.4 and suspended in buffer B, pH 7.4 containing the same component as in buffer A except $NaHCO_3$, glucose and in the presence of 0.1% aprotein. After homogenization, the homogenates were centrifuged at 2,000 x g 10 min. To the pellet, 5 ml of 0.3 M lactose and 1.0 M urea solution containing 0.1% aproteinin was added and incubated at 37°C for 30 min. Further low speed (10,000 x g) centrifugation was then carried out for 10 min. The supernatant was now subjected to high speed (100,000 x g, 1 hr) centrifugation. The pellet protein components were deoxycholate-(DOC, 0.5%)-solubilized at 30°C and filtered through GF/C filter. The filtrate was then dialyzed against buffer B in the presence of 0.05% DOC. Affinity chromatography on Octyl-Sepharose 4B and/or concanavalin A-Sepharose column were carried out according to the procedure published previously (11). Salt fractionation cut between the range of 30-45% $(NH_4)_2SO_4$ saturation was obtained as a crude extract for agglutination assay.

Agglutination Assay: The agglutination assays, using trypsinized human or sheep erythrocytes, were routinely carried out at room temperature by two different methods: 1) Clearance Assay: this assay was performed in a microtiter cluster dish (Costar). To each well in the dish, 25 μl of cell suspension of trypsinized erythrocytes were added and then the culture dish was shaken for 4 min and left to stand for 180 min before agglutination was evaluated. The end point of agglutination was taken as the highest dilution of LMF which agglutinates erythrocytes. Protein concentration was routinely determined by the method of Lowry et al., (12). Specific agglutination activity was defined as the hemoagglutination per μg of protein. 2) Hypochromicity assay: This assay was done according to Lis and Sharon (13). Hypochromic effect at 600 nm was measured as a function of various ligand concentration or dilution of the LMF.

Electrophoresis: The ribonucleoprotein components from affinity chromatograph were prepared for electrophoresis in 8% polyacrylamide sodium dodecyl sulfate gels as described previously (14). Molecular weight was estimated from the electrophoretic mobility relative to standard protein markers.

Induction Assay: Inductive assay was carried out following the lens cell developmental parameter against the control. These parameters includes cell aggregation, elongation, lentoid body or minilens formation. Two kinds of control were set up e.g., the mice lens cell without the addition of LMF and chicken embryo kidney cells with addition of the factor.

Microscopic Measurements: Inductive activity was done routinely by phase contrast microscopy. Scanning and transmission electron microscopic methods were also used for cytostructural analyses. The minilens were perfused with 1% glutaraldehyde-paraformaldehyde in phosphate-buffered fixative solutions. The thin section preparations were stained with uranyl acetate and lead citrate as described previously (15).

TABLE 1. MOLECULAR PROPERTIES OF LENMOFIN

MOLECULAR WEIGHT[1]	-SH[2]	RNA[3]	DNA[4]	TRYPTOPHAN[5]
52,000	+	1 - 2%	0	+

THE DETERMINATIONS WERE BASED ON THE METHODOLOGY OF 1, SDS GEL ELECTROPHORESIS; 2, (^{3}H)-IODOACETAMIDE METHOD; 3, MODIFIED ORCINOL OF MEJBAUM; 4, DIPHENYLAMINE AND 5, FLUORESCENCE SPECTRA.

RESULTS

Isolation and Characterization of Lens Morphogenic Factor (LMF) from Chick Neural Retina

LMF was isolated from freshly prepared chick embryo retina tissue or from 5 days old retinal cell culture. After affinity chromatography on octyl-Sepharose and salt fractionation, the deoxycholate-solubilized LMF migrates as a single discrete band on SDS/polyacrylamide gel electrophoresis with an apparent molecular weight of 52,000 (Fig. 2). Table I represents some molecular parameters of the purified LMF. Using modified orcinol method (16), we found that the LMF contains about 1-2% RNA. The presence of a RNA component was further substantiated by the RNaşe-digested abolishment of agglutination activity.

Alkylation of LMF with (^{3}H)-iodoacetamide before and after reduction yielded equivalent radiolabeling from the G-50 Sephadex column. These data suggest the presence of free -SH group. LMF binds to concanavalin A-Sepharose and can be eluted with methyl-α-mannoside (11) as desorbing agent suggesting a glycoprotein structure. The presence of tryptophan was demonstrated by measuring the typical emission spectrum at 345 mμ with excitation spectrum at 285 mμ. The possible involvement of tryptophan in agglutination activity was revealed by hyperchromicity change upon binding to various sugar ligands. (See Figure 3).

Determination of Agglutination Activity

The initial steps of cellular differentiation involves cell-to-cell interaction. One would expect therefore that the isolated LMF can induce cellular agglutination. We have employed red blood cell agglutination as our preliminary assay system. Figure 3A shows the agglutination activity of LMF isolated from different tissues as a function of dilution factor. It is clear

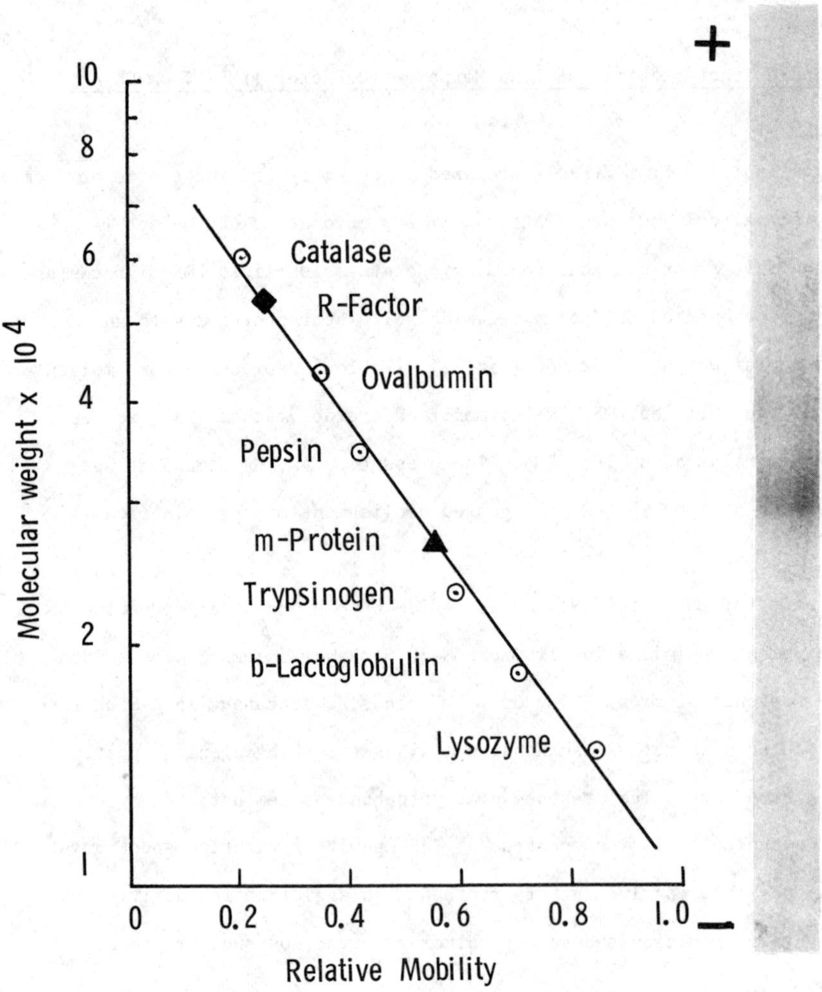

Figure 2. Sodium dodecyl sulfate-gel electrophoresis of Con-A Sepharose purified lenmofin. The left panel represents the calibration of standard protein markers run under the same condition. The molecular weights of the standards were: ovalbimun (45,000), catalase (60,000), pepsin (34,700), trypsinogen (23,500), β-lactoglobulin (18,400), lysozyme (14,300). ▲M-protein represents the enriched protein component isolated from the LMF induced normal culture system.

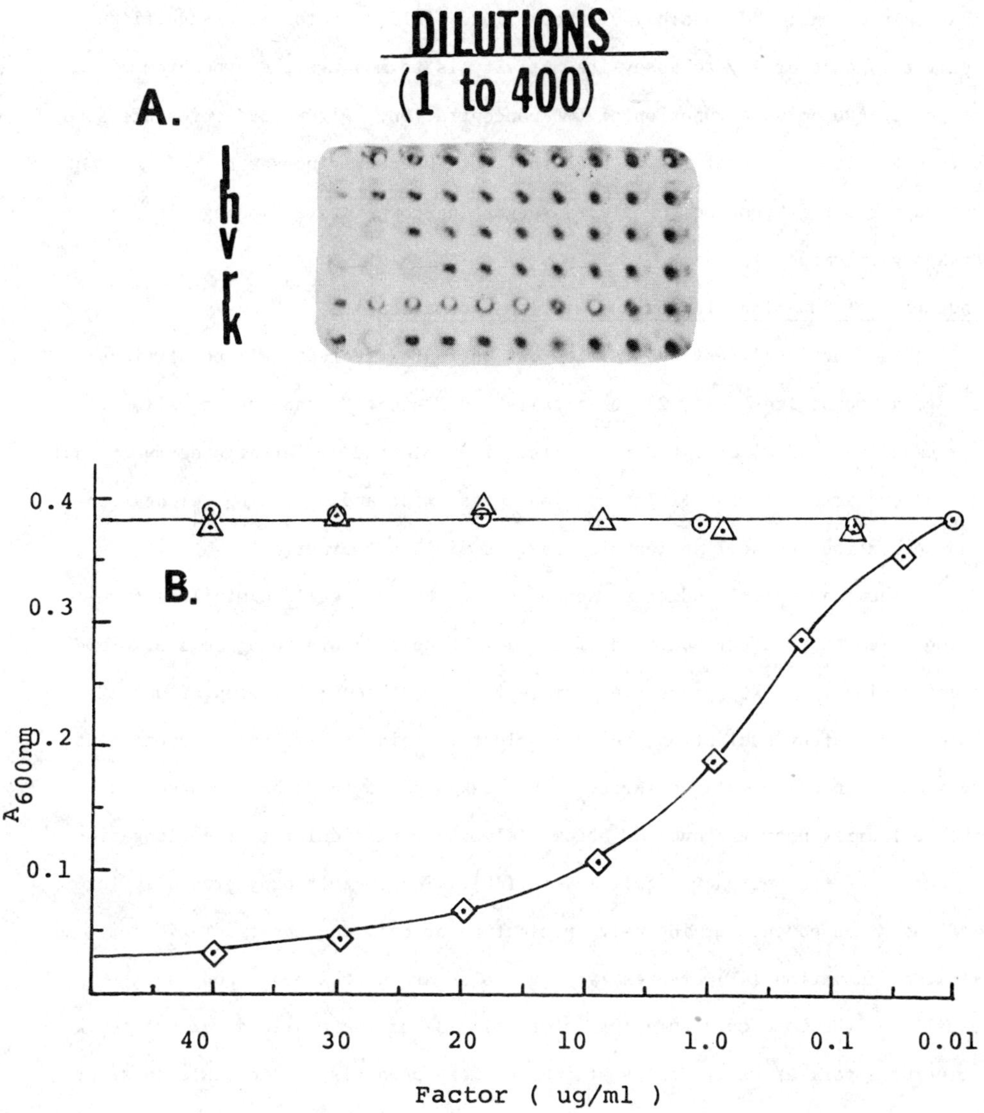

Figure 3. Hemoagglutination of trypsinized erythrocytes by lenmofin. (A) Activity of various factors isolated from: 1, (liver), h, (heart), V, (vitreous fluid), r, (retina) and K, (kidney). (B) hypochromic assay. The optical density at 600 nm of the erythrocytes suspension is plotted against the dilution of purified lenmofin. ◇, LMF alone; △, LMF + 8 mM mannose; ⊙, LMF + 8 mM N-acetylglucosamine.

that only retinal and possibly vitreous fluid yield a factor with significant activity. Another way to assay its activity is to measure the hypochromic effect at 600 nm as a function of LMF concentration. Figure 3B represents such a result. It is interesting to note that both N-acetylglucosamine and mannose can suppress the hypochromic effect, suggesting competative binding sites for both sugar components.

Induction of Superlentoid Body or Minilens Formation

This functional test was carried out on four different cell populations: 1) normal mouse lens cell, 2) genetically - deficient Nakano mutant cell, 3) human lens and 4) cAMP-treated cells. The selected differentiated functional parameters are cell-to-cell interaction, attachment and spreading, elongation and nucleation, as well as lentoid body and minilens formation.

Fig. 4 shows a typical inductive sequence from RF lens cell to minilens formation. Two hours after addition of LMF, the RF cells begin to aggregate. Sometimes nucleation takes place even before full length cellular elongation (4C). Immediately after nucleation, the cell cluster begins to form with concomitant reorientation of cellular polarity. This takes place in such a fashion that the cell array near the nucleus becomes almost preparticular to the elongation direction of the peripheral cell layers (4D). The lentoid body gradually forms (4E) as a consequence of intensive nucleation or cell condensation. The minilens formation (4F) creates large vacuoles around the dense body but usually at the more curvature (or anterior) side of the minilens. Zonules-like structures form at early stages of induced morphogenesis and remain even after the maturation of minilens structure.

So far this morphogenic process occurs only to the normal lens culture. The Nakano lens cell culture response different to this induction. The addition of LMF can only induce lentoid body-like structure (Fig. 5), as compared to the

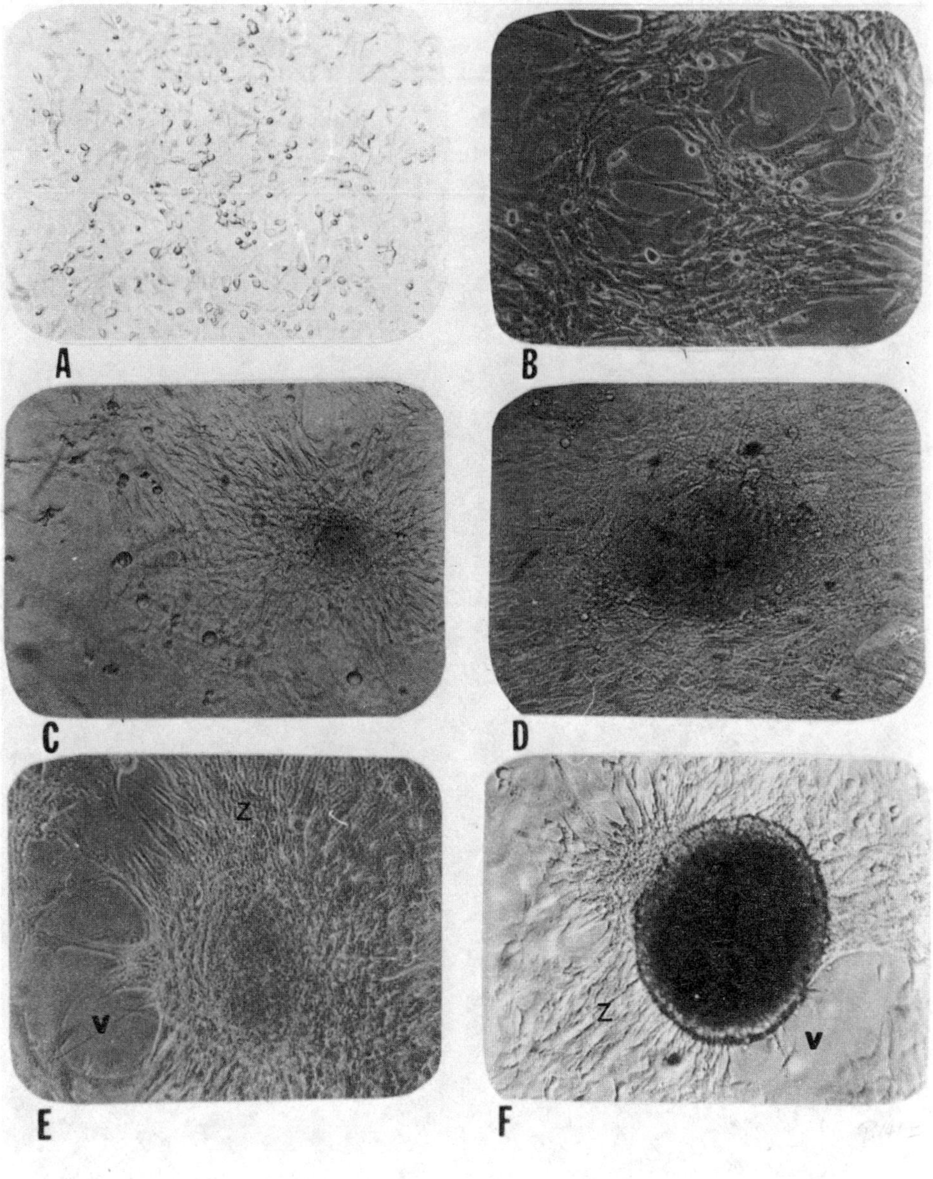

Figure 4. Phase contrast micrograph, showing the approximate sequences of minilens formation induced by lenmofin. A. starting RF normal lens cell culture. B. 12 hrs. after addition of lenmofin (30 μg/ml). C. nucleation. D. polarized condenzation. E. lentoid body production, and F. minilens formation v and z represent vacuoles and zonule-like structure respectively.

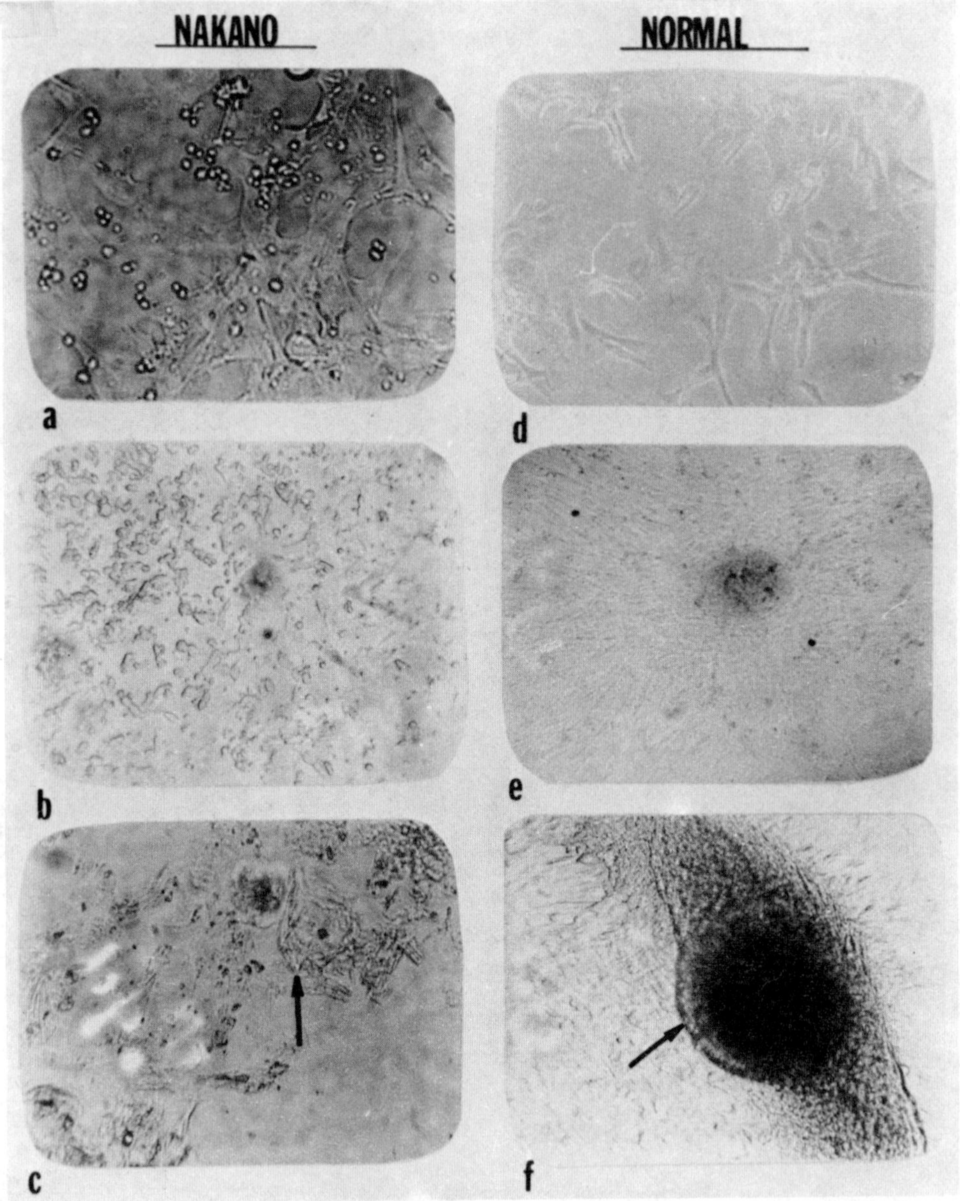

Figure 5. Comparison of lentoid body formation between normal and congenital caractous lens cells. Arrow in 5c points to the putative lentoid body formation of the lens cell from Nakano strain while in 5f the arrow indicates the super-lentoid body or minilens formation. a,b,c, and d,e,f represents the sequences of lentoid body formation for Nakno and normal cell culture respectively.

well-formed minilens from the normal culture system. Table 2 summarize the response of different cell populations to the addition of approximately equal quantities of LMF.

Evidence for Organized Structure

One important criterion for minilens formation is identification of its organized lens structure. Fig. 6 represents a histological cross-section of LMF induced minilens. The rounded form of epithelial cells is localized on the outperiphery layers while the elongated fiber-like cells layered around the nucleus which appears to be amorphous and hardening, resembling physiological lens topograph.

Examination by scanning electron microscopy reveals elongated fiber-like structure (Fig. 7). Transmission electron micrograph indicate the following structural features (1) microfilament in the elongated fiber-like and dividing cells (Fig. 8), indicating active cellular differentiation. (2) polyribosomes on the rough ER (Fig. 9), suggesting cellular proliferation and amplification of genomic activity. (3) desmosomes and defined gap functions (Fig. 10), indicating sites of maximum adhesion and interdigitation between the cell surfaces. All these cytoarchitectural data suggest that the lenmofin can sequentially induce lens morphogenesis.

Enzymatic Modifications and Molecular Hybridization

In order to determine the nature and composition of LMF, various enzymatic modifications were carried out. Trypsin digestion abolished all the binding activity, suggesting a protein moiety and possible cell-to-substratum interaction mechanism. Thrombin and neuramidase treatment reduce the aggretination capacity to 30 and 20% respectively, the latter indicating carbohydrate components.

Incubation with RNase T_1 and U_2 did not appear to effect LMF's activity while treatment with RNase A cause a loss of 80% agglutination. These data

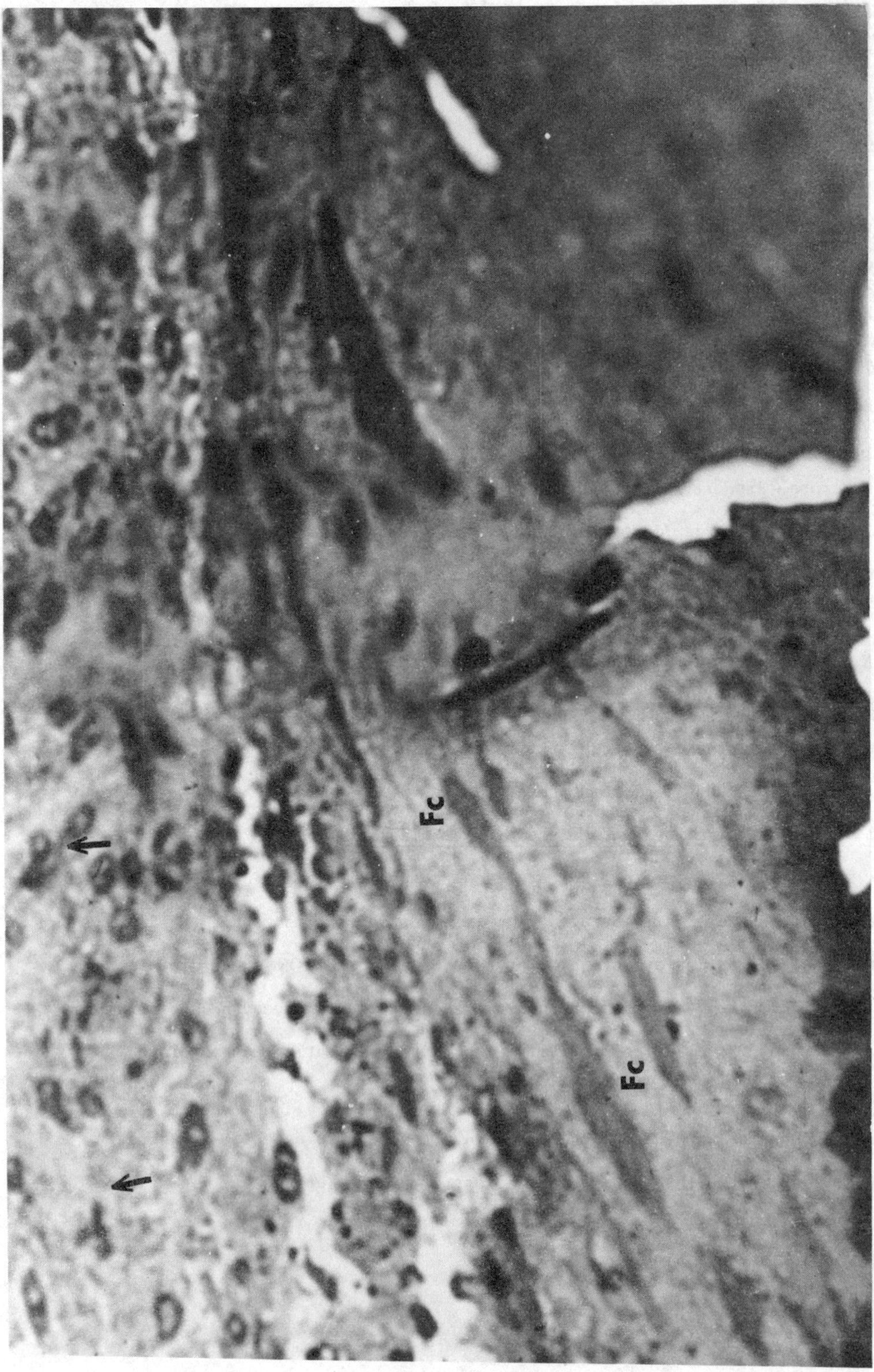

Fig. 6

Figure 6. Phase micrograph of a thin cross-section of minilens. The cells at the periphery are elongated while amorphouse shape are evident in the putative minilens nucleus phase. The disappearance of organelles and denucleation are apparent in the inner center region of the minilens. (x 1000, 1 μM Epon section, tabuline blue stain). The arrows at the upper left are directed toward the out-peripherial region of the minilens. Fc, fiber-like cells.
(Figure reduced to $\frac{8.5}{10}$ during printing.)

Table 2. Effect of Lenmofin on the Differentiated Function of Various Lens Cells

Function / Cell Type	Cell Aggregation	Spreading	Lentoid Body Formation	Minilens Formation
Normal Mouse Lens Cell	+	+	+	+
Nakano Mutant Lens Cell	+	+	+	-(?)
Human Lens Cell	+	±	-	-
c-AMP-Treated Lens Cell	+	-	-	-

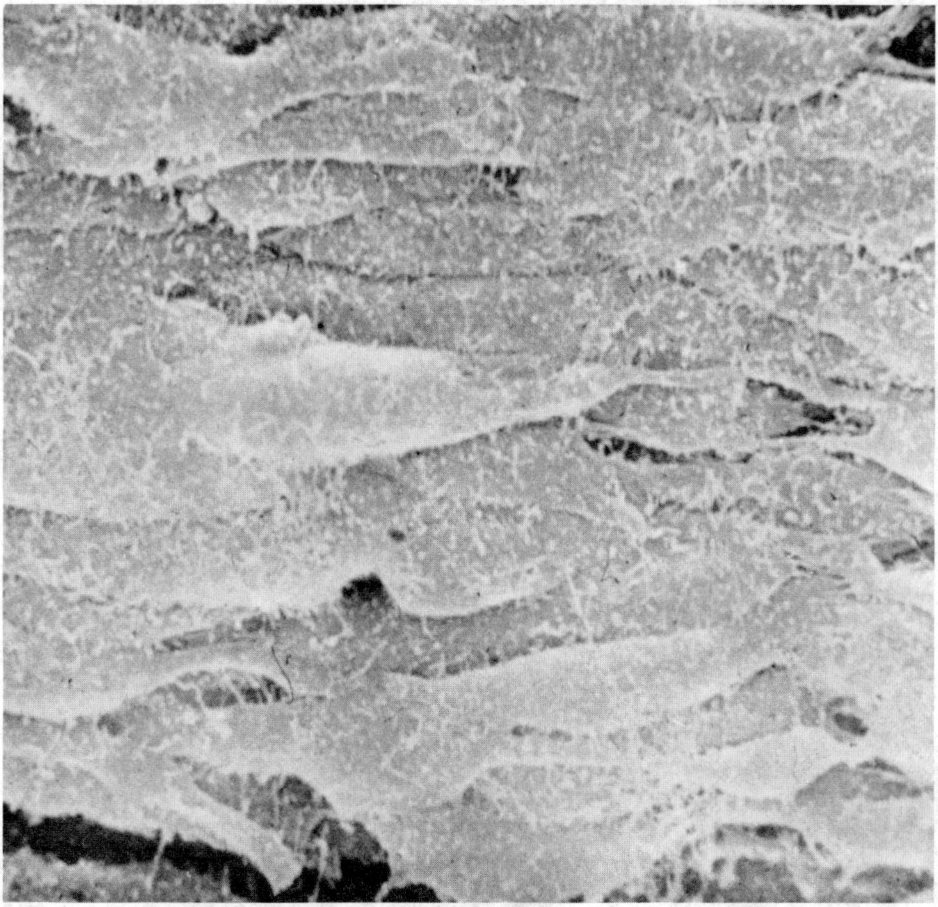

Figure 7. Scanning electron micrograph of a cross-section through the putative cortical fibers of the newly-formed minilens. The long axes of these elongated spindle-like cells are parallel to the surface. Adjacent cells are in close contact (x 5000).

(Figure reduced to $\frac{7}{10}$ during printing.)

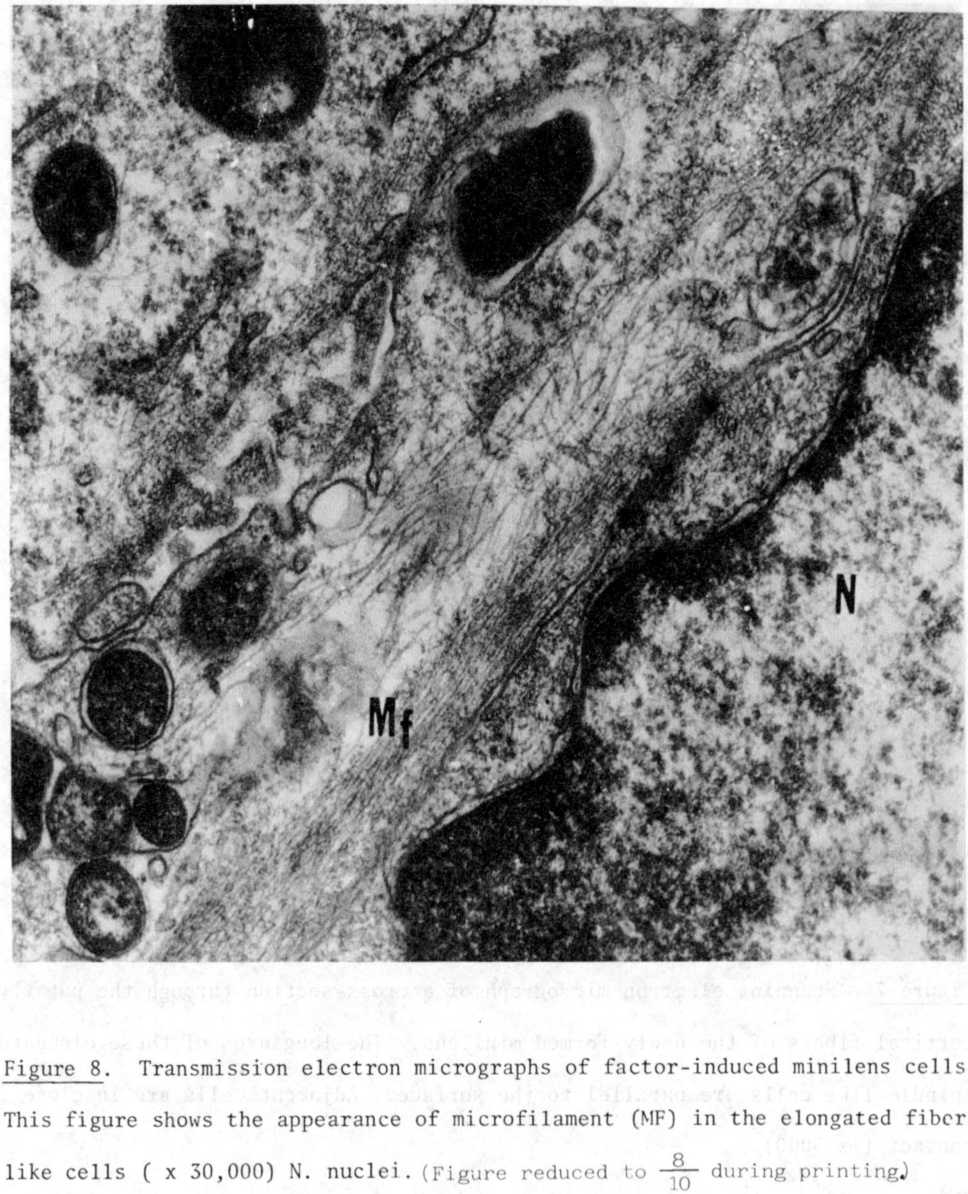

Figure 8. Transmission electron micrographs of factor-induced minilens cells. This figure shows the appearance of microfilament (MF) in the elongated fiber like cells (x 30,000) N. nuclei. (Figure reduced to $\frac{8}{10}$ during printing.)

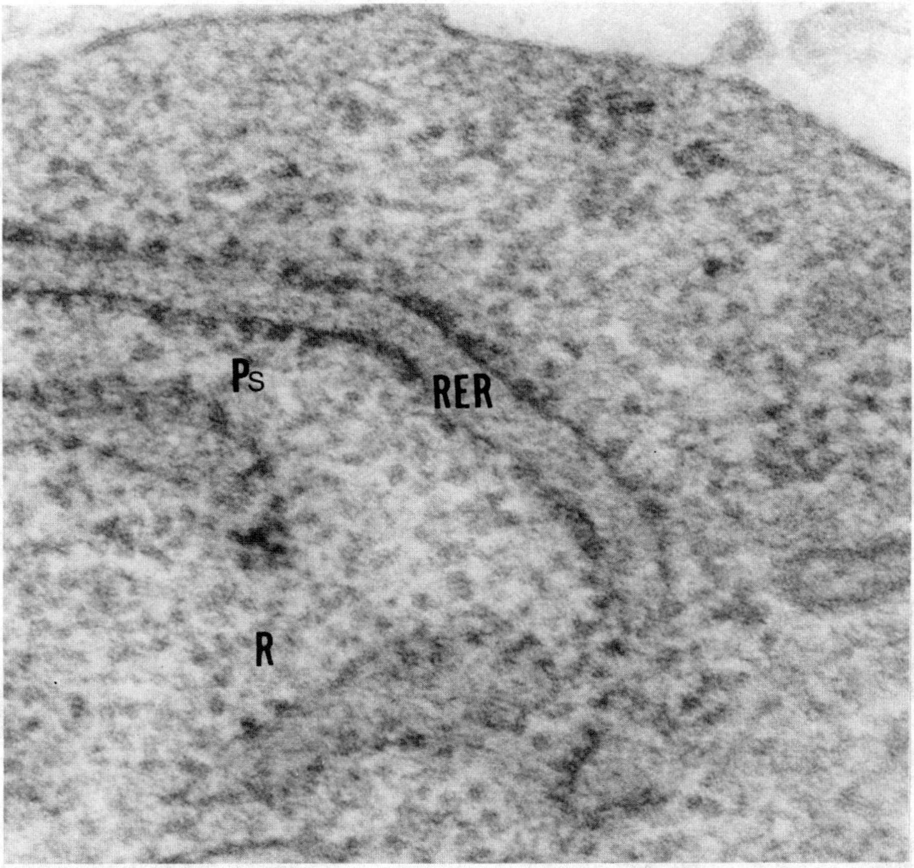

Figure 9. Transmission electron micrograph of a putative cortical lens fiber section of minilens. Polysomes (Ps) are clear evident. R, free ribosome. RER, rough endoplasmic reticulum (72,000).
(Figure reduced to $\frac{7.5}{10}$ during printing.)

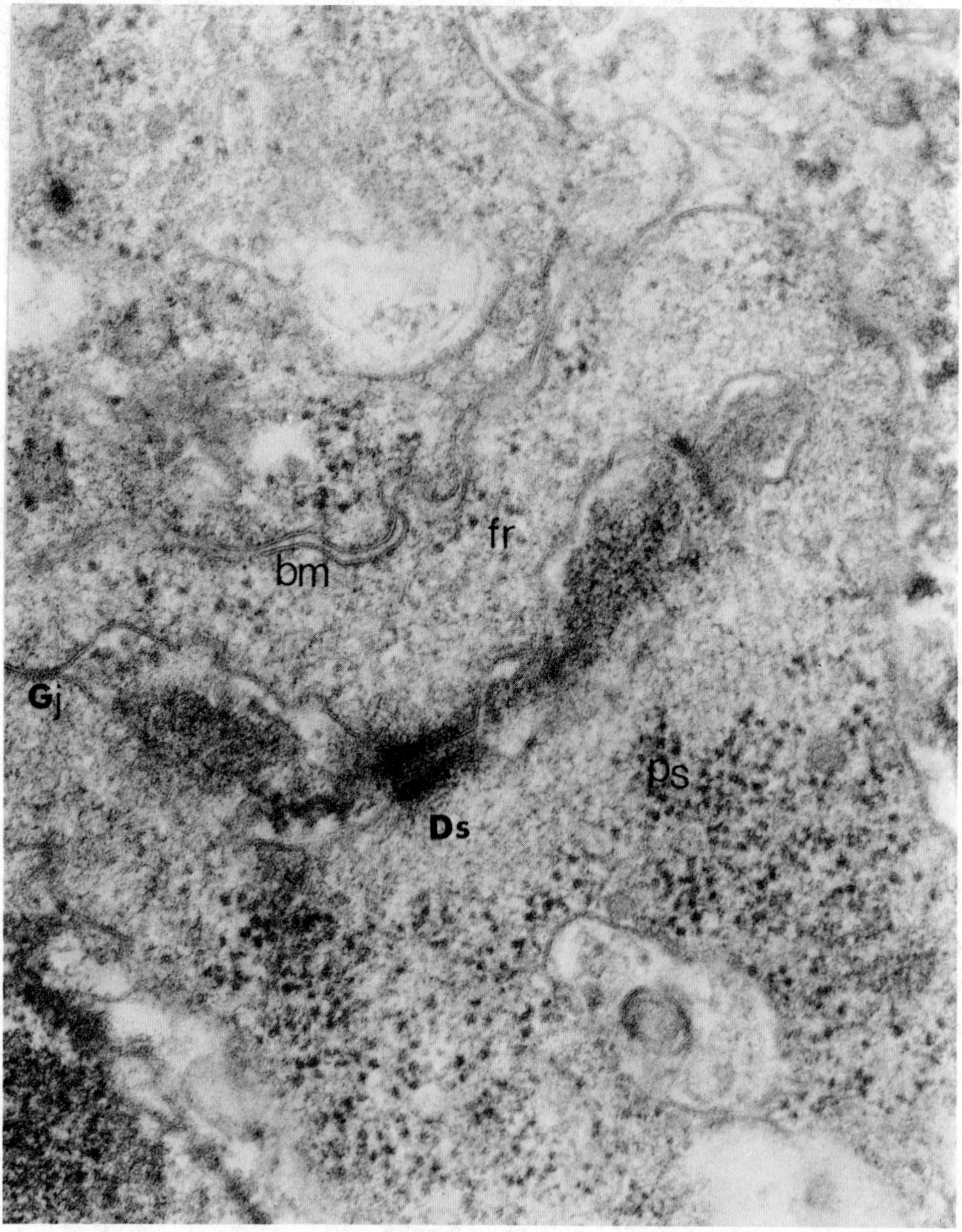

Figure 10. Transmission electron micrograph of minilens appearing in cultures of normal lens cells. Free ribosomes (Fr) and polysomes (Ps) are abundant. Desomosome (Ds) and bilayers of cell membrane (Bm) are clearly evident; possible gap-junction (Gj). This structure can be seen 48 hrs after the addition of lenmofin (x 50,000). (Figurc reduced to $\frac{8.5}{10}$ during printing.)

indicate that not only RNA is an integral component of the factor but uridylic and/or cytidylic residues are likely involved in maintenance of the LMF configuration for activity. The LMF is relatively heat stable since treatment at 100°C for 10 min does not effect it intracellular binding activity (Table 3).

A molecular hybridization test was performed for the detection of mRNA appearance as a function of differentiation stages. It can be seen in Fig. 11 that at the onset of minilens formation (2 days) the expression of crystallin mRNAs is almost 10 times the level found in 4 days after the lentoid body formation, while essentially nondetectable in the control. This result suggests that the mRNA appearance and minilens production is closely related. These data are also consistent with that of the transmission electron micrograph of the 2 day lens culture (see Fig. 9). It shows more polysome assemblies at the onset of minilens formation.

TABLE 3. EFFECT OF VARIOUS TREATMENT ON THE FACTOR'S INTERCELLULAR BINDING ACTIVITY

TREATMENT*	TEMP (100^oC, 10') (A)	VARIOUS DIGESTION: TRYPSIN (B)	THROMBIN (C)	RNASE A (D)	RNASE T_1 (E)	NEURAMINIDASE (F)	RNASE U_2 (G)
PERCENT OF ACTIVITY REMAINED	100	10	30	20	100	20	100

*VARIOUS TREATMENTS AND ENZYMATIC DIGESTIONS WERE CARRIED OUT ACCORDING TO THE FOLLOWING CONDITIONS. (A) 100^oC FOR 10 MIN (B) TRIS-HCL (0.5 M) BUFFER, PH 8.0 AT 25^oC FOR 20 MIN (C) 0.2 M TRIS-HCL AND 30 MM $CACL_2$, PH 8.1 AT 30^oC FOR 30 MIN. (D) 0.1 M TRIS-HCL IN 0.0001 M EDTA, PH 7.4 AT 37^oC FOR 60 MIN. (E) 0.01 M TRIS-HCL PH 7.4 AND 0.001 M EDTA AT 37^oC FOR 45 MIN. (F) 0.1 M ACETATE BUFFER, PH 5.0 IN THE PRESENCE OF 1% MUCIN AT 37^oC FOR 40 MIN. (G) 0.01 M TRIS-HCL, PH 7.2 AT 37^oC FOR 2 HRS.

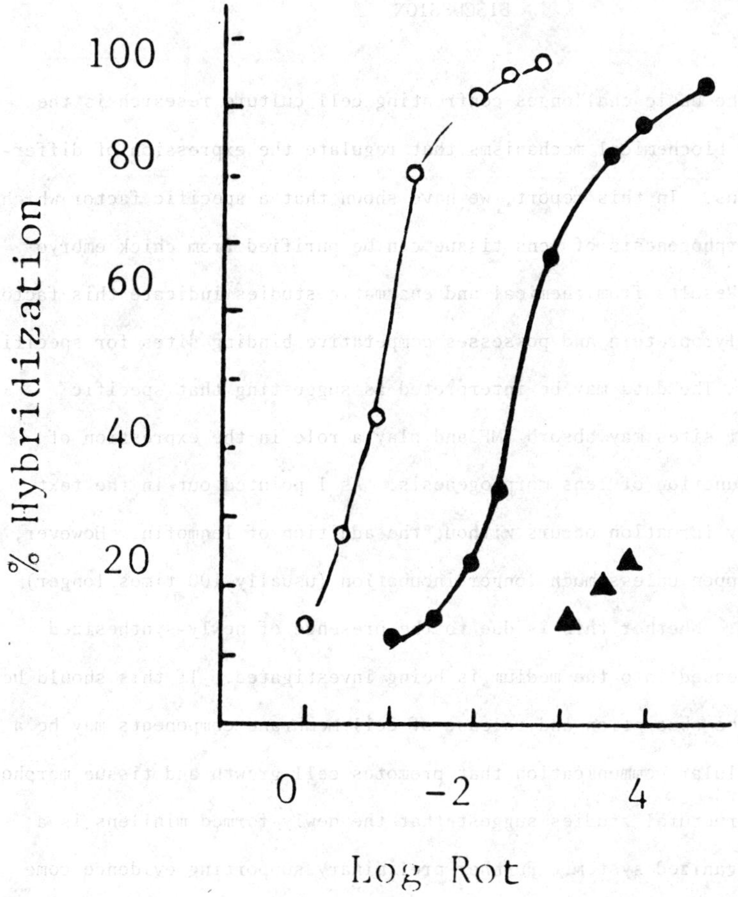

Figure 11. Hybridization kinetics of lenmofin-induced differentiating normal mice cell culture. 1 μg of total lens mRNA from 2 days (o) and 4 days (●) cell culture and 2 ng (^{3}H) cDNA transcribed from poly(A)$^+$ lens mRNA were annealed at 68°C in 1 ml 0.24 M sodium phosphate buffer pH 6.8, containing 0.05% SDS. At the Rot values shown, samples were removed and the percentage of cDNA present as hybride was determined by S_1 nuclease digestion. ▲ 2 day control without the addition of lenmofin.

DISCUSSION

Among the basic challenges confronting cell culture research is the understanding of biochemical mechanisms that regulate the expression of differentiated functions. In this report, we have shown that a specific factor which regulates the morphogenesis of lens tissue can be purified from chick embryo neural retina. Results from chemical and enzymatic studies indicate this factor is a ribonucleoglycoprotein and possesses competative binding sites for specific sugar molecules. The data may be interpreted as suggesting that specific membrane receptor sites may absorb LMF and play a role in the expression of differentiated function of lens morphogenesis. As I pointed out in the text that lentoid body formation occurs without the addition of lenmofin. However, this will not happen unless much longer incubation (usually 100 times longer) time was allowed. Whether this is due to the presence of newly-synthesized glycoprotein released into the medium is being investigated. If this should be the case, then the absorption and release of cell membrane components may be a form of intercellular communication that promotes cell growth and tissue morphogenesis. Cytostructural studies suggest that the newly-formed minilens is a proliferating organized system. Further preliminary supporting evidence come from the isolation and enrichment of a 26,000 dalton protein from the lenmofin induced lens culture system. (See Figure 2). Recent studies indicate that a similar molecular weight protein is associated with membrane structure (17) lens skeleton (18) and gap junction system (19).

During lens morphogenesis, we have observed huge vacuoles formation preferably at one side of the putative minilens. This phenomenum is consistent with the fact that during embryological development posterior chamber or vitreous chamber is always there even before lens formation (20)

It is known that cAMP can induce cellular elongation (21). We have tested and found that it can only cause cellular aggregation. Furthermore, the cAMP treated cell population fails to form lentoid body upon addition of LMF, even after several washes and transfers of the culture. This data suggest that there may be an non-competative sites overlap between cAMP and LMF. The response of the congenital cataractone strain to lenmofin induction is most interesting. The fact that the mutant cells can only form lentoid body, not minilens, raises the question of whether this is the crucial configuration that the mutated site(s) involved. It is conceivable and hoped that a series of mutant can be selected in order to determine the role of lenmofin played in this congenital disease. Most recently we have abserved that the LMF activity is graduatly diminishing from 14 days old chick retinal cells, suggesting a regulatory role of LMF in the differentiation of lens tissue.

ACKNOWLEDGMENTS

I am indebted to Dr. B. Worgul of Columbia University and Dr. S. Chen of New York University for their assistance in the E. M. work. I thank Dr. T. Fu, T. W. Chen for their technical assistance. I am also indebted to Dr. Y. S. Shih of the Institute of Developmental Biology, Academia Sinica for her participation in the initial phase of this work. The preparation of this manuscript has benefited from the helpful suggestions of Dr. Hodgins of New York University. The initial cell culture was kindly provided by Dr.P. Russell of National Eye Institute. This research has been supported by the National Eye Institute, NIH Grant EY 03173-03 and by Fight for Sight Inc.,Grant G-672.

REFERENCES

1. Papaconstantinou, J. (1967) Science 156 338-340.

2. Toivonen, S. (1945) Ann. Soc. Zool-bot. Fenn. Vanamo. 11 1-28.

3. Coulombre, A.J. (1969) Invest. Ophthalmol. 8 25-28.

4. Eguchi, G. and Okada, T.S. (1973) Proc. Natl. Acad. Sci. (USA) 70 1495-1499.

5. Russell, P., Fukui, H.N., Tsunematsu, Y., Huang, F.L. and Kinoshita, J.H. (1977) Invest. Ophthal. Visual Sci. 16 243-246.

6. Tsunematsu, Y., Fukui, H.N. and Kinoshita, J.H. (1978) Exp. Eye Res. 26 671-685.

7. Beebe, D., Feagans, D.E., and Jebens, H.A.N. (1980) Invest. Ophthal. Visual Sci. 52th 155.

8. Chen, J.H. and Spector, A. (1977) Biochemistry 16 499-505.

9. Chen, J.H., Lavers, G.C., Spector, A., Schutz, G. and Geigelson, P. (1974) Exp. Eye Res. 18 189-199.

10. Chen, J.H., Lavers, G.C. and Spector, A. (1973) Biochem. Biophys. Res. Commun. 52 767-773.

11. Eitle, E. and Gerish, G. (1977) Cell Diff. 6 339-346.

12. Lowry, O.H., Rosebrough, N.J., Farr, A.L. and Randall, R.J. (1951) J. Biol. Chem. 193 265-275.

13. Lis, H. and Sharon, N. (1972) in Methods in Enzymology, ed. Ginsburg, V. (Academic Press, New York), 28 360-365.

14. Chen, J.H., Lavers, G.C. and Spector, A. (1976) Biochem. Biophys. Acta 418 39-51.

15. Chen, S. and D.E. Hillman (1980) Brain Research 187 487-493.

16. Brown, A.H. (1946) Arch. Biochem. 11 269-271.

17. Brockhuyse, R.M. and Haklmann, E.D. (1978) Expt'l Eye Res. 26 305-320.

18. Ramaekers, F.C., Selteri-Versteegen, A.M. and Bloemandal, H. (1980) Biochem.

Biophys. Acta 596 (1) 57-63.

19. Goodenough, D.A. (1978) Pharmacol. Rev. 30 (4) 383-392.

20. Balinsky, B.I. (1966) in An Introduction to Embryology Chapter 13-2 (W.B. Saunders Co. Philadelphia) 377-379.

21 Pastan, I. and Johnson, G.S. (1974) in Advances in Cancer Research eds. Klein, G., Weinhouse, S. and Haddow, A. (Academic Press, N.Y.) pp 303-329.

DIFFERENTIAL TRANSCRIPTION AND DIFFERENTIAL STABILITY OF mRNA: BOTH PLAY A ROLE IN ADENOVIRUS GENE CONTROL

J. E. Darnell, Jr., M. Wilson, J. R. Nevins
and E. Ziff

The Rockefeller University, New York, N. Y.
10021 USA

Almost twenty years ago large nuclear RNA molecules were discovered in cultured mammalian cells. Some of these large molecules were shown to be precursors to rRNA. Because some of the nuclear RNA had characteristics of cytoplasmic mRNA it appeared also possible they might be precursors to mRNA. These early findings made the definition of transcriptional units—regions of DNA including the stop and start sites for RNA synthesis—a necessary first step in determining the level or levels at which gene control lies in eukaryotic cells (1-5). It was necessary to prove or reject the large nuclear RNA transcripts as mRNA precursors. If both short and long transcriptional units had existed simultaneously in the DNA encoding a messenger RNA, then the long transcription units might not have produced true messenger RNA precursors. The first large nuclear transcripts that were proved to be mRNA precursors were those formed late in adenovirus infection from the righthand two-thirds of the virus genome. No short primary transcripts arose from this region (2-5) and therefore mRNAs must be derived from the large primary transcript that covers more than 25,000 bases of the DNA strand copied in the rightward direction. The characterization of additional early adenovirus transcription units and cell transcription units followed. The definition of adenovirus transcription units involved both "nascent chain" analysis of pulse-labeled RNA (3) and UV transcription analysis (4, reviewed in 1). In addition to the adenovirus transcription units cell mRNA had a target size for UV inactivation that was approximately three to five times larger than the mRNA (6).

Together with the data on transcription unit size and the discoveries of capping (7), polyadenylation (reviewed in refs. 8, 9), methylation (7) and finally splicing (10-12), a common pathway for formation of most mRNA molecules could be suggested(1). The likelihood of extensive processing during mRNA formation has renewed the old conjectures (13-15) that gene regulation might include differential processing but now the hypotheses could be stated in more specific guise.

In this paper we will describe direct measurements on control of mRNA formation that take advantage of the transcriptional map of adenovirus type 2 (Ad-2). The experiments demonstrate various levels at which mRNA concentrations from a variety of Ad-2 transcriptional units are controlled. Both positive and negative control at the initiation of transcription appear to occur during the regulation of several different Ad-2 transcriptional units. At the other end of the spectrum of possible control points for mRNA concentrations is stabilization of mRNA in the cytoplasm. For at least one early transcriptional unit and perhaps for others an increase in one of two different spliced mRNAs appears to depend on lengthening the half-life of one of the two mRNAs. These early results illustrate that a variety of methods are used for gene control in the levels of Ad-2 mRNA concentration. Thus as was the case in studying steps in mRNA biosynthesis, HeLa cells infected with Ad-2 remain an attractive system to continue studies aimed at discovering the proteins and cell structures that presumably participate in controlling mRNA concentrations.

TRANSCRIPTION UNIT MAP

Phillipson and colleagues and Sharp, Flint and their colleagues (reviewed in ref. 16) used the mRNA from cells both early and late after infection to determine which regions of the adenovirus DNA, employed as labeled DNA, would hybridize to mRNA. Separated strands of various regions were prepared to allow determination of the direction of transcription as well. From these studies a "messenger RNA map" was produced. Pulse-labeled nuclear RNA obtained from isolated nuclei (3) or extracted from pulse-labeled cells (4, 5) was analyzed to establish which of the various Ad-2 mRNAs derive from independent transcriptional units. The apparent origin of transcription could also be located by determining which restriction fragments of the Ad-2 genome hybridized short RNA molecules after a pulse. The analysis of such pulse-labeled RNA coupled with the analysis of RNA synthesis after UV irradiation (4) (which damages promoter distal transcription most but leaves promoter proximal RNA synthesis less impaired) revealed the apparent start sites for several separate early promoters, two intermediate promoters and the major late promoters. Gelinas and Roberts (17) then identified a similar if not identical capped oligonucleotide on all the late mRNAs. The capped 5' end of the mRNA was then shown by electron microscopy (10,11) to be homologous to the approximate region where the RNA initiation site had been mapped (2-4). The existence of adenovirus mRNA with sequence homology to noncontiguous regions of the adenovirus genome coupled with the evidence that the transcription unit contained all the sequences present in the mRNA was the basis for concluding that RNA·RNA splicing occurs (1).

The time seemed ripe therefore to try to establish definitely the sequence of the start site for RNA synthesis late in Ad-2 infection. The DNA sequence of over 1000 bases was determined around 16, the site at which synthesis had been judged to begin by the earlier UV and pulse label studies. In addition, the exact sequence of the 5' eleven bases found on all mRNAs was determined (18). Only one region of DNA (16.45) between ~ 15 and 18 on the genome was found capable of encoding the 5' end of the late mRNAs. Analysis of nuclear RNA from cells or isolated nuclei showed no transcribed oligonucleotides "upstream" from the cap site, although all the oligonucleotides predicted from this sequence "downstream" could be found (19). Thus it appeared that the cap site was the start site for late Ad-2 RNA synthesis.

This general approach of determining the DNA sequences around a presumed promoter site and determining the sequence of the capped oligonucleotide of mRNAs derived from that promoter had identified the apparent start site for a series of early transcriptional units in addition to the large late transcriptional unit. A common hexa- or heptanucleotide sequence, TATAAA, occurs ~ 25 to 30 nucleotides "upstream" from all of the proposed

Ad-2 cap sites. This same sequence appears in the same place before a number of cap sites in cellular genes as well. Thus it appears that one portion of a common signal for RNA polymerase II exists ~ 35 to 30 nucleotides before the RNA initiation site (19).

Cap addition is then very quick after RNA chain initiation. Chains less than 100 nucleotides in length from the late Ad-2 promoter have already received a cap (Babich, Nevins and Darnell, unpublished). Finally, it should be noted that Roeder and his colleagues (20) have described an <u>in vitro</u> transcription initiation system that utilized Ad-2 DNA, polymerase II, and a crude cell extract from infected or uninfected cells. In this system initiation occurs at the same late promoter site as that which functions inside the cell and the transcripts are capped. Knowing the apparent start site for RNA synthesis has enabled us to determine whether transcriptional control in the nucleus can be strictly demonstrated during Ad-2 infection.

CONTROL OF PROTEIN IX TRANSCRIPTIONAL UNIT

Two early promoters were established at positions 1.4 and 4.5 and an intermediate promoter at 9.1 by identification of short UV-resistant promoter proximal RNA segments (21). These presumed starting regions were confirmed by sequence analysis of caps in mRNA and corresponding sites in DNA (19). The promoters at 1.4, termed 1A, and 4.5, termed 1B, produce mRNAs both early and late in infection and can be scored as active sites of transcription by hybridization of UV-resistant short ($<$500 bases) nuclear RNA to fragments from 0-3,3-4.4,4.4-8, and 8-11. The third promoter at ~ 9 produces the mRNA for the capsid protein IX beginning about 6 hours after infection (ref. 22; infection produced by high multiplicities of ~3000 particles per cell). Short UV-resistant promoter proximal RNA for the protein IX was not detected at earlier times. This is the most sensitive assay presently available for control at the site of initiation (21). Only a measurement of the synthesis rate of the exact capped oligonucleotide complementary to the 9-11 region would improve this conclusion. It therefore appears that initiating events for RNA synthesis are limited at ~ 9.1 early in infection even though a neighboring promoter at 4.5 is active. While it is not known in our experiments where high multiplicity has been used whether both 1B and protein IX promoters are used in the same DNA molecule this must be the case when single particles infect cells, since protein IX is formed without any DNA replication (16).

CONTROL AT 16.45, THE MAJOR LATE PROMOTER

The beginning of labeling of late mRNAs from the major late promoter at 16.45 occurs after 10 to 12 hours of infection. The rate of RNA synthesis from the rightward reading strand from the region of 16 was compared to that from the region between 3-11 which includes all of the transcriptional units described above (23). Compared to early in infection there was at least a 50-fold increase in rightward transcribed RNA complementary to the 11-18 region late in infection. The nature of the first RNA synthesized from the 16 region is of interest because there may be some changes in the products from this region during infection. Late in infection only about 1 in every 4 polymerases that starts RNA synthesis at 16.45 completes the transcriptional unit all the way to 98 (2,23,23). The other initiation events result in premature termination producing capped RNA chains that are from ~100 to ~2000 bases in length that correspond to the beginning of the transcriptional unit. No role for these premature stop sites has been indicated yet but it remains a possibility that as RNA synthesis increases at 16 a differential and controlled termination could occur.

SURVEY OF PROMOTER ACTIVITY DURING EARLY INFECTION

After the definition of the Ad-2 promoters, a survey of patterns of transcription unit function can be made by pulse-labeling cells for 2 to 5 minutes and hybridizing total nuclear RNA to DNA on filters containing the start sites for various early transcription units (25). First, the transcription of region 1A reaches a maximum, followed closely by regions 3 and 4. All these transcription units reach a maximum rate of transcription in 2 to 3 hours. Transcription of region 2 however only reaches a maximum after ~ 6 hours of infection. In the initial studies on the transcription of the various early region it appeared that transcription units in regions 2 and 4 and possibly region 3 underwent a substantial decline in transcriptional rate later in infection. This decline has now been confirmed with additional experiments (27). The decrease in transcriptional rate from region 4 as well as possibly region 2 is blocked by cycloheximide treatment, indicating a need for protein synthesis in order to decrease transcription rate from these regions. The protein that mediates the suppression of transcription in region 4 is suggested by studies of ts 125, a mutant virus that does not make viral DNA at $41^{o}C$ (26) and produces a temperature-sensitive version of the ~ 72,000 dalton protein that binds to single-stranded DNA. The mutant virus goes through the same program of control of early transcription as does wild type when grown at $32^{o}C$, but fails to properly shut off region 4 synthesis when grown at elevated temperature (27). Thus a temperature-sensitive transcriptionally active protein would appear to have been uncovered.

SUMMARY OF TRANSCRIPTIONAL CONTROLS

A summary of transcriptional controls thus far established for the Ad-2 infectious cycle can be made: 1) RNA synthesis from region 1A possibly slightly precedes synthesis from other regions. Evidence from virus mutants (28,29) implicates the RNA from 1A or its protein product in assisting the synthesis of perhaps all the early transcription units, including 1A itself. The function of region 1A however apparently can be carried out by the cell if high multiplicities of infection are used or if long times elapse between infection and assay for early transcription (Nevins and Shenk, unpublished observations). 2) Synthesis from regions 3 and 4 follows iA quickly but region 2 has a definite lag in maximal activation. 3) Regions 2 and 4 undergo a suppression of synthesis after 4 to 5 hours declining to as little as 10% of their maximal rates. A new protein, perhaps the 72K protein, is needed for the suppression of at least region 4. 4) Transcription of the protein IX mRNA early in infection is undetectable. Transcription begins at ~6 to 8 hours and continues. Thus transcriptional control at the initiation step appears to be almost absolute for this transcriptional unit. 5) There is at least a 50-fold increase in transcription from the start site at the late transcriptional unit at 16 hours compared to 5 hours. This increase includes the prematurely terminated transcripts and is not simply a relief of premature termination.

GENE CONTROL AT OTHER LEVELS

Since the discovery of poly (A) in mRNA and especially since the discovery of splicing, ideas about differential processing of primary transcripts have been discussed (5). Can a cell exercise a choice in conducting any of the necessary posttranscriptional steps and thereby regulate what mRNAs appear in the cytoplasm? Can a cell choose variably one of two or more possible poly (A) sites or splice sites in a transcriptional unit containing more than one of either of these sites? While it is difficult to rule out the possibility of occasionally using such variable choices the evidence is against such controls in the formation of Ad-2 mRNA as a major factor in determining individual mRNA levels. For example, Ad-2 late transcription appears to give rise to one mRNA each time the late transcript is synthesized (30). This single mRNA is chosen in two steps, first, the 3' end, the poly (A) site, and then the splice site (30). At the moment no variations in the choice of poly (A) sites or splice sites are proven. In the manufacture of cell mRNAs it is also true that while only about one in 3 to 4 transcripts has poly (A) added, those that do are largely conserved (Harpold, Wilson and Darnell, unpublished observations). The unconserved nuclear transcripts appear to differ qualitatively from the conserved transcripts (31). Much more work is required to support a firm opinion but it is at least possible that variable controlled processing occurs rarely if at all.

However, new evidence on Ad-2 mRNA formation shows quite clearly that all regulation of mRNA concentration is not at the level of transcription. Transcriptional unit 1B of adenovirus produces two mRNAs spliced differently to give products of $\sim$ 1200 and $\sim$ 2000 nucleotides, respectively (32,33). The ratio of these two forms of 1B mRNA varies in various circumstances. In transformed rat cells where presumably the same 1B transcriptional unit operates, the longer mRNA predominates in a molar ratio of at least 4 or 5 to 1 (32). During early lytic infection the ratio of the two RNAs is about equal (32,22). Late in infection a change occurs in the two mRNAs from 1B so that the relative amount of the shorter mRNA increases perhaps 5-fold (22). This result could be due either to: 1) more frequent production of 14S mRNA than 20S mRNA, or 2) late in infection the 14S mRNA could be more stable in the cytoplasm. Using the accumulation of labeled mRNA as the guide to mRNA half-lives, the turnover rate of the 14S 1B mRNA has been shown to be ~ 5 to 10 times longer late in infection compared to early (Wilson and Darnell, unpublished observations). A similar set of circumstances appears to be true for a small RNA from the 1A region. Thus differential mRNA stabilization, not either differential transcription or RNA processing, of the 14S mRNA accounts for its relative increase in concentration late in infection.

CONCLUSION

These early results on different types of control in the progression of Ad-2 through its infectious cycle suggests that Ad-2 may offer a legitimate model for a developmental program. Genes "come on" and "stay on" and "come on" and "go off" due to transcriptional controls. Other mRNAs change levels because of differential mRNA stabilization. The value of the adenovirus system aside from the necessary fact that such an array of controls exists, is that infected cells all progress through the same pathways uniformly. An additional most important advantage of studying Ad-2 is that infected cells synthesize adenovirus RNA faster than mRNA precursor synthesis from any other single transcriptional unit yet discribed. Thus it is not an unrealistic hope, knowing which of the possible regulatory points are utilized in controlling a transcriptional unit, that the proteins or cell structures involved in this control may be studied directly.

References

1. Darnell, J. E. 1979. Transcription units for mRNA production in eukaryotic cells and their DNA viruses. In Progress in Nucleic Acid Research and Molecular Biology. W. E. Cohn, Ed. Vol. 22. (Academic Press, New York) pp.327-353.

2. Darnell, J. E., R. Evans, N. Fraser, S. Goldberg, & J. Nevins. 1977. Cold Spring Harbor Symp. Quant. Biol. XLII:515-522.

3. Weber, J., W. Jelenik, & J. E. Darnell. 1977. Cell 10:611-616.

4. Goldberg, S., J. Weber & J. E. Darnell. 1977. Cell 10:617-621.

5. Evans, R., N. Fraser, E. Ziff, J. Weber & J. E. Darnell. 1977. Cell 12:733-740.

6. Goldberg, S., H. Schwartz & J. E. Darnell. 1977. Proc. Nat'l. Acad. Sci. USA 74:4520-4524.

7. Shatkin, A. J. 1976. Cell 9:645-653.

8. Edmonds, M. & M. A. Winters. 1976. Polyadenylate polymerases. In Progress in Nucleic Acid Research and Molecular Biology. W. E. Cohn, Ed. Vol. 17. (Academic Press, New York), pp. 149-179.

9. Brawerman, G. 1976. Characteristics and significance of the polyadenylate sequence in mammalian messenger RNA. In Progress in Nucleic Acid Research and Molecular Biology. W. E. Cohn, Ed. Vol. 17. (Academic Press, New York), pp. 117-148.

10. Berget, S. M., C. Moore & P. A. Sharp. 1977. Proc. Nat'l. Acad. Sci. USA 71:3171-3175.

11. Chow, L. T., R. Gelinas, T. R. Broker & R. J. Roberts. 1977. Cell 12:1-8.

12. Klessig, D. F. 1977. Cell 12:9-21.

13. Scherrer, K. & L. Marcaus. 1968. J. Cell. Physiol. 72 (Suppl. 1):181-212.

14. Georgiev, G. P. 1969. J. Theor. Biol. 24:473.

15. Darnell, J. E. 1968. Bact. Rev. 32:262-290.

16. Flint, J. 1977. Cell 10:153-166.

17. Gelinas, R. E. & R. J. Roberts. 1977. Cell 11:533-544.

18. Ziff, E. B. & R. M. Evans. 1978. Cell 15:1463-1475.

19. Baker, C. D. & E. B. Ziff. 1979. Cold Spring Harbor Symp. Quant. Biol. XLIV, in press.

20. Weil, P. A., D. S. Luse, J. Segall & R. G. Roeder. 1979. Cell 18:469-484.

21. Wilson, M. C., S. Sawicki, M. Salditt-Georgieff & J. E. Darnell. 1978. J. Virol. 25:97-103.

22. Spector, D. J., M. McGrogran & H. J. Raskas. 1978. J. Mol. Biol. 126;395-414.

23. Fraser, Nigel W., Pravinkumar B. Seghal & James E. Darnell, Jr. 1979. Proc. Nat'l. Acad. Sci. USA 76:2571-2575.

24. Evans, R., J. Weber, E. Ziff & J. E. Darnell. 1979. Nature 278:367-370.

25. Nevins, J. R., H. S. Ginsberg, J.-M. Blanchard, M. C. Wilson & J. E. Darnell. 1979. J. Virol. 32:727-733.

26. Ginsberg, H. S., Ul Lundholm & T. Linne. 1977. J. Virol. 23: 142-151.

27. Nevins, J. R. & J. Jensen-Winkler. 1980. Proc. Nat'l. Acad. Sci. USA, in press.

28. Berk, A. T., F. Lee, T. Harrison, J. Williams & P. A. Sharp. 1979. Cell 17:935-944.

29. Jones, N. & T. Shenk. 1979. Proc. Nat'l. Acad. Sci. USA 76: 3665-3669.

30. Nevins, Joseph R. & James E. Darnell, Jr. 1978. Cell 15: 1477-1493.

31. Harpold, Michael M., Ronald M. Evans, Marianne Salditt-Georgieff & James E. Darnell. 1979. Cell 17:1025-1035.

32. Wilson, Michael C., Nigel W. Fraser & James E. Darnell, Jr. 1979. Virology 94:175-184.

33. Berk, A. J. & P. A. Sharp. 1975. Cell 14:695-711.

CHARACTERIZATION OF ANTIGEN-SPECIFIC IMMUNE i-RNAs AND OF IMMUNE RNA PRESENTATION TO IMMUNOCOMPETENT CELLS

M. Fishman

Division of Immunology, St. Jude Childreh's Research Hospital, 332 North Lauderdale, P.O. Box 318, Memphis, Tennessee, 38101, USA.

P. Bilello and G. Koch

Physiologisch-chemisches Institut, Universitat Hamburg, Grindelallee 117, 2 Hamburg 13, West Germany.

Abstract The immune RNAs, both the RNA-antigen complex and the i-RNA, may have *in vivo* relevance in participating in the early events in antibody formation in which antigen-stimulated macrophages can interact with lymphocytes to bring about an expansion of antibody forming cells in the absence of lymphocyte proliferation. It is also known that i-RNA, in contrast to RNA-antigen complexes, (a) is a m-RNA of sufficient size to code for a complete immunoglobulin, (b) can be presented to lymphocytes via macrophages that are not histocompatible, (c) stimulates nonspecific antigen reactive lymphocytes, and (d) is synthesized by a unique subpopulation of macrophages that possess specific surface membrane receptors for the antigen.

(This research was supported by Cancer Center Support Grant CA21765-02, and by ALSAC.)

The role of macrophages in immunogenesis has been a controversial subject for many years. In recent years there has developed a strong trend away from the historical view that macrophages are destructive of antigen and therefore have a negative or at best a neutral role [1, 2]. Instead, macrophages are now thought to fill a productive and indeed essential role as helper cells in the induction of antibody formation [3]. The trend is indeed so strong that antigens that only recently were considered to be independent of a requirement for macrophage "help" are now said to require such help [4].

A positive role for macrophages as accessory cells in *in vivo* experiments can be deduced from circumstantial evidence with the observations on the phenomenon of peripolesis, indicating macrophage-lymphocyte interactions, being the most circumstantial. Recent studies have called attention to an important role for histocompatibility antigens in macrophage-lymphocyte interactions. Requirements for syngeneity of donors of macrophages and lymphocytes have been documented [5] and Shevach has shown [6] that macrophages express an Ir gene product (Ia antigens) which appears to be important in the activation of immune T cells. However, although the activation of T cells by histocompatible antigen-pulsed macrophages is antigen specific, attempts to prevent a successful interaction by either removing the antigen from the macrophage surface by proteolytic enzyme treatment or by quenching the antigen with specific antibody have failed [7, 8]. The nature of the antigen within the macrophage or the requirement of antigen for specific stimulation or activated T lymphocytes remains unclear. In contrast to the inability to block this activation with antibody directed against the antigen, successful interference of lymphocyte activation occurs when the antigen pulsed macrophages are pre-incubated with antibody against Ia antigens.

Genetic restriction also observed at the level of macrophage-lymphocyte binding [9] in which cluster formation (peripolesis) between antigen-pulsed macrophages and specifically activated lymphocytes requires histocompatibility of the two cell types. In a recent paper, Braendstrup _et al_. [10] showed that there was a unique clustering between macrophages and lymphocytes consisting of a macrophage attached to a central lymphocyte and an attachment of several peripheral lymphocytes to the central lymphocyte. Genetic restriction only occurred between the macrophage and the central lymphocyte whereas the peripheral lymphocytes were neither restrictive nor had to be derived from an animal immune to the antigen used to pulse the macrophage. Thus specific interactions between the macrophage and the central lymphocytes were antigen directed whereas the specific interactions between the central and peripheral lymphocytes were not. The initial stimulation of a central lymphocyte by antigen-pulsed macrophages is thus thought to be transmitted to the surrounding peripheral lymphocytes causing an increase in the number of antibody-forming cells without lymphocyte proliferation. This could explain the discrepancy between the generation time for antibody-forming cells (4 hours) and the doubling time for lymphocyte proliferation (8 hours) noted by Dutton and Mishell [11]. It could also be postulated that this lateral spread of information from macrophage to lymphocytes may be due to an immune RNA which is capable of initiating a specific immune response in the absence (i-RNA) of antigen.

To summarize the current status of immune RNA it is necessary to clearly distinguish the RNA-antigen complex from i-RNA both extracted from macrophages pre-incubated with antigen. Table I outlines only a few of the major differences.

Table I

Distinguishing Characteristics Between i-RNA and RNA-Ag Complexes

i-RNA	is free of demonstrable antigen.
	translates into 19S antibody possessing the allotypic specificity of the macrophage donor.
	stimulates lymphocytes with or without specific antigen receptors.
	is synthesized in macrophages exposed to either low or high input of antigen.
RNA-antigen Complex	activity is dependent on both the antigen and RNA moieties.
	initiates a 7S antibody response with the immunoglobulin allotype of the lymphocyte donor; is not translated in a cell-free system.
	is synthesized in macrophages exposed only to high input of antigen.
	stimulates only lymphocytes with specific antigen receptors.

This paper will further define the m-RNA activity of i-RNA and describe the role of genetic restriction in the presentation of immune RNA to lymphocytes by macrophages.

i-RNA

This immunogenic material was first described as a species of RNA of an estimated 8-16S, produced in a subpopulation of peritoneal exudate cells representing only 8% of the total cells. The RNA has been shown to be highly sensitive to degradation by pancreatic RNAse and to be resistant to DNAse and proteases. It was designated as informational i-RNA when it was shown to transfer to recipient cells the ability to make antibody specific for the antigen to which the RNA donor cells had been exposed and that such antibody had the allotype specificity (both on the a and b loci) for the donor rabbits immunoglobulin. In more recent work [12] it was possible to show the i-RNA fulfilled the criteria of m-RNA in several ways: It incited antibody synthesis in cells in the presence of Actinomycin D; it was present in the polysomal fraction of donor cells and displays increased specific activity upon purification on dT-cellulose columns and maybe most importantly, it led to synthesis of specific antibody carrying the donor rabbits allotype markers in a cell-free system derived from mouse L-cells.

Results illustrating the synthesis of specific anti-T2 antibody in a L cell-free system are shown in Table 2. Various RNA preparations were added to the cell-free system and the detection and allotype classification of the specific antibody among the translation products was determined at the same time by means of an anti-allotype amplification procedure. It

Table 2

Amplification of Neutralization of T2 Phage With Anti-allotype Serum

Material Added to L-cell-free System	% Neutralization[a] Anti-4	Anti-5
i-RNA[b]		
11/44 T2 i-RNA	37	0
dT-cellulose T2-i-RNA	35	9
Sucrose 16-18S T2 i-RNA	40	0
23/55 T2 i-RNA	2	30
Normal RNA	13	0
Endogenous (Control)	15	3

[a]A 1:8 dilution of the translation product was incubated with T2 phage for 30 minutes at 37°C, followed by incubation with anti-allotype serum for 30 minutes at 37°C and 1 hour at 4°C; neutralization of more than 25% is highly significant at the 1% level.

[b]i-RNA was prepared from peritoneal exudate (PE) cells pre-incubated with T2 phage at a ratio 1 T2 phage to 100 PE cells; 100 μg of T2-i-RNA and 10 μg of dT cellulose purified i-RNA were added to cell-free system.

can be seen that i-RNA extracted from peritoneal exudate cells from homozygous 44 or 55 rabbits was responsible for the synthesis of 44 or 55 immunoglobulin, capable of neutralizing T2 phage, respectively. From this and other experiments the effective dose range for synthesis of antibody was 20-600 μg/ml of total i-RNA or 2-20 μg of dT-cellulose purified i-RNA. The molecular size of i-RNA was 16-18S which is sufficient to code for a complete immunoglobulin molecule. The size of the product responsible for specific neutralization of T2 was shown, by means of sucrose gradient analysis, to be coincident with rabbit serum proteins having the 19S size of IgM. However, in attempting to demonstrate L and H chains in the cell-free supernatants by means of polyacrylamide gel electrophoresis, H chains but not L chains were detected. These results were in conflict with our observations that the anti-T2 neutralizing activities of such supernatants were consistently amplified with antiserum directed against a light chain allotype marker. Recent experiments were conducted to detect L chains in the cell-free supernatants by immunoprecipitation procedures. i-RNA extracted from antigen-incubated PE cells from homologous 44 rabbits was translated in an L-cell system and the presence of L-chain determinants among the translational products was measured by (a) co-precipitation with 4-anti-4 serum, and by (b) binding to a polymerized anti-4 column. Controls consisted of supernatants from cell-free systems in which globin m-RNA was translated and supernatants from L-cell systems to which no exogenous RNA was added. In addition, the amount of nonspecific binding was determined by co-precipitation with equivalent amounts of 5-anti-5 serum. From these preliminary studies, our results showed that approximately 2-3% of the radioactive TCA-precipitable material was specifically bound to the polymerized anti-4 column and that a similar amount was also specifically co-precipitated with 4-anti-4 serum. The results thus suggest that L-chain synthesis did occur but for reasons yet to be explained. These proteins could not be detected by acrylamide gel electrophoresis procedures.

The molecular size of i-RNA was further analyzed on sucrose-DMSO gradients to reduce distorting effects due to any secondary structures of the RNA. Results of such an experiment are presented in Table 3. i-RNA extracted from antigen-pulsed macrophages were placed on 5-20% sucrose gradients containing 80% DMSO. Fractions were collected after centrifugation at 58,000 rpm for 18 hours at 25°C and assayed for m-RNA activity in the L-cell-free system. Two active m-RNA fractions were detected with sedimentation coefficients of 16 and 28S which gave rise to the formation of specific antibody. The molecular size of such m-RNAs would be equivalent to 4-10 x 10^5 daltons of nucleic acid, enough to code for 40,000-100,000 daltons of protein and could thus code for immunoglobulin heavy and light chains which are 55,000 and 22,000 daltons, respectively. Similar results were obtained for dT-cellulose purified i-RNA. In light and modern nucleic acid biochemistry the larger m-RNA may represent a m-RNA precursor.

The entrance of exogenous T2 m-RNA into immunocompetent lymphocytes lacking specific T2 surface membrane receptors [12] has been shown in our laboratory [13] and suggests that the exogenous i-RNA may bring about changes in the cells synthetic activities that go beyond translation of message contained in such foreign RNA. Our observations suggested that IgM antibody molecules found in response to exogenous i-RNA were made of polypeptide chains that contain exclusively the allotype markers of the

Table 3

Fractionation of i-RNA in Sucrose-DMSO Gradients

dT-cellulose i-RNA		
Fraction No.		% Phage Neutralization (1:8 Dilution)
3		0
4	28S	37
5	28S	43
6		12
7	16S	25
8	16S	20
9		6

Total i-RNA		
Fraction No.		% Phage Neutralization (1:8 Dilution)
3	28S	42
4	28S	27
5	28S	10
6		6
7	16S	11
8	16S	34
9		0

Translation products tested for anti-T2 activity in the alltype amplified neutralization assay (see Table 2).

i-RNA donor. No evidence was obtained for the synthesis of "asymmetric" or hybrid molecules which could contain one or more chains specified by m-RNA of the recipient cell origin. We have, however, observed the formation of "hybrid" molecules in response to a mixture of exogenous i-RNA preparations in our spleen cell culture experiments. In these experiments, i-RNA was prepared from cells of a b4 rabbit which had been stimulated with T2 phage and another i-RNA preparation was made from cells of a b5 animal which had been incubated with SP82 phage (*B. subtilis* phage). Mixture of these two i-RNAs were then added to b6 spleen cells. The results presented in Table 4 clearly show that IgM antibodies against T2 and SP82 phage contained both b4 and b5 but not b6 allotype markers. Such hybrid molecules could be translational products of a processed m-RNA derived from the large precursor RNAs of both T2-RNA and SP82-RNA. This would involve a mechanism by which the HmRNAs are excised, spliced and ligated into the final form of m-RNA coding for such hybrid immunoglobulin molecules [13].

Genetic Restriction in Immune RNA Presentation to Immunocompetent Lymphocytes

The initiation of an immune response with immune RNAs *in vitro* was found to be dependent on the presence of specific antigen-reactive cells for the expression of RNA-antigen complex activity but not for i-RNA activity. However, both RNAs required the presence of an adherent population of cells (macrophages) for the RNAs to enter lymphocytes and subsequently bring about specific antibody formation [14]. This requirement for macrophages to "feed" RNA to surrounding lymphocytes allowed us to examine the question as to whether these two cell types had to be histocompatible. In these experiments spleen cells from inbred and noninbred

Table 4

Formation of Hybrid Antibody Molecules with a Mixture of Two Different Antigen-specific m-RNAs

i-RNA Added to 66 Spleen Cells[a]	% Neutralization			
	T2 Phage[b]		SP82 Phage	
	Anti-4	Anti-5	Anti-4	Anti-5
11/44 T2-i-RNA	48	5	2	4
33/55 SP82 i-RNA	8	6	2	70
11/44 T2-i-RNA + 33/55 SP82-i-RNA	42	42	41	53

[a]i-RNAs prepared from macrophages incubated with phage at a ratio of 100 macrophages:1 phage; 200 μg of i-RNA were added per culture of 2×10^7 spleen cells.

[b]% neutralization activity of tissue culture fluids diluted 1:8; 25% neutralization significant at the 1% level; no anti-phage activity was detected with anti-6 amplification.

rabbits were used as donors of adherent and nonadherent spleen cells. The adherent cells were incubated with immune RNA extracted from peritoneal exudate cells incubated with a high input of T2 (ratio of macrophages: T2 was 1:100). After incubation for 30 minutes, excess RNA was removed and various nonadherent cell populations including (a) autologous, (b) matched, and (c) mismatched cells were added back to the tissue culture plates containing the RNA-exposed adherent cells. Tissue culture fluids were removed 5 days after incubation and assayed for total anti-T2 activity as well as antibody activity associated with 19 and 7S immunoglobulins separated by appropriate sucrose gradient centrifugations. The i-RNA activity was monitored by the presence of 19S antibody activity and the RNA-antigen complex activity was noted by the presence of 7S antibody activity. The results of one such experiment are shown in Table 5. The need for histocompatible macrophages and lymphocytes was only seen with the formation of 7S antibody initiated by the RNA-antigen complex. 19S antibody formation due to the i-RNAs was found in all the cultures regardless of the genetic source of macrophages or lymphocytes. Thus, genetic restriction was operative with an immune RNA which contains antigen and could only stimulate specific antigen-reactive lymphocytes. Results from other laboratories [15] suggest that genetic restriction between antigen-pulsed macrophages and activated T cells was evident during the secondary immune response rather than in the primary response.

Characterization of the m-RNA Donor Cells

One of our major interests in the investigation of immune i-RNA is to determine the nature of the donor cell. The concept that a m-RNA for

Table 5

The Requirement for Histocompatible Macrophages for the Presentation of Immune RNA to Lymphocytes

Rabbit #			% Neutralization				
			Total Activity			Sucrose Gradient Fraction	
AD and	NA Cells + Immune RNA		1:5	1:10	1:20	19S	7S
1	1	autologous mixtures	64	42	23	+	+
2	2		58	27	21	+	+
3	3		46	36	19	+	+
1	2	histocompatible (matched)	46	36	17	+	+
2	1		49	34	26	+	+
2	3	nonhistocompatible (mismatched)	81	54	31	+	-
3	2		52	41	27	+	-
1	3		70	54	36	+	-
3	1		36	34	27	+	-
-	1		16	15	2		
-	2		3	0	2		
-	3		3	9	8		

immunoglobulin synthesis is present in a cell that is thought not to be immunocompetent was difficult to comprehend. There have been, however, recent reports indicating the presence of m-RNA in cells that do not synthesize its translational products. This is true for T lymphocytes [16] which contain m-RNA for kappa chain immunoglobulin and for MOPC · 104E cells which contain m-RNAs for kappa fragment synthesis and for lambda chain synthesis but only produce lambda light chain proteins (D. Baltimore, personal communication).

In our own studies we have identified the i-RNA producing peritoneal exudate cells as representing only 8% of the total cell population, a subpopulation that can be enriched by appropriate density gradient centrifugation. Evidence from two lines of experimentation has been obtained that suggests that the donor cells may possess surface membrane receptors that recognize the antigen used to stimulate the formation of specific i-RNA. One is the observation that i-RNA is synthesized by macrophages exposed to a low input of phage in which the ratio of cells to phage could be as low as 1000:1. This suggests that some recognition mechanism must exist between the i-RNA-producing cells and the antigen. The other line of evidence comes from suicide experiments in which macrophages were incubated with highly radioactive (I-125) T2 phage. These exposed PE cells were no longer able to synthesize i-RNA but were able to produce immune RNA-antigen complexes. Specific elimination of the T2-i-RNA-producing cells was evident in these experiments in that these cells were able to respond to SP82 stimulation and synthesize both immune i-RNA and RNA-antigen complexes (see Table 6).

Table 6

Suicide of i-RNA Producing Cells with ^{125}I-labeled Antigen

Source of PE Cell RNA		Antibody Response			
		19S		7S	
1st Incubation[a]	2nd Incubation[b]	Anti-T2	Anti-SP82	Anti-T2	Anti-SP82
PEC + T2	T2	+[c]	-[d]	+	-
PEC + T2	SP82	-	+	-	+
PEC + ^{125}I-T2	T2	-	-	+	-
PEC + ^{125}I-T2	SP82	-	+	-	+

[a] 1 x 10^9 PE cells incubated with 1 x 10^9 T2 or radioactive T2 (1 x 10^6 cpm) for 2.5 hours at 4°C.

[b] PE cells incubated with nonradioactive phage for 30 minutes at 37°C; ratio of PE cells to phage was 1:1000.

[c] Positive response: 1:8 dilution of tissue culture supernatants resulting in greater than 25% neutralization.

[d] Negative response: undiluted tissue culture supernatants resulting in less than 25% neutralization.

The nature of the i-RNA donor cell has been difficult to identify because of the presence of two cell types in the active peritoneal exudate cell population. One of these cells is a macrophage (70%) and the other cell has been designated as A cell (30%) which is microscopically distinct from macrophages and lymphocytes [17]. The A cells which lack Fc receptors were separated from macrophages that were rosetted with appropriate antibody-coated erythrocytes. Removing the rosetting cells by means of Ficoll hypaque density gradient centrifugation resulted in a preparation that consisted of 70% A cells and only 30% macrophages. Preliminary studies to determine the i-RNA synthesizing capacity of such cell preparations have yielded negative results. Work is, however, continuing to examine the "A" cells for i-RNA synthesis as well as any possible regulatory role they may have on macrophages in immune RNA formation.

In conclusion, the immune RNAs, both the RNA-antigen complex and the i-RNA, may have *in vivo* relevance in participating in the early events in antibody formation in which antigen-stimulated macrophages can interact with lymphocytes to bring about an expansion of antibody forming cells in the absence of lymphocyte proliferation. It is also known that i-RNA, in contrast to RNA-antigen complexes, (a) is a m-RNA of sufficient size to code for a complete immunoglobulin, (b) can be presented to lymphocytes via macrophages that are not histocompatible, (c) stimulates nonspecific antigen reactive lymphocytes, and (d) is synthesized by a unique subpopulation of macrophages that possess specific surface membrane receptors for the antigen.

REFERENCES

1. Burnet, J. M. CIBA Foundation Symposium on the Immunologically Competent Cell. Little, Brown and Company, Boston, p. 100, 1963.

2. Ehrenreich, B. A. and Cohn, Z. A. J. Exp. Med. 126:941, 1967.

3. Lee, K.-C., Shiozawa, C., Shaw, A. and Diener, E. Eur. J. Immunol. 6:63, 1976.

4. Chused, T. M., Kassan, S. S. and Mosier, D. E. J. Immunol. 116: 1579, 1976.

5. Rosenthal, A. S., Blake, J. T., Ellner, J. J., Greineder, D. K. and Lipsky, P. E. Immunobiology of the Macrophage, Ed. D. S. Nelson, Academic Press, New York, p. 131, 1976.

6. Shevach, E. M. J. Immunol. 116:1482, 1976.

7. Ellner, J. J., Lipsky, P. E. and Rosenthal, A. S. J. Immunol. 118: 2059, 1977.

8. Ben-Sasson, S. Z., Lipscomb, M. F., Tucker, T. F. and Uhr, J. W. J. Immunol. 119:1493, 1977.

9. Lipsky, P. E. and Rosenthal, A. S. J. Immunol. 115:440, 1975.

10. Braendstrup, O., Werdelin, O., Shevach, E. M. and Rosenthal, A. S. J. Immunol. 122:1608, 1979.

11. Dutton, R. W. and Mishell, R. I. Cold Spring Harbor Symposium on Quantitative Biology, Antibodies 32:407, 1967.

12. Bilello, P., Fishman, M. and Koch, G. Cell. Immunol. 23:309, 1976.

13. Williamson, B. Nature 270:295, 1977.

14. Schaefer, A. E., Fishman, M. and Adler, F. L. J. Immunol. 112:1981, 1974.

15. Pierce, C. W., Kapp, J. A. and Benacerraf, B. J. Exp. Med. 144:371, 1976.

16. Storb, U., Hager, L., Putman, D., Buck, L., Farin, F., and Clagett, J. Proc. Nat. Acad. Sci. USA 73:2467, 1976.

17. Rice, S. G. and Fishman, M. Cell. Immunol. 11:130, 1974.

RNA-DEPENDENT RNA POLYMERASES

Heinz Fraenkel-Conrat, M. Ikegami and S. Stein

Department of Molecular Biology
and Virus Lab
University of California, Berkeley

Abstract Plants, in contrast to bacteria and probably also animal, contain RNA dependent RNA polymerases. These enzymes are largely in the cytoplasm, when free, but became associated with membranes when they are bound to templates such as the RNA of infecting viruses. The enzymes are strictly plandspecific and yet appear to be able to replicate such viral RNAs.

The RNA dependent RNA polymerase of tobacco has been purified about a 30,000 fold. It shares the following properties with the enzymes from other plants: 1) great lability; 2) requivement for about 0.005 M Mg; 3) template dependence and inference but no template specificity; 4) about 170,000 molecular weight, consisting of subunits; 5) increase in amount but not nature of enzyme by RNA virus infection; 6) production of only the complementary strand of RNA or polynucleotide template; 7) production in vitro of only short polynucleotides.

It is nevertheless believed, but not proven that these enzymes can in vivo and in situ serve as viral RNA replicases.

The role of these enzymes in the plants' physiology remains uncertain. For lack of any other known function, they are proposed to serve in the production of specific double stranded RNA molecules that may serve regulatory roles. Such RNA molecules have been demonstrated in tobacco leaves.

As a consequence of the concept and formulation of the central dogma of molecular biology, cellular RNA serves primarily in ribosomal, messenger, and adaptor functions. No convincing evidence has been brought forth that these species of molecules are replicated, their only origin being through transcription from genomic DNA.

An exception to these normal cellular processes in represented by RNA virus- infected cells in which RNA must be both, transcribed and replicated. The more complex of these virus families, such as the viruses causing flu and measles, carry negative strand RNA. These viruses carry the enzymes necessary to transcribe and replicate their RNAs as the first step in the infection process. This is also true for the RNA tumor viruses although here a reverse transcriptase occurs in the virion, that transcribes the viral RNA to DNA.

The situation is different for the simple plus-strand RNA viruses of bacteria, animals and plants. These carry no enzymes in the virion. Instead RNA bacteriophages have been shown to carry the gene for a comparatively small protein (60,000 daltons) that is able to form a complex with 3 non-enzymatic bacterial proteins which is then active as an RNA-dependent RNA polymerase. The molecular weight of that enzyme-active complex is about 210,000.

Although the nature of the polymerases that replicate the RNA of the corresponding class of animal viruses, e.g.poliovirus, is less clear and no enzyme has been isolated from infected cells in pure from, nor characterized nearly as well as that of Qβ phage, it appears very probable that the RNA replicase of polio-infected cells is composed of both a virus-coded peptide chain or its proteolytically processed derivative of 77K and 63K daltons, respectively, and probably several not well-defined host components. No definitive evidence exists that non-virus-infected bacteria or animal cells can replicate RNA molecules, although negative results are never as definitive as positive findings.

Surprisingly, it has been demonstrated in 1971 that RNA-dependent RNA polymerase activity is present in extracts of plants, and specifically Chinese cabbage (1). This activity is increased by RNA virus infection such as turnip yellows mosaic virus which is usually grown in these plants (1). Therefore it was suspected that some of the control plants might have accidentally become virus-infected and that the observed enzyme activity was a consequence of such cryptic viral infection. This suspicion enabled the scientific community to avoid accepting the unorthodox conclusion that RNA dependent RNA polymerase existed in healthy cells, - not an unusual mode of reacting to new, unexpected and seemingly anti-dogma findings.

Further studies of other plants, such as tobacco (2-13), cowpea (9, 14), cauliflower (15) and now many other plants (8, 16, 17) have demonstrated the general presence of such enzymes in healthy plants with only one exception (cucumber cotyledons) (18). This has gradually led to an acceptance of the fact that there exist RNA-dependent RNA polymerases in healthy plants.

The purification of these enzymes has progressed slowly, largely because they were found very labile under all kinds of conditions and most of the activity was lost in the course of purification. Also the fact that upon virus infection most of the increased amounts of enzyme activity is bound to membranes, while it is free and soluble in the healthy plants, has hindered the purification and the recognition of the nature of the enzyme. These facts are true for all plant RNA-dependent RNA polymerases, as well as for bacterial ones with the exception of the Qβ -enzyme, and for the corresponding enzymes from virus-infected animal cells. Only in 50% glycerol at - 70°C can the RNA polymerase activity of the enzyme from tobacco be retained in purified preparations for several days.

Nevertheless we have reported methods of solubilizing and purifying the enzyme (3, 10) and have recently found an additional method that has represented a major advance in this regard. Our most highly purified samples, utilizing as a final step (step 5) a RNA cellulose affinity column, are about 40-fold more active in terms of specific activity than the best previously reported preparations and as active as the best RNA polymerases of other sources (see Figure 1 and Table 1). They incorporate about 2 nmole

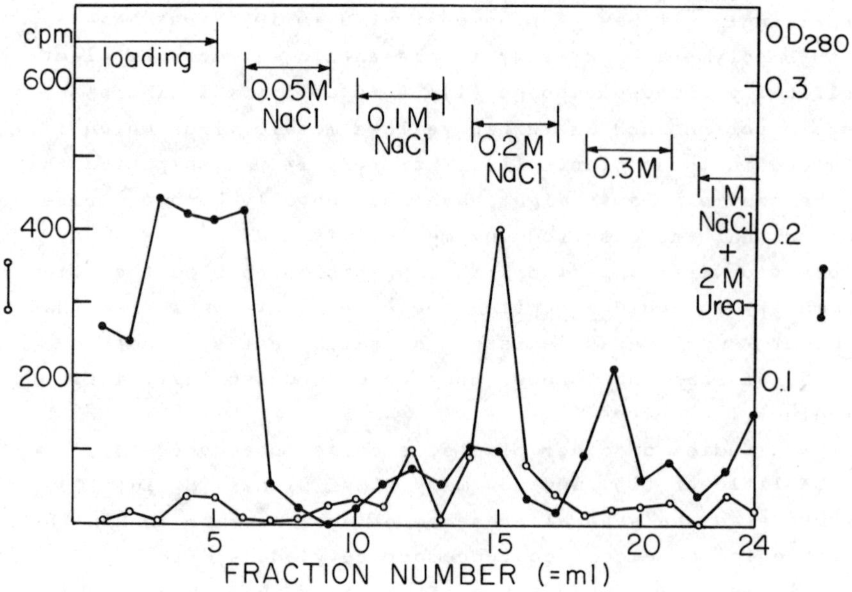

Figure 1. The 5'-terminal sequence of turnip yellow mosaic virus RNA, used to illustrate potential complementary base pairing, that in any RNA can represent obstacles to RNA polymerase action. This is schematically presented in the inset with the lines terminating in arrows indicating short transcription product sequences, and the arrows the polymerase. Note that the product is largely but not completely H-bonded to the template.

Table 1

Summary of the purification procedure for TNV stimulated Tobacco RNA dependent RNA polymerase from 125 g plant leaves

	Total Activity (cpm X 10^{-6})	Total Protein (mg)	Specific Activity (cpm/μg protein)	Purification
1. Crude extract	5.1	850	6	1
2. Low salt extract	21.1	700	31	5
3. PEG-Dextran two phase separation	16.1	24	676	113
4. DEAE-Sephadex	11.1	4	3083	514
5. RNA cellulose	0.63*	0.02	28,636	4,773

* Peak tube only

UTP per mg enzyme per minute, using viral RNA as template, and 5 fold move with poly U, C.

Almost all workers in this field have now accepted the fact that healthy plants contain such an enzyme, but no consensus has been reached whether this enzyme is responsible for the replication of the RNA of infecting viruses. To test the latter hypothesis we have compared many properties, both physical and enzymatic, of three not yet highly purified enzyme preparations from the same plant, namely from healthy, tobacco mosaic virus (TMV) infected, and tobacco necrosis virus (TNV) infected tobacco (7). We found no differences between these 3 preparations. We then compared enzyme preparations from two different plants, tobacco and cowpea, both either TNV-infected or healthy (9). Here we found some differences in the protein-chemical proportion between the enzymes from the different plants, be they infected or healthy. Our conclusion therefore is that the enzyme is host-specific, coded for by the host's gene, and not to any easily detected extent by viral genes. This conclusion is not yet generally accepted, although evidence favoring it is steadily increasing, at least as far as tobacco is concerned (8, 11, 17, 19).

The nature of the enzymatic action of the most highly purified preparations is in most respects the same as that reported by us and others for less active preparations (3, 4, 7 - 10, 20, 21). The criteria tested were sedimentation rate (mol. wt. ~170,000), divalent cation optima, kinetics, non-specificity but quantitative differences for different templates etc.

With turnip yellow mosaic viral (TYMV) RNA as template, the product is largely but not completely double-stranded with 30% ± 10% being usually digestible by RNase in high salt solution (2x SSC) (see Table 2). The undenatured template-product complex is quite large, about 1-2 x 10^6 molecular weight, but heterogeneous (see Figure 2). Upon denaturation the newly synthesized polynucleotides are also heterogeneous but quite small, mostly of the order of tRNA. These data suggest multiple initiation of transcription on the 2 x 10^6 TYMV template. The fact that the template remains large indicates the absence of ribonucleases from our "step 5" - preparations. When less purified polymerase was used, the template-product complex was of low molecular weight. This is in agreement with the finding of others. Various attempts

Table 2: Evidence that in vitro tobacco RNA polymerase products are completely hybridizable to template RNA and are in good part singlestranded.

RNA polymerase	Template mg/ml	Product Analyzed (cpm)	RNase-resistent (in 2x SSC) Melted %	Self-annealed %	Annealed with TYMV RNA % mg (RNA/ml)
Step 4, sp. act. 3000, 20 hrs	4	ds* (3578)	yes, 5	24	91, 97 (0.7, 1.4)
Step 5, sp. act. 65,000, 4 hrs	7.6	ds* (203)	yes, 9†	75	86, 99 (0.2, 1.0)
Step 5, sp. act. 65,000, 4 hrs	76	ds (972)	yes, 6	84	91, 100 (0.2, 1.0)
Step 5, sp. act. 48,000, 17 hrs	7.6	ss^Δ (159)	no, 57°		88, 97 (0.75, 1.5)
Step 5, sp. act. >18,000, 21 hrs	7.6	ss (22,405)	no, 51		87, 102 (0.7, 1.4)
Step 5, sp. act. 48,000, 17 hrs	76	ss (1107)	no, 63		92, 99 (0.75, 1.5)

* Product isolated by phenol extraction, RNase treatment in 2x SSC, protease K, phenol extraction, Sephadex G75 and ethanol precipitations

† Mock-annealed with yeast RNA (2 mg/ml)

Δ Product isolated without phenol and RNase treatment by Sephadex G75 and ethanol precipitations. This is thus native replicative intermediate, as indicated about 43% singlestranded, and fully able to anneal.

° These data represent a measure of the proportion of single- to doublestranded product made by highly active enzyme. The singlestranded fraction is always between 30 and 50%, similar from 15' to 21 hrs of enzyme action, and not affected by template concentration. The doublestranded component is not completely degraded by RNase at low salt concentration.

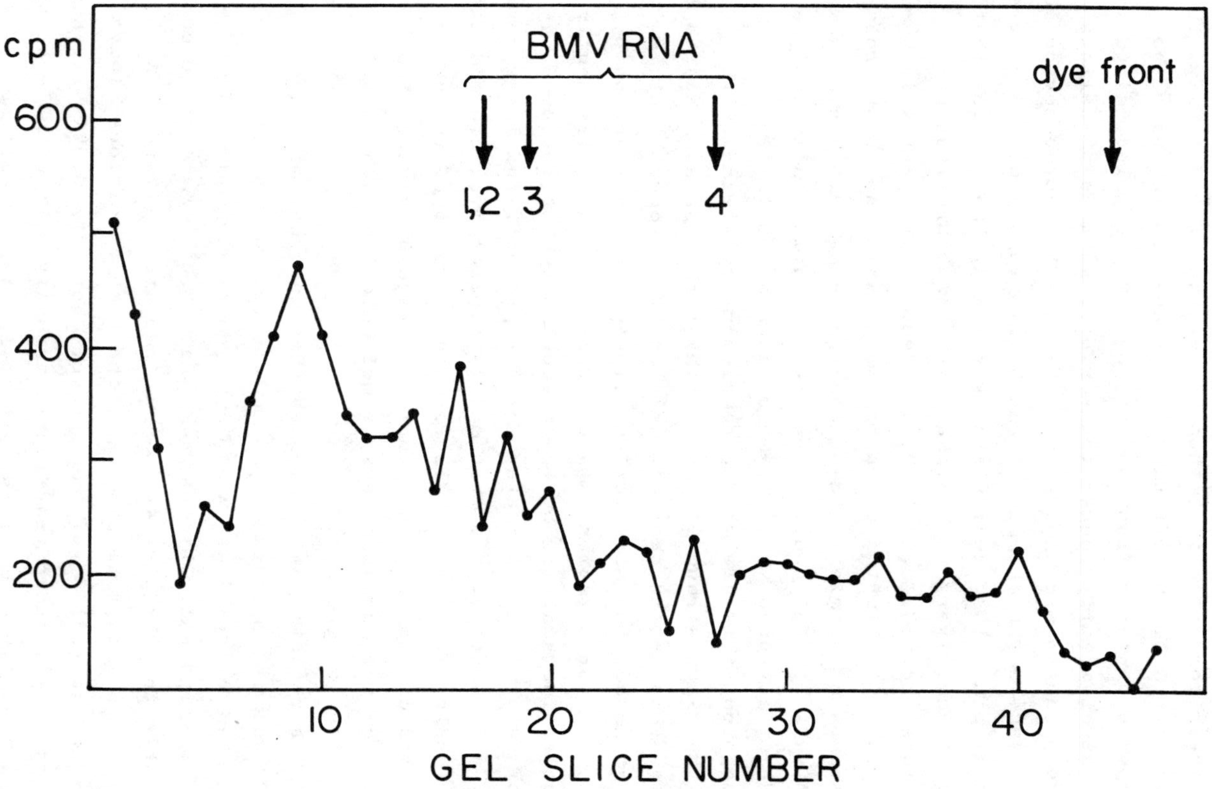
cpm
600
400
200
BMV RNA
1,2 3
4
dye front
10
20
30
40
GEL SLICE NUMBER

Fig. 2

to obtain large transcription products were not successful. Among these was the use of long incubation periods (17-21 hrs). We also utilized equal concentrations of the four nucleoside triphosphates rather than the customary method of using the radioactive one at a much lower concentration, thus higher specific activity. This slightly increased the efficiency of transcription when the triphosphate concentration was not too low, but also failed to produce significantly longer transcripts. We tend to believe that the products are always small because the enyme may be able, in vitro and in the absence of RNA binding proteins, to copy only the nonhairpin folded segments of viral RNA. The finding that incorporation is very much higher with structure-less polynucleotides (poly U, C, poly U, G,) seems to support this hypothesis. This is illustrated on Figure 3, which shows the relative amounts of single - and doublestrandedness in a small part of TYMV RNA (the 3' end), and presents schematically the presumed mode of action of the enzyme on such a template.

The product is always the complementary strand of the template, the minus strand if TMV or TYMV viral RNA are used as template. This is the case not only for the product that is double-stranded, but also for the singlestranded fraction. Besides the hybridization data on Table 2, we also frequently used nucleotide analysis by the nearest neighbor technique as a means of determining the nature of the product. The interpretation of all these data is that mostly replicative intermediate-type molecules are made by displacement of part of the newly synthesized chains from the template, presumably under the influence of the advancing enzyme molecule (see schematic presentation on Figure 3).

It has recently been reported that upon use of a great excess of enzyme as related to template (e. g. 1/40 of the customary amount of template) the formation of some plus strand material was indicated by the fact that about 30% of the product was not able to hybridize to excess template (+strand) RNA (13). In contrast, our experiments under the same conditions always show that very close to 100% of the product, both the double-stranded and the single-stranded fraction is actually hybridizable to the viral RNA (see Table 2). This is true for the product made with the most highly purified as well as with a somewhat cruder prepara-

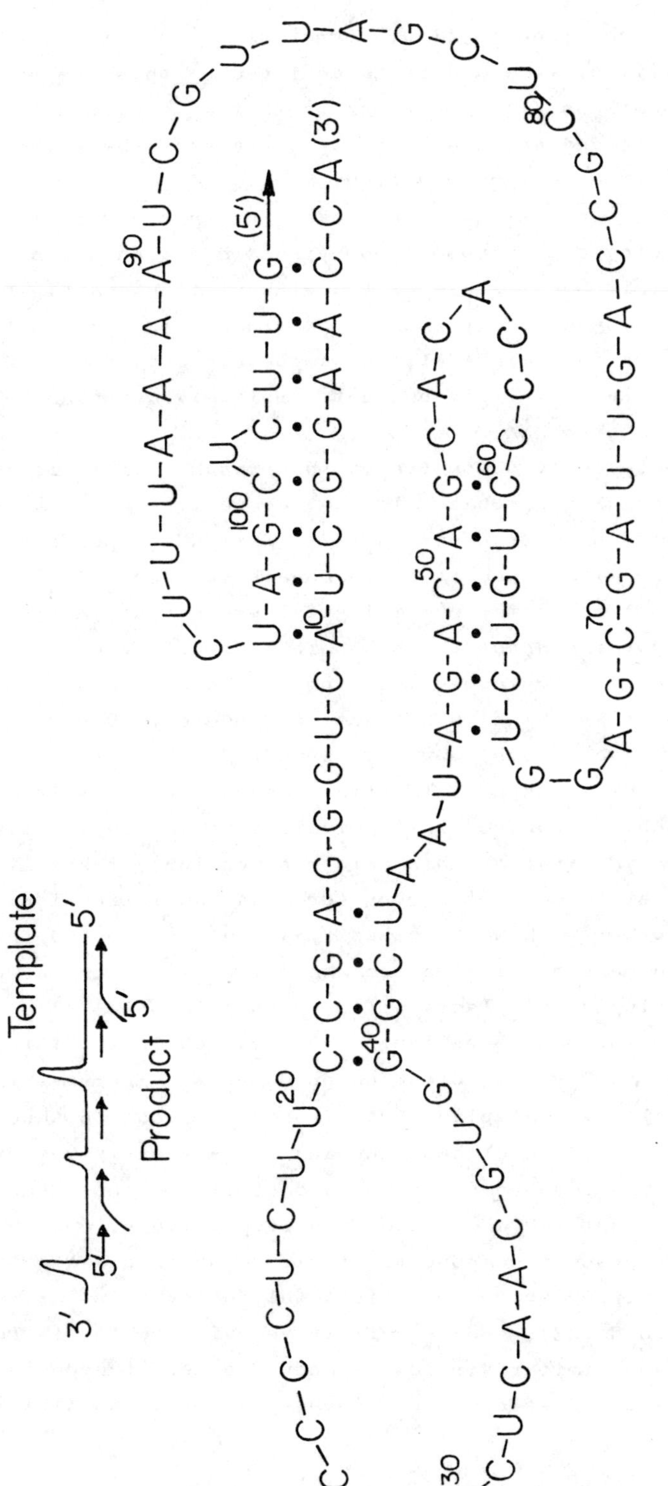

Fig. 3

tion. Thus we continue to find, regretfully, no evidence for true replication (i. e. +→ - -→+) being achieved by this enzyme under our iv vitro conditions. It appears probable that additional factors and/or the membrane-associated natural state of the enzyme are required for true replicase action.

The most important question regarding this enzyme is: What is its physiological purpose? Upon virus infection its production is increased; it becomes associated with viral RNA at membranous sites; and during active virus propagation it makes much viral plus strand RNA on a minus strand template. However, in uninfected cells the enzyme is largely but not completely cytoplasmic and not associated with RNA.

We have recently demonstrated the presence of small amounts of double-stranded RNA in tobacco leaves, as well as yet smaller amounts of partially double-stranded RNA (22). This RNA appears to largely be of a single kind since it self-anneals very quickly. It may well represent the normal physiological product of this enzyme. But we know nothing about the biological significance of that RNA.

The largely cytoplasmic location of the enzyme and the fact that its levels are high in not actively photosynthesizing tissues (e. g. cauliflower heads) makes it appear unlikely that it is involved in photosynthesis. The findings of very little of the partially double-stranded RNA (equivalent to the replicative intermediates in viral nucleic acid replication), makes it appear improbable that the enzyme serves for mRNA amplification. For lack of any other working hypothesis, we have suggested that its purpose is to make double-stranded RNA molecules which may play a regulatory role in the plant cell's physiology (22).

Another important question to be answered in the future is the mechanism by which viral infection produces an increase in enzyme production. All typical plant RNA viruses have one to three gene products of unknown function. One each is made early upon infection by TMV, and the bromo-viruses and similar groups; these early proteins are about 140 and 35k dalton respectively, and were believed to represent components of virus-specific RNA polymerases. There is now strong evidence against this hypothesis, as far as TMV is concerned (23), and no definitive evidence for it regarding the bromo - and similar viruses is known to me. It appears possible that the role of these and other such viral proteins lies in stimulating the production of the host's RNA dependent RNA polymerase in some yet unknown manner.

Summary

Further purification of the RNA dependent RNA polymerase of tobacco leaves is reported. The properties of this very unstable enzyme are no different than those of less purified preparations. The synthesized polynucleotides, appear large as long as they remain bound to a large template RNA which is the case when highly purified nuclease-free enzyme is used. The products are always in part singlestranded corresponding in this regard to the replicative intermediate of viral nucleic acid replication. The products appear to be entirely complementary to the template under all conditions tested. Upon melting the product polynucleotides become quite short and they always correspond to only a small fraction of the amount of template used. This inefficiency is attributed tentatively to the conformational inaccessibility of much of the template, a conclusion that is supported by the greater efficiency of less structured polynucleotides, e. g. poly (U, C), as templates.

References

(1) Astier - Manfacier, S. and Cornuet, P. (1971) Biochim. Biophys. Acta 232. 494-493

(2) Duda, C. T., Zaitlin, M. and Siegel, A. (1973) Biochim. Biophys. Acta 319, 62-71

(3) Fraenkel-Conrat, H. (1976) Virology 172, 23-32

(4) Stussi-Garaud, C., Lemius, J. and Fraenkel-Conrat, H. (1977) Virology 81, 224-236

(5) LeRoy, C., Stussi-Garaud, C. and Hirth, L. (1977) Virology 82, 48-62

(6) White, J. L. and Murakishi, H. H. (1977) J. Virol. 21, 484-492

(7) Ikegami, M. and Fraenkel-Conrat, H. (1978) Proc. Nat. Acad. Sci. U.S.A. 75, 2122-2124

(8) Romaine, C. P. and Zaitlin, M. (1978) Virology 86, 241-253

(9) Ikegami, M. and Fraenkel-Conrat, H. (1978) F E B S Lett. 96, 197-200

(10) Ikegami, M. and Fraenkel-Conrat, H. (1979) J. Biol. Chem. 254, 149-154

(11) Clerx, C. M. and Bol, J. F. Virology 91, 453-463 (1978)

(12) Duda, C. T. Virology 92, 180-189 (1979)

(13) Chifflot, S., Sommer, P., Hartmann, D., Stussi-Garaud, C., and Hirth, L. Virology 100, 91-100 (1980)

(14) White, J. L. and Dawson, W. (1978) Virology 88, 33-43

(15) Astier-Manifacier, S. and Cornnet, P. (1978) C. R. Seances Acad. Sci. Ser D. 287, 657

(16) Weening, C. J. and Bol., J. F. (1975) Virology 63,77-83

(17) Lazar, E., Walter, B., Stussi-Garaud, C., and Hirth, L. (1979) Virology 96, 553-563

(18) Kumarasamy, R. and Symons, R. H. (1979) Virology 96,622-632

(19) Linthorst, H. J. M., Bol, J. F. Jaspars, E. M. J. (1980) J. Gen. Virol. 46, 511-515

(20) Zaitlin, M., Duda, C. T. and Petti, M. A. (1973) Virology 53, 300-311

(21) Ikegami, M. and Fraenkel-Conrat, H. (1980) Virology 100, 185-188

(22) Ikegami, M. and Fraenkel-Conrat, H. (1979) Proc. Nat. Acad. Sci. U.S.A. 76, 3637-3640

(23) Scalla, R., Romaine, P., Asselin, A., Rigaud, J., and Zaitlin, M. (1978) Virology 91, 182-193

EFFECTS OF POLYAMINES ON MACROMOLECULAR METABOLISM AND DEVELOPMENTAL PROCESSES IN PLANT CELLS[1]

Arthur W. Galston and Ravindar Kaur-Sawhney

Department of Biology, Yale University, New Haven, Connecticut, USA

A. Introduction: Over the last several decades, considerable quantitative correlational evidence has linked high intracellular concentrations of the naturally-occurring polyamines (especially putrescine, spermidine and spermine) to the onset of cell division in microbes and animal cells (2, 6). More recent evidence indicates that polyamines may play an important regulatory role in plants as well (10, 20), their multiple effects on growth, mitosis and the prevention of senescence being mediated mainly through control of macromolecular metabolism, and partly through effects on the various membrane of the cell.

As polycationic substances at cellular pH's, the di-, tri- and tetra-amines form electrovalent linkages with the ionized phosphate groups of the backbone of DNA and RNA molecules (21); they can also associate with proteins. In many instances, the association results in altered conformation of the macromolecule, with altered activity, stability and resistance to degradative enzymes. In eukaryotic cells, spermine tends to be most abundant in the nucleus, where it associates with nuclear DNA and RNA; spermidine is often found in abundance on ribosomes and, together with putrescine, can also be found associated with t-RNA (and presumably m-RNA) of the cytosol. *In vitro*, virtually every step in transcription and translation can be stimulated by the addition of polyamines (2).

1 Aided by grants from the National Science Foundation, the U.S.-Israel BARD program, and Mr. Albert Nerken. We are grateful to Arie Altman, Yao-ren Dai and Hector Flores for collaboration and Liu-mei Shih (Yale), Nitsa Levin and Ra'anan Friedman (Hebrew University) for technical assistance.

The polyamines are formed from amino acids through the action of carboxylases and transferases. Arginine gives rise to putrescine (Pu) either by way of ornithine and a final decarboxylation catalyzed by ornithine decarboxylase (ODC), or through an arginine decarboxylase (ADC)-mediated decarboxylation, to yield agmatine, followed by loss of urea (Fig. 1). Methionine, activated by ATP to form S-adenosylmethionine (SAM), contributes aminopropyl groups that attach successively to putrescine to make spermidine (Spd) and spermine (Spm). This process is activated by SAM decarboxylase and aminopropyl transferase. Cadaverine (Cd) is formed by direct decarboxylation of lysine.

ODC is the most rapidly turning over enzyme in eukaryotes, with a half life of 10-12 min (2). When quiescent cells are stimulated to divide by growth hormone, injury, or carcinogen, the rise in ODC activity is usually the first measurable biochemical parameter. Inhibition of ODC by α-methyl-ornithine or SAMDC by methylglyoxal-bis-guanylhydrazone (MGBG) prevents the onset of polyamine synthesis and the chain of events leading to cell division. Putrescine addition to the medium frequently inhibits ODC activity through a feedback process involving formation of an "antizyme" of ODC (8). This may play a role in the rapid turnover of the enzyme.

B. Experiments with plant protoplasts

Plant protoplasts, cells whose walls have been enzymatically removed, have aroused considerable interest because of their ability, under proper conditions, to resume growth and reconstitute the entire organism (3). They have also found use in somatic genetics, since protoplasts of different genomes can be made to fuse *in vitro* and to produce genomic and cytoplasmic combinations not attainable in other ways (5, 18). While tobacco and other members of the *Solanaceae* regenerate readily, the ability to regenerate varies widely amongst the families of flowering plants, and except for a

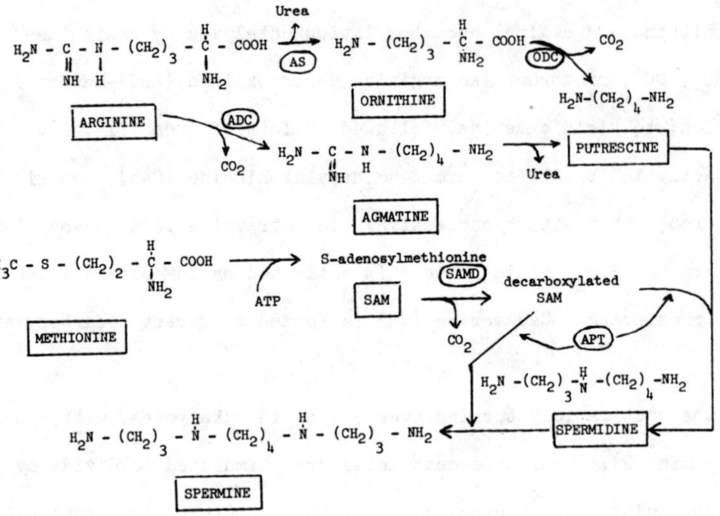

Figure 1. Pathways of biosynthesis of the major polyamines. The formation of cadaverine by decarboxylation of lysine is not shown. AS = arginase; ADC = arginine decarboxylase; ODC = ornithine decarboxylase, SAMD = S-adenosylmethionine decarboxylase; APT = aminopropyltransferase.

recent report (22) on pearl millet (Pennisetum) the large group of cereal grasses is entirely unrepresented in the list of plants capable of regenerating the entire organism from protoplasts. Our investigations began with attempts to cultivate the mesophyll protoplasts of oat (4), a cereal whose protoplasts had been investigated as to physiological characteristics (19).

We soon found that oat protoplasts were extremely fragile, quick to senesce and virtually devoid of growth potential. In agar media, they persisted for several weeks in a viable state, but lost characteristic chloroplast morphology and green color, developed a spherical brown droplet at one end of the cell, and showed clumped, rather than evely distributed cytoplasmic components. They did usually reform weak cell walls and occasionally went into nuclear division and cytokinesis, but never produced more than a "micro-callus" in culture (4). In hanging drop culture, protoplasts in B5 medium (11) deteriorated and lysed within 24 hours. We found that 1-10 mM arginine, a compound that is known to inhibit senescence in detached oat leaves (17) also preserved the integrity of isolated oat (Avena sativa) protoplasts (Fig. 2) (1, 13, 9). Since lysine was also effective, it was natural to test the activity of the diamines, putrescine and cadaverine, resulting from the decarboxylation of these two amino acids. When they proved even more effective than their amino acid presursors, we tried spermidine and spermine, and found them to be the most effective of all (1). In addition to preventing chloroplast clumping and protoplast lysis, the polyamines were found to promote the synthesis of RNA and protein (Fig. 3), to initiate some DNA synthesis, and to initiate some mitotic activity and the formation of binucleate protoplasts (Fig. 4) (16).

C. Experiments with excised leaf sections

Leaves detached from plants are known to undergo a rapid series of senescent changes including massive hydrolysis of protein and RNA and loss of

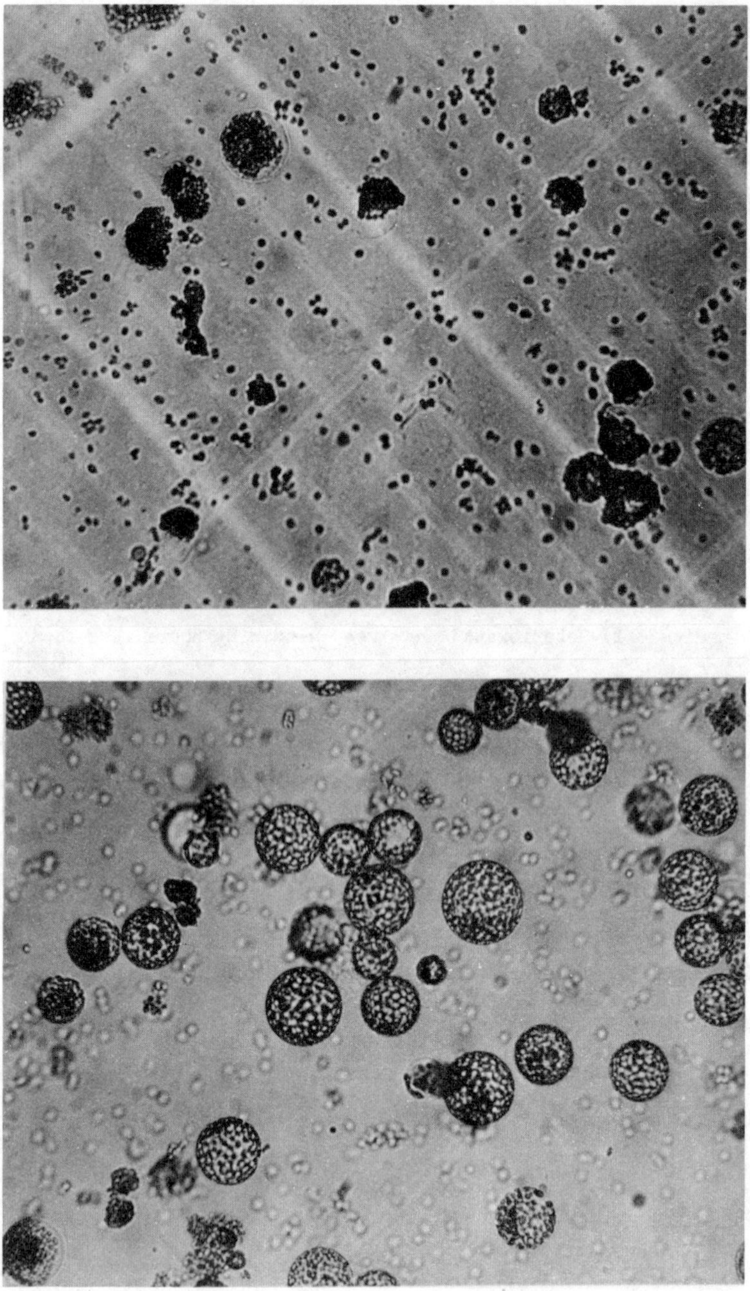

Figure 2. The stabilization of oat protoplasts in hanging drop culture by 10 mM arginine. Above, control; below + arginine. Lysine and the polyamines putrescine, cadaverine, spermidine and spermine are also effective.

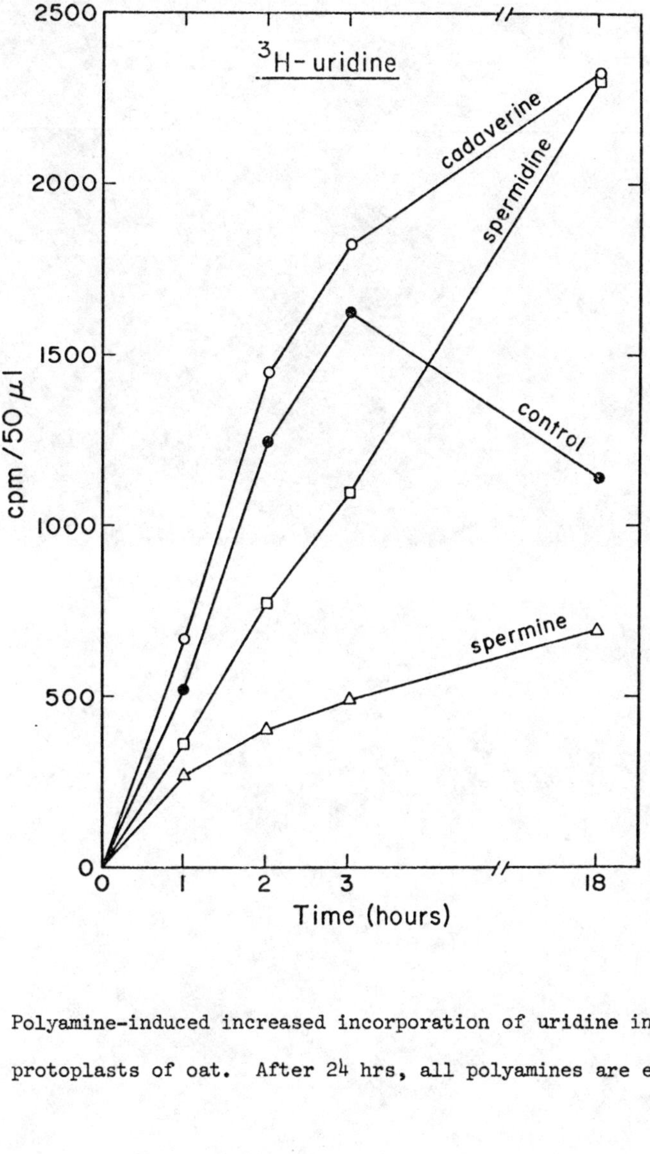

Figure 3. Polyamine-induced increased incorporation of uridine into RNA in protoplasts of oat. After 24 hrs, all polyamines are effective.

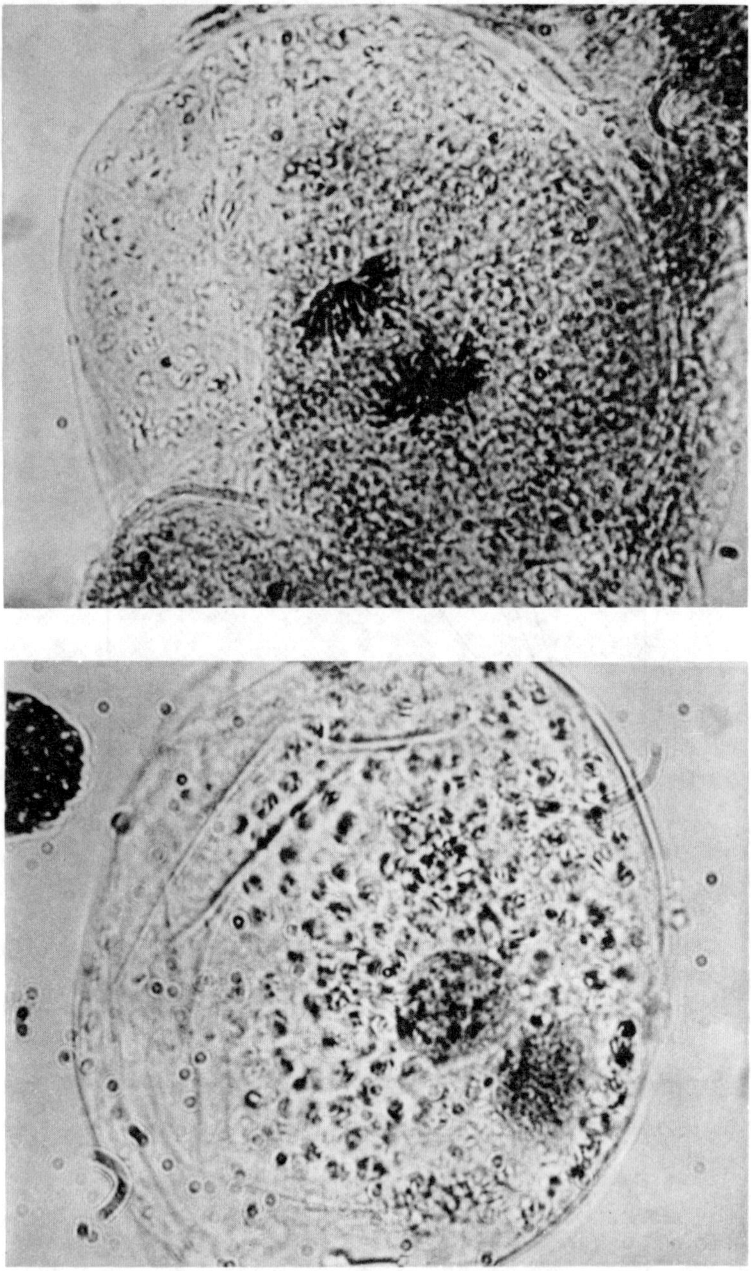

Figure 4. Anaphase figures (above) and binucleate protoplasts (below) in cultures treated with 1 mM spermidine. These occur rarely in controls, but their frequency is greatly enhanced by the polyamines.

chlorophyll. If peeled oat leaf segments are floated on 1-10 mM polyamines, this entire syndrome of senescent changes is delayed or arrested completely, and chlorophyll is retained for many days in the dark, under conditions where control leaves bleach completely (15). Spermine is generally the most effective compound at 1 mM, and spermidine at 10 mM and beyond, where spermine begins to be toxic. Even a 10-60 min dip into 1 mM spermine followed by vigorous rinsing suffices to arrest chlorophyll disappearance over the subsequent 48-72 hr dark period. The polyamines are also effective on peeled corn (*Zea mays*), wheat (*Triticum aestivum*) and barley (*Hordeum sativum*) leaf segments. If the leaves are only partially peeled, then polyamine action is limited to the peeled areas; this indicates that polyamines cannot penetrate the epidermis and do not move from cell to cell once they do penetrate. In dicots, the polyamines are effective on discs cut from leaves with a cork-borer. Among the species whose chlorophyll loss in the dark was delayed by exogenously applied polyamines, are rape (*Brassica napa*), bean (*Phaseolus vulgaris*), pea (*Pisum sativum*), and tobacco (*Nicotiana tabacum*). The last of these, the most frequently used species for protoplast cultivation, shows the lowest spontaneous chlorophyll loss and the least response to applied polyamines. This suggests that the reason for its successful regeneration from protoplasts may lie in a high endogenous polyamine titer.

Biochemical correlates of leaf senescence

In oat leaves, ribonuclease activity rises markedly within minutes of excision, doubles within 4-6 hrs, and attains treble its control value within 24-48 hrs. Acid protease activity starts to rise after about 6 hrs and continues to augment linearly thereafter for at least 48 hrs; chlorophyll disappearance starts only after 24-36 hrs, then proceeds rapidly (Fig. 5). The addition of 1 mM spermine to the medium in which the leaves are floating completely prevents these changes for at least 48 hrs (15). Addition of

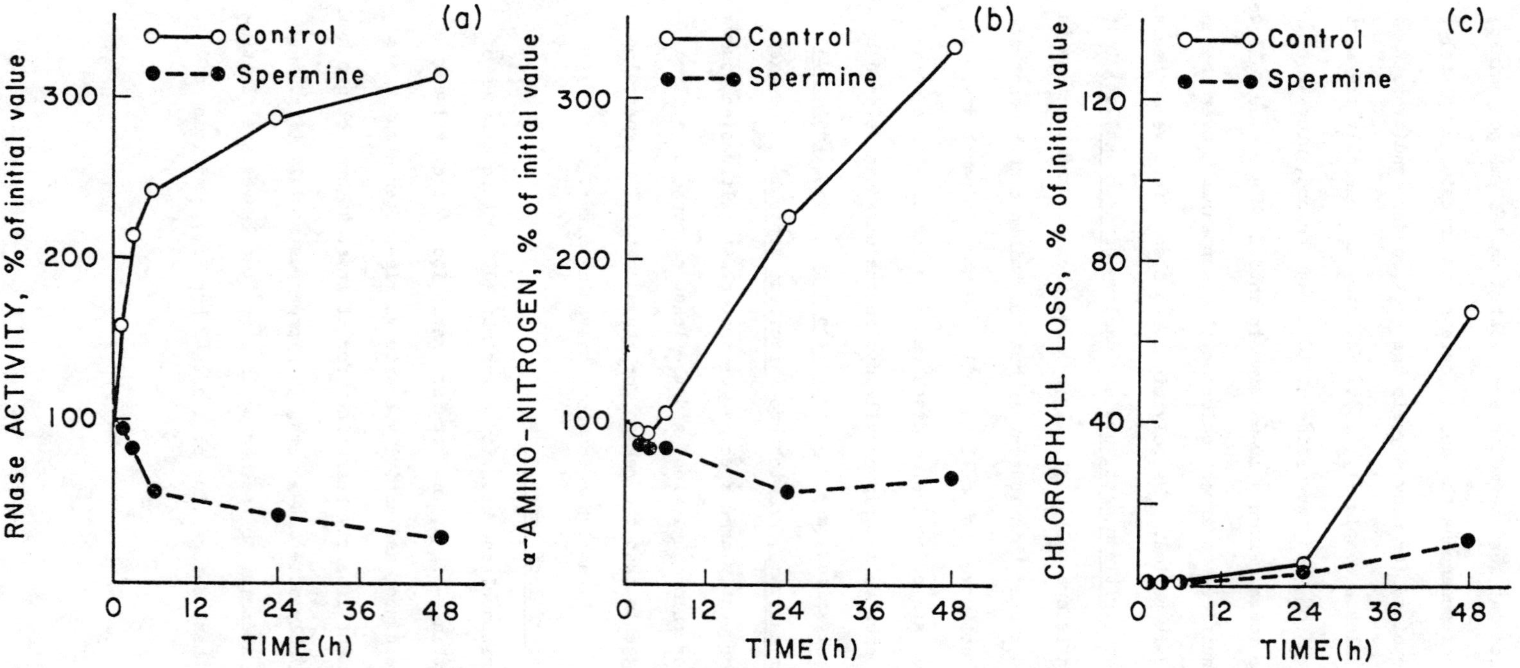

Figure 5. The effects of 1 mM spermine on changes related to senescence. Left: RNase activity; center: protease activity; right: chlorophyll degradation.

polyamines at various times after senescent changes have begun is partially effective; RNase levels decline, protease levels are stabilized, and chlorophyll loss is slower. If calcium ion is added together with the polyamines, the antisenescent action of the latter is partially prevented. Since Ca^{++} is not effective in senescence prevention on its own, it would appear that the inhibitory effect of Ca^{++} may result from competition for a common entry site at the membrane.

In the light, where excised oat leaf sections do not spontaneously lose chlorophyll over a 48-hr period, applied polyamines cause a photobleaching, while at the same time preventing the usual rise in RNase and protease activities.

In vitro effects of polyamines

It has long been known that polyamines, by virtue of their electrostatic association with RNA, can protect all species of RNA molecules against attack by RNase (2). Our investigations corroborated these reports, in showing that the RNase released from lysed oat mesophyll protoplasts could be inhibited by applied polyamines *in vitro* (14). Thus, polyamines lower RNase effectiveness in two ways: (a) by repressing its *de novo* formation *in vivo* and (b) by inhibiting the action of preformed RNase *in vitro*. The *in vivo* effect but not the *in vitro* effect, is also produced by arginine and lysine, the amino acid precursors of the polyamines.

We have also found that polyamines, which repress *de novo* protease formation (15) also inhibit the action of proteolytic enzymes *in vitro* (Fig. 6). This is true not only of the uncharacterized protease from oat mesophyll protoplasts, but also for crystalline trypsin and papain. With trypsin, inhibition develops following a time-dependent process, presumably involving association between polyamine and enzyme. Thus addition of polyamine at zero time, when enzyme and substrate are mixed, produces only slight inhibition of

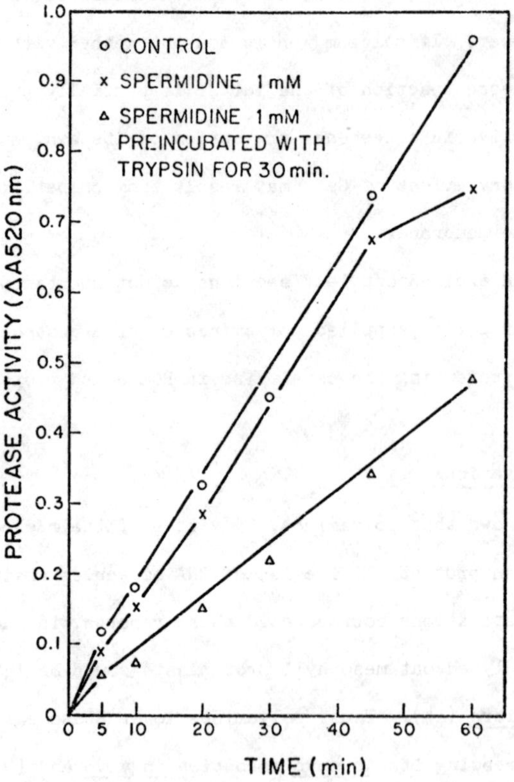

Figure 6. The inhibition of the action of crystalline trypsin on azocoll by spermidine. Note that there is a time-dependent inhibition which is greater when the polyamine is preincubated with the enzyme for 20 minutes.

trypsin activity, but a 30-minute preincubation of enzyme with polyamine induces a significant loss of activity. Once again, the amino acid precursors of the polyamines, which are effective in repression in vivo are ineffective in the in vitro inhibition.

Response of polyamine metabolism to light and hormones

A crucial test of the hypothesis that polyamines can control growth and development would be the demonstration that polyamine synthetic enzymes respond in meaningful ways to developmental triggers, like light and hormones, that exert differential control over the growth of various organs. It is thus significant that the activity of ADC is markedly increased in the buds and decreased in the epicotyls of etiolated pea plants that have previously been exposed to photomorphogenically active red light (7). Since red light absorbed by the P_r form of phytochrome promotes growth of etiolated pea buds and inhibits the growth of etiolated pea epicotyls (12), enzyme behavior parallels growth, as would be predicted if a rise in polyamine titer was a necessary prelude to increased growth. It is clear that the action is due to light absorption by phytochrome, since there is excellent mutual photoreversibility between red and far-red light. The altered enzyme activity is seen within two hours, and the rise in bud ADC activity is sensitive to cycloheximide, but not to actinomycin D or chloramphenicol. Thus, a rapid translational control of ADC synthesis appears to be induced by phytochrome conversion. Neither ODC nor SAMDC activities were much affected.

The application of the plant hormone gibberellin, known to reverse the effects of red light in many plant systems, is effective in negating effects of red on ADC activity, especially in buds. Similarly, in light grown mung-bean seedlings, a marked inhibition of ADC activity was seen in the terminal buds and subapical epicotyl sections of seedlings sprayed with 0.1 mM gibberellic acid several days before harvest. In this case, the decrease in

ADC activity is paradoxical, since both the buds and subapical epicotyl sections were promoted in their growth. We have considered that we must go to shorter time experiments, since an early promotion of ADC activity may have occurred, leading to putrescine accumulation and feedback repression, possibly through formation of an antizyme (8). It is significant that a marked GA-induced repression of ADC formation was also observed with excised epicotyl sections, under conditions in which the other plant hormones and analogs indoleacetic acid, kinetin and abscisic acid were only slightly effective or ineffective.

We have also examined the changes in activities of ODC, ADC and SAMD in tissues and leaves of different ages. ADC and ODC are highest in activity in the base of the oat leaf, where mitosis is still in progress, and lowest at the tip, which is physiologically the oldest. If mature leaves of different ages are compared, there is an inverse relation between age and ADC activity; ODC shows a lesser correlation. When leaves are detached and allowed to senesce in the dark, the activities of ADC and ODC decline markedly, while SAMD remains constant. In the light, where leaves do not senesce rapidly, ADC and ODC activities actually rise over a 7-day period. Thus, all analytical probes indicate a close relation between a probable high polyamine titer and the maintenance of vigorous macromolecular synthesis and cellular activity; conversely, a low polyamine titer seems conducive to senescence. Direct assays of polyamine content by TLC and HPLC are in progress to provide additional evidence on this question.

Summary

Fragile protoplasts of oat mesophyll cells are considerably stabilized by 1-10 mM arginine or lysine or the polyamines (putrescine, cadaverine, spermidine, spermine) formed from them. These polyamines also induce greater incorporation of ^{3}H-leucine, ^{3}H-uridine and ^{3}H-thymidine into protein, RNA and DNA, respectively, and can trigger a small fraction of the cells into mitosis.

Excised leaf sections senesce rapidly. The earliest detected senescent change is a rise in RNase activity (within 1 hr); later there is a rise in acid protease activity (6-8 hrs), and ultimately a loss of chlorophyll (24-36 hrs). All these changes are prevented by 1-10 mM polyamine. Four species of monocot leaves (oat, wheat, corn, barley) and four species of dicot leaves (pea, bean, tobacco, rape) respond to polyamines, but with differing effectiveness.

In addition to repressing formation or release of RNase and protease, polyamines inhibit the activities of the pure enzymes (pancreatic RNase, trypsin) in vitro. This inhibition involves a time-dependent combination of the polyamine with enzyme or substrate.

The enzymes of polyamine biosynthesis, especially ADC, respond to light and hormonal triggers in a manner that generally parallels effects of these agents on growth. Thus, phytochrome conversion to the P_{fr} form, which promotes bud growth and inhibits epicotyl elongation in etiolated peas, promotes and inhibits, respectively, the ADC activity in these organs. Differences are observable within 1-2 hrs and augment for at least 6 hrs after a red light flash. The differences are reversible by either far-red light or gibberellin. Effects on transcription are indicated on the basis of inhibitor studies.

The accumulated evidence indicates that polyamines play a role in plant growth regulation through effects on membranes and macromolecular metabolism.

References

1. Altman A, R Kaur-Sawhney, A W Galston Plant Physiol 60: 570-574, 1977
2. Bachrach U Function of Naturally Occurring Polyamines, Academic Press, 1973.
3. Bajaj Y P S Protoplast isolation, culture and somatic hybridization. in Plant Cell, Tissue and Organ Culture, J Reinert, Y P S Bajaj (eds), Springer-Verlag, Berlin, pp 467-496, 1977
4. Brenneman F, A W Galston Biochem Physiolog Pflanzen 168: 453-471, 1975
5. Carlson P E, H H Smith, R D Dearing Proc Natl Acad Sci (US) 69: 2292-2294, 1972
6. Cohen S S Introduction to the Polyamines, Prentice-Hall 1971
7. Dai Y-R, A W Galston Plant Physiol (Submitted for publication)
8. Fong W F, J S Heller, E S Canellakis Biochimica et Biophysica Acta 428: 456-465, 1976
9. Galston A W, A Altman, R Kaur-Sawhney Plant Sci Lett 11: 69-79, 1978
10. Galston A W, R Kaur-Sawhney What's New in Plant Physiology 11: 5-8, 1980
11. Gamborg O L, R A Miller, K Ojima Exp. Cell Res 50: 151-158, 1968
12. Goren R, A W Galston Plant Physiol 41: 1055-1064, 1966
13. Kaur-Sawhney R, W R Adams Jr, J Tsang, A W Galston Plant and Cell Physiol 18: 1309-1317, 1977
14. Kaur-Sawhney R, A Altman, A W Galston Plant Physiol 62: 158-160, 1978
15. Kaur-Sawhney R, A W Galston Plant, Cell and Environment 2: 189-196, 1979
16. Kaur-Sawhney R, H E Flores, A W Galston Plant Physiol 65: 368-371, 1980
17. Martin C, K V Thimann Plant Physiol 50: 432, 1972
18. Melchers G, M D Sacristan, A A Holder Carsberg Res Commun 43: 203-218, 1978
19. Ruesink A W, K V Thimann Proc Natl Acad Sci USA 54: 56, 1965
20. Smith T A Plant Amines, Encycl Plant Physiol 8 Chapter 10, 1978
21. Stevens L Biol Rev 45: 1-27, 1970
22. Vasil I K, V Vasil, D W R White, H R Berg Somatic hybridization and genetic manipulation in plants. in Plant Regulation and World Agriculture, T K Scott, ed, Plenum Publ Corp, N Y, pp 63-84, 1979

REGULATION OF GENE EXPRESSION IN AVIAN LIVER

By Robert Goldberger
National Institutes of Health
Bethesda, Maryland

ABSTRACT

Unlike many estrogen-responsive tissues, livers of egg-laying species contain abundant mRNAs encoded by both estrogen-responsive and constitutively expressed genes. We have constructed cDNA clones from three members of the abundant mRNA class of hen liver. One of these mRNA species was identified as serum albumin mRNA, another as vitellogenin mRNA, and the third as apoVLDLII mRNA. Hybridization analyses using cloned cDNA probes indicate that expression of the apoVLDLII gene in rooster liver, like that of the vitellogenin gene, is completely dependent upon the administration of estrogen, whereas the serum albumin gene is expressed constitutively. The dramatic induction of vitellogenin and apoVLDLII synthesis by estrogen is due to both an increased rate of transcription of the genes and to stabilization of the mRNAs.

A few years ago, my colleagues, Dr. Roger Deeley and Dr. Kathleen Mullinix, and I began to search for a system that would tell us something about regulation in eukaryotic cells. We were guided in this search by the following considerations. First of all, we did not want to work with a system in which cellular differentiation was itself an integral part of the regulatory response. We felt that we could obtain more clear cut information if we could study a response in a tissue that was already fully differentiated and in which the regulatory response did not require DNA synthesis. We recognized that a hormonally responsive system would be the easiest to manipulate, and for practical reasons we wanted a system in which the hormonal response was of very great magnitude. And then, too, we wanted the tissue we study also to contain a whole _group_ of genes that are hormonally responsive, so that we would be able to study not only the effect of hormone on a specific gene but also the mechanisms that coordinate the expression of the hormone-responsive domain of the tissue. And of course it would be nice if our hypothetical tissue would also have a gene expressed at a high level that is _not_ responsive to hormonal stimulation. Such a gene would provide an ideal control for _in vitro_ transcription studies, in which one must show that the specificity of hormonal responsiveness reflects that of the tissue _in vivo_. The system that we finally chose, and that actually does fulfill these criteria, is the avian liver and its response to estrogen.

It has been known for over 40 years that the liver is the site of synthesis of avian yolk proteins and that synthesis of these proteins is normally carried out in the female and can be induced in the male -- even in the mature male -- by administration of estrogen (1). Historically, egg yolk proteins were divided into two groups -- the low and the high density lipoproteins (2). The major egg yolk phosphoproteins are found among the high density lipoproteins. They are known as phosvitin (actually there are two different phosvitins) and lipovitellin. It had been shown that induction of these three proteins by estrogen was tightly coordinated, so it seemed to us that they could serve as a starting point for studies on the estrogen-responsive genes of a highly active, terminally differentiated tissue -- the liver -- in which the hormonal response must, of necessity, be superimposed on a background of functions that the liver fulfills before, during, and after the hormonal response. One of these constitutive hepatic functions is the synthesis of large amounts of serum albumin, the system we chose as our control.

When we began to study this system, we soon found that our selected group of estrogen-responsive genes was in fact only one gene, because the egg yolk phosphoproteins turned out to be products of cleavage of a single precursor protein, vitellogenin (3). Clearly we would have to look beyond vitellogenin to capture a group of estrogen-responsive genes. In this paper I will describe some of the work we have done with the vitellogenin system, how we have approached the problem of our control system -- serum albumin -- how we have begun to isolate cloned sequences representing various parts of the estrogen-responsive domain of avian liver, and finally, how we have started to investigate transcription of chromatin in vitro.

A schematic representation of avian vitellogenesis is shown in Figure 1. Vitellogenin is synthesized in the liver of the female as a huge polypeptide with a molecular weight of 240,000 (3). This polypeptide is subsequently phosphorylated, glycosylated, and associated with lipid. It is carried in the blood as a dimer (with a molecular weight of about 500,000) to the ovary, where it is taken up by the developing oocyte and cleaved specifically to form the egg yolk phosphoproteins, lipovitellin and phosvitin (4). The function of vitellogenin has never been elucidated, though most authors state (without any apparent hesitation) that it serves as a phosphate storage protein for the developing embryo or as a metal ion transport protein. Whatever the true function of vitellogenin may be, it is certainly a molecule to be reckoned with. Each egg contains about a gram of vitellogenin, which represents approximately 150 mg of high energy phosphate.

Figure 2 shows a preparation of pure vitellogenin mRNA (along with 18 and 28S ribosomal RNA markers) subjected to electrophoresis in agarose containing methylmercuric hydroxide as the denaturing agent (5). Its size, determined by gel electrophoresis, as shown here (but with many RNA standards of known size), and by contour analysis of electron photomicrographs, as shown in Figure 3, is 2.35 million daltons, or 7,000 nucleotides, approximately 600 nucleotides longer than required to specify the vitellogenin polypeptide (5).

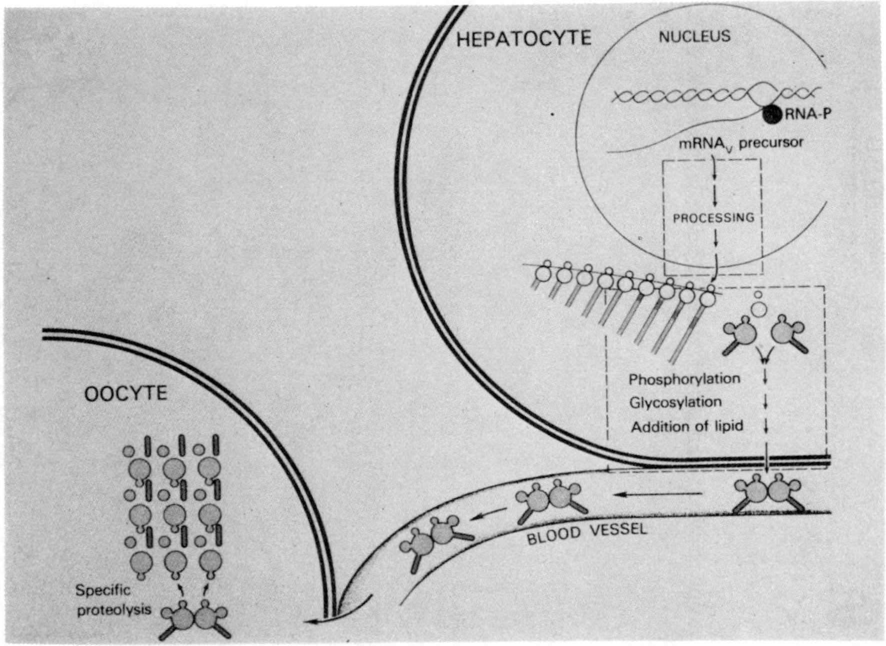

Figure 1: Schematic representation of avian vitellogenesis. See text for details.

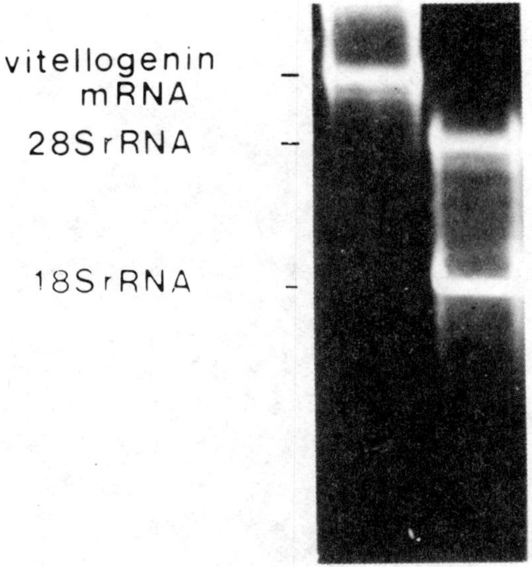

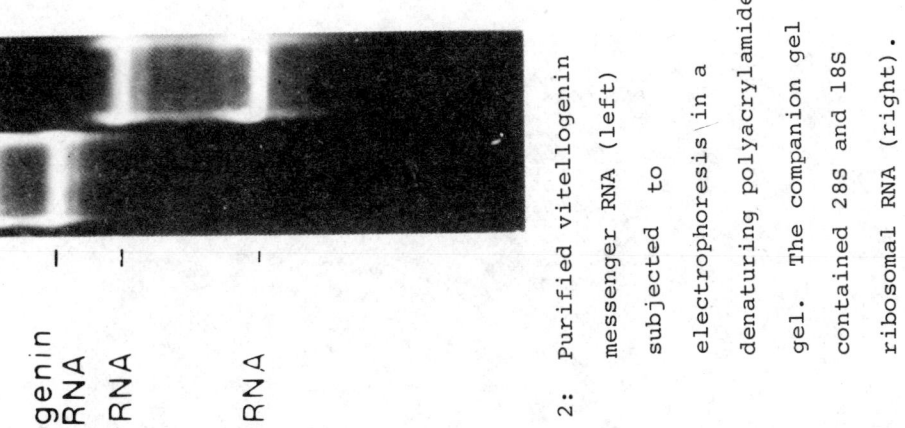

Figure 2: Purified vitellogenin messenger RNA (left) subjected to electrophoresis in a denaturing polyacrylamide gel. The companion gel contained 28S and 18S ribosomal RNA (right).

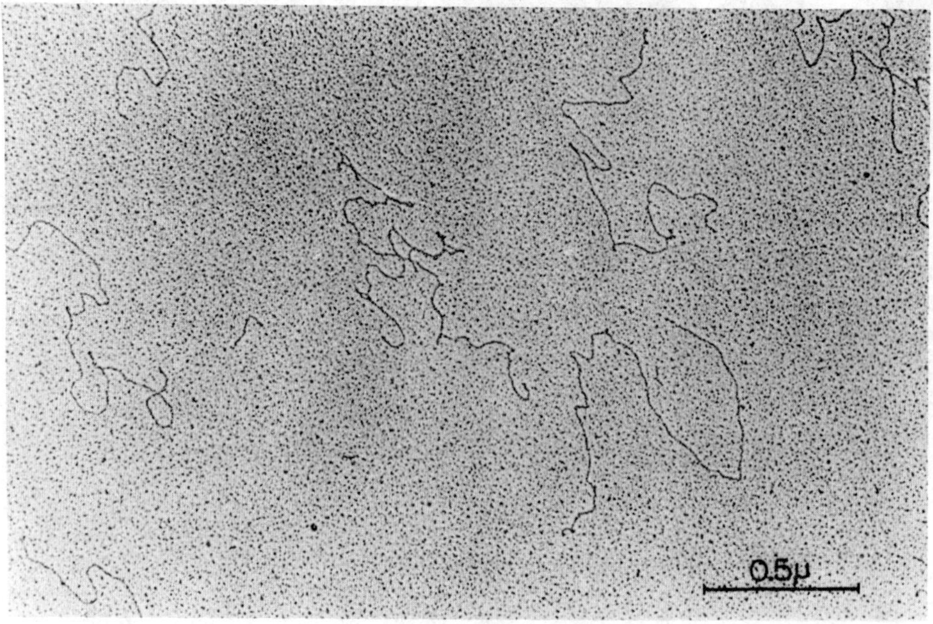

Figure 3: Electron photomicrograph of vitellogenin messenger RNA.

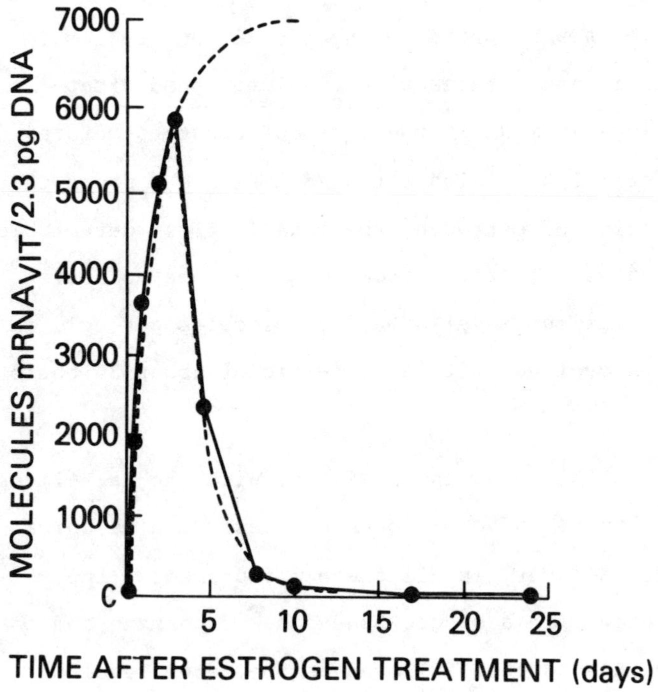

Figure 4: Kinetics of accumulation and disappearance of vitellogenin messenger RNA in rooster liver after a single large injection of 17β-estradiol. See text for explanation.

After purifying the vitellogenin mRNA, we then made cDNA and cloned this cDNA in *E. coli* (5,6). We were then in a position to have ample amounts of a pure probe for hybridization studies in which we could determine the rates of accumulation and disappearances of vitellogenin mRNA in the liver of roosters before, during, and after stimulation with estrogen. Figure 4 shows that after a large injection of estrogen, the mRNA is first detectable within about 30 minutes. It rises from less than one molecule per cell, reaches a maximum level of 6,000 molecules per cell in three days, and then declines with a half-life of about 30 hours (7).

This half-life of 30 hours is consistent with the half-lives of other abundant hormone-inducible mRNA species, such as chick ovalbumin mRNA (8). However, in all these measurements, the animals are stimulated with a single huge dose of hormone and the amount of specific mRNA is then followed as the serum level of hormone shoots up very high and then slowly decays, so that the withdrawal phase is very gradual. Recent experiments on mRNA stability in acutely withdrawn chick oviduct (9) indicate that in the total absence of hormone these abundant mRNAs become quite unstable. But there is an important criticism of these experiments in oviduct. The criticism is that the oviduct is a tissue that is itself *dependent* on hormone. When hormone is withdrawn this tissue regresses. Cells die. Is it any wonder that the hormone-dependent messages decay quickly in the dying cell? Since the liver does not regress -- it remains biosynthetically active after withdrawal from hormone treatment -- we were interested in determining whether or not a similar destabilization of vitellogenin mRNA would occur, and if so, whether the destabilization was characteristic of all abundant mRNAs or just those that are hormone-responsive.

Figure 5 shows what happens to the level of mRNA when the hormone is withdrawn very rapidly. This was done by using silastic implants to administer hormone, and then removing the implant, so that the hormone is withdrawn suddenly (10). We found that the vitellogenin mRNA disappeared from the liver very rapidly -- at a rate consistent with a half-life of only three hours. Remember that under conditions in which the hormone was withdrawn slowly, the half-life of the message was ten times as long. The open circles of Figure 5 represent the results with another estrogen-inducible message, that for apoVLDLII, for which we found the same thing: a marked destabilization of the mRNA when estrogen is not present. The open squares of Figure 5 show the results for serum albumin mRNA, which remains unaffected by removal of hormone, indicating the the destabilization of the mRNA for vitellogenin and apoVLDLII is specific for hormone-responsive mRNAs. The conclusion is that for abundant mRNAs that are induced by hormone the stability of these mRNAs is greatly increased in the presence of hormone and these messages are much less stable in the absence of hormone. And this effect is specific, since the stability of albumin mRNA (which is not hormone dependent) is not influenced by the presence or absence of hormone. And the effect has nothing to do with cell death,

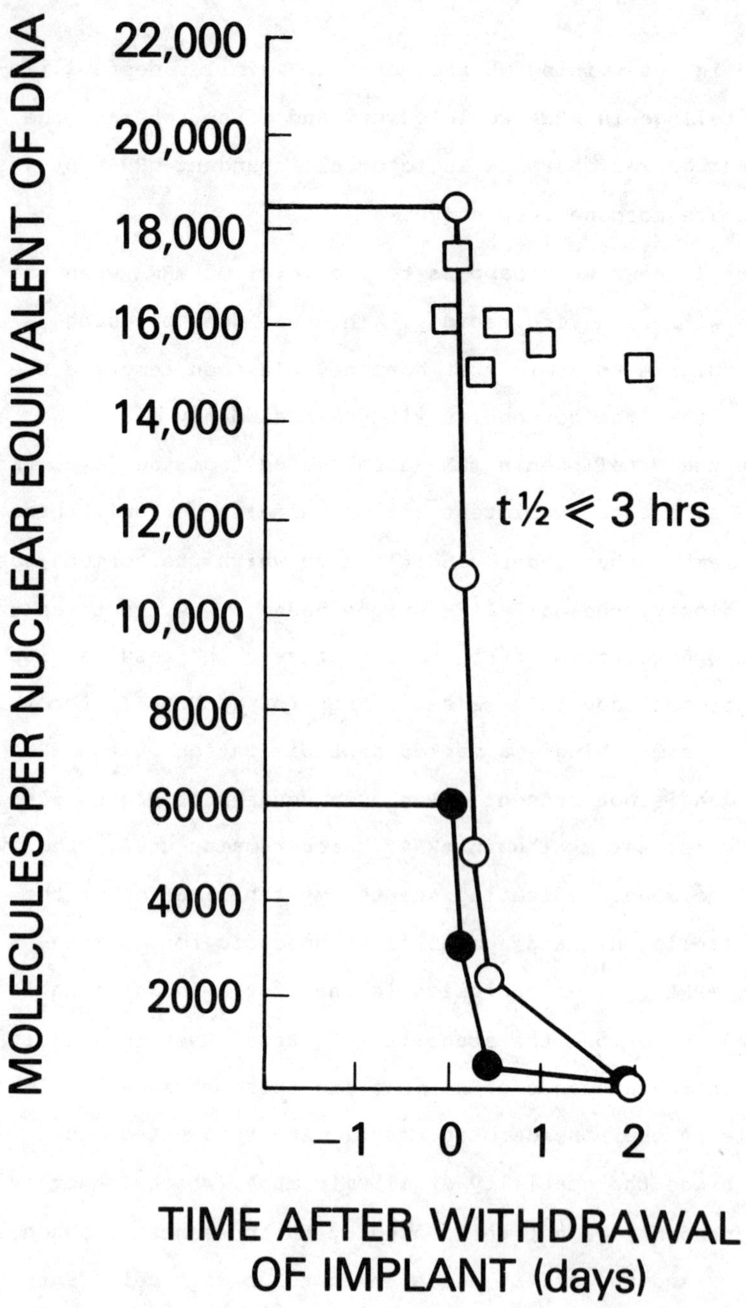

Figure 5: Kinetics of disappearance from rooster liver of three species of messenger RNA after sudden withdrawal of estrogen. The closed dots represent vitellogenin mRNA; the open dots, apoVLDLII mRNA; and the open squares, chicken serum albumin mRNA.

since it is observed in the liver where withdrawal of hormone does not result in regression of the tissue (10).

You may recall that in the beginning of this paper I enumerated the features of the system we hoped to find. Well, we have our tissue that is terminally differentiated and highly active metabolically; we have our vitellogenin gene that responds dramatically to estrogen stimulation without requiring DNA synthesis; and we have our control system -- serum albumin, which is truly a constitutive function of the liver. But we are still missing specific probes for other genes that respond to estrogen, which we need if we are to be able to study the coordinated response of the liver to stimulation by estrogen.

I will not describe how we have gone about isolating and cloning various parts of the estrogen-responsive domain of avian liver, but I would like to summarize some of our findings for you. First of all, we found that stimulation of the rooster with estrogen causes the expression of about one hundred genes not previously expressed. Secondly, we have studied the most abundant mRNAs and the most highly inducible mRNAs. In this group, there are only two mRNAs -- that corresponding to the vitellogenin gene and that corresponding to the apoVLDLII gene (6,10).

Contrary to many reports in the past, the apoVLDLII message is not present in normal rooster liver, and is induced by estrogen to about 18,000 molecules per cell.

Now I want to turn to a new topic -- *in vitro* transcription. We sought to develop an *in vitro* system that achieves faithful transcription by RNA polymerase II and that potentially can be manipulated in a controlled way. We used chromatin from the nuclei of rooster liver as the transcriptional material to study the level of transcription of the vitellogenin gene. This chromatin, isolated by a modification of the procedure of Marzluff and Huang (11), has high levels of endogenous RNA polymerase II activity, and all transcription I will be talking about has been carried out by the endogenous, chromatin-associated RNA polymerase II (12). The chicken serum albumin gene is an ideal control for experiments designed to study *in vitro* transcription of the vitellogenin gene since its transcription is constitutive and relatively unaffectd by estrogen.

In order to measure transcription of the vitellogenin and albumin genes in chromatin, we have analyzed the radioactive RNA made *in vitro* by hybridization to filter-bound cloned albumin cDNA and cloned vitellogenin cDNA, as well as other cloned cDNAs, under conditions of vast DNA excess (12). The raw data from one experiment in which we measured the transcription of the vitellogenin and albumin genes in chromatin from the livers of normal and estrogen-treated roosters are shown in Table I. In this experiment, the synthesis of vitellogenin RNA in chromatin from the liver of estrogen-stimulated roosters comprised 0.47% of the total RNA synthesized. Albumin RNA, synthesized in the same reaction, comprised 0.12% of the RNA. In contrast, there was very little vitellogenin RNA synthesized in chromatin from normal

TABLE I

SYNTHESIS OF VITELLOGENIN AND ALBUMIN RNA IN CHROMATIN FROM LIVERS OF NORMAL AND ESTROGEN-TREATED ROOSTERS

Chromatin	DNA on Filter	ΔCPM	RNA Input (CPM)	Percent of Transcripts Hybridized	Estrogen Treated/ Normal
Normal Rooster Liver	Vitellogenin	14	2.6×10^5	0.02	-
	Albumin	252	2.6×10^5	0.15	-
Estrogen-Treated Rooster Liver	Vitellogenin	855	6.0×10^5	0.47	23.5
	Albumin	438	5.6×10^5	0.12	0.80

roosters (0.02%). Albumin RNA, however, comprised approximately the same proportion (0.15%) as in the chromatin from the liver of estrogen-stimulated roosters. It should be noted that the absolute amount of albumin RNA synthesis in chromatin from liver of estrogen-stimulated roosters increased by 1.7-fold compared to normal roosters. This increase is commensurate with the overall increase in transcriptional activity produced by estrogen treatment. But the absolute level of vitellogenin RNA synthesis after estrogen treatment was 61-fold higher! The effect is really quite dramatic. We have done these experiments with cloned apoVLDLII DNA, and the results are extremely similar to those for vitellogenin DNA -- transcription is greatly stimulated by or even completely dependent upon prior stimulation by estrogen (12).

In summary, I have discussed the system we are studying -- the avian liver and its response to estrogen -- and have given our rationale for choosing a fully differentiated tissue in which one can study the response to a hormone against a background of constitutive cellular functions. I have described some of our studies on the vitellogenin system, and I have also described our work on another gene in avian liver -- serum albumin -- which we have utilized as a control for our investigation of hormonal responsiveness. I introduced our studies on the estrogen-responsive domain of avian liver, from which we could place an upper limit on the diversity of the hormonal response: there are no more than about one hundred genes whose expression is altered by exposure to estrogen. We have found that the vitellogenin and apoVLDLII genes are the only ones that are several thousand-fold induced by estrogen, resulting in the production of abundant mRNAs. Neither vitellogenin mRNA nor apoVLDLII mRNA is present at a significant level in the liver of the normal rooster, but the very high levels achieved after stimulation with estrogen are due to two effects. One is the specific stabilization of these mRNAs and the other is the increased transcription of the two genes.

One of our current interests is focused on how the expression of these two genes and of other groups of estrogen-responsive genes is so tightly coordinated.

REFERENCES

1. Clegg, R.E., Sanford, P.E., Hein, R.E., Andrews, A.C., Hughes, J.S., and Muetter, C.D. (1951) Science 114, 437

2. Bernardi, G., and Cook, W.H. (1960) Biochim. Biophys. Acta 44, 86

3. Deeley, R.G., Mullinix, K.P., Wetekam, W., Kronenberg, H.M., Meyers, M., Eldridge, J.D., and Goldberger, R.F. (1975) J. Biol. Chem. 250, 9060

4. Bergink, E.W., and Wallace, R.A. (1974) J. Biol. Chem. 249, 2897

5. Deeley, R.G., Gordon, J.I., Burns, A.T.H., Mullinix, K.P., Bina-Stein, M., and Goldberger, R.F. (1977) J. Biol. Chem. 252, 8310

6. King, C.R., Udell, D., and Deeley, R.G. (1979) J. Biol. Chem. 254, 6781

7. Burns, A.T.H., Deeley, R.G., Gordon, J.I., Udell, D.S., Mullinix, K.P., and Goldberger, R.F. (1978) Proc. Natl. Acad. Sci. USA 75, 1815

8. Palmiter, R.D. (1973) J. Biol. Chem. 248, 8260

9. Hynes, N.E., Groner, B., Sippel, A.E., Jeep, S., Wurtz, T., Nguyen-Huu, M.C., Giesecke, K., and Schutz, G. (1979) Biochemistry 18, 616

10. Wiskocil, R., Bensky, P., Dower, W., Goldberger, R.F., Gordon, J.I., and Deeley, R.G. (1980) Proc. Natl. Acad. Sci. USA, in press

11. Marzluff, W.F., and Huang, R.C.C. (1975) Proc. Natl. Acad. Sci. USA 72, 1082

12. Mullinix, K.P., Meyers, M.B., Christmann, J.L., Deeley, R.G., Gordon, J.I., and Goldberger, R.F. (1979) J. Biol. Chem. 254, 9860

ROLE OF DOUBLE-STRANDED RNA IN TRANSLATIONAL CONTROL

Haim Grosfeld* and Severo Ochoa+

Roche Institute of Molecular Biology

Nutley, New Jersey 07110 U.S.A.

ABSTRACT

Reticulocytes contain cyclic AMP-independent protein kinases which inhibit polypeptide chain initiation through phosphorylation of the α subunit of the chain initiation factor eIF-2. One of these kinases (heme-controlled translational inhibitor, HCI) is activated in the absence of free heme, the other (double-stranded RNA activated translational inhibitor, DAI) is activated by low concentrations of double-stranded RNA. In other cells, e.g., Erlich ascites tumor cells, mouse fibroblasts, DAI but not HCI, is induced by interferon. HCI and DAI differ in their localization in the cell, in molecular weight, and in the effect of N-ethylmaleimide which activates HCI but inhibits both DAI activation and activity. We have been able to purify reticulocyte pro-DAI only partially but have purified preactivated DAI to near homogenity. This enzyme, active in nanogram amounts, is highly unstable. Our results suggest that following interaction with double-stranded RNA, DAI undergoes self phosphorylation in the presence of ATP, and is thereby activated to phosphorylate the eIF-2 α subunit as well as histone.

---

*Present address: Israel Institute for Biological Research
P. O. Box 19, Ness-Ziona, Israel.

+To whom correspondence regarding this paper should be addressed.

Abbreviations: HCI, heme controlled translational inhibitor; DAI, double-stranded RNA activated translational inhibitor; dsRNA, double-stranded RNA; NEM, N-ethylmaleimide; eIF-2, eukaryotic initiation factor 2; BSA, bovine serum albumin.

Translational Control in Reticulocyte Lysates

In reticulocytes, globin synthesis is controlled by the level of heme, the prosthetic group of hemoglobin. Protein synthesis in reticulocyte lysates is but briefly maintained in the absence of added hemin (1-4). Gross and Rabinovitz (5) showed that hemin prevents the formation of an inhibitor of chain initiation from a proinhibitor of similar molecular weight. As seen in Fig. 1A (6) the synthesis of protein in a lysate of rabbit reticulocytes comes to a standstill in a few minutes unless hemin is added to the incubation medium. The translational inhibitor formed in the absence of hemin will be referred to as the heme controlled inhibitor (HCI).

Translational inhibition with the same kinetics observed in hemin deficiency is also obtained, in the presence of hemin, by treatment with N-ethylmaleimide (NEM) (5), oxidized glutathione (7,8), or (Fig. 1B) low concentrations of double-stranded RNA (dsRNA) (9-11). However, whereas NEM and oxidized glutathione activate HCI, dsRNA activates a second translational inhibitor. This will be referred to as the dsRNA activated inhibitor (DAI).

HCI and DAI are cyclic AMP independent protein kinases that catalyze the phosphorylation of the small, α subunit of the chain initiation factor eIF-2 (9,12-14). *In vitro* the two enzymes phosphorylate the same site(s) of α eIF-2 (15). HCI and DAI differ in a number of ways: (a) HCI is present in the postribosomal supernatant whereas DAI is ribosome associated; (b) the subunit molecular weight of HCI (about 90,000) is higher than that of DAI (about 68,000) (16, this paper); (c) HCI is activated by NEM (5) but DAI is inhibited (17, this paper)

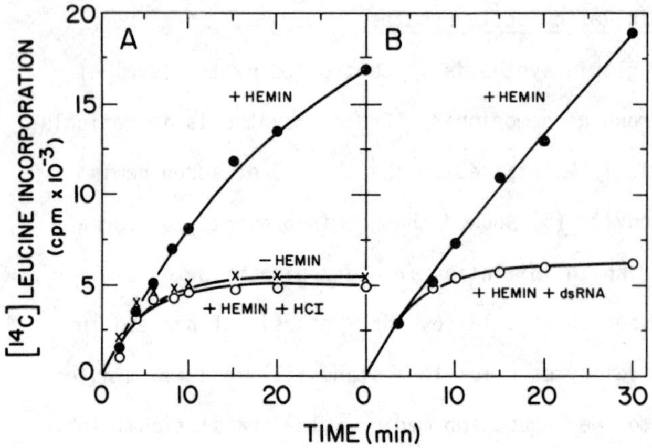

Fig. 1. Protein synthesis in the reticulocyte lysate in the presence of hemin, in the absence of hemin, or in the presence of both hemin and HCl (A) or in the presence of hemin without or with dsRNA (B). The concentration of hemin, when present, was 30 μm (A) or 20 μm (B). The concentration of (partially purified) HCI was 25 μg/50 μl, and that of dsRNA (P. chrysogenum mycophage) was 28 ng/ml. Samples were incubated at 30° and 5 μl aliquots withdrawn for assay of protein synthesis at the indicated times. Prepared from data of Clemens et al. (6) (panel A) and Levin and London (10) (panel B).

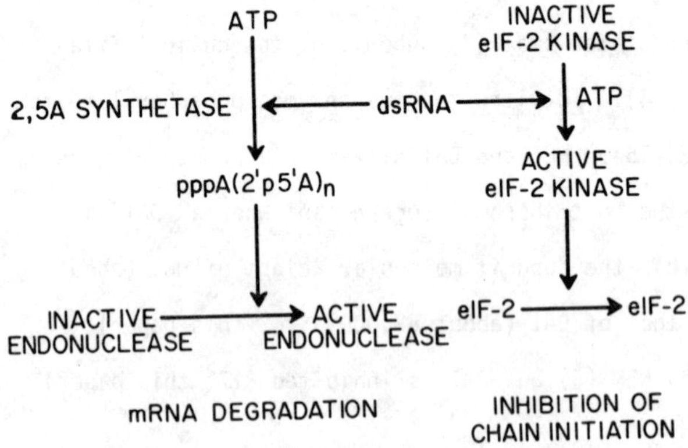

Fig. 2. Dual control of translation by dsRNA. The 2,5-A synthetase and the eIF-2 kinase are constitutive in reticulocytes; in other cells (e.g., Erlich ascites tumor cells, mouse fibroblasts) they are induced by interferon. The endonuclease appears to be constitutive in all cases (From Farrell et al. 30).

(d) HCI specifically phosphorylates α eIF-2 but DAI can also phosphorylate histone (18,19, this paper); and (e) HCI and DAI do not cross-react immunologically (20). Both HCI and DAI are present in reticulocytes but, unlike HCI, DAI is induced in other cells (Erlich ascites tumor cells, mouse fibroblasts) by interferon (18,19).

There is little doubt that the translational inhibition observed following activation of HCI or DAI is related to the phosphorylation of the eIF-2 α subunit (9,21,22). This effect is highly specific. Thus, phosphorylation of the eIF-2 β subunit by a cAMP-independent protein kinase (casein kinase) from reticulocyte lysates has no effect on translation (10,23,24). The way in which phosphorylation of the eIF-2 α subunit inhibits chain initiation has not been fully elucidated but it is known to interfere with stimulation of eIF-2 activity by ESP (25-29), whereas phosphorylation of the eIF-2 β subunit does not (27).

There is a dual mechanism of control of translation by dsRNA (Fig. 2) in both reticulocytes and interferon treated cells. Initiation can be inhibited by DAI whereas mRNA translation can be inhibited by an oligonucleotide (2,5-A) consisting of one 5'-ATP and two or more AMP residues linked through 2'-5' phosphodiester bonds (31). 2,5-A is a potent inhibitor of translation, active in nanomolar concentrations. It activates a latent, constitutive endonuclease that degrades mRNA (32-34). In the present paper we shall be concerned with the isolation and properties of DAI from reticulocyte lysates.

MATERIALS AND METHODS

Ternary Complex Formation Assay

Active DAI inhibits ternary complex formation and this reaction was used routinely for DAI assay. The assay was conducted in three stages as follows:

Stage 1. First step of DAI activation: incubation with dsRNA. Unless otherwise specified, samples (30 μl) containing 20 mM Hepes buffer, pH 7.6, 100 mM KCl, 1.5 mM $Mg(OAc)_2$, 1 mM dithiothreitol, DAI as indicated in the legends, and dsRNA [unless otherwise noted, 0.16 μg/ml of poly(I).poly(C)], were incubated for 5 min at 30°. Controls without dsRNA were run occasionally.

Stage 2. Second step (phosphorylation) of DAI activation and phosphorylation of eIF-2 α subunit. Samples (40 μl) containing 25 mM Hepes buffer, pH 7.6, 120 mM KCl, 2.5 mM $Mg(OAc)_2$, 1.25 mM dithiothreitol, about 4 pmol of eIF-2 CM-350 (or as specified in the legends), 0.05 to 0.125 mM ATP, and a suitable aliquot of the incubated stage 1 sample, were incubated for 5 min at 30°.

Stage 3. Ternary complex formation. The incubated stage 2 samples were supplemented with 1 pmol of [^{35}S]Met-$tRNA_i$ (about 160,000 cpm), 0.2 mM GTP, and 16 μg of CM-200 ESP, unless otherwise noted. The volume was made up to 50 μl and the samples were incubated for 5 min at 30°. The reaction was stopped by addition of 3 ml of ice-cold buffer [20 mM Tris-HCl, pH 7.6, 100 mM KCl, 2 mM $Mg(OAc)_2$]. The samples were filtered through nitrocellulose membranes and the retained radioactivity was measured. eIF-2 and ESP were prepared as described by de Haro and Ochoa (27). The two factors were eluted together from the DEAE-cellulose column by raising the concentration of KCl to 200 mM and were separated from each other by chromatography on carboxymethyl Sephadex. ESP was eluted with 200 mM KCl and eIF-2 with 350 mM KCl. These preparations are referred to as ESP CM-200 and eIF-2 CM-350. For some experiments the eIF-2 was further purified by chromatography on phosphocellulose (eIF-2 PC) as described (27). As judged by polyacrylamide-SDS gel electrophoresis eIF-2 CM-350 was 20-25% pure; the phosphocellulose purified eIF-2 (eIF-2 PC) was 70-75% pure.

Other assays

Occasionally DAI was assayed by measuring the phosphorylation of the eIF-2 α subunit or of histone IIa, following incubation with $\gamma[^{32}P]$ ATP. The incubation samples (25 μl) containing 20 mM Hepes buffer, pH 7.6, 100 mM KCl, 3.5 mM $Mg(OAc)_2$, 0.033 mM $[^{32}P]$ATP (2,000 cpm/pmol), about 7 pmol of eIF-2 PC, or 25 μg histone IIa, and an appropriate amount of preactivated DAI, were incubated for 7 min at 30°. After incubation, duplicate samples were subjected to polyacrylamide-SDS gel electrophoresis and stained with Coomassie blue. An autoradiogram was obtained from one of the gel slabs. The eIF-2 α band, or the histone band, was cut out from the other slab, the gel section dissolved in H_2O_2 (3 hr, 95°) and its radioactivity measured.

Protein was generally determined by the Lowry method (35) but, with very dilute protein solutions, the Bradford procedure (36) was used. Bovine serum albumin was used as the standard in both cases.

Polyacrylamide-SDS gel electrophoresis and autoradiography

Polyacrylamide-SDS gel electrophoresis was carried out essentially by Laemmli's procedure (37) in 0.1% sodium dodecylsulfate, 10% acrylamide, 0.1% N,N'-methylenebisacrylamide, pH 7.6, for 4 hr at 3.5 mA. Gels were stained with 0.2% Coomassie blue or by the more sensitive silver reduction method (38). For autoradiography the gels were dried, placed on Kodak X-Omat RP X-ray film and exposed for an appropriate length of time.

Preparation of DAI DE-80 fraction

DE-80 fraction was prepared from the ribosomal salt wash of reticulocyte lysates. All operations were conducted at 0-2°. One liter of rabbit reticulocytes (Pel-Freez) yielded about 400 ml of lysate. Ribosomal salt wash (640 mg of protein), prepared as described (25), was dialyzed overnight against buffer A [20 mM Hepes buffer, pH 7.6, 0.1 mM EDTA, 1 mM dithiothreitol, 5% (vol/vol) glycerol] containing 80 mM KCl. The dialyzed solution (20 ml) was applied to a DEAE-cellulose (Whatman DE-52) column (1.5 x 16 cm) equilibrated with buffer A containing 80 mM KCl and washed with the same buffer until the A_{280} nm decreased to below 0.2 (27). The bulk of the DAI activity was in the wash (22 ml, 550 mg of protein). This fraction was used for some experiments after precipitation with ammonium sulfate at 70% saturation, followed by dialysis against buffer A.

Preparation of Sepharose-histone column

50 g of Sepharose 4B were washed with water on a Büchner funnel and suspended in 100 ml of 2.0 M K_2CO_3 (pH 11.0) with mechanical stirring in the cold room. Ten ml of cyanogen bromide dissolved in acetonitrile (1.5 g/ml) were then added with continuous stirring for 90 seconds. The activated Sepharose was washed exhaustively with cold water and added to 150 mg of histone (Sigma IIa from calf thymus) suspended in 30 ml of 0.2 M $NaHCO_3$ containing 0.5 M KCl and stirred overnight at 0°.

RESULTS

DAI activation

One polypeptide (M_r about 68,000) is heavily labeled when DAI DE-80 is activated with dsRNA in the presence of $\gamma[^{32}P]$ATP (Fig. 3, cf. tracks 1 and 3). The faintly labeled peptide in track 3 (M_r about 100,000) is a contaminant which is removed in the last step of purification of preactivated DAI (see below). In the presence of dsRNA and eIF-2 an additional peptide (M_r about 38,000) is labeled. This is a small α subunit of eIF-2 (Fig. 3, cf. tracks 4 and 6). Neither the 68,000 nor the 38,000 daltons

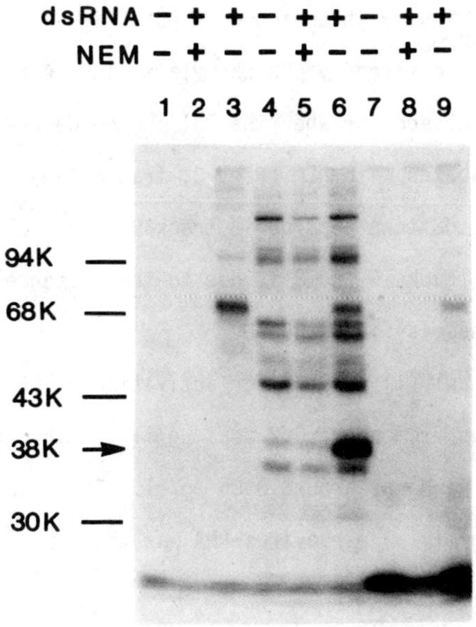

TRACKS 1-3, NO SUBSTRATE, 4-6, eIF-2,

7-9, HISTONE II

Fig. 3. Phosphorylation of 68 kilodaltons polypeptide, eIF-2 α subunit, and histone, upon activation of DAI:inhibition of DAI activation by NEM. The incubations were conducted in two steps. In step I the samples (25 μl) contained 20 mM Hepes buffer, pH 7.4, 100 mM KCl, 1.5 mM $Mg(OAc)_2$, 1 mM dithiothreitol, and 7.2 μg of pro-DAI DE-80 (specific activity after activation, about 75), without or with poly(I).poly(C) (1 μg/ml), and either without further additions or with addition of either 9 pmol (approximately) of eIF-2 CM-350 or 20 μg of histone IIa. The samples were incubated for 5 min at 30°. Duplicates of each of three samples (no substrate, eIF-2, histone) that were to receive poly(I).poly(C) were prepared by preincubating DAI (5 min, 30°) with NEM (10 mM), followed by neutralization of the unreacted NEM with dithiothreitol (20 mM), addition of the remaining components, and incubation as above. In step II, the samples were supplemented with $\gamma[^{32}P]$ATP (0.05 mM, specific radioactivity, 3,000 cpm/pmol) and incubated for 7 min at 30°. Polyacrylamide-SDS gel electrophoresis and autoradiography were performed as described in the text.

peptide is labeled when DAI is previously incubated with NEM (Fig. 3, tracks 2 and 5). Histone IIa is also strongly phosphorylated in the presence of dsRNA and less so in its absence or when the DAI in the dsRNA-containing sample has been preincubated with NEM (Fig. 3, tracks 7-9). Note phosphorylation of the 68,000 daltons peptide in track 9. The relatively intense phosphorylation in tracks 7 and 8 is due to the presence of non-dsRNA dependent histone kinase(s) in DAI DE-80.

The optimal concentration of poly(I).poly(C) for activation of DAI DE-80 varied with different batches from about 0.2 to 1 μg/ml. With reovirus RNA the optimal concentration was around 0.05 μg/ml. There was no activation at 0.001 or at 10 μg/ml. The reovirus RNA was the kind gift of Dr. A. Shatkin.

DAI like HCI (25-29) blocks the eIF-2-ESP interaction. Fig. 4 shows the time course of ternary complex formation and the pronounced stimulation by ESP which, by itself, binds no Met-tRNA$_i$. In the presence of DAI and ATP the stimulation of complex formation by ESP is markedly reduced.

Purification of Pro-DAI

DAI DE-80 (20 ml, 530 mg of protein) was dialyzed against a solution containing 20 mM potassium phosphate buffer, pH 6.7, 1 mM dithiothreitol, 25 mM KCl, and 10% (vol/vol) glycerol. The dialyzed solution (19 ml) was applied to a column (1.3 x 25 cm) of phosphocellulose (Whatman P-11) equilibrated with the same solution. The column was washed with the potassium phosphate solution until the A_{280} nm was below 0.01. Protein was then eluted with 500 ml of a linear gradient (100-700 mM) of KCl in the same buffer and fractions (6 ml) were collected. DAI was assayed by inhibition of ternary complex formation and by histone IIa phosphorylation. As seen in Fig. 5, coincident peaks of ternary complex inhibition

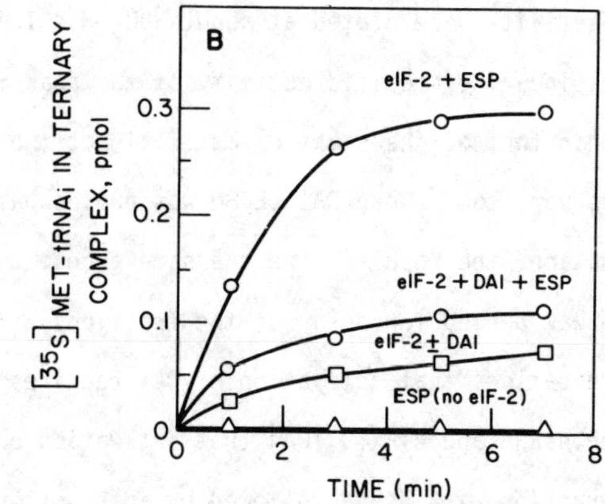

Fig. 4. Time course of ternary complex formation with and without ESP and DAI + ATP. The standard assay was used. Reactants were used in the following amounts: eIF-2 (CM 350), about 4.5 pmol; ESP (CM 200), 16 μg; preactivated DAI DE-80, 16 μg.

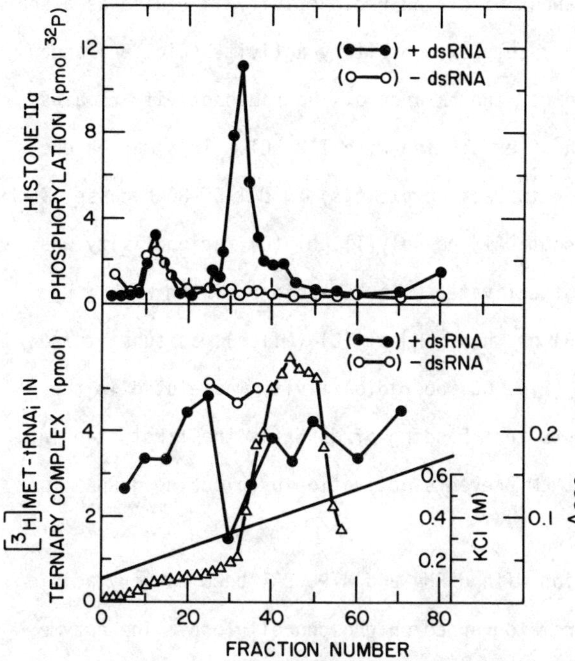

Fig. 5. Phosphocellulose chromatography of pro-DAI. DE-80 fraction, was chromatographed on phosphocellulose as described and assayed for inhibition of ternary complex formation (lower panel) or phosphorylation of histone IIa (upper panel) with (black circles) or without (open circles) dsRNA activation but after preincubation with ATP in all cases. Reovirus RNA was used for activation. The open triangles (lower panel) show the A_{280} nm of the fractions.

and histone phosphorylation activity were eluted at about 300 mM KCl just ahead of the bulk of the protein. The specific activity of the peak fractions was 3-4 times higher than that of the starting material but the yield of units (about 7%) was very low. When DAI DE-80 was preincubated with dsRNA prior to fractionation, the results were the same except that no further addition of dsRNA was needed for assay of the fractions. This suggests [despite earlier indications that activation of DAI requires simultaneous interaction with dsRNA and ATP (9,10)] that activation occurs in two separate steps: interaction with dsRNA followed by interaction with ATP. We have some indications that in the first step dsRNA binds to DAI. If one adds a small amount of [^{3}H]-labeled poly(I).poly(C) (Miles Laboratories) to DAI DE-80, and fractionates the enzyme on phosphocellulose as above, about 6% of the radioactivity is eluted as a sharp peak together with ternary complex inhibitory activity (Fig. 6A) ahead of the bulk of the protein. The balance of the radioactivity remains bound to the column even after elution with 1 M KCl. This may be due to binding of the polymer to basic protein(s) in DAI DE-80 because, in control experiments without DAI, no poly(I).poly(C) radioactivity was retained. When DAI was inactivated with NEM before mixing with [^{3}H] poly(I).poly(C) (Fig. 6B) or when [^{3}H]poly(C) (Miles) was substituted for [^{3}H]poly(I).poly(C) (Fig. 6C) no radioactivity was eluted in the DAI region. This suggests that binding of dsRNA is the first step in DAI activation and that NEM prevents activation by blocking dsRNA binding.

After full activation with dsRNA and ATP, DAI becomes very acidic and can no longer be chromatographed on phosphocellulose. The enzyme now binds strongly to DEAE-cellulose from which it can be eluted, with considerable purification, at high KCl concentrations.

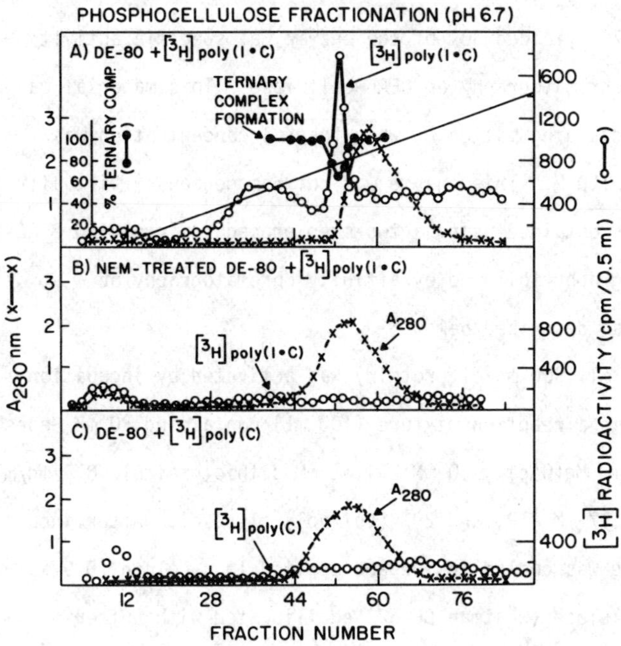

<u>Fig. 6</u>. Binding of [^{3}H]poly(I).poly(C) by DAI. DAI DE-80 (21 mg of protein) was incubated for 10 min at 30°, in a final volume of 7.5 ml, either with [^{3}H]poly(I).poly(C) (223,000 cpm, 38,000 cpm/μg) without (panel A) or with (panel B) prior treatment with NEM (see legend to Fig. 3) or with [^{3}H]poly(C) (201,000 cpm, 49,624 cpm/μg) (panel C). The samples were then chromatographed on phosphocellulose as described in the text for purification of pro-DAI. The A_{280} nm and radioactivity were measured in all gradient fractions from each experiment. The mid fractions in the experiment corresponding to panel A were also assayed for inhibition of ternary complex formation.

Purification and properties of preactivated DAI

Activated DAI DE-80 (specific activity in ternary complex inhibition assay, about 70) yielded 30% of the enzyme, at specific activity about 170,000, on chromatography on DEAE-cellulose. This material was eluted with KCl-containing buffer A, when the KCl concentration was raised from 0.4 to 1.0 M. This enzyme was not homogeneous. Virtually homogeneous DAI was obtained using a two-step procedure involving DEAE-cellulose chromatography followed by affinity chromatography on Sepharose-histone as described below.

DAI DE-80 (37 ml, 500 mg of protein) was activated by incubation for 10 min at 30° in a reaction mixture (100 ml) containing 20 mM Hepes buffer, pH 7.6, 2 mM $Mg(OAc)_2$, 30 mM KCl, 1 mM dithiothreitol, 0.6 mg/ml poly(I).poly(C), 0.17 mM ATP, and 20% (vol/vol) glycerol. After incubation, the mixture was cooled in ice and applied to a column (0.9 x 12 cm) of DEAE-cellulose (Whatman DE-52) equilibrated with buffer A containing 80 mM KCl. All operations were conducted at 0°-2°. The column was washed with the 80 mM KCl-containing buffer A and then with buffer A containing 200 mM KCl until the A_{280} nm of the effluent was below 0.02. More protein was then eluted with buffer A containing 800 mM KCl. The protein-containing fractions were pooled and the solution was diluted with buffer A to make the KCl concentration 500 mM. The DE-800 fraction (5 ml) was applied to a Sepharose-histone column (0.4 x 5 cm), previously equilibrated with buffer A containing 500 mM KCl, and the column washed with the same buffer until the A_{280} nm was below 0.01. DAI was then eluted with buffer A containing 1.0 M KCl. The peak activity fractions were pooled to yield 3.6 ml of solution containing 5 μg of protein/ml. We found highly purified DAI to be extremely unstable;

it did not withstand storage for more than a few hours under a variety of conditions.

A summary of the purification procedure is given in Table I. When assayed by ternary complex inhibition the Sepharose-histone DAI was purified 5,000-fold from the DE-80 fraction. When assayed by histone IIa phosphorylation the degree of purification appeared to be much less (about 1,400-fold). It should be noted, however, that the DE-80 fraction contains dsRNA-independent along with dsRNA-dependent histone kinase activity so that the histone phosphorylation assay gives abnormally high values for DAI in this fraction.

In order to estimate the degree of purification of DAI between lysate and DE-80 fraction, we measured translational inhibition produced in a standard assay with hemin-supplemented lysate [containing 15 μg/ml poly (I).poly(C) to prevent activation of endogenous DAI (9,10)] by the addition of known amounts of preactivated lysate or DAI DE-80. The results (not shown) indicated that the specific activity of DAI in the DE-80 fraction is about 50 times higher than in the lysate. If so, the overall purification of DAI from the lysate (ternary complex inhibition assay) would be 50 x 5,000 or 250,000-fold.

Fig. 7 shows polyacrylamide-SDS gel electrophoretic patterns at various purification stages of DAI. The nearly homogeneous enzyme (track 5) migrated slightly behind a 68,000 daltons marker. To see whether the homogeneous, active DAI is phosphorylated, DAI DE-80 was activated in the presence of $\gamma[^{32}P]$ATP and purified as described above. Fig. 8 shows that this was indeed the case. The stoichiometry of phosphorylation has not been determined.

Highly purified DAI phosphorylated both the eIF-2 α subunit and

Table I. Purification of preactivated DAI

Step	Protein (mg)	Ternary complex inhibition assay Units[1]	S.A.	Histone phosphorylation assay Units[2]	S.A.
DE-80 before rechromatography	500	38,000	76	160,000	320
DEAE-cellulose rechromatography, DE-200	426	2,600	6		
DEAE-cellulose rechromatography, DE-800	0.8	11,000	13,750	39,000	48,750
Sepharose-histone chromatography	0.02	7,600	380,000	9,000	450,000

[1]One unit, amount of enzyme causing 50% inhibition of ternary complex formation under standard assay conditions.

[2]One unit, amount of enzyme causing the transfer of 1 pmol of $[^{32}P]$ from $\gamma[^{32}P]$ATP to histone IIa under standard assay conditions.

S.A., specific activity: units/mg protein. Protein was determined by the Bradford procedure (36).

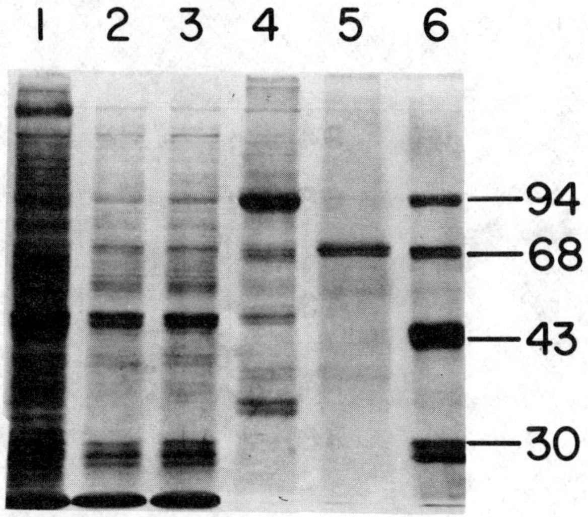

Fig. 7. Purification of preactivated DAI. Polyacrylamide-SDS gel electrophoresis of DAI at several steps of purification. Pro-DAI DE-80 was preactivated and purified by chromatography on DEAE-cellulose and affinity chromatography on Sepharose-histone as described in the text. Aliquots of various fractions were subjected to polyacrylamide-SDS gel electrophoresis. Proteins were stained by the silver reduction procedure (38). 1. Ribosomal salt wash, 130 μg of protein. 2. DE-80 before activation, 37 μg of protein. 3. DE-80 before activation, 50 μg of protein. 4. DE-80, 7.5 μg of protein. 5. Sepharose-histone, 4.3 μg of protein. The sample used for electrophoresis was previously concentrated by partial lyophilization after extensive dialysis against distilled water. This may have entailed loss of protein so that the amount actually applied on the gel may have been below the stated 4.3 μg. 6. Molecular weight (kilodaltons) markers.

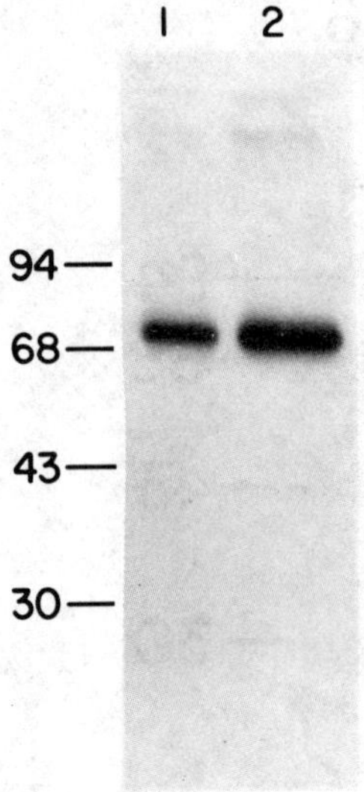

Fig, 8

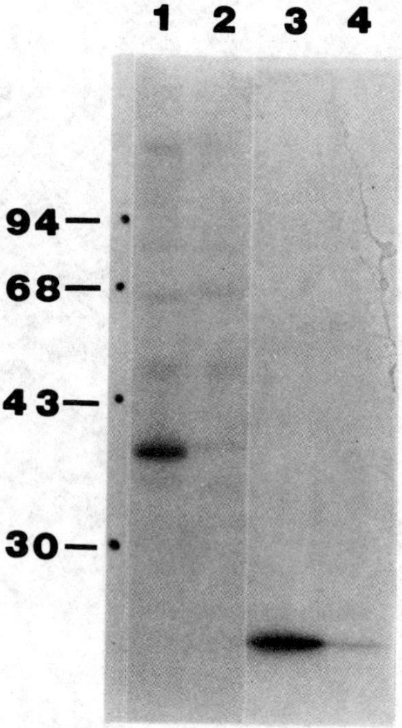

Fig, 9

Fig. 8. Autoradiogram of [^{32}P]DAI. Sepharose-histone DAI was prepared as described in the text from DAI DE-80 (84 mg of protein) that had been activated, in a final volume of 20 ml, with 0.6 μg/ml poly(I).poly(C) and 50 μM γ[^{32}P]ATP (1,200 cpm/pmol). The Sepharose-histone fraction was dialyzed and concentrated by partial lyophilization prior to polyacrylamide-SDS gel electrophoresis and autoradiography. The radioactivity of the protein applied to the gel was 3,600 cpm in track 1 and 7,200 cpm in track 2. The migration of molecular weight markers (kilodaltons) is shown at the left.

Fig. 9. Phosphorylation of the eIF-2 α subunit and histone IIa by highly purified DAI. Samples were prepared and incubated as described in the text under Other assays. The preactivated DAI (16 ng/sample) was a fraction obtained by DEAE-cellulose chromatography eluted on raising the concentration of KCl from 0.4 to 1.0 M; its specific activity (ternary complex inhibition assay) at the time of use was 120,000. After incubation the samples were subjected to polyacrylamide-SDS gel electrophoresis followed by autoradiography. 1 and 2, eIF-2 as phosphate acceptor without and with prior treatment with NEM, respectively. 3 and 4, histone IIa as phosphate acceptor without and with prior treatment with NEM, respectively.

histone IIa as shown in Fig. 9. The figure also shows that prior treatment of the enzyme with NEM abolished DAI activity. Previously we have seen (cf. Fig. 3) that NEM can also inhibit DAI activation. It may be noted that no phosphorylation of the 68 kilodaltons band is seen in Fig. 9. This is presumably because the DAI, that had been activated with non-labeled ATP, was already phosphorylated.

Since nearly homogeneous DAI can phosphorylate both eIF-2 α and histone (Table I) it appears that, unlike HCI DAI can utilize both substrates. This possibility is strengthened by the data of Table II. They show that histone IIa inhibits phosphorylation of the eIF-2 α subunit by DAI. Although this has not been ascertained, it is likely that eIF-2 and histone compete for an active site on the enzyme.

Table II. Inhibition of eIF-2 α subunit phosphorylation by histone

Control samples			Experimental samples		
Additions (μg)	^{32}P bound (cpm)	Inhibition (%)	Additions (μg)	^{32}P bound (cpm)	Inhibition (%)
None	3,500		None	3,500	
BSA (20)	3,490	0	Histone (20)	2,100	40
BSA (40)	3,530	0	Histone (30)	1,480	58
			Histone (40)	610	83

The incubation samples contained 25 mM Hepes buffer, pH 7.6, 4 mM $Mg(OAc)_2$, 0.04 mM $\gamma[^{32}P]$ATP (1,000 cpm/pmol), approximately 14 pmol of eIF-2 PC, 2 ng of preactivated DAI (specific activity, 120,000), and other additions as indicated, in a final volume of 28 μl. After incubation for 8 min at 30°, the samples were subjected to polyacrylamide-SDS gel electrophoresis as described in the text. The gels were stained with Coomassie blue and the α band of eIF-2 was cut off. The gel section was dissolved in 0.4 ml of H_2O_2 at 95° and the radioactivity measured in Hydrofluor. The DAI used here was the same as in Fig. 9.

DISCUSSION

We have been able to isolate preactivated DAI from reticulocyte lysate as a nearly homogeneous, highly unstable phosphoprotein with an apparent subunit mass of 68,000 daltons. Previously (39,40) only partial purification of this enzyme had been achieved. The molecular weight of pro-DAI or active DAI was estimated to be about 120,000 (40). Our work supports the view that DAI is a single enzyme that can catalyze the phosphorylation of both histone and the α subunit of eIF-2. This appears to be true also of interferon-induced DAI (18,19). The fact that virtually homogeneous, active DAI is phosphorylated (Fig. 8) confirms earlier suggestions that the enzyme is activated by phosphorylation (9,10,39,40) and is consistent with a model [reactions (1)-(3)] whereby, upon activation by

(1) $\text{DAI} + \text{dsRNA} \longrightarrow \text{active DAI}$

(2) $\text{DAI} + \text{ATP} \xrightarrow{\text{(active DAI)}} \text{DAI(P)} + \text{ADP}$

(3) $\text{eIF-2 (or histone)} + \text{ATP} \xrightarrow{\text{[DAI(P)]}} \text{eIF-2(P) [or histone(P)]} + \text{ADP}$

dsRNA, DAI phosphorylates itself and becomes active to phosphorylate its substrates, eIF-2 α and histone. Whether the self-phosphorylation is intramolecular, intermolecular, or both, remains an open question.

Our preliminary results with labeled poly(I).poly(C) suggest that the first step in DAI activation is dsRNA binding (Fig. 6). This is consistent with the observation that NEM blocks both DAI activation (Fig. 3) and dsRNA binding (Fig. 6).

REFERENCES

1. London, I. M., Clemens, M. J., Ranu, R. S., Levin, D. H., Cherbas, L. F. and Ernst, V. (1976) Fed. Proc. 35, 2218-2222.

2. Zucker, W. V. and Schulman, H. M. (1968) Proc. Natl. Acad. Sci. U.S.A. 59, 582-589.

3. Hunt, T., Vanderhoff, G. and London, I. M. (1972) J. Mol. Biol. 66, 471-481.

4. Mathews, M. B., Hunt, T. and Brayley, A. (1973) Nature New Biol. 243, 230-233.

5. Gross, M. and Rabinovitz, M. (1972) Biochim. Biophys. Acta 287, 340-352.

6. Clemens, M. J., Henshaw, E. C., Rahamimoff, H. and London, I. M. (1974) Proc. Natl. Acad. Sci. U.S.A. 71, 2946-2950.

7. Kosower, N. S., Vanderhoff, G. A., Benerofe, B., Hunt, T. and Kosower, E. M. (1971) Biochem. Biophys. Res. Commun. 45, 816-821.

8. Ernst, V., Levin, D. H. and London, I. M. (1978) Proc. Natl. Acad. Sci. U.S.A. 75, 4110-4114.

9. Farrell, P. J., Balkow, K., Hunt, T., Jackson, R. J. and Trachsel, H. (1977) Cell 11, 187-200.

10. Levin, D. H. and London, I. M. (1978) Proc. Natl. Acad. Sci. U.S.A. 75, 1121-1125.

11. Lenz, J. R. and Baglioni, C. (1978) J. Biol. Chem. 253, 4219-4223.

12. Levin, D. H., Ranu, R. S., Ernst, V. and London, I. M. (1976) Proc. Natl. Acad. Sci. U.S.A. 73, 3112-3116.

13. Kramer, G., Cimadevilla, J. M. and Hardesty, B. (1976) Proc. Natl. Acad. Sci. U.S.A. 73, 3078-3082.

14. Gross, M. and Mendelewski, J. (1977) Biochem. Biophys. Res. Commun. 74, 559-569.

15. Ernst, V., Levin, D. H., Leroux, A. and London, I. M. (1980) Proc. Natl. Acad. Sci. U.S.A. 77, 1286-1290.

16. Hunt, T. (1979). In: Miami Winter Symposium: From Gene to Protein, eds. Russell, T. R., Brew, K., Schultz, T. and Haber, H. (Academic Press, New York) Vol. 16, pp. 321-346.

17. Grosfeld, H. and Ochoa, S. (1979) Fed. Proc. 38, 781.

18. Sen, G. C., Taira, H. and Lengyel, P. (1978) J. Biol. Chem. 253, 5915-5921.

19. Kimchi, A., Zilberstein, A., Schmidt, A., Shulman, L. and Revel, M. (1979) J. Biol. Chem. 254, 9846-9853.

20. Trachsel, H., Ranu, R. S. and London, I. M. (1978) Proc. Natl. Acad. Sci. U.S.A. 75, 3654-3658.

21. Farrell, P. J., Hunt, T. and Jackson, R. J. (1978) Eur. J. Biochem. 89, 517-521.

22. Ernst, V., Levin, D. H. and London, I. M. (1979) Proc. Natl. Acad. Sci. U.S.A. 76, 2118-2122.

23. Benne, R., Edman, J., Traut, R. R. and Hershey, J.W.B. (1978) Proc. Natl. Acad. Sci. U.S.A. 75, 108-112.

24. Tahara, S. M., Traugh, J. A., Sharp, S. B., Lundak, T. S., Safer, B. and Merrick, W. C. (1978) Proc. Natl. Acad. Sci. U.S.A. 75, 789-793.

25. de Haro, C. and Ochoa, S. (1978) Proc. Natl. Acad. Sci. U.S.A. 75, 2713-2716.

26. Ranu, R. S. and London, I. M. (1979) Proc. Natl. Acad. Sci. U.S.A. 76, 1079-1083.

27. de Haro, C. and Ochoa, S. (1979) Proc. Natl. Acad. Sci. U.S.A. 76, 1741-1745.

28. de Haro, C. and Ochoa, S. (1979) Proc. Natl. Acad. Sci. U.S.A. 76, 2163-2164.

29. Das, A., Ralston, R. O., Grace, M., Roy, R., Ghosh-Dastidar, P., Das, H. K., Yaghmai, B., Palmieri, S. and Gupta, N. K. (1979) Proc. Natl. Acad. Sci. U.S.A. 76, 5076-5079.

30. Farrell, P. J., Sen, G. C., Dubois, M. F., Ratner, L., Slattery, E. and Lengyel, P. (1978) Proc. Natl. Acad. Sci. U.S.A. 75, 5893-5897.

31. Kerr, I. M. and Brown, R. E. (1978) Proc. Natl. Acad. Sci. U.S.A. 75, 256-260.

32. Vaquero, C. M. and Clemens, M. J. (1979) Eur. J. Biochem. 98, 245-252.

33. Slattery, E., Ghosh, N., Samanta, H. and Lengyel, P. (1979) Proc. Natl. Acad. Sci. U.S.A. 76, 4778-4782.

34. Schmidt, A., Chernajowsky, Y., Shulman, L., Federman, P., Berissi, H. and Revel, M. (1979) Proc. Natl. Acad. Sci. U.S.A. 76, 4788-4792.

35. Lowry, O. H., Rosebrough, N. J., Farr, A. L. and Randall, R. J. (1951) J. Biol. Chem. 193, 265-275.

36. Bradford, M. M. (1976) Anal. Biochem. 72, 248-254.

37. Laemmli, U. K. (1970) Nature 227, 680-685.

38. Switzer, R. C., III, Merril, C. R. and Shifrin, S. (1979) Anal. Biochem. 98, 231-237.

39. Levin, D. H., Petryshyn, R. and London, I. M. (1980) Proc. Natl. Acad. Sci. U.S.A. 77, 832-836.

40. Petryshyn, R., Levin, D. H. and London, I. M. (1980) Biochem. Biophys. Res. Commun. 94, 1190-1198.

INFORMATION FOR HISTONE SYNTHESIS AND CHROMATIN STRUCTURE IN DEVELOPMENT

Paul R. Gross* and Robert J. Arceci

Marine Biological Laboratory, Woods Hole, Mass., and Department of Biology, University of Rochester, Rochester, New York, U.S. A.

Abstract The messenger RNA molecules encoding histones are an abundant class of "maternal" messages, i.e., those provided by transcription during oogenesis, stored in the unfertilized egg, and used to direct a part of the protein synthesis during early embryonic development. In sea urchins, the transcription of histone genes begins during the first cleavage cycle, hence the messages directing histone synthesis can be of one or the other origin - maternal or embryonic. Early histone gene transcription takes place on the same genes as are represented in the maternal mRNA population, but as development proceeds there is a changeover to at least one other, quite distinct set of histone genes. Both the transcription products and the proteins made on them are distinguishable as "early" or "late" forms. During postgastrula development and probably much earlier, the chromatin is in consequence organized by two different groups of histones. The changeover from early to late histone gene expression is already programmed in the earliest determined blastomeres (at the 16-cell stage), and it takes place in progeny of those cells even in the absence of cell contact and normal embryonic topography. It can also be shown that utilization of mRNA from the histone-encoding pool is selective and preferential, with regulation somehow associated with the presence of replicating (but not necessarily transcribing) genomes. The early and late histones organize chromatin differently, as evidenced by a systematically changing internucleosomal repeat length.

---

*To whom correspondence about this paper should be sent, addressed to The Director, M. B. L., Woods Hole, MA 02543, U. S. A.

INTRODUCTION

In 1969, Kedes and Gross identified a complex of abundant, newly-transcribed (rapidly labeled) RNAs of sea urchin embryos as histone-coding genetic messages. The histones are, in turn, among the most abundant of newly-synthesized proteins in early development when, although there is no net increase of mass, there is an exponential increase in the number of cell nuclei and, of course, of chromatin. At that time a search was underway for means by which the "masked maternal mRNA" hypothesis, which had been proposed in essentially its final form in 1963 and 1964 (Gross and Cousineau), could be tested without the use of metabolic inhibitors such as actinomycin, against which there was then, as to a smaller extent now, some (not entirely justified) prejudice.

The means for testing came to hand in the early 1970's in the form of improved techniques of nucleic acid hybridization and cell-free translation in heterologous systems. Application of these methods to the analysis of RNA in unfertilized eggs yielded, in 1973, a useful result. There were found present in the egg many template-active RNAs, capable of directing *in vitro* the synthesis of specific proteins, among them all five kinds of histones, made normally in early development (Gross, *et al*., 1973). Skoultchi and Gross (1973) confirmed the presence in the maternal RNA of histone mRNA specifically, by means of quite independent hybridization studies, while Ruderman and Gross (1974) established that synthesis of all five histone classes begins during the first cleavage cycle.

These studies were the subject of a report from our laboratory to the first Conference on RNA in Reproduction and Development, in 1973. Attention was called there to the curious fact that histone mRNA is both a "maternal" species and a product of transcription during early development (i.e., an "embryonic" species). No clue was available then as to the significance of such a duplex system of provenance for histone-coding templates.

The maternal mRNA hypothesis had, nevertheless, been tested, and with the aid of studies on histone mRNA. Proof now available, certain derivative issues acquired a renewed interest. Among them were: the possibility that maternal mRNA is a component of all female gametes; the chemical mechanism by which their translation is prevented in the egg and then allowed in the embryo; the selectivity, if any, of that permitted translation; the homogeneity or heterogeneity of distribution, in relation to egg geometry, of the maternal mRNA population (with the implied chance that maternal mRNA might be identified with the classically-defined "morphogenetic determinants"); the information content of maternal mRNA; and finally, the significance of the duplex provision of templates for early developmental protein synthesis, especially of histones.

There has been, since the last Conference, an extraordinary production of research addressing these issues. It is fair to say that most of the questions implied and identified by them have been answered, at least in principle (see, for a review up to late 1975, Davidson, 1976). Contributions from other participants in the present Symposium address some of the remaining questions in a most satisfactory way, while a number of those not addressed here directly have nevertheless been dealt with, by appropriate methods, during the past few years. An example of the latter is the demonstration of regional asymmetry in distribution of the single-copy maternal RNA sequence set of early sea urchin embryos (Rodgers and Gross, 1978).

We therefore discuss here a restricted group of questions based upon the now generally accepted masked maternal mRNA hypothesis (for a recent, specific test of which, see Kaumeyer, Jenkins, and Raff, 1978); specifically, those maternal mRNAs coding for histones. While these questions, and the experiments designed to explore them, deal with but a single (and perhaps unusual) multigene family and its expression, there is reason to expect that some of the findings will prove representative of other genes as well.

In what follows, each question heads a section of the report, wherein the present situation is summarized, sometimes with the aid of representative data recently published or in press. There is no attempt, for economy of space, to make the summaries adequate reviews of the literature, nor to consider species other than sea urchins. The final section is an attempt to assign significance, without undue speculation, to all the findings so summarized.

ARE EMBRYONIC HISTONE mRNAs COPIED FROM THE SAME GENES AS THE MATERNAL mRNAs?

Soon after it became clear that all five classes of histones are present and being synthesized from the first cleavage onward (Ruderman and Gross, 1974), an earlier interest in the histone content of embryos, and in changes thereof, resumed (e.g., Seale and Aronson, 1973; Easton and Chalkley, 1972). The first and most obvious change to be reported, in the period of renewed activity, was that in the H1 histone of sea urchin embryos. In 1974, Ruderman and Gross and Ruderman, Baglioni, and Gross showed that the altered electrophoretic behavior of H1 histone at the midblastula-early gastrula stage was the result of synthesis of an _entirely_ _new_ polypeptide, directed by mRNA distinct from that functioning earlier.

Improved methods of electrophoretic separation of histones, employing the anionic detergent Triton X-100, permitted Cohen, Newrock, and Zweidler (1975), and Newrock, _et_ _al_. (1978) to arrive at the same conclusion respecting histones H2A and H2B, as well as to identify yet another set of histones (named "CS") whose synthesis is regulated during the period of cleavage in _Strongylocentrotus purpuratus_. Uniqueness of these different histone sets was established, ultimately, by amino acid analysis and partial sequencing of the proteins.

In due course, hybridization experiments employing cloned histone-coding DNA proved that the observed changes of histone synthesis pattern are the expression of distinct gene sets, each encoding an entire group of the five histone types (review: Kedes, 1979). The genes transcribed during the period of cleavage are the same as those transcribed for the maternal histone mRNAs, while the genes responsible for late histones are separate and different. Details of timing and of the number of genes thus employed differ from one species to another, but the general outcome (which appears to apply to many animals other than sea urchins) is that histone synthesis begins with an "early" set (henceforth "E") and is eventually changed to yield only a "late" set ("L").

In 1976, Arceci, Senger, and Gross showed by the method of cell-free translation that the mRNA for histone $H1_L$ is absent from the unfertilized egg. This was confirmed, subsequently, by hybridization methods, and appears likely to apply also to the mRNA for the other L-histones (e.g., Childs, Maxson, and Kedes, 1979). The transcription of late histone genes is probably initiated quite soon after fertilization, but their association with polysomes and the appearance of L-histones in detectable quantity is later, and considerably more abrupt. This suggests, but does not prove, that translation-level selectivity exists, as between E- and L-histone messages, at least in certain stages.

In short, there is a regulated transition of histone synthesis pattern from the beginning to the end of development, which is the consequence of a phased expression of different genes. The late histone genes are among the only ones known so far whose messages are used during development but not present in the maternal set. The resulting synthesis of the different classes of histones is coordinate. That, taken together with the finding that newly synthesized histones are associated quickly with chromatin and are exceptionally stable thereafter (Ruderman and Gross,

1974; Cohen, et al., 1975; Arceci and Gross, 1977; Brandt, et al., 1979), implies that the ratio of E- to L-histones in chromatin should decline according to a simple dilution function as development proceeds.

This was shown to be the case for H1 by Poccia and Hinegardner (1975), and it is, as shown in Figure 1, true for H2A and H2B as well. All variants follow the same dilution curve, as coordinate change predicts. The idea cannot be tested for the remaining two histone classes, H3 and H4, because there is no evidence for differences (to date) in the primary structures of E- and L-variants, but the different multigene families **do** contain genes for these histones, and their transcription products are distinguishable.

IS THERE SELECTIVITY, AT THE LEVEL OF TRANSLATION, FOR DIFFERENT MEMBERS OF THE MATERNAL mRNA SET?

An increasing body of evidence, earlier indirect and more recently direct and sequence-specific, supports the idea that translation of the maternal mRNA is selective in the non-trivial sense that it is not merely a result of differing affinities among members of the mRNA population for an unchanging set of translation factors. Selectivity has been demonstrated in a general way by cell-free translation experiments employing purified RNA from different stages of development (e.g., Rosenthal, Hunt, and Ruderman, 1980, for embryos of the clam, *Spisula*). It can be deduced from experiments on the quantitative course of histone synthesis in sea urchin and amphibian embryos (Kedes,

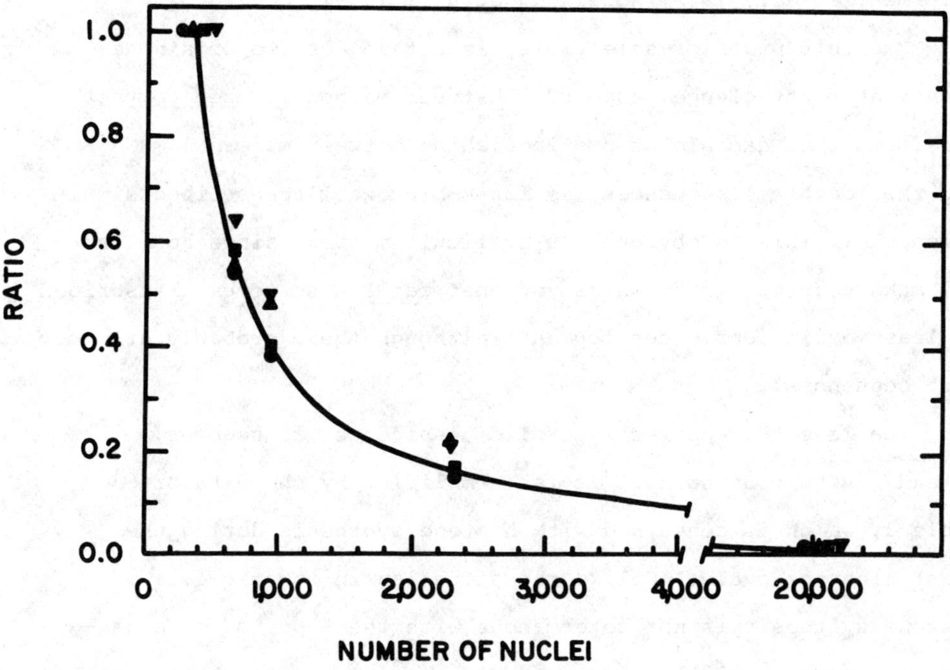

Figure 1. The ratio of number of nuclei at blastula to the number at later stages, and of $H1_E$, $H2A_E$, and $H2B_E$ to the L-variants of the same histones, as a function of nuclear number. The curve is the theoretical dilution function. From Arceci and Gross, 1980c.

_et al_., 1969; Moav and Nemer, 1971; Adamson and Woodland, 1977; Ruderman, et al., 1979; Woodland, _et al_., 1979).

Explicit proof of selectivity is not so easy to obtain as might, at first glance, appear. Whatever method is used to test for it must be capable of distinguishing between maternal mRNA and the identical sequences (as for E-histones) transcribed simultaneously. This is obviously a difficult matter, since no maternal mRNA sequence is known as yet that _fails_ also to be transcribed at least during early development (although there probably are such sequences).

The less than perfectly explicit evidence is, however, already quite convincing. It is exemplified by the data of table 1, which is concerned with histone synthesis during the first cleavage cycle. Unlike the situation in somatic cells, histone synthesis is not coterminous with the S-period of cleaving blastomeres. On the contrary, it continues into G_2, during which time the rate actually increases (Arceci and Gross, 1977).

Sea urchin embryos can be cultured in the presence of actinomycin D so that they proceed normally (although with a slight increase of the intermitotic time) through the cleavage cycles despite a complete suppression of histone mRNA (and nearly all other RNA) synthesis. Such a drug treatment has no detectable effect on the pattern of protein synthesis (see also Sargent and Raff, 1976), nor, in the short period in question, on the rate of protein synthesis. As is shown in table 1, the histone fraction of total protein synthesis rises sharply through the first cleavage cycle . The absolute rate of protein synthesis overall

Table 1

Histone Synthesis in Control and Actinomycin-Treated Embryos During the First Cleavage Cycle

Status	Histone Synthesis as per cent of total (^{3}H)-lysine incorporation into proteins* CONTROL	ACTINOMYCIN
first S	3.3	3.6
first G_2	4.2	5.1
second S	6.1	5.8
second G_2	8.6	8.8

*Histone radioactivity identified by high resolution electrophoresis and radiofluorography.

Table 2 Micrococcal Nuclease Digestion Kinetics of Nuclei from Different Developmental Stages.

Stage	K(min^{-1})	Final Fraction Digested (%)	Goodness of Fit	Rate of Digestion Relative to Sperm
Blastula	0.132±.019	82±3.2	0.044	4.55
Gastrula	0.106±.016	60±2.2	0.029	3.66
Pluteus	0.086±.008	47±1.3	0.014	3.00
11 Day Larva	0.046±.006	34±1.4	0.009	1.58
Sperm	0.029±.007	20±2.5	0.008	1.00

Values were calculated assuming first order kinetics by the nonlinear least square fitting program of Pearson, Davidson and Britten (1977), using no fixed parameters.

is also rising at this time (e.g., Fry and Gross, 1970a,b); hence the rate for histones is rising even faster. It is found that this happens whether or not transcription is allowed, i.e., with or without actinomycin.

Experiments of this kind can be done for longer periods of synchronous development, with the same result, up to the morula stage (ca. 60 - 100 cells), when non-transcribing embryos fail to sustain the rising rate of histone synthesis characteristic of the controls. That is also the stage at which (on the basis of other kinds of evidence) newly-transcribed histone mRNA contributes an already significant fraction of the total polysomal histone mRNA (Arceci and Gross, 1980a).

Thus histone mRNAs can be recruited selectively from among the members of the maternal sequence set, there being no reason whatsoever to believe that in the presence of actinomycin it is the amount or utilization of all other maternal messages that is declining. It would be surprising, therefore, if such a capacity for selective recruitment to the translational machinery were to be limited to the histone mRNAs and their controllers, among the multiple thousands of sequences making up the maternal set.

Preferential recruitment of histone mRNA from the maternal set seems, finally, to depend upon the presence of one or more nuclei capable of replicating DNA. Limitation of space prevents a proper description here of the complexly-controlled experiments (done with merogones, actinomycin, and hydroxyurea) that lead to such a conclusion, but their principle is simple enough to state briefly.

HIstone synthesis takes place even in the absence of a nucleus if the cytoplasm is from an egg, i.e., a merogone (Nishioka and Mazia, 1977). Artificially activated merogones show the same rise of protein synthesis rate (and for a time the same pattern of protein synthesis) as do normal zygotes. Whereas, however, the histones are a steadily increasing fraction of total protein synthesis normally, they are not so in the enucleate merogones.

Enucleates can be provided secondarily with a nucleus, by fertilizing them. When this is done, the normal, cycle-dependent increases of histone synthesis relative to the total are restored, and this happens, furthermore, even if transcription is blocked. All that is needed is for DNA synthesis to continue, as it does in actinomycin. The recruitment of histone mRNAs from the general maternal mRNA pool is therefore controlled in some way by events coupled to genomal replication. The mechanism is mysterious, but the control makes sense, in view of the weight-for-weight equivalence of histone and DNA in chromatin.

WHEN AND HOW IS THE SIGNAL FOR SWITCHING FROM ONE SET OF HISTONE GENES TO ANOTHER GIVEN IN EMBRYOGENESIS?

At the fourth cleavage, which is asymmetric, there are formed sixteen cells of three size classes: macro-, meso-, and micromeres. All cell types are at least partially "determined" at that time, i.e., their developmental potentials have been sharply and differentially restricted. The micromeres, for example, are progenitors only of mesenchyme that will form the larval skeleton, although on the way thereto they serve as the primary inducers of gastrulation.

At this stage only E-histones are synthesized (albeit with interesting quantitative disferences among the cell types; see Senger, Arceci, and Gross, 1978; Senger and Gross, 1978). In normal circumstances some thirty or forty hours must pass (for L. pictus and depending upon temperature) before histone synthesis on mRNA transcribed from the L-histone genes replaces the synthesis of E-histones completely.

Cells of the 16-cell embryo can be dissociated and cultured, where they divide about the normal number of times and appear otherwise to be in good physiological condition (Hynes, et al., 1972a,b). The separated blastomeres can be cultured with reaggregation either allowed or prevented. With it, embryoid clusters form and develop a complex morphology; without it, the cells remain single and fall apart into two when they divide.

Remarkably, all three blastomere types separated in bulk from 16-cell stage embryos, and carried in culture for a period equal to the normal interval for completion of the E-to-L-histone changeover, make the same changeover, and do so in the complete absence of cell-cell contacts. An example of data from which this conclusion is drawn is given in figure 2.

This does not answer the question: how is the switching signal given? But it does show quite clearly when it is given: at or prior to the fourth cleavage - a long time and many cell generations in advance of the first detectable expression of the late genes. The first determined blastomeres are, in other words,

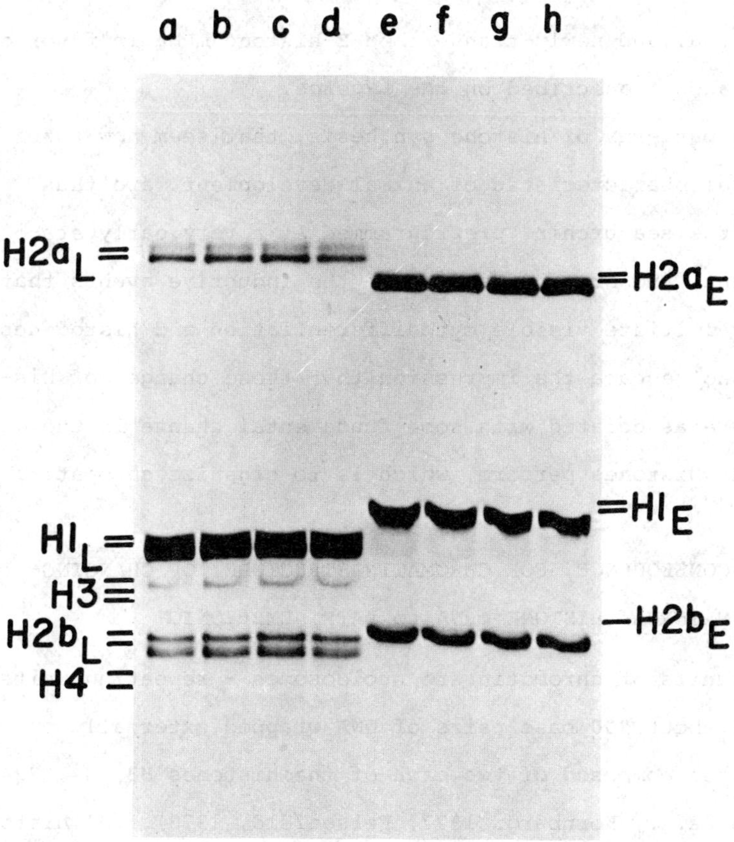

<u>Figure 2.</u> Histones synthesized by intact embryos and by progeny of isolated 16-cell stage blastomeres after 5 hours (e-h) and 40 hours (a-d) in culture. The 5-hour cultures all synthesize early histones, and the 40-hour cultures only the late variants. Fluorogram of a gel electrophoretic separation. Arceci and Gross, 1980b.

already committed to and programmed for all processes that, in the intact embryo, bring about the gradual decline and disappearance of maternal and newly-transcribed E-histone mRNA in favor of histone messages transcribed on the L-genes.

Altered patterns of histone synthesis, that seem now to be a very general characteristic of animal development, are thus, at least in the sea urchin, preprogrammed at a very early stage of development, and are independent of the inductive events that so typically initiate visible cytodifferentiation and histogenesis.

One cannot escape the impression that these changes of histone genes are associated with some fundamental change in the function that histones perform, which is to organize chromatin.

WHAT IS THE CONSEQUENCE, FOR CHROMATIN STRUCTURE, OF CHANGING FROM EARLY, MATERNAL HISTONE mRNA to LATE, EMBRYONIC?

The subunits of chromatin are nucleosomes - repeating units comprised of about 200 base pairs of DNA wrapped externally on octomeric cores composed of two each of the histones H3, H4, H2A, and H2B (see, e.g., Kornberg, 1977; Felsenfeld, 1978). H1 histone associates primarily with the DNA between particles, i.e., with the so-called "linker" DNA (VanHolde, *et al*., 1974).

When chromatin or a preparation of nuclei is attacked, in a medium of suitable composition, by micrococcal nuclease, the preferred initial cutting sites are in the linker regions. Such cuts cause the release of an oligomer series whose length increments are determined by the length of DNA in the nucleosome repeating unit (Hewish and Burgoyne, 1973; Noll, 1974). As digestion proceeds further, there accumulate relatively stable "core particles," whose associated DNA contains about 145 base pairs. This is a remarkably regular result for all the chromatins examined.

The nucleosome repeat length, on the other hand, is variable, and this seems in the main to be ascribable to the many variants of H1 histone, some of them primary structure variants and some produced by post-translational modifications (e.g., Morris, 1976). It is noteworthy, in particular, that the internucleosomal repeat length is variable with the state of cellular differentiation (e.g., Zongza and Mathias, 1979; Spadafora, et al., 1976).

We have found that (1) there is a systematic and stage-dependent change in the kinetics of nuclease digestion of nuclei from sea urchin embryos, and (2) this change is associated with a changing nucleosomal repeat length. Table 2 summarizes the results of experiments on the nuclei of blastulae, gastrulae, plutei, 11-day (feeding and growing) larvae, and spermatozoa. Values of the first-order rate constant and of the asymptotic fraction (final per cent digested), for the reaction under identical conditions of each kind of nuclear preparation with micrococcal nuclease, were estimated from the raw data with the aid of a nonlinear least squares curve-fitting computer program (Pearson, et al., 1977), and are given in the table.

Evidently there is a marked and progressive fall in the rate of digestion and in the final digested fraction with the developmental series: blastula, gastrula, pluteus, feeding larva, sperm. This change is very closely correlated with the fraction of total histone as L-histone (for which see Figure 1 and imagine its inverse). The correlation coefficients are: for rate of digestion vs. %

L-histone, r = 0.99; for final fraction digested vs.% L-histone, r = 0.98.

Electrophoretic separation (in this case on a 1.8% non-denaturing slab gel) of the purified DNA fragments is represented - from among the many methods used to study them (Arceci and Gross, 1980c) - in Figure 3. This includes, along with the raw data (fluorescence photograph of the gel, densitometric scans), some elements of the calibration system employed for very accurate sizing of the separated DNA fragments. When these sizes are plotted as a function of oligomer number, a series of straight lines is obtained from an experiment such as is shown in the Figure, the slopes of which give, with standard errors, the nucleosomal repeat lengths. Results of a comprehensive series of such experiments are summarized in table 3.

Clearly then the internucleosomal repeat length changes systematically in the course of development from the morula, a stage in which the chromatin contains no detectable L-histones, to the feeding larva, in which the residual E-histones have been so diluted as to be no longer detectable. Of interest therefore is the condition of chromatin in mid-development, i.e., in post-gastrula, pre-larval stages with the most intense differentiative and histogenetic activity.

This has been studied by analysis of the mean nucleosomal repeat lengths as a function of the extent of digestion, the working hypothesis being that if the kinetics of digestion change with repeat length (as is shown above), and if repeat length is correlated with histone type, then nuclei with definitely heterogeneous chromatin (in respect to histone type content) should

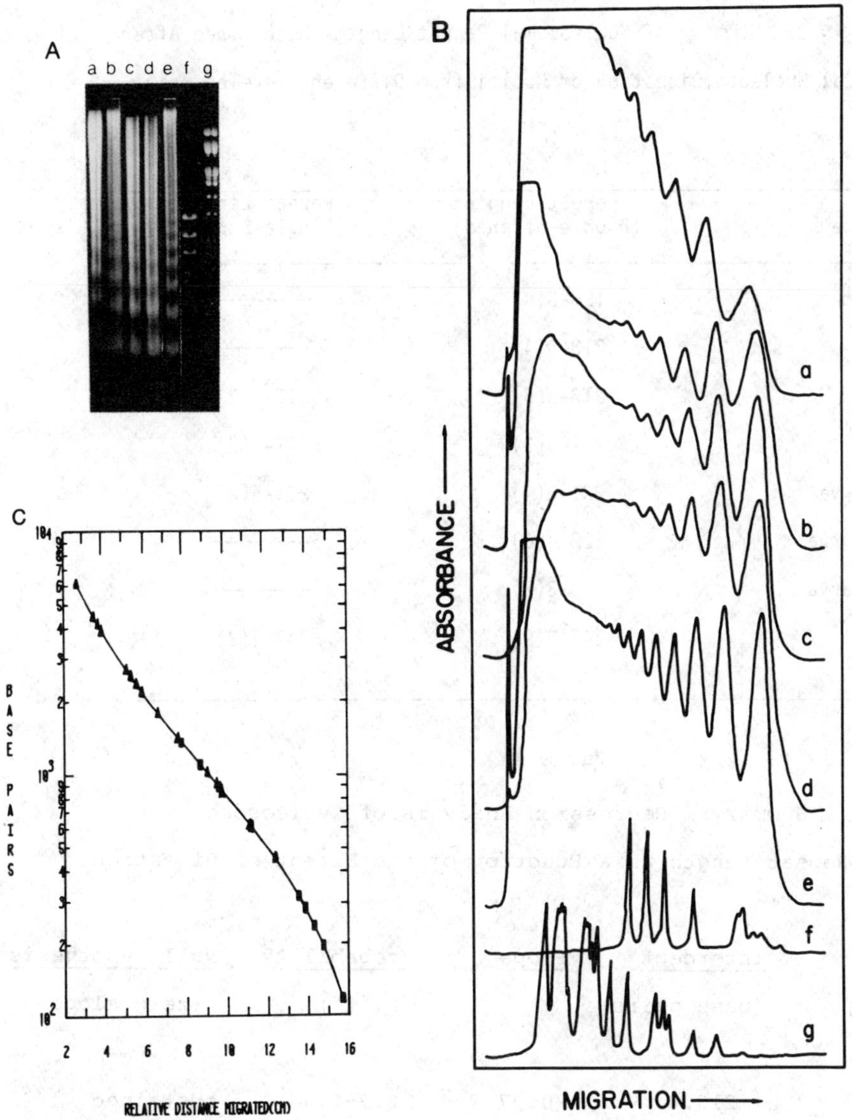

Figure 3. Analysis of products of micrococcal nuclease digestion of nuclei from different stages of development. A. Photograph of the original 1.8% agarose slab gel stained with ethidium bromide. (a) sperm, (b) 9-day larva, (c) late gastrula, (d) blastula, (e) calf thymus, (f) HAE III-cut ØX174RF DNA marker; (g) Hpa-cut wild-type T7 DNA marker. B. Densitometric scans of same. C. Calibration curve used for sizing the DNA fragments. Data from Arceci and Gross, 1980c.

Table 3. Summary of Nucleosomal Repeat Lengths Determined After Micrococcal Nuclease Digestion of Nuclei from Different Developmental Stages

Stage	Repeat Length (Double-Straned)	Repeat Length (Single-Straned)
Calf Thymus	196±3(4)	————
Morula	213±3(4)	————
Blastula	213±3(5)	212±3(2)
Gastrula	217±3(3)	217±3(2)
9 Day Larva	231±3(4)	231±5(2)
11 Day Larva	228±4(6)	————
16 Day Larva	230±3(2)	————
Sperm	247±3(9)	247±4(2)

Table 4

Summary: Regression Analysis of Nucleosome Repeat Length as a Function of the Extent of Digestion

Stage	Intercept (base pairs)	Slope	Probability ("t")	Null Hypothesis (zero slope)
Morula	213.8	-0.072	0.3-0.4	sustained
Pluteus	217.0	+0.233	<<0.001	rejected
11-day Larva	229.3	-0.066	0.2-0.3	sustained
Sperm	247.9	-0.174	0.01-0.02	?

display a change in nucleosomal repeat length with extent of digestion.

The expectation is realized in experiments such as are summarized in Figure 4, which gives the calculated mean repeat length at different points in digestion for morula, pluteus, 11-day larva, and sperm. Table 4 provides a statistical summary of the data. For stages in which either E- or L-variants predominate overwhelmingly (morula and larva, respectively), there is no significant regression of repeat length on extent of digestion. For sperm, which has (interestingly) the longest repeat length and also an unique set of histones, there is a marginally significant regression, and it is negative. For the pluteus, however, the stage with significant bulk amounts of both the E- and L-variant histones, the nucleosomal repeat length increases with digestion, and the positive slope of the fitted regression line is highly significant (P for the null hypothesis is $\ll 0.001$).

Such results, selected from a much more detailed analysis of the problem (Arceci and Gross, 1980c), imply what cannot be given exhaustive documentation here: that the organization of chromatin, as probed (coarsely!) in nuclease digestions, changes from the beginning to the end of development, with intermediate stages in which the chromatin is a mosaic of the new (long spacing) and old (short spacing) arrangements. We are diffident about offering such an implication without full documentation. Fortunately an account of the evidence has been published (Arceci and Gross, 1980c,d) and can be consulted by the interested reader.

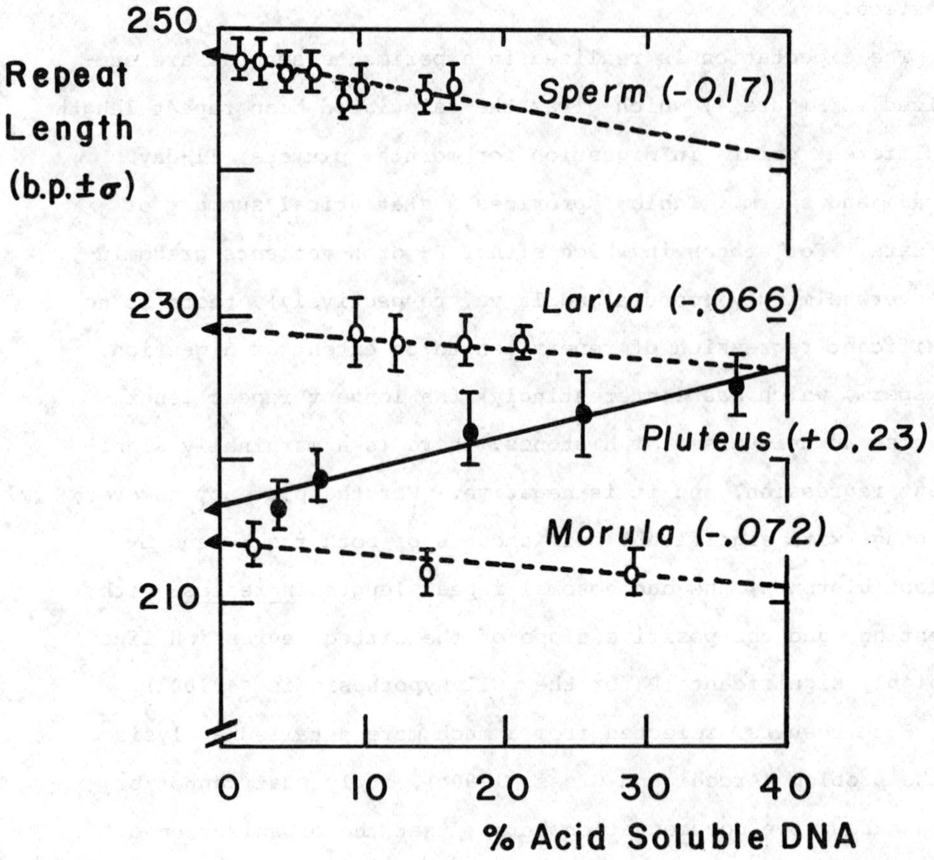

Figure 4. Regression analysis of calculated nucleosomal repeat lengths for chromatin of different stages, as a function of the extent of digestion. From Arceci and Gross, 1980c, to which the reader is referred for details. Tests of significance reported in Table 4.

That some extraordinary reorganization of chromatin must take place in the course of embryogenesis cannot, however, be denied. Even the appearances insist, as Figure 5 may help to recall. There are shown electron micrographic profiles, at the same magnification, of nuclei from early cleavage (4-cell stage) and from postgastrula ("prism" stage) embryos. As is clear from these transmission micrographs, some large-scale change of organization must take place in the chromatin: otherwise, how to account for the 40-fold reduction of nuclear volume from the early to the late stage, and for the massive clumping of nuclear contents into dense, fibrous blocks?

CONCLUSIONS

The information that directs histone synthesis has a multiple (or at least a double) origin: early histone genes and late histone genes. That condition seems to be characteristic of all animal embryonic life (so far as the question has been put), but not of somatic cellular life. It persists throughout development, including the periods of determination, cytodifferentiation, and organogenesis.

The provenance of RNA messages bearing this information is itself duplex: at least through the critical period in which cells and their clonal descendants are being determined, the messages directing synthesis of early histones have one of two origins - maternal or embryonic - and in a changing ratio of one to the other, while messages directing the synthesis of late histones are uniquely the products of recent transcription in the nucleus.

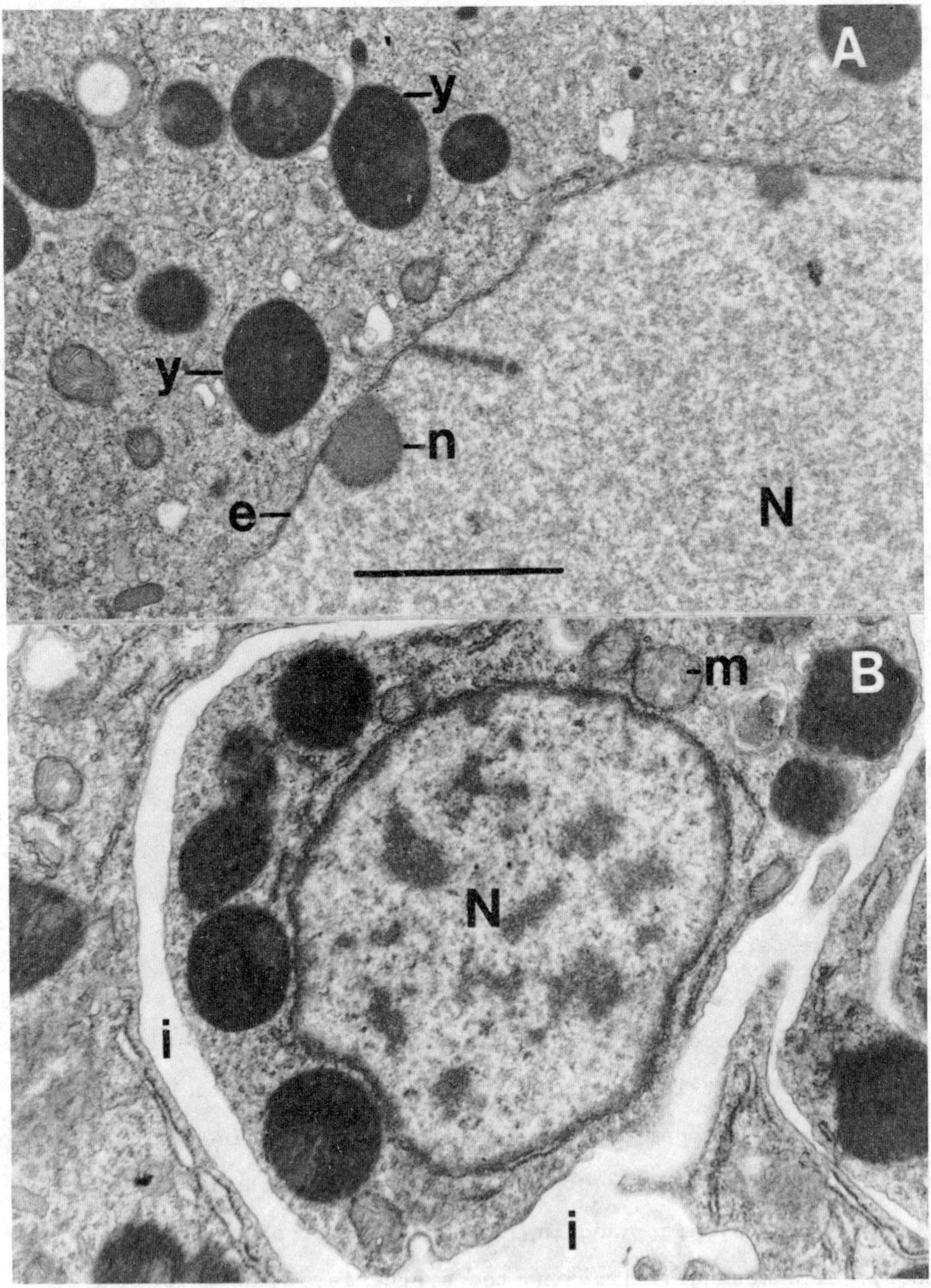

Figure 5. Normal transmission electron micrographs of sections of (A) 4-cell stage sea urchin embryo and (B) a "prism" larva (cell from the anterior epithelium shown). The solid line in A represents one µM, and the magnification is 32,000X for both micrographs. N, nucleus. e, nuclear envelope. n, an "agranular nucleolus, chracteristic of early cleavage stage nuclei. y, yolk particles. m, mitochondrion. i, intercellular space. (Figure reduced to $\frac{9}{10}$ during printing.

Accordingly, the ratio of early to late histones being made also varies, and eventually no more early ones are produced. The duration of development is sufficient, however, to insure that by the time adult behavior is achieved (as in the feeding and growing larva), early histones are a vanishingly small fraction (so far as the whole animal is concerned) of the total in chromatin. Note that this says nothing about individual cells. For those surviving without additional division from early development, when the synthesis of E-histones is active, the chromatin may well be heterogeneous even in the adult or larva.

By the same token cells in the course of determination and differentiation are mostly, if not entirely mosaic in respect to the histone content of their chromatin. That implies that they are also mosaic in such features of chromatin architecture as would be dependent upon histone composition. It has been shown here that histone composition does, in fact, determine - or is strongly correlated with - the internucleosomal spacing.

It would be surprising if such changes were not in some way related causally to the changes known to occur in the biochemistry and structure of developing systems, e.g., a sharp change in the rate of cell division, an elongation of the cell cycle, an altered size and morphology of nuclei and of their nuclear sap, changes in the genomal representation of HnRNA and in the rate of transcription, etc. Whatever the details of those correlations, they depend critically upon the existence in the egg of the distributed, spatially heterogeneous, secondary genome that is the maternal message population, and also upon the time-course and selectivity of utilization of that set of RNA molecules.

ACKNOWLEDGMENTS

We are grateful to many persons, institutions, and granting agencies for support of the work summarized here. In particular we thank our colleagues, Drs. Robert Angerer and Martin Gorovsky, for much advice and useful argument; the University of Rochester and the Marine Biological Laboratory for many forms of help in the maintenance of a collaboration after the two collaborators were separated by a change of positions of one of them; and the officials of granting agencies, such as the National Institute of Child Health and Human Development (grant HD 08652) and the Rockefeller Foundation (GA PD 7812), whose sensible administration of grants in support of our work optimized that collaboration. One of us (R.J.A.) was a trainee of the Medical Scientist Training Program at the University of Rochester during the course of the work.

REFERENCES

Adamson, E. D., and Woodland, H. R. (1977). Devel. Biol. 57: 136.

Arceci, R. J., and Gross, P. R. (1977). Proc. Nat. Acad. Sci. U.S.A. 74: 5016.

Arceci , R. J., and Gross, P. R. (1980a). In preparation.

Arceci, R. J., and Gross, P. R. (1980b). Science, In Press.

Arceci, R. J., and Gross, P. R. (1980c,d). Devel. Biol. In Press.

Arceci, R. J., Senger, D. R., and Gross, P. R. (1976). Cell 9: 171.

Brandt, W. F., et al. (1979). Eur. J. Biochem. 94: 1.

Childs, G., Maxson, R., and Kedes, L. H. (1979). Devel. Biol. 73: 153.

Cohen, L. H., Newrock, K. M., amd Zweidler, A. (1975). Science 190: 994.

Davidson, E. H. (1976). GENE ACTIVITY IN EARLY DEVELOPMENT. Academic Press, New York.

Easton, D., and Chalkley, R. (1972). Exp. Cell Res. 72: 502.

Felsenfeld, G. (1978). Nature 271: 115.

Fry, B. J., and Gross, P. R. (1970a,b). Devel. Biol. 21: 105,125.

Gross, P. R., and Cousineau, G. H. (1963). Biochem. Biophys. Res. Commun. 4: 321.

Gross, P. R., and Cousineau, G. H. (1964). Exp. Cell Res. 33: 368.

Hewish, D. R., and Burgoyne, L. A. (1973). Biochem. Biophys. Res. Commun. 52: 504.

Hynes, R. O., et al.(1972a,b). Devel. Biol. 27: 150, 457.

Kaumeyer, J. F., Jenkins, N. A., and Raff, R. A. (1978). Devel. Biol. 63: 266.

Kedes, L. H., and Gross, P. R. (1969). Nature 223: 1335.

Kedes, L. H., et al. (1969). J. Mol. Biol. 45: 337.

Kedes, L. H. (1979). Ann. Rev. Biochem. 48: 837.

Kornberg, R. D. (1977). Ann. Rev. Biochem. 46: 931.

Moav, B., and Nemer, M. (1971). Biochemistry 10: 881.

Morris, N. R. (1976). Cell 9: 627.

Newrock, K. M., et al. (1978). Cell 14: 327.

Noll, M. (1974). Nature 251: 249.

Nishioka, D., and Mazia, D. (1977). Cell Biol. Int. Repts. 1: 23.

Pearson, W. R., Davidson, E. H., and Britten, R. J. (1977). Nucleic Acids Res. 4: 1727.

Poccia, D. L., and Hinegardner, R. T. (1975). Devel. Biol. 45: 81.

Rodgers, W. H., and Gross, P. R. (1978). Cell 14: 279.

Rosenthal, E. T., Hunt, T., and Ruderman, J. V. (1980). In Press. (Cell)

Ruderman, J. V., Baglioni, C., and Gross, P. R. (1974). Nature 247: 36.

Ruderman, J. V., and Gross, P. R. (1974). Devel. Biol. 36: 286.

Ruderman, J. V., et al. (1979). Devel. Biol. 71: 71.

Sargent, T. D., and Raff, R. A. (1976). Devel. Biol. 48: 327.

Seale, R. L., and Aronson, A. I. (1973). J. Mol. Biol. 75: 647.

Senger, D. R., and Gross, P. R. (1978). Devel. Biol. 65: 404.

Senger, R. D., Arceci, R. J., and Gross, P. R. (1978). Devel. Biol. 65: 416.

Skoultchi, A., and Gross, P. R. (1973). Proc. Nat. Acad. Sci. 70: 2840.

Spadafora, C., Noviello, L., and Geraci, G. (1976). Cell Differ-antiation 5: 225.

Van Holde, K. E., et al. (1974). Nucleic Acids Res. 1: 1579.

Woodland, H. R., et al. (1979). Cell 18: 165.

Zongza, V., and Mathias, A. P. (1979). Biochem. J. 179: 291.

THE BIOLOGICAL ACTION OF EXOGENOUS CYTOPLASMIC RNAs OF ANIMAL LIVER ON CULTURED HUMAN HEPATOMA CELLS AND THEIR DISTRIBUTION IN VIVO

Gu Jian-ren, Guo Chan, Chen Yuan-ching,
Huang Zong-ji, Hu Lifu, Huang Le-hong

(Department of Biochemistry and Molecular Biology,
Shanghai Cancer Institute)

Abstract The total RNA and mRNA-riched cytoplasmic-RNA extracted from normal liver were: (1) incubated with mouse ascitic hepatoma cells in vitro, and then injected the ascitic cells into mouse peritoneum; (2) injected directly into peritoneal cavity of mice. The inhibition of tumor growth, the change of G-6-p metabolic enzyme patterns and the synthesis of specific proteins of hepatoma cells were observed. The results indicated that hepatoma cells after incubated with normal liver RNA could be redifferentiated possessing more normal phenotypic characters.

It was shown that labeled RNA distributed preferentially into mouse liver and transplanted liver tumor.

INTRODUCTION

Cancer cells are characterized by their abnormal differentiation and persistent, uncontrolled proliferation. Most of the special features of cancer cells are considered as aberrant manifestations of gene expression. If the mistakes or alterations in the mechanism of gene expression could be corrected, they would be expected to cause or force cancer cells redifferentiation. It would be possible to lead to new clues in the treatment of cancer.

In our laboratory the redifferentiation of hepatoma cells by RNAs extracted from normal liver has been studied since 1963[1]. The mouse ascitic hepatoma cells are completely deprived of their capability of tumorigenicity after incubation with normal liver

RNA in vitro, and are partially inhibited in growth after intraperitoneal injection of liver RNA. The hepatoma cells could be promoted in the synthesis of serum albumin, and their enzyme pattern in G-6-p metabolic pathway seems to shift towards that of normal liver cells. The synthesis of AFP is inhibited. The regulatory effect on host gene expression of cancer cells and the template activity of mRNA are both present in the exgenous cytoplasmic RNAs.

MATERIALS AND METHODS

1. Animal tumors and cell line

Swiss mice of 20-25gm are used in all the experiments. Mouse ascitic hepatoma is given by Institute Materia Medica, Chinese Academy of Science. The human liver cancer cell line 7402 and a transplantable rat hepatoma, originally induced by diethylnitrosamine, are derived from Institute of Cell Biology, Chinese Academy of Science.

2. RNA extraction and purification[2]

The aqueous layer RNAs used in the early experiments for incubation of ascitic hepatoma cells are total RNAs prepared from rat or mouse liver from aqueous phase according to Laskov, the phenolic layer RNA is extracted from interphase and phenolic phase by the method of Lu et al.

In later experiments, mRNA enriched cytoplasmic RNAs are prepared from liver of various kinds of animals(sheep, pig, calf, rabbit etc) essentially according to Brawerman[3] with some modification. The phenolic phase and interphase after phenol-chloroform extraction are repeated by extracted with pH 9.0 buffer. Polysaccharides are removed by ethylene monomethyl ether. RNAs are precipitated by CTA and recovered by repeated washings with 0.1M NaAC-75% ethanol.

3. Characterization of RNA

RNAs are analyzed by PAGE in 2.5% gel-0.5% agarose. The electrophoresis is carried out according to Maurer[4]. The gels are scanned by a densitometer.

The template activity of mRNA in RNA preparation is assayed by wheat embryo cell-free system as described by Roberts[5].

4. Iodination of RNA[6]

RNA was labeled as described by Getz, M. J. (1972). In 1ml mixture containing 6.25 x 10^{-5}M KI, 0.5mCi, ^{125}I-NaI or ^{131}-NaI (^{125}I SA 40mCi/ml, ^{131}I SA 100mCi/ml), 2.3 x 10^{-3}M $TiCl_3$ to remove free ^{125}I. The labeled RNA solution is added by 1mg un-labeled RNA carrier and then precipitated by 2 volume alcohol, washed, dissolved, passed through a column of Sephadex G-25. The fractions containing ^{125}I-RNA are pooled. The labeled rate is about 30% with specific radio-activity of 75 μCi/mg RNA.

Iodination of nucleotide is under taken by alkaline hydrolysis of ^{125}I-RNA.

5. Inhibitory effect of RNA on tumorigenicity and growth of ascitic hepatoma cells(1)

Mouse ascitic hepatoma cells are harvested and washed with normal saline 2-3 times, then are incubated with aqueous layer RNA(8mg/ml) or phenolic layer(less than 2mg/ml) at 0-4°C for 12-16 hours. After being counted with 0.5% nigrosis, 6 x 10^6 viable tumor cells are injected intraperitoneally per mouse. In general, cell viability is about 70-95%. Ascitic tumor cells are incubated with normal saline, liver RNA treated with RNase, RNase alone and extracted from other tissues at same concentration and inoculated intraperitoneally as control groups. After 8-10 days, the mice are sacrificed and tumor cells are washed out from the peritoneal cavity with normal saline and counted. Since no ascitic tumor formation is observed in the liver RNA incubation group, some animals are sacrificed at the end of 3 months after tumor cell inoculation.

In vivo experiments, cytoplasmic RNA, containing PVS at 0.15mg/mouse, are injected intraperitoneally 6 to 12 hours after intraperitoneal inoculation of 6 x 10^6 ascitic hepatoma cells. Two further injections of RNA are given on day 2 and 3 after inoculation. The mouse are sacrificed on the 8th to 10th day, the tumor cells are collected and counted. PVS alone and saline are injected by the same time schedule as control.

6. The distribution of ^{125}I or ^{131}I labeled RNA in mice(7)

^{125}I (or ^{131}I)-RNA were injected intravenously into normal and implanted hepatocarcinoma-bearing mice(5 Ci/100 μg RNA, 20 μCi/400 μg RNA/0.5ml). At the same time the mice underwent whole body scintillation scanning in order to reveal the distribution in various organs. As well as made the autoradiograph.

7. Enzyme assay of G-6-p metabolic pathway in RNA-treated mouse ascitic hepatoma cells[1]

Mouse ascitic hepatoma cells are harvested on day 4-5 after inoculation, washed with 50 volumes of saline at 4°C 4-5 times, and incubated in Tris-HCl buffer pH 7.8, containing aqueous layer RNA at 50 A_{260} units/ml or phenolic layer RNA at 10-15 A_{260} units/ml at 0-4°C.

After incubation the cells are centrifuged. The cells are suspended in Rabinovitz medium containing 20 amino acids and RNA with the same concentration as described above at 37°C for 2 hours. Tumor cells are then collected by centrifuge, washed, counted and subjected to enzyme assay.

Tumor cells are homogenized in Dounce homogenizer in distilled water and adjusted to isotonicity immediately by addition of KCl. Enzyme assay is carried out with 10,000g supernatant of the homogenate. Glucose-6-phosphatase(G-6-pase) activity is measured by the method of Swanson[8], G-6-p dehydrogenase(G-6-PDH) according to Glock and Mclean[9], phosphoglucomutase(PGM) by the method of Najjar[10], phosphohexose isomerase (PHI) by the method of Bruns and Hinsberg[11], fructose diphosphatase(FDP) according to McGilvery[12]. The protein of the supernatant is measured by Lowry's method. The specific activity of enzymes is expressed in enzyme activity units per mg protein of nitrogen.

8. Synthesis of mouse and human serum albumin by human hepatocarcinoma cell lines 7402

Human and mouse serum albumins are prepared by ammonium sulfate fractionation and purified by starch gel column electrophoresis. Rabbit antisera against corresponding serum albumin are prepared with a titer of 1:32 and 1:64 respectively. No cross reactivity between human and mouse serum albumin can be demonstrated in double diffusion or counter-immunoelectrophoresis.

Human hepatocarcinoma cells(7402) are grown in RPMI 1640 10% calf serum. Cells are harvested on day 4 after subculture, then are washed and transferred to leucine free 1640 medium containing RNA(6mg/4ml) and ^{14}C-leucine(5-10 μCi/4ml, 15mCi/mM) for 24 hours. The viability of cells is checked by vital stain with nigrosin. Cells are washed and lyzed by repeatedly freezing and thawing. Aliquots of supernatant are subjected to counter-electrophoresis with corresponding antisera after the addition of human or mouse serum albumin as carrier.

The precipitation bands are cut after thoroughly washing the agar plate. The incorporation of ^{14}C-leucine into corresponding serum albumin is counted by liquid scintillation spectrometer.

Aliquots of supernatant of each group are precipitated with TCA after the addition of serum albumin carrier. The precipitates are washed with cold and hot TCA, and counted as total protein synthesis.

Actinomycin D is added in some control groups to block the transcription of cancer cells at a concentration indicated in legend of table 4.

9. Regulatory effect of exogenous liver RNA on synthesis of serum albumin and AFP in rat hepatoma cells

The culture medium and the incubation condition are similar to those in the experiment described above for synthesis of human and mouse serum albumin. Rat transplantable hepatoma solid nodules are dissected from peritoneal cavity of rats. Hepatoma cells are dispersed and isolated, then inoculated at a density of 10^8 cells per bottle. Tumor cells are incubated with normal mouse liver RNA. Cells are harvested at different time intervals. After thoroughly washing, cells are lysed. The supernatant is subjected to counter immunoelectrophoresis with anti-AFP and anti-albumin antisera respectively. The procedures of measurement of incorporation of ^{14}C-leucine into AFP, serum albumin and total protein are the same as above.

RESULTS

1. The inhibitory effect of normal liver RNA on tumorigenicity of mouse hepatoma cells

After incubation of mouse hepatoma cells in vitro with normal rat or mouse liver RNA (at 8mg/ml for aqueous layer RNA or 2mg/ml for phenolic layer RNA) at 0-4°C for 12-16 hours, the hepatoma cells are completely deprived of their tumorigenicity after inoculation i.p. into mice (Table 1). After treatment with RAase, the inhibitory effect of RNA is abolished. No inhibitory

Table 1 The inhibition of tumorigenicity of mouse ascitic hepatoma cells after incubation with normal mouse liver RNA

Incubation of Hepatoma Cells	RNA Concentration (mg/ml)	no.of Mice	Experimental Results	
			No.of mice with ascitic tumor	No. of mice without ascitic tumor
Normal mouse liver RNA				
Aqueous layer RNA	8.0	69	0	69
	4.0	16	2	14
	1.0	8	8	0
Phenolic layer RNA	2.0	25	0	25
Normal saline	---	60	60	0
RNase 0.4 mg	---	10	10	0
0.8 mg	---	5	5	0
*Normal mouse liver aqueous layer RNA previously treated with RNase	4.0	10	1	9
	8.0	8	0	8

* 8.0 mg/ml RNA are preincubated with 0.8 mg RNase/ml at 37°C 30 minutes; 4.0 mg/ml RNA are treated in same condition with 0.4 mg/ml RNase.

2. The inhibitory effect of exogenous liver RNA on the growth of mouse ascitic cells in vivo.

Table 2

Group	No. of mice	No. of cancer cells	rate of inhibition
Normal saline	10	4.39×10^8	-----
P.V.S. control	20	4.20×10^8	4.3 %
RNA + PVS	10	2.35×10^8	46.4 % ($P < 0.05$)

Six million hepatoma cells were inoculated intraperitoneally per mouse and RNA, P.V.S., saline were injected intraperitoneally per day three days.

activity on tumor growth is found by RNase, or RNA extracted from hepatoma cells, kidney, spleen, heart and muscle. These experiments indicate that the inhibitory effect of exogenous normal liver RNA must be attributed to the RNA macromolecules themselves. Since RNAs extracted from tissues other than normal liver are ineffective and RNAs from liver of various kinds of animals such as mouse, rat, pig, calf, rabbit, turtle, tortois etc, are all effective, the inhibitory activity of RNA seems to be organspecific, but not speciesspecific.

Hepatoma cells remain dye-resistant to nigrosin or trypan blue staining after incubation with RNA. Some animals inoculated with 6 x 10^6 hepatoma cells treated with normal liver RNA were examined at different time intervals till the end of 3 months after inoculation. On the 12th day after inoculation, there were still 1-2 x 10^6 viable cells in peritoneal cavity. They are gradually degenerated and disappeared in 1 month. These results suggested that the inhibitory effect of exogenous RNA could not be attributed to the cytotoxic effect of RNA.

The time of first injection was six hours after tumor inoculation. Tumor cells were counted on the day of sacrifice. Polyvinylsufate (PVS) is a potent RNase inhibitor. In the RNA + PVS group the rate of inhibition is 46.4%. RNA alone has no degradation of RNA by RNase in vivo.

Similar to the experimental results in vitro, the inhibitory effect of normal liver RNA on growth of hepatoma cells in vivo is organ specific, but not species specific. Normal liver RNA from goats, pigs, calfs, turtles were all effective (Table 2).

2. The regulatory effect of exogenous mouse liver RNA on enzyme pattern of G-6-p metabolic pathway of mouse hepatoma cells.

Table 3. It is shown that in the G-6-p metabolic enzymes system, there are remarkable differences between normal liver cells and hepatoma cells. For example, the decrease of the activities of glucose-6-phosphatase, fructose-diphosphatase, phosphoglucomutase is closely correlated with the degree of malignancy and the rate of growth of the tumor.

On the other hand, the increase of activity of glucose-6-phospho-dehydrogenase is parallel to malignancy and rate of growth

Table 3 Effect of exogenous RNA on G-6-P metabolic enzyme patterns of hepatoma cells

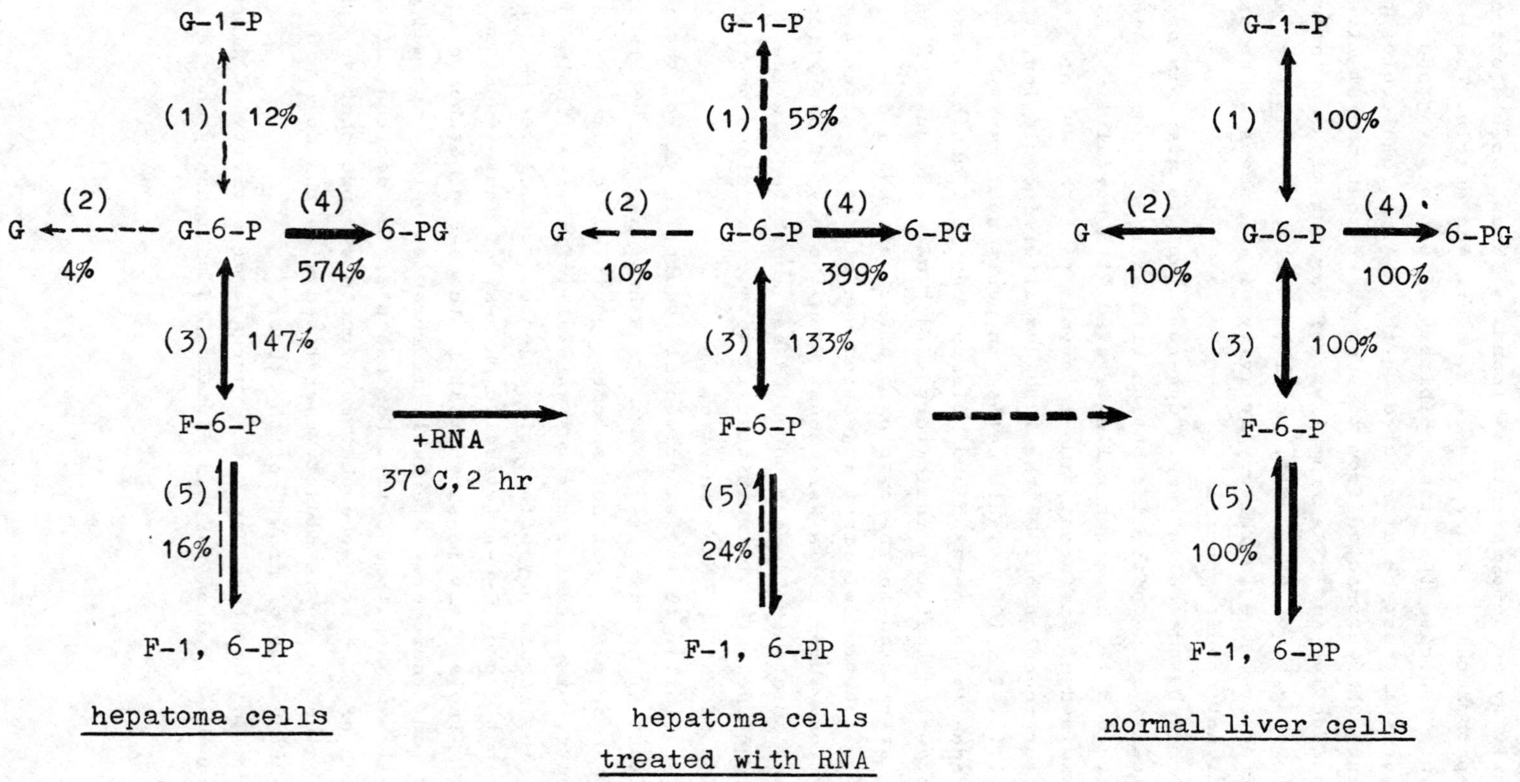

Table 4

The effct of mouse liver RNA on the synthsis of serum albumin of human hepatocarcinoma (7402) cells

No. of experiment	materials	synthesis of human serum albumin			synthesis of mouse serum albumin		
		incorporation (c.p.m)	net stimulated incorporation	%	incorporation (c.p.m)	net stimulated incorporation	%
(1)	7402 cells	84			48		
	7402 cells +RNA	207	123	150	183	135	280
(2)	7402 cells	106			58		
	7402 cells +RNA	379	273	260	192	134	230
	7402 cells +D	38	-68	-63	34	-24	-41
	7402 cells +D+RNA	91	-15	-15	281	223	385
(3)	7402 cells	124			100		
	7402 cells +RNA	359	235	188	270	170	170
	7402 cells +D	89	-35	-28	59	-41	-41
	7402 cells +D+RNA	192	69	54	213	113	113

D actinomycin D 1.60 g/ml (1) RNA 0.8mg/ml (2) RNA 1.25mg/ml (3) RNA 1.25mg/ml

Figure 1,

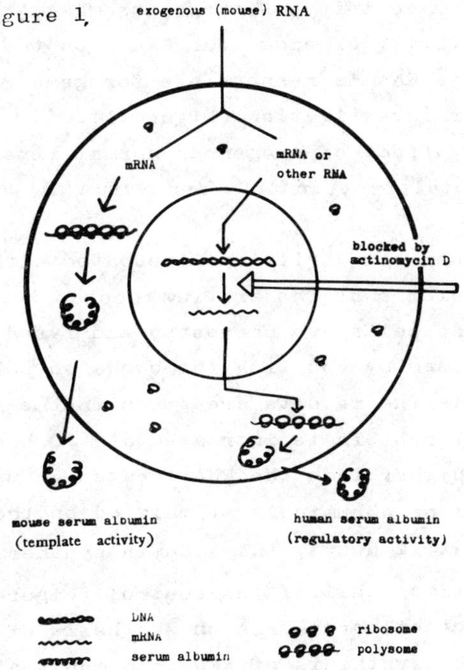

of tumor. Therefore, such enzyme systems could be used as markers for the degree of differentiation of cancer cells.

3. The regulatory effect of exogenous mouse normal liver RNA on the synthesis of serum albumin of human hepatocarcinoma cell line 7402.

Serum albumin is a liver specific protein, which is also an indicator of differentiation of normal liver cells. In liver cancer cells, the synthesis of serum albumin is usually suppresed or completely abolished. After normal mouse liver RNA incubated, the synthesis of serum albumin increases and could synthesize mouse albumin as well as human serum albumin.

The synthesis of human serum albumin was stimulated 3 fold and such a stimulating effect could be blocked by Actinomycin-D. Therefore there might be two ways to affect of exogenous RNA. One is the templete activity of exogenous mRNA, the other is the regulatory activity of exogenous RNA's on the gene expression of host cells whose RNA is responsible for gene regulatory activity are under investigation (Figure 1).

4. The regulatory effect of exogenous normal liver RNA on the synthesis of total protein AFP and serum albumin of rat hepatoma cells

(1) The rat hepatoma cells are incubated with exogenous mouse liver RNA. The medium contains tritium labeled leucine. At different time intervals, the cells are harvested and lysed. The albumin synthesis is analysed by counting the bands on polyacrylamids gel electrophoresis the results are given in the figure 2. The synthesis of total protein is increased at 12 hours but the control group is higher than the RNA treated group.

The synthesis of albumin is stimulated on the course of treatment by RNA at 12 hours, the albumin synthesis in RNA-treat group is 6 times that of the control (Figure 3).

(2) The effect of normal RNA on synthesis of AFP of rat hepatoma cells. The synthesis of AFP is demonstrated by counter-electrophoresis, the precipitated bands are cut and counted by liquid scintillation spectrometry. The synthesis of AFP is inhibited to a certain extent when the cells are incubated at different time intervals. Both in the control and experimental group, level of AFP synthesis drops at the first 6 hours interval.

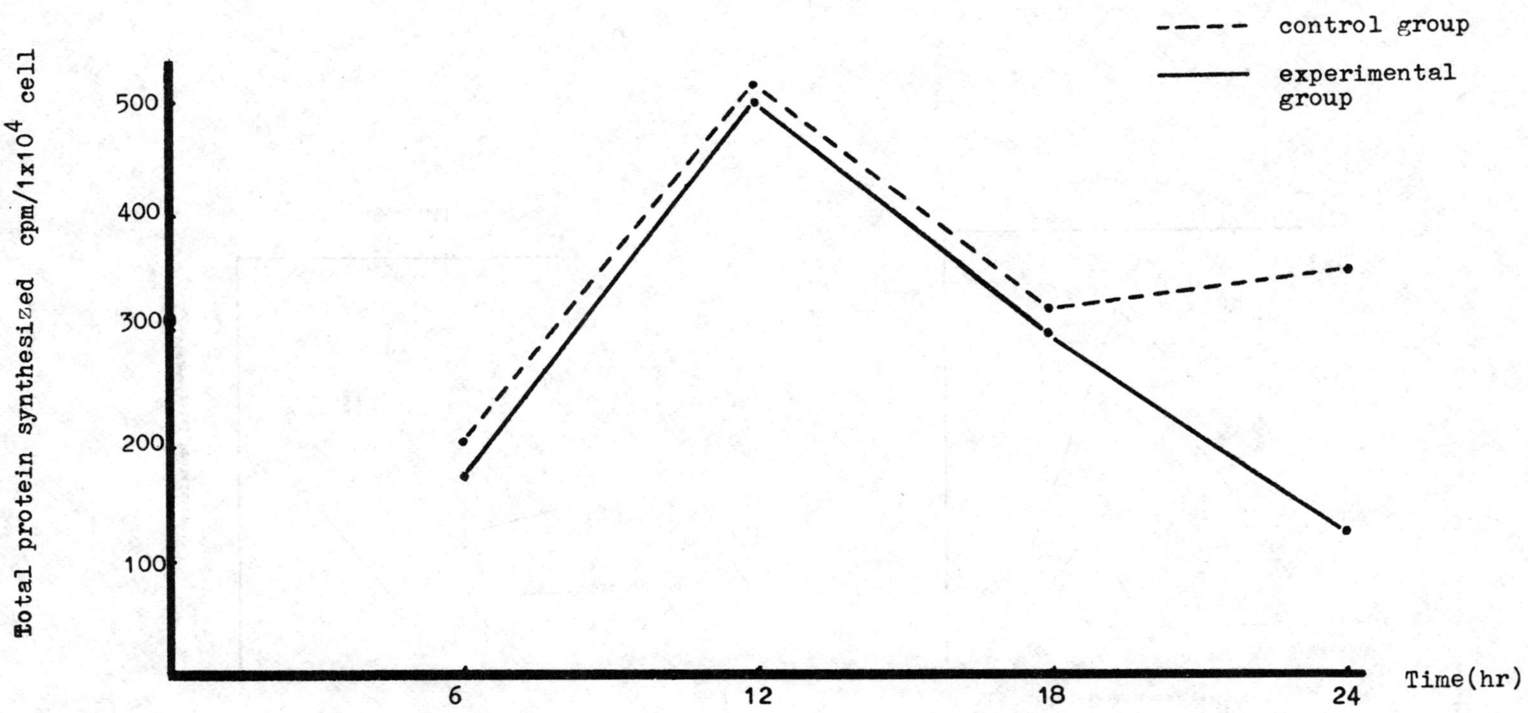

Figure 2. The effect of normal RNA on total protein of hepatocarcinoma cells

Figure 3

The effect of normal RNA on Albumin synthesis of rat hepatocarcinoma cells

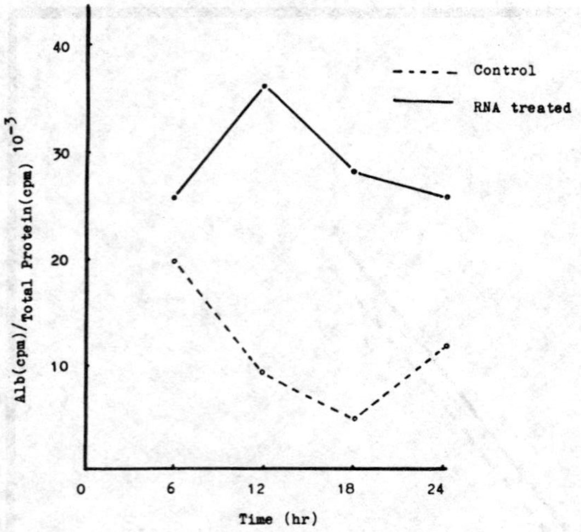

Figure 4

The effect of normal RNA on AFP synthesis of rat hepatocarcinoma cells

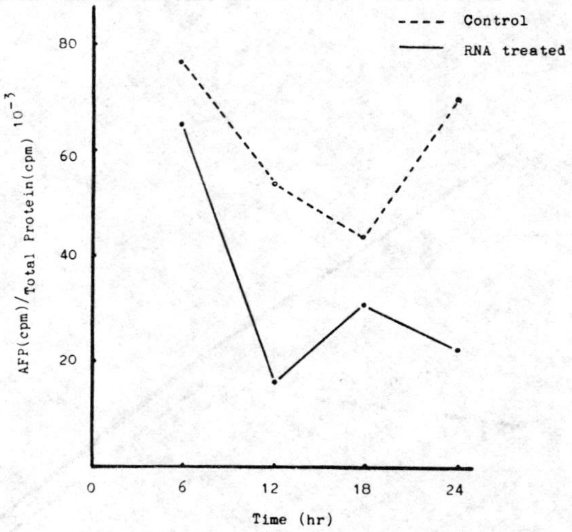

It may be explained by the fact that hepatoma cells are primary culture cells, which have not adapted to the enviroment in culture (Figure 4).

5. The distribution of ^{125}I and ^{131}I-labeled exogenous liver RNA in mice.

The normal liver RNAs were labeled with ^{131}Iodine and ^{125}Iodine. The ascitic hepatoma cells were inoculated intrahepatically and used as a tumor model. After intravenous injection of labeled RNA the distribution of isotopes was examined by total body scanning and determination of radioactivity of individual organs was done by cell autoradiography. The two figures show radioactive scanning of the whole body, thirty minutes after injection of five to ten microcuries of labeled RNAs (Figure 5). In autoradiography, 15 minutes after injection of labeled RNA, silver granules were demonstrated in hepatic parenchymal cells and hepatoma cells. This indicated that some exogenous RNA molecules may enter liver cells and hepatoma cells in vivo (Figure 6).

DISCUSSION

1. The inhibition of the growth and the induction of the redifferentiation of hepatoma cells by exogenous normal liver RNA. From the in vitro incubation and the intraperitoneal injection experiment of exogenous RNA in vivo, it is evident that the tumorigenicity and the growth of ascitic hepatoma cells can be definitely inhibited by RNAs extracted from normal animal liver. The inhibitory effect of RNAs has been proved to be tissue specific rather than species-specific. The inhibitory effect must be attributed to RNA macromolecules themselves as revealed by control experiments, in which no inhibitory effect could be demonstrated in RNase-treated RNA of RNA extracted from other tissues than normal liver.

The loss of tumorigenicity of hepatoma cells after incubation with normal liver RNA is not a cytotoxic effect of RNA, since the tumor cells are still viable after incubation as shown by vital stain. It is evident that the hepatoma cells, after incubation with liver RNA, can not only synthesize serum

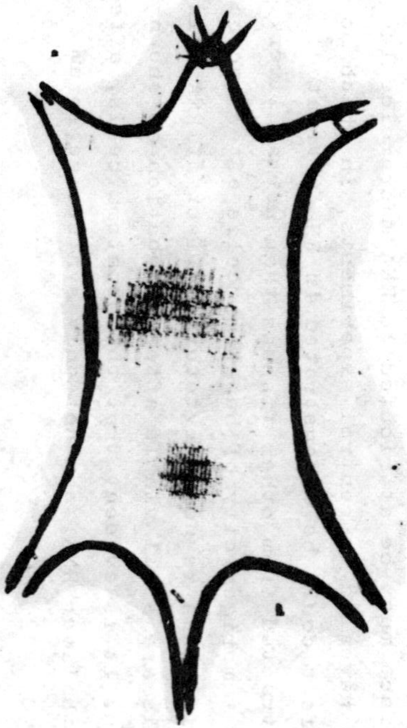

Total body scanning of mouse 30' after ^{125}I-RNA iv.

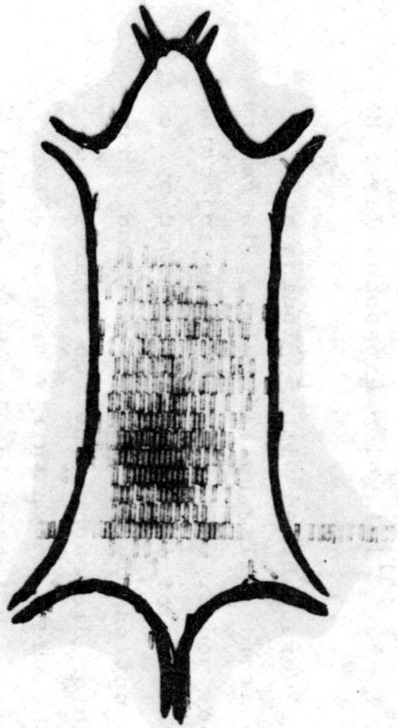

Total body scanning of mouse 30' after ^{125}I-NT iv.

Figure 5. The distribution of exogenous RNA in vivo

Figure 6. The distribution of exogenous RNA in hepatoma cells and normal liver cells.

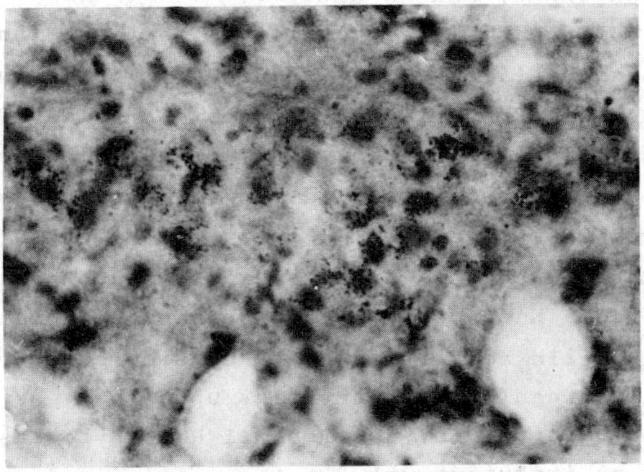

There are silver granules demonstrated in hepatoma cells in vivo.

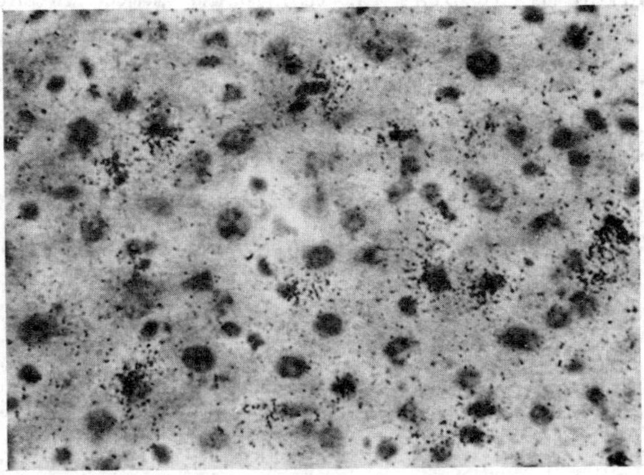

15' after injection of ^{125}I-RNA there are silver granules demonstrated in hepatic parenchymal cells in vivo.

albumin, but also resume their enzyme activities which are intensively depressed in untreated cancer cells.

2. As the exogenous RNA can promote the synthesis of serum albumin which is a liver specific protein and decreased in hepatocarcinoma cells, and induce the shift of enzyme pattern of G-6-P metabolic pathway, it indicates that the inhibitory effect of exogenous liver RNA on hepatoma cells is intimately related to the redifferentiation of hepatoma cells. It is postulated that the loss of malignancy or tumorigenicity may be the result of the redifferentiation of liver cancer cells induced by exogenous liver RNA.

The regulatory mechanism of gene expression of eukaryotes remains unsolved. Many cellular components, especially the nuclear or chromatin constituents are considered to responsible for the positive regulatory activities upon the gene expression, such as nonhistone protein, chromosomal RNA. Cytoplasmic mRNA may also play an important role in the regulation of gene expression(12), and in the alteration of heritable gene alterations(14). In our experiments, the stimulatory effect of the synthesis of serum albumin of human hepatocarcinoma cells by mouse cytoplasmic RNA and the blockage of this effect by Actinomycin-D may denote that there is a factor in our RNA preparation which might play an important part in the regulation of cancer cell gene expression. Furthermore, the shift of enzyme pattern of G-6-p metabolic pathway and the increase in the synthesis of serum albumin parallel to the decrease of the synthesis of AFP may also support the view that the phenotypic changes of cancer cells after treatment with normal liver RNA are the result of an alteration in gene expression of cancer cells. It will be of great importance to determine whether the regulatory effect on gene expression is the activity of mRNA itself or is due to some protien regulatory molecules translated by exogenous mRNA or the presence of some specific regulatory RNA molecules in cytoplasm. This work is in progress.

3. In vivo distribution of exogenous RNA

^{125}I or ^{131}I labeled RNAs are prepared and their distribution is traced in vivo by whole body scanning and cell autoradiography. The results suggest that labeled RNA could be taken up preferentially in liver. The silver grains are demonstratrated in hepatoma cells as well as in liver parenchymal cells after 15 minutes pulse labeling. Though the possibility of RNase-degradation of injected RNA macromolecules could not yet be ruled out, it is possible that some fragments or even a certain amount of intact molecules surviving enzyme degradation might be taken up by the liver and hepatoma cells. However, the search for protective agents for RNA against the attack of RNase is of extreme importance. The research on how to entrap RNA molecules with liposome or other vehicles is now under way in our laboratory.

References

1. Ku Tien-Jien, Yan Gen-Bao, Li Ming-lie, Xue Ru-zhen, Xu Kai-li, Xu Xiu-lan, Xu Shi-kang, Chen Yuan-ching, Sun Shu-ping, Lia Law-zhen.: The biological action of RNA extracted from normal cells on cancer cells. Chinese Med. J. 52 (4): 209-223, 1966
2. Lu Chia-hung, Ku Chien-jen, Wang Pei-yu, and Chu Ying-Shu.: Ribonucleic acid in the silk gland of silkworm. Shih Yen Sheng Wu Hsuch Pao 9 (3): 300-310, 1964
3. Brawerman, G.: The isolation of messenger RNA from mammalian cells. In Moldave, K. Grossman, L. (Eds) Methods in enzymology. Academia press New York and London Vol. XXX part F p.605, 1974.
4. Maurer, HR.: Disc electrophoresis and related techniques of polyacrylamide gel electrophoresis. 2nd Ed.: Walter de Kruyter. New York, 1971.
5. Robert, BE.: Efficient translation of TWV RNA and rabbit globin 9S RNA in a cell-free system from commercial wheat germ. Proc. Natl. Acad. Sci. USA. 70:2330, 1970.
6. Getz, MJ., Altenburg, LC. and Saunders, GF.: The use of RNA labeled in vitro with ^{125}Iodine in molecular hybridization experiments. Biochim. Biophys. Acta. 287:485, 1972.
7. Section of Biochemistry. Shanghai Cancer Institute: The biological action of exogenous cytoplasmic RNA's of animal liver on cultured human hepatocarcinoma cells and their distribution in Vitro. Acta. Biochim. Biophys. Sin. 9 (3): 237-248, 1977.

8. Swanson, MA.: Glucose-6-phosphatase from liver. In Colowick, SP. and Kaplan, NO. (Eds). Method in Enzymology, Academic press INC. publishers New York Vol. II p.541, 1955
9. Glock, GE. and Mclean, P.: Properties and assay of glucose-6-phosphate dehydrogenase and 6-phosphogluconate dehydrogenase of rat liver. Biochem. J. 55:406-408, 1953
10. Najjar, VA.: Phosphoglucomutase from muscle. In Colowick, SP. and Kaplan, NO. (Eds): Methods in Enzymology, Academic press INC. publishers New York Vol. I p.294, 1955
11. Bruns, FH., Jacob, W. and Weverinck, F.: Phosphohexoisomerase, phosphoribisomerase and lactic acid dehydrogenase in cerebrospinal fluid. Clin. Chim. Acta. 1:63-66, 1956.
12. McGilvery, RM.: Fractose-1,6-diphosphatase from liver. In Colowick, SP. and Eaplan, No. (Eds): Methods in Enzymology. Academic press INC. publishers New York Vol. 2:543, 1955.
13. Yang, SF. and Niu, MC.: Albumin synthesis in mouse uterus in response to liver mRNA. Proc. Natl. Acad. Sci. USA. 74 (5): 1894-1898, 1978.
14. Tung, TC. and Niu, MC.: Nucleic acid-induced transformation in goldfish. Scientia Sinica 3:377, 1973
15. Ku Tsien-jien et al: The preparation and biological characteristics of lipid vesicles and liposomes containing RNA, and the incorporation of the entrapped RNA into cancer cells. Acta. Biochimica et Biophysica. In press. 1979

AN ACIDIC ENDOGENOUS DNA POLYMERASE FROM MOUSE ASCITES TUMOR CELLS

Shu-hsi Hsiao

Institute of Developmental Biology
The Chinese Academy of Sciences

ABSTRACT

An acidic endogenous DNA polymerase was detected in mouse Ehrlich ascites tumor cells. This enzyme was found to exist as a complex with its endogenous template, and no exogenous template was required for enzyme activity. Through DEAE-cellulose chromatography, the enzyme-template complex was resolved into enzyme protein and nucleic acid fractions which showed no enzyme activity separately, but became active when combined. The enzyme-template complex, as well as the hollow enzyme without its endogenous template, showed maximum enzyme activity around pH 6.0 when tested with two buffer systems. The enzyme required all four deoxynucleotide triphosphates for activity, and incorporated either ^{3}H-dTTP or ^{3}H dGTP into acid insoluble products. Manganese ion at low concentrations was required as the divalent cation, while magnesium ion was less effective. The presence of sulfhydryl agent and KCl increased the enzyme activity. The enzyme-template complex was found to be very stable, while the separated hollow enzyme without its endogenous template lost activity rapidly. In addition to its own endogenous template, the hollow enzyme could also use activated calf thymus DNA, native DNA, as well as synthetic poly(dA)-oligo(dT) as template, but could not copy poly(rA)-oligo(dT) or messenger RNA. The enzyme-template complex was found sensitive to N-ethylmaleimide, actinomycin D and detergent treatments. Gel electrophoresis showed that the size of the enzyme-template complex was very large, and the separated endogenous template was quite small.

INTRODUCTION

Three main forms of DNA polymerases (α, β and γ) have been isolated and characterized in eukaryotic cells (1-5). Polymerase α has been isolated from both the cytoplasm and nucleus of cells. It has optimal enzyme activity around pH 7.5, requires Mg ion for activity, and is sensitive to high salt concentrations. Polymerase β is a low molecular weight enzyme isolated from the nucleus of cells. It is an alkaline enzyme with optimal enzyme activity around pH 9.3. It requires either Mn or Mg ion for enzyme activity and is resistant to sulfhydryl group inhibitors, such as N-ethylmaleimide. Polymerase γ has been shown to be identical to the mitochondrial DNA polymerase (6-9). Its optimal pH for enzyme activity is around pH 8.0 and requires Mn ion for activity. It differs from the other two DNA polymerases in its ability to copy the synthetic template, poly(rA)-oligo(dT), at a faster rate than it copies activated DNA, although polymerase β under certain conditions also copies poly(rA)-oligo(dT) (10).

In this paper, we present a preliminary report on an acidic endogenous DNA polymerase isolated from mouse Ehrlich ascites tumor cells. This enzyme differs from other DNA polymerases reported in the literature mainly in three respects: (1) the pH optimum for enzyme activity is different, (2) exogenous template is not required for enzyme activity, and the enzyme was found to exist as a complex with its endogenous template, and (3) the enzyme-template complex appeared to be very stable.

EXPERIMENTAL PROCEDURE

Ehrlich ascites tumor cells were collected from mice 7 days after inoculation and washed with buffer (20mM KPO_4, pH 6.8, 0.14M KCl, 1 mM EDTA). The cells were sonicated in glycerol-phosphate buffer (10% glycerol, 20mM KPO_4, pH 6.8, 2mM mercaptoethanol) and centrifuged at 27,000g for 15 minutes at $4^{o}C$ to remove nuclear and mitochondrial fractions. The supernatant was layered on a discontinuous glycerol gradient consisted of (from top) 5ml of sample, 3.6ml of 15% glycerol, 1.8ml of 30% glycerol, and 1.8ml of 60% glycerol (these gradient solutions were made up in buffers containing 20mM KPO_4, pH 6.8 and 2mM mercaptoethanol). After centrifugation at 100,000g for 16 hours at 4°C, fractions of 0.6ml were collected from the bottom of the tube and assayed. Fractions containing the enzyme-template complex activity were pooled and applied on a DEAE-cellulose column equilibrated with 20mM KPO_4, pH 6.8, 2mM mercaptoethanol, and 10% glycerol. The column was developed with either gradient elution (from 0 to 1M KCl in equilibrating buffer) or stepwise elution (0.3M and 1.0M KCl in starting buffer), and fractions of 4ml were collected and assayed.

The assay mixture consisted of 50mM KPO_4, pH 6.0, 0.5mM $MnCl_2$ or $MnSO_4$, 2mM DTT, 0.1M KCl, 0.5mg/ml BSA, 80-100µM unlabeled dNTP, and ^{3}H-dTTP (47-50 Ci/mmol, 1µCi/assay, 1pmol=50,000cpm) or ^{3}H-dGTP (9.3 Ci/mmol, 1µCi/assay, 1 pmol=9,300 cpm). Enzyme fractions (5-20µl) were added to 100µl of the assay mixture and incubated at 37°C for 15-60 min. Acid insoluble material was precipitated with 10% TCA containing 0.02M sodium pyrophosphate, collected on filters of 0.45µm pore size, washed with 5% TCA and counted in a liquid scintillation counter.

RESULTS AND DISCUSSION

Isolation of the enzyme-template complex

Discontinuous glycerol gradient sedimentation was used to isolate the enzyme-template complex (Fig.1). After centrifugation, fractions were assayed without the addition of template, and the endogenous enzyme activities were located in the 30% glycerol zone, but not in the sample zone. Thus the enzyme appeared to exist as a complex with its endogenous template sedimenting through the 15% glycerol zone and accumulating in the 30% glycerol zone. In this way, a large volume of the sample could be applied in each centrifuge tube resulting in a 4 to 5-fold concentration of the complex from the post-mitochondrial supernatant.

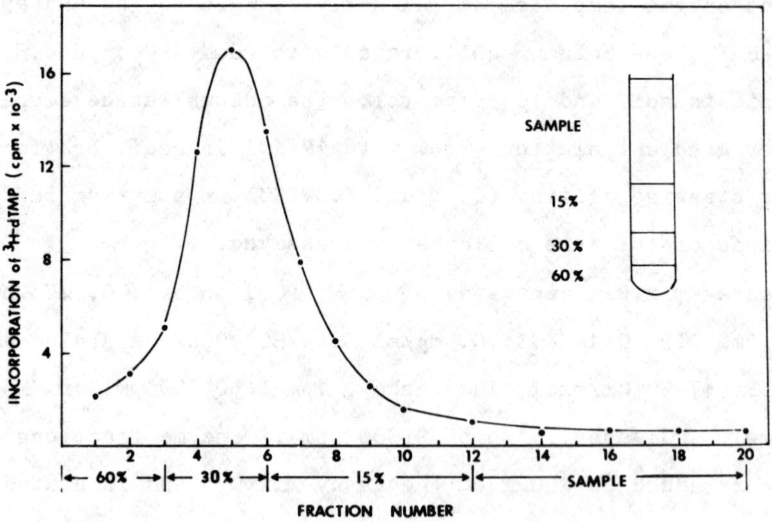

Fig. 1 Isolation of the enzyme-template complex by discontinuous glycerol gradient centrifugation. The composition of the discontinuous gradient is shown in the right. The enzyme activity profile shows the location of the enzyme-template complex at the 30% glycerol zone.

Separation of the endogenous template from the complex

The enzyme-template complex isolated from discontinuous glycerol gradient sedimentation was resolved into the template-free enzyme (the hollow enzyme) and nucleic acid template fractions by DEAE-cellulose chromatography (Fig. 2). At this stage, the hollow enzyme (Fraction A) alone could not incorporate ^{3}H-dTTP or ^{3}H-dGTP into acid insoluble products. Exogenous templates, such as activated DNA, had to be added to detect the enzyme activity and to locate the position of the enzyme peak. The fractions eluted with high salt buffer contained nucleic acids (Fraction B) which were sensitive to DNase treatment but resistant to RNase (data not shown). Neither Fraction A nor Fraction B alone could incorporate ^{3}H-dTTP or ^{3}H-dGTP into acid insoluble products, but when Fraction B was added back to Fraction A, activity of the hollow enzyme could be restored.

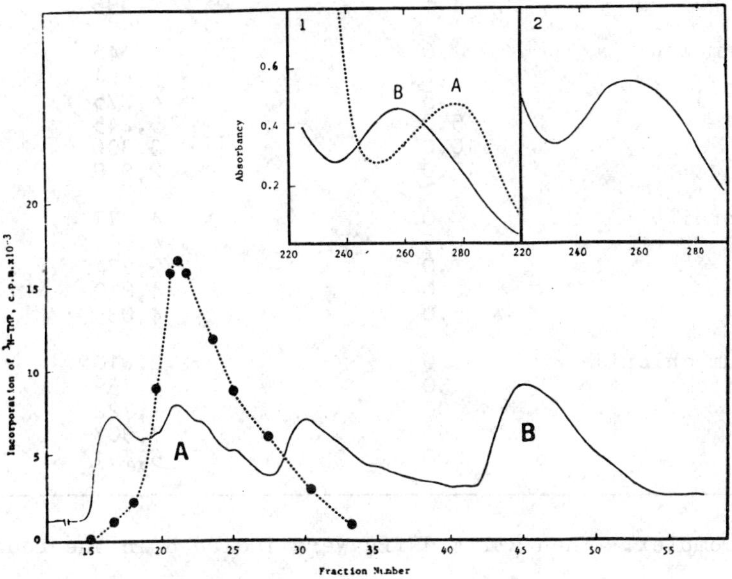

Fig.2 Separation of the complex into enzyme and template fractions by DEAE-cellulose chromatography. The post-mitochondrial supernatant was applied to a DEAE-cellulose column and eluted with increasing concentrations of KCl (0-1 M) in the starting buffer. The solid line shows the 280nm absorption profile of the fractions. The dotted line shows enzyme activities (Fraction A) assayed with activated calf thymus DNA as template.
Insert 1 shows UV absorption spectra of Fractions A and B.
Insert 2 shows the UV absorption spectrum of calf thymus DNA.

Requirements for enzyme activity

Table 1 shows the requirements for the activity of the enzyme-

Table 1. Requirements for Enzyme Activity

Addition	Concentration (mM)	Incorporation of ^{3}H-dTMP (cpm)
Deoxynucleotide triphosphates		
Complete		6,507
-dATP		2,818
-dCTP		2,782
-dGTP		2,960
Manganese ion	0	148
	0.1	6,533
	0.25	8,020
	0.5	11,249
	1.0	8,578
	1.5	7,396
Magnesium ion	0	148
	1.25	1,684
	2.5	4,275
	5.0	5,245
	10.0	3,306
	15.0	2,038
Dithiothreitol	0	4,377
	0.5	4,804
	1.0	5,274
	2.0	4,810
	5.0	4,085
Potassium chloride	0	2,610
	50	5,459
	100	7,722
	150	7,907
	200	6,791

template complex. When all 4 dNTPs were included in the reaction mixture, maximum enzyme activity was obtained. Since either ^{3}H-dTTP or ^{3}H-dGTP could be incorporated into acid insoluble products, the enzyme preparation did not seem to contain terminal transferase activity. The product of the enzyme reaction was found to be sensitive to DNase treatment, but resistant to RNase or dilute alkali treatment, hence the product appeared to be DNA (data not shown).

Divalent cations were found necessary for enzyme activity. In the absence of divalent cations, enzyme activity was very low. Manganese ion at 0.05mM was found much better than magnesium ion in restoring the enzyme activity. Monovalent cations (K^+ or Na^+) were found to have an activating effect on the enzyme activity. The effect of sulfhydryl agents was not very clear, although dithiothreitol was usually included in the assay mixture, while mercaptoethanol was added to the buffer for enzyme isolation.

Effects of templates

As mentioned before, when the crude enzyme extract or the enzyme-template complex was assayed, no exogenous template was required. However, after DEAE-cellulose chromatography the hollow enzyme (Fraction A) without its endogenous template required the addition of exogenous template for enzyme activity. Results of using various DNAs, RNAs, synthetic polynucleotides, as well as Fraction B (the endogenous nucleic acid fraction separated from the enzyme-template complex) as templates for Fraction A are shown in Table 2.

Table 2. Effects of Various Templates

Template	Quantity	Incorporation of ^{3}H-dTMP (cpm)
None		206
Activated DNA	4 µg	15,555
Native DNA	4 µg	18,578
Poly(rA)-oligo(dT)	2 µg	498
Poly(dA)-oligo(dT)	2 µg	12,466
mRNA(rat liver)+oligo dT	4 µg	356
mRNA(carp egg)+oligo dT	4 µg	293
Fraction B	0 µl	335
(concentrated, A_{260}= 10)	1 µl	3,510
	3 µl	8,245
	6 µl	11,910
	10 µl	12,548

It is clear from these results that mRNAs and synthetic poly(rA)-oligo(dT) were not copied by the enzyme in Fraction A, whereas native or activated calf thymus DNA, synthetic poly(dA)-oligo(dT), as well as Fraction B could all be used as templates for the enzyme.

Optimal pH for enzyme activity

The pH curve for this enzyme was quite unusual (Fig. 3). It showed maximum enzyme activity peak around pH 6.0, and at pH 7-9 where other DNA polymerases were generally assayed, the activity of this enzyme was very low. The same characteristic pH curves were obtained not only with the crude extract, but also with the isolated enzyme-template complex, and the recombined Fractions A and B from DEAE-cellulose chromatography. Moreover, two buffer systems were employed to determine the pH curves. In addition to phosphate buffer, sodium maleate buffer were also used, and similar pH curves for the enzyme were obtained.

In order to compare the pH curves of this enzyme with those of other DNA polymerases, the activities of the following enzymes were determined in the same buffer systems. These were E. coli DNA polymerase, sea urchin DNA polymerase α, lymphocyte DNA polymerase β, HeLa cell DNA polymerase γ, and AMV reverse transcriptase (these polymerases were kindly provided by Dr. L. Loeb of the Institute for Cancer Research, Fox Chase, Philadelphia). Results indicated that polymerase β did not show any enzyme activity below pH 7. The other DNA polymerases showed very low enzyme activities around pH 6, and their activities increased as the pH values of the buffer became higher (unpublished data). None of these DNA polymerases showed a maximum enzyme activity peak below pH 7, whereas the endogenous DNA polymerase from ascites tumor cells seemed unique, showing its maximum enzyme activity peak around pH 6.0.

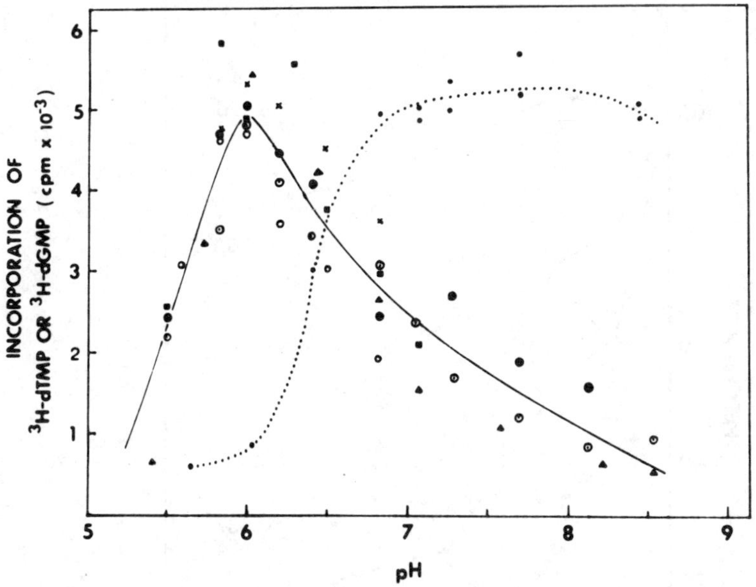

Fig.3 Optimal pH curves. The solid line represents the endogenous enzyme activities of the crude extract and the isolated enzyme-template complex in assay mixtures containing Mn ion and phosphate buffer. The dotted line shows the activity of the same crude extract assayed in Tris buffer and in the presence of Mg ion and exogenous template. Those experimental points on the figure represent the activities of different enzyme fractions obtained during various steps of isolation (including the hollow enzyme without its own template, Fraction A), and assayed in either phosphate or maleate buffer systems.

Stability of the enzyme-template complex

The crude extract and the isolated enzyme-template complex were found to be very stable. The crude extract, or the post-mitochondrial supernatant, could be stored in sterile 20% glycerol phosphate buffer at -20°C for more than six months without losing any enzyme activity. Even at refrigerator temperature, the crude extract was stable for 2-3 months. After that, the enzyme activity gradually decreased but was still detectable for another 2 months.

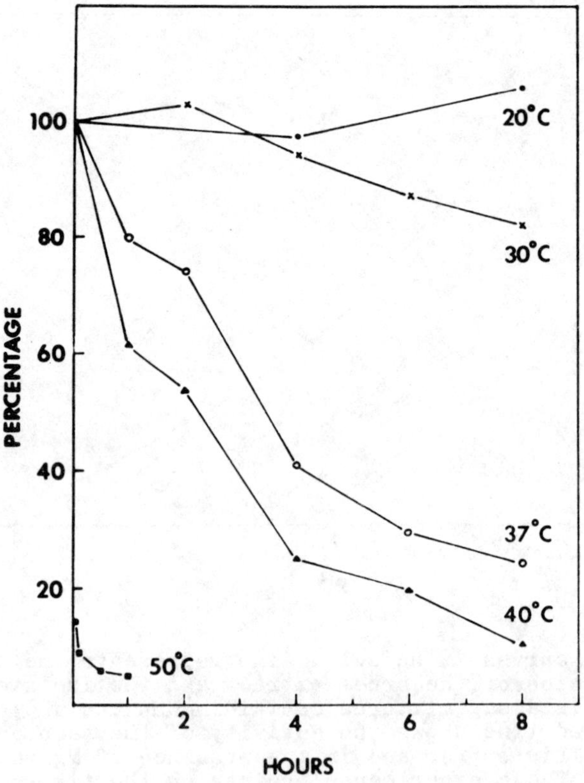

Fig.4 Stability of the enzyme-template complex at different temperatures.

When the crude extract or the isolated enzyme-template complex was kept at different temperatures and aliquots removed at various time intervals for enzyme assay, the template protected enzyme was again shown to be stable at room temperature (Fig. 4). After standing for 8 hours at 20°C, the enzyme did not show appreciable loss of activity. After 8 hours at 30°C, 80% of the enzyme activity could still be detected. Based on these results, the enzyme-template complex could be isolated at room temperature without losing much of its activity. At higher temperatures, however, the enzyme lost its activity rapidly. For exemple, 10 minutes at 50°C, 90% of the enzyme activity were lost. In contrast, the hollow enzyme without its template (Fraction A) was less stable and lost its activity rapidly at 4-5°C.

Effects of various agents

The effects of various agents on the activity of the enzyme-template complex are shown in Table 3. The sulfhydryl inhibiting agent, N-ethylmaleimide, showed little effect at low concentrations, but at 10mM, about 75% of the enzyme activity was inhibited. Actinomycin D, as expected, inhibited the enzyme activity. The

Table 3. Effects of Various Agents

Agents	Concentration	Incorporation of ^{3}H-dTMP (cpm)	%
N-ethylmaleimide	0 mM	6,951	100
	0.5 mM	8,074	116
	1.0 mM	6,813	98
	5.0 mM	5,990	86
	10.0 mM	1,679	24
Actinomycin D	0 µg/ml	4,529	100
	10 µg/ml	2,652	59
	25 µg/ml	1,945	43
	50 µg/ml	1,455	32
	100 µg/ml	1,041	23
Triton X-100	0 %	4,279	100
	0.01 %	2,235	52
	0.025 %	1,369	32
	0.05 %	1,176	28
	0.10 %	946	22
Na dodecylsulfate	0 %	5,175	100
	0.002 %	2,385	46
	0.005 %	799	16
	0.01 %	631	12
	0.02 %	453	9

enzyme was found very sensitive to detergent treatment even at low concentrations. Triton X-100 at 0.01% and sodium dodecylsulfate at 0.002% reduced the enzyme activity to 50%. These results differ from many viral DNA polymerases which require detergent treatment to release the enzyme from the virion.

Polyacrylamide gel electrophoresis

The size and purity of the enzyme-template complex and of Fractions A and B were studied with polyacrylamide gel electrophoresis (Fig. 5). The post-mitochondrial supernatant showed many bands, whereas the enzyme-template complex showed bands only on the upper half of the gel, indicating that smaller protein molecules did not enter the 30% glycerol zone and remained either in the soluble fraction or in the 15% glycerol zone.

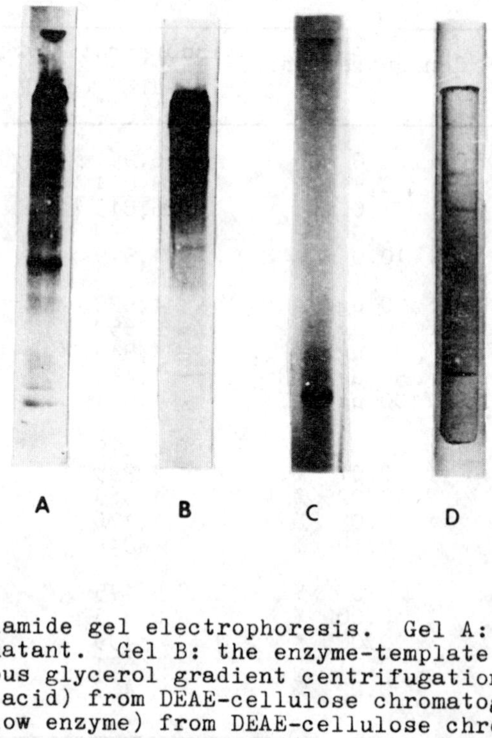

Fig.5 Polyacrylamide gel electrophoresis. Gel A: the post-mitochondrial supernatant. Gel B: the enzyme-template complex isolated from discontinuous glycerol gradient centrifugation. Gel C: Fraction B (nucleic acid) from DEAE-cellulose chromatography. Gel D: Fraction A (hollow enzyme) from DEAE-cellulose chromatography.

The concentrated Fraction B was detected as a major band migrating to the anode, corresponding to the position of the 4-5S

RNA of rat liver in a parallel run. Sucrose density sedimentation of the concentrated Fraction B also showed a single 260nm absorption peak sedimenting around 10% sucrose density similar to the position of the 5S RNA marker in a parallel run (data not shown). The concentrated Fraction A still showed a few bands, indicating that at this stage the enzyme was only partially purified. Glycerol gradient centrifugation showed that the concentrated Fraction A sedimented at a position containing 25% glycerol (unpublished data). Thus the size of this hollow enzyme appeared still quite large.

From these experimental results, the endogenous DNA polymerase from mouse Ehrlich ascites tumor cells appeared to be different from the known DNA polymerases in that the former differs mainly in its low pH optimum for enzyme activity and the presence of endogenous template. In addition to these two points, polymerase α has been found sensitive to 50mM KPO_4 and high salt concentrations (5), whereas the ascites tumor enzyme is activated by them; polymerase β is a low molecular enzyme and is resistant to sulfhydryl inhibitors, whereas the latter enzyme exists as a large molecular enzyme-template complex and is sensitive to N-ethylmaleimide; polymerase γ can copy synthetic polyribonucleotides, whereas the latter enzyme can only copy natural and synthetic poly-deoxyribonucleotides (5). Moreover, the ascites tumor enzyme also differs from viral reverse transcriptase and the avian particle enzyme (11-13) which copy both natural RNA and synthetic polyribonucleotides and require detergent treatment to release the enzyme from the virion. To further distinguish this enzyme from other DNA polymerases reported in the literature, it would be necessary to compare the effects of specific DNA polymerase inhibitors such as aphidicolin, dideoxy-TTP, ara-CTP (14-16), etc. on these enzymes.

ACKNOWLEDGEMENT

The author wishes to thank Dr. M.C. Niu of Temple University for supporting the initial phase of the work carried out in his laboratory in 1975, and to thank Mr. Chao Wei for his expert assistance for the part of work carried out in China since 1979.

REFERENCES

(1) Weissbach, A., Baltimore, D., Bollum, E., and Gallo, R. (1975) Science 190 401-402.

(2) Weissbach,A. (1977) Annu. Rev. Biochem. 46 25-47.

(3) Sarngadharan, M.G., Robert-Guroff, M., and Gallo, R.C. (1978) Biochim. Biophys. Acta 516 419-487.

(4) Sheinen,R., and Humbert, J. (1978) Annu. Rev. Biochem. 47 277-316.

(5) Knopf, K.W., Yamada, M., and Weissbach, A. (1976) Biochem. 15 4540-4548.

(6) Hubscher, U., Kuenzle, C.C., and Spadari, S. (1977) Eur. J. Biochem. 81 249-258.

(7) Bolden, A., Pedrali-Noy, G., and Weissbach, A. (1977) J. Biol. Chem. 252 3351-3356.

(8) Bertazzoni, U., Scovassi, A.I., and Brun, G.M. (1977) Eur. J. Biochem. 81 237-248.

(9) Fujisawa, T., Tanaka, S., Kobayashi, M.,and Koike, K. (1977) Biochim. Biophys. Acta 475 611-622.

(10) Weissbach, A. (1975) Cell 5 101-108.

(11) Bauer, G., and Hofschneider, P.H. (1976) Proc. Natl. Acad. Sci. USA 73 3025-3029.

(12) Bauer, G., Friis, R.R., Jilek, G., and Hofschneider, P.H. (1978) Biochim. Biophys. Acta 518 125-137.

(13) Bauer, G., and Temin, H.M. (1979) J. Virol. 29 1006-1013.

(14) Ohashi, M., Taguchi, T., and Ikegami, S. (1978) Biochem. Biophys. Res. Commun. 82 1084-1090.

(15) Wist, E., and Prydz, H. (1979) Nucleic Acid Res. 6 (4), 1583-1590.

(16) Wist, E. (1979) Biochim. Biophys. Acta 562 62-69.

THE MOLECULAR BIOLOGY OF VESICULAR STOMATITIS VIRUS AND ITS DEFECTIVE INTERFERING PARTICLES

Alice S. Huang, Ph.D.

Department of Microbiology and Molecular Genetics
Harvard Medical School
and
Division of Infectious Diseases
Children's Hospital Medical Center
300 Longwood Avenue
Boston, Massachusetts 02115

Vesicular stomatitis virus (VSV) is a large, enveloped animal virus (see Wagner, 1975). Its genome consists of a single strand of RNA, completely covalently linked and made up of approximately 12 kilobases. Most of this RNA contains coding sequences for 5 proteins, L, G, N, NS, and M (Rose et al., 1975; Freeman et al., 1977). The non-transcribed regions consist of 2 bases found between each intergenic sequence and about 50 bases at the termini of the genome (McGeoch, 1979; Semler et al., 1979; Keene et al., 1980; McGeoch et al., 1980). These sequences at the termini are believed to control transcription and replication of the genome as well as to provide nucleation sites for encapsidation. At the beginning of each coding region the sequences are 3' UUGUC--UAG 5' (Rose, 1975). At the end of each coding region there is the sequence 3' AUAC 5' followed by U_7 (J.K. Rose, unpublished observations). How these sequences control transcription is at present unknown, but it is likely that the polymerase recognizes transcriptional stop and start sequences in the synthesis of multiple complementary monocistronic mRNA.

All five of the proteins coded for by the viral RNA genome are found as structural proteins in the virion. The genome is completely covered by the RNA-binding N protein (49,000 daltons) which protects the RNA from digestion by ribonuclease. Associated with this nucleocapsid structure is an RNA-dependent RNA polymerase, the L protein (190,000 daltons) which binds to the nucleocapsid via a phosphorylated ligand, the NS protein (31,700 daltons). The RNA genome and these 3 proteins consitute the core of the virus particle. This core is further surrounded by a matrix (M) protein (23,000 daltons) and then a lipid bilayer which is associated with a transmembrane glycoprotein (G protein, 67,000 daltons). This G protein is the surface antigen specific to VSV and determines the host range as well as the serotype of the virus.

The growth cycle of VSV begins by attachment to and penetration into susceptible cells (see Wagner, 1975). This virus has an extraordinarily large host range. Once inside the cell, the virus uncoats to the core stage. This consists of the RNA genome protected by N protein in association with the polymerase (L) protein and NS protein. Transcription is initiated with the synthesis of all 5 complementary mRNAs. These, like other eucaryotic mRNA, are capped at the 5' end and contain poly(A) at the 3' end. Protein synthesis then proceeds. Replication of the genome is dependent on protein synthesis. The large genomic 40 S RNA is copied into a completely complementary RNA which is also protected by N protein and acts as template for the synthesis of progeny minus strand RNA. These progeny strands can participate as templates for transcription and replication or they associate with other VSV structural proteins to form progeny virions. All the VSV synthetic steps occur in the cytoplasm. Virions are formed by budding out of the plasma membrane.

To the molecular biologist, VSV offers serveral interesting aspects for study. (1) The RNA genome cannot function as a template without the RNA-binding N protein. In this way it acts as chromatin. The interaction of the N protein with RNA may help to elucidate how proteins regulate gene expression. (2) The polymerase poses several interesting questions regarding recognition, promotion of transcription and the details of synthesis of individual mRNA. (3) Both of the phosphorylated proteins, NS and M, interact with the polymerase and in some way regulate transcription versus replication of the genome. The M protein is also a membrane-associated, matrix protein which determines the bullet-like morphology of VSV and interacts with the G protein, presumably to help anchor it in the lipid bilayer. (4) The viral glycoprotein G, is readily labeled and its synthesis can be followed from various intracellular membranes until it reaches the cell surface and either forms part of budding progeny virions or is shed as a soluble antigen from the surface of the cell.

Our laboratory has been concerned with VSV RNA synthesis and the role of defective interfering (DI) particles in attenuating the cytopathic effect of VSV of the host (Huang, 1973; Huang and Baltimore, 1977; Huang, 1977).

Interference with the growth of standard VSV by DI particles occurs primarily during the step involving genome replication. To understand how this competition occurs we have characterized the genomes of standard virus and DI particles by oligonucleotide mapping and sequence analysis (Huang, 1977; Huang et al., 1980). Although specific sequences are recognized by the polymerase, competition between 2 different DI particles suggests that the extent of self-complementary sequences at the termini of the genomes determines which genome is preferentially replicated (Huang et al., 1980).

VSV RNA species are divided into two groups, those that are

synthesized by standard VSV and those that are specific to DI particles (Stampfer et al., 1969; Huang, 1975). RNA specific to standard virus are the 5 mRNA coding for the 5 viral proteins as well as the negative strand genomic RNA (40-42S) and its complement. Leader or small RNAs which 46-48 nucleotides long and contain sequences that are identical to the 5' termini of the genome RNA and its complement are also found in minute quantities during RNA synthesis directed by standard virus (Leppert et al., 1979). RNA specific to DI particles are limited to the minus strand genome RNA of the DI particle, its complement and a small RNA, the sequence of which is represented at the 5' termini of both DI genomic RNA and its complement (Rao and Huang, 1979).

When cells are co-infected with DI particles and standard VSV, transcription is reduced to that carried out by input standard virus. Replication of the genome of standard VSV is inhibited and the majority of the RNA species synthesized are specific to DI particles (Huang and Manders, 1972; Perrault and Holland, 1972). It is possible to infect cells with 2 different DI particles together with the helper standard virus and show that during a one cycle growth, one of the DI particles can virtually outcompete both the standard virus and the other DI particle. This ability to establish a genome as the one to be replicated preferentially in the presence of other RNA genomes is determined within 2 hr after infection of cells. Once decided, the advantage conferred onto an RNA molecule appears to be irreversible (Huang and Wagner, 1966; Rao, unpublished observations).

To understand the molecular basis of this competition we have isolated 2 DI particles, each independently derived from the same parent (Huang et al., 198Ø). DI-T particles have genomes one-third the length of standard VSV, whereas, DI Ø.52 has about half the genome of standard VSV. Both particles fail to code for any VSV specific proteins and they each compete effectively against standard VSV, resulting in complete inhibition of the replication of the standard genome. When cells are triply infected, DI Ø.52 particles wins over both standard and DI-T genomes.

Oligonucleotide mapping the RNAs of DI Ø.52 and DI-T particles indicates that they are derived from the region of the L gene at the 5' end of the VSV genome (Huang et al., 198Ø). Extragenomic oligonucleotides derived from the RNAs of these DI particles suggest that they are unlikely to be simple deletion mutants. Sequence analysis of these extragenomic oligonucleotides shows that single base changes exist as well as the acquisition of complementary sequences (F.S. Hagen, unpublished observations). DI Ø.52 particles contain about 6Ø nucleotides at the 3' and 5' termini which are complementary. These sequences can be annealed, digested with ribonuclease and isolated as small duplex molecules (D.D. Rao, unpublished observations). DI-T particles contain about 5Ø nucleotides at the termini of its genome which are complementary. In contrast, standard virus genomes contain 18 nucleotides at the termini which are mostly complementary with two base mismatches (Keene et al., 1979; Freeman, unpublished observations). Because the first 46 bases at the termini of the RNAs from DI-T and DI Ø.52 particles appear to be identical by sequencing, the extra length of complementarity in DI Ø.52 RNA must confer an advantage in replication over the RNA of DI-T particles.

How could such complementarity contribute to an advantage during RNA replication? Little is known about the molecular mechanisms involved during VSV RNA replication. Replication appears to be semi-conservative, where the genome RNA is copied into a complete complementary piece which in turn acts as template for the synthesis of more genomic RNA. Transcription, in distinction to replication, copies the genomic RNA into multiple smaller complementary mRNA species. Therefore, the earliest decision concerning replication occurs during the sysnthesis of the large complementary strand.

It is possible that the degree of intramolecular self-complementarity becomes the decisive factor for the synthesis of the first large complementary RNA. The ability of the genome to circularize in a stable form may mark it out as a template for replication as opposed to transcription. Along these lines, examination of nucleocapsid template structures in the electron microscope has shown that some of those implicated in replication are in a circular form (Naeve et al., 198Ø).

ACKNOWLEDGMENTS

This work is supported by research grants from the U.S. Public Health Service (National Institute of Allergy and Infectious Diseases) and from the American Cancer Society. Dr. Huang is Professor of Microbiology and Molecular Genetics, Harvard Medical School, and Director, Laboratories of Infectious Diseases, Children's Hospital Medical Center, Boston, MA, USA.

REFERENCES

Freeman, G.J., Rose, J.K., Clinton, G.M., and Huang, A.S. 1977. RNA synthesis of vesicular stomatitis virus. VII. Complete separation of the mRNA's of vesicular stomatitis virus by duplex formation. J. Virol. 21: 1094-1104.

Huang, A.S. 1973. Defective interfering viruses. Ann. Rev. Microbiol. 27: 101-117.

Huang, A.S. 1975. Ribonucleic acid synthesis of vesicular stomatitis virus. In NEGATIVE STRAND VIRUSES, eds. R.D. Barry and B.W.J. Mahy. Academic Press, N.Y., pp. 353-360.

Huang, A.S. 1977. Viral pathogenesis and molecular biology. Bacteriol. Rev. 41: 811-821.

Huang, A.S., and Baltimore, D. 1977. Defective interfering animal viruses. In COMPREHENSIVE VIROLOGY, Vol. 10, eds. H. Fraenkel-Conrat and R.R. Wagner, Plenum Publishing Corp., N.Y., pp. 73-116.

Huang, A.S., and Manders, E.K. 1972. Ribonucleic acid synthesis of vesicular stomatitis virus. IV. Transcription by standard virus in the presence of defective interfering particles. J. Virol. 9: 909-916.

Huang, A.S., Rao, D.D., and Lanman, G. 1980. Defective interfering particles of vesicular stomatitis virus: Structure-function relationships. Ann. N.Y. Acad. Sci. (in press).

Keene, J.D., Schubert, M., and Lazzarini, R.A. 1979. Terminal sequences of vesicular stomatitis virus RNA are both complementary and conserved. J. Virol. 32: 167-174.

Keene, J.D., Schubert, M., and Lazzarini, R.A. 1980. The intervening sequence between the leader region and the nucleocapsid cistron of vesicular stomatitis virus RNA. J. Virol. (in press).

Leppert, M., Rittenhouse, L., Perrault, J., Summers, D.F., and Kolakofsky, D., 1979. Plus and minus strand leader RNAs in negative-strand virus-infected cells. Cell 18: 735-748.

McGeoch, D. 1979. Structure of the gene N:gene NS intercistronic junction in the genome of vesciular stomatitis virus. Cell 17: 673-681.

McGeoch, D., Dolan, A., and Pringle, C.R. 1980. Comparisons of nucleotide sequences in the genomes of the New Jersey and Indiana serotypes of vesicular stomatitis virus. J. Virol. 33: 69-77.

Naeve, C.W., Kolakofsky, C.M., and Summers, D.F. 1980. Comparison of vesicular stomatitis virus intracellular and virion ribonucleoptroteins. J. Virol. 33: 856-865.

Perrault, J., and Holland, J.J. 1972. Absence of transcriptase activity and transcription-inhibiting ability in defective interfering particles of vesicular stomatitis virus. Virology 50: 159-170.

Rao, D.D., and Huang, A.S. 1979. Synthesis of a small RNA in cells coinfected by standard and defective interfering particles of vesicular stomatitis virus. Proc. Nat. Acad. Sci. USA 76: 3742-3745.

Rose, J.K. 1975. Heterogeneous 5'-terminal structures occur on vesicular stomatitis virus mRNAs. J. Biol. Chem. 250: 8098-8101.

Rose, J.K., and Knipe, D. 1975. Nucleotide sequence complexities, molecular weights and poly(A) content of the vesicular stomatitis virus mRNA species. J. Virol. 15: 994-1003.

Semler, B.L., Perrault, J., and Holland, J.J. 1979. The nucleotide sequence of the 5' terminus of vesicular stomatitis virus RNA. Nucleic Acids Res. 6: 3923-3930.

Stampfer, M., Baltimore, D., and Huang, A.S. 1969. Ribonucleic acid synthesis of vesicular stomatitis virus I. Species of ribonucleic acid found in Chinese hamster ovary cells infected with plaque-forming and defective particles. J. Virol. 4: 154-161.

Wagner, R.R. 1975. Reproduction of Rhabdoviruses. In COMPREHENSIVE VIROLOGY, Vol. 4, eds. H. Frankel-Conrat and R.R. Wagner, Plenum Press, N.Y., pp. 1-94.

STUDIES ON ANTICANCER EFFECT OF RNA

Huang Hua-zhang

Experimental Cancer Research Laboratory, Tianjin Medical College, Tianjin, China

Yang Qi-lun

Department of Biology, Nan-Kai University, Tianjin, China

Yu Jian-kang

Institute of Developmental Biology, Academia Sinica, Beijing, China

Li Wei-lian

Oncology Department, the Second Central Hospital, Tianjin, China

Zhang Zhi-ping

Immunological Laboratory, Tianjin People's Hospital, Tianjin, China

Wang Hui-yuan

Department of Surgery, The First Central Hospital, Tianjin, China

Abstract We report that RNA-rich preparations, obtained by cold phenol method from liver of rabbits and rams immunized with breast cancer of TA-2 mice, showed an inhibitory effect on the early growth of the experimental tumor.

Further more, the preliminary results in short-period-improvement of immunotherapy of human cancer patients with RNA-rich extracts from ram livers immunized with the same histological type thuman cancer tissue is also obtained.

In order to enhance the efficacy of RNA therapy, instead of using syngeneic spleen cells preincubated with immune RNA *in vitro*, the skin bleb fluid incubation method (SBFI method an *in vivo* incubation system for the transfer of nucleic acid information and/or stimuli) is used to treat cancer patients. The results of our initial clinical therapeutic trials of RNA rich preparations by SBFI method on a total of 65 cancer patients with grossly detectable and measurable metastatic diseases showed 30 cases improved, 11 stable, and 24 failures, according to Pilch's criteria.

The present investigation shows that RNA-rich extracts prepared by cold phenol method from liver of rabbits and rams immunized with breast cancer of TA-2 mice have an inhibitory effect on the early growth of the tumor. (TA-2 mice is an inbred strain raised over 40 generations in our laboratory, according to Jackson Memorial Laboratory, Bar Harbor, Maine, USA. The average survival time of the tumor bearing mice was about 3 weeks.)

The RNA-rich extracts from liver have a typical UV absorption ratio $A_{260nm}/A_{280nm} \geqslant 2$, $A_{260nm}/A_{230nm} \geqslant 2$. Protein contamination is less than 2% in general (by Lowry's method[1]. The poly-A containing RNA constitutes about 1-2% of the total RNA when fractionated on Sigmacell 38 columes by the method of Schutz[2].

As seen in table 1, only immunized rabbit liver RNA causes a retarded growth at early stage, showing a 64.5% inhibition in average tumor weight.

Table 2 shows a similar result.

Table 3 shows the inhibition on the growth of tumor isografts.

Table 4 gives the same results on 5-day tumors (palpable tumor) by immunized sheep liver RNA-rich extrats and a lower inhibition rate (34.5%) by a Schlager[3] type treatment.

In table 5, the inhibitory effect is reduced by RNase.

After the above mentioned favorable results of animal experiments, human cancer tissue was used to immunize rams from which liver RNA-rich extracts were prepared for clinical therapeutic

Table 1. Increased Resistance of TA-2 Mice to a Transplantable Breast Cancer by Repeated Injections of RNA-Rich Extracts

Group	Mode of Tumor Administration	Injections (day 1,3,5, 7 after challenge)	Number of Tumor Takes/Number of Tumor Challenged Mice Day after challenge								Average Tumor Wt. (gm)	% Inhibition	P Value
			9	10	11	12	13	14	15	16			
Control	$5x10^5$ tumor cells (S.C.)	4x0.5ml Saline (S.C.)	5/8	5/8	6/8	6/8	6/8	7/8	7/8	7/8	0.567		
Freund's Rabbit Liver RNA	$5x10^5$ tumor cells (S.C.)	4x1mg RNA (S.C.)	4/5	4/5	5/5	5/5	5/5	5/5	5/5	5/5	0.870	(-)	$P > 0.1$
Immunized Rabbit Liver RNA	$5x10^5$ tumor cells (S.C.)	4x1mg RNA (S.C.)	0/6	0/6	1/6	2/6	2/6	2/6	4/6	4/6	0.169	64.5%	$P < 0.05$

Table 2. Increased Resistance of TA-2 Mice to Transplantable Breast Cancer by Repeated Injections of RNA-Rich Extracts

Group	Mode of Tumor Administration	Injections (day 1,3,5,7 after challenge	Number of Tumor Takes/number of Tumor challenged Mice Day after challenge					Average Tumor Weight (gm)	% Inhibition	P Value
			13	14	15	16	17			
Control	$2.5x10^5$ tumor cells (s.C.)	4x0.5ml saline (S.C.)	3/4	3/4	4/4	4/4	4/4	0.505		
Normal Rabbit Liver RNA	$2.5x10^5$ tumor cells (S.C.)	4x1mg RNA (S.C.)	0/5	2/5	4/5	5/5	5/5	0.318	37.4%	P > 0.1
Immunized Rabbit Liver RNA	$2.5x10^5$ tumor cells (S.C.)	4x1mg RNA (S.C.)	3/5	3/5	3/5	4/5	5/5	0.260	48.5%	P < 0.02

Table 3. The Inhibitory Effect on Transplantable Breast Cancer of TA-2 Mice at Early Growth by Immunized Rabbit Liver RNA-Rich Extracts

Exp.No.	Group	Number of Mice	Mode of Tumor Administration	Injections (Day 1,3,5,7 after Isograft)	Average Tumor Wt. (gm.)	% Inhibition	P Value
III	Control	9	fragments (S.C.)	4x0.5ml Saline (S.C.)	0.718		
	Immunized Rabbit Liver RNA	15	fragments (S.C.)	4x1mg RNA (S.C.)	0.157	78.1%	$P < 0.02$
IV	Control	10	fragments (S.C.)	4x0.5ml Saline (S.C.)	1.05		
	Immunized Rabbit Liver RNA	11	fragments (S.C.)	4x1mg RNA (S.C.)	0.32	69.5%	$P < 0.01$

Table 4. The Inhibitory Effect of Immunized Sheep Liver RNA-Rich Extracts on the Early Growth of Transplantable Breast Cancer of TA-2 Mice

Group	Number of Mice	Mode of Tumor Administration	Injections	Average Tumor Volume* CM^3 (% Inhibition) Day after Isograft			
				12	15	18	22
Control	9	fragments (S.C.)	6x0.5ml saline (Day 5,7,9,11, 13,15 after isograft.)	0.278	0.927	2.031	2.791
Immunized sheep Liver RNA	8	fragments (S.C.)	6x1mg RNA (Day 5,7,9,11, 13,15 after isograft)	0.223 (20.1%)	0.635 (31.5%)	0.992 (51.2%)	1.130 (59.7% $P < 0.01$)
Immunized Sheep Liver RNA (Schlager treatment)	10	fragments (S.C.)	$5.5x10^6$ PEC 5mg RNA 1mg Antigen (Day 5 after isograft)	0.218 (21.66%)	0.700 (24.5%)	1.35 (33.5%)	1.830' (34.5% $P < 0.05$)

* Tumor volume by $V = \frac{4}{3}\pi r^3 \quad = 4/3\ \pi r_1 . r_2 . r_3 \quad = \pi/6\ D_1 . D_2 . D_3$

Table 5. The Inhibitory Effect of Immunized Sheep Liver RNA-Rich Extracts on the Early Growth of Transplantable Breast Cancer of TA-2 Mice

Group	Number of Mice	Mode of Tumor Administration	Injections (Day 6,8,10,12, 14 after isograft)	Average Tumor Volume* CM^3 Day after isograft 10	14	16
Control	9	fragments (S.C.)	5x0.2ml saline (S.C.)	0.1823	0.5426	0.8739
Immunized Sheep Liver RNA	7	fragments (S.C.)	5x1mg RNA (S.C.)	0.1060	0.3562	0.4951 (43.35% inhibition $P < 0.01$)
Immunized Sheep Liver RNA+RNase	7	fragments (S.C.)	5x1mg RNA (S.C.)	0.0973	0.5704	0.7732
Immunized Sheep Liver RNA+RNase	7	fragments (S.C.)	1mg RNA i.v. (Day 5 after isograft)	0.1026	0.4665	0.7057

* Tumor volume by $V = \frac{\pi}{6} D_1 D_2 D_3$

studies of histologically similar cancer patients. In order to enhance the efficacy of RNA therapy, instead of using syngeneic spleen cells preincubated in vitro with immune RNA, Chinese cantharis skin bleb is used to take the place of spleen cells in cancer patients. This is the so called skin bleb fluid incubation method—an in vivo incubation system for the transfer of RNA information. The procedure is as follows:

A. The production of the skin bleb (method of Tsin-Sen Yao and You-Huai Chang of the Cancer research Institute of Chinese Academy of Medical Science, Beijing[4]).
   1. Chinese cantharis tincture was prepared by soaking 10 grams of Chinese Cantharis (Cicindela chinensis) in 100 ml of 95% ethyl alcohol for 15 days.
   2. Two layers of filter paper (2 x 2 cm) soaked with cantharis tincture were laid on the inner side of the forearm, covered with a piece of 2 x 2 x 0.2 cm glass, and fixed with adhesive tape for 4 hours.
   3. The paper and glass were removed and the area was covered with a protecting frame to allow the bleb to form.
   4. After 24 hours, a bleb of about 2 cm in diameter was produced, containing 1-4 ml of bleb fluid (Fig. 1).

B. Incubation with RNA. (Skin bleb incubation method — an in vivo incubation system)
   1. The RNA (1-2 mg) was dissolved in 0.5 ml of water for injection.
   2. The dissolved RNA was injected from the base of the bleb with the needle turned up to reach the inside of the bleb (Fig. 2). After injection, the bleb was again covered with the frame for 24 hours (or 1 hour for urgent need).

C. Therapeutic injection
   1. After 1-24 hours incubation the bleb fluid was collected with a syringe for injection.
   2. Route of injection:- (1) intratumoral and paratumoral, (2) intravenous, (3) subcutaneous, (4) intra- and paralymph node, and (5) intralymphatic. Different ways of injection could be used singly or in combination.

The above mentioned mode of administration was named skin-bleb-incubated-injection-method which differs from the currently

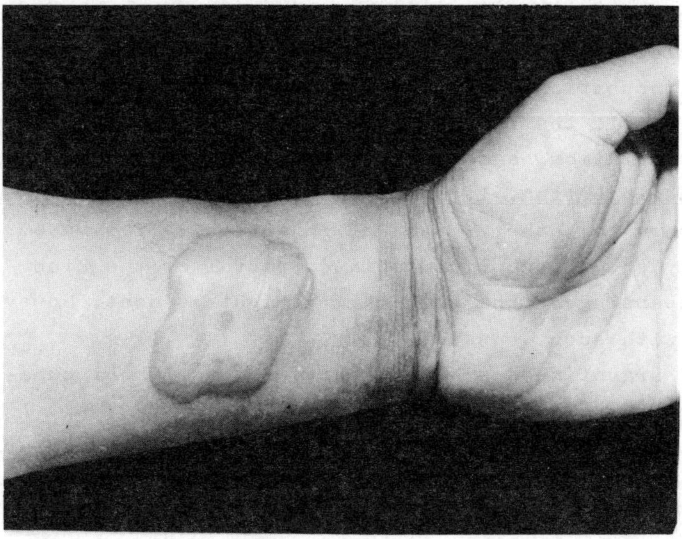

Fig.1 A 24 hours skin bleb on forearm
produced by Chinese cantharis tincture

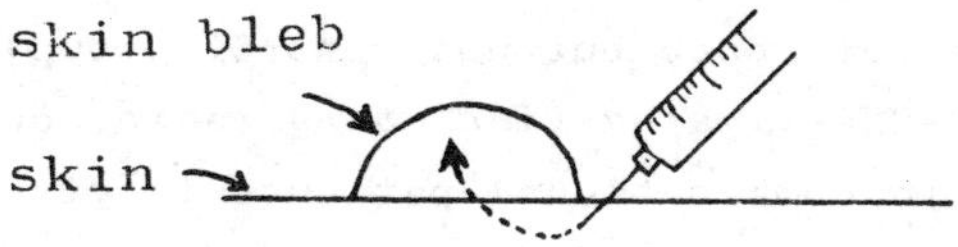

Fig.2 The injection of RNA solution into
the skin bleb.

used subcutaneous injection method.

The rate of RNA uptake is not known. The opinion of Jachertz was in milliseconds[5], and others used 15-30 minutes or longer to obtain satisfactory results[5]. For the purpose of safety, we lengthened the incubation period to 1-24 hours to allow the patient enough time to handle the foreign material— to transfer and amplify the RNA information, and have enough time to observe the safety after the injection. If safety is ensured, then we can turn to intravenous or other routes of injection. (If symptom of hypersensitivity should occur, immediate removal of the bleb content, and cleansing of the bleb cavity might be done. However, we haven't met with such cases.)

In case of urgent need or during hot weather, the incubation time can be shortened.

In unexpected situations, fluid from blebs of short duration was directly mixed with RNA in the ampoule and used for subcutaneous injection, but intravenous injection was not allowed, for fear of the presence of trace of foreign protein.

Skin bleb incubated injection was usually given to a patient every 2 weeks to one month but can be used weekly when necessary.

Therapeutic trials on advanced and earlier cancer patients was carried out, and therapeutic effect was estimated by tumor remaining stable or reduced in size in a short period of time (according to Pilch's standard[6]) or even occassionally complete regression of the tumor.

Table 6 summarizes the results of our initial clinical trials of RNA therapy on cancer patients with grossly detectable and measurable metastatic diseases according to Pilch's criteria.

30 patients were considered as improved, 11 as stable, 24 as failure in a total of 65 patients.

The comparison (as reference but not control group) between the RNA-treated and non-RNA-treated (including radio, chemo, Chinese drug therapy singly or in combinations) patients is given in table 7.

The RNA-treated patients (16 cases of advanced adenocarcinoma of stomach and 11 cases of advanced hepatoma among a total of 65 cancer patients) were from the Oncology Department of Second Central Hospital, Tianjin, China.

Table 6. Initial Clinical Trials of RNA Therapy
Summary of Clinical Results: Gross Disease

Tumor Type	Improved	Stable	Failure	Total
Adenocarcinoma of Stomach	8	4	11	23
Malignant Lymphosarcoma	6	4	5	15
Hepatoma	6	1	4	11
Malignant Melanoma	3	0	2	5
Kidney Cancer	5	1	1	7
Breast Carcinoma	2	1	1	4
Total	30	11	24	65

Table 7. The comparison of the Effect of RNA and Non-RNA Therapy of Advanced Adenocarcinoma of stomach and Hepatoma Patients

Tumor Type	Treatment	Improved	Stable	Failure	Total
Adenocarcinoma of stomach	RNA therapy	6	2	8	16
	Non-RNA therapy	1	0	17	18
Hepatoma	RNA therapy	6	1	4	11
	Non-RNA therapy	0	0	13	13

The non-RNA-treated patients were also from the same Oncology Department in the same period of time, including all the cases of inoperable advanced adenocarcinoma of stomach (18 cases) and hepatoma (13 cases).

Illustrative cases:

Case 1. A 65-year-old male, gastric cancer, no history of surgical operation, nor other treatment had been received. Under examination, Octobar 13, 1976, 6 x 4 cm sized tumor was palpable in middle upper abdomen, a left subclavian nodule 1.7 cm in size being present. One mg of RNA-rich extract was injected subcutaneously monthly by skin bleb incubation method (begining Oct. 16). In addition, 1 mg of RNA per week was also injected subcutaneously. On Oct. 30, the abdominal tumor was reduced in size to 2 x 1.5 cm, and the subclavian node to 0.7 cm. X-ray of the upper G. I. tract confirmed regression of the tumor (Fig. 3).

Case 2. A 62-year-old male, a case of hepatoma, with pain, enlarged liver 10 cm below the inferior margin of the right chest, ascites, and edema of lower extremities. Colloidal ^{198}Au scanning showed localization in the liver. α FP was positive. No treatment was given before RNA administration (June 1978). After admission to the hospital, the same scheme of RNA treatment as in case 1 was given. And the patient was discharged and improved after 6 weeks of hospitalization. He resumed light physical work thereafter. Reexamination on Feb. 1979, showed that liver enlargement, ascites, and edema of lower extremities all disappeared, appetite being good (taking 0.85 to 0.9 kg of rice per day).

Case 3. A 52-year-old male, with a tentative diagnosis of lymphosarcoma on left side of the neck, (10 x 9 x 3 cm) connected with a small node and another node (2 x 1 cm) on the right side of the neck. Spleen was enlarged. No treatment had been received. RNA treatment began in April, 1978, 3 mg twice weekly of RNA from ram liver immunized with lymphosarcoma injected subcutaneously. The skin bleb incubation method was used in the 1st and the 10th injections given intratumorally and paratumorally.

Half a month later, the main tumor decreased in size. After 30th injection (15 wks), all tumors regressed, including the tumor on the opposite side of the neck which received no injection of RNA. Spleen was reduced to normal size. Patient resumed work. (Fig. 4 a.b.c.)

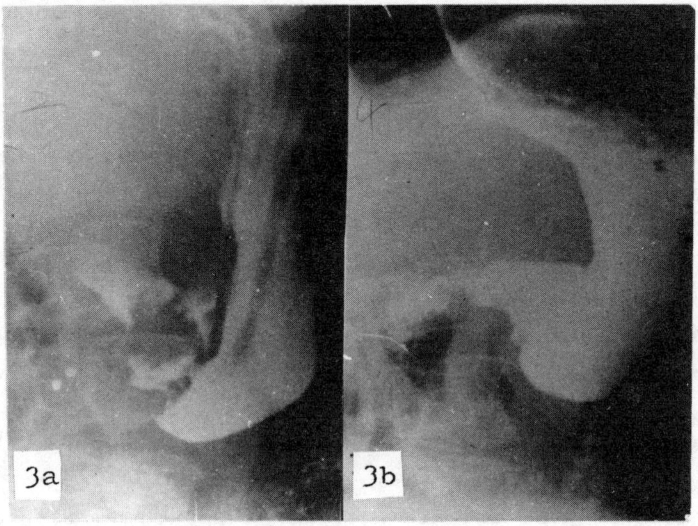

Fig.3 X-ray of stomach of case 1. a. Before RNA treatment, b. After RNA treatment.

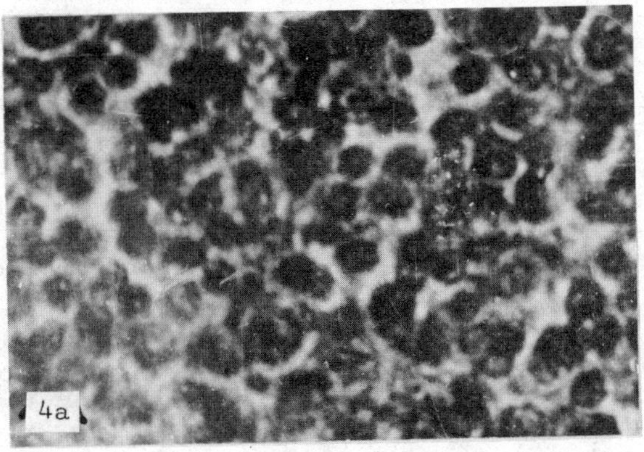

Fig.4a Pathological section of Case 3 (high power).

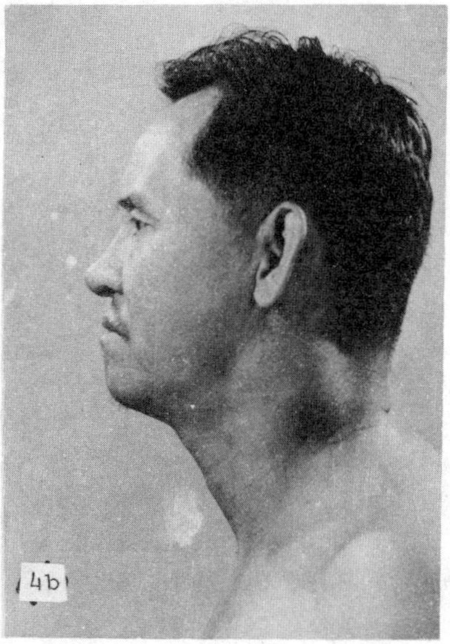

Fig.4b Case 3 before RNA treatment.

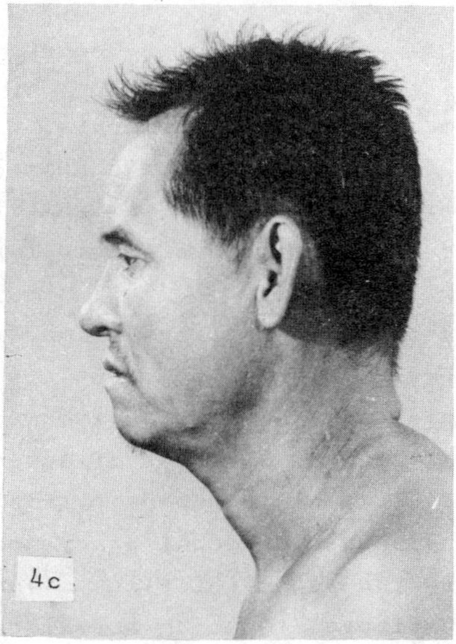

Fig.4c Case 3 after RNA treatment (tumor completely regressed).

Case 4. An 18-year-old girl was diagnosed lymphosarcoma on left side of the neck, tumor being 4 x 3 x 0.5 cm, without any treatment until April 28, 1978. RNA treatment was carried out with 2 mg RNA twice per week subcutaneously, and the same skin bleb incubation method was used as in case 3. Half a month later, tumor began to regress. On May 29, 1978, the 10th injection of RNA was given, and complete tumor regression was observed (Fig. 5 a. b. c.)

Case 5. A 65-year-old male was diagnosed adenocarcinoma of stomach with ascites, and multiple peritoneal matastases. Gastrojejunostomy had been performed.

Weekly skin bleb incubated RNA injection began July 1977, for 1 year. Following the treatment, the tumor gradually regressed from 7 x 4 cm to 6 x 4 cm, to 4 x 3 cm, and finally to 3 x 2cm. Ascites completely disapeared, hepato-renal function, blood routine, immunoglobin IgG, IgM, IgA, rosette test etc. became normal. Body weight increased by 8 kg. Patient went back to work (Fig. 6 a.b.c.).

DISCUSSION

From our animal experiments, we consider that RNA-rich extracts of liver from tumor immunized animals (rabbits, rams, and tumor bearing mice) have a certain inhibitory effect on the early growth of tumors of the same histologic type. This agreed with Garvey's results(7).

This inhibitory effect is manifested by an apparent retardation of early tumor growth, but a better therapeutic value remains to be explored by more extentive research.

In a total of 65 cases of advanced and earlier cancer patients, clinical therapeutic trials of RNA-rich extract produced not a single case of local or systemic toxicity. These results are similar to the report of Pilch(6) and to the report by Shanghai Cancer Institute(8) using normal animal liver RNA. All this preliminary therapeutic trials on tumors by RNAs appeared to be quite safe and encouraging; an extented period of follow-up observation is of course desirable.

In case 3, RNA injected into the tumor on the left side of the neck was followed by complete regression not only of the tumor

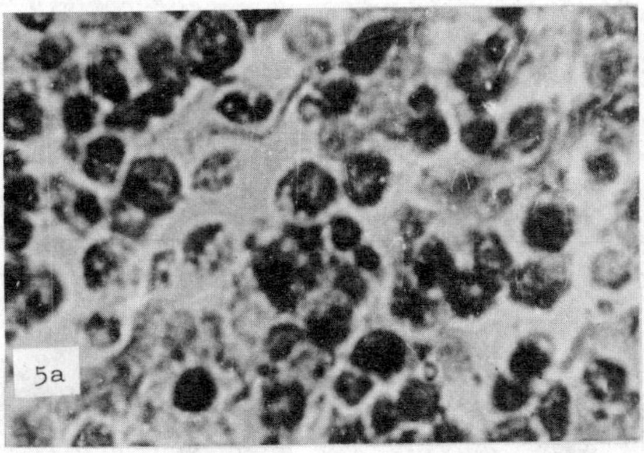

Fig.5a Pathological section of Case 4
(high power)

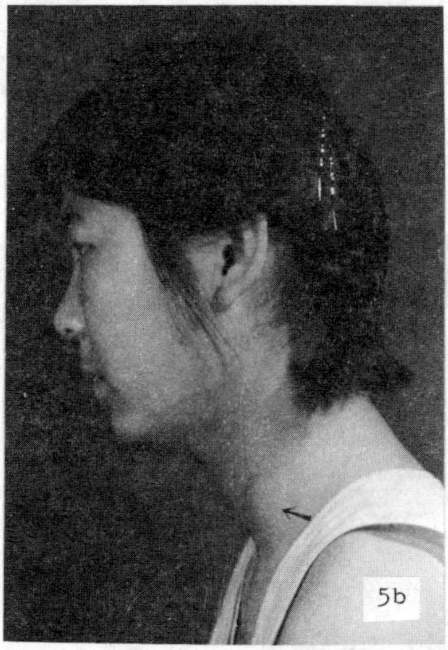

Fig.5b Case 4 before RNA treatment (The arrow indicates the position of tumor)

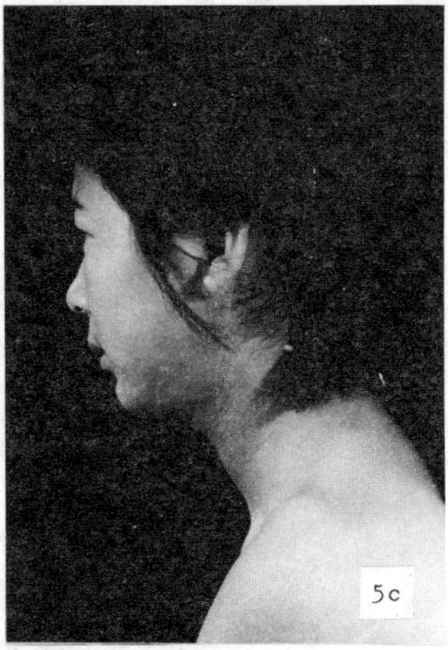

Fig.5c Case 4 after RNA treatment (complete regression)

Fig.6 G.I. X-ray of Case 5

6a. Before treatment (pyloro-duodeno-passage obstructed)

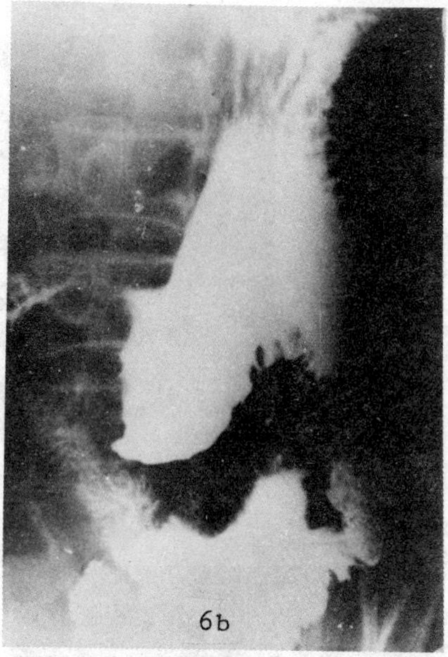

6b. After gastrojejunostomy (phyloro-duodeno-passage remained obstructed as before)

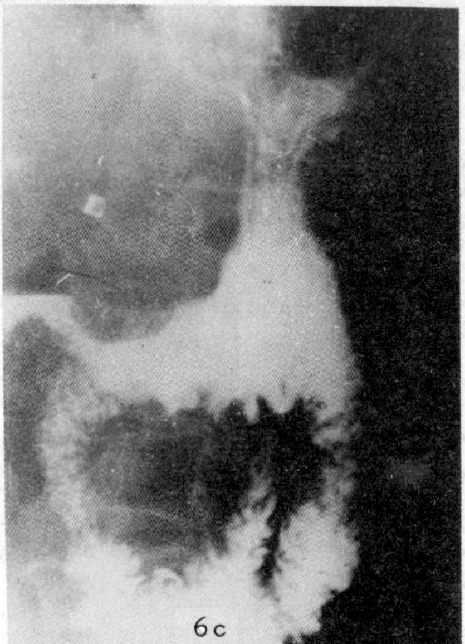

6c. After RNA treatment, pyloro-duodeno-passage became patent. It coinsided with cancer regression.

on the left side but also that on the right side. It highly suggests that is being a result of systemic immuno therapy. To explain this result one may recall the skin bleb incubated injection method.

Futher more, when newly sythesized ^{3}H-uridine labeled RNA of mice is incubated with cancer patient's skin bleb cells (in which RNA synthesis has been blocked by actinomycin D) the silver granules can be seen in the cells by radioautography (Fig. 7), and the TCA precipitate of the cells also showed radioactivity. These suggest that large molecules of RNA (TCA precipitated portion) can enter the skin bleb cells. (Huang and Mu-Zhen Yu: unpublished work). Therefore, skin bleb incubation method can be used to transfer nucleic acid information and/or stimuli to a large number of skin bleb cells (about 5-10 x 10^6 cells per ml of bleb fluid, including about 50% monocytes, 10% lymphocytes, 10% eosinophiles, 30% neutrophiles) and thus enhance their immunoactivity.

The advantages of this method are: (1) it is simple and easy to perform, (2) no complicated instruments are needed, (3) the isolation of white blood cells is avoided, and (4) since patient's own cells are used, the risk of hypersensitivity to foreign cells can be avoied.

Longer period of follow-up observation and more strictly controlled experimental as well as clinical work are needed, in order to understand the exact machanism of the present apparent therapeutic success, even though the success is only partial.

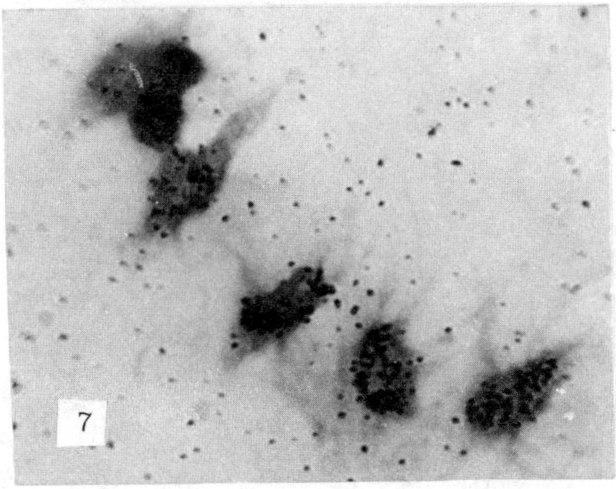

Fig. 7 The radioautography of skin bleb cells of cancer patient incubated with ^{3}H-RNA.

REFERENCES

(1) Lowry, O. H. et al.: Jour. Biol. Chem. 193, 265 (1951)

(2) Schutz, G. et al.: Biochem. Biophys. Res. Commun. 49, 680 (1972)

(3) Schlager, S. I. et al.: Cancer Res. 35, 1907 (1975)

(4) 中国 医学科学院肿瘤防治研究所等：中华医学杂志1 9 7 6年第4期2 2 9頁

(5) Adler, F. L.: Ann. New York Acad. Sci. 207, 489 (1973)

(6) Pilch, H. Y. et al.: Ann. N. Y. Acad. Sci. 277,592 (1976)

(7) Linker-Israeli, M. and Gravey, J. S.: ibid. 207, 481 (1973)

(8) 上海市肿瘤研究所生化組：上海医学，1 9 7 8年第一期6 － 1 0頁。

STUDIES ON THE SPLEEN RNA OF TUMOR-BEARING DONORS

Jiang Ximing, Ding Renrui, Zhang Tongwen
Fang Genlai, Cha Shijin

(Department of Biology, Hangzhou University)

Hong Changfu, Yang Bainan, Yu Hui

(Laboratory of Microbiology and Immunology,
Zhejiang People's Academy of
Experimental Hygiene)

Zhang Minchuan, Wang Tingyu, Li Yonglun

(Department of Oncology and Cancer Laboratory,
Ningbo Second Municipal Hospital)

Most of the immune RNA used for the immunotherapy of malignant tumors so far reported were derived from the lymphoid organs of animals immunized with the appropriate tumor cells. Whether the RNA extracted from the lymphoid organs of animals bearing a progressive tumor would retain the same transferring ability for tumor immunity were still obscure. Pennline et al 1979[1] had reported that the RNA extracted from the spleen of tumor-bearing mice was able to transfer to normal lymphocytes the abilities to produce suppressive factors for tumor-cell cytotoxicity as well as to mediated specific antibodydependent cell cytotoxicity, while that of the tumor-immune mice transferred only the "positive" immunological responsiveness. At the mean while, suppressive cells had

---

Liu Xianglin from Zhejiang Medical University had joined a part of the experiment.

been demonstrated in spleen of tumor-bearing mice in several laboratories,[2,3,4] that they would inhibit the proliferation of lymphocytes as well as the immunological responsiveness. Hence the immunological status of the tumor-bearing hosts were more complicated than that of the immunized hosts. How would be the transferring ability of this spleen RNA of tumor-bearing donor was a problem of both immunological and therapeutical significance. This problem would be evaluated here as follows.

I. THE IN VITRO TRANSFER OF CELL-MEDIATED IMMUNITY WITH THE SPLEEN RNA OF TUMOR-BEARING MICE

The spleen-RNA was extracted from the spleen of mice bearing Sarcoma 180 by the hot phenol method after Yang et al, 1978[5]. The transferring ability of this spleen-RNA was assayed with mixed lymphocyte tumor cells reaction according to Deckers et al, 1974[6] with slight modifications, in which the leukocytes in whole blood of normal mice were incubated with spleen-RNA and then cultured with Mitomycin-C treated S_{180} cells. Lymphoblastogenesis, as indicated by the incorporation of ^{3}H-thymidine, in the mixed culture system were choosen as the indication of cell-mediated immunity, which would be expressed by the stimulation index calculated from the following formula.

$$S\ I = \frac{\text{cpm of leukocytes with tumor cells} - \text{cpm of leukocytes without tumor cells}}{\text{cpm of leukocytes without tumor cells}}$$

The results of 8 in vitro transfer experiments were summarized in table 1. The stimulation index of the spleen-RNA groups were significantly different from that of the leukocyte treated with normal spleen-RNA or spleen RNA treated with RNase. While the leukocytes of S_{180} tumor-bearing mice manifested also a high SI in mixed culture system. These results indicated that a conversion of normal leukocyte into an immunocompetent cell was achieved through the incubation with spleen-RNA of tumor-bearing mice, and thus the transferring ability of this spleen-RNA were

Table 1. Summary of the Experiments Showing In Vitro Transfer of Cell-Mediated Immunity With Spleen RNA of S_{180}-Bearing Mice

Cells Tested	Radioactivity of Transformed Leukocytes Pulse/min (cpm)		SI	P*
	Cell Culture Without Tumor Cells	Cell Culture With Tumor Cells		
Normal Leukocytes	340.94±259.40	292.55±196.17	-0.1419	
Normal Leukocytes + Spleen RNA of S_{180}-Bearing Mice	301.67±167.30	698.27±522.97	1.3147	< 0.001
Normal Leukocytes + Normal Spleen RNA	260.67±192.11	340.25±186.33	0.3053	> 0.3
Normal Leukocytes + Spleen RNA of S_{180}-Bearing Mice Treated with RNase	433.25±158.40	681.50±352.82	0.5730	> 0.1
Leukocytes From S_{180}-Bearing Mice	273.25±97.76	1100.60±617.87	3.0278	< 0.01

*Significance of the radioactivity in experimental groups as compare with that in normal leukocytes was determined by t test.

demonstrated in the mixed lymphocyte-tumor cell reaction by the lymphocyte transformation. The transferred host leukocytes seemed to be in the similar immunocompetent status as the lymphocytes of the S_{180} tumor bearing mice as indicated by their significant stimulation index.

After sucrose density gradient centrifugation, the profile of the O. D. in 260mμ had shown 3 main peaks with different sedimentation values (Fig. 1). After treatment with ribonuclease, the density gradient profile remained only 1 peak of lighter components (Fig. 2). At the same time, the transferring activity of the spleen-RNA were abolished after the treatment with ribonuclease.

The available data from these in vitro experiments demonstrated obviously that the spleen RNA from tumor-bearing mice possessed some properties as same as that of the immune RNA drived from the immunized animals cited in previous literatures. These included the transfer of the cell-mediated immunity as revealed by the stimulation of lymphoblastogenesis of the normal lymphocytes after incubation with this spleen RNA, and the ingradient of RNA which would be inactivated by the treatment of ribonuclease. Thus we could consider the spleen RNA of tumor-bearing mice to be equivalent to the immune RNA extracted from the lymphoid organs of immunized animals at least in the sense of transferring of immunological responsiveness.

2. THE TUMOR INHIBITORY EFFECT OF SPLEEN RNA OF TUMOR-BEARING DONORS IN ANIMAL EXPERIMENTS

In vivo experiments were carried out in mice to evaluate the tumor inhibitory effect of the spleen RNA from the tumor-bearing mice. In the experiments, mice were inoculated with S_{180} cells and administrated subsequently with 3 subcutaneous injections of spleen RNA in every other days. Physiological saline was used instead of spleen RNA in the control groups. The tumor inhibition effect of the spleen RNA were measured by the tumor weight diminished in the experimental groups as compared with that of the control in 14 or 21 days of experimentation.

Table 2 were a cumulative data of 9 experiments conducted within 3 years. Although the tumor cell quantity inoculated and the dosage of the spleen RNA administrated were different from experi-

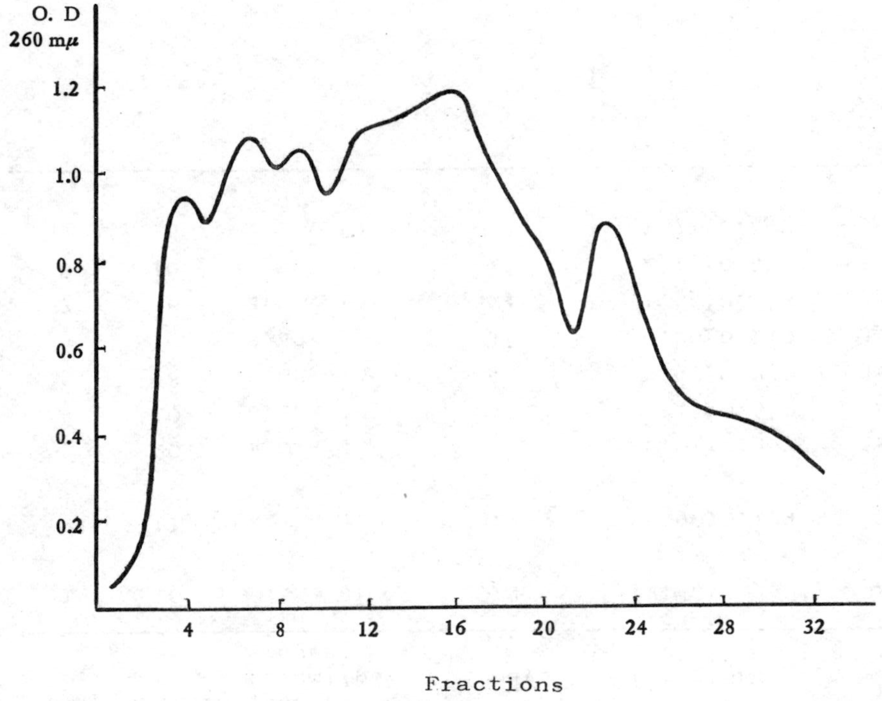

Fig. 1 Sucrose Density Gradient Profile of Spleen RNA from Mice Bearing S_{180} Tumor

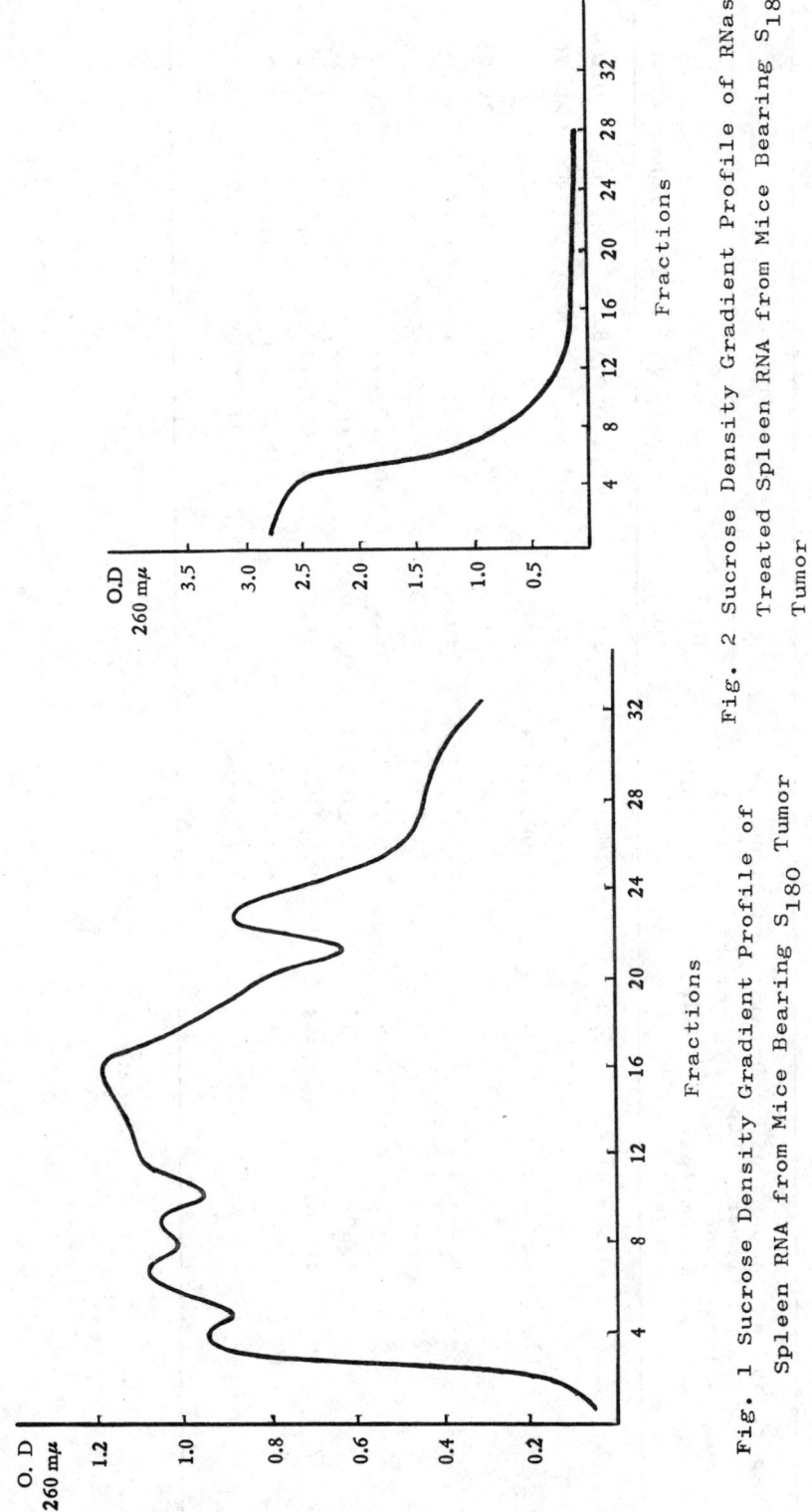

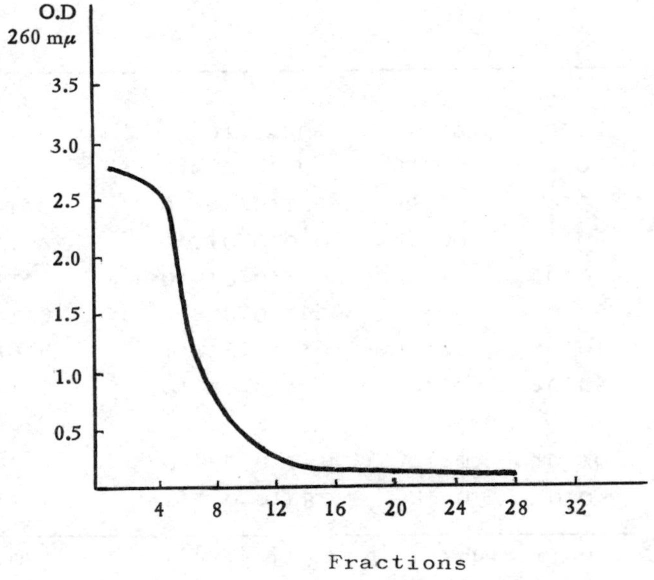

Fig. 2 Sucrose Density Gradient Profile of RNase Treated Spleen RNA from Mice Bearing S_{180} Tumor

Table 2. Tumor Inhibitory Effect of Spleen RNA from S_{180} -Bearing Mice

Exper. No.	No. of Animals	Amount of Tumor Cells Inoculated (0.2ml/per mouse)	Duration of Experiment (days)	Tumor Weight of Controls (g ± SE)	Experimentals Dose (mg)	Experimentals Tumor Weight (g ± SE)	% of Inhibition	P
1	10	3.6×10^6	14	2.42±0.540	3x1.5	1.150±0.282	52.40	< 0.05
					3x0.5	0.560±0.151	37.70	> 0.10
2	11	4.0×10^6	14	0.90±0.174				
					3x1.5	0.520±0.182	42.20	> 0.05
3	20	4.0×10^6	14	2.01±0.367	5x0.4	1.173±0.110	41.75	< 0.01
4	10	0.88×10^6	14	2.14±0.340	3x1.5	1.430±0.190	33.00	< 0.05
5	8	0.88×10^5	14	0.97±0.250	3x1.5	0.610±0.150	37.10	< 0.05
6	10	0.80×10^5	14	0.20±0.110	3x1.7	0.110±0.020	45.00	< 0.05
7	18	2.0×10^5	21	1.39±0.448	3x1.0	0.463±0.108	66.92	< 0.02
8	10	2.0×10^5	21	2.13±0.470	3x1.0	1.030±0.380	52.60	> 0.05
9	15	2.0×10^5	21	1.71±0.297	3x1.0	0.897±0.209	48.07	< 0.05

ment to experiment, all of them had a tumor inhibitory effect in extent from 33% to 67%, in which 7 of 9 experiments were statistically significant.

In experiments 7 and 9 (Table 3), tumor inhibitory effects of spleen RNA from tumor-bearing mice was compared with that of spleen RNA treated with DNase, RNase and the spleen RNA from normal mice. Spleen RNA treated with DNase exhibited the same tumor inhibitory effect in S_{180} inoculated mice as intact spleen RNA from tumor-bearing mice, while the spleen RNA treated with RNase and spleen RNA from normal mice exerted no demonstrable tumor inhibitory effect as compared to the untreated controls. These results indicated that the active principle which induced the performance of anti-tumor effect were integrated RNA in the spleen cells of tumor-bearing mice. This spleen RNA was distinct from spleen RNA of normal mice as revealed in the experiment, and thus a tumor specificity had been elucidated in the spleen RNA from tumor-bearing mice.

3. THE THERAPEUTIC EFFECT OF SPLEEN RNA FROM TUMOR-BEARING PATIENTS

Human spleen RNA were prepared from the resected spleen of patients with adenocarcinoma of stomach. 17 patients of gastric cancer with tumor of same histological type as the RNA donor, who had been operated by subtotal gastrictomy, received spleen RNA from tumor-bearing patients as an immunotherapeutic treatment. The dosage was 4 mg per week in total of 88 mg spleen RNA. Another 22 patients with matched gastric cancer and similar operation were choosen as controls.

The results of the follow up survey within 3 years on the cancer patients of therapeutic and control groups were shown in the Table 4. The curative effect of the therapeutic groups as indicated by the survival rate was somewhat more encouraging in the first 2 years after the RNA treatment. At the end of the second year, the survival rate of the therapeutic group was 82%, while that of the control was 50% only.

The 3 year survival rate was 52.94% in the therapeutic group and 36.4% in the control. Here would be mention that there were

Table 3. The Anti-Tumor Specificity of Spleen RNA of S_{180} -Bearing Mice

		Controls	Experimentals			
			Anti-S_{180} Spleen-RNA	Spleen-RNA +DNase	Spleen-RNA +RNase	Normal Spleen RNA
Exper. 7.	Tumor Weight (g ± SE)	1.397±0.488	0.463±0.108	0.389±0.162	2.037±0.450	0.904±0.317
	% of Inhibition		66.92	71.80	-45.81	35.28
	No. of Tumor-Bearing Mice	9/9	11/18	5/9	9/9	9/10
Exper. 9	Tumor Weight (g ± SE)	1.715±0.297	0.890±0.209		1.382±0.272	1.374±0.158
	% of Inhibition		48.07		19.36	19.88
	No.of Tumor-Bearing Mice	15/15	11/14		13/13	12/12

Tabe 4. The Results of Immunotherapy with Spleen RNA From Cancer Patients

Time After Surgery	Therapeuticals NO. Survival	Survival Rate (%)	Controls No. Survival	Survival Rate (%)
1 st. Year	16/17	94.10	16/22	72.7
2 nd. Year	14/17	82.40	11/22	50.0
3 rd. Year	9/17	52.94	8/22*	36.4

* Including 3 cases with critical relapse or metastasis.

3 patients in the control group, who had suffered from critical relapse or metastasis at the time of visiting survey. They would reasonably not be included in the healthy survivals. When these 3 patients were discounted from the survivals of the control group, the 3 year survival rate of the control would be reduced to 22.73%, thus made the difference from the 52.94% of the therapeutic group more significantly. It appeared that the spleen RNA of tumor-bearing patients might exert some anti-tumor effect at least in the case of gastric cancers. The estimation of some immunological indexes of these patients before and after the therapeutic treatment with spleen RNA had demonstrated a widespread improvement of the immunological status in the treated patients (Table 5). It was conceivable that these improved immunological responsiveness, which would be transferred by the spleen-RNA administrated, would played a certain role in the tumor inhibitory effect in the therapeutic patients.

From the available data, it was conceivable that the spleen RNA of the tumor-bearing donors had demonstrated a similar immunological potentialities as the immune RNA from immunized animals. The transfer of the cell-mediated immunity had been demonstrated by the mixed lymphocyte-tumor cell reaction in the in vitro experiment. In the in vivo tests, tumor inhibitory effects by the introducing of spleen RNA from tumor-bearing donors were achieved both in mice and human tumor experiments, while the RNA from spleen of normal unimmunized animal failed to do so. All of these results indicated that the spleen RNA of tumor-bearing donors could transfer the "positive" immunological responsiveness against the tumor which had borne in the donor. But the immunological status was more complicated in the tumor bearing donors than the immunized donors. Besides the immunocompetent cells present in the spleen and other lymphoid organs in the tumor-bearing host, immune suppressive cells had also been found in the spleen of tumor-bearing mice by various investigators[2,3,4]. Nelson et al 1975[7] reported that the culture supernatants of spleen cell from tumor-bearing mice inhibited (blocked) specific cellmediated cytotoxicity to the tumor borne by the spleen donor, and they also mediated specific antiserumdependent cytotoxicity with control lymphoid cells. Such dual characters in the spleen of tumor-bearing animal had

Table 5. The Immunological Indexes of Patients of Gastric Cancer Before and After Treatment With Spleen RNA From Tumor-Bearing Patients

Immunological Index	No. of Patients	Before Therapy (Average)	After Therapy (Average)
Lymphocyte Transformation (%)	17	38.18	53.41 (16)*
Rosette Formation (%)	17	38.88	51.41 (16)
IgG (mg/ml. serum)	17	7.85	12.33 (13)

* The figure in the parenthesis indicates the number of patients with increasing index.

also been demonstrated in several laboratories[1,8]. From the results of our experiments, only the "positive" immunological responsiveness of anti-tumor effect had been detected in the hosts which had received the spleen RNA of tumor-bearing donors. No suppressive effect on the tumor immunity had been detected so far with the methods used in our experiments, although the suppressor cells had been identified in the spleen of the tumor-bearing mice[1,2]. These problems would remain to be clarified in the further investigations.

References

(1) Pennline, K.J., Evans, S.B., Nawrocki, J.F., Rees, J.C., Johson, C.S. Vallera, D.A., & Dodd, M.C. Release of soluble "blocking" and "suppressor" factors from normal lymphocytes treated with RNA from spleen of tumor-bearing mice. Brit. J. Cancer, 39:247-258, 1979.

(2) Kirchner, H., Muchmore, A. V., Chused, T.M., Holdern, H. T., and Herberman, R.B. Inhibition of proliferation of lymphoma cells and T lymphocytes by suppressor cells from spleen of tumor-bearing mice. J. Immunol., 114:206-210, 1975.

(3) Fujimoto, S., Greene, M.I. and Sehan, A.H. Regulation of the immune response to tumor antigen. I. Immunosuppressor cells in tumor-bearing hosts. J. Immunol., 116:791-799, 1976.

(4) Takei, F., Levy, J. G. and Klburn, D.G. In vitro induction of cytotoxicity against syngeneic mastocytoma and its suppression by spleen and thymus cells from tumor-bearing mice. J. Immunology, 116: 228-293, 1976.

(5) Yang, S.Y., Wang, G.D., Chang, I. ZL. and Ye, Q.W. Immune cytolysis of human liver cancer cells mediated by xenogeneic "immune" RNA. Acta Biol. Exper. Sinica, 11:189-198, 1978.

(6) Deckers, P.J., Wang, B.S. and Mannick, J.A. A quantitative in vitro assay of immune RNA. Surg. Forum, 25:114-116, 1974.

(7) Nelson, K., Pollack, J.B. and Hellstrom, K.E. Specific anti-tumor responses by cultured immune spleen cells. I. In vitro culture method and initial characterization of factors which block immune cell-mediated cytotoxicity in vitro. Int. J. Cancer, 15:806, 1975.

(8) Chang, Z.L., Wang, J., Wang, G.D. and Yao, Z. The immune status of tumor-bearing host. III. Observations on the dynamics of the cytostatic effect of spleen cell of tumor-bearing mice on tumor cell in vitro. (^{125}I UdR uptake inhibition) Acta Biol. Exp. Sinica 12:13-27, 1979.

STUDIES ON THE BIOCHEMISTRY OF THE INTERFERON SYSTEM*

P. Lengyel, R. Broeze, B. Jayaram, J. Pichon,
E. Slattery, J. Dougherty, H. Samanta,
and G. Floyd-Smith

(Department of Molecular Biophysics and Biochemistry
Yale University, New Haven, Connecticut 06511 U.S.A.)

Interferons were discovered in 1957 as antiviral agents.[1] Studies during the last 23 years however, revealed their involvement in the control of a large variety of seemingly diverse physiological processes. These include e. g.: cell proliferation, antibody response, graft rejection, delayed hypersensitivity, histocompatibility antigen expression, natural killer cell recruitment, and macrophage activation.[2-7] It appears to be in line with this multiplicity of effects that the biochemistry of interferon synthesis and interferon action are also complex.

Interferons are proteins synthesized in a variety of cells from vertebrates (in organisms or in cell culture) upon viral infection or treatment with other inducers of interferon. One of the most effective inducers is double-stranded ds RNA (e.g, poly (I)·poly (C))[3]. dsRNAs (or at least complementary RNA strands) are formed in the course of the replication of some viruses presumably as intermediates or side-products[8,9] and at least in some cases viral dsRNA was reported to be a mediator in the induction of interferons by a virus.[10]

---

* Supported by NIH research grants AI12320 and CA 16038
abbreviations: ds, double-stranded
EAT, Ehrlich Ascites tumor cells

One organism, or even a single cell, may produce several types of interferons.[11] Three types of human interferons were designated as: Interferon α (leukocyte type), interferon β (fibroblast type), and interferon γ (immune interferon).[12,13] The types of human interferons synthesized depend on the nautre of both the inducer and of the producing cell.[11] Human interferon genes were reported to be located on three different chromosomes.[14,15]

The interferons are excreted from the producing cells. They interact with other cells and alter the biochemical and immnuological characteristics of these in various ways. At least some of these alterations require the presence of the nucleus, RNA synthesis and protein synthesis.[3-7] After the removal of the interferons from the medium the alterations gradually disappear. Thus, cells exposed to interferons at a concentration at which virus replication is impaired usually do not suffer lasting damage, though in the presence of interferons the rate of cell division may be decreased. Interferons are in general host specific i. e., they are the most active in cells of the species in which they were produced.[3,5] They are remarkably potent; in cell culture 3×10^{-14} M interferon may impair virus replication.[16]

The role of the interferon system in the defense against viral infections and malignancies was revealed in animal experiments involving the use of antisera to interferons.[17,18] Clinical studies with exogeneous interferons have been impeded by the inadequate supply of human interferons.[3]

We have been examining various aspects of the biochemistry of the interferon system primarily in mouse cells.

ISOLATION AND CHARACTERIZATION OF VARIOUS MOUSE INTERFERONS

We induce interferon formation in mouse Ehrlich ascites tumor (EAT) cells in culture by infection with live Newcastle disease virus and use classical methods of protein purification to isolate the interferons.[19,20] The specific activity of our purified interferons is about 2×10^9 NIH mouse reference standard units/mg protein. Polyacrylamide gel electrophoresis in the presence of sodium dodecyl sulfate reveals interferons of three size classes: A. 35,000 to 40,000 daltons, B. 26,000 to 30,000 daltons, and C. about 20,000 daltons. The tryptic peptide patterns of interferons A and B are very similar[19] and the sequences of their first 24 amino terminal amino acids (number 17 has not been identified) are identical.[21] It is conceivable that the broadness of the molecular weight distributions of interferons A and B is due to nonuniform substitution with carbohydrates. The tryptic peptide pattern of interferon C is different from those of A and B and the sequence of the first 20 amino terminal amino acids of interferon C differs from those of A (or B) in 18 position[21]. Moreover, an antiserum to interferon A does not neutralize interferon C (ref[22]). There is some homology in amino terminal sequence between mouse interferons A (or B) and a human fibroblast interferon(β type) and a more pronounced homology between mouse interferon C and human lymphoblastoid interferon (α type).[21] These homologies serve as a basis for classifying our mouse interferons A and B as β type and our mouse interferon C as α type.[13]

BIOCHEMISTRY OF INTERFERON ACTION

Interferon, dsRNA and mRNA Cleavage

In the course of investigating the mechanism of inhibition of reovirus replication in mouse L cells by interferon we established that the early steps of replication e. g., virus binding to cells, penetration and partial uncoating, are unaffected whereas viral protein and RNA accumulation are decreased.[23,24] These observations stimulated us to compare the rate of reovirus mRNA degradation in an extract prepared from cells treated with interferon to that in an extract from untreated control cells. Reovirus mRNAs which had been prepared in vitro and were contaminated with genomic dsRNAs of the virus were degraded faster in the extract from interferon-treated cells than in the extract from control cells when incubated under conditions of cell-free protein synthesis. Remarkably, the difference in the rate of degradation was not observed in experiments with purified reovirus mRNA preparations from which the contaminating ds reovirus RNA had been removed. However, added ds reovirus RNA (or poly (I). poly (C)) greatly enhanced the rate of degradation of the purified reovirus mRNAs in the extract from interferon-treated cells whereas it affected this rate in the extract from control cells barely if at all.[25] In addition to dsRNA, ATP was also required to accelerate reovirus mRNA degradation in the extract from interferon-treated cells. This dsRNA and ATP dependent endonuclease activity in extracts from interferon-treated cells was designated as $endonuclease_{INT}$[26].

The $endonuclease_{INT}$ system consists of at least two complementary enzyme fractions. One of these when incubated with dsRNA and ATP gives rise to a heat stable, low molecular weight substance. This substance activates the second enzyme fraction a latent endoribonuclease (designated as RNaseL).[27,28] The substance is identical with a series of (2'-5') linked oligoadenylates $(2'-5')(A)_n$. These were originally discovered by Kerr and his colleagues as inhibitors of protein synthesis.[29]

$(2'-5')(A)_n$ Synthetase

We isolated the enzyme synthesizing $(2'-5')(A)_n$ (designated as $(2'-5')(A)_n$ synthetase) from mouse EAT cells as a homogeneous protein.[30] The treatment of these cells with interferon increases the level of the enzyme 10 to 100 fold. In the presence of dsRNA the purified enzyme can convert the large majority (over 97%) of the ATP added to $(2'-5')(A)_n$ (where n extends from 2 to about 15) and pyrophosphate, though it does not cleave the pyrophosphate.[30] The stoichiometry of the reaction can be formulated as:

$$(n+1)\ ATP \longrightarrow (2'-5')\,pppA(pA)_n + n\ pyrophosphate$$

($(2'-5')pppA(pA)_n$ stands for the same series of compounds that we abbreviate otherwise as $(2'-5')(A)_n$. The different abbreviation is used here to indicate that the 5' terminus of $(2'-5')(A)_n$ is a triphosphate). The extent of the reverse reaction. i. e., the pyrophosphorolysis of $(2'-5')(A)_n$ was below the level of detection in our conditions. The affinity of the enzyme for ATP is low; the rate of the reaction increases when the ATP concentration is increased from 5 mM to 10 mM.[31]

It is remarkable that both interferon A and interferon C, which, as noted, have clearly different N-terminal sequences (the rest of the molecules have not been sequenced) are similarly effective in inducing $(2'-5')(A)_n$ synthetase in mouse cells. (Interferon B was not tested).[22]

RNase L

This latent endoribonuclease which can be activated by $(2'-5')(A)_n$ was purified several hundred fold from EAT cells.[28] The activation of RNase L by $(2'-5')(A)_n$ is reversible. Upon removal of $(2'-5')(A)_n$ from the activated enzyme (e. g., by gel filtration) the enzyme reverts to the latent state; readdition of $(2'-5')(A)_n$ reactivates the enzyme. The activation does not seem to result in a large size change of the enzyme. This and other data make it unlikely that the activation should involve the binding or the release of a protein.[28] The partially purified RNase L preparation, if activated, cleaves a large variety of single-stranded RNAs (but not poly (C), poly (G), and poly (A), only very slowly) and not dsRNA (G. Floyd-Smith, to be published).

The following observation is in line with the possible involvement of the $(2'-5')(A)_n$ synthetase, RNase L system, in mediating at least some of the effects of interferon action: the replication of reovirus in mouse L929 cells is inhibited by interferon and reovirus messenger RNA is degraded much faster in interferon-treated L929 cells than in L929 cells not treated with interferon.[32]

Interferon, dsRNA and Protein Phosphorylation

The need for dsRNA and ATP for the activation of the endonuclease$_{INT}$ system[26], the knowledge that several proteins are activated or inactivated by phosphorylation[33], and other considerations prompted us to test the effect of dsRNA on protein phosphorylation in extracts of interferon-treated and control cells. We found that the addition of dsRNA (from reovirus or (poly(I). poly(C)) to an extract from interferon-treated EAT cells (but not, or only to a much lesser extent to an extract from control cells) results in the phosphorylation of at least two proteins: P_1 (67,000 daltons) and P_2 (37,000 daltons).[34,35,36,37] It is in accord with these findings that dsRNA added to intact, interferon-treated cells (but not to control cells) also results in the phosphorylation of P_1. We purified the dsRNA-activatable protein kinase system several thousand fold from EAT cells treated with interferon. Interferon A and interferon C are similarly effective as inducers.[22] The treatment with interferon increases the level of the system several fold. The purified enzyme preparation is free of dsRNA independent kinase activity and is similar to a dsRNA dependent kinase in reticulocyte lysates.[38] (2' -5')$(A)_n$ is not synthesized by the activated kinase and (2' - 5')$(A)_n$ does not substitute for dsRNA in activating the enzyme.[39] The P_2 protein that can be phosphorylated by the kinase appears to be the small subunit of the peptide chain initiation factor eIF-2[39]. The identity of P_1 has not been definitely established. However, P_1 copurifies with the dsRNA-activatable protein kinase system throughout a several thousand fold purification. The most highly purified kinase preparation we have at present consists of P_1 as the major protein band and two minor bands (as tested by labeling with ^{125}Iodine).[40] Studies are underway to test whether or not P_1 is identical with the protein kinase. The purified enzyme if activated can also phosphorylate some histones.

The addition of the activated protein kinase preparation to a cell-free protein synthesizing system from EAT cells or reticulocyte lysates results in the inhibition of peptide chain initiation.The inhibition can be overcome by the addition of further eIF-2 and is apparently the consequence of the phosphorylation of eIF-2.[39]

Two Disinct Mechanisms for Inhibiting Protein Synthesis in Extracts From Interferon-Treated Cells

As discussed in the preceding sections interferon treatment of cells increases the level of at least two enzyme systems. These impair mRNA translation in the presence of dsRNA by two distinct mechanism; a) by accelerating mRNA cleavage and b) by inhibiting peptide chain initiation. The inhibition of peptide chain initiation can be reversed by the action of an enzyme dephosphorylating the phosphorylated peptide chain initiation factor eIF-2.[41] RNase L that had been activated by (2' - 5')$(A)_n$ can be reverted to the latent state upon enzymatic degradation of (2' - 5')$(A)_n$[42]

dsRNA Does Not Have to be "Free" to Activate the Latent Enzymes

The presence of dsRNA (or at least of complementary RNA strands) has been established in vaccinia virus-infected cells[8,10] and in cells infected with viruses with a dsRNA genome,

e. g., reovirus. In the latter case, however, the dsRNA is apparently always packaged in a protein coat.[43]

This prompted us to test whether or not dsRNA has to be "free" to activate the endonuclease$_{INT}$ system and the protein kinase. We found that reovirions, reovirus cores (i. e., reovirions partially uncoated by treatment with chymotrypsin) and reovirus subviral particles (formed from reovirions in infected cells by cleavage and removal of some outer coat proteins) purified by centrifugation through CsCl gradients can replace dsRNA in promoting RNA degradation in an extract from interferon-treated cells. However, in enhancing RNA cleavage the reovirions are only 4 per cent as efficient on a per mg dsRNA basis as free dsRNA. Furthermore reovirions treated with RNase III (an enzyme specific for cleaving dsRNA), under conditions in which most free dsRNA but little or no dsRNA in virions is degraded, are no longer able to promote RNA cleavage. This suggests that dsRNA on the surface of the reovirions may be responsible for the activation of the endonuclease$_{INT}$ system.[33,44]

It should also be noted that reovirus subviral particles isolated by CsCl density gradient centrifugation from an extract of cells which had been treated with a partially purified interferon preparation manifest an endonuclease activity and synthesize _in vitro_ many reo mRNAs shorter than full size. Reovirus subviral particles isolated from untreated cells have, however, little or no endonuclease activity and synthesize full size reo mRNAs.[23]

Possible Rationale for the Multiple Role of dsRNA in Interferon Induction and Action

It remains to be understood why dsRNA has become an important modulator (e. g., both inducer and enzyme activator) of the interferon system. It is conceivable that dsRNA in some form (i. e., "free" or partially coated) is a side product or intermediate in the replication of many viruses. If that is the case it might serve as a signal revealing the presence of replicating viruses in the cell.

As noted in previous sections the treatment of cells with interferon induces, among others, two enzyme systems both of which are latent: endonuclease$_{INT}$ (consisting of $(2' - 5')(A)_n$ synthetase and RNaseL) and the protein kinase. The activation of these latent enzyme systems requires dsRNA in some form. This puzzling complexity might be rationalized in the following hypothesis. Localized virus infection (perhaps if accompanied by some formation of dsRNA) results in the synthesis of interferon in the infected cells and its secretion and spreading in the body. In the cells exposed to interferon the two enzyme systems are induced. Since they are latent they do not impair cell metabolism. When the cells previously exposed to interferon become infected with a virus, however, this might result (in the formation of some dsRNA and perhaps thereby) in the activation of the two enzyme systems. These in turn impair protein synthesis in the virus-infected cells. It remains to be verified that the physiological activator of these two enzyme systems in intact cells is dsRNA.

It should be noted that at least two of the biochemical phenomena elicited by interferon i. e. the impairment of mRNA methylation in vitro[45,46] and the accelerated inactivation of some transfer RNA species in vitro[47] are manifested even in the absence of dsRNA. Our studies concerning these phenomena have been summarized elsewhere.[32]

mRNAs and Proteins Induced by Interferon Treatment

The alterations in biochemical characteristics upon exposure to interferons prompted us to compare the protein and mRNA compositions of interferon-treated and control cells. We found that the treatment of EAT cells with a highly purified interferon preparation results in the enhancement of the accumulation of a particular mRNA and the corresponding protein (145,000 daltons). The correspondence was established by translating the mRNA into the protein in question *in vitro*.[48] Treatment of mouse L cells with a mouse interferon preparation and human cells (HeLa) with a human interferon preparation was found to induce the formation of several mRNAs[49] and proteins.[49-51] The function of the proteins induced in mouse cells and human cells by interferons remains to be elucidated.

CONCLUSIONS

The isolation and characterization of pure human[52,53] and mouse interferons,[19,21,54,55] the cloning of several human interferon genes,[56-58] the production of human interferons in E. coli harboring interferon gene-containing plasmids,[59] the recognition of the central role of interferons in controlling a large veriety of immunological processes,[5] the starting of clinical studies with interferons on cancer patients[60] and also the beginning of the understanding of the unexpectedly complex biochemistry of interferon action, are among the highlights of interferon research in the last 5 years.

However, there are still large gaps in our knowledge of the interferon system and specifically of its biochemistry. We do not know if interferon acts only at the cell surface, only within the cell, or in both locations.[61] The roles of the recently discovered enzyme systems (e. g. $(2' - 5')(A)_n$ synthetase, RNase L, and the dsRNA activatable protein kinase) induced by interferons in mediating the various actions of interferons remain to be established. The identities and functions of several proteins induced by interferon treatment have to be elucidated. Genetic approaches e. g. selection of mutants impaired in one or another of the activities affected by interferons may facilitate the answering of these questions. It is most probable that the increased knowledge of biochemistry of the interferon system will ultimately benefit clinical medicine.

REFERENCES

1. Isaacs, A. and Lindenmann, J. (1957). Proc. Roy. Soc. Ser. B. 147, 258-267.

2. Finter, N.B.(editor). (1973). Interferons and interferon inducers. North-Holland, Amsterdam.

3. Stewart, W.E., II. (1979). The interferon system. Springer Verlag, New York.

4. Gresser, I. (1977) Cellular Immun. 34, 406-415 .

5. DeMaeyer, E.,and DeMaeyer-Guignard, J. (1979) Comprehensive Virology 15, 205-260 (Fraenkel-Conrat and Wagner, R. editors) Plenum, New York.

6. Baron, S., and Dianzani, F., (editors). (1977). Tex. Rep. Biol. Med. 35, 1-573. Plenum, New York

7. Vilcek, J., (editor). (1980) Regulatory functions of interferons. Ann. New York Acad. Sci (in the press)

8. Colby, C., and Duesberg, P.H. (1969) Nature (London) 222, 940-944.

9. Marcus, P., and Sekellick, M.J. (1977) Nature (London) 266, 815-819.

10. Boone, R.F., Parr, R.P., and Moss, B. (1979) J. Virol. 30, 365-374.

11. Havell, E.A., Hayes, T.G., and Vilcek, J. (1978) Virology 89, 330-334.

12. Youngner, J.S. (1977) Tex. Rep. Biol. Med. 35, 17-22.

13. Stewart, W.E. II, Nature (London) 286, 110.

14. Slate, D., and Ruddle, F.H. (1979) Pharmac. Ther. 4, 221-230

15. Meager, A., Graves, H.E., Walker, J.R., Burke, D.C., Swallow, D.M., and Westerveld, A. (1979) J. gen. Virol. 45, 309-322.

16. Kawakita, M., Cabrer, B., Taira, H., Rebello, M., Slattery, E., Weideli, H., and Lengyel, P. (1978) J. Biol. Chem. 253, 598-602.

17. Gresser, I., and Tovey, M.G. 1978. Biochim. Biophys. Acta 516, 231-247.

18. Gresser, I., Tovey, M.G., Bandu M.-T., Maury, C., and Brouty-Boye, D. (1976) J. Exp. Med. 144, 1305-1315.

19. Cabrer, B., Taira, H., Broeze, R.J., Kempe, T.D., Williams, K., Slattery, E., Konigsberg, W.H., and Lengyel, P. (1979) J. Biol. Chem. 254, 3681-3684.

20. Taira, H., Broeze, R.J., Slattery, E., and Lengyel, P. (1980) J. gen Virol. (in the press).

21. Taira, H., Broeze, R.J., Jayaram, B.M., and Lengyel, P. (1980) Science 207, 528-530.

22. Broeze, R.J., Dougherty, J.P., and Lengyel, P.(1980) Fed. Proc. 39, 2205.

23. Galster, R.J., and Lengyel, P. (1976) Nuc. Acids Res. 3, 581-598.

24. Gupta, S.L., Graziadei, W.D.III., Weideli, H., Sopori, M.L., and Lengyel, P. (1974) Virology 57, 49-63.

25. Brown, G.E., Lebleu, B., Kawakita, M., Shaila, S., Sen, G.C., and Lengyel, P. (1976). Biochem. Biophys. Res. Commun. 69, 114-122.

26. Sen, G.C., Lebleu, B., Brown, G.E., Kawakita, M., Slattery, E., Lengyel, P.(1976) Nature (London) 264, 370-373.

27. Ratner, L., Wiegand, R., Farrell, P., Sen, G.C., Cabrer, B., and Lengyel, P. (1978) Biochem. Biophys. Res. Commun. 81, 947-954.

28. Slattery, E., Ghosh, N., Samanta, H., and Lengyel,P. (1979) Proc. Natl. Acad. Sci. USA 76, 4778-4782.

29. Kerr, I.M., and Brown, R.E. (1978) Proc. Natl. Acad. Sci. USA 75, 256-260.

30. Dougherty J.P., Samanta, H., Farrell, P.J., and Lengyel, P. (1980) J. Biol. Chem. 255, 3813-3816.

31. Samanta, H., Dougherty, J.P., and Lengyel, P. (1980) J. Biol. Chem. (in the press)

32. Lengyel, P., Desrosiers, R., Broeze, R., Slattery, E., Taira, H., Dougherty, J., Samanta, H., Pichon, J., Farrell, P. Ratner, L., and Sen, G. (1980) In Microbiology 1980, 219-225.(Schlessinger, D., editor), American Society for Microbiology, Washington, D.C.

33. Rubin, C.S., and Rosen, O.M. (1975) Ann. Rev. Biochem. 44, 831-887.

34. Lebleu, B., Sen, G.C., Shaila, S., Cabrer, B., and Lengyel, P. (1976) Proc. Natl. Acad. Sci. USA 73, 3107-3111.

35. Roberts, W.K., Hovanessian, A., Brown, G.E., Clemens, M.J., and Kerr, I.M. (1976) Nature (London) 264, 477-480.

36. Shaila, S., Lebleu, B., Brown, G.E., Sen, G.C., and Lengyel, P. (1977) J. gen. Virol. 37, 535-546.

37. Zilberstein, A., Federman, P., Shulman, L., and Revel, M. (1976) FEBS Lett. 68, 119-124.

38. Sen, G.C., Taira, H, and Lengyel, P. (1978) J. Biol. Chem. 253, 5915-5921.

39. Farrell, P.J., Sen, G.C., Dubois, M.-F., Ratner, L., Slattery, E., and Lengyel, P. (1978) Proc. Natl. Acad. Sci. USA 75, 5893-5897.

40. Lengyel, P., Samanta, H., Pichon, J., Dougherty, J., Slattery, E., and Farrell, P. (1980) Ann. New York Acad. Sci. (in the press)

41. Zilberstein, A., Kimchi, A., Schmidt, A. and Revel, M. (1978) Proc. Natl. Acad. Sci USA 75, 4734-4738.

42. Schmidt, A., Chernajovsky, Y., Schulman., L., Federman, A., Berissi, H., and Revel, M. (1979) Proc. Natl. Acad. Sci. USA 76, 4788-4792.

43. Joklik, W.K. (1974) Comprehensive Virology 1, 231-234 (Fraenkel-Conrat, H., and Wagner, R., editors) Plenum Press, New York.

44. Ratner, L. (1979) PhD. Thesis. Yale University.

45. Sen, G.C., Lebleu, B., Brown, G.E., Kawakita, M., Slattery, E., and Lengyel, P. (1976) Nature (London) 264, 370-373.

46. Sen, G.C., Shaila, S., Lebleu, B., Brown, G.E., Desrosiers, R.C., and Lengyel, P. (1977) J. Virol. 21, 69-83.

47. Sen, G.C., Gupta, S.L., Brown, G.E., Lebleu, B., Rebello, M.A., and Lengyel, P. (1976) J. Virol. 17, 191-203.

48. Farrell, P.J., Broeze, R.J., and Lengyel, P. (1979) Nature (London) 279, 523-525.

49. Farrell, P.J., Broeze, R.J. and Lengyel, P. (1979) Nature Ann. New York Acad. Sci. (in the press).

50. Gupta, S.L., Rubin, B.Y., and Holmes, S.L. (1979) Proc. Natl. Acad. Sci. USA 76, 4817-4821.

51. Knight, E., and Korant, B. (1979) Proc. Natl. Acad. Sci. USA 76, 1824-1827.

52. Zoon, K.C., Smith, M.E., Bridgen, P.J., Anfinsen, C.B., Hunkapiller, M.W., and Hood, L.E. (1980) Science 207, 527-528.

53. Knight, E., Hunkapiller, M.W., Korant, B.D., Hardy, R.W.F., and Hood, L.E. (1980) Science 207, 525-526.

54. DeMaeyer-Guignard, J., Tovey, M.G., Gresser, I., and DeMaeyer, E. (1978) Nature (London) 271, 622-625.

55. Iwakura, Y., Yonehera, S., and Kawade, Y., (1978) J. Biol. Chem 253, 5074-5079.

56. Mantei, N., Schwarzstein, M., Streuli, M., Panem, S., Nagata, S., and Weissmann, C. (1980) Gene 10, 1-10.

57. Taniguchi, T., Ohno, S., Fuji-Kuriyama, Y., and Muramatsu, M. (1980) Gene 10, 11-15.

58. Derynk, R., Content, J., DeClercq, E., Volckaert, G., Tavernier, J., Devos, R., and Fiers, W. (1980) Nature (London) 285, 542-547.

59. Nagata, S., Taira, H., Hall, A., Johnsrud, L., Streuli, M., Ecsodi, J., Boll, W., Cantell, K., and Weissmann, C. (1980) Nature (London) 284, 316-320.

60. Cantell, K., (1979) Interferon I pp. 2-25. (Gresser, I., editor). Academic Press. New York.

61. Friedman, R.M. (1979) Interferon I. pp. 53-72. (Gresser, I., editor) Academic Press, New York.

RNA METABOLISM IN THE EARLY GERMINATING SEED EMBRYO*

Abraham Marcus, Shirley Rodaway, and Ketaki Datta

The Institute for Cancer Research 7701 Burholme Avenue Philadelphia, Pennsylvania 19111

Abstract Germination is the process by which the developmentally arrested dry seed resumes growth when exposed to water. At the level of the embryo, the process occurs in 3 phases a period of rapid water uptake, a quiescent period of several hours and a period of sustained growth RNA synthesis, both of rRNA and mRNA, begins as soon as the tissue is hydrated. A substantial increase in the rate of both transcription and processing occurs during the quiescent period preparatory to growth. The regulation of these processes is not brought about by increased methylating capacity, gene amplification, or the levels of nucleotide triphosphates.

Early embryo growth seems not to require the synthesis of rRNA but except for the events when the first hour after germination, embryo growth is sensitive to inhibition of mRNA synthesis. Calculation of the maximal contribution of new mRNA to the total mRNA pool indicates however, a maximal contribution of 20%, assuming a functional equivalence between preformed and newly synthesized mRNAs. This suggests that the requirement may be for a select class of mRNAs that have an unusually high turnover rate. Experiments with azaUrd, studying the time course of inhibition of protein synthesis, are consistent with this idea. The inhibition which seems to be related to the formation of azaUrd-RNA lags about 1.5hr. beheind the peak formation of its RNA.

*This work was supported by Grant PCM 79-00268 from the National Science Foundation; by Grants CA-06927 and RR-05539 from the National Institutes of Health; and by an appropriation from the Commonwealth of Pennsylvania.

The tentative hypothesis is that a protein component involved in ribosomal transit is depleted within 3 hr. The mRNA for its component would also have a similarly high turnover such that within 3 hr. It would contain a state of azaUrd. A direct test of this idea is possible by an in vitro determination of protein synthetic capacity in extracts of the azaUrd treated cells.

Germination of a seed embryo is best considered as a triphasic process in which there is: a) an initially rapid uptake of water bringing the fresh weight of the embryo to somewhat more than twice the initial weight, b) a period of several hr that is quiescent with respect to growth, and finally c) a growth period in which the fresh weight of the embryos increases in a sustained manner. Fig. 1 illustrates this phenomenon for wheat embryos isolated mechanically from wheat seed and separated from endosperm by flotation in $CHCl_3$-CCl_4. Fig. 2 presents similar data for axes of soybean seed isolated manually. The major difference between the two systems is the greater length of the quiescent period in the soybean axes.

In studying RNA metabolism in the embryonic axis, we have maintained a focus on these three phases. We consider, on one hand, the status of RNA synthesis at the different stages in germination and, on the other hand, the extent to which such synthesis of RNA is necessary for an embryo to accomplish the germination process.

RNA synthesis during germination. Analysis of RNA synthesis in the germinating embryo is approachable by straightforward methodology and has been the subject of considerable study. Initially, a number of laboratories reported that RNA synthesis, in particular the synthesis of poly A(+) RNA, was delayed considerably relative to protein synthesis but subsequent studies failed to support this contention. It appears that an essentially concomitant synthesis of 4-5S RNA, rRNAs, and poly A(+) RNA begins within the first hr of imbibition (Sen et al., 1975; Spiegel et al., 1975; Payne, 1977; Delseny et al., 1977). With intact seeds there is a substantial delay in the incorporation of radioactive precursors into RNA, and autoradiographic studies have shown that synthesis begins at the peripheral areas of the seed (Payne et al., 1978).

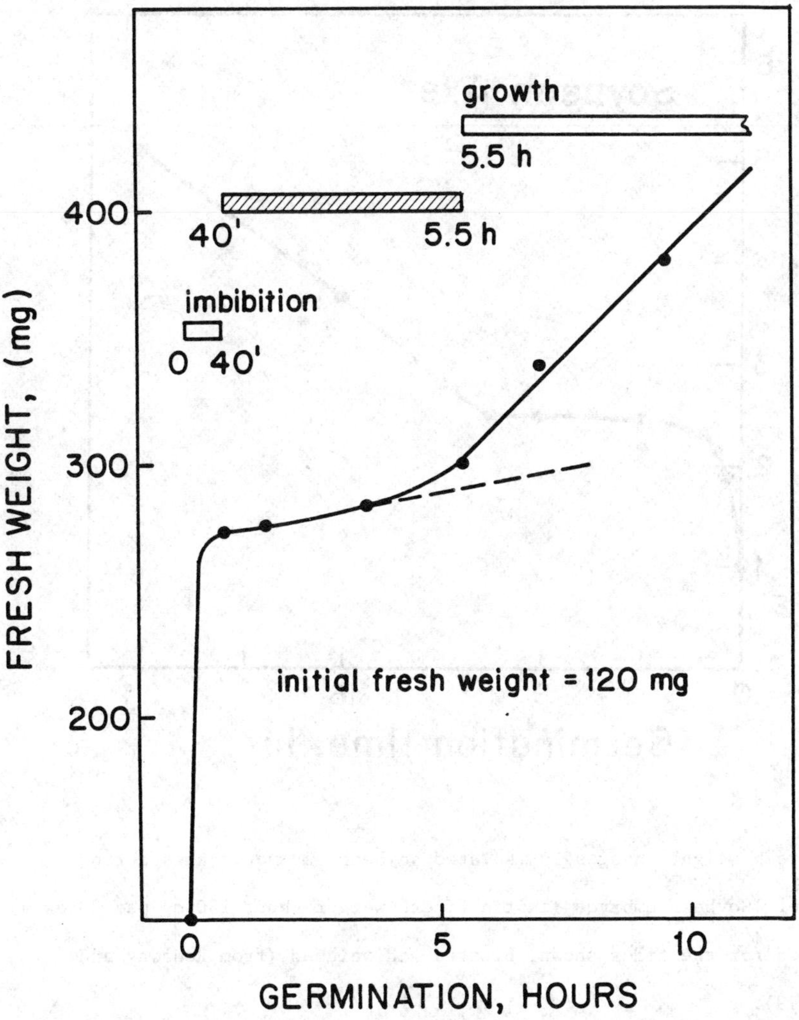

Fig. 1. Fresh weight changes in wheat embryonic axes during early imbibition. Wheat embryos were imbibed in H_2O for the times shown, blotted and weighed.

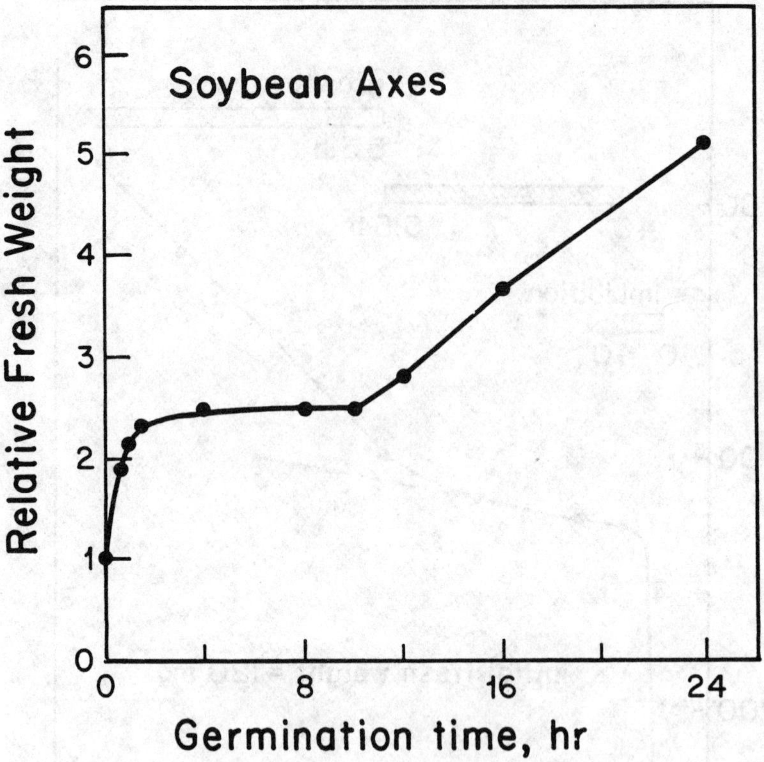

Fig. 2. Fresh weight changes in isolated soybean embryonic axes during germination. Soybean embryos (initial fresh weight about 130 mg per 30 axes) were imbibed for the times shown, blotted and weighed (from Rodaway and Marcus, 1979).

Since hydration proceeds slowly with intact seeds, this result is almost certainly a consequence of the time course of water uptake and not a manifestation of a delay in transcriptional capacity.

On a quantitative level there appear to be changes, both in the rate of transcription of the different RNAs, and in the rate of processing, at least of the ribosomal precursor RNA (Huang et al., 1980). Table I presents data from an experiment with wheat embryos measuring the incorporation of radioactive uridine into cytoplasmic high molecular weight poly A(+) RNA (mRNA) and poly A(-) RNA (primarily rRNA). Between 1.5 and 5.5 hr, there is a two-fold increase in the rate of mRNA synthesis and about a three-fold increase in the rate of rRNA synthesis with most of the increase occurring up to 3.5 hr. Table II presents data for soybean axes comparing the incorporation into polysomal and ribosomal fractions. The relative changes in the rate of incorporation is similar in both fractions increasing about 5.5 fold between 1.5 and 9.5 hr. Approximately 25% of this change occurs by 4 hr.

An important requirement for a comparative analysis using the incorporation of a radioactive precursor is that the precursor either be labeled to the same extent in both situations or that the data be corrected for the specific activity of the precursor. Another approach to looking at changes in RNA synthesis is the analysis of the rate of processing of rRNA. This type of comparison can be made without concern for the precursor specific activity so long as the rate of change of precursor specific activity is similar in the different time periods. Experimentally, the relative rates of processing are ascertained by following the ratio of radioactivity appearing in the newly synthesized 26S and 18S rRNAs. The rationale in this approach is that to the extent that processing is delayed, there is a delay

Table I. Rates of RNA Synthesis in Germinating Wheat Embryos

Period	UTP Sp. Act.	Incorporation into High Mol Wt RNA (nmol UMP)			
hr	cpm/nmol x 10^{-4}	nuclear	cytoplasmic		total
			A(+)RNA	A(-)RNA	
1-2	9.0	2.7	0.44	0.55	3.7
3-4	11.5	3.6	0.69	1.9	6.2
5-6	15.5	4.6	0.77	2.0	7.4
17-18	21.0	4.5	0.62	1.4	6.5

75 mg wheat embryos var. Norana 1974, were incubated with 100 μC_i ^{3}H-uridine at 10^{-5}M. RNA from cytoplasmic ribosomes was extracted with phenol-chloroform in the presence of SDS and separated into poly A(+) and poly A(-) RNAs on oligo(dT) cellulose. UTP was determined from trichloroacetic acid extracts of embryos incubated under identical conditions. Complete details are given in Huang et al. (1980).

Table II. Rates of RNA Synthesis in Germinating Soybean Axes

Period	UTP Sp. Act	Incorporation into High Mol Wt RNA (nmol UMP)			
hr	cpm/nmol x 10^{-4}	nuclear	polysomal	ribosomal	total
1.0-2.33	1.5	0.9	0.5	0.5	1.9
3.5-4.87	2.1	1.6	1.4	1.0	4.0
9.0-10.33	3.6	4.1	3.2	2.5	9.8

30 soybean axes, var. Kent, were labeled with 200 μC_i ^{3}H-uridine at 0.5 x 10^{-5}M for 80 min. The remainder of the procedure is described in Table I.

in the appearance of 26S rRNA relative to 18S rRNA in the cytoplasm. Since the specific radioactivity of the precursor UTP is increasing with time, the newly synthesized 18S rRNA will reflect the higher specific activity. Fig. 3 presents an analysis of the poly A(-) RNA from both the polysomal and postpolysomal fractions of soybeans axes incubated with ^{3}H-uridine for 80 min at 1.5, 4, and 9 hr. Major changes in the rate of processing occur at the different time periods. The radioactive RNA evinces the $\underline{A}260$ absorbance pattern typical of rRNA only in the 9 hr samples. The 26S/18S ratios for the radioactive 4 hr and 9 hr samples are 0.44 and 1.08, respectively, while that of the 1.5 hr sample is indeterminable. Similar, though less striking, results were obtained in experiments with wheat embryos.

What are the mechanisms whereby transcription and processing are regulated? Methylation is an important step in maintaining the normal rate of processing of pre-rRNA (Vaughn et al., 1967). Nevertheless, when the methylating capacity of cells is inhibited, ribosomes are formed that accumulate undermethylated rRNA (Caboche and Rocchelerie, 1977). We, therefore, considered that if methylating potential were regulating rRNA synthesis during early embryo germination, we would find undermethylated rRNA in the newly synthesized ribosomes. Analysis of the rRNAs in soybean axes, indicated however, that the rRNA made at 3.5 hr is methylated to the same extent as that made at 9 hr (Huang et al, 1980). Thus the methylating capacity of the embryos, at the least, keeps pace with the capacity for rRNA transcription.

A second possible explanation for the change in the rate of RNA synthesis is the amplification of the rRNA genes, a process known to occur during oogenesis in amphibians and in the development of drug-resistance in

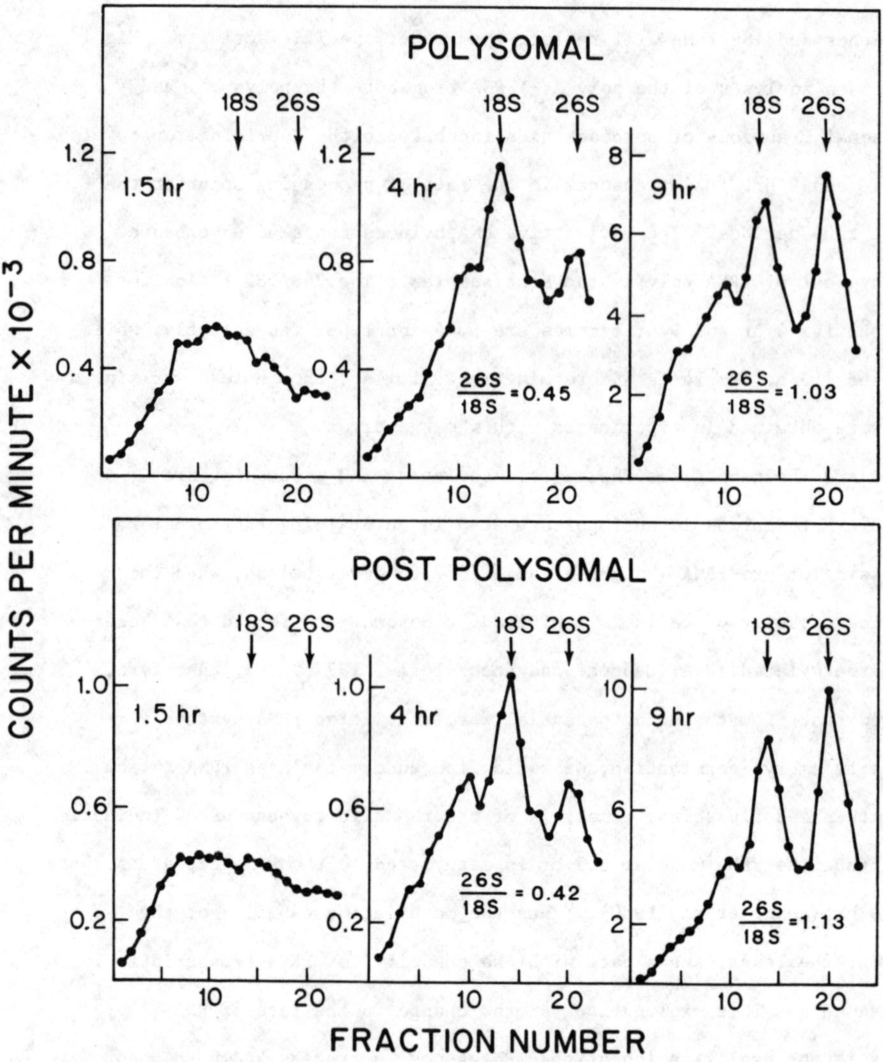

Fig. 3. Sucrose Gradient Analysis of Radioactive poly A(-) RNA of Soybean Axes. Axes were incubated in ^{3}H-uridine for 80 min at 1.67, 4.0 and 9.0 hr and the RNA was extracted from the polysomal and postpolysomal fractions. After separation on oligo(dT) cellulose, the poly A(-) RNA was analyzed on a sucrose gradient (see Huang et al. 1980 for complete details).

mammalian cell lines. In an early report (Chen and Osborne, 1970) such amplification was reported to occur early in the germination of wheat embryo. Subsequently, more careful analysis (Ingle and Sinclair, 1972) showed that the fraction of DNA capable of hybridizing with either 26S or 18S rRNA remains the same throughout both embryo development and early germination, i.e., there is no amplification of the ribosomal genes.

Another possibility that has been considered for a role in regulating RNA synthesis is the level of the nucleoside triphosphates. Dry embryos have only minimal levels of nucleoside triphosphates, and these increase dramatically very early during embryo imbibition (Brooker et al., 1977, Cheung and Suhadolnik, 1978). Based on the findings with wheat embryos that UTP and CTP increased with a somewhat greater delay than ATP and GTP, Cheung and Suhadolnik (1978) suggested that the levels of these nucleotides might be regulating RNA synthesis. Careful analysis of UTP levels in soybean axes (Table III) showed, however, that 80% of the increase in UTP that occurs throughout the 9.5 hr period is achieved prior to 4 hr. Since 75% of the increase in the rate of RNA synthesis occurs after 4 hr, it seems unlikely that the UTP level is regulatory. Finally, one other obvious component necessary for RNA transcription are the polymerases. Polymerase II, the enzyme responsible for transcribing DNA into HnRNA and pre-mRNA, is found in abundant quantities in dry soybean axes and unimbibed wheat embryos. A lower molecular weight form (IIB) increases steadily after inhibition and it has been suggested that this is the form of polymerase active in transcription (Guilfoyle and Jendrisak, 1978). Such a change could explain the differences in rates of A(+) RNA synthesis. Another possible explanation is based on the finding that nuclei and chromatin fractions isolated from ungerminated axes contain low levels of polymerases whereas comparable preparations from

Table III. Levels of ATP and UTP in Germinating Soybean Axes

Time	ATP	UTP
hr	nmol/axis	
1.67	4.9	0.9
4.13	6.9	2.3
9.67	7.3	2.5

Trichloroacetic acid extracts of the embryos were analyzed for ATP and UTP by enzymatic procedures (see Huang et al. 1980).

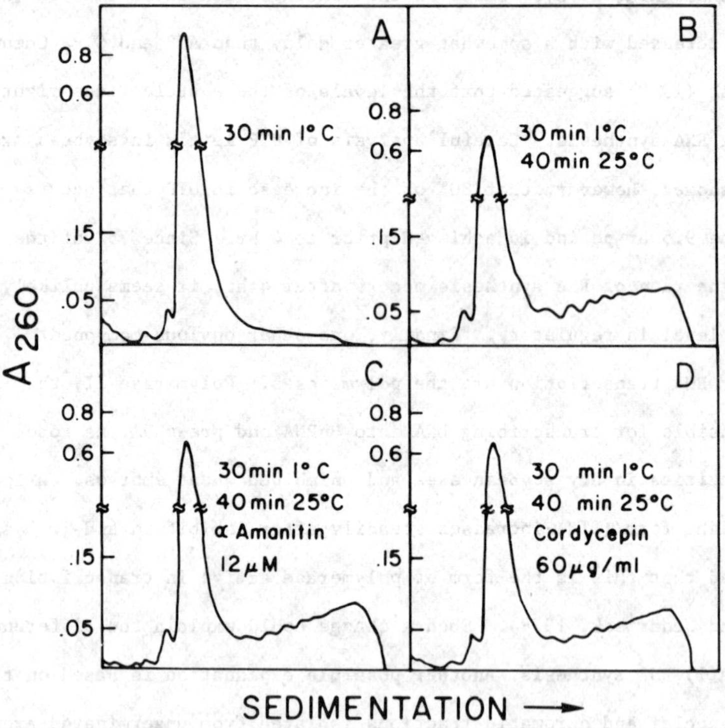

Fig. 4. Polysome Formation During Early Germination of Wheat Embryos and the Effect of RNA Synthesis Inhibitors. Embryos were imbibed A) in H_2O for 0 min, B) in H_2O for 20 min at 1°C then 40 min at 25°C, C) as B but in 12 μM α-amanitin, D) as B but in 250 μM cordycepin. (see Spiegel and Marcus, 1975 for complete details).

germinated axes are several-fold higher in activity (Guilfoyle, 1981). Thus, with the onset of germination, the enzymes may be mobilized to the nucleus and perhaps modified so as to increase transcriptional activity. With regard to the change in rate of rRNA processing, similar ideas would be appropriate although no experimental support is yet available.

Requirement for RNAs during the process of embryo germination. The question of RNA requirement is best considered in two categories, RNAs pre-existent in the dry embryos and RNAs newly synthesized during germination. It is a well established observation that mature dry seed embryos have an abundant supply of ribosomes that are potentially active in protein synthesis (Marcus et al., 1968; Bewley and Larsen, 1979; Peumans et al., 1980). Nevertheless, as described earlier in detail, there is considerable synthesis of new rRNA throughout early germination. Based on uridine incorporation data into cytoplasmic poly A(-) RNA, we estimate a maximal net synthesis of 15 nmol of UMP residues in rRNA for 75 mg embryo in 7.5 hr. Given an endogenous level of 1000 nmol of UMP residues in rRNA, it is clear that a distinction between replacement of preformed ribosomes and supplementation cannot be made by direct analysis. One observation that we have made is that one can inhibit the synthesis of new rRNA by about 70% by adding 1.5 mM 5-fluorouridine at 1 hr after the start of imbibition, without effect either on protein synthesis or on embryo growth. Thus, most of the newly synthesized rRNAs are not necessary for germination.

A more equivocal question is that concerning mRNA. A key observation relevant to this question is shown in Fig. 4. During the first 40 min after imbibition, there is a substantial formation of polysomes, unaffected by the presence of either α-amanitin or cordycepin, treatments that inhibit A(+) RNA synthesis 94 and 83%, respectively (Spiegel and Marcus, 1975). These results indicate that the early increase in protein synthesis is brought about by a substantial mobilization of preformed mRNAs into polysomes, and that the synthesis of new mRNA is unnecessary for this early increase in protein synthesis. The presence of preformed mRNAs can be directly demonstrated in wheat embryos (Spiegel and Marcus, 1975; Brooker et al., 1978), and in several other plant species (Payne, 1976).

What about the requirement for A(+) RNA synthesis in the later periods of germination? Experiments of Spiegel (unpublished results) showed that the same treatments wherein embryos were imbibed in α-amanitin or cordycepin without effect at 1 hr, resulted in substantial inhibition of the rate of protein synthesis by 3 hr. Such data has recently been published for wheat embryos by Cheung et al. (1979) and by Caers et al. (1979). Fig. 5 presents recent data from our laboratory showing that presence of cordycepin can result in the inhibition of protein synthesis to an extent of 70-80%. In general, this early insensitivity to inhibitors of RNA synthesis followed by gradually increasing sensitivity has been interpreted to indicate a gradually increasing demand for newly synthesized mRNAs (see Bewley, 1981).

What newly synthesized mRNAs might be required? Based on rates of synthesis for 1 hr periods (Table I), there is a net synthesis of 4.5 nmol UMP residues of mRNA per 75 mg embryo in 7.5 hr. The steady state level of mRNA (approximately 2% of the total cytoplasmic high molecular weight RNA) is 25 nmol UMP residues of mRNA per 75 mg. Thus, assuming functional equivalence between newly synthesized and stored mRNAs, the maximal possible contribution of newly synthesized mRNAs to the total mRNA pool by 7.5 hr is less than 20% of the total. Since protein synthesis can be inhibited to an extent of 70-80% by inhibiting mRNA synthesis (see fig. 5), it would seem that it is not the overall loss of mRNA that is affecting protein synthesis,

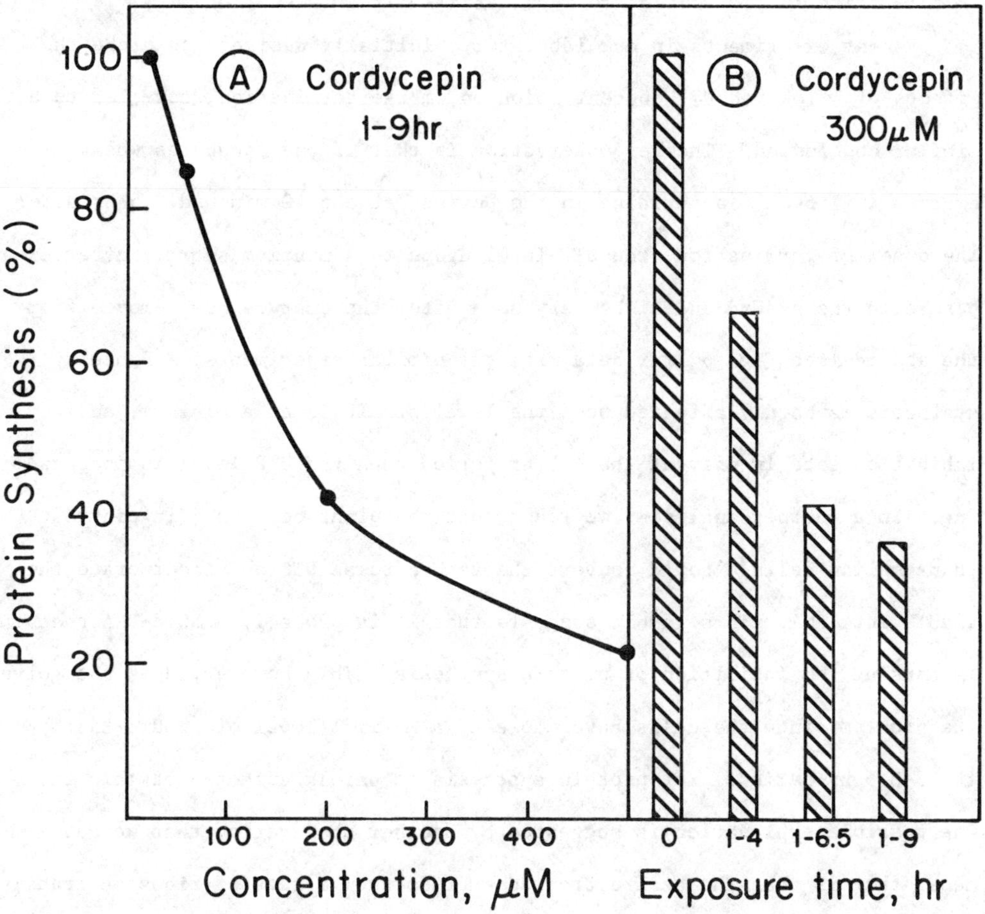

Fig. 5. Inhibition of Protein Synthesis in Wheat Embryos by Cordycepin (Rodaway, unpublished data). 75 mg wheat embryos were imbibed for 1 hr in H_2O and then transferred either to varying concentrations of cordycepin (A) or to 300 µM cordycepin for different periods of time (B). At the times indicated, the samples were assayed for protein synthesis in a 12 min assay with 8 x 10^{-3}M ^{14}C-leucine (Brooker et al., 1977). In (B), controls in water were assayed at the same time points.

but rather the decay of a select class of mRNAs that has a turnover rate at least 4 times that of the total mRNA of the germinating embryo.

Recent experiments in our laboratory, initially designed to probe the effects of a lowered UTP concentration on embryo germination, have led to a similar conclusion. The key observation is that if one incubates wheat embryos in 7.5 mM 6-azauridine in the period between 40 min and 2.5 hr after the onset of germination, the UTP level drops to a minimum shortly after exposure to the azaUrd and rises very soon after the embryos are removed from the azaUrd (see Fig. 6 for data from two similar experiments). Protein synthesis is hardly affected when the level of UTP is at a minimum, and inhibition sets in only at the 4-5 hr period when the UTP level is increasing. In a further analysis, we found that the plant embryos (in contrast to mammalian cells) could convert the azaUrd to azaUTP and incorporate the azaUTP into RNA. This result suggests that it is probably azaUrd-RNA that is causing the inhibition of protein synthesis. This however, does not solve the problem since the cells have close to a maximal level of azaUrd-RNA by the 3-3.5 hr period. Yet protein synthesis is barely affected at this time. One possible explanation is suggested by another observation that we have made; that is that the azaUrd treated embryos show a rate of ribosome transit 40% of that of control embryos. This result suggests that by the 4-5 hr period, when the cells have had azaUrd-RNA for 3 hr, there is a depletion of a protein component(s) functioning in ribosome transit. The mRNA for this component(s) would have a high turnover such that by the 4-5 hr period, a significant part of this mRNA would contain azaUrd, and as such would result in the synthesis of an inactive translational component. A test of this idea is in progress using appropriate lysates and determining their _in vitro_ protein synthesizing capacity.

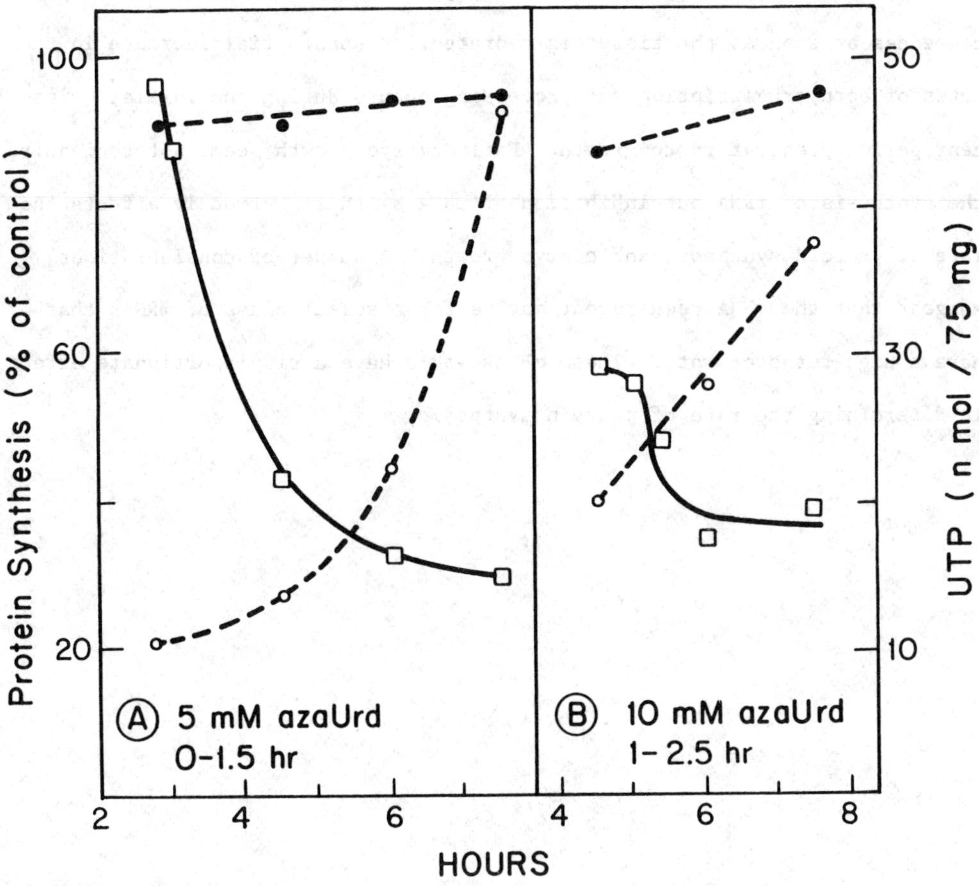

Fig. 6. Effect of 6-azaUridine in Lowering Cellular UTP and In Inhibiting Protein Synthesis. 75 mg wheat embryos were exposed to azaUrd for the periods shown and then transferred to water. At the various time points, the embryos were either assayed for protein synthesis (□--□) in a 12 min incorporation with 8 x 10^{-3} M ^{14}C-leucine (Brooker et al., 1977) or trichloroacetic acid extracts were made and analyzed for UTP (control ●--●; azaUrd ○--○) (Huang et al., 1980).

SUMMARY

RNA synthesis, both of mRNA and rRNA, in the germinating seed embryo, commences as soon as the tissue is hydrated. A substantial increase in the rates of both transcription and processing occurs during the initial quiescent period preparatory to growth. Early embryo growth seems not to require the synthesis of rRNA but inhibition of mRNA synthesis strongly affects the rate of protein synthesis and embryo growth. A number of considerations suggest that the mRNA requirement may be for a select class of mRNAs that have a high turnover rate. These mRNAs would have a disproportionate effect in determining the rate of protein synthesis.

REFERENCES

Bewley, J. D. (1981). in Encyclopedia of Plant Physiology, in press.

Bewley, J. D. and Larsen, K. M. (1979). Phytochem. 18, 1617-19.

Brooker, J. D., Cheung, C. P., and Marcus, A. (1977). In Protein Synthesis and seed Germination (A. Khan, ed.) pp. 347-356, Elsevier Press, New York.

Brooker, J. D., Tomaszewski, M., and Marcus, A. (1978). Pl. Physiol. 61, 145-9.

Caboche, M. and Bacchelerie, J. P. (1977). Eur. J. Biochem. 74, 19-29.

Caers, L. I., Peumans, W. J., and Carlier, A. R. (1979). Planta 144, 491-6.

Chen, D. and Osborne, D. J. (1970). Nature 226, 1157-60.

Cheung, C. P. and Suhadolnik, R. J. (1978). Nature 271, 357-8.

Cheung, C. P., Wu, J., and Suhadolnik, R. J. (1979). Nature 277, 66-7.

Delseny, M., Aspart, L., and Guitton, Y. (1977). Planta 135, 125-28.

Guilfoyle, T. J. in Biochemistry of Plants, Vol. VI. Proteins and Nucleic Acids (A. Marcus, ed.). (1981). In press.

Guilfoyle, T. J. and Sendrisak, J. J. (1978). Biochem. 17, 1860-66.

Huang, B. F., Rodaway, S., Wood, A., and Marcus, A. (1980). Plant Physiol., in press.

Ingle, J. and Sinclair, J. (1972). Nature 235, 30-32.

Marcus, A., Luginbill, B., and Feeley, J. (1968). Proc. Natl. Acad. Sci. USA 56, 1770-77.

Payne, P. I. (1976). Biol. Revs. 51, 329-363.

Payne, P. I. (1977). Phytochem. 16, 431-34.

Payne, P. I., Dobrzanska, M., Barlow, P. W., and Gordon, M. E. (1978). J. Exp. Botany 29, 77-88.

Peumans, W. J., Carber, A. R., and Schreurs, J. (1980). Planta 147, 302-306.

Rodaway, S. and Marcus, A. (1979). Plant Physiol. 64, 975-981.

Sen, S., Payne, P. I., and Osborne, D. J. (1975). Biochem. J. 148, 381-387.

Spiegel, S. and Marcus, A. (1975). Nature 256, 228-230.

Spiegel, S., Obendorf, R., and Marcus, A. (1975). Pl. Physiol. 56, 502-507.

Vaughn, M. H., Soeiro, R., Warner, J. R., and Darnell, J. E. (1967). Proc. Natl. Acad. Sci. USA 58, 1527-1537.

ACCUMULATION OF HISTONE GENE TRANSCRIPTS DURING OOGENESIS AND EARLY DEVELOPMENT OF THE SEA URCHIN STRONGYLOCENTROTUS PURPURATUS

Alex Mauron, Shoshana Levy and Laurence H. Kedes

The Howard Hughes Medical Institute Laboratory
and Department of Medicine
Stanford University School of Medicine
and Veterans Administration Hospital
Palo Alto, California 94304

Introduction

Gene expression in sea urchin embryos begins as early as the first zygotic cleavage division. The transcription of histone genes and proteins has been studied extensively. As development proceeds there is an increase in the relative amounts of histone protein synthesized on polysomes. The histone template RNAs utilized in this process are made up of both RNAs accumulated and stored during oogenesis (maternal mRNAs) and newly transcribed messengers. It has been estimated that at mid-cleavage (200-400 cells) as much as one-third of the histone templates are still maternal in origin (Skoultchi and Gross, 1973).

The histone genes transcribed during the rapid cleavage stage of development are members of a moderately repetitive family of tandemly arrayed genetic units each of which contains coding sequences for the five classes of histone mRNA interspersed with spacer DNA (reviewed in Kedes, 1979). It is not yet clear whether transcription is polycistronic. If it is, then there should be clear evidence that transcripts from the spacer DNA exist in cellular RNA, even if for a brief period. The purpose of the experiments reported here is to quantify the number of histone mRNA molecules per egg and the number of histone mRNA molecules that accumulate at several stages of early development. In addition we provide evidence that there is no transcription of spacer DNA and that, hence, transcription is not polycistronic.

Experimental Design

The method we have employed to determine numbers of mRNA molecules depends on titration of cellular RNA with sequence specific complementary probes. Such probes were constructed by isolation of DNA coding fragments from the histone genes of *Strongylocentrotus purpuratus* contained in recombinant plasmids (reviewed in Kedes, 1979). We radiolabelled the mRNA specific probes with I^{125} and then separated the strands by denaturation and electrophoresis on polyacrylamide gels. After the separated strands were electroeluted from the gel, any residual contamination of opposite strand was removed by allowing the DNA to undergo self reassociation. The samples were then twice passed over hydroxy-apatite columns to remove double-stranded material. The specific activity of the probes was carefully determined by back hybridization of the separated strands and comparison of the rate of reassociation to that of a known standard.

The concentration of a specific RNA molecule in a population was determined by hybridizing increasing amounts of RNA with a constant but excess amount of probe. The hybrids were scored by S1 nuclease resistance or by hydroxy-apatite column separation of single stranded and double stranded material. The initial slope of the saturation curve provides a direct measure of the number of hybridizing RNA molecules per unit mass of total RNA.

Figure 1, 2 and 3 provide the DNA sequence and codon translations of the probes used in these experiments (Sures, Lowry and Kedes, 1978). Three kinds of probes were used: probes corresponding to DNA segments transcribed into H2b and H3 mRNA, their complementary strands, and probes constructed from a region of spacer DNA that lies just

Figure 1

p2b-3 H2b CODING PROBE

```
met val ile met asn ser             asp ile phe glu arg ile ala gly glu ser ser
ATG GTC ATC ATG AAC AGC TNN GTN NAC GAC ATC TTC GAG CGA ATT GCC GGC GAA TCT TCC

arg leu ala gln tyr asn lys lys             ser ser arg glu ile gln thr ala val
CGC CTC GCT CAG TAC AAC AAA AAG TNN ACN NTC AGC AGT CGC GAG ATT CAG ACC GCC GTC

arg leu ile leu pro gly glu leu ala lys his ala val ser glu gly thr lys ala val
CGC CTC ATT CTC CCC GGA GAG CTG GCA AAG CAC GCT GTG AGC GAG GGT ACC AAG GCA GTG

thr lys tyr thr thr ser lys end
ACG AAA TAC ACT ACC TCC AAG tag ACAGGTCATATCCTGCTCTAATTGGACATAATACANNGG
```

FIGURE 2

pH3-1 H3 CODING PROBE

```
ile arg arg tyr gln lys ser thr glu leu leu ile arg lys leu pro phe gln arg leu
ATT CGC CGC TAC CAG AAG AGC ACT GAG CTT CTC ATC CGA AAA CTG CCA TTC CAG CGT CTA

val arg glu ile ala gln asp phe lys thr glu leu arg phe gln ser ser ala val met
GTG CGT GAG ATT GCA CAG GAC TTC AAG ACA GAG CTA CGT TTC CAG AGT TCC GCT GTG ATG

ala leu gln glu ala ser glu ala tyr leu val gly             asp thr asn leu cys
GCC CTT CAA GAA GCC AGC GAG GCA TAC CTA GTT GGC CNN NNN NGA GAC ACC AAC CTG TGT

ala ile his ala lys arg val thr ile met pro lys asp ile gln leu ala arg arg ile
GCC ATC CAC GCC AAG AGG GTT ACC ATC ATG CCC AAA GAC ATC CAG CTC GCC CGT CGa atc

arg gly glu arg ala end
cgc gga gaa cgc gcc tag
```

FIGURE 3

p2b-1 H2b SPACER PROBE

```
        10         20         30         40         50
CCTTTAGATA TATTATATAG GTTCCCTTTT TCATTTGCTG TAACCAAATA

        60         70         80         90        100
ATTTTGAAAT ACAATGTCAA AGATATTTCA TTGCTCATTT CTGTTATATC

       110        120        130        140        150
ATTACGTTTA TTATTATTGT CTATCATTGT GTATATTGTG TGAGAAGAAA

       160        170        180        190        200
TGAAAAATAA ATTCGCTCTT TACATATAGA GAGAGCTACT ACTACTACTA

       210        220        230        240        250
TATAATTATA TATCTCCGAG ATAATGTCAA AAGTAGGAGT GANNNNNNNN

       260        270        280        290        300
NNNNNNNTGC AAGTATGTCT TGGATAAAAA GTCTCGACAT GTTCCATATT

       310        320        330        340        350
CCCATCAAAA TTCATCGTCT TCTTCAACTT TTTCACATTT CTCATTCTTT

       360        370        380        390        400
GGGGATTGAA TTGAACAATG CAGACAGANN NNNNNNNNTG CCCGTATGAT

       410        420        430        440        450
CACTATGTCG CCATCTCTAG GCAGGGGATG GAACAGGCAC TAANCTGCGA

       460        470        480        490        500
CGCCTAAGAC CAATGAAAGG ATCGAGACCG AGGCTCATTT GCATACGGAC

       510        520        530        540        550
CGCAGCATAC Ggatccggcc ccgtgtataa aaaggaaagg ttctcgctgg

       560        570        580        590        600
ccattcacag tatccaaaga atatttgctt gacatactcg tttcgctgca

       610        620        630
tctttacaga ccagaaaacc tcaattcatc atg
```

upstream from the mRNA map position (Levy, Sures and Kedes 1979). In each of the figures, bases in capital letters are in the probe and those in lowercase are not. The H2b coding probe (Figure 2) includes 39 nucleotides present in the 3' end of the mRNA but not translated. The relation of the spacer probe to the H2b mRNA is seen in Figure 3 where the junction between nucleotides in the probe (upper case) and those in adjacent to the probe (lower case) is 9 bases 5' to the TATAAAAA box. The location of the capping box (Sures, Levy and Kedes, 1980) and the initiation codon are designated by underlining.

Methods

Methods for the culture of sea urchin embryos and RNA extraction have been published elsewhere (Childs, Maxson and Kedes, 1979). Preparation of previtellogenic oocytes and oocyte RNA was performed as described by Hough-Evans *et al* (Hough-Evans, Wold, Ernst, Britten and Davidson, 1977). Titration of specific RNA sequence concentration was performed essentially as described by Scheller *et al* (Scheller, Costantini, Kozlowski, Britten and Davidson (1978) except that the probe was labelled with I^{125} (Chan, Ruyechan and Wetmur, 1976). hybridizations using the H3 and H2b coding region clones were scored with S1 nuclease (Wickens, Buell, and Schimke, 1978) in the presence of 5 per cent glycerol and 12 mM mercaptoethanol.

The DNA fragments used in these experiments were derived from subclones of pSp2 and pSp17 (Cohn, Lowry and Kedes, 1976). The methods of bacterial growth, plasmid extraction and isolation of restriction fragments have been described elsewhere (Cohn and Kedes, 1979).

Results

Histone mRNA Accumulation In Oogenesis

The number of mRNA molecules were measured in RNA from both previtellogenic oocytes and mature, post-meiotic eggs. Radiolabelled, single stranded probes corresponding to both transcribed and non-transcribed coding regions of H2b and H3 genes were hybridized in solution to increasing amounts of whole gamete RNA. The probes were in molar excess over the hybridizing complementary sequences. Similar experiments were performed with probes isolated from adjacent spacer region DNA. The hybrids were scored by S1 nuclease digestion resistance or by their ability to bind to hydroxyapatite.

The results are presented in Table I. We detect no histone mRNA molecules in previtellogenic oocytes. Nor are there any sequences which hybridize to the non-transcribed strand of the histone gene coding region probe. Mature eggs, on the other hand have accumulated large numbers of histone mRNA molecules. There are about 10^6 molecules of both H3 and H2b messengers in each egg. From the known molecular length and sequence of each RNA, we calculate that this is equivalent to 29.5 fg of H2b mRNA and 27.4 fg of H3 mRNA per egg. Since the amount of RNA per *S. purpuratus* egg is approximately 2.9 nanograms, the two histone mRNAs each make up nearly 0.01 per cent of egg RNA.

TABLE I

NUMBER OF RNA MOLECULES HYBRIDIZING TO HISTONE DNA

	Spacer		H2b		H3	
	+*(a)*	-	+	-	+	-
oocyte*(b)*	<5×10^9(c)	<5×10^9(c)	<1×10^7(c)	<1×10^7(c)		
egg	<40	<40	1.1×10^6	<5×10^3	0.97×10^6	
5.75 hr	<40	1700	1.1×10^6		1.0×10^6	
9 hr	1300	6100	1.4×10^6		3.4×10^6	
11 hr	3500	1800	8.6×10^6	<3000 *(c)*	9.9×10^6	<3000 *(c)*
13 hr	2300	500	11.3×10^6		8.3×10^6	
16 hr	1200	300	2.2×10^6		3.5×10^6	
19 hr			1.3×10^6		3.5×10^6	
22 hr	1200	800	1.2×10^6		4.4×10^6	

(a) Probes labelled "+" are derived from the strand that hybridizes to mRNA.

(b) Pre-vitellogenic oocytes; the amount of RNA per cell is not known and the results are expressed as the molar fraction of hybridizing RNA.

(c) These numbers represent calculated maximum values derived from the RNA/DNA ratio of the hybridization reaction. In no case was any hybridization actually detected.

In strong contrast, there are no detectable egg RNA sequences that hybridize to probes from the non-coding strand of the mRNA region.

Histone mRNA Accumulation During Embryogenesis

Dramatic changes in histone mRNA content begin at about 5-6 hours after fertilization. An accelerating accumulation reaches its peak at about 12 hours (200 cells) for both mRNAs examined when there are about 10×10^6 molecules of each of H2b and H3. The absolute numbers of each RNA at each time point are probably within experimental error. Thus the number of mRNA molecules per embryo rises 10 fold and in concert over a brief 6 hour period of development. The number of nuclei per embryo during this same period rises only some 2-3 fold.

An abrupt decrease in both sets of mRNAs occurs between 15 and 20 hours. The decay of H2b sequences has a half-lifetime (T 1/2) of 1.5 hours. This measurement must be the maximum half-lifetime since our calculation does not take into account that transcription of histone mRNAs is still proceeding.

At no point in development are transcripts of the non-coding strand detected in embryo RNA for either the H3 or H2b probes.

Spacer Transcripts

The data in Table I also demonstrate that a few transcripts accumulate that do hybridize to the pre-H2b spacer probes. In experiments not reported here (A. Mauron and L. Kedes, unpublished) we have used an analogous pre-H3 spacer probe that also detects a few transcripts. Both sets of spacer transcripts are present in few copies relative to the mRNAs transcribed from adjacent DNA sequences. Neither the rise nor fall of the complementary H2b spacer transcripts is coincidental with each other or with that of the mRNAs. Most importantly, more than 95 per cent of the hybrids formed with spacer probe denature at temperatures 12-15 ^{0}C below that of homologous hybrids. Thus we conclude that there is essentially no transcription of the H2b and H3 spacer regions from which the probes were derived.

Discussion

Accumulation of mRNA During Oogenesis

Maturation of female gametes from previtellogenic oocytes to eggs may take as long as 5-6 months (Piatigorsky, 1975). During this interval a number of gene products accumulate in the growing oocyte. Among them are the elements essential for carrying out the metabolically active events that occur after fertilization. Thus it is no surprise that DNA polymerase, RNA polymerase, ribosomes and the other elements of protein synthesis all accumulate and are stored in egg cytoplasm during oogenesis. The presence of storage form mRNA has been well established. Two biological features may account for the fact that histone mRNAs make up such a large fraction of egg RNA in the sea urchin. First, there is no appreciable pool of preformed histone proteins stored in the egg. Secondly, DNA (and chromatin) replication is so rapid following fertilization that large amounts of histone protein are required to keep up with the rapid rate of cell division. Thus, the evolutionary strategy devised for the urchin seems to be the accumulation of enough histone mRNAs to provide for the histone proteins needed by the first few cleavage divisions.

The histone mRNAs stored during oogenesis appear to be identical to those newly synthesized during cleavage stage of subsequent development (Childs, Maxson and Kedes, 1979). Thus it appears likely that the same genes active in transcribing histone mRNAs during oogenesis are the same genes responsible for the transcription of the histone mRNAs that are synthesized and accumulate during embryogenesis. However, the formal possibility remains that, given the heterogeneity of histone genes in general and within a single species in particular, there may be a specialized subset of gene(s) that are active in oogenesis and a different subset active in embryogenesis. Each would code for identical or nearly identical mRNAs (and proteins).

Even two (diploid) single copy genes in an oocyte nucleus, working continuously for a 5-6 month period of maturation, are capable of generating 10^6 copies of an histone mRNA. Experiments in progress in collaboration with B. Hough-Evans and E. Davidson will attempt to determine if accumulation is continuous during oogenesis or whether it is confined to a particular stage of maturation.

Histone mRNA Accumulation During Cleavage

Although histone mRNA synthesis is detectable from the earliest stages of embryogenesis (Childs, Maxson and Kedes, 1979; Maxson, 1978), our experiments fail to detect significant increments of accumulation per embryo until about 8 hours after fertilization (150-200 cells). This data is consistent with the evidence that the rate of histone gene transcription rises 4-8 fold beginning at about 6-7 hours (Maxson, 1978).

We believe it is no coincidence that the two mRNAs examined here have essentially identical stoichiometries as pertains to both their storage pool in eggs and their absolute accumulation during oogenesis. The 1:1 stoichiometry of H2b and H3 proteins in chromatin is well established. There are several possible mechanisms whereby such stoichiometry might be regulated. First there might be close regulation of mRNA synthesis. If different amounts of RNA were made, then differential stability or efficiency of translation might also account for the protein products being stoichiometric. Our data now allow us to formally distinguish these possibilities.

The absolute numbers of the two mRNAs per embryo are essentially identical. The two proteins are produced at identical rates and are stable. Thus we conclude that each mRNA must be translated with the same efficiency. Since the rates of transcription (Maxson, 1978) and accumulation (this work) of the two mRNAs are essentially identical, then their stabilities (half-lifetimes) must also be the same. We believe that these two linked genes, and perhaps all histone mRNA genes, are coordinately regulated and subject to the same set of metabolic regulatory pathways as regards their transcription, the stability and storage of their mRNA products and the efficiency of their translation. Since the H1 protein apparently accumulates with one-half stoichiometry, we await with interest the results of our experiments on H1 gene transcription and mRNA accumulation.

No Spacer Transcripts

The basic structure of the linked coding regions immediately suggests an underlying regulatory mechanism. Co-transcription has been put forward as a likely possibility (reviewed in Kedes, 1979) and high molecular weight molecules that hybridize to histone gene probes have been described in both Hela cells and sea urchins (Melli et al, 1978; Spinelli et al 1980). However, experiments from our laboratory have been unable to detect high molecular weight RNA molecules using spacer DNA probes hybridized to RNA separated on sucrose or acrylamide gels (unpublished data of G. Childs and S. Levy). However, such an approach might miss rare molecules that only occur in a few copies per cell. Thus we were particularly interested to bring the DNA probe excess titration method to bear on this question. The data presented in Table I do demonstrate the presence of RNA-DNA hybrids formed by spacer probes derived from DNA segments adjacent to the 5' end of the H2b mRNA capping site. The total number of such RNA molecules accumulated is more than 10^3 fewer at its peak than the number of mRNA molecules accumulated.

However, when the spacer hybrid homology is examined by thermal sensitivity and scored by hydroxyapatite binding, the Tm of the RNA-DNA hybrids is 12-15 ^{0}C less then the homologous DNA-DNA hybrids. A similar result obtained using probes from the H3 spacer region (data not shown). Thus we conclude that there are no detectable transcripts of the spacer DNA immediately upstream from the two histone genes we have examined to date. We can detect as few as one such transcript per 10 cells.

Kinetic considerations preclude the possibility that the histone mRNA sequences accumulate as the end product of a high molecular weight transcription event that is then followed by rapid turnover of 5' leader sequences. If histone mRNAs were in fact processed from high molecular weight precursors, then one can estimate the number of such sequences produced per unit time. Nuclear RNA is transcribed at about 5 nucleotides per second (Aronson and Chen, 1977) and histone mRNAs appear to be synthesized at about the same rate (Maxson, 1978). If we assume that each histone mRNA that accumulates is derived from a 1000 base transcript, then each molecule has a transit life-time of 3.3 minutes. Our data demonstrate that between 5 and 12 hours of development, each cell of a 100-200 cell embryo accumulates about 10^4 molecules per hour of each histone mRNA. Thus at least .055 of the 10^4 molecules (550 molecules) per cell are in process of being synthesized at any moment. These numbers are far in excess of our lower detection limit of 1 molecule per 10 cells. We conclude that between fertilization and gastrulation, there is essentially no transcription from these histone gene spacer regions and, as a correlary, histone gene transcripts are probably no larger than the histone mRNAs themselves.

The spacer sequences of histone genes do not have a random base composition or sequence. They contain small repeating elements. It is likely that some portions of the spacer probes bear limited homology with many regions of the genome, some of which are transcribed. Thus it is not clear that the RNA-DNA hybrids formed with the spacer probes are anything other than random associations.

No Transcripts of Non-coding Strand

No RNA sample contained sequences that were capable of hybridizing to probes constructed from the non-coding histone mRNA strand. Thus we conclude that all RNA transcripts from the histone genes are strand specific. It has been noted that in the sea urchin embryo, almost 40 per cent of the DNA sequences are transcribed to some extent (Hough *et al*, 1975). This means that nearly 80 per cent of the DNA regions are transcribed. When strand separated probes to moderately repetitive DNA are used to asses transcription, both strands are found to be transcribed. Our results suggest that this is not the case for the histone genes.

Acknowledgements

The authors are grateful to Barbara Hough-Evans for provision of previtellogenic RNA and to her and Eric Davidson for seminal discussions and hospitality. This work was supported in part by grants from the National Institutes of Health, The American Cancer Society, and the Veterans Administration. A.M. is a Fellow of the Swiss National Research Fund. L.H.K. is an Investigator of the Howard Hughes Medical Institute.

References

Aronson, A. and K. Chen 1977. *Develop. Biol.* 59:39-48

Cohn, R.H. and L.H. Kedes 1979. *Cell* 18:843-53

Cohn, R.H., J. Lowry and L.H. Kedes 1976. *Cell* 9:147-61

Chan, H.C., W. Ruychan and J.G. Wetmur 1976. *Biochemistry* 15:5487-90

Childs, G., R. Maxson and L.H. Kedes 1979. *Develop. Biol.* 73:153-73

Hough, B.R., M. Smith, R.J. Britten and E.H. Davidson 1975. *Cell* 5:291-99

Hough-Evans, B.R., B. Wold, S. Ernst, R.J. Britten and E.H. Davidson 1977. *Develop. Biol.* 60:258-77

Kedes, L.H. 1979. *Ann Rev. Biochem.* 48:837-70

Levy, S., I. Sures and L.H. Kedes 1979. *Nature* 279:737-39

Maxson, R., 1978. *Ph.D. Thesis* Univ. of Calif., Berkeley

Melli, L., G. Spinelli, H. Wyssling, and E. Arnold 1977. *Cell* 11:651-61

Piatigorsky, J. 1975. In *The Sea Urchin Embryo* G. Czihak *Ed.* Springer-Verlag, New York

Scheller, R., F. Costantini, M. Kozlowski, R.J. Britten and E.H. Davidson 1978. *Cell* 15:189-203

Skoultchi, A. and P.R. Gross 1973. *Proc. Nat. Acad. Sci., USA* 70:2840-44

Spinelli, G., M. Melli, E. Arnold, C. Casano, F. Gianguzza, and M. Ciaccio 1980. *J. Mol. Biol.* in press

Sures, I., J. Lowry, and L.H. Kedes 1978. *Cell* 15:1033-44

Sures, I., S. Levy and L.H. Kedes 1980. *Proc. Nat. Acad. Sci., USA* in press.

Wickens, M., G. Buell and R.T. Schimke 1978. *J. Biol. Chem.* 253:2483-95

NUCLEIC ACID MEDIATED GENETIC CHANGES IN *NEUROSPORA CRASSA*

N. C. Mishra

Dept. of Biology, University of South
Carolina Columbia, S. C. 29208, U.S.A.

Abstract An unequivocal proof for transformation of Neurospora is provided. A mutant strain with a temperature sensitive (*ts*) inositol requirement for growth was used as a donor in the transformation of another mutant recipient; latter required inositol for growth at any temperature. The recipient was found to yield temperature sensitive transformants following treatment with the donor (*ts*) nucleic acid preparation. The characteristics of the *Neurospora* transformation system is discussed.

Introduction

A study of genetic transformation can provide an insight into the molecular mechanism of DNA repair, recombination and gene expression. Transformation (as mediated by naked DNA molecules or isolated chromosomes) has now become an established method for the transfer of genetic information from one cell type to another both in prokaryotes and eukaryotes. Recently such a role of isolated nucleic acid molecules in the genetic transformation of *Neurospora crassa* has been described (Mishra Szabo and Tatum 1973, Mishra and Tatum 1973, Mishra 1976, Mishra 1979). Our original findings regarding the transformation of *Neurospora* by DNA molecules have been recently confirmed (Schablik *et al* 1976, Case *et al* 1979, Szabo and Schablik 1980). Mishra has independently provided an unambiguous evidence for transformation of *Neurospora* by the transfer of a temperature sensitive (ts) character (requirement for inositol) from a donor to a recipient strain (Mishra 1979).

A further unequivocal proof of transformation is provided in this paper. Data are presented to show that transformation of a young recipient culture can be performed routinely and that the transformants have either a temperature sensitive (*ts inl*) or a wildtype (non *ts* inl^{+}) phenotype for inositol requirement depending on the nature of donor DNA used (i.e. whether the recipient is treated with DNA from a temperature sensitive (*ts* inl^{-}) or from a wildtype (non *ts* inl^{+}) strain).

Materials and Methods

A multiple marked strain 2506 (A *rg*$^-$, *arg*$^-$ *pan*$^-$) of *Neurospora crassa* carrying an *inl*$^-$ allele (89601) was used as the recipient. The allo-DNA was prepared from the wild-type strain (RL 3-8A) or from an allelic temperature sensitive strain (RL3-8A-2) which required inositol for growth at 37° C but not at 25° C. The iso-DNA was prepared from the strain (2506) used as recipient in the transformation experiments. These strains have been described elsewhere (Mishra 1979). DNA was prepared as described by Marmur (1961); mRNA was prepared as described by Mishra, Niu and Tatum (1974). DNase and RNase treatments were performed as described in an earlier paper by Mishra, Szabo and Tatum (1973). The transformation experiments were carried out by treatment of a young recipient culture (20 - 30 hr old) with Nucleic acid (50 μg/ml) as described by Mishra (1979). Presumptive transformants were selected by plating nucleic acid treated culture on a medium without inositol (Mishra, Szabo and Tatum 1973) and were then examined for their temperature sensitive (*ts inl*) phenotype with respect to inositol requirement (Mishra 1979).

Results

Data presented in Table 1 show that allo-DNA mediated transformation of *inl* recipient can be performed reproducibly with the use of a young culture. In such an experiment, the iso-DNA was found to be ineffective in producing transformants; also a pretreatment of allo-DNA by DNase was found to abolish

TABLE 1

Transformation of inl⁻ recipient

TREATMENT	TOTAL NUMBER OF COLONIES EXAMINED	NUMBER OF inl^+ TRANSFORMATION SCORED	TRANSFORMATION FREQUENCY ($\times 10^6$)
1. None	8.4×10^6	0	
	9.2×10^6	1	
	6.3×10^6	0	0.042
2. Allo-DNA	3.3×10^6	5	
	0.5×10^6	1	1.62
	4.2×10^6	7	
3. Allo-DNA treated with DNase	4.6×10^6	0	
	10×10^6	0	0
	3.8×10^6	0	
4. Iso-DNA	4.2×10^6	0	
	8.7×10^6	1	0.048
	7.6×10^6	0	

A young culture (20-30 hours old) of the recipient strain (inl^-, arg^-, pan^-, rg^-) was treated with 50 ug/ml of allo-DNA at 30°C for 40-48 hours and then scored for inl^+ colonies on medium without inositol.

TABLE 2

Specific effect of Allo-DNA in transformation of inl^- recipient

Source of Allo-DNA used in the transformation experiment	No. of transformants examined	No. of transformants of each genotype: inl ts	inl^+ (non ts)
1. Wild type strain (ie non-ts inl^+)	135	0	135
2. Temperature sensitive mutant strain (ie ts inl^-)	98	77	21

A young culture of the recipient strain (inl^-; 89601) was treated with Allo-DNA preprations (from the wild type RL3-8A or from a temperature sensitive strain, (RL3-8-2A), colonies growing on medium without inositol at 25°C were examined for their ts or non-ts phenotype (with respect to inositol requirements for growth). ts strains required inositol for growth at 35° C but not at 25°C; non-ts inl^+ strains did not require inositol for growth at any temperture.

its transforming activity. In three parallel experiments only the allo-DNA was found to yield inl$^+$ transformants; the frequency of transformation was 1.6 x 10^{-6}; this represents a 50-100 folds incrase over the reversion frequency (0.02 - 0.04 x 10^{-6}) (see Table 1). In experiments using recipient culture older than 30 hrs, the frequency of transformation was found to be inconsistent (data not shown).

Allo-DNA prepared from the wildtype and from a temperature sensitive strain was used in order to investigate the specificity of donor DNA in the transformation of an inl recipient. The data presented in Table 2 clearly show that the phenotype of a transformant is a function of a donor DNA used. All 135 transformants obtained after treatment of the inl recipient with a wildtype DNA preparation were found to possess wild-type phenotype (i.e. did not require inositol for growth at any temperature). A similar treatment of the inl recipient with allo-DNA from a temperature sensitive donor yielded 77 temperature sensitive transformants out of a total of 98 transformants examined. Thus the data in Table 2 provide an unequivocal proof for the donor DNA specific genetic transformation in *Neurospora*.

The data presented in Table 3 show the inducing effect of allo-mRNA in the reversion of inl recipient. The frequency of inl$^+$ reversion mediated by allo-mRNA is comparable to the value reported earlier (Mishra, Niu and Tatum 1974). The effect of allo-mRNA was specific since a pretreatment with RNase was found to abolish its activity (see Table 3).

TABLE 3

Effects of RNA treatment on the genetic reversion of inl⁻ locus in Neurospera Crassa

TREATMENT	TOTAL NO. OF COLONIES EXAMINED	REVERTANTS TOTAL NO.	REVERTANTS FREQUENCY (x 10^6)
1. None	6.1×10^6	0	0
2. Allo-mRNA	14.7×10^6	6	0.406
3. RNase treated Allo-mRNA	5.6×10^6	0	0

A young culture of the recipient strain was treated with mRNA and then scored for the inl^+ colonies on a medium without inositol.

TABLE 4

Transmission pattern of inl^+ character by the transformants/revertants among their sexual progency

TREATMENT	TRANSMISSION PATTERN[1]	COMMENTS
Allo-DNA	Mendelian & Non-Mendelian	Transformation with or without intergation of donor genetics information.
Allo-mRNA	Mendelian	Transformation with intergration of donor genetic information.
None	Mendelian	Genetic reversion only.

[1] A segregation pattern of 4 inl^+ and 4 inl^- spores in an ascus (from a cross inl^+ x inl^-) was considered as Mendelian Transmission whereas any departure from this ratio (commonly 0 inl^+ : 8 inl^-) was considered as non-Mendelian.

Discussion

The data presented in this paper provide additional unequivocal proof for genetic transformation in *Neurospora crassa*. The occurrence of genetic transformation mediated by total DNA preparation has been recently confirmed (Schablik *et al* 1976, Case *et al* 1979, Szabo and Schablik 1980). The transformation of *inl* recipient using a hybrid plasmid DNA (as reported for mutants of Quinic acid metabolism in Neurospora by Case *et al* (1979)) is not technically feasible due to the non-availability of inl^- mutation in *Escherichia coli*. At least some of the transformants reported by Case *et al* (1979) may be considered to arise as a result of mutation induced by donor DNA since the authors have not examined the effect of plasmid DNA alone on the reversion of *qa* alleles (Case *et al* 1979). The likelihood of mutagenesis by donor DNA as the cause of genetic reversion is further supported by their observation that the transformation of *qa* 4 gene was accompanied by the mutation of an adjoining gene ($qa\text{-}3^+$)(Case *et al* 1979). Such mutagenic effect of donor DNA in *Neurospora* has been reported earlier (Mishra 1976, Szabo and Schablik 1980).

The characteristics of the *Neurospora* transformants have been described earlier (Mishra and Tatum 1973, Mishra 1976). The *Neurospora* transformants are essentially of two kinds as based on their ability to transmit the transformed character to their sexual progeny. Both Mendelian and non-Mendelian transmission patterns of the transformed character (as summarized in Table 4) have been described (Mishra and Tatum 1973, Mishra Niu and Tatum 1974, Szabo and Schablik 1980). The involvement of a nuclease gene (nuc-1) in the donor DNA uptake and transformation of _inl_ recipient is described (Mishra 1979).

Summary

An unequivocal proof for transformation of _Neurospora_ is provided. A mutant strain with a temperature sensitive (_ts_) inositol requirement for growth was used as a donor in the transformation of another mutant recipient; latter required inositol for growth at any temperature. The recipient was found to yield temperature sensitive transformants following treatment with the donor (_ts_) nucleic acid preparation. The characteristics of the _Neurospora_ transformation system is discussed.

Acknowledgement

This work was supported by a grant from the Dept. of Energy (contract No. E.P.-78-S-09-1071).

References

Case, M.E., Schweizer, M., Kushner, S.P. and Giles, N.H. (1979) Proc.Natl.Acad.Sci (U.S.A.) 76, 5259-5263.

Marmur, J. (1961) Jour.Mol.Biol. 3, 208-218.

Mishra, N.C. (1979) Jour.General.Microbiology 113, 255-259.

Mishra, N.C. (1977) Genetical Res. 29, 9-19.

Mishra, N.C. (1976) Nature 264, 251-253.

Mishra, N.C., Niu, M.C. and Tatum E.L. (1974) Proc.Natl.Acad.Sci. (U.S.A.)

Mishra, N.C., Szabo, G. and Tatum E.L. (1973) in The role of RNA in Development and Reproduction (Ed. Niu, M.C. and S.J.Segal North Holland Publishing Co.). 261-273.

Mishra, N.C. and Tatum E.L. (1973) Proc.Natl.Acad.Sci. (U.S.A.) 70, 3875-3879.

Schablik, M., Szabolcs, M., Kiss, A., Aradi, J., Zsindley, A. and Szabo, G. (1977) Acta Biologia Academae Scientiarum Hungaricae 28, 273-279.

Szabo, G. and Schablik, M. (1980) Acta Biologia Academae Scientarum Hungaricae (in press).

GENETIC MANIPULATION IN HIGHER ORGANISMS. II. THE mRNA-MEDIATED TRANSFER OR ERYTHROCYTE MESSAGE FROM RABBIT TO GOLDFISH+

Lillian Chang Niu*, Xue Guo-xiang** and M. C. Niu*
(*Department of Biology, Temple University, Philadelphia, PA 19117
and **Institute of Developmental Biology, Academic Sinica, Beijing)

ABSTRACT. Subsequent to the finding that carp liver mRNAs play a role in the formation of the liver specific lactate dehydrogenase isozymes ($LDH-C_4$)/(17), we have investigated the role of globin mRNA in goldfish development. The mRNA we used was a gift from Dr. Jerry Lingrel, University of Cincinnati College of Medicine, Ohio, and was isolated from reticulocytes of anemic rabbit. It was injected into fertilized goldfish eggs. The goldfish that developed look normal. Blood from 7 of the mature goldfish was withdrawn from the caudal artery 4 times at monthly intervals. After washing with physiological saline 3 times, red blood cells (RBC) from each withdrawal were homogenized with a phosphate buffer (0.1 M, pH 7) and centrifuged. The supernatant was subjected to the Ouchterlony double diffusion test, starch gel electrophoresis for LDH and starch gel electrophoresis for glucose-6-phosphate dehydrogenase (G-6-PD). The RBC in 2 of the 7 goldfish were found to possess (1) rabbit Hb, (2) an additional LDH band corresponding to one of rabbit-RBC LDH and (3) additional band corresponding to one of rabbit G-6-PD (this phenotypic expression is less distinct). None of the three was found in controls.

INTRODUCTION

During oogenesis, genomic readouts of developmental information are released into the cytoplasm (1,2). Upon fertilization, they are segregated into recognizable particles; for example, germ plasm (3) and gray crescent (4) in amphibians, anteriorizing factors in Smittia (5), polar granules in Drosophila (6), polar lobe in Dentalium and Ilyanassa (7), etc. These particles contain RNA and are sensitive to ultraviolet light irradiation. The damage caused by irradiation can be repaired by microinjection of cytoplasm from non-irradiated eggs (1, 3, 8). Similarly, the lethal genetical defect in axolotl (1, 9) and in deep orange embryos of Drosophila (10) has been corrected by microinjection of cytoplasm from normal eggs. The chemical nature of the active component in cytoplasm has not been defined. Some authors found evidence for protein or acidic protein (9, 11, 12) while others for RNA or polynucleotide (5, 13). However, mRNAs have been isolated from the eggs of sea urchin, amphibian and fish. Identifiable mRNAs from sea urchin and amphibian are of two kinds: histone and tubulin-mRNAs (14, 15, 16). They encode the synthesis of the "house keeping proteins" required for cleavage or cell division. In contrast, fish eggs contain mRNAs with the potential to prime the synthesis of albumin and globin (see Yu & Niu, in this volume, pp. 893). Furthermore, the livers from goldfish separately injected with carp egg- and carp liver-mRNAs possess the same additional intermediate band between the 2 cathodally migrated bands of the

---

+An abstract of this work appeared in Am.Zool., 19(3):978,1979.
This work was supported in part by Rockefeller Foundation, New York, N.Y.

LDH-C^4 (17). None of the other organs in the injected goldfish had altered LDH patterns. These results lead to the conclusion that egg mRNAs contain a liver-forming subfraction.

The translation capacity of mRNA has been tested proficiently in Xenopus oocytes (18). When rabbit and mouse globin mRNAs were separately injected into fertilized Xenopus eggs, rabbit/mouse globins were detected in tissues from all parts of the young embryo (19). However, the authors had not specified where in tissues of Xenopus the alien globin was. There is good reason to believe that rabbit/mouse globins are localized in RBC. In order to test this possibility, rabbit reticulocyte (RE)-mRNAs were injected into fertilized goldfish eggs. The amount injected was 0.05 - 0.1ul of Niu-Twitty solution with and without the Re-mRNA (2ug/ul). If injected mRNAs participate in the development of red blood cells, genotypic characters of rabbit RBC would be detected in goldfish RBC.

MATERIALS AND METHODS

Reticulocytes were collected from anemic rabbits produced by 5 daily subcutaneous injections of 2.5% neutralized phenylhydrazine hydrochloride. Polysomal RNAs were prepared from washed reticulocytes according to the procedure of Evans and Lingrel (20). Reticulocyte Poly-A attached RNAs were obtained from polysomal RNAs by affinity chromatography on oligo (dT) 12-18 cellulose using the procedure of Aviv & Leder (21).

Polyacrylamide gel electrophoresis of the Poly-A attached RNAs, twice chromatographed, is shown in Fig. 1. The appearance of a single band may

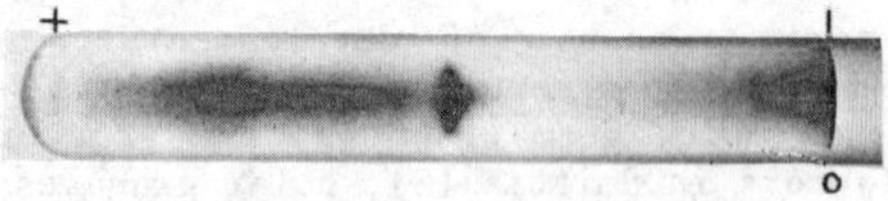

Fig. 1 Polyacrylamide gel electrophoresis of carp egg mRNAs. Note the presence of a single band - heterogenous mRNAs.

mean that the reticulocyte Poly-A attached RNAs are globin mRNA (as early workers have designated) or that reticulocyte Poly-A attached RNAs are heterogenous containing a number of species. They were tested for translation activity using a cell-free wheat germ translational system. Figure 2 shows the time course of incorporation of H^3-leucine into acid insoluble materials. Forty five minutes of incubation was optimal. The optimal amount of Poly-A-attached RNAs was 5 ug/50ul of the incubation mixture (Fig. 3).

The reaction products were analyzed using goat anti-rabbit globin serum (purchased from Cappel Laboratories, Pa.). The immunoprecipitates were collected by centrifugation and washed twice with a phosphate buffer containing Triton x-100 and leucine. They were dissolved in a phosphate buffer containing SDS (2%) and 2-mercaptoethanol (2%). The solution was heated to 100°C for 1-1/2 minutes to help dissociate the immuno-precipitates. After adding

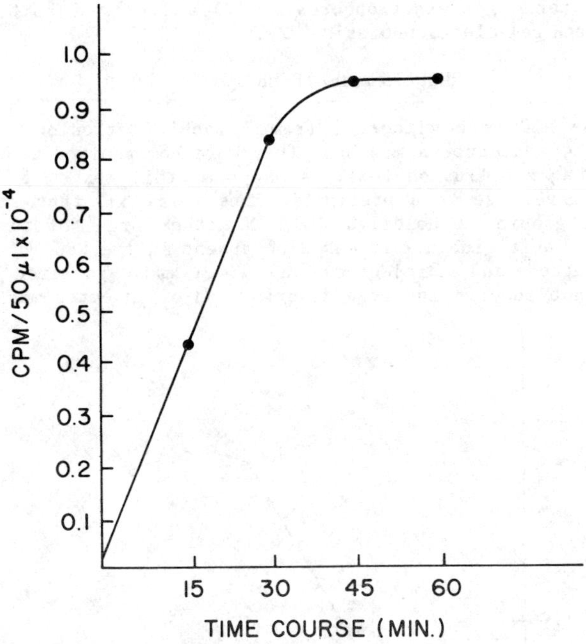

Fig. 2 Time course of carp egg mRNA mediated incorporation of H^3-leucine into acid - insoluble materials (Protein). Forty-five minutes were optimal.

some rabbit hemoglobin as carrier, the mixture was electrophoresed on polyacrylamide gel. One gel was stained and the other sliced into 1-2mm sections. Each slice was solubilized at 60°C in a vial containing 1 ml of a mixture (60% perchloric acid/30% H_2O_2, 2:1). The UV absorption at 280 was recorded with a Gilford spectrophotometer and radioactivity counted with a Beckman 9,000 scintillation counter. The results are shown in Fig. 4. Here you can see that the amount of rabbit globin is illustrated by a dotted line and the newly synthesized H^3- labled protein by a solid line. Both lines are peaked at a point where the band is localized on the gel. In other words, the newly synthesized protein and the authentic rabbit globin are immunologically and electrophoretically indistinguishable from each other. Therefore, the reticulocyte Poly-A attached RNAs we used in the present study are mRNAs.

The reticulocyte mRNA were injected into fertilized goldfish eggs. We obtained some 80 larval fishes. Their external morphology does not differ from the saline injected and liver mRNA injected goldfish. Half of the larval fish were sacrificed for blood in the Fall, 1978. Those blood samples were unfortunately lost by an accident. Only 7 of the rest matured in June, 1979.

Blood samples were collected from the caudal artery in June, July, August and September. All samples were washed, lysed and analysed for the presence of (1) Rabbit globin using Ouchterlony double diffusion test, (2) Rabbit component of LDH using starch gel electrophoresis (17) and (3). Rabbit component of G-6-PD using starch gel electrophoresis (22).

RESULTS AND DISCUSSION

Rabbit globin: The RBC were withdrawn from 7 rabbit reticulocyte mRNA injected goldfish on 4 consecutive months. They were homogenized and subjected to the Ouchterlony agar diffusion test. Among 4 monthly tests of each fish, only Nos. 1 and 7 gave rise to a precipitin line (Fig. 5), thus showing the presence of rabbit globin in goldfish RBC. No other organ of the injected goldfish contained rabbit globin. It was also absent in the RBC of both control (saline- injected and uninjected) and liver mRNA injected goldfish. These findings do not support the even distribution of injected mRNA and sub-

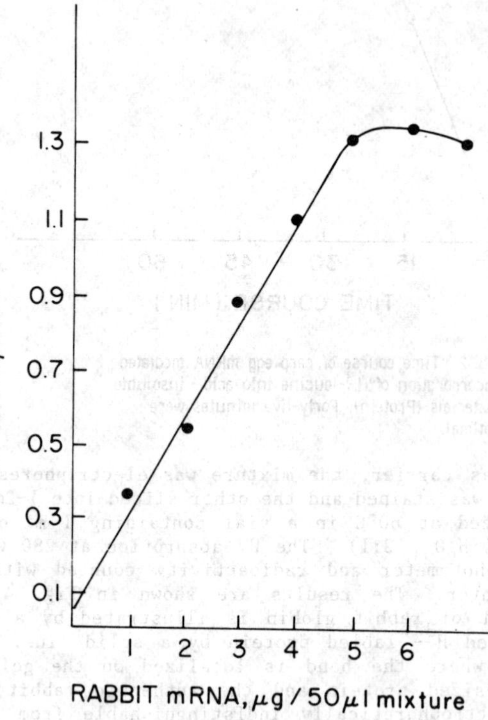

Fig. 3 The effect of mRNA concentration on the rate of the incorporation of H^3-leucine into protein.

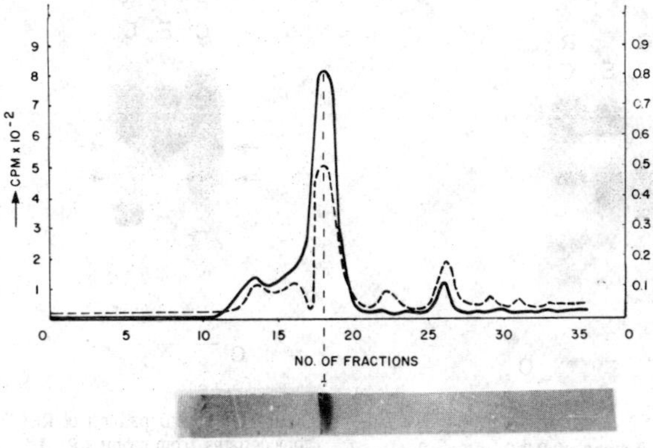

Fig. 4 Sodium dodecyl sulfate / polyacrylamide gel electrophoresis of the anti-rabbit hemoglobin serum precipitable material from the reaction mixture. Aliquots of the immunoprecipitates were dissociated by 2% SDS and 2% 2-mercaptoethanol in phosphate buffer, and then subjected to polyacrylamide gel electrophorsis. One gel was fixed and stained and another was sliced into 1-2mm sections and solubilized at 60°C in a vial containing a mixture (60% perchloric acid and 30% H_2O_2, 2 : 1). Both UV absorption at 280 and radio activity were recorded.

sequent translation of hemoglobin (19). On the contrary, they showed mRNA mediated the acquisition of rabbit globin in goldfish RBC . The proportion between goldfish and rabbit globins in the RBC had not been determined. However, we are now studying the physiology of the composite RBC with particular reference to the significance in evolution.

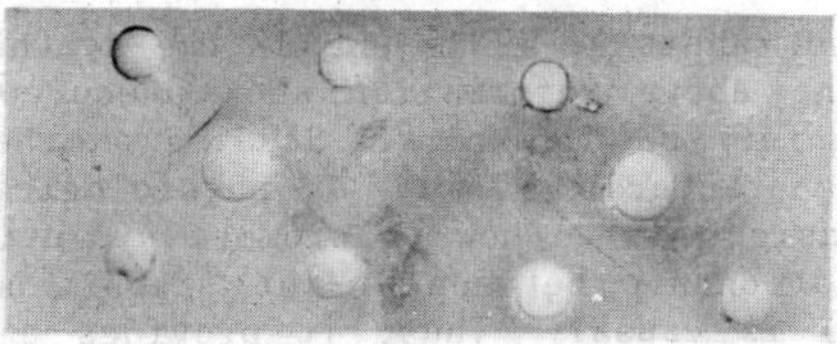

Fig. 5 Ochterlony agar diffusion test: Left plate-Center well contains goat antirabbit globin serum. Precipitation line occurs in-between central well and the well with homogenates from the injected goldfish #1 and right plate has precipitation line in-between central well and the well with homogenates from the injected goldfish #7. Other holes contain homogenates from two controls (saline injected and un-injected) and liver mRNA injected goldfish.

Lactate dehydrogenase isozymes (LDH): Starch gel electrophoresis of the homogenates showed that No. 1 goldfish gained the far anodal band of rabbit reticulocyte LDH (Fig. 6). Apparently this is a hybrid pattern of LDH between rabbit and goldfish RBC and, thus, reveals a genetic alteration of goldfish RBC. Nos. 5 and 7 also had an additional band, but not striking as in No. 1.

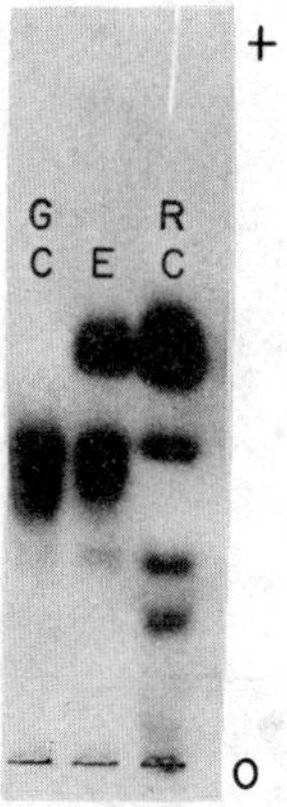

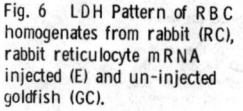

Fig. 6 LDH Pattern of RBC homogenates from rabbit (RC), rabbit reticulocyte mRNA injected (E) and un-injected goldfish (GC).

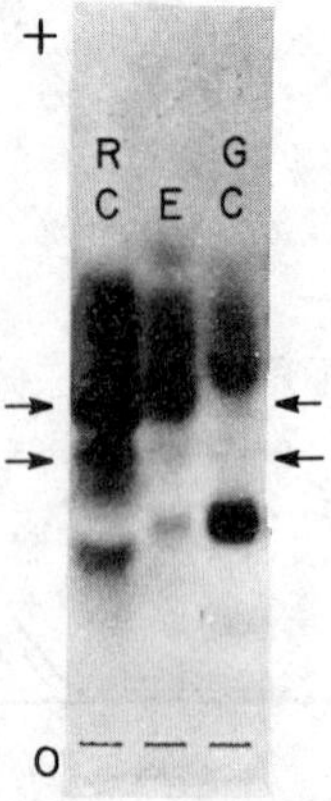

Fig. 7 G-6-PD pattern of RBC homogenates from rabbit (RC), reticulocyte mRNA injected (E) and un-injected goldfish (GC). Arrows point to where the differences are in the three.

Glucose-6-phosphate-dehydrogenase (G-6-PD): Both Nos. 1 and 7 goldfish possessed RBC with a faint band (lower arrow) of rabbit RBC's G-6-PD (Fig. 7). It can be seen that the band in E is less conspicuous than in RC. The extra band of E was present in the RBC obtained from 4 consecutive months. Using a similar procedure as we did in this paper, the reticulocyte mRNAs prepared by early workers (e.g., 18, 19, 20, 21 and Miles Laboratories, Inc.), have been designated as globin mRNA. On account of the 3 markers just described, it would seem proper to change globin-mRNA to reticulocyte-mRNAs.

Both carp egg mRNAs and rabbit Re-mRNAs encode the synthesis of globin. These data coupled with the above information show unequivocally that egg mRNAs contain a subfraction for RBC formation. Studies on the mRNA mediated acquisition of alien RBC characters as well as physiological significance may perhaps lead to some insight as to how to promote evolution in the laboratory. Besides, the relationship between in-put (rabbit mRNAs) and output (rabbit RBC's proteins) is causal. Thus, it provides a strong support for the use of mRNAs in the genetic manipulation of higher organisms.

The mode of action of the injected mRNAs is under active investigation. There are several theories and one of them is that the injected macromolecules act as the template for the synthesis of copy DNA (cDNA). The crux of the theme is availability of the activity of the enzyme (reverse transcriptase) in eggs and developing tissues. Recently, one of our co-workers has found the presence of an enzyme in fish eggs utilizing mRNA as the code for DNA synthesis (Yan et.al.,unpublished). The cDNA, thus obtained, would integrate with the genome in a like manner as the insertion of complimentary DNA into normal cells infected with oncogenic viral RNA (23,24).

LITERATURE CITED

1) Briggs, R. & G. Cassens (1966). Accumulation in the oocyte nucleus of a gene product essential for embryonic development beyond gastrulation. Proc. Nat. Acad. Sci. (U.S.A.), 55:1103.

2) Davidson, E. (1976). Gene activity in early development. 2nd edition (1976), Academic Press, New York and London.

3) Smith, L. D. and M. A. Williams (1975). Germinal plasm and determination of the primordial germ cells. In "The developmental biology and reproduction." eds. Markert & Papaconstantinou, P.3, Academic Press, New York.

4) Grand, P. and J. F. Wacaster (1972). The amphibian gray crescent region- a site of developmental information. Dev. Biol., 28:454.

5) Kardler-Singer, I. and K. Kalthoff (1976). RNase sensitivity of an anterior morphogenetic determinant in an insect egg. Proc. Nat. Acad. Sci. (U.S.A.), 73:3739.

6) Illmensee, K. and A. P. Mahowald (1974), Transplantation of posterior polar plasm in Drosphila. Induction of germ cells at the anterior pole of the egg. Proc. Nat. Acad. Sci. 71:1016.

7) Newrock, K. M. and R. A. Raff (1975). Polar lobe specific regulation of Ilyanassa obsoleta. Dev. Biol. 42:242.

8) Chung, H. M., and G. M. Malacinski (1975). Repair of U.V. irradiation damage to a cytoplasmic component required for neural induction in the amphibian egg. Proc. Nat. Acad. Sci. 72:1235.

9) Brigge, R. (1972). Further studies on the maternal effect of the O gene in the Mexican axolotl. J. Exp. Zool., 181:271.

10) Garen, A. and W. Gehring (1972). Repair of the lethal development defect in deep orange embryos of Drosophila by injection of normal egg cytoplasm. Proc. Nat. Acad. Sci. 69:2982.

11) Gurdon, J. B. (1977). Egg cytoplasm and gene control in development. Proc. Royal Soc. (B), 198:211.

12) Brothers, A. J. (1976). Stable nuclear activation dependent on a protein synthesized during oogenesis. Nature. 260:112.

13) Okada, M., I. A. Kleinman and H. A. Schneiderman (1974). Repair of a genetically caused defect in oogenesis in Drosophila melanogaster by transplantation of cytoplasm from wild type eggs and by injection of pyrimidine nucleotides. Dev. Biol. 37:55.

14) Gross, K. W., J. Ruderman, M. Jacobs-Lorena, C. Baglion and P. R. Gross (1973). Cell-free synthesis of histones directed by messenger RNA from sea urchin embryos. Nature. (London), New Biol. 241:272.

15) Raff, R. A., H. V. Colot, S. E. Selvig and P. Gross (1971). Oogenetic origin of mRNA for embryonic synthesis of microtubale proteins. Nature. 235:211.

16) Adamson, E. D. and H. R. Woodland (1977). Changes in the rate of histone synthesis during oocyte maturation and very early development of Xenopus laevis. Dev. Biol. 57:136 and Woodland and Adamson (1977). The synthesis and storage of histones during the oogenesis of Xenopus. Dev. Biol., 57:118.

17) Niu, M. C. and T. C. Tung (1977). Genetic manipulation in higher organisms. I. Goldfish ova as materials of operation: - mRNA mediated alteration of the liver specific isozymes. Scientia Sinica, 20:803.

18) Gurdon, J. B., C. D. Lane, H. R. Woodland and Marbaix (1971). Use of frog eggs and oocytes for the study of mRNA and its translation in living cells. Nature. 233:177.

19) Gurdon, J. B., H. R. Woodland and J. B. Lingre (1974). The translation of mammalian globin mRNA injected into fertilized eggs of Xenopus laevis. I, message stability in development, Dev. Biol. 39:125.

20) Evans, M. J. and J. B. Lingral (1969). Hemoglobin mRNA. Synthesis of 9S and ribosomal RNA during erythroid cell development. Biochemistry 8: 3000.

21) Aviv, H. and P. Leder (1972). Purification of biologically active globin mRNA by chromatography on oligo(dT)-cellulose. Proc. Nat. Acad. Sci. 69:1408.

22) Shaw, R. C. and R. Prssad (1969). Starch gel electrophoresis of enzymes-A compilation of receipts. Biochem. Genetics, 4:297.

23) Temin, H. M. and S. Mizutani (1970). RNA-dependent DNA polymerase in virions of Rous sarcoma virus, Nature. 226:1211.

24) Battimore, D. (1970). RNA-dependent DNA polymerase in virions of RNA tumor viruses. Nature. 226:1209.

THE EFFECT OF mRNA ON NUCLEAR ACTIVITY IN DEVELOPING SYSTEMS

M. C. Niu*

(Department of Biology, Temple University, Philadelphia, PA 19122
and
Institute of Developmental Biology, Academia Sinica, Beijing)

ABSTRACT

This report will discuss what we have done in the past few years. The word "we" means my group at Temple University and the collaborative team at the Institute of Developmental Biology. The central theme of our research is mRNAs in growth and differentiation. Back in 1973, we started a series of experiments dealing with the use of mRNAs in the analysis of organ formation in normal development. Accumulated data show clearly that goldfish eggs contain mRNAs with subfractions for tail and liver development. Recently reticulocyte-forming mRNAs were found also in fish egg mRNAs. These organ- or cell-forming mRNAs are highly specific and thus can be used as the basis for genetic manipulation of embryogenesis in animal as well as in plants. However, the mode of mRNA action is not clearly understood in the developing eggs. The equipotentiality of DNA and egg mRNAs in the mediation of tail transformation leads to the consideration that mRNAs may act as template for DNA synthesis. This is supported by recent finding of RNA dependent DNA polymerase in fish eggs. Other possibilities are that mRNAs may undergo replication or that mRNAs may transform into DNAs by a coupled processes of de-oxygenation and methylation of uridines.

Each phenotypic expression of an organism is derived from the dominance, co-dominance or recessiveness of the gene. The genes for tail of crucian and carp are dominant while that of goldfish are recessive. The fact that crucian- and carp-egg mRNAs mediated the tail transformation of goldfish and conversely goldfish egg mRNAs did not induce carp's tail change have led to the hypothesis that the tail-forming component of carp egg mRNAs is dominant and that of goldfish egg mRNAs recessive. Similarly, the rat liver mRNA mediated change of goldfish liver-specific C_4 bands of LDH (lactate dehydrogenase isozymes) can also be explained by the co-dominance of the LDH genes.

In adult tissue, mRNAs mediate differential gene activation. This is shown by functional study of liver mRNAs on uterine wall of young mice or castrated mice. Uterine wall is incapable of synthesizing albumin. However, injection of rat, calf or chicken liver mRNAs into the lumen resulted in the synthesis of two species of albumins: mouse and alien; rat, bovine or chicken. The production of mouse albumin is rich and of longer duration while that of rat, bovine or chicken albumin poor and shorter duration. Puromycin stopped the syntheses of both albumins. Dactinomycin inhibited the synthesis of mouse albumin and had no effect on the production of calf albumin. It appears, therefore, that exogenous mRNAs have a major function of activating the genome and a minor one of encoding protein synthesis.

In early embryogenesis of vertebrate animals there is dependent and independent differentiation. The type of dependent differentiation is shown by primary induction in which chordamesoderm induces the overlying ectoderm to develop into nervous tissue and sense organs. Independent differentiation deals with self differentiation of the presumptive organ-forming areas as illustrated by Vogt (1). If, therefore, the presumptive area for heart formation is excised from a urodele gastrula, in Hoftfreter solution it will develop into a beating heart (2). Similarly, extirpation of vegetal plasm in eggs and embryos of Discoglissus Pictus results in the development of a sterile animal (3). The agents for neural formation and germ cell production have been traced respectively to the cortex of gray crescent (4) and germ plasm (3,5,22) of the fertilized egg. It appears that cytoplasm of mature eggs contains agents for both dependent as well as independent differentiation. Cytoplasmic control of nuclear activity has long been known (6). Crucial tests came from homoplastic nuclear transplantation in Rana Pipiens (7) and in Xenopus laevis (8). However, it is xenoplastic nuclear transplantation that shows not only cytoplasmic effects on the implanted nucleus, but also species characters of cytoplasm. In a recent paper directed by the late Professor T. C. Tung (9), nuclei from carp blastula (Cyprinus carpio L.) were transplanted into the enucleated eggs of a crucian (Carassius auretus L.). The developed fish possess barbs (carp character), pharyngeal teeth (3-1-1 of carp instead of 4 as in crucian), vertebrae (26-30 crucian instead of 32-36 in carp), and lateral line scales (33-34 intermediate between 36 carp and 28 crucian) (Fig. 1). Among the four taxonomic characters examined, it may be said at first glance that the genes for barbs and pharyngeal teeth formation are being activated and that the genes for the number of vertebrae and lateral line scales altered. I shall come to these points in a moment.

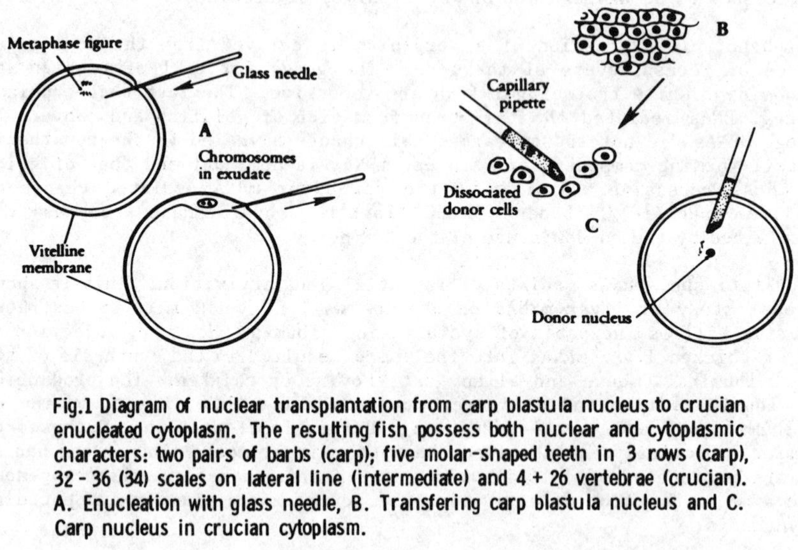

Fig.1 Diagram of nuclear transplantation from carp blastula nucleus to crucian enucleated cytoplasm. The resulting fish possess both nuclear and cytoplasmic characters: two pairs of barbs (carp); five molar-shaped teeth in 3 rows (carp), 32 - 36 (34) scales on lateral line (intermediate) and 4 + 26 vertebrae (crucian). A. Enucleation with glass needle, B. Transfering carp blastula nucleus and C. Carp nucleus in crucian cytoplasm.

Studies on the active agents of egg cytoplasm are a subject of current interest. One of the most likely candidates is the preformed or stored mRNAs. Poly A attached RNAs were partially purified from whole egg RNAs or from polysomal RNAs by column chromatography with Sigma cell #38 (#50) cellulose (economical) or oligo (dT) 12-18 cellulose. The partially purified egg poly A attached RNAs were heterogenous and were analyzed using color reactions with orcinol for RNA, Dische reagent for DNA and Lowry procedure for protein (10). In the past two years, RNA preparation was treated with RNase-free DNase (Worthington). DNA contamination was less than 0.1% and protein less than 1%. The UV absorption profile was typical of a nucleic acid. A single band was found on polyacrylamide gel electrophoresis. The heterogenicity of the single banded poly A attached RNAs was revealed by the ability to encode the syntheses of albumin and globin (see Yu and Niu in this volume, p.893). Poly A attached RNAs were also prepared by the same procedure from carp and rat livers, and rat ascites tumor cells. Both carp and rat liver poly A attached RNAs were capable of encoding albumin synthesis in the cell-free wheat germ translational system. Ascites tumor cell RNAs were not tested. Therefore, egg and liver poly A attached RNAs used in this paper are mRNAs.

To the filtrate of the above column chromatography, NaCl was added up to 1 M in the cold (4°C), and mixed well. The precipitates were collected by centrifugation. They are ribosomal RNAs (rRNAs).

Egg mRNAs were injected into fertilized goldfish eggs according to the procedure published elsewhere (11). Before presenting our results, I would like to show an eight minute color movie about our team work on mRNA mediated transformation of goldfish tails. It was made during the summer before the death of Professor T. C. Tung, Vice President of Academia Sinica. In the movie, you saw 14 members of our team at work and in discussions, doing RNA isolation and identification, artificial fertilization of goldfish eggs, micro-injection of mRNA into goldfish eggs, and goldfish with veil-shaped (control) and fork-shaped (transformed) tails (Fig. 2).

Table 1 shows the results of an early series of egg mRNA injected- and two series of control-goldfish. It can be seen that the rate of single tail forma-

Table 1

The effect of Crucian egg mRNAs on goldfish tail formation

MATERIALS INJECTED	EGGS INJECTED	LARVAL FISH	TYPE OF TAILS	
			DOUBLE	SINGLE
CRUCIAN OVARY mRNAs, 100A per ML HOLTFRETER SOLU'N	(100%) 454	(71%) 320	214 (67%)	106 (33%)
HOLTFRETER SOLUTION	(100%) 1065	(71%) 765	753 (98%)	12 (2%)
UN-INJECTED		426 (100%)	413 (97%)	13 (3%)

Fig. 2 Top - normal goldfish with double tail and bottom - experimental goldfish with single tail.

tion in crucian egg mRNA injected goldfish is 33% versus 2-3% for the two controls. Similar frequency of single tail formation in the saline injected and noninjected goldfish shows that the trauma caused by injection has no effect on normal tail formation. The lack of effect in the series injected with liver mRNAs, tumor poly A attached RNAs and rRNAs (Table 2) illustrates the specificity of egg mRNAs. The effect is statistically significant ($P < 0.001$).

TABLE II

THE EFFECT OF mRNA AND rRNA FROM CARP's MATURED OOCYTES, LIVERS, RAT's WALKER ASCITES TUMOUR CELLS ON TAIL FORMATION OF GOLDFISH

Experiments	Kinds of Nucleic Acid	Experimental				Control (uninjected)		
		# Eggs Injected	Larval Fish					
			Total #	Double Caudal Fin	Single Caudal Fin	Total #	Double Caudal Fin	Single Caudal Fin
1	Carp ovary mRNA (100 O.D./ml)	454	70.5% 320 100%	214 66.9%	106 33.1%	370	358 96.8%	12 3.2%
2	Carp ovary rRNA (50 O.D./ml)	147	16.4% 24 100%	24 100%	0	24	24 100%	0
3	Carp liver mRNA (100 O.D./ml)	215	25.1% 54 100%	54 100%	0	56	56 100%	0
4	Rat's Walker Ascites tumour cell mRNA (50 O.D./ml)	69	42% 29 100%	29 100%	0	33	31 93.9%	2 6.1%

Deoxyribonucleic acid (DNA) was prepared according to the procedure of Mirsky and Pollister (12) from crucian liver and ovary. It was treated with RNase and then pronase. The effect on tail transformation was about the same as egg mRNAs (Table 3). When urodele DNA prepared by the same procedure was

TABLE III

THE EFFECT OF CRUCIAN DNA ON GOLDFISH TAIL FORMATION

Kinds of DNA	Experimental				Control (uninjected)		
	# Eggs	Larval Fish			Larval Fish		
	Injected	Total #	Double Tail	Single Tail	Total #	Double Tail	Single Tail
From Ovary	100% 667	34.9% 234 100%	180 74.1%	63 25.9%	657 100%	646 98.3%	11 1.7%
From Liver	100% 231 100%	39.8% 92 100%	59 64.2%	33 28.7%	119 100%	114 95.8%	5 4.2%
Total	100% 898	37.3% 335 100%	237 71.3%	96 28.7%	776	760 97.9%	16 2.1%

injected into goldfish eggs, there was tail transformation. In addition, one balancer (larval character of some urodele) was found at the right or left postero-lateral part of the mouth (Fig. 3). The frequency was about 1% (4 in 365 injected fish). Its resemblance to the urodele balancer was reported elsewhere (13). This experiment was repeated in 1979 using DNA prepared from Triturus pyrrhogaster. Out of 235 larvae, 14 developed one balancer (about 6%). On account of the reproducibility, we are tempting to breed them and see if this DNA induced acquisition (the balancer) can be passed on to the next generation.

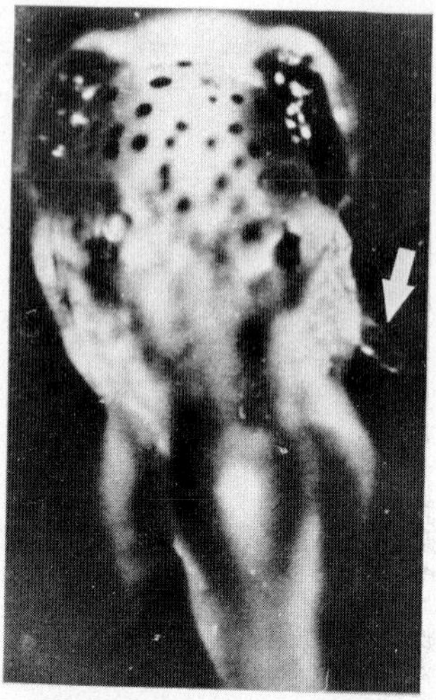

Fig. 3 Anterior portion of a goldfish with a balancer on right side of the mouth.

Goldfish are mutants of crucian, obtained by domestication and selection. Carp belongs to a different genus. The effect of crucian egg mRNAs was repeated with carp egg mRNAs on eggs from the breed Red Dragon Eye (strain from Institute of Developmental Biology) mixed Red Dragon Eye (fish from Shenyang, Nanning and Beijing) and Wen Yu (imported from Japan). Experimental results are shown in Table 4. Apparently carp egg mRNAs act in the eggs from the three groups of goldfish in a manner similar to crucian egg mRNAs. The frequency of induced tail change varies according to the donor strain of eggs (14). Tail reversion of Red Dragon Eye is lowest (0.1%). Its response to egg mRNAs is also lowest. In contrast, Wen Yu has the highest rate of tail reversion and its egg's response to egg mRNAs is also the highest. This relationship would indicate that the response of egg to egg mRNAs is genetically controlled.

TABLE IV

Distribution of Tail Pattern in Three Groups of Experimental and Control Fish
Group I, local Red Dragon Eye; Group II, mixed Red Dragon Eye, from 3 localities; Group III, imported Wenyü

	Experimental			Control			P Value
	Total Fish	Double Tail	Single Tail	Total Fish	Double Tail	Single Tail	
Group I	761	665 87.4%	96 12.6%	762	761 99.9%	1 0.1%	<0.001
Group II	1303	1008 77.4%	295 22.6%	1649	1551 94.1%	98 5.9%	<0.001
Group III	826	573 67.4%	253 30.6%	924	833 90.2%	91 9.8%	<0.001
Total	2890	2246 77.7%	644 22.3%	3335	3145 94.3%	190 5.7%	<0.001

Different batches of carp egg mRNAs were tested for transforming activity. The activity of carp egg mRNAs varied from batch to batch (Table 5). However, none of them has ever failed in our hands to show transforming capacity.

TABLE V

THE EFFECT OF FOUR BATCHES OF EGG mRNA ON TAIL FORMATION OF THE GOLDFISH FROM GROUP 1 OF TABLE 4

# of mRNA	Experimental			Control		
	Total	Double Tail	Single Tail	Total	Double Tail	Single Tail
3	220	173 78.6%	47 21.4%	320	319 99.7%	1 0.3%
524	43	34 79.1%	9 20.9%	34	34	0
525	282	261 92.6%	21 7.4%	231	231	0
531	216	197 91.2%	19 8.8%	177	177	0
Total	716	665 87.4%	96 12.6%	762	761 99.9%	1 0.1%

Mating between crucian or carp and goldfish produces offspring with a single tail. It appears that crucian or carp tail is dominant and goldfish tail recessive (15). The exact number of genes controlling tail formation has not been determined, however.

Egg mRNA mediated tail transformation in goldfish is supported by the finding that injection of cytoplasm of carp eggs also resulted in tail change (Tung et al., unpublished). Conversely, however, injection of egg mRNAs or cytoplasm from goldfish into carp fertilized eggs did not produce tail change (over 1,000 larval fishes). A likely explanation is that the tail-forming cytoplasm or the tail-forming mRNAs of carp egg is dominant and of goldfish recessive. This fits well with the correction of "0" lethal gene of an axolotl mutant by injection of a bit of normal egg's cytoplasm (16).

The stability of the nucleic-acid-induced single-tailed goldfish was tested by mating. The type of tails observed in this study varied according to the experimental series (17). Among the offspring of common Red Dragon Eye, there were 96% double tail vs. 2% single tail and variation between double and single was 2% (Table 6). By contrast, the offspring of crucian egg-mRNA (Table 7) and

TABLE VI

TAIL PATTERNS IN THE OFFSPRING FROM CROSSES OF DOUBLE-TAIL-FINNED GOLDFISH (THE CONTROL)

# of the Pairs	# of Larval Fish	Types and Number (%) of the Tail Fin			
		1	2	3 Varieties of Single Tail Fin	
		Double	Single	A 3-Lobed	B Split-Lobe(s)
1	195	195(100%)			
2	26	25(96.2%)			1(3.8%)
3	103	95(92.2%)	5(4.9%)	3(2.9%)	
4	102	95(93.1%)	4(3.9%)	2(2.0%)	1(1.0%)
Total	426	410(96.2%)	9(2.1%)	5(1.2%)	2(0.5%)

TABLE VII

TAIL PATTERNS IN THE OFFSPRING FROM CROSSES OF THE mRNA-INDUCED SINGLE-TAIL-FINNED GOLDFISH

# of the Pairs	# of Larval Fish	Types and Number (%) of the Tail Fin					
		1	2	3 Varieties of Single Tail Fin		4 Abnormal	5 Rudimentary
		Double	Single	A 3-lobed	B Split-Lobe (s)		
1	359	210(58.5%)	45(12.5%)	86(24.0%)	4(1.1%)		14(3.9%)
2	68	26(38.2%)	24(35.3%)	14(20.6%)	4(5.9%)		
3	167	116(69.5%)	21(12.6%)	10(6.0%)	20(11.9%)		
4	86	44(51.1%)	23(26.7%)	14(16.3%)	4(4.7%)	1(1.2%)	
Total	680	396(58.2%)	113(16.6%)	124(18.3%)	32(4.7%)	1.(0.1%)	14(2.1%)

of crucian DNA-induced single-tailed-goldfish (Table 8) had single tails within the range of 17% and a variety of forms (25%) in-between the double and single. Compared with 33-35% of single tailed goldfish in crucian egg-mNRA- or DNA-injected series (parent), F_1 (offspring of the mating) had a greatly reduced rate of single tail formation (17%). This finding indicates that there are more than one gene for tail formation. Experiments designed to ascertain the number of genes controlling goldfish and carp tail formation are in progress.

TABLE VIII

TAIL PATTERNS IN THE OFFSPRING FROM CROSSES OF THE DNA-INDUCED SINGLE-TAIL FINNED GOLDFISH

# of the Pairs	# of Larval Fish	Types and Number (%) of the Tail Fin			
		1	2	3 Varieties of Single Tail Fin	
		Double	Single	A 3-Lobed	B Split-Lobe(s)
1	11	5(45.4%)	3(27.3%)	3(27.3%)	
2	6	2(33.3%)		1(1.67%)	3(50.0%)
3	39	26(66.6%)	5(12.8%)	3(7.7%)	5(12.8%)
4	25	14(56.0%)	6(24.0%)	3(12.0%)	2(8.0%)
Total	81	47(58.0%)	14(17.8%)	10(12.3%)	10(12.3%)

The transmission of a nucleic-acid-induced character (tail) from one generation to another leads to the examination of changes in visceral organs. Liver was first studied. The rationale for choosing liver is based on its specific lactate dehydrogenase isozymes (LDH). Among all tissues studied with starch gel electrophoresis, goldfish liver is the only organ with cathodally migrating C_4 (18) (Fig. 4). The C_4 bands of carp liver stained very faintly (hardly reproduced in photography) and migrated slightly toward the cathode (19). Hybridization between goldfish and carp produced offspring whose liver specific LDH-C_4 migrated toward cathode. The stained bands vary slightly according to the individual fish and, by and large, there are five recognizable bands. Liver extracts from 35 carp egg-mRNA injected goldfish were subjected to starch gel electrophoresis, and 15 of them showed an additional narrow band between the proximal and distal C_4 bands (Fig. 5). Apparently, the genesis of the additional band in liver of injected goldfish is due to the presence of injected mRNAs. This implies that carp egg-mRNAs contain a liver-forming subfraction. Experimental support for this idea comes from two studies: (1) Template activity of egg-mRNAs. In collaboration with Yu (see Yu and Niu in this volume, p. 893), carp egg-mRNAs were tested in the cell-free wheat germ translational system. Among the reaction products, both carp albumin and globin were detected by immunoprecipitation tests and identified respectively by polyacrylamide gel electrophoresis of the immunoprecipitates and peptide mapping

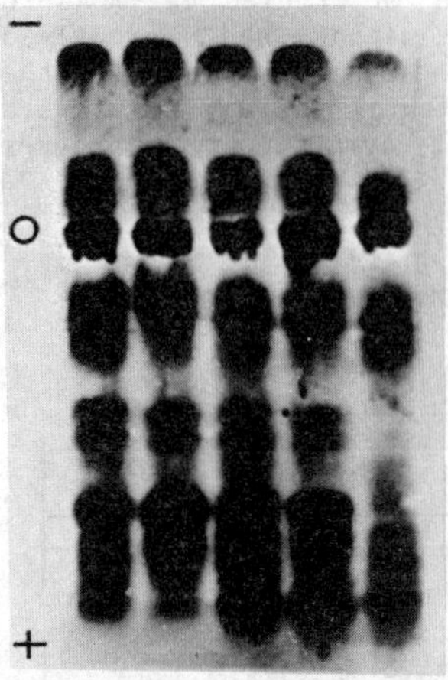

Fig. 4 Starch gel electrophoresis of five goldfish's liver specific LDH-C_4 isozymes --two cathodally migrated bands.

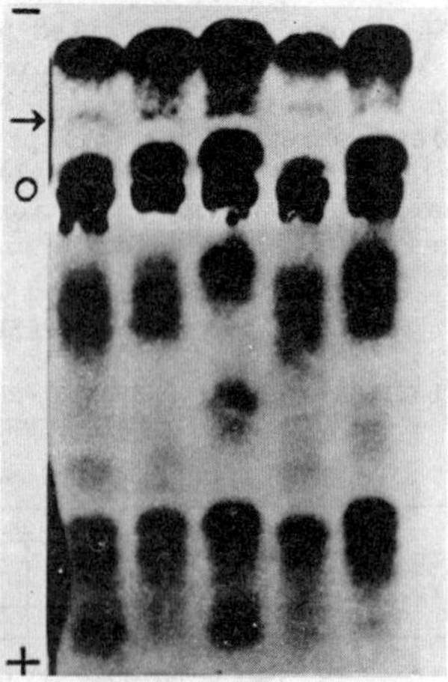

Fig. 5 Starch gel electrophoresis of the liver specific LDH - C_4 isozymes from carp egg mRNA injected goldfish. Note the presence of a narrow band in-between the two distinct cathodally migrated bands of goldfish.

analysis. Therefore, a subfraction of egg-mRNAs acts as liver mRNAs to encode the synthesis of a liver specific protein, albumin. (2) Effect of carp liver-mRNAs in developing eggs. For this, carp liver-mRNAs were injected into fertilized eggs. Starch gel electrophoresis of the liver extracts from the liver-mRNA injected goldfish revealed the presence of an additional diffuse band (Fig. 6). The migration distance was the same as the one found in liver from carp egg-mRNA injected goldfish (20). In order to show species specificity of liver-mRNAs, rat livers (Fig. 7) were used to prepare mRNAs. They were injected into goldfish eggs. The liver of the developed goldfish was characterized by broad proximal and broad distal C_4 bands. No intermediate band was seen between the two broad bands (Fig. 8).

In a separate report (see Niu, Xue and Niu in this volume, p.407), data were presented to show that carp egg-mRNAs contain another "morphogen," a RBC forming subfraction that specifically causes the development of RBC. The causal relationship between the input (mRNAs) and the output (organ or tissue formation) provides the rationale for genetic manipulation in higher animals and plants. The manner by which macromoleculate mRNAs act on the genome, we believe, is similar to that of the RNA-rich particles, germ plasm, in amphibian eggs (5) or polar granules in insects (21). During the process of cleavage, those cells receiving germ plasm transform into primordial germ cells. Replacement of a region with primordial germ cells by the same region from another subspecies made the gonad produce sperm or egg of the donor subspecies rather than

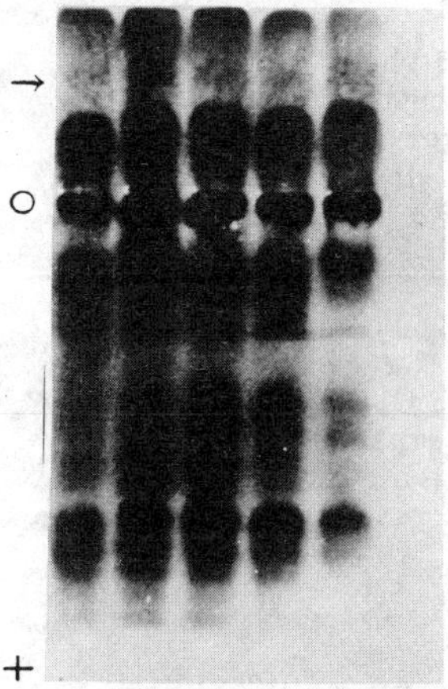

Fig. 6 Starch gel electrophoresis of the liver specific LDH - C_4 isozymes from carp liver mRNA injected goldfish. Note the presence of a diffuse band in-between the two cathodally migrated bands.

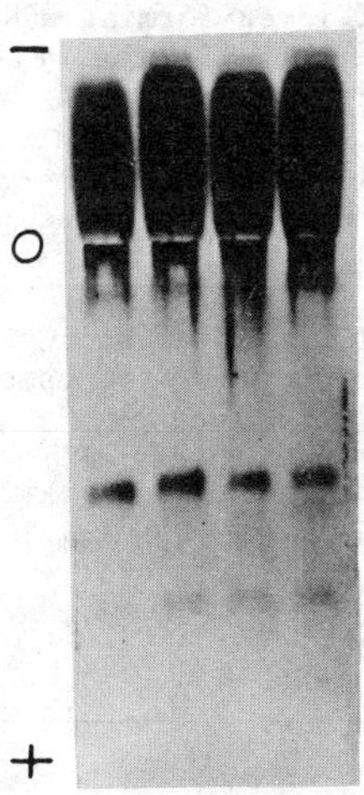

Fig. 7 Starch gel electrophoresis of the liver specific LDH - isozymes from rat. Note the long cathodally migrated bands.

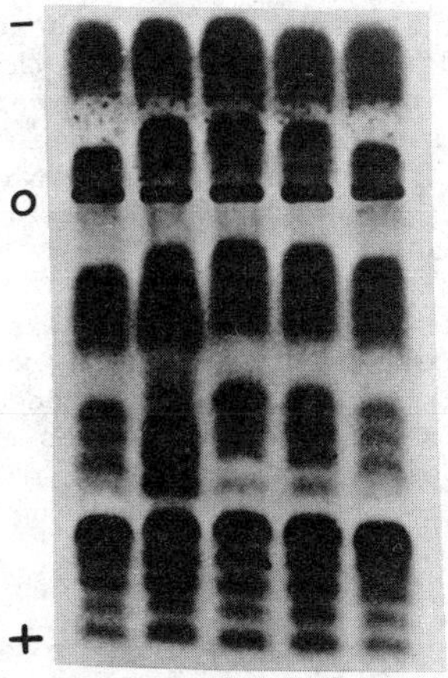

Fig. 8 Starch gel electrophoresis of the liver specific LDH - C_4 isozymes from rat liver mRNA injected goldfish. Note the mRNA caused intensification of the distal cathodal band and compare with Fig. 4.

of the host (22). Thus, the germ plasm induced cellular commitment; primordial germ cells would migrate into gonad of another species for the genesis of ova or spermatozoa. The stage of cellular commitment appears very early (23), and probably coincides with the recognizability of presumptive organ-forming areas in the period covering late blastula to gastrula. It seems likely that germ plasm, liver-forming mRNAs, RBC forming mRNAs, tail forming mRNAs and other organ forming mRNAs may be confined to cells of respective areas in the presumptive map.

The manner by which the active component of germ plasm or injected mRNA acts upon the developing egg is a subject of great interest. DNA is the material basis of inheritance. The fact that both egg-mRNA and crucian DNA were capable of mediating the transformation of goldfish tails would imply that injected egg-mRNA may somehow act in the same way as DNA in the developing eggs. The oncogenic viral RNA induced transformation of normal cells into cancerous tissue is a good example. As a matter of fact, it was Temin (24) who proposed the role of reverse transcriptase in embryonic differentiation. The proposal demands in the developing system the availability of (1) tissue or organ-forming mRNAs and (2) activity of the enzyme reverse transcriptase. The best developing systems are fertilized eggs in which there are stored or preformed mRNAs. Among them, histone mRNAs and tubulin mRNA are well known in sea urchins (see Gross in this volume, p.254) and amphibian eggs (25). Both histone- and tubulin-mRNAs encode the synthesis of histone and tubulin respectively for the needs of cleavage. The presence of liver-forming mRNAs in egg mRNAs has already been demonstrated early in this report. In the cell-free wheat germ translational system, as reported earlier, carp egg mRNAs were found to encode globin synthesis (see Yu and Niu in this volume, p.893). Moreover, the mRNA induced transfer of rabbit RBC characters to goldfish (see Niu, Niu and Xue in this volume, p.407) provides powerful evidence for the presence of globin mRNA in egg-mRNAs.

Reverse transcriptase was found in eucaryotic cells (26) and normal embryonic cells (27). Goldfish oocytes or ova were used for the extraction of an enzyme according to the procedure of Bolden et al. (26)(Fig. 9). The activity of this crude enzyme was tested using carp egg mRNA as a template. It was proportional to the amount of mRNA added. In 50 ul of reaction mixture, 10 ug of mRNA was optimal. With 10 ug of mRNA in each sample, the rate of ^{3}H-thymidine

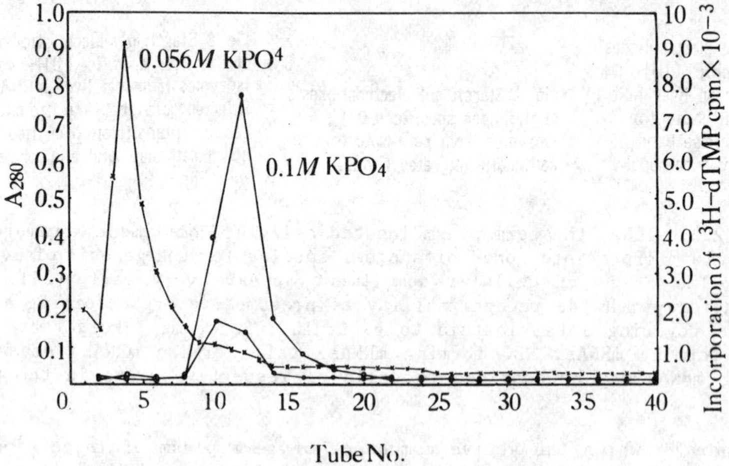

Fig. 9 Elution profile of the crude enzyme from deae-cellulose column and the enzyme activity in each fraction. Goldfish egg extract was prepared and chromatographied on DEAE-52 cellulose column according to Bolden et al. (26) with slight modification. Twelve ml of egg extract (100 A_{280} units of protein) were applied to the column and eluted with 6 volumes of a linear gradient from 0.02 M KPO_4 to 0.3 KPO_4 containing 0.5 M DTT (pH 7.5). The flow rate was 30 ml/hour. Fractions of 2.5 ml were collected and absorption at 280 determined (x—x—x). Each fraction was tested for the activity to encode DNA synthesis (cDNA). Tubes 10-12 and 11-13 are respectively active in egg mRNA (•—•—•) and activated thymus DNA directed synthesis of cDNA (o—o—o).

incorporation into acid insoluble material was linear and sensitive to dactinomycin and RNase. The acid insoluble material was sensitive to RNase-free DNase, but resistant to RNase. Both samples were treated with Dische reagent and turned blue. All data suggest that the ^{3}H-labelled, acid insoluble material is ^{3}H-cDNA (Niu et. al., unpublished). Details of the study on the activity of RNA dependent DNA polymerase will be published elsewhere. It appears that exogenous mRNAs in coordination with the counterpart of egg mRNAs are catalyzed automatically by an endogenous enzyme for the synthesis of DNA. This complimentary DNA (cDNA) would then be inserted at the locus(i) of the genome of primordial cells. Thus they are committed for the development into an organ of the mRNA tissue source, e.g. liver and RBC. The phenotypic expression of an organ or tissue has to be considered in view of the dominance, co-dominance or recessiveness of the gene(s) involved.

The effect of mRNA on somatic tissue of living animals is different. Castration is known to abolish sexual drive and to make secondary sexual organs atrophy. Injection of uterine RNA into the lumen of atrophied uterus resulted in the resumption of its morphological and enzymatic entity (28,29). Both saline and calf liver RNA had no observable effect on uterine morphology. A relevant question is whether or not liver mRNA can act on the uterine epithelial cells. To analyze this problem, C^{14}-leucine was co-injected with liver mRNA into the lumen of atrophied uterus (18 mice per experiment). At different intervals of time, two uteri were excised from 2 injected mice, washed with saline and homogenized. Rabbit antisera against mouse-, rat-, bovine-and chicken-albumins were used selectively to precipitate the respective albumin in the homogenates. It should be noted here that all antisera used in these experiments were properly absorbed. Rabbit anti-mouse albumin serum does not react with rat-, bovine-, or chick-albumin; nor anti-rat serum with mouse albumin; nor anti-bovine serum with mouse albumin; nor anti-chicken serum with mouse albumin. The radioactivity (CPM) of the immunoprecipitates was plotted against time (Fig. 10, upper curve: mouse albumin, lower: alien albumins). Apparently, mouse albumin synthesis reached the peak one hour after injecting RNA into 4 experimental series. Parallel to mouse albumin synthesis, alien albumins (rat and bovine) were also peaked 1 hour after injection (29).

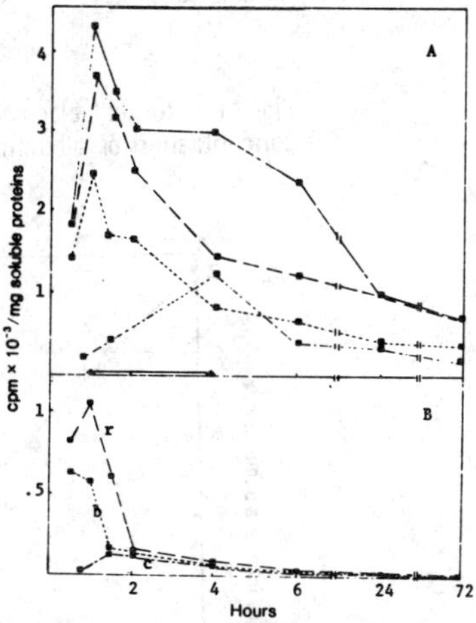

FIG. 10 Synthesis of albumin in the uterus primed by liver mRNA. Five one-hundreth milliliter of mouse (■—■), rat (■- - -■), calf (■- - -■), or chicken (■·······■) liver mRNA (200 A_{260} units of mRNA containing 20 μCi of [^{14}C]leucine per ml of saline) or saline with [^{14}C]leucine alone (▲—▲) was injected into the right uterine horn. At various times after injection, the uteri were excised and washed for preparation of protein extract. The albumin immunoprecipitates from the extract were prepared. The radioactivity of the immunoprecipitates was determined in a toluene scintillation fluid. (*A*) Mouse albumin; (*B*) alien albumin. r, rat; b, bovine; c, chicken.

A graded concentration of albumin was used for "rocket" immunoelectrophoresis (Fig. 11). The amount of albumin was plotted against the migration distance, giving a standard curve (Fig. 12). The content of albumin in each experimental and control series was estimated, and is shown in Fig. 13. The albumin of the control was derived from the serum left in uterine tissue at the time of excision. Figures 10 and 13 show respectively that there are two kinds of albumin synthesized by uterine wall in response to the injection of liver mRNAs. Each liver mRNA of the four species (mouse, rat, bovine and chicken) was capable of mediating the synthesis of mouse albumin. The duration of this process was a number of days. In contrast, the synthesis of alien albumin was short and the amount small. The authenticity of the newly synthesized albumin (C^{14}-labelled) was further tested by polyacrylamide gel electrophoresis of the dissociated immunoprecipitates. Figure 14 shows that the C^{14}-albumin and carrier albumin have the same electrophoretic migration distance.

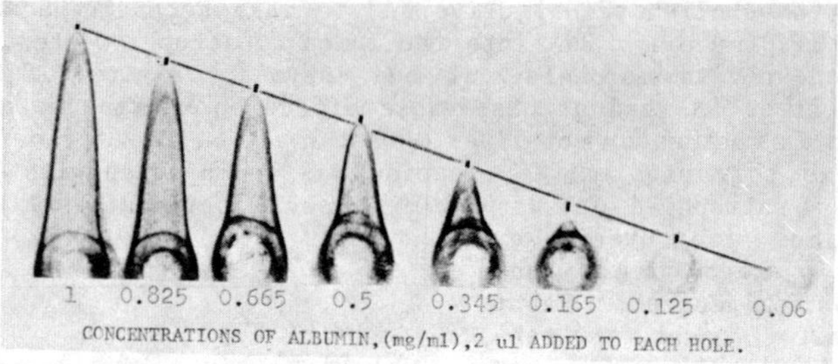

Fig. 11 "Rocket" electroimmunodiffusion of a graded concentration of albumin 30 .

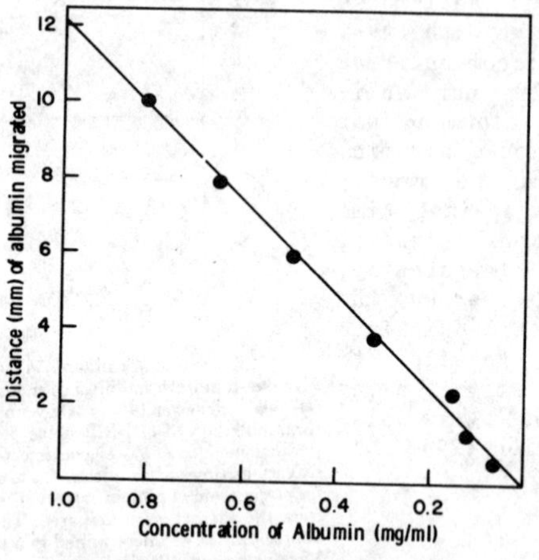

Fig. 12 The relation between albumin concentration and distance of migration.

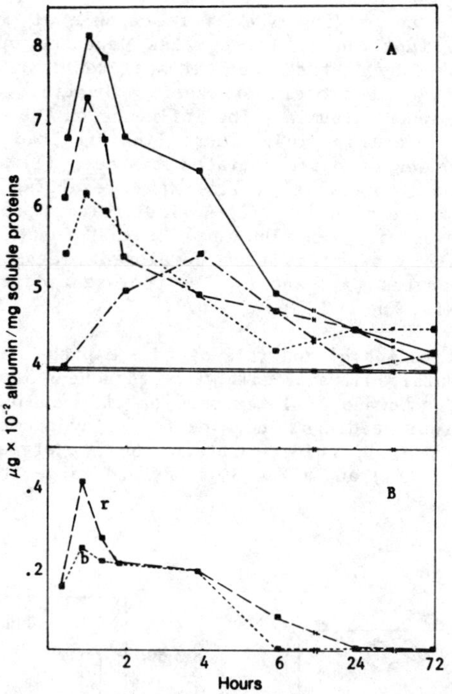

FIG. 13 Time course of albumin accumulation in uterus treated with liver mRNA. The amount of albumin was determined by comparing the height of the precipitation cones in antibody–agar-coated slides with a standard curve by electroimmunodiffusion technique (26). (*A*) Mouse albumin; (*B*) alien albumin. See legend of Fig. 10 for other explanation.

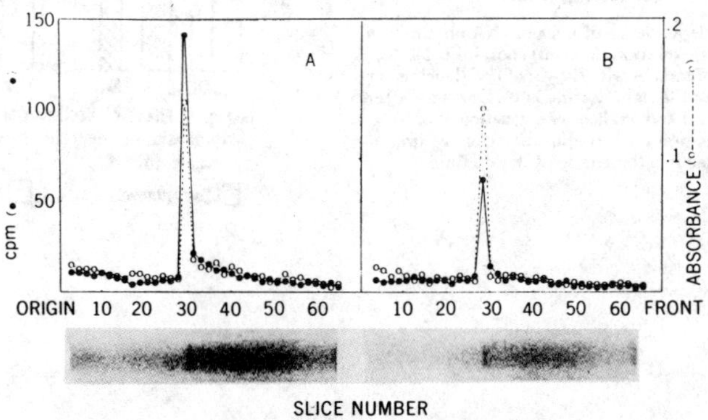

FIG. 14 Sodium dodecyl sulfate/polyacrylamide gel electrophoresis of anti-albumin serum immunoprecipitable material from extract of uterus treated with rat liver mRNA. Aliquots of the immunoprecipitates (antigen–antibody complex) from the experiments of Fig. 3 were dissociated by 2% sodium dodecyl sulfate and 2% 2-mercaptoethanol in phosphate buffer, and then subjected to polyacrylamide gel electrophoresis. The gels were then fixed and stained. After scanning, the gel was sliced into 1-mm sections and digested in a gel solubilizer. A_{280} was measured and radioactivity was determined in liquid scintillation fluid. (*A*) Mouse albumin; (*B*) rat albumin.

Neither uterus mRNAs nor kidney mRNAs were capable of mediating the synthesis of albumin. The function of liver mRNAs depended upon concentration (Fig. 15) and was abolished by pretreatment with pancreatic RNAse and inhibited by puromycin, an inhibitor of protein synthesis. Dactinomycin was found to inhibit the synthesis of mouse albumin. Its influence on the synthesis of alien albumin was practically none (Fig. 16). These findings lead to the conclusion of a dual function of mRNA in differentiated tissue: (1) Activation of the nuclear activity of mouse epithelial cells: the resulting products (mRNAs) encode the synthesis of mouse albumin. (2) A minor role in protein translation: this is shown by production of alien albumin. The last function of mRNA is well known. Exploration of this type has been developed elegantly by Gurdon and associates in Xenopus oocytes (31) and is also represented in this volume (see Yu and Niu, p. 893 , and Niu, Xue and Niu, p. 407).

Studies on gene activation or controls of gene expression have come a long way. The first experimental analysis dealt with ecdysone-stimulated chromosonal puffing in insects (32). Nowadays, literature on this subject is voluminous. They are grouped under four headings: hormone (e.g., hydrocortisone, see Goldberger in this volume, p. 215), acidic protein (33), acetylated histone (see Allfrey in this volume, p. 1) and mRNA (34). By and large, the mode of action is still not clearly defined.

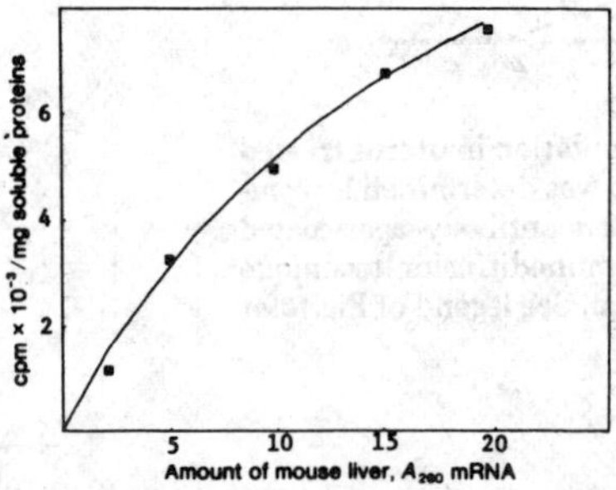

FIG. 15 Concentration dependence of liver mRNA on albumin synthesis. Mouse liver mRNA solution (0.05 ml) containing 50, 100, 200, 300, or 400 A_{260} units of mRNA with 20 μCi of [^{14}C]leucine per ml of saline was injected into the right uterine horn. One hour after injection the uteri were excised and washed for extraction of protein. The immunoprecipitate was prepared. Radioactivity of the immunoprecipitate was determined in a toluene scintillation fluid.

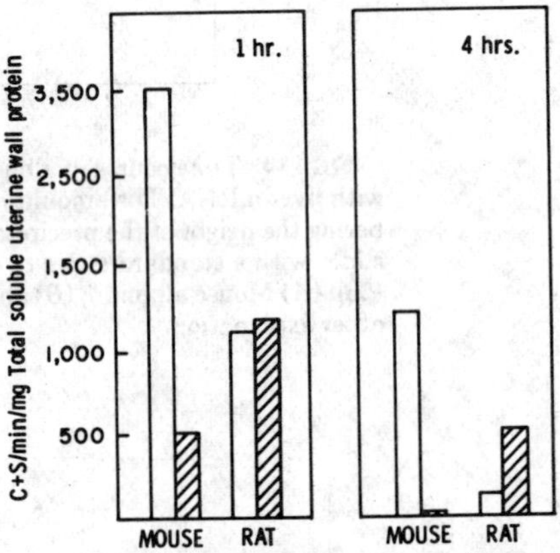

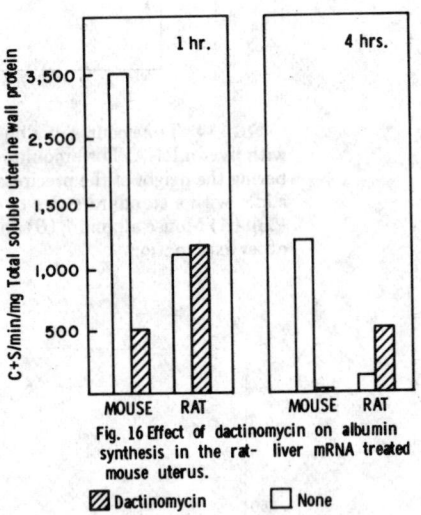

Fig. 16 Effect of dactinomycin on albumin synthesis in the rat- liver mRNA treated mouse uterus.

Dactinomycin   None

*The author wishes to thank all of his co-workers from the laboratories in the Department of Biology, Temple University and the Institute of Developmental Biology, Academia Sinica. This research was supported in part by a grant from the Rockefeller Foundation, New York.

LITERATURE CITED

1. Vogt, W. (1929) Gestaltungsanalyse am Amphibienkein mit artlicher Vitalfarbung. 2. Gastrulation und Mesodermbildung bei Urodelen und Anuran. Arch. f. Ent. mech. 120:384.

2. Bacon, R. L. (1945) Self-differentiation and induction in the Heart of *Ambystoma*. J. Exp. Zool., 98:87.

3. Librera, E. (1964) Effects on gonad differentiation of the removal of vegetal plasm in eggs and embryos of *Discoglissus pictus*. Acta Embryol. Morph. Exptl., 7:217.

4. Curtis, A. S. G. (1960) Cortical grafting in *Xenopus laevis*. J. Embry. Exp. Zool., 8:163.

5. Smith, D. L. and M. A. Williams (1975) Germinal plasm and determination of the primordial germ cells. 33rd Symposium Soc. Devel. Biol., Academic Press, N.Y., pp. 3-21.

6. Morgan, T. H. (1934) Embryology and Genetics, Columbia University Press.

7. Briggs, R. and King, T. J. (1959) Nucleocytoplasmic interaction in eggs and embryos. In "The Cell" (Brachet and Mirsky, eds.) vol. 1, p. 537. Academic Press, New York.

8. Gurdon, J. B. (1977) In "Control of Gene Expression Animal Development," Harvard University Press, Cambridge, MA.

9. Tung, T. C. *et al*. (1980) Nuclear transplantation in teleosts: 1. hybrid fish from the nucleus of carp and the cytoplasm of crucian. Scientia Sinica, 23:517.

10. Lowry, D. H., N. J. Rosebrough, A. L. Farr and R. J. Randall (1951) Protein measurement with the folin phenol reagent. J. Biol. Chem., 193:265.

11. Tung, T. C. and M. C. Niu (1973) Nucelic acid-induced transformation in goldfish. Scientia Sinica, 16:377.

12. Mirsky, A. E. and A. E. Pollister (1946) Chromosin, a deoxyribose nucleoprotein complex of the cell nucleus. J. Gen. Physiol., 30:117.

13. Tung, T. C. and M. C. Niu (1977) Organ formation caused by nucleic acid from a different class. I. Urodele DNA mediated balancer formation in goldfish. Scientia Sinica, 20:56.

14. Tung, T. C. and M. C. Niu (1977) The effect of carp egg-mRNA on the transformation of goldfish tail. Scientia Sinica, 20:59.

15. Matsui, Y. (1934) Genetical studies on goldfish of Japan. J. Imperial Fisheries Institute, 30:1.

16. Briggs, R. (1972) Further studies on the maternal effect of the gene in the Mexican axolotl. J. Exp. Zool., 181:271.

17. Tung, T. C. and M. C. Niu (1975) Transmission of the nucleic acid-induced character, caudal tail, to the offspring in goldfish. Scientia Sinica, 18:223.

18. Wilson, F. P., G. C. Whitt and C. L. Prosser (1973) Lactate dehydrogenase and malate dehydrogenase isozyme patterns in tissue of temperature-acclimated goldfish. Comp. Biochem. physiol., 46B:105.

19. Shaklee, J. B., D. L. Kepes, and G. S. Whitee (1973) Specialized lactate dehydrogenase isozymes: the molecular and genetic basis of the unique eye and liver LDH's of teleost fishes. J. Exp. Zool., 185:217.

20. Niu, M. C. and T. C. Tung (1977) The mRNA mediated alteration of the liver located dehydrogenase isozymes. Scientia Sinica, 20:803-806.

21. Mahowald, A. P., et al. (1979) Germ plasm and pole cells of Drosophila. Symp. Soc. Dev. Biol., 37:127.

22. Blackler, A. B. and M. Fishberg (1961) Transfer of primordial germ-cells in Xenopus laevis. J. Embryol. Exp. Morph., 9:634.

23. Brothers, A. J. (1976) Stable nuclear activation dependent on a protein synthesized during oogenesis. Nature, 260:112.

24. Temin, H. M. (1971) The protuvirus hypothesis: speculations on the significance of RNA-directed DNA synthesis for normal development and for carcinogenesis. J. Nat. Cancer Inst., 46(2):III-VII.

25. Adamson, E. D. and H. R. Woodland (1977) Changes in the rate of histone synthesis during oocyte maturation and very early development of Xenopus laevis. Devel. Biol., 57:136. Ruderman, J. V. and M. L. Pardue (197) In vitro translation analysis of mRNA in echinoderm and amphibian early development. Devel. Biol., 60:48.

26. Bolden, A., M. Fry, R. Muller, R. Citarella and A. Weisssbach (1972) The presence of a polyriboadenylic acid-dependent DNA polymerase in eucaryotic cells. Arch. Biochem. Biophy., 153:26-33.

27. Bauer, G. and P. H. Hofschneider (1976) An RNA dependent DNA polymerase, different from the known viral reverse transcriptase, in the chicken system. Proc. Nat. Acad. Sci., 73:3025.

28. Mansour, A. M. and M. C. Niu (1965) Functional studies with uterine RNA. Proc. Nat. Acad. Sci., 53:764-770.

29. Yang, S. F. and M. C. Niu (1977) Albumin synthesis in mouse uterus in response to liver mRNA. Proc. Nat. Acad. Sci., 74:1894-98.

30. Laurell, C. B. (1966) Quantitative estimation of proteins by electrophoresis in agarose gel containing antibodies. Anal. Biochem., 15:45-52.

31. Gurdon, J. B., A. R. Woodland and J. B. Lingrel (1974) The translation of mammalian globin mRNA injected into fertilized eggs of Xenopus laevis. 1. Message stability in development. Devel. Biol., 39:125-133.

32. Clever, U. (1966) Gene activity patterns and cellular differentiation. Am. Zool., 6:33.

33. Wang, T. Y. and N. C. Kostraba (1973) Non-histone proteins as gene derepressor moleculer. In "The role of RNA in reproduction and development" eds. M. C. Niu and S. J. Segal. Am. Elsevier Publishing Co., New York.

34. Frenster, J. H. and P. R. Herstein (1973) RNA in gene derepression. In "The role of RNA in reproduction and development," eds. M. C. Niu and Sheldon Segal. Am. Elsevier Publishing Co., New York.

CHARACTERIZATION OF LYMPHOID CELL RNA WHICH MODULATES SPECIFIC CELLULAR IMMUNITY

Ronald E. Paque[1], Donald P. Braun[2] and Sheldon Dray[3]

[1]Department of Microbiology
The University of Texas Health Science Center
San Antonio, Texas

[2]Department of Internal Medicine
Rush Medical Center
Chicago, Illinois

[3]Department of Microbiology and Immunology
University of Illinois at the Medical Center
Chicago, Illinois

Abstract Mediation of cellular immunity by hot-cold phenol extracts of RNA occurs with molecular species derived by different RNA separation methodologies. A sucrose-derived "B" fraction with a molecular weight ranging from 6-12 s separates on a sucrose gradient into two distinct species "B_1" and "B_2". B_1 (the lighter fraction) transfers tumor specific and chemically defined antigenic sensitivities to immunologically naive lymphoid cells. Occasional transfer of cellular sensitivity also occurs with the heavier B_2 frac-

Abbreviations used in this paper include: MIF, migration inhibitory factor; PPD, purified protein derivative; KLH, keyhole limpet hemocyanin; Ln-1, line 1; ln-10, line 10; DTH, delayed type hypersensitivity; ARSNAT, mono(p-azobenzene-arsonate)-N-chloroacetyl-L-tyrosine; PEC, peritoneal exudate cells; RNase, ribonuclease; poly-A, polyadenylic acid sequences; mRNA, messenger ribonucleic acid; r-RNA, ribosomal ribonucleic acid; I-RNA, immune ribonucleic acid; t-RNA, transfer ribonucleic acid; SDS, sodium dodecyl sulphate; PVS, polyvinyl sulphate.

tion. Oligo (dT) affinity chromatography of whole, hot-cold phenol extracts results in the presence of two fractions designated as I and II. The bound fraction II containing polyadenylic acid sequences also transfers various antigenic sensitivities. Assessment of the B fraction by affinity chromatography results in a poly-A containing peak which is apparently absent from sucrose derived peaks, A, C, D, and E. Centrifugation of fraction II on sucrose gradients results in the appearance of two peaks with similar S values and fractionation displacement volumes as the "B" fraction. Atomic absorption spectroscopy of whole RNA-ARSNAT extracts results in an inability to detect arsenic atoms, a chemical marker for the ARSNAT antigen,at a level of 0.1 ng sensitivity. The poly (A) containing fraction II and the B fractions possess similar physical, chemical, and immunobiological characteristies. RNA extracts protectively entrapped in liposomes also transfer various antigenic sensitivities in vitro. Multiple species of RNA in the B fractions, or fraction II could be required for the observed transfer of lymphoid cell specific antigenic sensitivities. Alternatively, isolation and separation of homogeneous RNA speices will be required.

A. INTRODUCTION: Transfer of cell-mediated immunity with RNA extracted from lymphoid cells.

Immunological reactions classified as cell-mediated can be transferred to a suitable host by pre-sensitized donor lymphoid cells but not by serum antibodies of any known class. The type of cells mediating the transfer are thymus-derived lymphocytes (T cells) that demonstrate specific reactivity to the sensitizing antigen and appear to be an obligatory component in the delayed-type hypersensitivity (DTH) skin test reaction, rejection of histo-incompatible grafts, hapten-induced contact sensitivity, graft-vs-host reaction, production of many lymphokine mediators, and rejection of tumors. Since 1967, our laboratories have studied the transfer of cell-mediated immunity with RNA extracted from various lymphoid tissues of immunized donors. The active subcellular components in the extracts, generally denoted as "Immune-RNA" (I-RNA) is obtained by the hot/cold phenol extraction procedure of Scherrer and Darnell (1962) with some modifications of our own (Thor and Dray, 1968; Paque and Dray, 1970). Our studies have utilized I-RNA from humans, monkeys, guinea pigs, and mice in syngeneic, allogeneic, and xenogeneic transfers of specific cell-mediated immunity (Table 1). Both *in vitro* and *in vivo* transfer methodology have been developed utilizing antigen-specific lymphokine release as an *in vitro* correlate of cell-mediated immunity (MIF assay), and tumor regression as an *in vivo* correlate of cell-mediated immunity. The antigenic specificities that we have used include the mycobacterial derived antigen (PPD), fungal antigens (histoplasmin or coccidioidin), a guinea pig brain antigen, keyhole limpet hemocyonin (KLH), chemically-defined antigens (alpha- and epsilon-DNP-oligolysines and ARSNAT) and tumor antigens (Ln-1, Ln-10 and MOPC-315). The results of these studies demonstrate that transfer of cell-mediated immunity by I-RNA

TABLE 1. Transfer of Cellular Immunity with Immune RNA Extracts

Exp.	Donor RNA Tissue	Donor RNA Species	Recipient Cells Tissue	Recipient Cells Species	Antigen	Assay	Reference
ALLOGENEIC TRANSFERS							
1	LN	Man	LN	Man	PPD	MIF	Thor (34); Thor & Dray (35)
2	LN	Man	LN	Man	Histoplasmin	MIF	Thor (34); Thor & Dray (36)
3[a]	LN	Man	LN	Man	PPD	MIF	Thor & Dray (34)
4[b]	LN	Man	LN	Man	Histoplasmin	MIF	Thor & Dray (35)
5	LN+Sp	GP	PEC	GP	PPD	MIF	Jureziz et al. (11)
6	LN+Sp	GP	PEC	GP	Coccidiodin	MIF	Jureziz et al. (11)
7	LN+Sp	GP	PEC	GP	ARSNAT	MIF	Schlager et al. (31)
8	LN+Sp	GP	PEC	GP-S13	PPD	MIF	Paque et al. (19)
9	LN+Sp	GP	PEC	GP-S13[b]	α, DNP-OL	MIF	Paque et al. (21)
10	LN+Sp	GP	PEC	GP-S13[b]	ε, DNP-OL	MIF	Paque et al. (21)
SYNGENEIC TRANSFERS							
11	LN[e]	GP	PEC	GP	KLH	MIF	Paque (23)
12	LN+Sp[e]	GP	PEC	GP	ARSNAT	MIF	Paque & Nealon (27)
13	LN+Sp[e]	GP	PEC	GP	KLH, ARSNAT	MIF	Paque & Nealon (24)
14	LN+Sp[e]	GP	PEC	GP	KLH, ARSNAT	MIF	Paque (26)
15	LN	GP-S13	PEC	GP-S13	PPD	DHSR	Jureziz et al. (11)
16	LN	GP-S13	PEC	GP-S13	Brain	DHSR	Jureziz et al. (12)
17	LN	GP-S2	PEC	GP-S2	α, DNP-OL	DHSR	Jureziz et al. (12)
18	LN	GP-S2	PEC	GP-S2	ε, DNP-OL	DHSR	Jureziz et al. (12)
19	LN+Sp	GP-S2	PEC	GP-S2	Line-1 TSA	MIF	Paque et al. (19)
20	LN+Sp	GP-S2	PEC	GP-S2	Line-10 TSA	MIF	Schlager et al. (31)
21	LN+Sp	GP-S2	PEC	GP-S2	Line-1 TSA	MIF	Paque (25)
22	LN+Sp	GP-S2	PEC	GP-S2	Line-10 TSA	MIF	Paque (25)
23	LN+Sp	GP-S2	PEC	GP-S2	Line-10 TSA	Tu Reg	Schlager et al. (31) Schlager et al. (32)
24	Sp	BALB/c	PEC	BALB/c	MOPC-300 TSA	MIF	Braun & Dray (2,3)
25	Sp	BALB/c	PEC	BALB/c	MOPC-315 TSA	MIF	Braun & Dray (2,3)
26	Sp	BALB/c	PEC[c]	BALB/c[d]	MOPC-315 TSA	MIF	Braun & Dray (2,3)
XENOGENEIC TRANSFERS							
27	LN+Sp	Monkey	PEC	GP	PPD	MIF	Paque & Dray (16, 17, 18)
28	LN+Sp	Monkey	PEC	GP	Coccidiodin	MIF	Paque & Dray (16, 17, 18)
29	LN+Sp	Monkey	PEC	GP-S2	Line-10 TSA	MIF	Schlager et al. (31)
30	LN+Sp	Monkey	PEC	GP-S2	Line-10 TSA	Tu Reg	Schlager et al. (31) Schlager et al. (32, 33)
31	LN+Sp	Monkey	LN	Man	Coccidiodin	MIF	Paque & Dray (17, 18)
32	LN+Sp	Monkey	LN	Man	KLH	MIF	Paque & Dray (17, 18)
33	LN+Sp	Monkey	LN	Man	KLH	MIF	Paque & Dray (17)

[a] 5-13 S fractions of RNA extracts used

[b] Strain 13 guinea pigs are nonresponders to α, or ε, DNP-oligolysines

[c] BALB/c bearing 5 day tumor whose PEC respond to tumor specific antigen in MIF test

[d] BALB/c bearing 14 day tumor whose PEC are unresponsive to challenge with tumor specific antigen in MIF test

[e] RNA containing poly-A sequences

is apparently specific for the antigen used to sensitize/immunize the RNA donor.

This paper presents our experience in isolating and characterizing the biologically active molecular species of I-RNA preparations and in developing methods for incorporation of RNA into lymphoid cells.

B. Methods of Isolating Active I-RNA preparations.

1. Whole RNA extracts.

The importance of high-grade, freshly distilled, impurity-free phenol cannot be overemphasized for use in RNA extraction, since its purpose is to degrade and precipitate proteins. Generally, no more than 2 liters of phenol is distilled at any one time to minimize the possibility of oxidation during continued storage. A 2-liter florence flask containing about 50 glass beads and fitted with a glass sidearm and a high-temperature thermometer is connected to a glass condenser. This unit is put inside a large heating mantle. The flask is filled with 2 liters of reagent-grade phenol (Fisher Chemical) and heated to 168°C. The impurities and water distilling from the phenol before the 168°C boiling point are discarded and appear as a cloudy precipitate in the collecting bottle. The phenol is distilled into a clean brown reagent bottle containing an amount of sodium acetate buffer calculated to be 12% of the total volume of phenol that will be distilled. This is because most commercial supplies of phenol are generally only approximately 88% phenol. For example, for distilling 1 liter of phenol, 120 ml of .01M sodium acetate buffer is added to the storage and collecting vessel before distillation. Distillation is carried out in a chemical fume hood. It is best not to connect the condenser to a hot- or cold-water supply, because the distilling phenol will quickly crystallize in the condenser opening, causing stoppage of the flow of phenol from the condenser. After the distillation is complete, 0.1% by total volume of 8-hydroxyquinoline is added to the phenol as an antioxidant and preservative. The distilled phenol is stored in the dark and is used for RNA extractions generally within a 60-day period.

The standard buffer utilized in all our RNA extractions is a 0.01 M sodium acetate buffer adjusted to pH 5.0 with acetic acid. The buffer we have continually utilized incorporates reagent-grade 0.01 M sodium acetate (Fisher Chemicals) buffer containing 0.5% reagent-grade SDS and 2 or 4 μg/ml of PVS. We have on occasion utilized this buffer incorporating 1.0 or 2.0 mg/ml of thoroughly washed bentonite. Our experience indicates that the addition or omission of bentonite in the extraction buffer is not crucial to the biological or biochemical quality of the RNA preparations. One of the major critical factors relevant to the preparation of buffers for RNA extraction is the source of water. We use distilled tap water that has been filtered and deionized and redistilled in a Bellco glass water still (Bellco Glass, Vineland, N.J.) containing a quartz heating element. Metal distilling elements such as those of copper or steel should not be used. The water is collected and stored in a clean, sterile glass carboy. Before preparing the sodium acetate buffer, the water is sterilized in clean, clear-glass, 1-liter reagent bottles.

The extraction of sensitized lymph nodes, spleens, or PEC for RNA follows the same basic procedure regardless of the species of animal used. The method of extraction is essentially described by Scherrer and Darnell (1962) with the modifications of Thor and Dray (1968) and Paque and Dray (1972). The procedure has been that described as the hot-cold phenol method. Essentially, a 150-ml Virtis homogenizing flask (Virtis Instrument Co., Gardiner, N.Y.) is filled with chilled, fresh, buffer-saturated phenol at a concentration of 12 ml of phenol per gram of tissue. The tissue is weighed prior to extraction, utilizing a forceps dipped in buffer to handle the frozen tissue. After weighing, the frozen tissue is immediately plunged into the cold phenol and rapidly homogenized in a Virtis tissue homogenizer for 15 min at 4°C. The sodium acetate buffer used in the extraction process is always adjusted to pH 5.1, the optimal pH for the isolation of RNA (Steel and Busch, 1963).

We generally pack crushed ice around the Plexiglas shield surrounding the homogenizing flask to maintain a chilled environment while homogenization takes place. While the tissue is homogenizing, an equal volume of 0.01 M sodium acetate buffer is saturated with phenol in a 150-ml separatory funnel. The phenol appears as a yellowish, milky precipitate when shaken with the buffer, but upon standing for a few minutes the two liquids separate into two distinct phases. The phenol is drawn off, leaving the phenol-saturated buffer in the separatory funnel. After homogenization, an equal volume of the phenol-saturated sodium acetate buffer is added to the tissue homogenate and vigorously stirred, and the homogenate is dispensed into 50-ml polypropylene tubes previously soaked and drained with buffer. The homogenate is then mixed on a high-speed vortex mixer, the top of the tube being temporarily capped with a tight-fitting rubber stopper, and heated to 60°C in a water bath with frequent stirring. After reaching 60°C the tubes are rapidly cooled to 4°C in an ice-water bath and then centrifuged at 4°C at 40,000 g in a refrigerated high-speed centrifuge for 12 min.

After centrifugation, the aqueous phase containing the RNA is carefully removed with a sterile Pasteur pipet and deposited in clean 50-ml polypropylene centrifuge tubes soaked earlier in 0.1 M sodium acetate buffer. An equal amount of buffer-saturated phenol is added to the aqueous phase, and the extraction procedure described above is repeated three times or until the interface appears clear. The quality of the extraction procedure can be tested by reading the A_{260}/A_{280} ratio of an appropriate dilution (usually 1:50) of the aqueous phase. It should be approximately ≤ 2.0. After the last extraction, the volume of the aqueous phase is measured and divided equally into clean, sterile, Corex high-speed glass centrifuge tubes (Corning Glass Co., N.Y.). Next, 3.0 M sodium acetate buffer, pH 5.0, is added to each volume of aqueous phase so that the final concentration is approximately 0.3 M. At least 5 volumes of cold 95% reagent-grade ethanol is added, and the RNA is permitted to precipitate at -20° C overnight. After precipitation overnight, the RNA is centrifuged at 4°C at 40,000 g for 12 min. The pellet is dissolved in sterile 0.3 M sodium acetate buffer and reprecipitated with 95% chilled ethanol. This procedure is repeated at least three times before using the RNA in any of the *in vitro* or *in vivo* tests for biological activity.

2. Isolation of RNA Fractions by sucrose density gradient centrifugation.

Many methods are employed in the separation of RNA molecules by sedimentation characteristics in surcose density gradients. We have routinely used analytical density gradient analysis for monitoring the quality of RNA extractions and employ a similar method in the preparation of RNA fractions by density gradient analysis. We present in detail a method utilized in our laboratory for both analytical and preparative surcrose density gradient separation of RNA fractions.

Linear sucrose density gradients are first prepared in a Buchler density gradient maker (Buchler Instruments Co., Fort Lee, N.J.). An ultrapure, RNase-free grade of sucrose (Schwarz-Mann, Orangeburg, N.J.) is prepared in a 0.11 M sodium acetate buffer containing 0.001 M EDTA adjusted to pH 5.1 with sterile acetic acid. Solutions of sucrose, 6 and 40%, are dissolved in the buffer in a total volume of 20 ml for each sucrose solution. It is important that the containers used for mixing density gradient solutions are clean and sterile in order to minimize and eliminate residual RNase activity. After the sucrose has been dissolved, the mixing chambers in the density gradient maker are filled. The first chamber is carefully filled with 19 ml of 40% sucrose, with the stopcock between the mixing chambers remaining closed. Next, 19 ml of a 6% sucrose solution is added to the other chamber. The stopcock is carefully opened to permit the escape of air bubbles between the mixing chambers. We use Beckman 1.6 X 10.2 cm cellulose nitrate tubes (Beckman Instruments, Fullerton, Calif.) for the preparative runs and 1.4 X 9.5 cm tubes for the analytical runs. The lines running from the gradient maker to the gradient tubes are carefully evacuated of air, utilizing a 5.0-ml syringe connected to a 28-gauge needle inserted into each line and slowly withdrawing the air. Mixing of the sucrose solutions is accomplished by the use of a small mixer inserted into the chamber containing the 40% sucrose solution. It is best if the chambers are inclined slightly forward to compensate for the tendency of the heavier 40% sucrose to run into the opposite chamber.

The gradients are usually made 2 or 3 hours before starting the centrifuge run and can be stored briefly in a test tube rack in the cold at 4°C. RNA to be analyzed or prepared is first centrifuged at 40,000 g, the pellet is dissolved in 1.5-2.0 ml of 0.11 M sodium acetate buffer, and the absorbance is measured at 260 nm in a Gilford Model 240 spectrophotometer. After obtaining the absorbance for the RNA preparation, the concentration of RNA in the solution is calculated and 125-150 μg is carefully layered on the analytical gradients with a sterile Pasteur pipet. In removing the aliquots of RNA from the solution, sterile, chilled glass pipets should be used. The aliquot is then deposited in a sterile test tube cap and drawn into the Pasteur pipet for gradient layering. Larger quantities of RNA can be utilized for preparing fractions to be tested for immunobiological activity <u>in vitro</u>. For the preparative runs, we commonly layer approximately 0.8-1.2 mg of RNA on the density gradient. Next, the preparative gradients are centrifuged at 5°C for 18 hours at 27,000 rpm in a prechilled Spinco Model SW-27.1 Ti rotor in a Beckman Model L5-50 ultracentrifuge. Analytical gradients are centrifuged for 18 hours at 28,000 rpm in a Beckman SW-40 rotor at 5°C or 40,000 rpm for 6 hours.

After centrifugation, the gradients are fractionated utilizing an ISCO Model 640 density gradient fractionator (Instrument Specialties Co., Lincoln, Nebr.) connected to a Model UA-5 ultraviolet analyzer set at 254 nm. The cellulose nitrate tubes are punctured at the bottom, and a solution of 70% glycerol is used to float the fractions of RNA upward. An expanded-scale recorder, ISCO Model 613, is utilized to record the RNA peaks as they pass through the optical unit. The syringe drive is stopped at appropriate points indicated by the peaks traced on the recorder, and each RNA fraction is collected in chilled, clean, sterile Corex glass centrifuge tubes. The collected fractions are immediately precipitated with cold ethanol and stored at -20°C until used in the *in vitro* assays to test transfer activity.

3. Isolation of RNA fractions by affinity chromotography on oligo(dT) cellulose.

The presence of poly A in almost all mRNA and its absence from tRNA or rRNA permits one to differentiate among these various species of RNA. Poly A can bind with a complementary base (thymidylic acid) and is the basis for the utilization of affinity chromatography in isolating and separating mRNA from other species of RNA.

The procedure we describe here is used in our laboratory and in other laboratories for isolating poly-A-containing mRNA from a variety of sources. Oligo(dT)cellulose (type T3) can be obtained from Collaborative Research, (Waltham, Mass.). The columns to be utilized for binding poly-A-containing RNA are sterile Pasteur pipets usually packed with 0.5-1.0 gm of oligo(dT) cellulose. The weighed quantity of cellulose is first washed three times in a high-salt buffer consisting of 0.5 M KCl and 0.01 M Tris-HCl adjusted to a pH of 7.5. A very small piece of autoclaved cotton or fiberglass is gently packed in the bottom junction of the pipet, and a small sterile piece of 1-mm tubing is connected to the small end of the pipet. The washed slurry of oligo(dT) cellulose is pipetted into the Pasteur pipet column and allowed to settle for 5 minutes. After packing the column, the oligo(dT)cellulose is overlaid with high-salt buffer, the tubing is clamped, and the column is stored at $4^{o}C$. After using the column several times, the flow rate decreases substantially. It is best in this case to remove the oligo(dT)cellulose from the column and wash with the high-salt buffer. Oligo(dT)cellulose can be regenerated for additional use by washing three times with 0.1 M NaOH with centrifugation at 150 g each time. After preparing the column, the RNA extract to be fractionated is centrifuged at 41,000 g for 10 min at $4^{o}C$, and the ethanol removed with a suction flask.

The RNA pellet is dissolved in approximately 0.8-1.0 ml of high-salt buffer and carefully drawn into a chilled, sterile Pasteur pipet. One should not try to dissolve over 1 mg of RNA for application in this particular column setup, as we have found that very sluggish flow rates and undissolved RNA can clog the oligo(dT) column. The dissolved RNA solution is carefully run down the sides of the column, gently applying it on the surface of the high-salt buffer in the column. The tubing at the bottom of the column is connected to a 0.2-cm flow cell in a Gilford Model 250 recording spectrophotometer with the wavelength set at 254 mμ. The high-salt buffer RNA solution is permitted to pass through the flow cell in the spectrophotometer. The first recorded peak, presumably free of poly A, is eluted first and has been designated fraction I. The column is then rinsed three times with high-salt buffer. After rinsing, the column is filled with the low-salt buffer containing 0.1 Tris-HCl adjusted to pH 7.5. The low-salt buffer is then permitted to flow through the column, eluting the second peak of RNA bound to the oligo(dT) cellulose and containing the poly-A sequences. Each fraction of RNA is collected in sterile, chilled Corex glass centrifuge tubes and immediately precipitated with cold 95% ethanol and stored at -20°C until needed. Typical peaks of fractions I and II separated by oligo(dT) chromatography are shown in Figure 4. Different lots of oligo(dT) cellulose vary in their ability to bind poly A. We have generally used T3-type oligo(dT) cellulose (Collaborative Research, Waltham, Mass.) which is reported to bind 40-150 absorbance units per gram of oligo(dT)cellulose. It is best to store oligo(dT)cellulose at -20°C in the powdered form, and a wet slurry of the material can be stored at 4°.

4. Experimental design for dual isolation of RNA fractions by sucrose density gradients and affinity chromatography.

The experimental design of the comparison studies in isolation of the various RNA fractions previously shown to mediate delayed hypersensitivity are shown in Figure 1. Hot-cold phenol extracts of "whole" lymphoid cell RNA (2-3 mg) are divided equally into 1.0 to 1.5 mg quantities. Each equal portion of the "whole" extract is subjected to (a) preparative sucrose density gradient fractionation or (b) affinity chromatography on oligo(dT) columns (Fig. 1). After oligo(dT) isolation of Fractions I and II, each fraction is subjected to ultracentrifugation on sucrose density gradients and fractionated (Fig. 1). In contrast, the B fraction derived from the preparative sucrose density gradient, as well as A, C, D, and E, are subjected to oligo(dT) chromatography to assess the presence of polyadenylic acid sequences binding to the column. The oligo(dT) fractions derived from sucrose density gradient fractions and

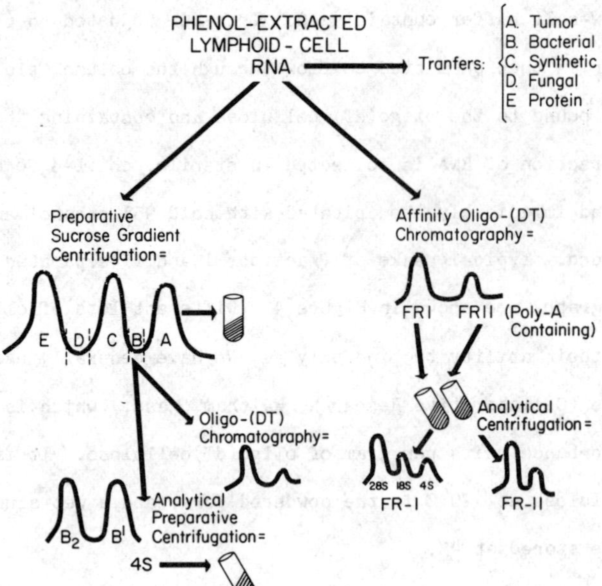

Figure 1. Experimental design of the fractionation and isolation of the various RNA fractions described herein. RNA extracted from the spleen or lymph nodes of ARSNAT sensitized guinea pigs is divided equal and subjected to sucrose density gradient fractionation or oligo(dT) affinity chromatography. Fraction II [oligo(dT) poly(A) containing] is then applied to sucrose density gradient centrifugation while the B fraction is subjected to oligo(dT) chromatography. The low-molecular-weight 4 S peak is to the right of the B subfractions. From Paque and Nealon, 1979, Cell. Immunol. 43, 48-61.

sucrose density gradient fractions derived from oligo(dT) chromatography are each tested for their ability to transfer cell-mediated immunity *in vitro*.

C. Characterization of Active RNA Preparations

1. Sucrose Density Gradient Analysis.

To localize the biological activity in the RNA extracts, RNA was prepared from guinea pigs immunized against line 10 tumors or various other antigens. A total of 8 mg of RNA-rich extracts were layered on individual sucrose density gradients so that each gradient contained approximately 800 to 1000 µg of RNA. The gradients were centrifuged until they demonstrated the characteristic 4 S, 18 S, and 28 S peaks, and 5 separate fractions were cut and collected from each gradient. Each fraction was then assessed for its ability to transfer line 10 tumor-specific sensitivity to GP-PEC in the cell-migration-inhibition assay.

The fractions designated as A, B, C, D, and E correspond to the cuts shown on a typical preparative sucrose density gradient in Figure 2. In all experiments, the 2nd fraction, designated as "B", transferred specific tumor sensitivity to nonsensitized GP-PEC. On the other hand, all other fractions corresponding approximately to 4 S, 18 S, 22 S, and 28 S failed to transfer line 10 sensitivity to nonsensitized GP-PEC. Similar results in regard to the "active" Fraction B are observed utilizing "immune" RNA prepared from animals immunized against chemically defined, fungal or bacterial antigens.

To further resolve and isolate the species of RNA capable of transferring line 10 or line 1 tumor-specific sensitivity, various biologically active fractions that we have designated as Fraction B, prepared from line 1 or line 10 lymphoid-cell RNA extracts, were recentrifuged on linear sucrose density gradients. Recentrifugation of line 10 or line 1 I-RNA fraction B results in 2 distinct peaks, which we have designated as B1 and B2 subfractions, and are

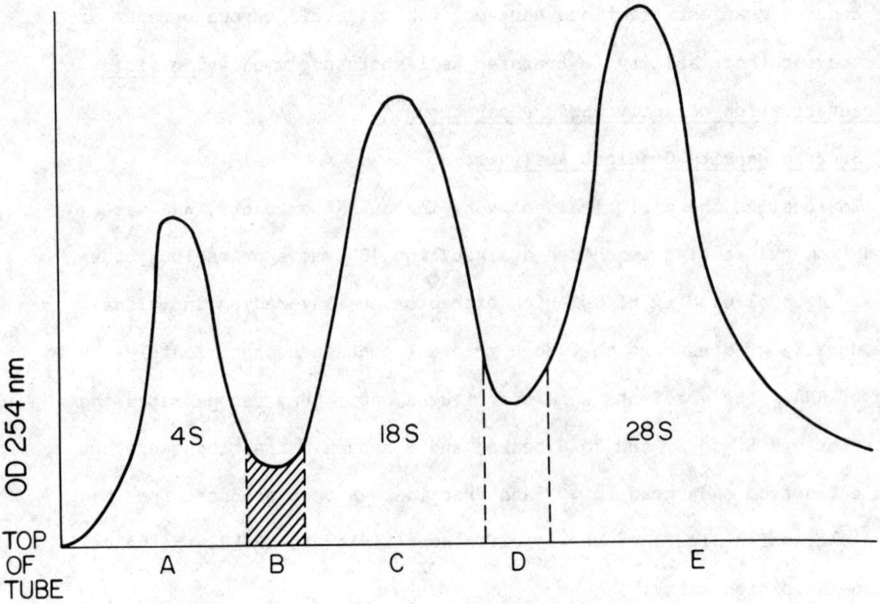

Figure 2. A typical sucrose density gradient pattern of RNA-rich extracts prepared from the lymphoid issues of strain 2 guinea pigs imunized to line 1 or line 10 hepatomas. -----, approximate "cut" of each fraction; shaded area, approximate position of Fraction B. From Paque, R.E., 1976. Cancer Res. 36,4530-4536.

shown in Figure 3. These peaks were cut and collected from the biologically active isolated Fraction B and assessed for their ability to transfer tumor-specific sensitivity to nonsensitized GP-PEC in the cell migration-inhibition assay. Fraction B1, in the line 10 system, transferred line 10 sensitivity to GP-PEC reproducibly in most instances. In addition, Fraction B2 occasionally transferred line 10 sensitivity, but generally, in most experiments, the B2 fraction has been found to be inactive. The occasional transfer of sensitivity found with the B2 fraction may reflect technical limitations present in the fractionation procedure. In a total of 14 experiments, where 30 B1 or B2 fractions were tested for their ability to transfer line 10 or line 1 sensitivity to GP-PEC, transfer of sensitivity for line 10 or line 1 with the B1 fraction occurred in 10 experiments.

2. Oligo(dT) affinity chromatography analysis.

RNA extracted from the lymph nodes or spleens of guinea pigs skin test sensitive to KLH or ARSNAT were separated on oligo (dT)-columns. Preparations of RNA extracts prepared by the hot-cold phenol method were each dissolved in high salt buffer and were applied to oligo (dT)-columns connected to a flow cell in a Gilford 250 spectrophotometer. After collecting the unbound RNA fraction eluted in the void volume with high salt buffer (Fraction I), the second peak of RNA bound to the column was eluted with the low salt buffer. Typical peaks eluted from the oligo (dT)-columns are shown in Figure 4. The unbound RNA, presumably containing none or few polyadenylic acid sequences is designated as Fraction I (Fig. 4). The RNA extract bound to the column, and presumably rich in poly-A sequences, is designated as Fraction II (Fig. 4). A total of 30 fractions (Fraction I and II) from four separate pools of phenol-extracted RNA were tested for their ability to transfer KLH sensitivity to nonsensitized GP-PEC. Fraction I was unable to transfer KLH sensitivity in 15

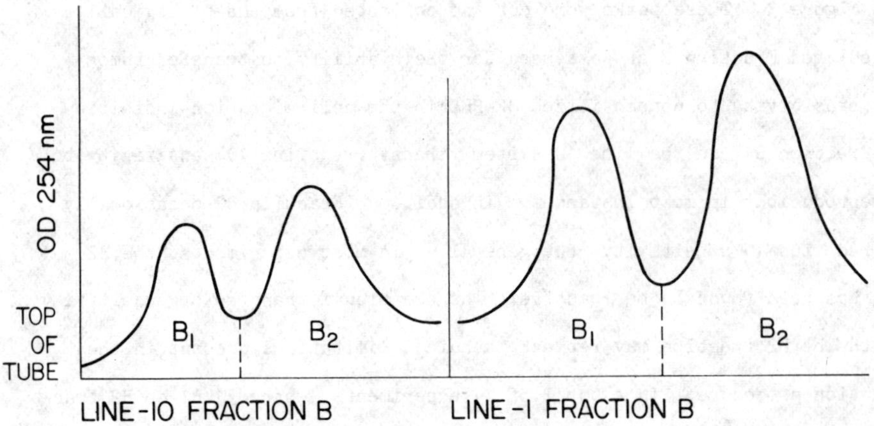

Figure 3. Sucrose density gradient patterns of subfractions of recentrifuged Fraction B (50 to 150 μg) prepared from RNA extracts of lymphoid tissues of animals immunized against line 10 or line 1 hepatomas. From Paque, R.E., 1976. Cancer Res. 36, 4530-4536.

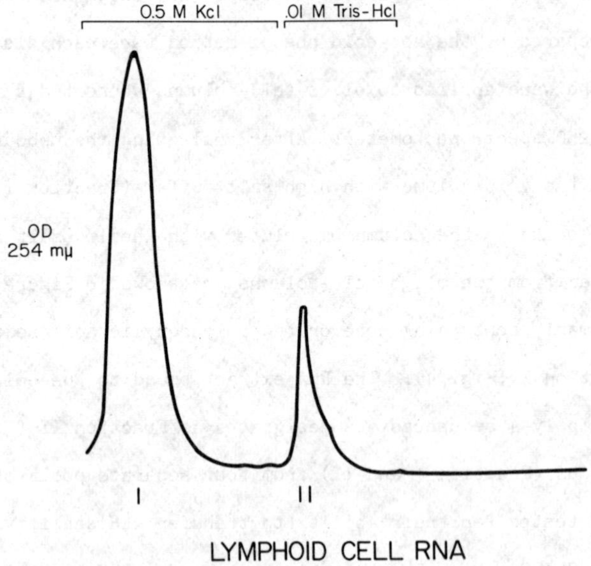

Figure 4. Hot-cold phenol extracted lymphoid cell RNA peaks obtained after elution from oligo(dT)-cellulose column. Fraction I, high salt (0.5 M KCl) buffer eluted; Fraction II, low salt buffer eluted (0.01 M Tris-HCl). From Paque, R.E., 1977. Cell. Immunol. 30, 332.

experiments. On the other hand, in the same number of experiments, Fraction II containing poly-A sequences routinely transferred KLH sensitivity to GP-PEC. Similar results were obtained utilizing Fraction II prepared from guinea pigs immunized against a chemically defined antigen, ARSNAT.

GP-PEC were collected 3 days after injection of light mineral oil and divided into three sterile 12-ml conical centrifuge tubes. Each preparation of nonsensitized cells was incubated with oligo (dT) Fraction I, Fraction II, or alone. After 15 min of incubation with the RNA fractions, the GP-PEC were centrifuged and immediately set up in the agarose droplet cell-migration-inhibition assay. The results of 10 experiments utilizing three separate pools of RNA with a total of 20 fractions, were tested for their ability to transfer ARSNAT sensitivity to GP-PEC. Fraction I, obtained from spleens or lymph nodes of ARSNAT sensitized guinea pigs, was unable to convert GP-PEC to specific ARSNAT sensitivity. On the other hand, Fraction II was able to transfer specific ARSNAT sensitivity in all experiments. Neither Fraction I nor Fraction II was able to inhibit the GP-PEC alone and Fraction I or II incubated with GP-PEC and an unrelated antigen, KLH in this case, also failed to inhibit the migration of GP-PEC. When Fraction II was incubated with GP-PEC and PPD, transfer of specific PPD sensitivity occurred in three experiments because the guinea pigs are immunized against ARSNAT which is emulsified in Freund's complete adjuvant containing *Mycobacterium butyricum*. Yet, presumably because the sensitization of the guinea pigs is not optimal, the transfer of PPD sensitivity occurs irregularly. To be sure that our GP-PEC were indeed "non-sensitive", GP-PEC were incubated with each of the antigens alone; ARSNAT, PPD, or KLH and these antigens alone failed to inhibit the migration of GP-PEC.

3. Sucrose Density Gradient Analysis of Fractions I and II of RNA Extracts Derived from Oligo (dT) Chromatography.

RNA extracted from the spleens or lymph nodes of ARSNAT-sensitized guinea pigs were dissolved in high salt buffer prior to affinity chromatography. The high salt buffer-dissolved RNA preparations were applied to an oligo (dT) column, and RNA Fraction I and II were eluted with high or low salt buffers, respectively. After isolation of Fractions I or II, each fraction was centrifuged on linear sucrose density gradients to assess the composition of each (Fig. 5). High salt buffer-eluted Fraction I had a sucrose density gradient profile similar to "whole" RNA extracts having 4 S, 18 S, and 28 S components. On the other hand, Fraction II possessed two peaks and a density gradient profile similar to those peaks obtained after recentrifugation of Fraction "B" from density gradient prepared RNA (Fig. 5). Each of the density gradient fractions shown in Fig. 5 derived from oligo (dT) affinity chromatography were collected and immediately precipitated with cold ethanol. Later, the fractions were tested for their ability to transfer ARSNAT sensitivity to GP-PEC.

Three days after the injection of light mineral oil, "nonsensitized" GP-PEC were collected and divided equally into three sterile 12-ml conical glass centrifuge tubes. Each tube of cells was centrifuged and incubated with the following reagents: GP-PEC incubated with gradient-prepared B1 subfraction (Fraction II derived); GP-PEC incubated with gradient-prepared B2 subfraction (Fraction II derived); or GP-PEC incubated alone. After incubation with the RNA for 15 min at 37°C the cells were centrifuged and set up in the agarose-droplet cell-migration-inhibition assay.

A total of 10 Fraction II-derived "B" subfractions were tested for their ability to transfer ARSNAT sensitivity. Subfraction B1 (Fraction II derived) was able to transfer ARSNAT sensitivity to GP-PEC in five experiments. On the

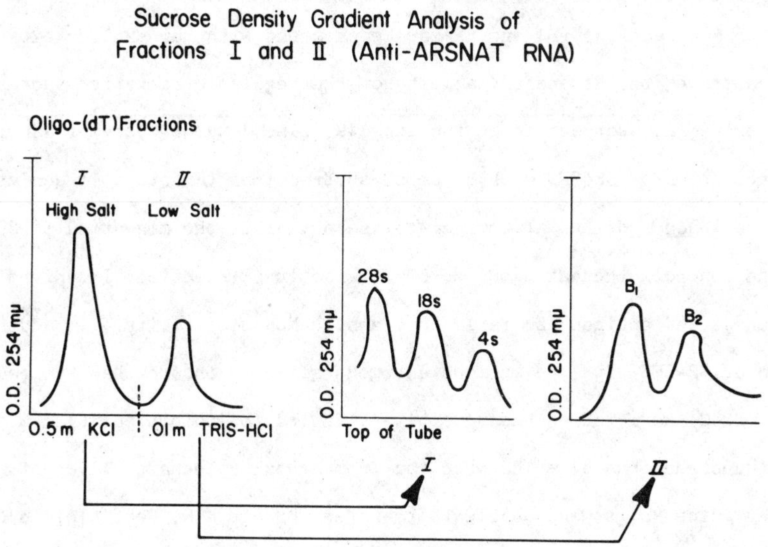

Figure 5. Phenol-extracted lymphoid cell RNA subjected to oligo(dT) chromatography followed by sucrose density gradient fractionation of Fraction I (high salt buffer eluted) and Fraction II (low salt buffer eluted, poly-A containing). The density gradient sedimentation patterns of Fractions I and II are shown on the right hand side of the figure. From Paque, R.E. and T. Nealon, 1979, Cell. Immunol. 43, 48-61.

other hand, the B2 (Fraction II derived) subfraction was able to transfer ARSNAT sensitivity to GP-PEC only in one experiment. The results with Fraction II derived B subfractions (B1 and B2) are consistent with our earlier data which demonstrated the "transfer" ability of the regular preparative sucrose gradient-derived B1 subfraction in the line 10, line 1 tumor system (Paque, 1976). Incubation of GP-PEC with B1 or B2 subfractions (Fraction II derived) alone in the absence of ARSNAT antigen fails to inhibit the migration of GP-PEC and routine controls incubating B1 or B2 subfractions (Fraction II derived) with the unrelated antigen KLH fails to transfer KLH sensitivity. In addition, incubation of GP-PEC with the various antigens alone to insure that the recipient cells (GPEC) were indeed "nonsensitive", failed to inhibit cellular migration. RNA subfractions B1 or B2 prepared from normal spleens or liver of unimmunized guinea pigs were unable to transfer ARSNAT, KLH, Coccidioidin or PPD sensitivity to nonsensitized PEC in these experiments. Furthermore, incubation of 15 μg of pancreatic RNase with either B1 or B2 subfraction derived from oligo (dT) Fraction II completely abolishes the ability to transfer ARSNAT sensitivity to nonsensitized GP-PEC.

4. Oligo (dT) Chromatography analysis of the RNA "B" Fraction Derived from Sucrose Density Gradient Centrifugation.

RNA was extracted from lymph nodes or spleens of guinea pigs sensitized against ARSNAT and PPD. The RNA (1-2 mg) was layered on 6 to 40% preparative, linear sucrose density gradients and centrifuged for 18 to 24 hr. After preparative density gradient centrifugation, the characteristic 4 S, 18 S, and 28 S peaks of "whole" lymphoid cell RNA extract can be seen (Fig. 6). After centrifugation, the RNA was fractionated and the biologically active B fraction was collected and precipitated with cold 95% alcohol. The B fraction was then applied to an oligo (dT) column to determine if polyadenylic acid sequences

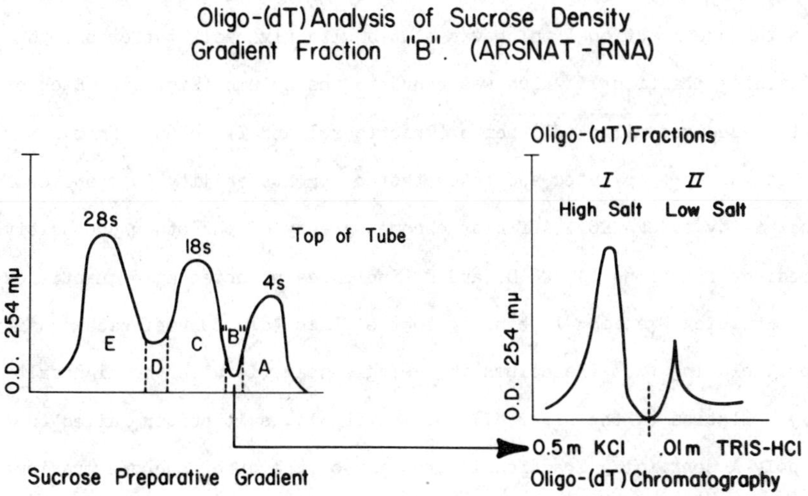

Figure 6. Phenol-extracted lymphoid cell RNA separated by sucrose density gradient centrifugation followed by oligo (dT) chromatography of the B fraction. A typical affinity chromatography pattern with elution of a poly-A containing Fraction II (B derived) is shown at the right. Affinity chromatography of fractions A, C, D, and E results in single peaks eluted with high salt buffer only. From Paque, R.E. and T. Nealon, 1979, Cell. Immunol. 43, 48.

were present. Application to an oligo (dT) column of the "active" sucrose gradient-derived B fraction results in the presence of two separate peaks. These peaks include Fraction I which was eluted with high salt buffer and the poly-A containing Fraction II which was bound to the column (Fig. 6). Each of the B fraction-derived oligo (dT) peaks (Fractions I and II) eluted from the column was ethanol precipitated and later tested for its ability to transfer ARSNAT sensitivity to GP-PEC. Affinity chromatography of the other preparative density gradient fractions (A, C, D, and E) indicates no detectable quantities of poly-A containing Fraction II binding to the oligo (dT) column; each density gradient peak eluting from the column in a single peak (I) with the high salt buffer only. Elution of the oligo(dT) column with low salt buffer failed to yield any poly-A containing Fraction II from any of the other sucrose gradient peaks (A, C, D, and E).

Nonsensitized GP-PEC were collected 3 days after injection of mineral oil and divided equally into three separate pools. Each preparation of "nonsensitized" GP-PEC was incubated with Fraction I (B-derived), Fraction II (B-derived), or in the absence of RNA fractions. After 15-min incubation at 37°C, the GP-PEC were centrifuged and immediately set up in the agarose droplet cell-migration-inhibition assay in the presence or absence of the various antigens. In six experiments, testing a total of 12 B-derived fractions, GP-PEC incubated with Fraction I (B-derived) failed to demonstrate transfer of ARSNAT sensitivity to GP-PEC. On the other hand, Fraction II (B-derived) binding to the oligo (dT) column was able to transfer specific ARSNAT sensitivity to GP-PEC as evidenced by specific inhibition of cell migration. This B-derived Fraction II presumably contains polyadenylic acid sequences as evidenced by its ability to bind an oligo(dT) cellulose column. GP-PEC incubated with Fractions I or II (B-derived) in the absence of the ARSNAT antigen failed to demonstrate inhibition of cell migration. In addition, GP-PEC incubated with Fraction I (B-derived) in the presence of PPD or coccidioidin fails to transfer specific sensitivity and GP-PEC incubated with oligo (dT) binding Fraction II (B-derived) failed to transfer coccidioidin sensitivity. Fraction II (B-derived) also transferred PPD sensitivity to GP-PEC. The transfer of PPD sensitivity with Fraction II (B-derived) would be expected since the guinea pigs were immunized against 150 μg of ARSNAT antigen incorporated in Freund's complete adjuvant containing *Mycobacterium butyricum*. Incubation of 15 μg of pancreatic RNase with oligo (dT) Fraction II (B-derived) destroys the ability of the RNA fraction to transfer ARSNAT sensitivity to GP-PEC. In addition, Fraction II (B-derived) prepared from the skeletal muscle or liver of ARSNAT sensitized guinea pigs failed to transfer ARSNAT sensitivity to nonsentized GP-PEC.

D. Presence of Ag in RNA preparations.

The presence or absence of antigen in I-RNA is a central question in the study of I-RNA-mediated transfer of immunity and raises the issue of the mechanism responsible for transfer. Two distinct hypotheses predominate the literature on this subject. Some investigators favor a super antigen concept in which RNA condenses with antigenic determinants to form a potent immunogen that can directly sensitize lymphoid cells. These RNA-antigen complexes and their role in I-RNA mediated transfer of immune reactivity have been reviewed extensively by Gottlieb and Schwartz (1972) and therefore, will not be considered here. Another hypothesis advanced to explain the action of RNA in transferring immunity proposes an informational role for the I-RNA. Presumably, informational I-RNA would not require the presence of antigen for transfer of reactivity. In our work, we have been unable to detect antigen contamination in our I-RNA preparations by immuno- logical or chemical tests. For example,

Jureziz, Thor and Dray (1970) tested the ability of I-RNA from DNP-oligolysine immune guinea pigs to sensitize normal animals. Guinea pigs that recieved up to 2.1 mg of I-RNA intraperitoneally, an amount of RNA 10 fold greater than the amount of material needed to convert nonsensitive lymphoid cells to a state of reactivity, did not develop positive DTH skin test upon challenge with DNP-oligolysine. In addition, normal spleen cells that had been treated with 25 µg of DNP-oligolysine were unable to transfer DTH skin test reactivity to normal guinea pigs. We have also attempted to release MIF from PPD-sensitized guinea pig lymphocytes with I-RNA prepared from PPD-sensitized hosts (Paque and Dray, 1970). Whereas the sensitized PEC were inhibited in their migration by as little as 10 µg/ml of PPD antigen, these same PEC were not inhibited in their migration by up to 3 mg of PPD-I-RNA. Still, these experiments do not exclude the presence of minute amounts of antigenic material which may be sequestered in the active I-RNA preparations.

For these reasons, we turned to chemical techniques for detecting antigen in our I-RNA preparations. One such antigen system which lent itself to the problem is the low MW (486) chemically-defined antigen ARSNAT. Approximately 15.4% of the ARSNAT antigen consists of arsenic which is detectable in trace amounts by known chemical and instrumental methods. Thus, arsenic can serve as a chemical marker for the presence of antigen in I-RNA capable of transferring ARSNAT sensitivity. RNA extracts from ARSNAT sensitive donors known to transfer sensitivity to the antigen were assessed for arsenic content by atomic absorption spectroscopy (AAS) (Schlager, Dray and Paque, 1974). This technique assays the arsenic directly, has a sensitivity greater than any biological or chemical assay, and since the sample is exhaustively incinerated at 2700°C, 'buried antigen' cannot escape detection. The absorbance at 1937 A is directly proportional to the quantity of arsenic in each sample. The peaks of absorbance in testing the positive controls of a standard arsenic solution and the ARSNAT samples can be readily observed (Fig. 7). When 13 and 23 ng of ARSNAT antigen were tested, the absorbance pattern was identical to that of the 2- and 5-ng standard arsenic samples. Testing large samples of RNA (5 or 10 mg) indicated nonspecific background absorbance due to the substantial quantities of RNA being assayed. The background absorption peaks are evidenced by first injecting 250 µg of an RNA sample and following with a standard 2-ng arsenic solution. The resulting absorbance pattern indicates two distinct peaks, one for nonspecific background absorbance (RNA) and a second corresponding to the 2 ng arsenic sample injected (Fig. 7). This control must be used when performing these studies to show that the instrument's ability to detect arsenic is not lost or concealed by the background absorbance of the RNA.

Because of the background absorbance initially noted with the RNA samples, an additional control must be used to ensure that the peaks of absorbance noted in Fig. 7 are indeed background. First, the wavelength of the apparatus is changed to 1942 A, the wavelength of the mercury ion line and a wavelength where arsenic atoms do not absorb light but which will indicate background absorbance. Assessing standard arsenic solutions and 5- and 10-mg quantities of ARSNAT, I-RNA extracts, the absorbance pattern of the water blank was identical to that of the standard 2- and 5-ng arsenic solutions; and the background absorbance for 5 and 10 mg of RNA was still present at 1942 A and identical to the peaks observed at 1937 A (Fig. 7). The inability to detect arsenic as an ARSNAT marker in several pools of RNA extracts transferring ARSNAT sensitivity suggests of its absence in the RNA extracts.

AAS, with a sensitivity of 0.1 ng, failed to detect arsenic in 250 µg to 10 mg of ARSNAT I-RNA. Therefore, if arsenic is associated with I-RNA material,

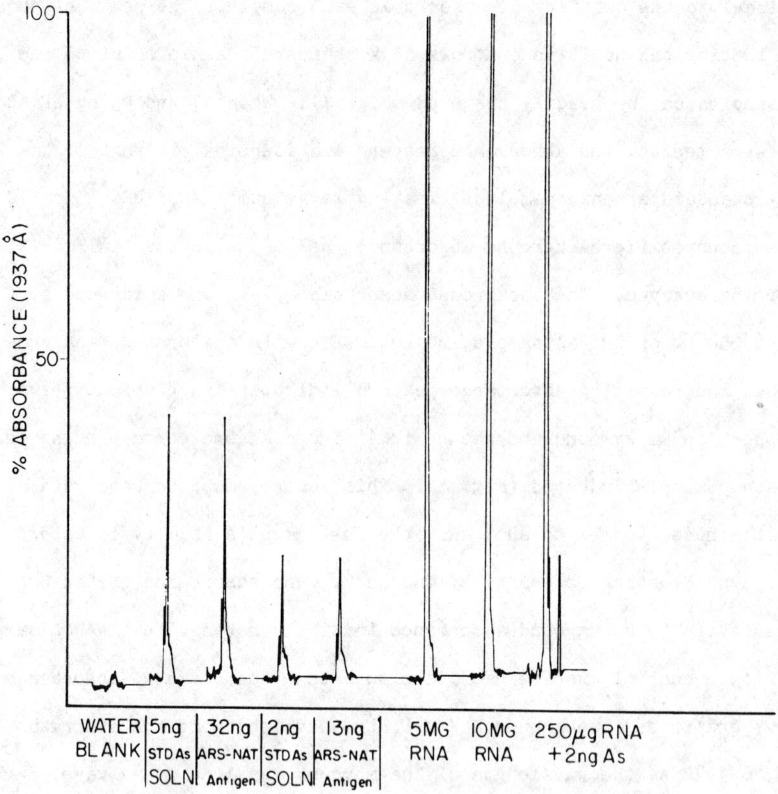

Figure 7. Atomic spectroscopic absorbance peaks at 1937 A of 5- and 2-ng standard arsenic solutions and control soutions of 13- and 32-ng samples of ARSNAT, RNA extract 5- and 10-mg samples were from guinea pigs sensitized to the ARSNAT. The peak on the extreme right is a localization control of background RNA and arsenic absorbance. From Schlager, Dray and Paque, 1974, Cell. Immunol. 14, 105-122.

it could be present in an amount of no more than 5 pg in 500 μg of RNA, the amount of RNA usually used for transfer of immunity. This corresponds to less than 0.0000065% ARSNAT antigen in the RNA extract.

Although, such a low level of possible contamination is strong indication that only RNA is involved, Crouch (1976) has stated that even this low level of possible contamination does not exclude the involvement of antigen, and has made the following calculations.

Starting with 1 mg of total RNA we then exclude ribosomal and tRNA giving us 50 μg of RNA. Estimates of the size of immune RNA, as determined by sucrose gradient sedimentation are on the order of 14-18 S or about 2,000 nucleotides. If we ascribe all 50 μg of RNA as being immune RNA, the following calculation can be made:

$$\frac{50 \times 10^{-6} \text{ g RNA}}{2 \times 10^{3} \text{ nucleotides } 0.320 \text{ g/mole nucleotide}} = 7.81 \times 10^{-11} \text{ moles immune RNA}$$

$$6.5 \times 10^{8} \text{g ARSNAT/g RNA or } 6.5 \times 10^{8} \text{mg ARSNAT/mg RNA}$$

$$\frac{6.5 \times 10^{-11} \text{ g ARSNAT}}{486 \text{ g/mole ARSNAT}} = 1.11 \times 10^{-13} \text{moles ARSNAT}$$

$$\text{giving } \frac{11.1 \times 1o^{-14} \text{ moles ARNSAT}}{7.81 \times 10^{-11} \text{ moles immune RNA}} = 1.04 \times 10^{-3} \frac{\text{moles ARSNAT}}{\text{moles immune RNA}}$$

or 1 molecule of ARSNAT per 1000 molecules of immune RNA.

In our view, there are certain limitations to this calculation, since: (1) Not all molecules are I-RNA specific for ARSNAT; (2) most RNA is degraded during incubation with cells and no selective uptake of message species has been reported; (3) the amount of ARSNAT used in the calculations represents the maximum amount possible and would therefore give an overestimate of the amount of ARSNAT possibly present. In addition, when we transferred line 10 or line 1 sensitivity with the B fraction, or ARSNAT sensitivity with Fraction II, the amount of RNA has ranged from 10 to 50 μg; amounts which are 50 to 10 fold smaller than the amount (500 μg) used to calculate the values above. Thus, although the complete absence of antigen contamination in I-RNA may be impossible to prove, we believe it is improbable that the activity of ARSNAT I-RNA is due to a super-antigen.

E. Incorporation of RNA into Lymphoid Cells.

The incubation of lymphoid cells with I-RNA is a crucial step in the transfer of biological activity and is often fraught with technical difficulties. Predominant among these is the extreme sensitivity of I-RNA to ribonucleases which appear to be ubiquitous to most biological systems. Any RNA transfer, be it *in vitro* or *in vivo*, must attempt to deal with the problem of ribonucleases. Most attempts have centered upon inhibiting the activity of these enzymes with sodium dextran sulphate or polyvinyl sulphate. The success of this approach, particularly for RNA injected *in vivo* is questionable since Enesco (1966) found that intravenously injected C^{14} labeled RNA is rapidly broken down to free bases and ribose, the bases contributing to the free nucleotide pool and the ribose fraction being utilized in carbohydrate metabolism or exhaled as CO_2. Another aspect of this problem is the limited capacity of cells to incorporate biologically active I-RNA. This difficulty may be a manifestation of both the influence of ribonucleases in the incubation environment as well as the competence of the cells to take up the I-RNA. Several lines of evidence suggest that the ability of cells to incorporate I-RNA is subject to biological restrictions. First, it is usually observed that cells incubated with I-RNA do not express the transferred activity unless they are subsequently stimulated with the immunizing antigen. This has been interpreted to indicate that the number of cells which are actually converted to antigenic sensitivity by I-RNA is relatively small. This was seen by Schlager, Paque and Dray (1975) in the Line-10 heptoma system who observed that guinea pigs treated with I-RNA and syngeneic nonsensitive PEC alone exhibited weak anti-tumor reactivity whereas the addition of extracted tumor antigens to the therapy regimen resulted in complete tumor regression. Apparently, the additional stimulation of converted cells by antigen led to the expansion of the antigen reactive cells to a level sufficient to cause tumor regression. In this study, however, the participation of additional host cells might also have been responsible for the differences seen in the presence or absence of antigen.

Aside from the observations suggesting that only a limited number of cells are actually converted by I-RNA, the tacit assumption that lymphoid cells can incorporate I-RNA has only rarely been tested. To directly test this assumption, Wang, Giacomoni and Dray (1973) studied the fate of the I-RNA incorporated by rabbit lymphoid cells and found that rabbit spleen cells can incorporate a small but measurable amount of radioactively-labeled rabbit lymph node RNA. Approximately 4×10^{10} daltons of RNA which was demonstrably resistant to RNase was incorporated per cell at saturation, representing approximately 0.2% of the amount of input RNA used. The RNA incorporated was found in all 3 subcellular fractions analyzed and had a nucleotide composition similar to that of the input RNA. Interestingly, the amount of bacterial RNA incorporated by rabbit spleen cells was substantially lower than the level of rabbit RNA incorporated and rapidly degraded intracellulary. Also, it was found that nonincorporated RNA was rapidly and completely degraded, presumably by cell-associated nucleases, within the first 15 minutes of incubation (Giacomoni, Wang and Dray, 1973). With all of these considerations in mind, it becomes difficult to envision how, under such circumstances a precise amount of RNA molecules can be delivered into the proper recipient cells in a reproducible manner. Thus, the ability of cells to incorporate I-RNA is subject to several severe restrictions and may well be the limiting factor which determines the success or failure of the transfer of biological activity.

Several procedures have been employed to facilitate the cellular incorporation of administered RNA, with the most utilized being the microinjection of specific messages into Xenopus oocytes (Graessmann, Graessman and Mueller, 1977). In this instance the production message-specific translation product has been reported. However, since the frog oocyte is an undifferentiated cell, data obtained in this system may not accurately reflect processes in differentiated eukaryotic cells. Graessmann et al (1977) have overcome this problem by the microinjection of RNA into differentiated eukaryotic cells. However, this method is not practical for studies requiring the administration of messenger RNA to a large number of cultured cells. A preliminary report by Anderson and Krueger (1976) has been published which documents the development of a technique which entails the encapsulation of globin mRNA within erythrocyte ghosts and the transfer of the sequestered message into hamster cells by the addition of sendai virus. While this procedure may facilitate the transfer of message to cultured cells, the addition of a viral fusogen virtually eliminates any potential clinical application.

The dual problem of the protection of RNA from ribonuclease degradation and the insertion of RNA into a large number of cells without the addition of a viral fusogen may be resolved by the sequestration of RNA within liposomes. In order to utilize liposomes as a vehicle for transporting messenger RNA into cells, two criteria must be fulfilled: (1) the liposomal aqueous space must be of sufficient volume so as to allow the sequestration of large numbers of molecules of very high molecular weight, (2) the structure of the liposomes must permit the efficient transfer of the sequestered RNA into the cytoplasm of the recipient cell. The latter condition would suggest the utilization of unilamellar vesicles since all but the contents of the outermost aqueous space of multilamellar vesicles would be processed by the lysosomal apparatus (Cohen _et al_, 1976).

Deamer and Bangham (1976) have devised a technique for the production of large unilamellar liposomes (1 diameter) by ether infusion. Ostro, Giacomoni and Dray (1977) have employed this technique to sequester high molecular weight RNA (4 S to 23 S) within liposomes and have demonstrated that the encapsulated RNA is resistant to external ribonuclease degradation.

In a subsequent report Ostro _et al_ (1978) demonstrated that cells treated with liposomally encapsulated globin mRNA are stimulated to produce a globin-like protein. Application of a similar approach by Magee, Cronenberger and Thor (1978) to the guinea pig Ln-10 hepatoma tumor system led to a marked stimulation of lymphocyte-mediated attack on tumor cells that was 12 times greater than the activity seen with naked I-RNA. This approach has also been utilized to encapsulate I-RNA prepared from guinea pigs immunized against the chemically defined antigen, ARSNAT. Paque, Magee and Nealon (1979), using positively charged multilamellar liposomes containing I-RNA, have successfully converted GP-PEC to ARSNAT sensitivity after _in vitro_ incubation with these liposome preparations. So far, the use of these liposome-I-RNA preparations in attempts to transfer skin test reactivity of normal guinea pigs has been unsuccessful.

F. DISCUSSION

The transfer of cell-mediated immunity via I-RNA is subject to a variety of technical difficulties. RNA transfer methodology entails immunization of

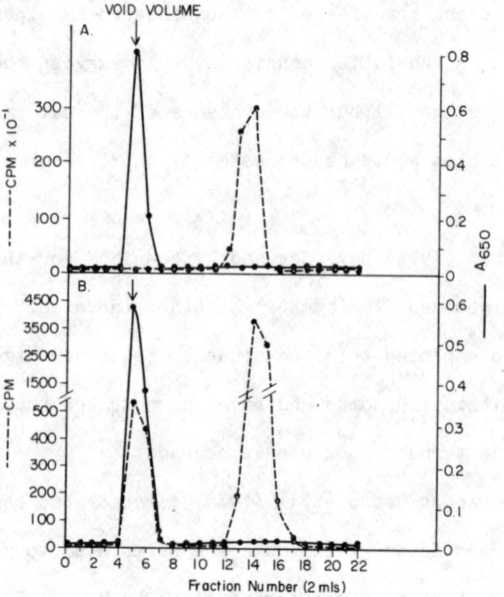

Figure 8. Elution profile derived from a 1.5 X 30 cm Sepharose 4B column equilibrated and run in 5 mM Hepes buffer, 0.145M KCl-NaCl, pH 7.µ. (A) Elution pattern obtained using 2 ml of ribonuclease-treated suspension of empty liosomes and exogenously added E. coli [^{3}H] RNA. B) Elution pattern obtained using a ribonuclease-treated 2 ml suspension of E. coli [^{3}H] RNA containing liposomes. From Ostro, Giacmoni and Dray, 1977, Biochem. Biophys. Res. Commun. 76, 836-842.

donors, extraction of I-RNA, incubation of I-RNA with recipient cells, and assessment of transfer activity by _in vivo_ and _in vitro_ tests. Some of the difficulties pertain to biological variability of cellular reactivity, incomplete delineation of the reaction mechanisms involved and the heterogeneity of lymphoid cell populations participating in the development and expression of cell-mediated immunity. These difficulties are compounded by the lack of quantitatively reproducible assays for measuring cell-mediated reactivity. But aside from these considerations, RNA transfers suffer from some special problems which can lead to difficulties in reproducing some phenomena. In our experience, most failures in attempts at I-RNA-mediated transfer of cellular immunity are due to the following: (1) the donor lymphoid cells may not contain a sufficient amount of I-RNA; (2) the I-RNA preparations are exquisitely sensitive to RNase and are degraded during its isolation or are toxic to the cells; (3) some antigens used to stimulate I-RNA-treated cells, particularly in tumor antigen systems, are poorly defined, impure and have not been quantitated; (4) heterogeneity of RNA molecules in active RNA preparations. Another major problem relates to the inability of I-RNA to convert immunologically naive recipients to specific skin test reactivity _in vivo_. Conversions of this nature have been sporadic and many times, non-reproducible. Studies focused on encapsulation and protection of I-RNA in liposomes may offer a new approach to successful and reproducible _in vivo_ transfer of sensitivity. The future application of I-RNA to the study of cell biology and particularly to the therapy of human disease states might well depend on the ability to deal with these problems.

The immunization of animals serving as donors of I-RNA is aimed at eliciting a hypersensitive state in the donor. Generally, animals that have marginal or weak reactivity to subsequent antigenic challenge are poor donors of I-RNA.

In the MOPC-315 plasmacytoma system, the ability of tumor-bearing mice to provide I-RNA with transfer activity for tumor antigen responsiveness is restricted to the early stage of lethal tumor growth which is characterized by host PEC responsiveness to tumor antigens. These observations suggest that the level of I-RNA synthesis that occurs in donor lymphoid cells is correlated with the degree of sensitivity possessed by the donor. If it can be assumed that the amount of I-RNA contained in the donor lymphoid tissue is correlated with the number of antigen-sensitive cells in the tissue, then the inability of marginally sensitized animals or terminal, unresponsive plasmacytoma-bearing mice to serve as good sources of I-RNA probably reflects an insufficiency in the number of antigen-sensitive cells in these hosts. This consideration raises the question of which cell type(s) are able to synthesize I-RNA. Early studies by Adler, Fishman and Dray (1966) and Fishman and Adler (1967) suggested that macrophage-derived RNA was able to transfer antibody synthesizing ability to lymphocytes. These reports suggested that the macrophage RNA might function as a modulator of lymphocyte responsiveness capable of signaling either antibody synthesis or lymphokine production upon antigenic challenge. Furthermore, Wang, Onikul and Mannick (1978) have presented evidence that macrophage RNA has transfer activity for anti-tumor cytotoxic responses in a murine tumor system. In contrast, Kern, Chow and Pilch (1978) have reported that I-RNA with transfer activity for anti-tumor cytotoxicity is synthesized by the T lymphocyte. In this study, both the cell synthesizing I-RNA and the cell which apparently incorporated the I-RNA was the T lymphocyte. It becomes apparent that there is considerable controversy concerning the nature of the cells involved in synthesizing I-RNA. Still, the identification of the I-RNA producing cell types(s) would aid greatly in attempts to enrich the activity of I-RNA. On the assumption that the number of cells with antigen reactivity even in a hypersensitized population is relatively small, one enrichment approach that might be exploited is the expansion of the I-RNA-producing cells either by cloning techniques or by the recently developed hybridoma technique. Hybridomas consisting of murine plasmacytoma cells and lymphocytes have been produced by Kohler and Milstein (1975). If this technology can be applied to obtain hybridomas synthesizing homogeneous I-RNA specific for a single antigen, it would represent a significant advancement in the technology for obtaining large quantities of relatively pure and biologically active I-RNA.

A variety of methods have been employed for preparing biologically active I-RNA. In our own experience, the hot/cold phenol method has been the most successful since the material recovered is usually undegraded and of low viscosity. The low viscosity is essential for effective incubation of RNA with cells. Two different approaches have been utilized to enrich the transfer activity of hot/cold extracted preparations. One method utilizes sucrose density gradient fractionation of the starting material to obtain the 5-18 S molecular weight RNA species (Paque, 1976). Most or all of the transfer activity contained in the initial preparation was localized to this fraction. Another approach at enrichment is by the isolation of poly-A-containing RNA species in the I-RNA preparation by oligo(dT) column affinity chromatography (Paque, 1977; Paque and Nealon, 1977). Both of these methods led to a 10-100 fold enhancement of transfer activity. Still, it is not yet clear whether transfer activity reflects the activity of a single species or of multiple species of RNA moieties since both of these fractions contain heterogeneous RNA species.

Clearly, the ability of I-RNA-treated lymphoid cells to respond to antigenic stimuli not specified by their genotype would argue strongly in favor of an informational or regulatory role for I-RNA. This is supported by evidence demonstrating the transfer of allotypic (Bell and Dray, 1969) and idiotypic (Giacomoni et al, 1974) immunoglobulin specificities in humoral systems and by the transfer of reactivity to the chemically defined antigen, DNP-oligolysine, to genetically nonresponder Strain-13 guinea pigs (Paque, Ali and Dray, 1975). In one series of experiments, I-RNA from DNP-oligolysine-immunized, responder Strain-2 guinea pigs was found to confer antigen-specific reactivity on non-responder Strain-13 guinea pig PEC as assessed by the MIF assay. These findings indicate that RNA-mediated transfer of cell mediated immunity can overcome a genetic inability of the recipient cells to respond to some chemically defined antigens.

Indeed, if I-RNA does function as a vehicle for the transfer of genetically-specified immune reactivity, it would suggest that a messenger RNA species might be responsible for this activity. Several lines of evidence are compatible with the possibility that the transfer activity of I-RNA is due to mRNA species. Estimates of the molecular size of the RNA species with transfer activity were first studied by Thor and Dray (1968) who reported the transfer of PPD reactivity with the 8-12 S fraction of a preparation of human I-RNA obtained from lymph nodes of a PPD-sensitive donor. Transfer activity of the separated 8-12 S RNA species was unaffected by DNase or trypsin but was abolished by RNase treatment. These findings were extended to a monkey I-RNA system by Paque and Dray (1974) who showed that transfer activity was confined to the RNA species located between the 4 S and 18 S regions. Subsequently, the size of the I-RNA species active in transferring anti-tumor reactivity in the guinea pig Line-10 hepatoma system was estimated to be 5-12 S (Paque, 1976). These sedimentation values are well within the range of the size of mRNA species determined in other systems. Another line of evidence which supports the idea that the transfer activity of I-RNA might be due to an mRNA species is the finding that transfer activity is associated with fractions of I-RNA having polyadenylic acid sequences characteristic of mRNA. Thus, Paque (1977) and Paque and Nealon (1977) have reported the transfer of PPD, KLH, and ARSNAT sensitivity to nonsensitive guinea pig PEC as assessed by the MIF assay with Poly-A containing I-RNA from hyperimmune donors. The amount of Poly-A containing I-RNA needed for transfer was approximately 10 fold less than the amount of whole I-RNA extracts required. Furthermore, the transfer activity of Poly-A containing I-RNA exhibited the same degree of specificity for the sensitizing antigen as seen for whole I-RNA preparations and resembled recentrifuged 5-12 S active fractions derived by sucrose gradients.

One difficulty in elucidating the mechanism of transfer in a cell-mediated immune system is the lack of knowledge of the receptors responsible for antigen recognition and activation of T lymphocytes. If an antibody-like molecule is indeed the receptor for antigen on the T cell, the size of the RNA species with transfer activity is sufficient to code for a light chain or idiotypic specificities but not for a complete immunoglobulin molecule. Other receptors dissimilar to antibodies have also been suggested, however, so that speculation as to the amount of RNA needed to code for a portion of a receptor sufficient for antigen-specific recognition is tenuous at best. It is conceivable that the transfer of antigen-specific recognition to an unsensitized cell that is genetically capable of a response could occur by depression of the host genome. The results of this event would presumably lead to the expression of genes encoding membrane receptors for antigen and to cellular activation upon subsequent antigenic exposure.

In our view, the goal of any future studies concerning I-RNA should focus, at least in part, on developing still better methods to (1) obtain active I-RNA by improved isolation from a homogeneous cell source; (2) incorporate I-RNA into cells via liposomes; (3) quantitiate the activity of I-RNA-treated cells; (4) biochemically characterize the active I-RNA specie. In terms of I-RNA procurement, a knowledge of the cell type(s) which synthesize I-RNA, coupled with methods to expand the I-RNA producing cell population, would greatly facilitate the isolation and characterization of the most active I-RNA moieties. The ability to incorporate large amounts of biologically active I-RNA into cells through the use of liposomes or other vehicles would allow for better defined experiments which are more reproducible. The development of a more precise means of assessing the biological activities of I-RNA would permit a rational study of the molecular mechanism(s) responsible for I-RNA-mediated immunological transfer. Acquisition of a homogeneous RNA species capable of transferring specific sensitivity would permit nucleotide sequencing and allow for some insight into the biochemical differences of receptor molecules from lymphoid cells exposed to different antigens. With improved I-RNA technology, it should become possible to develop the molecular biology of cell-mediated immune reactions and to modulate the immune responsiveness of lymphoid cells in a predictable manner for application to a variety of problems in both basic and clinical sciences.

REFERENCES

1. Anderson, W.F. and Krueger, L.J. (1976). Cell. Biol. 70, 185a.
2. Braun, D.P. and Dray, S. (1976). Proc. Am. Assoc. Cancer Res. 17, 7.
3. Braun, D.P. and Dray, S. (1977). Cancer Res. 37, 4138.
4. Cohen, G.M., Weissmann, G., Hoffstein, S., Awasthi, Y.C. and Srivastava, S.K. (1976). Biochem. 15, 452.
5. Crouch, R.J. (1976). In "Immune RNA in Neoplasia" (M. Fink, ed.), 279, Academic Press, New York.
6. Deamer, D. and Bangham, A.D. (1976). Biochim. Biophys. Acta. 443, 629.
7. Enesco, H.E. (1966). Exp. Cell. Res. 42, 640.
8. Giacomoni, D., Wang, S.R. and Dray, S. (1973). Ann. N.Y. Acad. Sci. 207, 251.
9. Gottlieb, A. and Schwartz, R.H. (1972). Cell Immunol. 5, 341.
10. Graessmann, A., Graessmann, M. and Mueller, C. (1977). Proc. Natl. Acad. Sci. USA 74, 4831.
11. Jureziz, R.E., Thor, D.E. and Dray, S. (1968). J. Immunol. 101, 823.
12. Jureziz, R.E., Thor, D.E. and Dray, S. (1970). J. Immunol. 105, 1313.
13. Magee, W.E., Cronenberger, J.H. and Thor, D.E. (1978). Cancer Res. 38, 1173.
14. Ostro, M.J., Giacomoni, D. and Dray, S. (1977). Biochem. Biophys. Res. Commun. 76, 836.
15. Ostro, M.J., Giacomoni, D., Lavelle, D., Paxton, W. and Dray, S. (1978). Nature 274, 921.
16. Paque, R.E. and Dray, S. (1970). J. Immunol. 105, 1334.
17. Paque, R.E. and Dray, S. (1972). Cell. Immunol. 5, 30.
18. Paque, R.E. and Dray, S. (1973). Ann. N.Y. Acad. Sci. 207, 369.
19. Paque, R.E., Meltzer, M.S., Zbar, B., Rapp, H.J. and Dray, S. (1973). Cancer Res. 33, 3165.

20. Paque, R.E. and Dray, S. (1974). Transpl. Proc. 6, 203.

21. Paque, R.E., Ali, M. and Dray, S. (1975). Cell. Immunol. 16, 261.

22. Paque, R.E. (1976). In "Immune RNA in Neoplasia" (M. Fink, ed.), p. 235, Academic Press, New York.

23. Paque, R.E. (1977). Cell. Immunol. 30, 332.

24. Paque, R.E. and Nealon, T. (1977). Cell. Immunol. 34, 279.

25. Paque, R.E. (1978). In "Methods in Cancer Research" (H. Busch, ed.), v. 15, p. 279, Academic Press, New York.

26. Paque, R.E. (1978). J. Reticuloendothelial Soc. 24, 403.

27. Paque, R.E. and Nealon, T. (1979). Cell. Immunol. 43, 48.

28. Paque, R.E., Magee, W. and Nealon, T. (1979). Fed. Proc. 38, 4581.

29. Scherrer, K. and Darnell, J.E. (1962). Biochem. Biophys. Res. Commun. 7, 486.

30. Schlager, S.I., Dray, S. and Paque, R.E. (1974). Cell. Immunol. 14, 104.

31. Schlager, S.I., Paque, R.E. and Dray, S. (1975). Cancer Res. 35, 1907.

32. Schlager, S.I. and Dray, S. (1975). Proc. Natl. Acad. Sci. USA 72, 3680.

33. Schlager, S.I. and Dray, S. (1976). Israel J. Med. Sci. 12, 344.

34. Thor, D.E. (1967). Science 157, 1567.

35. Thor, D.E. and Dray, S. (1973). Ann. N.Y. Acad. Sci. 207, 369.

36. Thor, D.E. and Dray, S. (1968). J. Immunol. 101, 469.

37. Wang, S.R., Giacomoni, D. and Dray, S. (1973). Exp. Cell. Res. 78, 15.

THE RIBONUCLEOPROTEIN STRUCTURE OF HETEROGENEOUS NUCLEAR RNA AND ITS RELATIONSHIP TO mRNA PROCESSING

Thoru Pederson
Cell Biology Group
Worcester Foundation for Experimental Biology
Shrewsbury, Massachusetts 01545 USA

Abstract: To explore the relationships between transcription, mRNA processing and nuclear structure, ribonucleoprotein particles containing heterogeneous nuclear RNA (hnRNP) have been purified from globin-producing mouse Friend erythroleukemia cells. These nuclear hnRNP particles sediment at 50 to 200S and contain, in addition to high molecular weight hnRNA, a specific set of nuclear proteins predominated by a major component of approximately 38,000 molecular weight. The hnRNP particles are free of histones and ribosomal structural proteins, indicating their purification from the two other major nucleoprotein components of the nucleus: chromatin and nucleolar ribosomal precursor RNP particles. The authenticity of the Friend cell hnRNP particles is demonstrated by the results of reconstruction experiments with deproteinized hnRNA, and by the resistance of the particles to dissociation during isopycnic banding in Cs_2SO_4 gradients without prior aldehyde fixation. Hybridization analysis with cloned mouse β-globin DNA demonstrates that hnRNP particles from induced Friend cells contain newly-synthesized transcripts of the β-globin gene. Agarose gel electrophoresis of hnRNP particle-derived RNA denatured in glyoxal followed by "Northern" transfer to diazobenzyloxymethyl-paper and hybridization with ^{32}P-labeled cloned mouse β-globin DNA reveals the presence in hnRNP of two size classes of β-globin gene transcripts, the larger of which corresponds to the pre-spliced 15S β-globin mRNA precursor previously identified in whole nuclear RNA, and the smaller of which corresponds to completely processed 9S β-globin mRNA. These results establish, for the first time, that the nuclear transcripts of a specific, well-defined eukaryotic structural gene can be isolated in a ribonucleoprotein particle form, and that their ribonucleoprotein structure persists throughout mRNA splicing.

An important question regarding both the structure and function of these nuclear particles is whether the proteins are uniformly or non-uniformly distributed on the hnRNA. To explore this issue, the nucleoprotein organization of globin mRNA-homologous sequences in Friend cell hnRNP particles has been examined in nuclease protection experiments. After mild (10-15%) pancreatic RNase digestion of particles, hybridization of the protected hnRNA with a cDNA probe reveals that globin mRNA sequences in hnRNP are preferentially attacked. Digestion of only 16% of the total hnRNA in nuclear RNP results in a 64% depletion of globin-mRNA-homologous sequences. This preferential nuclease sensitivity is not observed when deproteinized hnRNA is digested similarly, showing that it is related to the nucleoprotein structure of hnRNP. Additional experiments reveal that the nuclease-sensitive globin mRNA sequences in nuclear hnRNP particles may be as completely unprotected as naked RNA, suggesting that during at least the early stages of mRNA processing the hnRNP proteins may be clustered on non-messenger regions of globin gene transcripts such as intervening sequences. These results demonstrate that the organization of proteins in hnRNP is non-random, and they also indicate that hnRNP structure is related to mRNA processing.

Introduction:

This paper will deal with an important chapter in modern research on the structure and function of the cell nucleus: the biosynthesis of messenger RNA. The main theme that will be developed here is that the production of functional messenger RNA takes place in special nuclear particles, in which the RNA is intimately complexed with a unique set of proteins. These "ribonucleoprotein" particles were first discovered by cytologists engaged in microscopic studies of large chromosomes. Over the past few years, we have developed techniques for isolating these particles from the cell nucleus (Pederson, 1974a) and, employing a variety of physical and biochemical tools, my colleagues and I have begun to dissect their molecular structure and biological function (Kish and Pederson, 1975; Firtel and Pederson, 1975; Kumar and Pederson, 1975; Calvet and Pederson, 1977; Kish and Pederson, 1977; Calvet and Pederson, 1978, 1979a, 1979b). Our central conclusion is that these RNA:protein particles, not "naked" RNA molecules, are the native vehicles for mRNA biosynthesis in the living cell. The purpose of this paper is to describe how this conclusion was reached, and to also draw attention to the experimental promise these particles hold for gaining further insights into the cell biology of messenger RNA biosynthesis.

Historical perspective:

An important new concept on the landscape of chromosome research emerged during the period between 1955 and 1960, when it became apparent that there is a relationship between chromosome structure and gene action. It was Joseph Gall who noted that nascent RNA on amphibian oocyte chromosomes undergoing meiosis was in the form of ribonucleoprotein (Gall, 1955). In the oocytes of many Amphibia (especially the Urodeles: salamanders and newts), cartilaginous fish (elasmobranchs) and many other animals (probably most), the bivalent meiotic prophase chromosomes present a dramatic morphology in which segments of DNA in each of the two sister chromatids are looped out from the main chromosomal axis. Since these loops are bilateral, the chromosomes have the general appearance of brushes. Such chromosomes were termed "lampbrush chromosomes" by analogy with the instruments our forebears employed to brush out the carbon soot from kerosene lamps. The discovery by Gall of ribonucleoprotein on the lampbrush chromosome lateral loops, the comprehensive extension of these observations by Callan, and the demonstration by Gall and Callan working together that the ribonucleoprotein contained newly-synthesized RNA, were the key steps leading to the concept that the extended loops and their associated ribonucleoprotein represent genes in action. This work therefore provided the first clear evidence that the nuclear products of gene transcription are not simply "naked" RNA, but RNA:protein complexes.

Isolation of hnRNA:protein complexes from Friend erythroleukemia cells

We have developed methods for isolating nuclear hnRNA:protein complexes, or hnRNP particles, from a variety of eukaryotic cells, including slime mold, *Drosophila*, rat liver and HeLa cells (Pederson, 1974a, 1974b; Firtel and Pederson, 1975; Kish and Pederson, 1978). These particles are 50-250S, RNAase-sensitive, EDTA-resistant structures that contain high molecular weight hnRNA, as defined by its labeling kinetics, characteristic heterodisperse sedimentation properties, sensitivity to selective inhibitors of RNA polymerase II, and its content of specific marker sequences characteristic of hnRNA such as 3'-OH poly(A), intramolecular double-stranded regions arising by transcription of inverted-repeat DNA sequences, and oligo(A) and oligo(U) sequences (Kish and Pederson, 1975, 1977; Calvet and Pederson, 1977, 1978, 1979a). However, in the present experiments we have focused on the presence of messenger RNA sequences in hnRNP. For this purpose we have chosen to investigate hnRNP particles from Friend erythroleukemia cells with particular reference to their content of transcripts of the β-globin gene.

Our previously developed methods for purifying nuclear RNP particles are based upon the gentle disruption of nuclei by controlled, brief sonication to release hnRNP, which is normally anchored to elements of the nuclear structure. The released hnRNP particles are then separated from nucleoli and chromatin by successive sucrose gradient fractionations. Induced Friend erythroleukemia cells were labeled for 24 hours with ^{14}C-thymidine and hnRNA was then selectively pulse-labeled with ^{3}H-uridine in the presence of a low dose of actinomycin. Cells were disrupted by Dounce homogenization in a hypotonic buffer and the amounts of 10% trichloroacetic acid-precipitable ^{3}H and ^{14}C radioactivity in the homogenate were taken as 100%. 100% of the initial hnRNA and DNA was recovered in the first nuclear pellet. After disruption of the nuclei by mild sonication, all of the nuclear hnRNA and DNA is recovered. Sedimentation of the nuclear sonicate on 30% sucrose at 4500 x g produces a tightly packed, opalescent band at the 0:30% sucrose interface; this band contians 70% of the initial hnRNA, but only 12% of the DNA.

In the initial isolation method developed for HeLa cell hnRNP particles, the 4500 x g sedimentation step was designed to selectively pellet nucleoli through the 30% sucrose, as initially shown by Maggio, Siekevitz and Palade (1963). With HeLa cells, this nucleolar pellet contains only about 5% of the chromatin, with 84% remaining at the 0:30% sucrose interface along with the hnRNP particles and other small nuclear elements. The same general distribution of chromatin during the 4500 x g 30% sucrose step is observed with rat liver nuclei and nuclei from the slime mold Dictyostelium; in both cases the great majority of the chromatin remains at the 0:30% sucrose interface. However, as described above, the result for induced Friend cell nuclei is different. In this case, only 12% of the chromatin remains in the 0:30% sucrose band, rather than the usual values of 85-90% obtained with the aforementioned cell systems. This may be due to the highly condensed state of chromatin in the fully-induced Friend erythroleukemia cell nucleus. Thus, the 4500 x g 30% sucrose sedimentation step, originally designed to deplete the nuclear sonicate of specifically nucleoli, also serves in the case of the induced Friend cell to remove the majority of the chromatin as well.

To further purify the hnRNP particles from the small amount of remaining chromatin, the 0:30% sucrose band was removed and sedimented on linear 15-45% sucrose gradients over a 60% sucrose cushion. Figure 1 shows the typical results of this gradient fractionation. Friend cell hnRNP particles display a heterodisperse sedimentation profile between approximately 50 and 200S. The sedimentation behavior of Friend cell hnRNP (Figure 1) is dependent upon the integrity of the RNA, as shown by the marked reduction in S values observed if particles are subjected to mild pancreatic RNAase digestion (0.1 μg per ml, 15 minutes, 4°C) prior to gradient analysis (data not shown). It can also be seen in Figure 1 that, in contrast to the heterodisperse hnRNP which is mainly confined to the 50-200S region of the gradient, the chromatin (as ^{14}C-DNA) is present only as a flat background across the gradient. 77% of the chromatin present in the 0:30% sucrose band (12% of the total chromatin) fractionates in the gradient pellet and 60% sucrose cushion, while 83% of the hnRNP particles remain in the gradient.

hnRNP proteins

Fractions 6-30 from gradients such as that illustrated in Figure 1 were

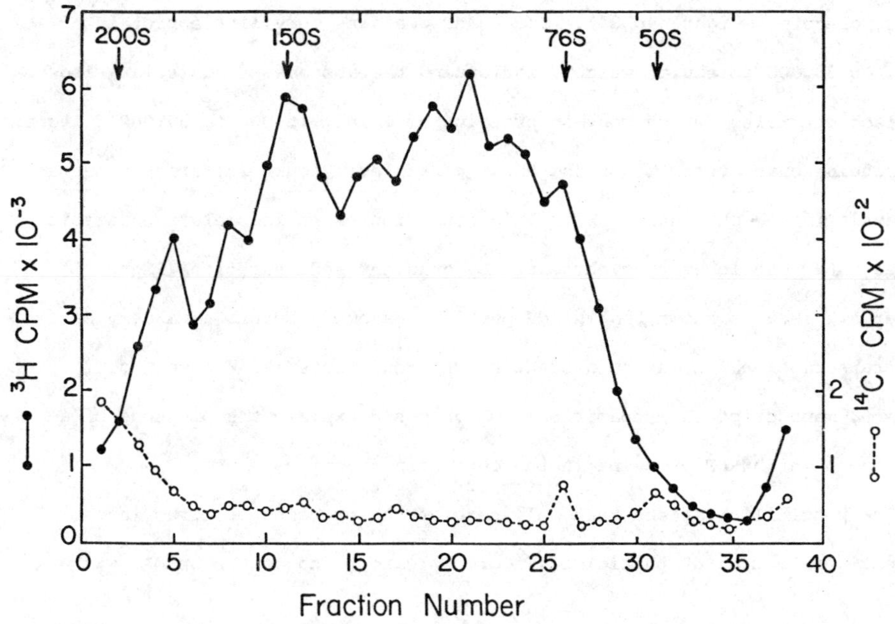

Figure 1. Sucrose gradient separation of hnNRP particles from chromatin. hnRNP particles were isolated from Friend cells that had been labeled for 24 hours with ^{14}C-thymidine and then pulse-labeled for 20 minutes with ^{3}H-uridine. The hnRNP particles were centrifuged in a 28 ml 15-45% sucrose gradient over a 4 ml 60% sucrose cushion in a Beckman SW27 rotor for 17 hours at 12,000 rpm (4°C); the direction of sedimentation is from right to left. The gradient was collected in 1 ml fractions and the amount of trichloroacetic acid-insoluble ^{14}C and ^{3}H radioactivity in each fraction was determined by liquid scintillation counting. ●——●: ^{3}H-uridine; o---o: ^{14}C-thymidine.

pooled and the particles pelleted by ultracentrifugation. The pellets from the preparative hnRNP gradients were also recovered for analysis of chromatin proteins. Analysis of these two nuclear fractions by SDS-polyacrylamide gel electrophoresis revealed major differences in their protein constituents. Histones dominate the chromatin fraction as expected (Figure 2B), but are present in only very low concentrations in the hnRNP particles (Figure 2A). The Friend cell hnRNP proteins are complex but, as in all the other eukaryotic cells previously investigated, there is a major component of approximately

38,000 molecular weight ("p38"). The hnRNP contains very little protein of less than 38,000 molecular weight, indicating the absence of nucleolar ribosomal precursor particles, which contain proteins of mainly 15,000 to 55,000 daltons. The proteins characteristic of the hnRNP particles (Figure 2A) are present at low levels in the chromatin (2B). This could represent incomplete separation of free hnRNP particles from chromatin by gradient sedimentation (Figure 1). However, it is also known that hnRNP particle assembly occurs as a very early post-transcriptional event on nascent hnRNA transcripts (E. Wieben and T. Pederson, manuscript in preparation), which could explain the presence of a small amount of hnRNP proteins in the chromatin fraction.

The possibility arises that only some of the proteins observed in the gradient-purified hnRNP particles (Figure 2A) are bound to the hnRNA, with

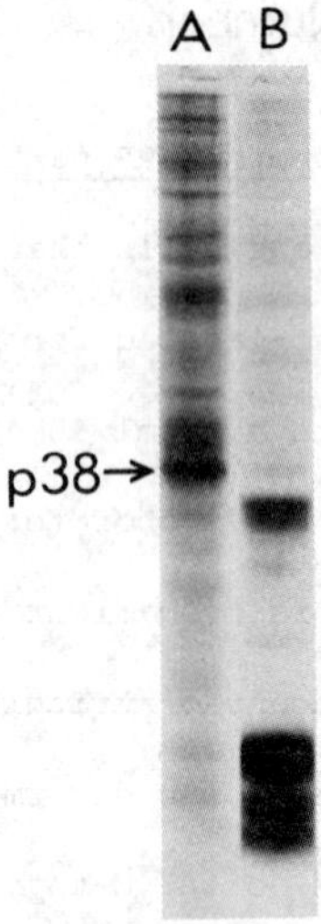

Figure 2. Distinct proteins of hnRNP and chromatin. hnRNP was recovered by high-speed centrifugation from fractions 6-30 of gradients such as shown in Figure 1, and chromatin was obtained as the gradient pellet. Samples were dissolved in SDS, electrophoresed and stained with Coomassie blue as previously described (Bhorjee and Pederson, 1972). A: hnRNP proteins; B: chromatin proteins. The molecular weight of the protein labeled "p38" in gel A was determined on three separate hnRNP samples, and the average value was 37,866 daltons (standard deviation = ± 986 daltons).

others being derived from co-sedimenting, non-ribonucleoprotein nuclear structures. This is ruled out by the fact that prior treatment of the pooled gradient fractions with pancreatic RNAase (500 μg/ml, 30 minutes, 37°C) followed by collection of any RNAase-resistant structures by high speed sedimentation results in a total elimination of all the protein components normally observed (not shown). This is not due to proteolysis during the RNAase digestion, since hnRNP mock-digested at 37°C without RNAase retains its usual protein complement.

Authenticity and stability of hnRNP

To investigate the possibility that hnRNP particles are formed as the result of non-specific RNA:protein interactions during nuclear fractionation, which had been ruled out in earlier studies on hnRNP isolated from other cells (Pederson, 1974a; Firtel and Pederson, 1975; Wilt et al., 1973), reconstruction experiments were performed in which deproteinized, ^{3}H-labeled Friend cell hnRNA was added to nuclei prior to their disruption. After isolation of the 0:30% sucrose band, which contained 88% of the added ^{3}H-hnRNA, this fraction was subjected to controlled pancreatic RNAase digestion under conditions in which previous studies had defined a significant differential in the digestion kinetics of naked hnRNA versus hnRNP. As shown in Figure 3, the added ^{3}H-hnRNA was digested much more rapidly than endogenous hnRNP (see legend for details), indicating that the addition of a naked hnRNA probe to nuclei does not result in the formation of material that is as nuclease-protected as the endogenous hnRNP particles. Similar results are obtained if the naked hnRNA probe is added to nuclei after sonication, and in both cases the digestion of the probe is very similar to that of naked RNA assayed in buffer alone. Thus, non-specific RNA:protein associations during nuclear fractionation cannot be a major factor in the formation of the observed hnRNP particles. These nuclease protection studies confirm the results of earlier reconstruction experiments in which non-specific hnRNA:protein interactions were ruled out using a velocity sedimentation analysis (Pederson, 1974a; Firtel and Pederson, 1975).

As shown in Figure 4, the gradient-purified hnRNP particles (Figure 1) are resistant to dissociation when banded in a preformed Cs_2SO_4 density gradient, even without prior fixation. This rather remarkable stability of hnRNP, which

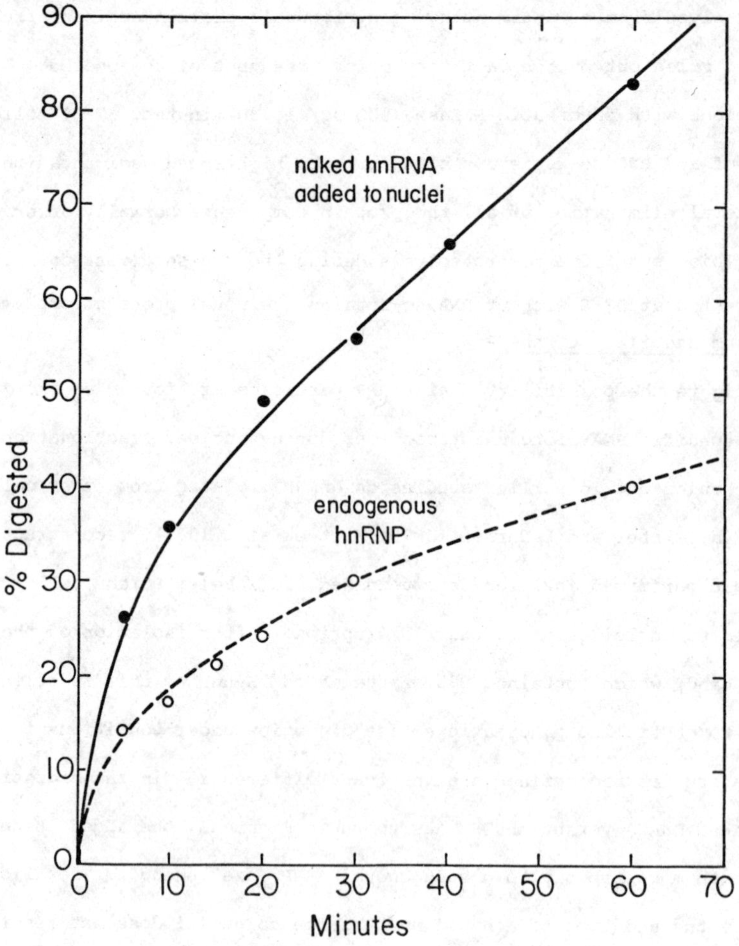

Figure 3. Reconstruction experiment. ^{3}H-labeled deproteinized hnRNA was added to 10^{8} isolated Friend cell nuclei in a final volume of 2.5 ml of RSB. The endogenous (unlabeled) hnRNA concentration was 11 μg/ml and the endogenous hnRNA:exogenous ^{3}H-hnRNA probe mass ratio was approximately 100:1. After nuclear fractionation, the hnRNP fraction was digested with 0.1 μg of pancreatic RNAase per ml at 4°C, and aliquots were removed at the indicated times and the digestion quenched by the immediate addition of mercaptoethanol (final concentration = 0.75mM, which is a 100,000-fold molar excess over the pancreatic RNAase). The amounts of ^{3}H-hnRNA probe remaining trichloroacetic acid-insoluble were measured (●——●). In parallel, a separate ^{3}H-labeled endogenous hnRNP preparation (o---o) was digested under identical conditions, including the same enzyme:endogenous hnRNA ratio as in the reconstruction experiment.

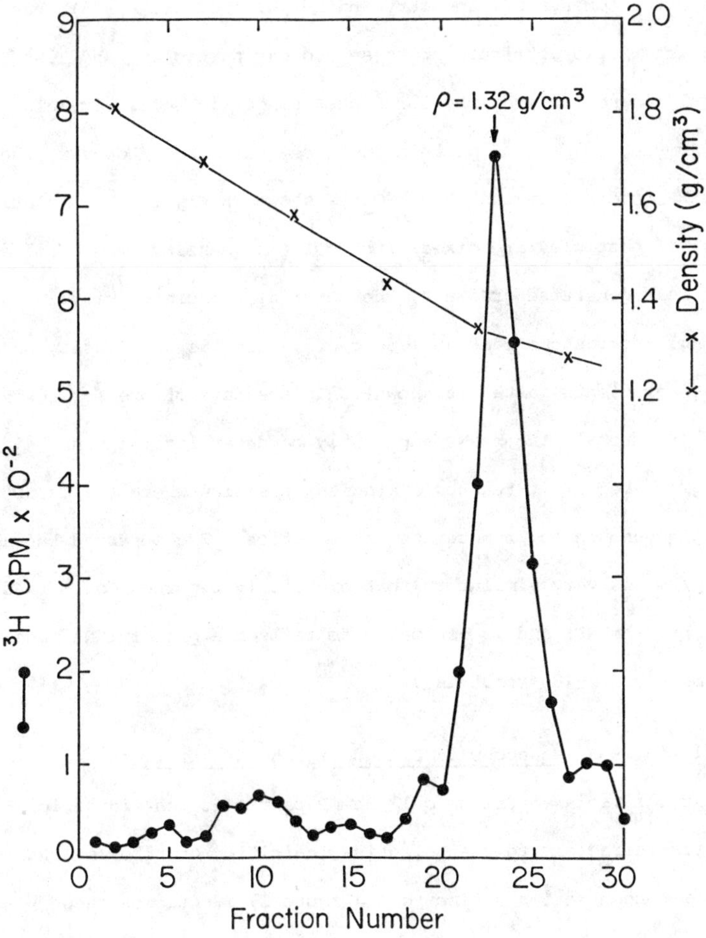

Figure 4. Stability of hnRNP particles in Cs_2SO_4. 0.50 ml of hnRNP was taken from fraction #20 of a preparative gradient such as shown in Figure 1 and layered on a preformed 4.5 ml Cs_2SO_4 gradient having an initial density range of 1.25 to 1.75 g/cm^3. The gradient was centrifuged in a Beckman SW50.1 rotor at 35,000 rpm for 65.5 hours (19°C). ●—●: trichloroacetic acid-insoluble ^{3}H-radioactivity, x—x: density.

was initially reported for sea urchin embryo hnRNP (Wilt et al., 1973), has emerged as a characteristic attribute of these nuclear particles, and also serves as a useful diagnostic property for hnRNP, since ribosomal particles are stripped almost completely free of protein in these gradients. However, the resistance of hnRNP to dissociation in Cs_2SO_4 as shown in Figure 4 is relative not absolute, for if particles are mixed with Cs_2SO_4 of density 1.50 g/cm^3 and then banded in a self-generated gradient, the ^{3}H-uridine radioactivity is observed as a single homogenous peak at 1.66 g/cm^3, which is the density of naked RNA in these gradients (data not shown). In the case of pre-formed Cs_2SO_4 gradients (Figure 4), the hnRNP apparently bands at its isopycnic density of 1.32 g/cm^3 before it reaches a gradient position where the Cs_2SO_4 ion activity is high enough to promote its dissociation. The observed buoyant density of 1.32 g/cm^3 is very similar to that previously reported for HeLa cell and sea urchin embryo hnRNP, and is estimated to reflect a protein:RNA mass ratio of approximately 4:1 (80% protein).

Detection of β-globin mRNA sequences in Friend cell hnRNP particles

The fact that the isolated Friend cell hnRNP particles contain rapidly-labeled RNA that is refractory to the selective inhibition of ribosomal RNA synthesis by a low concentration actinomycin (Figure 1) identifies these particles as containing hnRNA. This is confirmed by sucrose gradient analysis of the hnRNP-derived RNA after thermal denaturation in DMSO (Figure 5), which demonstrates this RNA to have the sedimentation behavior characteristic of mammalian hnRNA after disruption of intermolecular aggregates. While it is never possible to know how much endonucleolytic degradation has occurred in such hnRNA preparations relative to their native size in vivo (because the size of gene-specific RNA sequences is not visualized), the sedimentation profile in

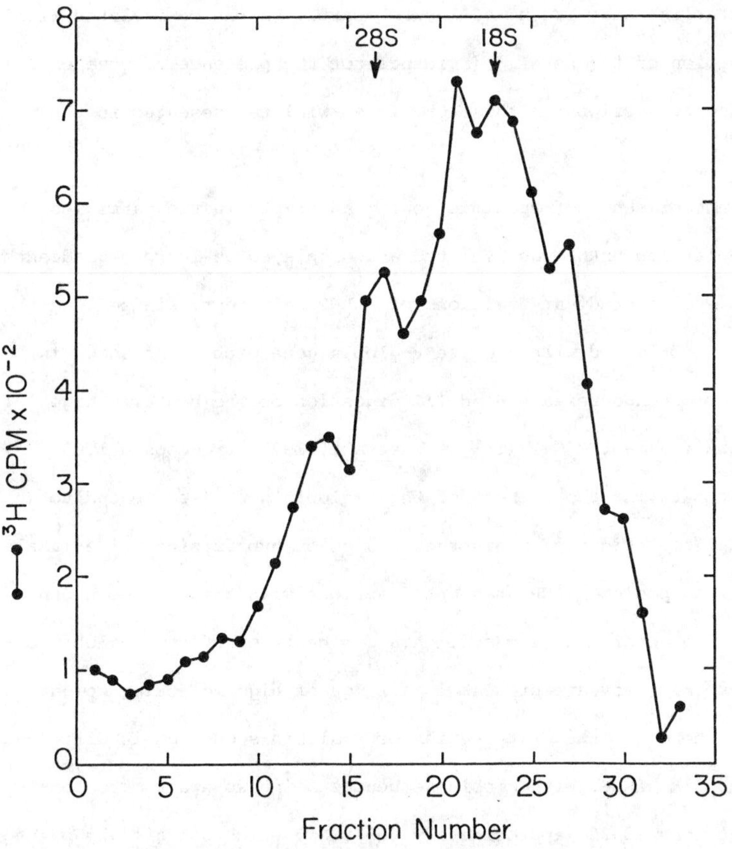

Figure 5. Size of hnRNP particle-derived RNA after denaturation in DMSO. RNA was extracted from ^{3}H-uridine-labeled hnRNP and denatured in 80% dimethylsulfoxide at 65°C. A 0.50 ml sample was layered on a 17 ml 15-30% sucrose-SDS gradient and centrifuged in a SW27.1 rotor at 23,500 rpm for 14.25 hours (20°C). The gradients contained 0.1M NaCl, 10mM EDTA, 10mM Tris-HCl, pH 7.0 and 0.5% SDS. ●——●: trichloroacetic acid-precipitable ^{3}H-radioactivity. Arrows denote the positions of 18S and 28S ribosomal RNA markers run in a parallel gradient.

Figure 5 does establish that Friend cell hnRNP particles can be isolated without gross degradation of the hnRNA. (Evidence for the presence of covalently intact globin gene transcripts in these particles will be presented in the following section.)

To examine the presence of specific gene transcripts in the particles, ^{3}H-uridine pulse-labeled hnRNP was isolated and displayed on sucrose gradients as in Figure 1. Pooled gradient fractions of hnRNP were deproteinized and the labeled hnRNA was hybridized with a mouse β-globin gene probe. As shown in Figure 6, restriction endonuclease Hind III digestion of the bacteriophage λ-cloned mouse DNA fragment MβG-2 (Tilghman et al., 1977) produces a 1.05 kilobase fragment carrying the 5' half of the β-globin gene (see legend to Figure 6 for details). Figure 7 shows the results of hybridizing pulse-labeled hnRNA retrieved from pooled hnRNP gradient fractions with this β-globin gene-specific probe. The heterodisperse sedimentation behavior of the β-globin sequences indicates that they are present in a range of high molecular weight hnRNP particles. However, the distribution of β-globin sequences is different from that of the bulk hnRNP, with globin sequences being somewhat more concentrated in fractions 18-30 (approximately 60-120S), which contain 52% of the β-globin sequences but only about 25% of the total hnRNP. Hybridization of parallel gradient fractions with total mouse DNA revealed a pattern that closely followed the total hnRNP profile (not shown), indicating that, in contrast to β-globin sequences, repetitive DNA sequence transcripts are present at a rather uniform level throughout the different hnRNP size classes. These data (Figure 7) show that newly-synthesized transcripts of the β-globin gene are present in high molecular weight hnRNP particles. The covalent integrity of these globin RNA sequences in hnRNP is now considered.

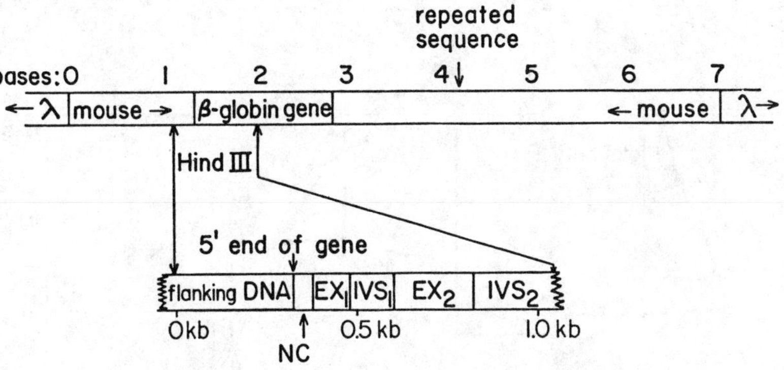

Figure 6. Isolation of β-globin DNA probe. The recombinant phage λgtWES·MβG-2 carries a 7 kilobase insert of mouse DNA containing the β-globin$_{major}$ gene (Tilghman et al., 1977). Hind III digestion of λgtWES·MβG-2 yields a 1.05 kb fragment containing the 5'-half of the β-globin gene, including a 52 nucleotide 5' non-coding leader sequence ("NC"), the two most 5'-ward coding regions of the gene ("ex_1" and "ex_2" respectively), the entire smaller intervening sequence ("IVS_1") and approximately one-third of the larger intervening sequence ("IVS_2"). The arrow labeled "repeated sequence" indicates the approximate location of a transcribed repetitive DNA sequence which hybridizes extensively with Friend cell hnRNA unless first removed by Hind III digestion (N.G. Davis and T. Pederson, unpublished results). The 1.05 kb Hind III fragment was used as a hybridization probe for β-globin gene transcripts in hnRNP (Figures 7 and 8).

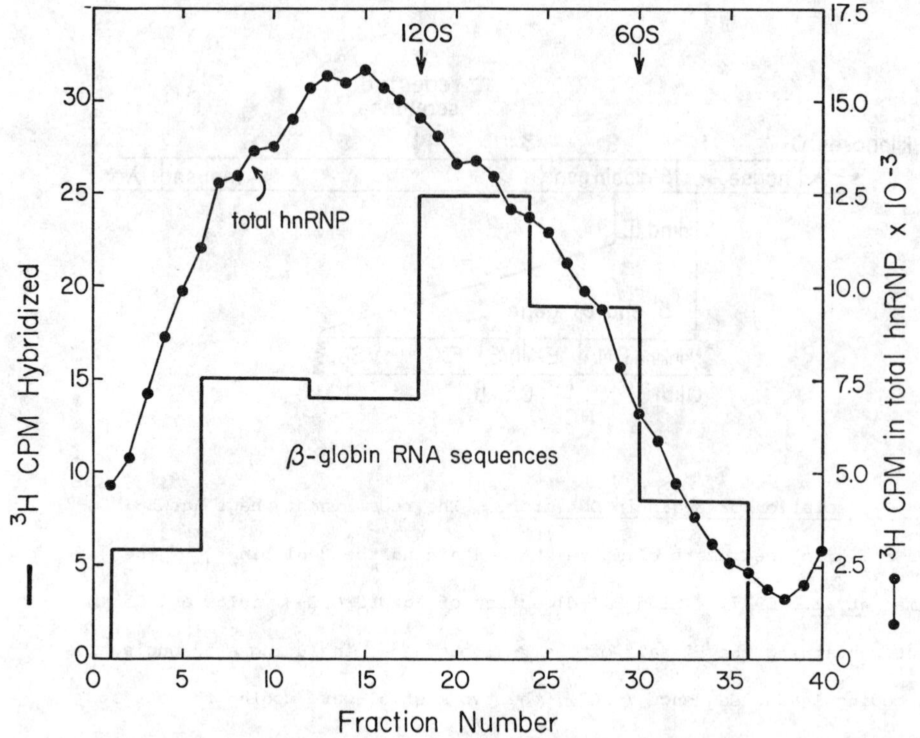

Figure 7. Detection of newly-synthesized β-globin gene transcripts in Friend cell hnRNP particles. ^{3}H-uridine pulse-labeled hnRNP was isolated and centrifuged in 15-45% sucrose gradients as in Figure 1. RNA was extracted from pooled gradient fractions as shown and hybridized to nitrocellulose filters containing cloned mouse β-globin DNA (see Figure 6). ●——●: trichloroacetic acid-precipitable ^{3}H radioactivity in total hnRNP, monitored in small aliquots of each gradient fraction. The histogram shows the total amount of β-globin RNA sequence present in each pool of gradient fractions.

Identification of covalently intact β-globin mRNA precursors in hnRNP

Unlabeled hnRNP was isolated from induced cells as usual and the RNA was deproteinized and subjected to agarose gel electrophoresis following thermal denaturation in the presence of glyoxal. The RNA was then transferred covalently onto diazobenzyloxymethyl-paper (Alwine et al., 1977) and hybridized with the 1.05 kb Hind III β-globin DNA fragment which had been ^{32}P-labeled by nick-translation. To calibrate the gel positions of the known mouse globin RNA species, total Friend cell poly(A)$^+$ cytoplasmic and nuclear RNA's were electrophoresed and analyzed in parallel. Lanes A and B of Figure 8 show the typical results obtained with the standards of poly(A)$^+$ cytoplasmic RNA (lane A) and poly(A)$^+$ nuclear RNA (lane B). The cytoplasmic RNA contains a single component reacting with the β-globin probe. Based upon the electrophoretic migration of RNA standards of known molecular weight, the cytoplasmic species reacting with globin DNA in Figure 8A is estimated to have a molecular weight of 280,000 daltons. For convenience, this cytoplasmic globin RNA will be hereafter referred to as "9S". As shown in lane B, the poly(A)$^+$ nuclear RNA contains two β-globin sequence components, one of which reproducibly co-migrates with the cytoplasmic 9S globin mRNA and a second, larger species which has a measured molecular weight of 600,000 daltons, corresponding to the "15S" mouse β-globin mRNA precursor previously described (Ross and Knecht, 1978). As can be seen in lane C, the RNA recovered from the purified nuclear hnRNP particles contains both the 15S and 9S globin sequences, and in approximately the same relative proportions as observed in the total nuclear RNA (lane B). These results therefore demonstrate that 15S, pre-spliced β-globin mRNA precursors (Kinniburgh et al., 1978, Smith and Lingrel, 1978; Tilghman et al., 1978) have a ribonucleoprotein structure in the Friend cell nucleus, and that these mRNA precursors

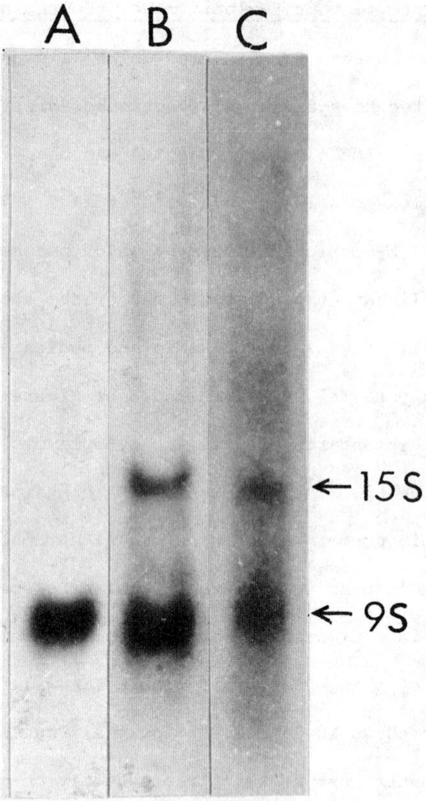

Figure 8. Northern blot hybridization of β-globin gene transcripts in hnRNP. RNA was recovered from nuclei, cytoplasm or purified hnRNP particles by Proteinase K-phenol deproteinization and electrophoresed in agarose gels after denaturation in glyoxal. After transfer to DBM-paper (Alwine *et al.*, 1977), the RNA was hybridized with nick-translated (^{32}P) β-globin DNA (see Figure 6). Lane A: total poly(A)$^{+}$ cytoplasmic RNA; lane B: total poly(A)$^{+}$ nuclear RNA; lane C: RNA from purified hnRNP particles. See text for description of molecular weight calibrations of the 15S and 9S β-globin RNA components.

can be isolated as covalently intact RNA molecules in ribonucleoprotein particles.

Altered nucleoprotein structure of mRNA sequences in hnRNP

Although very little is known about the functional significance of these hnRNA:protein interactions, it seems likely that the ribonucleoprotein organization of hnRNA is important for accurate mRNA processing. For example, there is evidence that hnRNP particle proteins may influence hnRNA secondary structure (Calvet and Pederson, 1977). These proteins might also fold the hnRNA transcript into specific tertiary structures that permit or facilitate post-transcriptional modifications such as splicing, and it is likely that the enzymes which catalyze hnRNA processing events are themselves particle-bound hnRNP proteins. One could also envision that hnRNA-associated proteins might specifically mask certain hnRNA sequences, leaving others relatively unprotected. This latter viewpoint predicts that the nucleoprotein organization of hnRNA is not merely a uniform protein "covering", but consists of a highly ordered and perhaps sequence-restricted set of RNA-protein interactions.

One way of testing this idea is to investigate the nucleoprotein structure of specific RNA sequences, or classes of sequences, in hnRNP particles. For example, we have shown that intramolecular double-stranded regions of HeLa cell hnRNA (Calvet and Pederson, 1971) have an accessible, relatively protein-free configuration within the hnRNP particle, as revealed by nuclease digestion experiments, analyses of dsRNA-bound protein and psoralen cross-linking studies (Calvet and Pederson, 1978, 1979a, 1979b). We now take up this issue for another class of hnRNA sequences: those which are homologous to cytoplasmic messenger RNA. Limited pancreatic RNAase digestion of Friend erythroleukemia cell hnRNP particles reveals that globin mRNA-homologous sequences are preferentially attacked, indicating that these sequences possess an altered nucleoprotein structure in hnRNP.

Absence of protein translocation during RNAase digestion of hnRNP particles

The experiments to be presented now involve the use of controlled nuclease digestion to seek information regarding the nucleoprotein organization of globin messenger RNA-homologous sequences in hnRNP particles from mouse erythroleukemia cells. This approach requires at the outset a demonstration that RNA:protein redistribution does not occur under the experimental conditions used for nuclease digestion. It can be seen from the results shown in Figure 9 that when deproteinized hnRNA (^{14}C) is added to a preparation of mouse erythroleukemia cell hnRNP particles (^{3}H), the kinetics of pancreatic RNAase digestion of the naked hnRNA proceed much faster than that of the hnRNP. This result confirms the previous observation that the hnRNP proteins confer upon hnRNA a substantial degree of protection from nuclease attack (Calvet and Pederson, 1978), and also corroborates the earlier mixing experiment (Figure 3) in which it was shown that the addition of naked RNA to nuclei does not lead to the formation of non-specific particles. However, the more important point in Figure 9 for the present investigation is that there is no detectable difference between the digestion kinetics of naked ^{14}C-hnRNA measured in the presence of hnRNP particles and those of naked hnRNA digested in buffer alone. This shows that under these conditions, Friend cell hnRNP proteins do not trnaslocate to the naked RNA probe to an extent measurable as nuclease protection. The failure to detect hnRNP protein translocation during RNAase digestion of the particles suggests also that movement of proteins within a given hnRNP particle does not occur either, although it should be recognized that this is only an inference. Like all other such measurements (e.g. Clark and Felsenfeld,

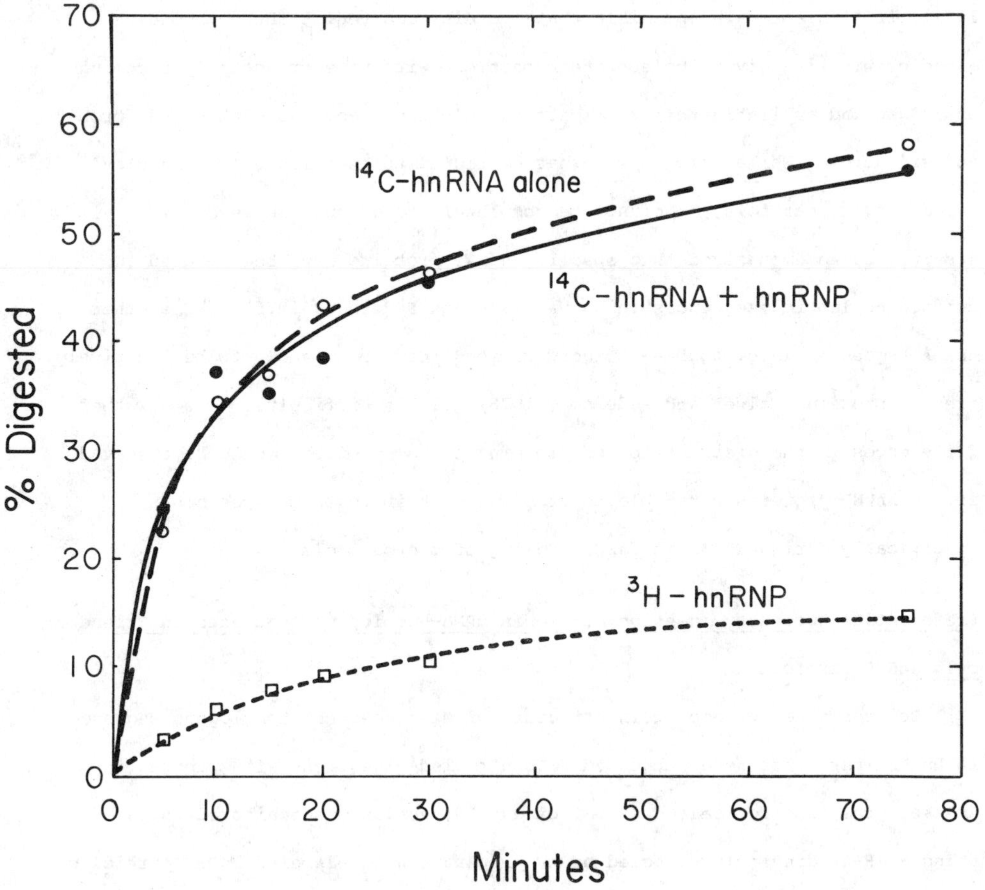

Figure 9. Absence of protein translocation during nuclease digestion of hnRNP. ^{3}H-uridine-labeled Friend cell hnRNP partilces were mixed with ^{14}C-uridine-labeled, deproteinized hnRNA, and the mixture was digested at 4°C with 0.05 μg of pancreatic RNAase per ml. At selected times samples were withdrawn and the digestion was quenched by the addition of a 400,000-fold molar excess of mercaptoethanol. The amounts of trichloroacetic acid-insoluble ^{3}H and ^{14}C radioactivity remaining were then determined. In parallel, a sample of the ^{14}C-hnRNA was digested alone. The concentration of ^{14}C-hnRNA was the same in both samples, and the hnRNP concentration was the same as that used in all the other nuclease digestion experiments. □: ^{3}H-hnRNP in the hnRNP-hnRNA mixture; •: ^{14}C-hnRNA in the mixture; o: ^{14}C-hnRNA in buffer alone.

1971), it always remains possible that proteins can redistribute themselves along or within a given nucleoprotein particle without ever actually becoming dislodged and available to participate in solution reactions with naked "probe" nucleic acids. While this is somewhat difficult to envision over long distances in the particle, it cannot be completely ruled out. However, it deserves to be emphasized that specific hnRNP proteins have been mapped to defined nucleotide sequences in hnRNA (Kish and Pederson, 1975), while other hnRNA sequences (e.g. double-stranded regions) have been shown to be relatively free of protein (Calvet and Pederson, 1978). It is difficult to see how these differences in the distribution of protein in the particles could be observed if the hnRNP proteins were able to freely redistribute themselves non-specifically during isolation and handling of the particles.

Preferential nuclease digestion of globin mRNA-homologous sequences in Friend cell hnRNP particles

To probe the nucleoprotein structure of messenger RNA-homologous sequences in hnRNP, particles were subjected to controlled digestions with pancreatic RNAase. The time course and extent of each digestion was monitored in parallel using a ^{3}H-uridine pulse-labeled hnRNP preparation. Unlabeled hnRNP particles were digested simultaneously under precisely the same conditions as the labeled "monitor" sample, including constant enzyme:substrate ratio.

In the experiment shown in Figure 10, ^{3}H-uridine-labeled hnRNP particles were digested with 0.1 μg of pancreatic RNAase per ml at 4°C until 16% of the ^{3}H-radioactivity had been hydrolyzed to acid-soluble material. When RNA from unlabeled hnRNP digested in parallel was hybridized with globin cDNA, a selective depletion of globin mRNA sequences was detected, relative to RNA from undigested hnRNP. As shown in Figure 10, RNA from mild RNAase-digested hnRNP

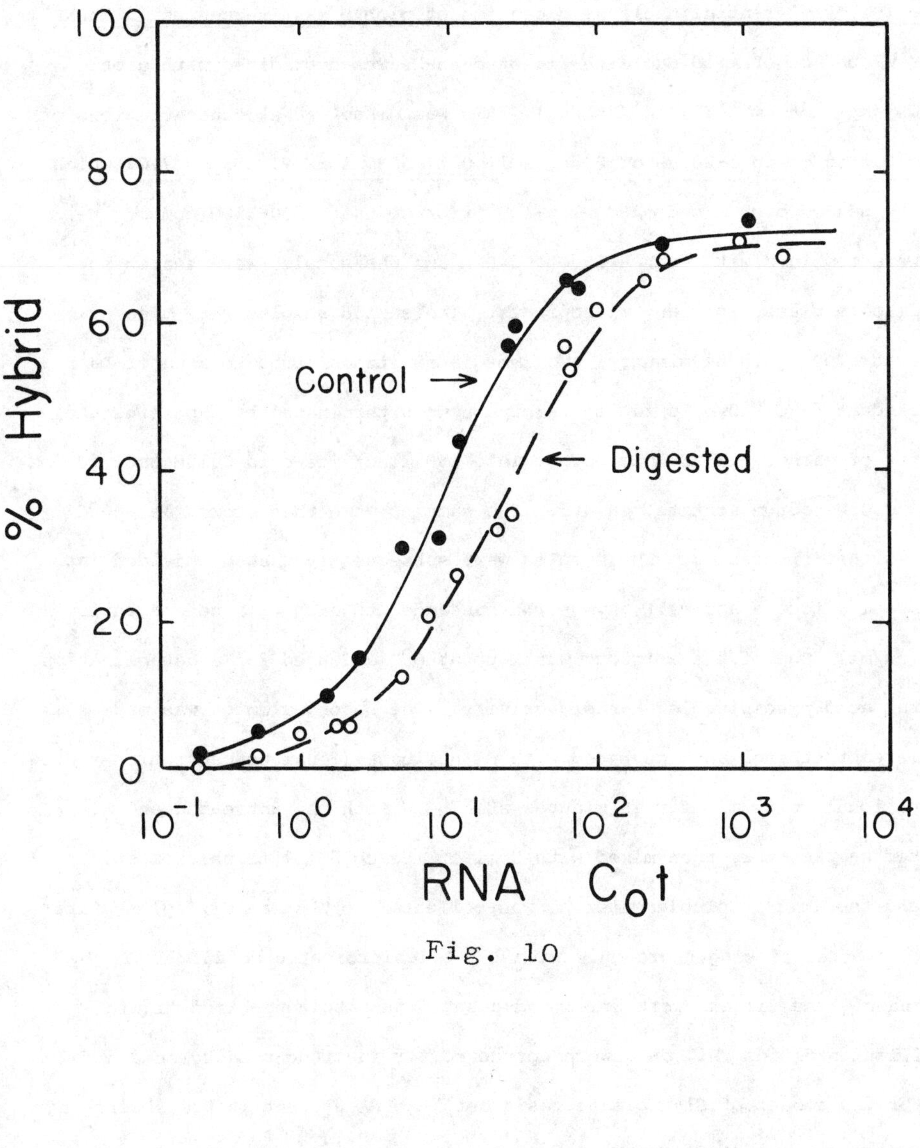
100
80
60
40
20
0
% Hybrid
Control
Digested
10⁻¹
10⁰
10¹
10²
10³
10⁴
RNA C₀t

Fig. 10

Figure 10. Preferential nuclease digestion of globin mRNA sequences in hnRNP. RNA from control of mild nuclease-digested hnRNP was hybridized with globin ^{3}H-cDNA (sp. act. = 3-5 x 10^7 dpm/μg). RNA was dissolved at concentrations of 0.1 to 2.2 mg/ml in 1-20 μL of 0.6M NaCl, 0.1% SDS, 10mM PIPES, pH 7.0, which had been passed over a bed of Chelex-100 prior to use. 500-1500 cpm of ^{3}H-cDNA were combined with each aliquot of RNA and the samples were sealed in borosilicate glass capillary micropipets. The sealed samples were then denatured (100°C, three minutes) and immediately transferred to a water bath regulated at 66°C. Hybridization reactions were terminated by expelling the contents of each capillary micropipet into 1.0 ml of ice-cold 0.15M NaCl, 0.2mM $ZnSO_4$, 0.05M sodium acetate, pH 4.5. The samples were then stored at -20°C until further processing. The samples were subsequently thawed, divided into two 450 μl aliquots and calf thymus DNA was added to each to a concentration of 50 μg/ml. One of the aliquots was kept at 0°C and used for a determination of total acid-precipitable ^{3}H-radioactivity. The second aliquot was made 25mM in 2-mercaptoethanol and incubated with S1 nuclease (final concentration = 0.24 International Units/ml) for 45 minutes at 37°C. Both the untreated and S1-digested samples were then mixed with 8 ml of ice-cold 10% trichloroacetic acid and the acid-insoluble material was collected on Whatman GF/C glass fiber filters. After rinsing thoroughly with 10% trichloroacetic acid followed by 95% ethanol, the filters were dried and counted in a toluene-based liquid scintillation fluid. All data were corrected for instrument background and also for the amount of S1 nuclease-resistant ^{3}H-cDNA present in the absence of any hybridizing RNA (1-3% of the total acid-insoluble ^{3}H-cDNA, depending upon experiment). All RNA C_ot values were corrected to "equivalent" C_ot's at 0.18M Na^+ (Britten *et al.*, 1974). Recent data (W.E. Hahn, personal communication) indicate that the empirical relationship between Na^+ concentration and the second-order rate constant for DNA reassociation (Britten *et al.*, 1974) is not applicable to RNA-driven hybridization reactions. However, this is not a factor in the present case, since all the hybridizations were consistently performed at the same Na^+ concentration (0.6M).

hybridizes with globin cDNA with kinetics which are about three times slower ($C_ot_{1/2}$ = 22 M·sec) than those of RNA from the undigested hnRNP ($C_ot_{1/2}$ = 8 M·sec). Both reactions display characteristic pseudo-first-order kinetics and reach identical termination plateaus. The difference between the two reactions indicates that globin mRNA-homologous sequences are preferentially diluted by mild RNAase digestion of hnRNP, relative to other sequences. Thus, digestion of only 16% of the RNA in hnRNP reduces the concentration of globin mRNA sequences by about 64% ($C_ot_{1/2}$ of 8 M·sec → 22 M·sec).

Data from several experiments like the one shown in Figure 10 are collated in Figure 11. It can be seen that digestion of only 10% of the total hnRNP results in a 40% depletion of globin mRNA sequences. Increasing the extent of hnRNP digestion to 16% leads to a further reduction in the concentration of globin mRNA sequences, to a level corresponding to only about 35% of the initial concentration. However, it can be seen that further digestion does not reduce the remaining globin mRNA sequences below this level. Thus, some of the globin mRNA sequences in hnRNP may be relatively nuclease-resistant, but the majority (65%) are more nuclease-sensitive than total hnRNP. It is possible that these nuclease-sensitive and more nuclease-resistant sequences represent different stages of globin mRNA processing in hnRNP (see Discussion).

The fact that most of the globin mRNA sequences are preferentially attacked during mild nuclease digestion of hnRNP (Figures 10 and 11) obviously means that a substantial fraction of the RNA must be more nuclease-resistant. This protected RNA could either be non-mRNA sequences in hnRNA or, alternatively, some (hypothetical) non-hnRNA component of the hnRNP particle preparation that is more nuclease-resistant than hnRNA altogether. However, the use of a short pulse label time (20 minutes) in the ^{3}H-uridine-labeled "monitor" hnRNP

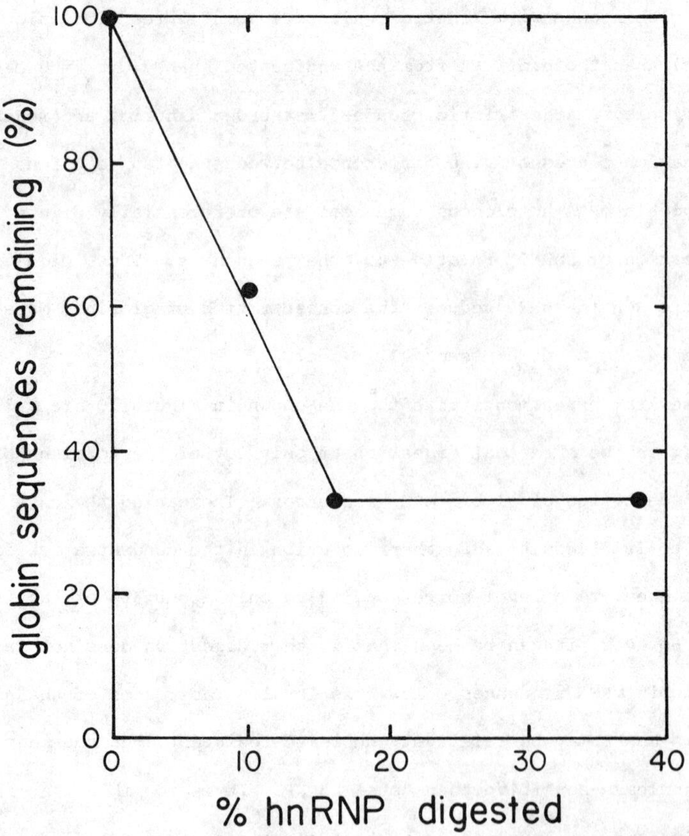

Figure 11. Preferential digestion of globin mRNA sequences as a function of % hnRNP digested. The points are derived from the RNA $C_ot_{1/2}$ values in Figures 10 and several other comparable experiments.

preparations and the selective inhibition of ribosomal RNA labeling by a low concentration of actinomycin (Perry *et al.*, 1970; Pederson, 1974a) insures that the measurement of ^{3}H-uridine radioactivity during RNAase digestion reflects specifically the behavior of hnRNA. Since, by this criterion, digestion of only 16% of the hnRNA in hnRNP is necessary to elicit a 64% reduction in the concentration of globin mRNA sequences, it follows that the presence of putative nuclease-resistant, non-hnRNA species could not generate such a result. It is therefore concluded that the more nuclease-resistant RNA sequences are components of hnRNA as we know it. The specific nature of these nuclease-protected hnRNP sequences is not revealed by these experiments, but they obviously could include other (*i.e.* non-globin) mRNA-homologous sequences or nucleus-restricted hnRNA sequences such as intervening DNA sequence transcripts.

Globin mRNA sequences are not preferentially nuclease-sensitive in deproteinized hnRNA

The observed preferential depletion of globin mRNA sequences upon partial RNAase digestion of hnRNP might be due to some inherent nuclease sensitivity of globin mRNA sequences themselves, perhaps related to their base composition or secondary structure, rather than being a reflection of their nucleoprotein structure in the hnRNP particle. To examine this remote possibility, hnRNP was deproteinized and the resulting hnRNA was digested with RNAase under conditions where a parallel monitor of ^{3}H-labeled hnRNA was digested 10%. As shown in Figure 12, the globin cDNA hybridization kinetics of the digested hnRNA were identical to those of undigested, control hnRNA. This shows that the preferential digestion of globin mRNA sequences observed previously (Figures 10 and 11) is related to the nucleoprotein organization of the hnRNP particle.

The control experiment with deproteinized hnRNA (Figure 12) provides

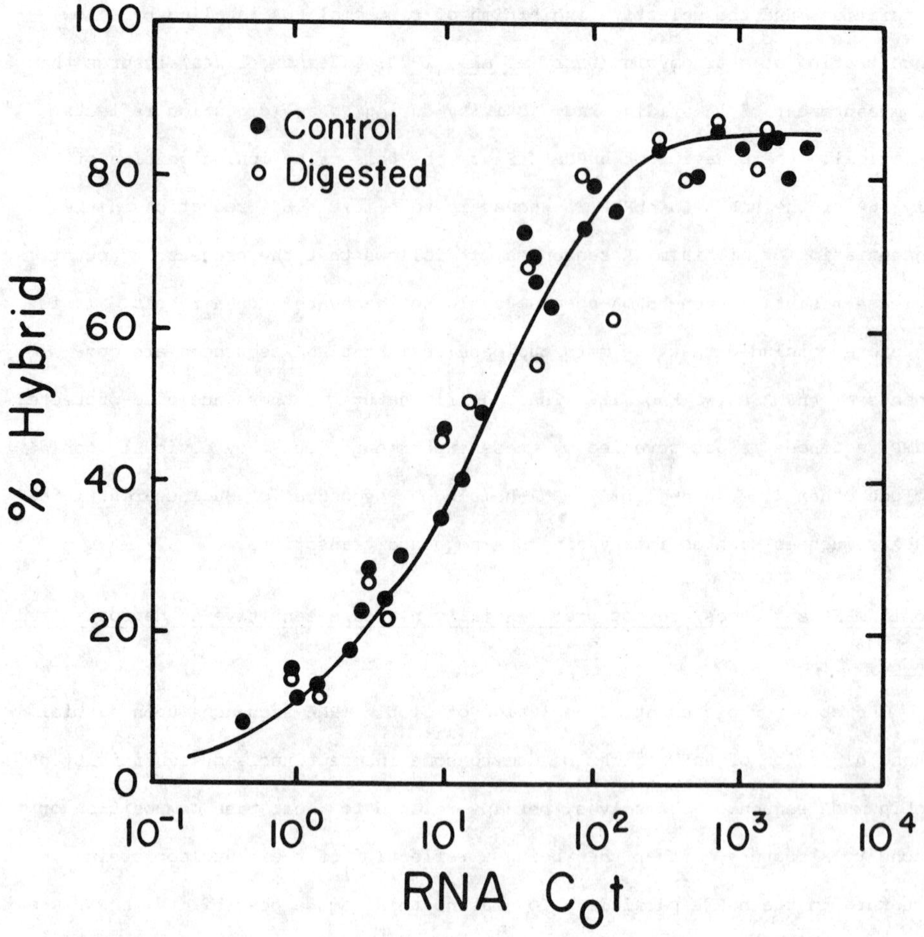

Figure 12. Absence of preferential digestion of globin mRNA sequences in deproteinized hnRNA. Deproteinized Friend cell hnRNA was digested 10% with pancreatic RNAase, and the control and digested RNA's were then hybridized with globin cDNA as detailed in Figure 10. (o): hybridization of control (undigested) hnRNA; (o): hybridization of nuclease-digested hnRNA.

assurance that the reduced rate of hybridization observed with RNA from partially digested hnRNP (Figures 10 and 11) is not an effect of overall RNA driver size on the cDNA hybridization kinetics. The hnRNA digested as naked RNA should, if anything, be smaller than the RNA from similarly digested hnRNP. Yet it hybridizes faster, indeed as fast as RNA recovered from intact hnRNP particles (Figure 12).

Globin mRNA sequences in hnRNP have an uncovered conformation

The preceding results indicate that globin mRNA-homologous sequences in hnRNP are more nuclease-sensitive than total hnRNP. However, they do not establish whether the globin mRNA sequences are simply complexed with somewhat less protein on average than total hnRNP or, alternatively, exist as uncovered, essentially protein-free RNA tracts in the hnRNP particle. To investigate this issue, the nuclease sensitivity of naked globin mRNA was determined as a point of reference against which to assess the nuclease sensitivity of globin mRNA sequences in hnRNP. To permit such a measurement, reconstruction experiments were performed in which pure, deproteinized mouse α,β-globin mRNA was added to a preparation of hnRNP particles from HeLa cells. This experiment is possible because (1) there is not a sufficient amount of endogenous globin mRNA sequences in HeLa cell hnRNP to obscure the behavior of the added mouse globin mRNA (Calvet, Holland and Pederson, unpublished data), and (2) naked RNA added to hnRNP does not become complexed with proteins to an extent sufficient to increase its nuclease resistance relative to naked RNA itself (Figure 9).

Deproteinized mouse α,β-globin mRNA was added to HeLa cell hnRNP in amounts that approximate the level of endogenous globin mRNA in Friend cell hnRNP, which was determined to be about 0.02% of the total hnRNA mass. This value was derived from a comparison of the RNA $C_ot_{1/2}$ for the reaction of Friend cell hnRNA with α,β-globin cDNA (i.e. 8 to 13 M·sec) with that for the back-hybridization of the cDNA with its own α,β-globin mRNA template (RNA $C_ot_{1/2} = 2 \times 10^{-3}$ M·sec).

When the HeLa cell hnRNP particle-mouse globin mRNA mixture was digested under precisely the same conditions used to effect a 15-20% digestion of Friend cell hnRNP, the naked globin mRNA (RNA $C_ot_{1/2}$ = 26 M·sec) was reduced about two-fold in its abundance relative to that in an undigested control ($C_ot_{1/2}$ = 13 M·sec). The $C_ot_{1/2}$ of the control reaction (13 M·sec) is the same as that observed for the hybridization kinetics of undigested Friend cell hnRNA ($C_ot_{1/2}$ = 8-13 M·sec). This shows that the mouse globin mRNA concentration in the HeLa hnRNP mixture was accurately reconstructed to mimic the actual situation in Friend cell hnRNP. Since the two-fold reduction in the cDNA hybridization kinetics observed after a 15-20% digestion of the HeLa hnRNP-mouse globin mRNA mixture is quantitatively very similar to the effect observed when Friend cell hnRNP is digested similarly (Figure 10), the nuclease sensitivity of the endogenous globin mRNA sequences in Friend cell hnRNP appear to be as high as that of protein-free globin mRNA itself. This indicates that globin mRNA sequences in hnRNP are not just more nuclease sensitive than total hnRNP but, in addition, are not complexed with hnRNP particle proteins to an extent sufficient to confer a detectable nuclease protection relative to naked RNA. Since hnRNP particles as a whole do show substantially reduced nuclease sensitivity relative to naked RNA (Figure 1), and since deproteinization of hnRNP abolishes the relative nuclease sensitivity of globin mRNA sequences (Figure 12), it follows that the nucleoprotein structure of hnRNA is hierarchical, and that globin mRNA sequences occupy uncovered domains in hnRNP.

Discussion

It is now firmly established that hnRNA exists in a ribonucleoprotein form in the eukaryotic cell nucleus. This fact was first suggested by light microscopic studies of nascent RNA on the lateral loops of amphibian oocyte lampbrush chromosomes (Gall, 1955), and has been more dramatically demonstrated by the electron microscopic identification of hnRNA:ribonucleoprotein particles *in situ* and on unfolded chromatin fibers spread from nuclei or cells by detergent lysis. Initial attempts to extract these hnRNP particles from nuclei resulted in the isolation of degraded, 40S RNP complexes (Samarina *et al.*, 1968), due to the action of endogenous nucleases. While these early degraded RNP preparations were of value for some purposes, such as enumerating the minimal set of proteins present in these so-called "core" particles, the fact remains that the 40S RNP's contain highly degraded hnRNA fragments and therefore cannot be regarded as native structures.

We have developed an alternative approach for purifying hnRNP that leads to the isolation of larger particles that contain high molecular weight hnRNA (Pederson, 1974a). In addition, it has been possible to demonstrate that these latter hnRNP preparations contain specific proteins bound to defined nucleotide sequences, such as poly(A) (Kish and Pederson, 1975; Firtel and Pederson, 1975), and that they contain different domains of nucleoprotein structure in which some RNA regions are less complexed with protein than others (*e.g.* double-stranded regions (Calvet and Pederson, 1978). In some cases, these same domains can be demonstrated in hnRNP *in vivo* (Calvet and Pederson, 1979). All of these facts suggest, but of course do not prove, that the isolated hnRNP particles represent native structures. However, it has not been possible in our previous studies to examine these particles in terms of specific, intact mRNA precursor sequences, due to the lack of gene-specific probes in the particular cellular systems being used, notably HeLa cells. In the present investigation, we have embarked upon the isolation of high molecular weight hnRNP particles from DMSO-induced mouse erythroleukemia cells that are synthesizing substantial quantities of globin messenger RNA, as a step toward the analysis of defined gene transcripts in hnRNP. The results of this study demonstrate that Friend cell hnRNP particles can be isolated in good purity by slight modifications of our original methods. The physical and biochemical properties of the Friend cell particles - sedimentation behavior, buoyant density, protein content, and nuclease sensitivity (Figures 1-4) - all recapitulate those described previously for HeLa cell hnRNP. More importantly, the present results establish that the purified Friend cell particles contain high molecular weight, rapidly-labeled hnRNA (Figure 5), including covalently intact nuclear precursors of β-globin messenger RNA (Figures 7 and 8).

In any isolation procedure for nucleoprotein, it is essential to show that the isolated material does not arise through non-specific nucleic acid:protein interactions during cell fractionation. Indeed, the original studies of eukaryotic <u>cytoplasmic</u> messenger RNP particles were compromised by just such artifacts, which were subsequently brought to light through specific reconstruction studies and more critical experimental work (Girard and Baltimore, 1966). These latter studies revealed that cytoplasmic extracts of mammalian cells contain a large concentration of RNA-binding proteins that rapidly associate non-specifically with added exogenous RNA's. The existence of these RNA-binding proteins does not of course eliminate the possibility that there is really such a thing as cytoplasmic mRNP (nor does it prove that there is), but it does argue for caution in its isolation and characterization. In contrast to the results for cytoplasmic fractions, reconstruction experiments involving the addition of naked RNA to nuclei or nuclear extracts have consistently failed to generate non-specific RNP complexes (Wilt *et al.*, 1973; Pederson, 1974a; Firtel and Pederson, 1975; and the present study). In these experiments the amounts of added probe RNA are very low relative to the endogenous nuclear protein mass (see legend to Figure 3), a situation that should be optimal for detecting non-specific interactions of proteins with the probe. Thus, the consistent negative outcome of these *nuclear* reconstruction experiments argues against the existence of a soluble pool of the major hnRNA-binding proteins in the cell nucleus. Apparently the hnRNP proteins are delivered from the cytoplasm to the nuclear interior at the correct stoichiometry to support the ongoing tempo of transcription and hnRNP assembly. This situation constitutes an intriguing case of intracellular coordination that is richly deserving of further study.

The present identification of pulse-labeled transcripts of the β-globin gene in Friend cell hnRNP constitutes the first demonstration of a specific, well-defined mRNA sequence in high molecular weight hnRNP particles. In a sense this demonstration comes as no surprise since β-globin sequences are abundant components of induced Friend cell hnRNA, and there was no reason to necessarily suspect *a priori* that hnRNP would prove to be depleted of mRNA-homologous sequences. The more important point however is that the present study demonstrates that covalently intact, 15S β-globin mRNA precursor can be isolated as hnRNP, which immediately opens the door to the use of these hnRNP particles to address mRNA processing, especially processing.

An important conclusion emerging from the experiments reported in this paper is that the ribonucleoprotein structure of hnRNA is non-random. hnRNP is organized differentially into distinct nucleoprotein domains, rather than merely consisting of a uniformly-distributed layer of protein covering all of each hnRNA transcript. That the structure of hnRNP is not this simple was suggested initially by our observation that intramolecular double-stranded hnRNA regions are relatively free of protein compared to total hnRNP (Calvet and Pederson, 1978). The present results now extend the concept of a hierarchical ribonucleoprotein organization of hnRNA to include at least 65% of the globin mRNA-homologous sequences in the hnRNP particles of mouse erythroleukemia cells, which are shown to be more nuclease-sensitive than total hnRNP and preferentially depleted during mild RNAase digestion of the particles. Additional experiments indicate that the preferentially digested globin mRNA-homologous hnRNP sequences have a nucleoprotein conformation that is as unprotected and nuclease-sensitive as naked RNA.

One interpretation of these results is that the nucleoprotein structure of hnRNA is somehow related to sequential stages of mRNA processing. There might be substantially different ribonucleoprotein structures before and after splicing or other post-transcriptional hnRNA modifications, so that each set of specific gene transcripts in hnRNP might consist of a mixed population of nuclease-sensitive and relatively nuclease-resistant structures. Thus, the fraction of nuclear globin mRNA-homologous sequences that is nuclease-sensitive in hnRNP, approximately 65% (Figures 10 and 11), might represent a specific stage or stages of globin mRNA processing. Another possibility is that hnRNP proteins may be clustered on non-messenger sequences in some globin gene transcripts. The known interspersed arrangement of coding and interveing sequences in primary transcripts of the mouse β-globin gene (Kinniburgh et al., 1978; Smith and Lingrel, 1978; Tilghman et al., 1978) provides a physical basis for such an alternating nucleoprotein organization, in which the hnRNP proteins might be clustered on intervening sequences, rendering the mRNA-homologous regions more nuclease-sensitive. This idea is also compatible with recent ultrastructural analyses of nascent hnRNP on chromatin spread from *Drosophila* embryos, which indicate hnRNP proteins are disposed in a non-uniform, interspersed pattern along the nascent hnRNA (Beyer *et al.*, 1979).

Interestingly, the present observation that globin mRNA-homologous regions of hnRNP are as nuclease-sensitive as naked RNA is the opposite of what might have been expected *a priori*, namely that mRNA-coding regions in hnRNP would be highly protected as a way of shielding them from endonuclease attack *in vivo*. That this is not the case raises the question of why, despite their being unprotected, the mRNA-coding regions in hnRNP are nevertheless the very sequences that survive during processing. One simple possibility is that the splicing endonucleases and ligase are so highly specific for sequence signals at intron: exon borders that a "protection" of the exons by hnRNP proteins is unnecessary. This view envisions the splicing enzymes to be integral, anchored components of hnRNP particle structure.

The use of a cDNA probe specific for globin mRNA-homologous sequences in hnRNP restricts the conclusions of the present study to this particular set of transcripts. Thus, we do not know whether the altered nucleoprotein structure detected here for globin mRNA sequences in hnRNP is representative of the situation for all mRNA-homologous sequence components of hnRNA. However, there is cytological evidence that the ribonucleoprotein structure of hnRNP may be gene locus-specific (discussed in Pederson, 1974a). It is therefore possible that certain hnRNP's might be considerably more nuclease-sensitive or resistant than others, depending upon their sequence-specific ribonucleoprotein structure. Experiments similar to those reported in this paper are therefore now underway using probes for other defined mRNA sequences in hnRNP.

Acknowledgments

Supported by NIH grant GM21595. I gratefully acknowledge the participation of my colleagues Nicholas Davis and Stephen Munroe in these experiments. We are also indebted to Eileen Falvey for capable assistance, Philip Leder for λgtWES·MβG-2, Arthur Skoultchi for a starter culture of mouse erythroleukemia cells and Jerry Lingrel for a generous gift of purified mouse α,β-globin mRNA.

References

Alwine, J.C., D.J. Kemp and G.R. Stark. 1977. Method for detection of specific RNAs in agarose gels by transfer to diazobenzyloxymethyl-paper and hybridization with DNA probes. Proc. Nat. Acad. Sci. U.S.A. 74:5350-5354.

Beyer, A., S. McKnight and O.L. Miller, Jr. 1979. Sequence-specific ribonucleoprotein structure. J. Cell Biol. 83:191 (abstract).

Bhorjee, J.S. and T. Pederson. 1972. Non-histone chromosomal proteins in synchronized HeLa cells. Proc. Nat. Acad. Sci. U.S.A. 69:3345-3349.

Britten, R.J., D.E. Graham and B.R. Neufeld. 1974. Analysis of repeating DNA sequences by reassociation. Methods in Enzymology XXIX; part E:363-418.

Clark, R.J. and G. Felsenfeld. 1971. Structure of chromatin. Nature New Biol. 229:101-106.

Calvet, J.P. and T. Pederson. 1977. Secondary structure of heterogeneous nuclear RNA: Two classes of double-stranded RNA in native ribonucleoprotein. Proc. Nat. Acad. Sci. U.S.A. 74:3705-3709.

Calvet, J.P. and T. Pederson. 1978. Nucleoprotein organization of inverted repeat DNA transcripts in heterogeneous nuclear RNA-ribonucleoprotein particles from HeLa cells. J. Mol. Biol. 122:361-378.

Calvet, J.P. and T. Pederson. 1979. Heterogeneous nuclear RNA double-stranded regions probed in living HeLa cells by cross-linking with the psoralen derivative aminomethyltrioxsalen. Proc. Nat. Acad. Sci. U.S.A. 76:755-759.

Calvet, J.P. and T. Pederson. 1979. Photochemical cross-linking of secondary structure in HeLa cell heterogeneous nuclear RNA in situ. Nucleic Acids Res. 6:1993-2001.

Firtel, R.A. and T. Pederson. 1975. Ribonucleoprotein particles containing heterogeneous nuclear RNA in the cellular slime mold, Dictyostelium discoideum. Proc. Nat. Acad. Sci. U.S.A. 72:301-305.

Gall, J.G. 1955. On the submicroscopic structure of chromosomes. Brookhaven Symp. Biol. 8:17-32.

Girard, M. and D. Baltimore. 1966. The effect of HeLa cell cytoplasm on the rate of sedimentation of RNA. Proc. Nat. Acad. Sci. U.S.A. 56:999-1002.

Kinniburgh, A.J., J.E. Mertz and J. Ross. 1978. The precursor of mouse β-globin messenger RNA contains two intervening RNA sequences. Cell 14:681-693.

Kish, V. and T. Pederson. 1975. Ribonucleoprotein organization of polyadenylate sequences in HeLa cell heterogeneous nuclear RNA. J. Mol. Biol. 95:227-238.

Kish, V. and T. Pederson. 1977. Heterogeneous nuclear RNA secondary structure: Oligo(U) sequences base-paired with poly(A) and their possible role as binding sites for heterogeneous nuclear RNA-specific proteins. Proc. Nat. Acad. Sci. U.S.A. 74:1426-1430.

Kish, V. and T. Pederson. 1978. Isolation and characterization of ribonucleoprotein particles containing heterogeneous nuclear RNA. Methods in Cell Biology 17:377-399.

Kumar, A. and T. Pederson. 1975. Comparison of proteins bound to heterogeneous nuclear RNA and messenger RNA in HeLa cells. J. Mol. Biol. 96:353-365.

Maggio, R., P. Siekevitz and G.E. Palade. 1963. Studies on isolated nuclei. II. Isolation and chemical characterization of nucleolar and nucleoplasmic subfractions. J. Cell Biol. 18:293-312.

Pederson, T. 1974a. Proteins associated with heterogeneous nuclear RNA in eukaryotic cells. J. Mol. Biol. 83:163-183.

Pederson, T. 1974b. Gene activation in eukaryotes: Are nuclear acidic proteins the cause or the effect? Proc. Nat. Acad. Sci. U.S.A. 71:617-621.

Perry, R.P., T-Y. Cheng, J.J. Freed, J.R. Greenberg, D.E. Kelley and K.D. Tartof. 1970. Evolution of the transcription unit of ribosomal RNA. Proc. Nat. Acad. Sci. U.S.A. 65:609-616.

Ross, J. and D.A. Knecht. 1978. Precursors of alpha and beta globin messenger RNAs. J. Mol. Biol. 119:1-20.

Samarina, O.P., E.M. Lukanidin, J. Molnar and G.P. Georgiev. 1968. Structural organization of nuclear complexes containing DNA-like RNA. J. Mol. Biol. 33: 251-263.

Smith, K. and J.B. Lingrel. 1978. Sequence organization of the β-globin mRNA precursor. Nucleic Acids Res. 5:3295-3301.

Tilghman, S.M., D.C. Tiemeier, F. Polsky, M.H. Edgell, J.G. Seidman, A. Leder, L.W. Enquist, B. Norman and P. Leder. 1977. Cloning specific segments of the mammalian genome: Bacteriophage λ containing mouse globin and surrounding gene sequences. Proc. Nat. Acad. Sci. U.S.A. 74:4406-4410.

Tilghman, S.M., P.J. Curtis, D.C. Tiemeier, P. Leder and C. Weissmann. 1978. The intervening sequence of a mouse β-globin gene is transcribed within the 15S β-globin mRNA precursor. Proc. Nat. Acad. Sci. U.S.A. 75:1309-1313.

Wilt, F., M. Anderson and E. Ekenberg. 1973. Centrifugation of nuclear ribonucleoprotein particles of sea urchin embryos in cesium sulfate. Biochem. 12: 959-966.

ANTI-TUMOR IMMUNE RESPONSES MEDIATED BY IMMUNE RNA

Yosef H. Pilch, M.D. and Shankar K. Nayak, Ph.D.

Surgical Oncology Service, Department of Surgery
University of California, San Diego, School of
Medicine, San Diego, California, 92103, U.S.A.

ABSTRACT

Preparations of ribonucleic acid extracted from lymphoid organs of animals immunized with tumor cells has been shown to transfer antitumor immune responses mediating rejection of rodent tumor isografts and cellular cytotoxic immune reactions against human and rodent tumor cells in vitro. Immune RNA has also been shown to induce antitumor antibody synthesis in rodent spleen cells in vitro. These immune responses are specific for the tumor associated antigens of the tumor cells used to immunize the animal from whose lymphoid organs the RNA is extracted. Preparations of Immune RNA (I-RNA) have been utilized in the immunotherapy of a spontaneously metastasizing rat mammary carcinoma and in preliminary studies of immunotherapy of human malignancies (melanoma, colon carcinoma and renal cell carcinoma).

Introduction:

Ribonucleic Acid (RNA) preparations extracted from lymphoid tissues of animals immunized with tumor cells has been shown to transfer anti-tumor immunological responsiveness by us (3,4,9-17,20-24, 26-28), as well as by several other investigators (1,2,5,6,8,18,19,29). Such immunoreactive RNA extracts of lymphoid cells have been termed "Immune" RNA (I-RNA) and have been shown to mediate tumor-specific immune responses, in vitro and in vivo, both in human (1,12) and in animal (9,10,14-17) tumor systems. In animal studies, both xenogeneic and syngeneic sources of I-RNA have been used. In human studies, xenogeneic and allogeneic sources of I-RNA have been employed. In this paper, we summarize our experience in the mediation of anti-tumor immune responses by I-RNA both in vitro and in vivo.

Rejection of Rodent Tumor Isografts Mediated by I-RNA:

In 1970-1971, we reported that immune rejection of tumor isografts in inbred mice could be mediated by syngeneic spleen cells pre-incubated, in vitro, with xenogeneic I-RNA extracted from the lymphoid organs of guinea pigs immunized with the specific tumor that was to be treated (23, 26,27). This immunity was manifested when such RNA-incubated spleen cells were injected intraperitoneally into syngeneic mice. This response was specific for the particular tumor used to immunize the guinea pig from whose lymphoid organs the I-RNA was prepared. Thus, spleen cells incubated with I-RNA against one tumor inhibited the growth of isografts of that tumor, but not isografts of a syngeneic but antigenically different tumor. Administration of spleen cells incubated with I-RNA against the second tumor inhibited the growth of isografts of that tumor but not isografts of the first tumor. RNA from guinea pigs immunized with Freund's adjuvant only was without effect. Likewise, injections of untreated syngeneic spleen cells (not incubated with RNA) did not inhibit the growth of tumor isografts. Most important was the finding that I-RNA from the lymphoid organs of guinea pigs immunized with syngeneic normal mouse tissues failed to mediate an anti-tumor response. This, together with the findings of the specificity of the I-RNA for the immunizing tumor, strongly suggested that the immunity expressed by lymphoid cells pre-incubated with anti-tumor I-RNA was directed against tumor-associated antigens.

We then sought a more direct and simpler method of utilizing I-RNA in mediating anti-tumor immune responses. Such a method would be the direct systemic administration of the I-RNA itself rather than lymphoid cells pre-incubated with I-RNA. In considering how best to achieve a transfer of immunity, in vivo, by the direct, systemic administration of I-RNA, our attention was drawn to the problem of the possible degradation of I-RNA by plasma and interstitial ribonucleases. Ribonuclease is known to inactivate preparations of I-RNA and, indeed, when the anti-tumor I-RNA preparations previously studied in our laboratory had been treated with RNase all anti-tumor immunoreactivity was abolished. These facts led to the initial assumption that the direct systemic administration of xenogeneic I-RNA without lymphoid cells would be ineffective due to enzymatic degradation and inactivation of I-RNA by interstitial and plasma ribonucleases. We attempted to eliminate this problem by administering our I-RNA in a soluation containing the potent RNase inhibitor, sodium dextran sulfate. When such mixtures of xenogeneic I-RNA and dextran sulfate were administered subcutaneously to mice or, by footpad injection, to rats every other day for 10 days, such animals were rendered specifically resistant to isografts of the particular tumor used to immunize the I-RNA donor (4,20,22).

These experiments with xenogeneic I-RNA provided convincing, but <u>indirect</u> evidence of the specificity of anti-tumor I-RNA for the tumor-associated antigens of these chemically-induced murine tumor. Because of certain obvious difficulties in the interpretation of results obtained in xenogeneic models, and so that <u>direct</u> evidence could be obtained that I-RNA could mediate immune responses to TAA, it was desirable to study a syngeneic system in which the tumor cells, the lymphoid tissues from which I-RNA was extracted, the spleen cells incubated with the I-RNA, and the recipient host were all from members of a single, inbred strain. This would eliminate the possibility that the inhibition of tumor isograft development observed might be due, in part, to immune responses directed against normal transplantation antigens. For this model, a syngeneic tumor-host system in female Fischer 344/N rats was used. Syngeneic I-RNA was prepared from the spleens of Fischer rats immunized to this tumor by the excision of growing tumor transplants. Control RNA preparations were extracted from the spleens of unimmunized Fischer rats.

Rats challenged with tumor cells synchronous with the first of three daily intravenous injections of syngeneic spleen cells pre-incubated with syngeneic anti-tumor I-RNA extracted from the spleens of immunized rats evidenced a much lower incidence of tumor isograft development than did control rats which recieved injections of syngeneic spleen cells alone (3). Furthermore, rats receiving spleen cells incubated with the same active I-RNA, treated with RNase prior to incubation with spleen cells, manifested a tumor incidence similar to that of control animals. Moreover, RNA from the spleens of unimmunized rats was ineffective in mediating an anti-tumor immune response.

Immunotherapy of a Spontaneously Metastasizing Rat Mammary Carcinoma with I-RNA:

We recently reported the immunotherapy of a transplantable, spontaneously metastasizing mammary adenocarcinoma (13762) in female Fischer 344/N rats with xenogeneic and syngeneic I-RNA (21). In this model, when growing primary tumor transplants are excised 18 days after transplantation, local recurrence does not occur, but all animals go on to die of distant metastases within approximately 95 days, if no additional treatment is given. Experiments were performed in which immunotherapy with I-RNA was utilized as an adjunct to surgical excision of primary tumor transplants. With excision of the primary tumor transplant alone, all animals died of metastases within 96 days. Xenogeneic I-RNA was extracted from the lymph nodes and spleens of Hartley guinea pigs immunized with mammary adenocarcinoma tissue. This I-RNA was administered to a group of rats at a dose of 1.0 mg in 1 ml of buffer containing 10 mg/ml of sodium dextran sulfate (molecular weight 500,000) (0.25 ml of this mixture was administered into each of the four foot pads of a group of rats every other day for 10 doses beginning 10 days prior to surgical excision of the primary tumor transplant). Of the animals so treated, 80% survived and remained free of disease for 180 days. When xenogeneic I-RNA was administered at a dose of 1 mg every other day for 10 doses beginning at the day of surgical excision of the primary tumor, 67% of animals so treated survived and remained free of disease for 180 days. Syngeneic I-RNA was extracted from the spleens of Fischer rats bearing growing tumor transplants and administered, at the same dosage schedule, pre- and postoperatively to another group of rats. Of the animals so treated, 50% survived and remained free of disease for 180 days.

Mediation of Cytotoxic Immune Reactions against Rodent Tumor Cells, In Vitro, by I-RNA:

Utilizing a microcytotoxicity assay for cell mediated immunity, we reported, in 1974, that xenogeneic I-RNA extracted from the lymphoid organs of guinea pigs immunized with murine tumor cells mediated tumor-specific cytotoxic immune responses, in vitro (9). I-RNA was extracted from the lymph nodes and spleens of guinea pigs which had been immunized with either of two chemically-induced C3H mouse sarcomas (MC-1 and BP-1) in complete Freund's adjuvant (CFA). Control guinea pigs were immunized with a pool of normal C3H tissues. Spleen cells from normal C3H mice were incubated with one or the other of the I-RNA preparations, or with the control (normal tissue) RNA. The cytotoxicity of these spleen cells, following incubation with I-RNA for either MC-1 or BP-1 tumor target cells was then measured.

Spleen cells pre-incubated with anti-MC-1 I-RNA were found to be markedly cytotoxic for MC-1 target cells but not for BP-1 target cells. Spleen cells, pre-incubated with anti-BP-1 I-RNA were cytotoxic for BP-1 target cells but not for MC-1 target cells. Again, RNA extracted from guinea pigs immunized with normal C3H tissues were inactive. When the active I-RNA preparations were treated with RNase, immune responses were abrogated but treatment with DNase or pronase did not effect activity. Wang and Deckers later confirmed these results independently using a different chemically-induced C3H sarcoma (31).

Using a somewhat different microcytotoxicity assay, similar results were obtained with syngeneic I-RNA in the rat tumor model previously described. In these experiments, the tumor used, the lymphoid tissues from which I-RNA was extracted, and the spleen cells incubated with the I-RNA were all obtained from members of a single inbred strain of rat (10). I-RNA was prepared from spleens of Fischer rats bearing growing tumor transplants of a chemically-induced tumor (MC3-R). Control RNA preparations were also extracted from the spleens of Fischer rats bearing a different chemically-induced tumor, BP1-R. Normal Fischer rat spleen cells, pre-incubated with I-RNA extracted from the spleens of rats bearing growing tumor transplants became significantly cytotoxic to target cells of the same tumor. When spleen cells were incubated with RNA from the spleens of rats bearing transplants of a syngeneic but antigenically different tumor, no significant cytotoxicity resulted. RNase treatment abolished the activity of the I-RNA preparations, but DNase or pronase treatment did not.

In 1976, we have described the intracellular localization of I-RNA in the lymphoid cells of RNA donor animals (13). RNA was extracted separately from homogenates of whole spleen cells, from nuclei alone, and from nuclei-free cytoplasmic fractions of spleen cells from rats sensitized to MC3-R tumor cells. The immunologically active components were confined mainly to RNA from cytoplasmic fractions. RNA from nuclear fractions was immunologically inactive.

In order to further characterize the anti-tumor I-RNA preparations, we studied the kinetics of I-RNA synthesis in the lymphoid tissues immunized rodents (14). As measured in a micro-cytotoxicity assay for cell mediated immunity, cytotoxic immune reactivity of syngeneic I-RNA was maximal when extracted from the spleens of Fischer rats 21-28 days following inoculation with 10^6 syngeneic MC3-R tumor cells. Maximum immunoreactivity of xenogeneic I-RNA extracted from the lymphoid organs of guinea pigs immunized with MC-1 mouse tumor cells was reached 14 days after immunization.

Both syngeneic and xenogeneic anti-tumor I-RNA were fractionated in preparative sucrose density gradients. The highest immune reactivity was consistantly obtained from I-RNA fractions with sedimentation values of 12-16S. The immunologically active I-RNA comprised only 5-7% of the total RNA extracted from the lymphoid tissues of immunized animals.

We then investigated the lymphocyte populations participating in cellular anti-tumor immune responses mediated by I-RNA (16). Xenogeneic I-RNA was extracted from the lymphoid organs of guinea pigs immunized with either MC-1 or BP-1 (chemically-induced tumors of C3H mice). Syngeneic I-RNA was prepared from spleens of Fischer rats bearing growing transplants of the chemically-induced sarcoma MC3-R. In both the mouse and the rat models, normal syngeneic spleen cells were used as effector cells. By using various commonly employed cell separation techniques, we performed experiments to determine the lymphoid cell of origin of I-RNA and the effector cell type responsible for the cytotoxic immunoreactivity mediated by I-RNA. T-lymphocytes were found to be intimately involved in cellular anti-tumor immune responses mediated by both syngeneic and xenogeneic I-RNA's. Syngeneic I-RNA was extracted from unfractionated spleen cell suspensions and from various subpopulations of spleen cells obtained from inbred Fischer 344/N rats bearing growing transplants of the MC3-R sarcoma. These I-RNA preparations were incubated separately with non-immune syngeneic spleen cells, and their cytotoxicity for MC3-R target cells was determined. Active I-RNA was extracted from populations of donor lymphoid cells enriched with respect to T-lymphoictes, while RNA from cells adherent to erythrocyte-antibody (EA) monolayers (B-cells) was inactive.

The effector cells acted on by I-RNA also appeared to be T-cells. Cells that formed EA rosettes (mostly B-cells) showed no increase in cytotoxicity for MC3-R target cells following incubation with I-RNA whereas cells not forming EA rosettes and nonadherent to nylon fiber (mostly T-cells) became more cytotoxic after incubation with I-RNA. After incubation with xenogeneic I-RNA extracted from the lymphoid organs of guinea pigs immunized with C3H tumor cells, the cytotoxicity of unfractionated C3H spleen cells for tumor target cells was increased. Removal of cells adherent to petri dishes did not change the cytotoxicity. Lymphocytes eluted from nylon fiber columns increased in cytotoxicity after being exposed to I-RNA. When these cells were treated with anti θ serum and complement, this response was abrogated. Therefore, the effector cells mediating cytotoxic anti-tumor reactions after treatment with I-RNA appeared to be T-lymphocytes.

<u>Induction of Anti-Tumor Antibody Synthesis by Mouse Spleen Cells, In Vitro, with Xenogeneic I-RNA:</u>

I-RNA has also been shown to induce in vitro anti-tumor humoral immune responses (17). Globulins derived from culture medium and cell lysates following incubation of normal mouse spleen cells with xenogeneic I-RNA were shown to contain complement-dependent cytotoxic anti-tumor antibody activity. Globulins derived from spleen cells incubated with I-RNA extracted from the lymphoid organs of guinea pigs that had been immunized with either of two cross-reacting chemically induced murine sarcomas reacted against target cells of both tumors, but not against target cells derived from a spontaneously arising syngeneic osteosarcoma. Globulins obtained from spleen cells that had been incubated with I-RNA from guinea pigs immunized either with osteosarcoma cells or cells from a spontaneous, syngeneic mammary tumor were not cytotoxic for target cells from either of the two chemically induced sarcomas, although globulins from spleen cells that had been incubated with anti-osteosarcoma I-RNA were cytotoxic for osteosarcoma target cells. Globulins derived from spleen cells incubated with I-RNA extracted from guinea pigs that had been immunized with complete Freund's adjuvant only or with a pool of syngeneic normal tissues, exhibited no cytotoxic activity against any of the three target cells tested. The anti-tumor antibodies whose synthesis was induced by xenogeneic I-RNA were apparently specific of the TAA of the particular tumor used to immunize the I-RNA donor.

Mediation of Cytotoxic Immune Reactions against Human Tumor Cells, In Vitro, by Xenogeneic I-RNA:

In 1974, we demonstrated that normal human peripheral blood buffy coat leukocytes pre-incubated with xenogeneic I-RNA extracted from the lymphoid organs of either sheep or guinea pigs immunized with human tumor cells became markedly more cytotoxic, in vitro, to human tumor cells of the same histologic type used to immunize the I-RNA donor. RNA from the lymphoid organs of animals immunized with Freund's adjuvant only was inactive (24,28). When purified mononuclear cell populations, prepared on Ficoll-isopaque gradients, were incubated with the I-RNA, the magnitude of the cytotoxicity observed was increased When the I-RNA preparations were treated with DNase or with pronase and then incubated with lymphocytes, the resultant cytotoxicity was not significantly altered. However, RNase treatment of the I-RNA abrogated the response.

In all these experiments, it was obvious that, since the effector lymphocytes were allogeneic with respect to the tumor target cells, the cytotoxic immune responses mediated by the I-RNA preparations were probably directed primarily against the normal transplantation antigens (HL-A antigens) on the surface of the tumor cells although immune responses aganst tumor associated antigen (TAA) might also occur. A critical experiment was then performed in an autologous system using autologous lymphocytes obtained from the same patient, E.D., from whose tumor tissue a melanoma target cell line (ED-H) originated (11). These lymphocytes were incubated, in vitro, with an anti-melanoma I-RNA extracted from the lymphoid organs of a sheep immunized with fresh melanoma cells and reacted with ED-H cells in vitro. The patient's lymphocytes incubated without RNA served as a control. When incubated with no RNA, the patient's lymphocytes were slightly cytotoxic to his tumor cells with a cytotoxicity index (CI) of 0.22 ± 0.05. However, when this patients' lymphocytes were incubated with anti-melanoma I-RNA the cytotoxicity of his lymphocytes tested against his own tumor cells more than doubled (CI+ 0.45 ± 0.02). This increase was significant at $p < 0.005$.

To derive additional evidence that immune responses directed specifically against human tumor-associated antigens could be mediated by I-RNA, a second experiment was performed which included additional controls (11). I-RNA was extracted separately from the lymphoid organs of two groups of guinea pigs. One group has been immunized with E.D.'s melanoma cells (anti-ED-Mel RNA), while the second had been immunized with E.D.'s autologous skin fibroblasts (anti-ED-Fib RNA). Each I-RNA preparation was incubated separately with normal, allogeneic lymphocytes (from a normal healthy donor) and with E.D.'s own autologous lymphocytes. Following incubation, each lymphocyte sample was tested for cytotoxic activity upon E.D.'s own melanoma target cells (ED-H cells). When effector cells which were allogeneic with respect to the target cells, were used, both anti-ED-Fib RNA and anti-Mel RNA were equally effective in mediating cytotoxic immune responses to ED-H melanoma target cells. However, when autologous lymphocytes were used as effector cells, only anti-ED-Mel RNA was active, and anti-ED-Fib RNA did not cause any increase in the cytotoxic activity of E.D.'s lymphocytes for autologous melanoma cells. We interpreted these results as follows: Both anti-ED-Mel RNA and anti-ED-Fib RNA induced allogeneic lymphocytes to effect immune responses directed against normal transplantation antigens (primarily HLA antigens). When these I-RNA preprations were incubated with autologous lymphocytes, I-RNA's directed against normal transplantation antigens were recognized as self by the lymphocytes and no immune responses against these antigens were elicited. Therefore, anti-ED-Fib RNA was inactive when incubated with autologous lymphocytes. When incubated with patient E.D.'s own autologous lymphocytes, only I-RNA's directed against foreign antigens (i.e. TAA) were capable of eliciting cytotoxic immune responses directed against autologous melanoma cells. I-RNA extracted from lymphoid organs of animals immunized with E.D.'s own normal fibroblasts, when incubated with E.D.'s autologous lymphocytes, failed to mediate cytotoxic immune responses against ED melanoma target cells, indicating a failure to elicit immune responses against self cell surface antigens.

Recently, in our laboratory, additional experiments were carried out which included normal fibroblast cell lines as target cells (15). These important controls, because of technical reasons, could not be included in our earlier experiments which had been performed exclusively with established tumor cell lines as target cells. Additional evidence that xenogeneic I-RNA can mediate immune response specifically against human TAA was obtained from these studies in two autologous systems. In these systems, malignant melanoma target cells, matching normal fibroblastic target cells, lymphocyte effector cells, and melanoma and normal skin tissue used to immunize RNA donor animals were derived from the same hosts. The target cells employed were from two malignant melanoma cell lines and their companion normal fibroblast cell lines which were established in our laboratory: (1) CA-H, a malignant melanoma explanted from a surgical specimen from patient C.A.; (2) CA-F, a culture of fibroblasts grown from biopsy of normal skin from patient C.A.; (3) RA-H, a malignant melanoma explanted from a surgical specimen from patient R.A., and (4) RA-F, fibroblasts grown from a biopsy of normal skin from patient R.A.

Target cell lines were established from both tumor and normal fibroblasts obtained from melanoma patient C.A. RNA donor animals were immunized either with melanoma tissue or with normal skin from this same patient. Patient C.A.'s peripheral blood lymphocytes were collected after his recovery from surgery and stored in liquid nitrogen.

When normal allogeneic lymphocytes from a healthy donor were incubated with I-RNA extracted from the lymphoid organs of a sheep immunized with C.A.'s melanoma tissue the CI of these lymphocytes for CA-H melanoma target cells more than doubled (from 0.25 to 0.55) and the CI of these lymphocytes for CA-F normal fibroblasts also doubled (from 0.30 to 0.65). The cytotoxicity of normal lymphocytes for CA-H melanoma target cells was also increased by incubation either with I-RNA from sheep immunized with tumor tissue or with RNA from guinea pigs immunized with normal skin from patient C.A. However, when autologous lymphocytes from patient C.A. were incubated with these same two RNA's, only I-RNA from the animal immunized with melanoma mediated immune reactions against CA-H melanoma cells. Aliquots of these autologous lymphocytes were tested, at the same time, on CA-F fibroblasts. Normal allogeneic lymphocytes became markedly more cytotoxic to CA-F target cells following incubation with RNA extracted either from animals immunized with melanoma or from animals immunized with normal tissue, whereas autologous lymphocytes from patient C.A. evidenced no increase in cytotoxicity for CA-F target cells following incubation with either of these RNA's. Incubation with RNA extracted from the lymphoid organs of guinea pigs immunized with complete Freund's adjuvant (CFA) only did not increase the cytotoxicity of normal lymphocytes for either CA-H or CA-F target cells.

Similar experiments were conducted in a second melanoma system when normal allogeneic lymphocytes were incubated with I-RNA extracted from guinea pigs immunized with normal skin tissue from patient R.A., the CI of these lymphocytes for RA-H melanoma target cells was increased from 0.08 to 0.42. When incubated with RNA from a sheep immunized with R.A.'s melanoma, the CI of these normal allogeneic lymphocytes increased to 0.45. In a second experiment, the cytotoxicity of autologous lymphocytes from patient R.A. against RA-H melanoma target cells was increased (from 0.30 to 0.56) following incubation with I-RNA from guinea pigs immunized with R.A.;s tumor tissue, but not by incubation with RNA from guinea pigs immunized with R.A.'s normal skin tissue.

In another experiment, the cytotoxicity of normal, allogeneic lymphocyte RA-H target cells was increased following incubation with RNA from guinea pigs immunized either with melanoma tissue or normal skin tissue from patient R.A. However, only RNA from animals immunized with melanoma induced autologous lymphocytes to display an increased cytotoxicity for RA-H melanoma target cells. When aliquots of these same lymphocytes were tested on RA-F fibroblasts, it was found that normal, allogeneic lymphocytes became markedly more cytotoxic for RA-F cells following incubation with RNA from guinea pigs immunized either with tumor tissue or with normal skin tissue from patient R.A. On the other hand, neither of these RNA;s converted autologous lymphocytes from patient R.A. to become cytotoxic for RA-F fibroblasts. It should be noted that RNA extracted from guinea pigs immunized with CFA only was inactive in all experiments.

The results of these studies provided corroborating evidence supporting our previous report (11) that xenogeneic anti-tumor I-RNA can mediate cytotoxic immune reactions directed specifically against human TAA. In two autologous systems it was shown that, when incubated with autologous lymphocytes, xenogeneic I-RNA extracted from the lymphoid organs of animals immunized with autologous melanoma tissue mediated cytotoxic immune reactions against autologous melanoma target cells but not normal, autologous fibroblasts. I-RNA from animals immunized with normal skin tissue from the autochthonous host did not increase the cytotoxicity of autologous lymphocytes for autologous tumor cells or fibroblasts.

It is important to note that, while the target cells for these assays were grown in medium containing fetal calf serum, the RNA donor animals were never immunized with tumor cells which had been exposed to fetal calf serum. (Fresh frozen tumor tissue or tumor cells grown in agamma human serum were used.) Thus, immune reactivity against neo-antigens related to fetal calf serum was ruled out.

Mediation of Cytotoxic Immune Reactions against Human Tumor Cells, in vitro, by Allogeneic I-RNA:

Additional evidence to strengthen our contention that I-RNA could mediate immune responses directed specifically against unique antigens associateed with human tumor cells was derived from yet another series of experiments involving a different approach to the problem of specificity (12). Allogeneic I-RNA was extracted from the peripheral blood lymphocytes of each of 5 cured melanoma patients (collected on the continuous-flow blood cell separator), 1 patient with colon carcinoma, 1 patient with hypernephroma, and 5 normal volunteers. Each I-RNA preparation was incubated with aliquots of a single sample of normal lymphocytes collected on the continuous-flow blood-cell separator from another healthy volunteer. Each aliquot was then tested for cytotoxic activity against ED-H melanoma target cells. All of the 5 I-RNA preparations extracted from lymphocytes of melanoma patients were effective in mediating cytotoxic immune responses against melanoma target cells. RNA from the lymphocytes of normal volunteers was invariably inactive, as was RNA's from the lymphocytes of the colon cancer patient and the hypernephroma patient. Presumably, the melanoma patients had been sensitised to melanoma-associated antigens in vivo, but not to normal histocompatibility antigens (all were male and none had been multiply transfused).

A similar experiment was carried out using HT-29M colon cancer cells as target cells. I-RNA's extracted from the peripheral blood lymphocytes of 3 out of 4 colon cancer patients incubated with normal allogeneic human lymphocytes effected cytotoxic immune reactions against colon cancer target cells. RNA's extracted from the peripheral blood lymphocytes of normal healthy volunteers were inactive. More importantly I-RNA's extracted from lymphocytes of patients with cancers other than colon cancer failed to elicit immune reactions directed against colon cancer target cells. These I-RNA's included two specimens from melanoma patients which had been found to be active in mediating immune responses against melanoma target cells. One of the I-RNA's from a colon cancer patient which was active against colon cancer target cells had previously been discovered to be inactive against melanoma cells.

I-RNA's extracted from the peripheral blood lymphocytes of two breast carcinoma patients, incubated with normal lymphocytes from a healthy volunteer, mediated immune reactions against BT-20 breast cancer target cells. RNA's extracted from the peripheral blood lymphocytes of normal donors, or from patients with cancers other than breast carcinoma (including a colon cancer I-RNA active against colon cancer and two melanoma I-RNA's active against melanoma), failed to mediate significant cytotoxic immune responses against breast cancer target cells. RNA extracted from the lymphocytes of a third breast cancer patient was inactive in this assay. A sucrose density gradient profile indicated that this particular RNA was almost completely degraded and was probably inactive as a result.

None of the cancer patients from whose peripheral blood lymphocytes these I-RNA's were extracted had been multiply transfused. All were male except for the breast cancer patients and none of the breast cancer patients were multiparous. It, therefore, seems unlikely that they had been sensitized to HL-A antigens or other normal human transplantation antigens except as cross reactivity might exist between human histocompatibility antigens and ubiquitous (e.g., bacterial) foreign antigens. We observed that only I-RNA's extracted from peripheral blood lymphocytes of melanoma patients mediated cytotoxic immune reactions against melanoma cells. Similarly, only I-RNA's extracted from the lymphocytes of colon cancer patients mediated cytotoxic immune responses against colon cancer target cells, and only I-RNA's from the lymphocytes of breast cancer patients mediated immune reactions against breast cancer target cells. In other words, cytotoxic immune reactions mediated by allogeneic I-RNA, extracted from the peripheral blood lymphocytes of cancer patients, were shown to be directed only against tumor target cells of the same histologic tumor type as the I-RNA donor. It appears that allogeneic I-RNA extracted from the peripheral blood lymphocytes of cancer patients, under the conditions of our experiments, mediates cytotoxic immune reactions which are directed against common tumor-associated antigens specific for the tumor type of the RNA donor.

Initial Clinical Trials of Immunotherapy with I-RNA:

The successful transfer of cell-mediated immune responses to human tumor-associated antigens in vitro by xenogeneic I-RNA provided a logical basis for trials of immunotherapy of human cancer with I-RNA. This form of immunotherapy offers several specific theoretical advantages First, large quantities of I-RNA can be produced without dependence upon human donors. Secondly, since histologically similar human tumors appear to share common, group-specific tumor-associated antigens, many patients with the same tumor type could be treated with I-RNA from an animal immunized with a single patient's tumor.

Because the accumulated laboratory studies strongly suggested that I-RNA could induce anti-tumor immune responses, and because of the many potential advantages of this form of immunotherapy, preliminary clinical trials were undertaken to examine the feasibility of I-RNA immunotherapy of human malignancy. Our clinical trials of adjuvant immunotherapy with I-RNA in patients with malignant melanoma and colorectal cancer were begun in September 1975. In our first clinical trial, 27 patients with malignant melanoma, at high risk for recurrence (Stage II or Stage I, Clark's Levels III, IV, V) following potentially curative surgical resection of all clinically detectable tumor (usually wide excision plus regional lymph node dissection) were included. Immunotherapy was initiated within 8 weeks of operation. All patients received 4mg of I-RNA weekly intradermally in multiple wheels of 0.1 ml near lymph node bearing areas (groins or axillae). Whenever possible, each patient received I-RNA from a sheep that had been immunized with his own, autologous tumor cells. Patients for whom autologous tumor tissue was not available, received I-RNA from a sheep immunized with allogeneic melanoma cells. If progression of disease was noted, I-RNA therapy was discontinued and alternative therapy instituted. If no recurrence developed, treatment was continued for two years and was then stopped. These patients were followed at 3 month intervals, for periods ranging between 20 and 66 months. The mean follow-up period is 49 months and the median follow-up period is 49 months. Fifteen patients eventually developed recurrent and/or metastatic disease. One patient developed a local recurrence at 10 months, at which time I-RNA treatment was discontinued. The recurrent tumor was resected and the patient is free of disease at 62 months. One patient had a local recurrence resected at 26 months and failed to return for follow-up. One patient developed metastases to the right axilla at 20 months. The patient underwent an axillary dissection and is free of disease at 43 months. Nine patients remain free of disease at 33,35,36,38,39,47, (2 patients), 49, and 63 months. The median survival rate for these patients is 29 months. In this pilot study, we do not have no treatment (control) group for statistical comparison. However, these patients did not fare significantly better than would have expected had they not received adjuvant therapy.

In a prospectively randomized controlled trial of adjuvant immunotherapy with I-RNA, 24 patients with Duke's B_2 or Duke's C colorectal cancer were studied following potentially curative surgery. Patients were eligible for admission to the study if, as determined by pathologic evaluation of resected specimen, their tumor involved all layers of the bowel, with or without positive nodes, or if their tumor did not penetrate through all layers of the bowel but one or more positive nodes were found in the resected specimen. Those patients with metastatic disease or in whom residual tumor was known to remain following surgery, were excluded from the study.

Twenty three patients were prospectively randomized into the study: 14 patients in the treatment group (surgery + I-RNA) and 9 patient in the control group (surgery only). In the treatment group, one protocol violation was discovered and hence excluded from the study two months after initiation of therapy. This patient had pulmonary metastases evident on pre-operative chest x-rays. Patients in the treatment group received 4 mg of I-RNA at weekly intervals, intradermally, in multiple wheels of 0.1 ml near lymph node bearing areas (groins or axillae). Treatment was initiated within 8 weeks of surgery and continued for 2 years unless recurrence or metastases developed, at which time I-RNA treatment was discontinued and appropriate alternate therapy instituted. All patients were followed at 3 month intervals. The follow-up period for patients in treatment group ranged from 23 months and 50 months. The mean and median follow-up periods are 36 and 34 months respectively. Eight patients receiving I-RNA treatment remain free of disease at 22,25,27,30,33,41, and 42 months. The median survival rate is 29 months.

Of the 13 determinate patients receiving I-RNA therapy, 3 received I-RNA from sheep immunized with autologous tumor tissue, while the others received I-RNA from sheep immunized with allogeneic colorectal carcinoma tissue. Four of these patients eventually developed recurrent and/or metastatic disease and died at 11,20,28, and 34 months. A fifth patient developed a local recurrence at 8 months, at which time I-RNA treatment was discontinued. The recurrent tumor was resected and the patient is free of disease at 27 months.

The follow-up period for the 9 patients in the control group ranges between 26 and 49 months. The mean and median follow-up period are 30 and 34 months respectively. Three of these patients developed metastases and died at 2,23, and 34 months following surgery. Six patients remain free of disease at 5,14,21,29,30, and 45 months. The median survival rate for the control group of patients is 23 months.

Even though there is a numerical difference in the median survival rates of patients in the treatment group (29 months) and the control group (23 months) this difference is not statistically significant when analyzed by the Mantel-Cox and Breslow tests.

Toxicity of the I-RNA has been absent or minimal. Total doses of up to 600 mg (over 36 months) and single doses as high as 60 mg have not resulted in significant side effects. A few patients reported mild, transient malaise and anorexia and 1 patient developed a transient low grade fever. No patient has experienced any local skin reactions and no allergic or anaphylactoid reactions have occurred. It is clear that sheep I-RNA, when prepared and administered intradermally in the doses and schedules described, is completely free of significant local or systemic toxicity and is very well tolerated. Many nonspecific immunoadjuvants currently utilized for the immunotherapy of cancer have significant local and/or systemic toxicity (e.g. BCG and Corynebacterium parvum), I-RNA, if proven effective, would constitute an attractive, nontoxic alternative agent for the immunotherapy of cancer.

Ramming and DeKernion (25) have treated 20 patients with metastatic hypernephroma (renal cell carcinoma) with I-RNA. This I-RNA was obtained from sheep which had been immunized with allogeneic human hypernephroma tissue. Weekly intradermal doses of 4-8 mg of I-RNA were given. Twelve patients had pulmonary metastases only while 5 had extrapulmonary metastases as well (to bone, brain or liver). One patient had a large mass of metastatic hypernephroma in superior mediastinal lymph nodes, proven by biopsy. Eight of 20 patients presented with metastases at the time of original diagnosis and underwent palliative nephrectomy prior to receiving I-RNA treatment. Eighty-six consecutive patients treated for metastatic hypernephroma at the same institution served as historical controls. Survival between subpopulations in each group, matched by computor according to extent and location of metastases, age, sex, and interval between nephrectomy and occurence of metastases, were compared by life table analysis. Survival was significantly greater in I-RNA-treated patients ($p < 0.05$) who had multiple metastases limited to the lungs, when compared with matched controls. I-RNA treatment did not influence survival of patients with metastases to other sites or multiple organ involvement. This group has recently initiated a prospectively randomized clinical trial of adjuvant immunotherapy with I-RNA in patients with primary renal cell carcinoma, who, following nephrectomy, are judged to have a high risk of recurrence (e.g., renal vein invasion or involvement of perinephric fat (Ramming, personal communication).

Steele, et al (29) have recently reported the results of I-RNA immunotherapy in patients with metastatic renal cell carcinoma. Their study group included 6 patients with metastatic renal cell carcinoma and a control patient with advanced malignant melanoma. Autologous lymphocytes, harvested from the patients by leukophoreses, were incubated immunized with autogenous tumor tissue, and were reinfused. Each patient received five such treatments every other day for 10 days. The control melanoma patient also received autologous lymphocytes incubated with renal cell carcinoma I-RNA. One renal cell carcinoma patient with biopsy proven pulmonary metastases had a complete remission which lasted 18 months, at which time he relapsed. He is presently alive over 2 years following treatment. Two patients evidenced partial regression of pulmonary metastases but subsequently relapsed, one at 4 months, and the other at 6 months. Stabilization of previously growing pulmonary lesions were noticed in two other patients. The sixth patient received only one treatment before serious morbidity from brain metastases precluded further therapy. All patients' lymphocytes showed increased in vitro cell mediated immune response only against renal cell carcinoma target cells.

Ishikawa, et al (7) have treated 14 patients undergoing "non-curative" resections for gastric carcinoma with ABO-Rh matched allogeneic leukocytes pre-incubated with I-RNA extracted from the lymphoid organs of rabbits immunized with a crude antigen extract prepared for autologous tumor tissue. The RNA-incubated leukocytes (0.6×10^9 - 1.2×10^9 cells) were infused intravenously 3 times at 3 day intervals. Two of the patients who were so treated survived free of disease for more than 5 years and are presumed cured.

Shanfang, et al (28) recently reported treating 17 patients with advanced lung cancer with xenogeneic Immune RNA extracted from the lymphoid organs of sheep immunized with autogenous lung carcinoma cells. The I-RNA was incubated with autologous exudate cells from chemically-induced skin blisters and the mixture injected intradermally into the same patients from whom the tumor cells used to immunize the sheep had been obtained. They reported that this therapy was "effective" in 10 of the 17 patients, "uncertain" in 4 patients who received other forms of treatment concurrently, and "ineffective" in 3 cases. These authors do not specify their criteria for an "effective" response but note that one of their patients had a 50% reduction in the size of his tumor which had been sustained for more than 14 months.

Thus far, based on the observations of Ramming, et al., (25) and Steele, et al, (29) immunotherapy with I-RNA, appears most promising in the treatment of patients with metastatic renal cell carcinoma. Results in this group of patients seem encouraging. Controlled clinical trials are necessary in order to determine whether any form of I-RNA immunotherapy can be proven to be established clinical benefit.

References:

1. Alexander, P., Delorme, E.J., Hamilton, L.D., and Hall, J.G.: Effect of nucleic acids from immune lymphocytes on rat sarcomata. Nature 213: 569-572 (1967).
2. Deckers, P.J., Wang, B.S., Stuart, P.A. and Mannick, J.A.: The augmentation of tumor-specific immunity with immune RNA. Transplantation Proc. 7: 259-263 (1975).
3. Deckers, P.J. and Pilch, Y.H.: Mediation of immunity to tumor specific transplantation antigens by RNA: Inhibition of isograft growth in rats. Cancer Res. 32: 839-846 (1972).
4. Deckers, P.J., and Pilch, Y.H.: Transfer of immunity to tumor isografts by the systemic administration of xenogeneic 'immune' RNA. Nature new Biol. 231: 181-183 (1971).
5. Fukushima, M., Machida, S., Hokama, A., Kojika, M., Nishikawa, T., Kikuchi, A., and Ishikawa, Y.: Passive transfer of the resistance to tumor with RNA. Tohuku J. Exp. Med. 112: 115-163 (1974).
6. Greenup, C.J., Vallera, D.A., Pennline, K.J., Kolodziej, B.J. and Dodd, M.C.: Anti-tumor cytotoxicity of poly(A)-containing messenger RNA isolated from tumor-specific immunogeneic RNA. Br. J. Cancer 38: 55-63 (1978).
7. Ishikawa, Y., Fukushima, M., Machida, S., Kakuta, H., Nishikawa, T., and Fujuda, S.: Specific cancer immunotherapy adjunctive to surgery for advanced gastric cancer: A report of six long term survivors. Gastroenterologia Japonica 13: 54-60 (1978).

8. Kennedy, C.T.C., Cater, D.B., and Hartveit, F.,: Protection of C3H mice against BP-8 tumor by RNA extracted from lymph nodes and spleens of specifically sensitized mice. Acta Pathol. Microbiol. Scand. 77: 796-799 (1969).

9. Kern, D.H. and Pilch, Y.H.: Immune cytolysis of murine tumor cells mediated by xenogeneic Immune RNA. Int. J. Cancer 13: 679-688 (1974).

10. Kern, D.H., Drogemuller, C.R. and Pilch, Y.H.: Immune cytolysis of rat tumor cells mediated by syngeneic Immune RNA. J. Natl. Cancer Inst., 52: 299-302 (1974).

11. Kern, D.H., Fritze, D., Drogemuller, C.R., and Pilch, Y.H.: Mediation of cytotoxic immune responses against human tumor-associated antigens by xenogeneic immune RNA. J. Natl. Cancer Inst. 57: 97-103 (1976).

12. Kern, D.H., Fritze, D., Schick, P.M., Chow, N., and Pilch, Y.H.: Mediation of cytotoxic immune responses against human tumor-associated antigens by allogeneic immune RNA. J. Natl. Cancer Inst. 57: 105-109 (1976).

13. Kern, D.H., deKernion, J.B. and Pilch, Y.H.: Intracellular localization of anti-tumor "Immune" RNA. Cell. Immunol. 22: 11-18, (1976).

14. Kern, D.H., Chow, N. and Pilch, Y.H.: Kinetics of synthesis and immunologically active fraction of anti-tumor Immune RNA. Cell. Immunol. 24: 58-68 (1976).

15. Kern, D.H., Drogemuller, C.R., Chow, N., and Pilch, Y.H.: Specifici
of antitumor immune reactions mediated by xenogeneic immune RNA. J. Natl. Cancer Inst. 58: 117-121 (1977).

16. Kern, D.H., Chow, N., and Pilch, Y.H.: Lymphocyte populations participating in cellular antitumor immune responses mediated by immune RNA. J. Natl. Cancer Inst. 60: 335-344 (1978).

17. Kern, D.H., and Pilch, Y.H.: Induction of anti-tumor antibody synthesis in vitro by xenogeneic Immune RNA. J. Natl. Cancer Inst. 60: 599-503, (1978).

18. Londner, M.V., Morini, J.C., Font, M.T. and Rabasa, S.L.: RNA-induced immunity against a rat sarcoma. Experentia 24: 598-599 (1968).

19. Mannick, J.A. and Edgahl, R.H.: Transfer of heightened immunity to skin homographs by lymphoid RNA. J. Clin. Invest. 43: 2166-2177 (1964).

20. Pilch, Y.H., Fritze, D., and Kern, D.H.: Mediation of immune respo
to human tumor antigens with 'immune' RNA. Clinical tumor immunolo
pp 169-190, Pergamon Press, Oxford (1976).

21. Pilch, Y.H., Fritze, D., DeKernion, J.B., Ramming, K.P. and D.H.: Immunotherapy of cancer with immune RNA in sanimal models and cance
patients. Ann. N.Y. Acad. Sci 277: 592-608 (1976).

22. Pilch, Y.H., Fritze, D., Walsman, S.R., and Kern, D.H.: Transfer of anti-tumor immunity by 'immune' RNA. Curr. Top. Microbiol. Immu
72: 157-190 (1975).

23. Pilch, Y.H. and Ramming, K.P.: Transfer of tumor immunity with ribonucleic acid. Cancer 26: 630-637 (1970).

24. Pilch, Y.H., Veltman, L.L., and Kern, D.H.: Immune cytolysis of human tumor cells mediated by xenogeneic 'immune' RNA: Implications for immunotherapy. Surgery, St. Louis 76: 23-24 (1974).

25. Ramming, K.P. and Dekernion, J.B.: Immune RNA therapy for renal cell carcinoma: Survival and immunologic monitoring. Ann. Surg. 186: 459-467 (1977).

26. Ramming, K.P., and Pilch, Y.H.: Mediation of immunity to tumor isografts in mice by heterologous ribonucleic acid. Science 168: 492-493 (1970).

27. Ramming, K.P. and Pilch, Y.H.: Transfer of tumor-specific immunity with RNA: Inhibition of growth of murine tumor isografts. J. Natl. Cancer Inst. 45: 735-750 (1971).

28. Shanfang, W., Meiping, Y., Huiming, T. and Ching, Y.: Clinical trial of the treatment of lung cancer with xenogeneic Immune RNA. Chinese J. Cancer 1: 150-155, 1979.

29. Steele, G., Jr., Wang, B.S., Richie, J.P., Erivin, T., Yankee, R. and Mannick, J.A.: Results of xenogeneic I-RNA therapy in patients with metastatic renal cell carcinoma. Cancer, in press.

30. Veltman, L.L., Kern, D.H. and Pilch, Y.H.: Immune cytolysis of human tumor cells mediated by xenogeneic "Immune" RNA. Cell. Immunol. 13: 367-377, (1974).

31. Wang, B.S. and Deckers, P.J.: Augmentation of concomitant tumor immunity with RNA. J. Surg. Res. 20: 183-194 (1976).

BIOLOGICAL AND PHYSICO-CHEMICAL PROPERTIES OF ANTITUMOR RNA*

R. Pottathil
Department of Pediatrics,
Duke University Medical Center,
Durham, North Carolina 27710
H.G. Bedigian , L. Shultz , and H. Meier
The Jackson Laboratory
Bar Harbor, Maine 04609

__Abstract__ It has been nearly two decades since the first experiments were reported that subcellular extracts derived from lymphoid tissues of immunized animals may transfer immune responsiveness to non-immune recipients (1-3). Also, another moiety called "dialyzable transfer factor" isolatable from lymphoid cells of individuals possessing immunity against certain antigens have been shown to transfer immunity to immunodeficient individuals (4,5). Subsequently, RNA _per se_ or RNA protein complexes have been found involved in several humoral and cell mediated immune reactions including tumor immunity (6-9).

We have previously reported the stimulation of lymphocytes in vivo as well as regression of a transplantable chemically induced fibrosarcoma (in Swiss/ICR mice) following intravenous administration of tumor cell extracts (10-12). Subsequent studies have shown that similar antitumor activities were also present in embryo cell extracts and that the active factor is an RNA-rich dialyzable fraction of cellular cytosol (13). Also, this fraction was embryotoxic when injected into pregnant mice.

---

* This work was supported in part by Training Grant CA 05013 from The National Cancer Institute and Research Contract NO1 CP 33255 Within The Virus Cancer Program of The National Cancer Institute. The work was done at the Jackson Laboratory, Bar Harbor, Maine. The Jackson Laboratory is fully accrediated by The American Association for Laboratory Animal Care.

In this paper we shall try to summarized the source and procedure for isolation and purification as well as some of the biological and physicochemical characteristics of the antitumor RNA.

Certain dialyzable RNAs isolated from murine tumors and embryos were shown to have source and strain-specific antitumor and embryo-toxic effects in vivo. A single i.v. injection of 15-30 μg of this RNA induced necrosis, hemorrhages and in some cases regression of solid tumors in the strain of origin. Its presence in both tumors and embryos suggests that it may result from re-expression of embryonic genes in tumor cells. Also, the RNA was shown to be immunogenic and blastogenic in its strain of origin. Preliminary studies indicate that the active moiety might be a double stranded RNA with approximate molecular weight of 6000. The exact mechanism by which these molecules bring about biological effects is still unknown.

INTRODUCTION:

It has been nearly two decades since the first experiments were reported that subcellular extracts derived from lymphoid tissues of immunized animals may transfer immune responsiveness to non-immune recipients (1-3). Also, another moiety called "dialyzable transfer factor" isolatable from lymphoid cells of individuals possessing immunity against certain antigens have been shown to transfer immunity to immunodeficient individuals (4,5). Subsequently, RNA per se or RNA protein complexes have been found involved in several humoral and cell mediated immune reactions including tumor immunity (6-9).

We have previously reported the stimulation of lymphocytes in vivo as well as regression of a transplantable chemically induced fibrosarcoma (in Swiss/ICR mice) following intravenous administration of tumor cell extracts (10-12). Subsequent studies have shown that similar antitumor activities were also present in embryo cell extracts and that the active factor is an RNA-rich dialyzable fraction of cellular cytosol (13). Also, this fraction was embryotoxic when injected into pregnant mice.

In this paper we shall try to summarize the source and procedure for isolation and purification as well as some of the biological and physicochemical characteristics of the antitumor RNA.

SOURCE OF RNA:

RNA was isolated from both spontaneous and transplanted tumors and embryonal tissues of inbred strains of A/J, AKR/J, C3H/HeJ, CBA/J, C57L/J and DBA/2J mice. In some experiments we also used a rabbit Whilm's tumor as the source of RNA. We used the spontaneous thymic leukemia of AKR/J mice and mammary adenocarcinoma of C3H/HeJ females as both RNA source and test system. Among transplanted tumors there were several different histological types i.e., the

BW 7756 hepatocellular carcinoma of C57L/J, BW10139 rhabdomyosarcoma of CBA/J, CaD2 mammary adenocarcinoma of DBA/2J, 6C3HED lymphosarcoma of C3H/HeJ and Sal fibrosarcoma of A/J mice. These tumors were obtained from the Jackson Laboratory Tumor Bank as subcutaneous transplants in 5-6 week-old female mice. Fetal tissues were obtained from timed-pregnant mice, 10-12 days postcoitus. For comparison similar preparations were also made from normal liver tissues of these strains of mice.

Procedure for isolation of antitumor moieties:

Steps involved in the isolation are summarized in Fig. 1. All isolation procedures were done at 4°C unless otherwise stated. For a standard preparation 10 gms of tumor tissue or 1 gm of embryonic tissue were homogenized with glass-teflon grinder in 10 ml of RPMI 1640 medium with a motor driven pestle (40 strokes average). The homogenate was then sonically disrupted and centrifuged at 10,000xg in a Beckman ultracentrifuge with a Type 30 rotor. The supernatant fluid was then centrifuged at 105,000xg for 2 hours with a Type 30 or Type 40 rotor. The supernatant was then dialyzed against RPMI 1640 medium, Tris-HCl buffer buffer pH 8.1 or TEAB buffer pH 7.0 for 24 hours using 12,000 to 14,000 molecular weight cutt-off membrane. The dialysate was then concentrated to a final volume equivalent to 1 gm tissue to 1 ml using an Amicon 8MC ultrafilter attached with UM2 membrane (1000 M.W. cut-off) or lypophilized and reconstituted in the case of TEAB buffer (evaporating buffer). The concentrated dialysate was then fractionated on a Sephadex G 25 column (3x60 cm) calibrated with Tris:HCl buffer pH 8.1 or TEAB pH 7.0. At a flow rate of 6-7 ml/hour, 100 drops fractions (6.5 ml) were collected and monitored for U.V. absorption at 254nm and 280nm in a Beckman Model ACTA III spectrophotometer. Molecular weight markers included Dextran 2000 (200,000 dalton), RNase B (13,500 dalton), RNase Tl (11,500 daltons) polyglycine (6000 dalton) and Phenol red (350 daltons).

PROCEDURE FOR ISOLATION

TUMOR TISSUE / EMBRYOS

↓

HOMOGENATE (1 gm : 10 ml PRMI 1640)

↓

CENTRIFUGED AT 10,000xg FOR 60 MINUTES

SUPERNATANT ← → SEDIMENT (discard)

CENTRIFUGED AT 105,000xg FOR 2 HOURS

SUPERNATANT ← → SEDIMENT (discard)

↓

DIALYSED 24 HOURS AGAINST RPMI 1640 using 12,000 MW cut-off membrane

↓

DIALYSATE CONCENTRATED TO 1/50th VOLUME IN ULTRAFILTER (1000 MW cut-off)

↓

CHROMATOGRAPHY ON SEPHADEX G 25- eluted with 50mM Tris:HCl,pH 8.1.

↓

FRACTIONS WITH MAXIMUM ABSORPTION AT 254nm COLLECTED AND CONCENTRATED

↓

DIGEST WITH PRONASE FOR 6 HOURS

↓

RNA EXTRACTED BY THE COLD PHENOL METHOD

↓

ETHANOL PRECIPITATED ——— RECONSTITUTED-RNase A DIGESTED FOR 6 HOURS

↓

RE-EXTRACTED WITH PHENOL - ETHANOL PRECIPITATED - RECONSTITUTED

↓

ANTITUMOR FACTOR (RNA?)

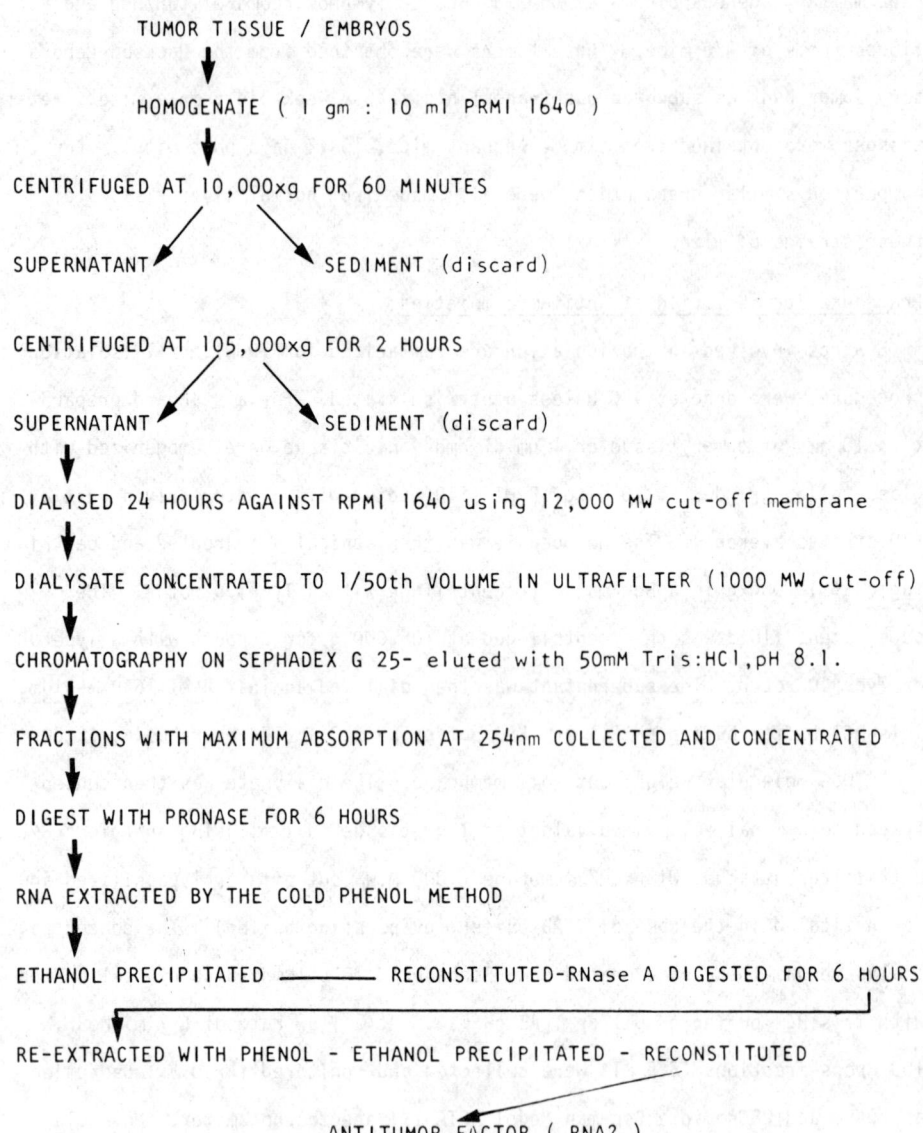

Fig. 1 Protocol for the isolation of antitumor RNA

Chromatographic profile of dialysates whether from tumors or embryos were similar (Fig. 2,3). The fractions 13-15 generally had the highest U.V. absorption, however, only fractions 11 and 12 yielded antitumor activity, all other fractions were ineffective (Table 1). In comparison with molecular weight markers, the approximate molecular weight of the active moiety was estimated to be around 6000 daltons.

The active fractions were further concentrated in an Amicon ultrafilter in the case of Tris-HCl buffer or lyophilized in the case of TEAB buffer and were digested with Proteinase K (100 μg/ml for 6 hours at 37°C) and RNA was extracted by cold phenol extraction procedure. Briefly, protease digested fractions were reconstituted in 2 ml 10 mM Tris-HCl-0.3 M NaCl- 1 mM EDTA pH 7.6. 1 ml of phenol-10% mCrysol mixture was added, followed by the addition of 1 ml of 24:1 chloroform -isoamyl alcohol. After 5 minutes vigorous mixing, aqueous phase was recovered by low-speed centrifugation. The aqueous phase was re-extracted three times as above and was extracted with 5 ml of diethyl ether. After removing the ether layer, the excess ether was evaporated by nitrogen bubbling. The procedure was repeated three times to ensure the complete elimination of possible traces of phenol. RNA in the aqueous phase was precipitated with three volumes of 95% pre-chilled ethyl alcohol for 18 hours at -20°C. The precipitate was recovered by centrifugation at 15,000xg for 60 minutes at 4°C. The precipitate was reconstituted in PBS or RPMI 1640 medium and stored frozen until tested for further analysis or biological activity. The amount of RNA was determined spectrophotometrically or spectroflyorometrically with yeast RNA as standard (14). 1 gm of tumor tissue gave a yield of approximately 0.0005%.

BIOASSAY OF ANTITUMOR RNA:

Spontaneous or transplanted tumor bearing mice were given single or multiple intravenous injections (15-30 μg of RNA in 0.2 ml RPMI 1640 medium/mice/

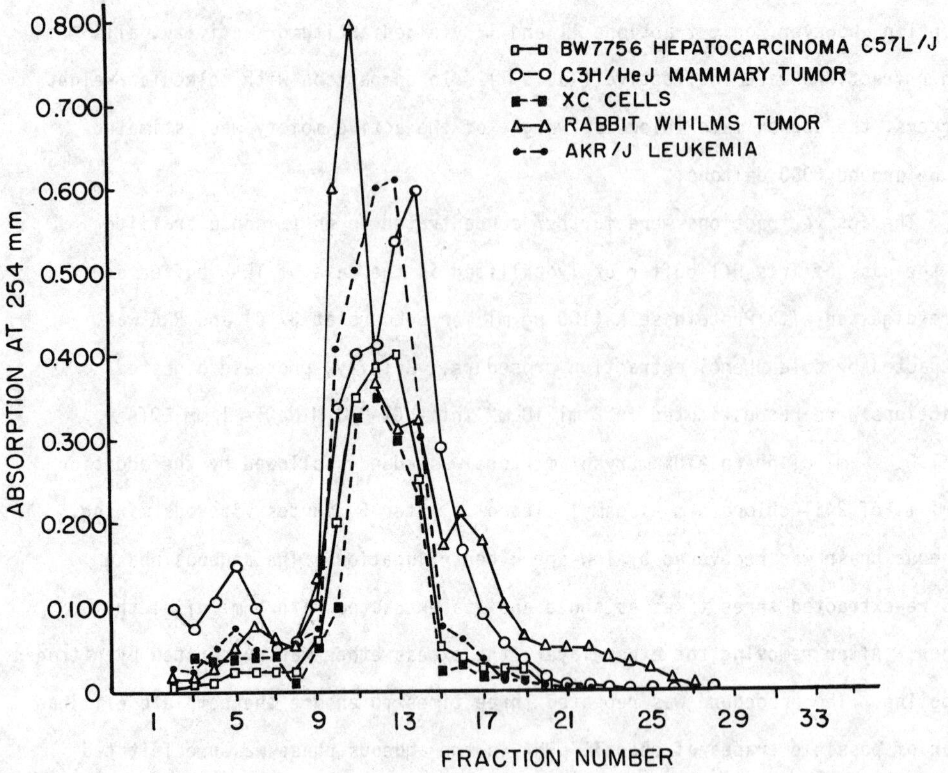

Fig. 2 Sephadex G25 column chromatographic profile of various cell dialysates. The columns were calibrated and eluted with Tris-HCl, pH 8.1. The flow rate was 6.5 ml/hour, each fraction contained 6.5 ml.

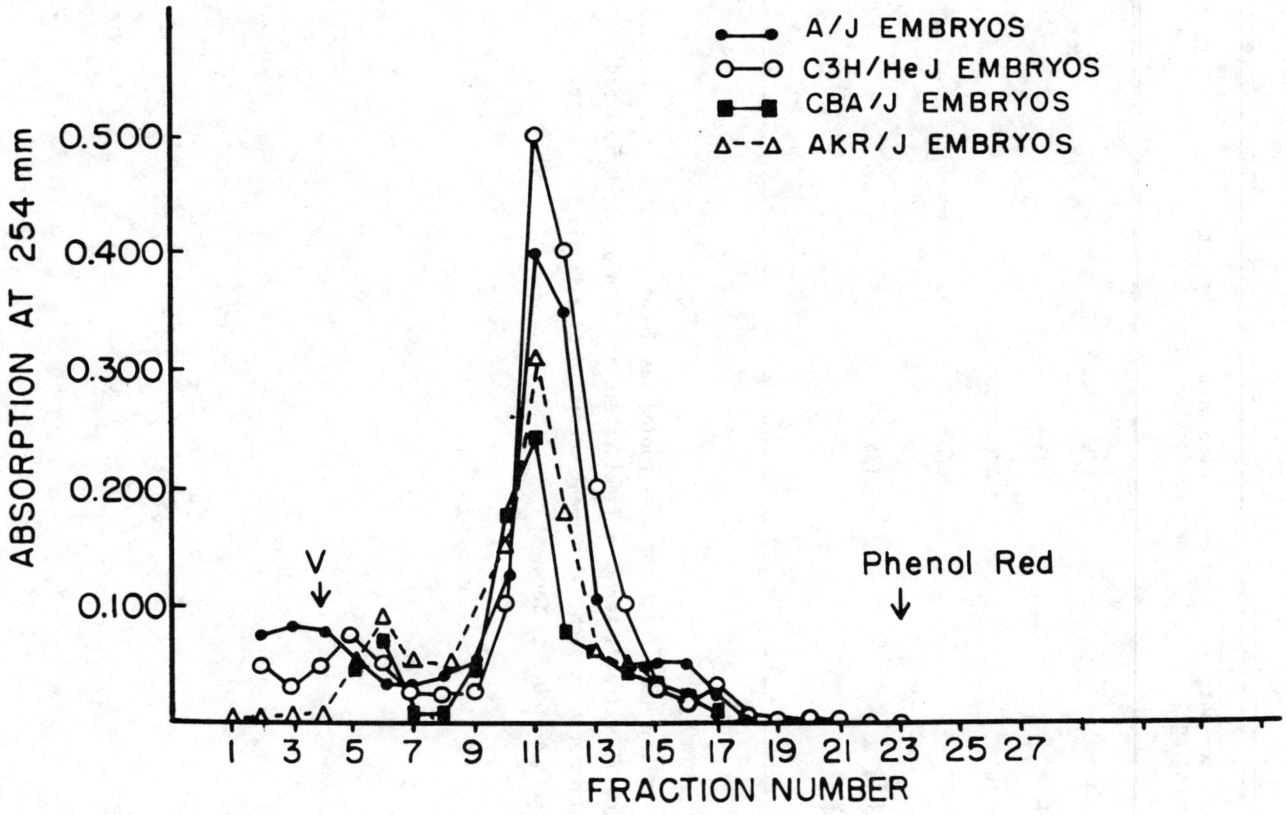

Fig. 3 Sephadex G25 column chromatographic profile of dialysates from embryos. The details of filtration were exactly the same as described for Fig. 2.

TABLE 1

BIOLOGICAL TESTING FOR ANTITUMOR ACTIVITY OF CHROMATOGRAPHIC FRACTIONS FROM SPONTANEOUS C3H/HeJ MAMMARY TUMORS

Chromatographic fraction	Number of mice with tumor regression	Results
1-4	0/6	- -
4-8	0/6	- -
9+10	0/6	- -
11+12	20/20	Extensive evidence of tumor regression
13-15	1/6	- -
24-28	0/6	- -

*Chromatographic fractions were pooled (as shown above) and concentrated using UM2 ultra filter to a final volume of 100 drops to 0.5 ml and administered as such. Treatment consisted of a single intravenous injection of 0.2 ml factor.

injection) of RNA preparations from tumors of the same origin (homologous), or unrelated origin (heterologous), embryos of same strain or unrelated origin, similar preparations from normal liver tissues of same or different strain or 0.2 ml of RPMI 1640 medium. The samples were coded in most of the experiments. Tumor growth was monitored by caliper measurements. Some mice were sacrificed periodically and tumor and visceral organs were processed for histopathological evaluation. In order to obtain unbiased data, histological preparations were always coded.

Spontaneous Tumors:

Leukemia in AKR/J mice:

Leukemic AKR/J mice were selected from a large colony of retired breeders and given intravenous injection of 15 μg of RNA prepared from AKR/J leukemic tissues, AKR/J embryos, AKR/J normal liver, C3H/HeJ mammary tumor, CeH/HeJ embryos or 0.2 ml of RPMI 1640 medium. In an attempt to enhance the biological activity some mice were also given normal spleen cells (intraperitoneal) together with RNA preparations.

Various degrees of necrosis, necrobiosis, karyolysis and hemorrhages were induced in target organs with RNA from leukemic tissues of AKR/J embryos, whereas RNA's from other sources were ineffective (Fig. Ia, Table 2). Survival curves of retired frmale AKR/J breeders treated with AKR/J leukemic cell RNA is given in Fig. 4.

Spontaneous mammary tumor in C3H/HeJ:

C3H/HeJ retired breeders bearing spontaneous mammary tumor of at least 1 cm mean diameter were given single intravenous injection of 15 μg of RNA prepared from C3H/HeJ mammary tumor, C3H/HeJ embryos, C3H/HeJ liver, AKR/J leukemic tissue, A/J embryos or 0.2 ml of RPMI 1640. Treatment of mice with RNA isolated from spontaneous mammary tumor tissues of C3H/HeJ embryos resulted in reduction

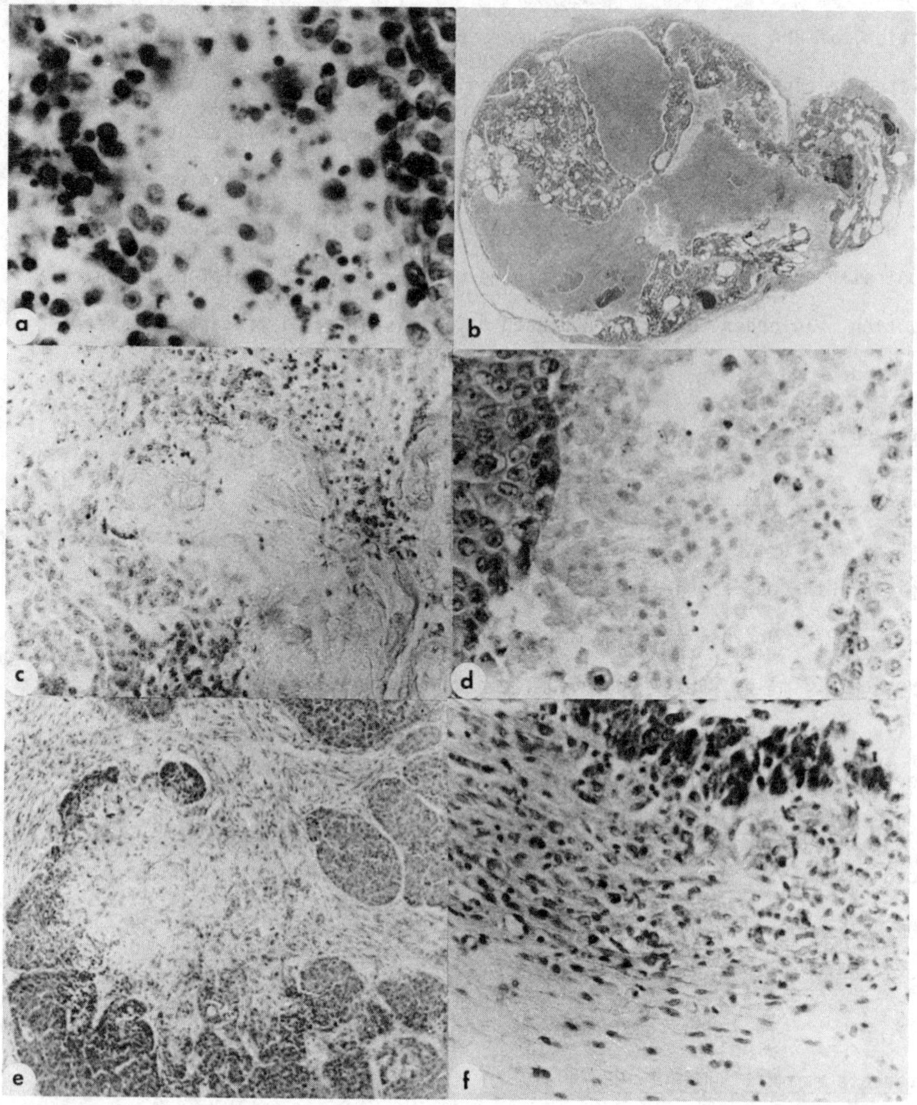

Fig. I 1a. Lymphnode from AKR/J leukemic mouse treated with homologous tumor RNA. Cytolysis, karyolysis and karyorhexis are evident 24 hours post-treatment. H&E, X 800.

1b. Spontaneous mammary adenocarcinoma in C3H/HeJ mouse, 24 hours after treatment with 15 μg of homologous tumor RNA. The tumor is blood filled and hemorrhagic. H&E, X 350.

1c. Tumor cell necrosis in spontaneous mammary adenocarcinoma bearing C3H/HeJ mouse, 24 hours after treatment with homologous tůmor RNA. H&E, X 900.

1d.-f. Stages of fibrosis and scar tissue formation in C3H/HeJ mouse bearing mammary daenocarcinoma treated with homologous tumor RNA, 7 days post-treatment. H&E, (d) X 800,,(e) X 150, ,(f) X 450.

(Figure reduced to $\frac{7}{10}$ during printing.)

TABLE 2

EFFECT OF LEUKEMIC CELL DIALYSATES ON AKR/J LEUKEMIA[1]

Source of extract	Number of treatments[2]	Number of mice with histological evidence of tumor suppression[4] (Days after first injection of extract) 1-10	10-30	30-40	40-60	60-80	>80
AKR/J leukemic lymphoid organs	1	17/17	5/5	6/8	-	-	-
AKR/J leukemic lymphoid organs	5	-	4/5	4/6	2/8	3/12	0/9
AKR/J leukemic lymphoid organs plus normal spleen cells[3]	5	-	5/5	-	4/8	-	-
AKR/J embryos	5	13/13	6/7	-	-	-	-
AKR/J normal liver	5	0/15	0/15	-	-	-	-
C3H/HeJ mammary tumor	5	0/12	0/8	-	-	-	-
C3H/HeJ embryos	5	-	0/10	0/8	-	-	-
RPMI 1640	5	-	0/10	-	-	-	-

[1] All mice in this experiment were selected from a large colony of AKR/J retired breeders and had obvious signs of leukemia.

[2] Each treatment consisted of microsomal dialysate estimated to contain 15 μg RNA/0.2 ml.

[3] Cells from a whole spleen were given per mouse as intraperitoneal injection.

[4] All specimens were coded before histological observation.

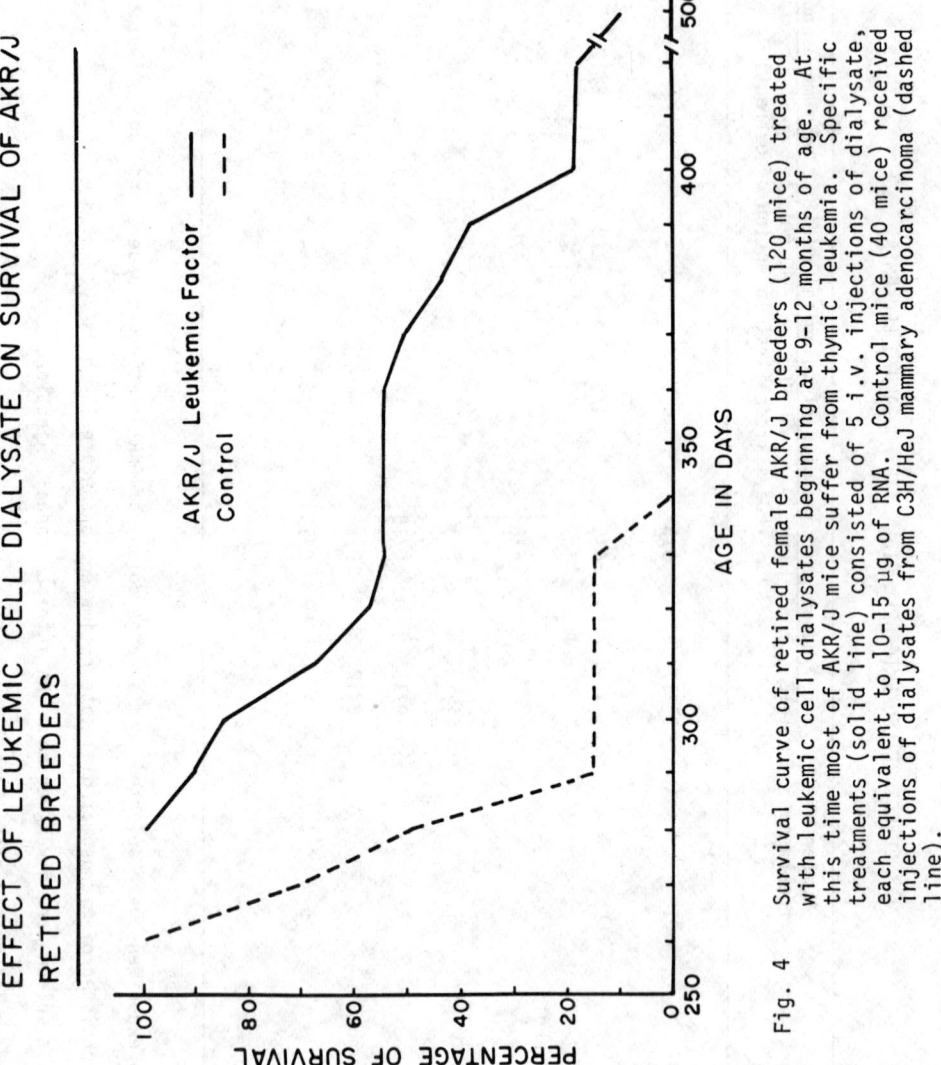

EFFECT OF LEUKEMIC CELL DIALYSATE ON SURVIVAL OF AKR/J RETIRED BREEDERS

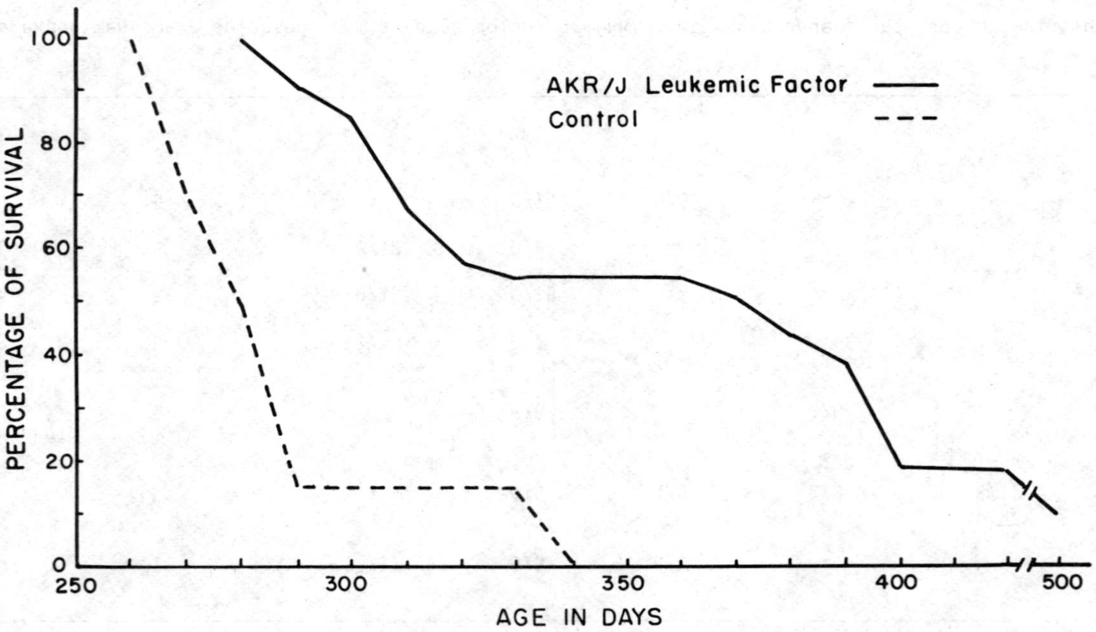

Fig. 4 Survival curve of retired female AKR/J breeders (120 mice) treated with leukemic cell dialysates beginning at 9-12 months of age. At this time most of AKR/J mice suffer from thymic leukemia. Specific treatments (solid line) consisted of 5 i.v. injections of dialysate, each equivalent to 10-15 μg of RNA. Control mice (40 mice) received injections of dialysates from C3H/HeJ mammary adenocarcinoma (dashed line).

of tumor sizes and in some cases complete tumor regression. Treatments were less effective if the tumor at the outset was larger than 1.0 cu cm (Fig. 5). Tumor necrosis and hemorrhages occurred as early as 24 hours after treatment (Fig. I b to d) within one week fibrosis and scar tissue formation appeared (Fig. I e-f). Treatments with RNA preparations from homologous liver or heterologous tumor or embryo sources were without effect (Table 3).

Transplanted tumors:

Hepatocellular carcinoma in C57L/J: Five/six-week-old C57L/J mice after five days of subcutaneous transplant of BW 7756 tumor were given 30 μg of RNA prepared from BW 7756 tumor, C3H/HeJ mammary tumor or 0.2 ml of RPMI 1640. Treatment with RNA from homologous tumor RNA but not with control preparations resulted in extensive hemorrhage and reduction of transplanted tumor size (Table 4).

Rhabdomyosarcoma in CBA/J: CBA mice bearing BW 10139 rhabdomyosarcoma were given 30 μg of RNA prepared from BW 10139 tumor, AKR/J leukemic tissues or RPMI 1640 seven days after tumor transplantation. Treatments with RNA prepared from BW 10139 resulted in complete regression of the transplant. Control preparations did not induce tumor regression (Table 4, Figs. 6 and 7).

Mammary adenocarcinoma in DBA/2J: Transplanted CaD2 adenocarcinoma bearing DBA/2J mice were given 30 μg of RNA prepared from CaD2, DBA/2J embryos, BW 10139 tumor or 0.2 ml of RPMI 1640 medium seven days after transplantation. Treatments with CaD2 tumor-RNA as well as DBA/2J embryo-RNA resulted in significant reduction in the growth of tumor transplant. Also, the survival of these mice were significantly prolonged as compared to heterologous RNA treated or untreated controls (Table 4, Fig. 8).

Lymphosarcoma in C3H/He/J: Transplanted lymphosarcoma (6C3HED) bearing C3H/HeJ mice were treated with 30 μg of RNA prepared from 6C3HED tumor, C3H/HeJ embryos, CaD2 mammary adenocarcinoma, C3H/HeJ mammary tumor or 0.2 ml of RPMI 1640 medium. Selected mice were sacrificed periodically to weigh the tumor mass

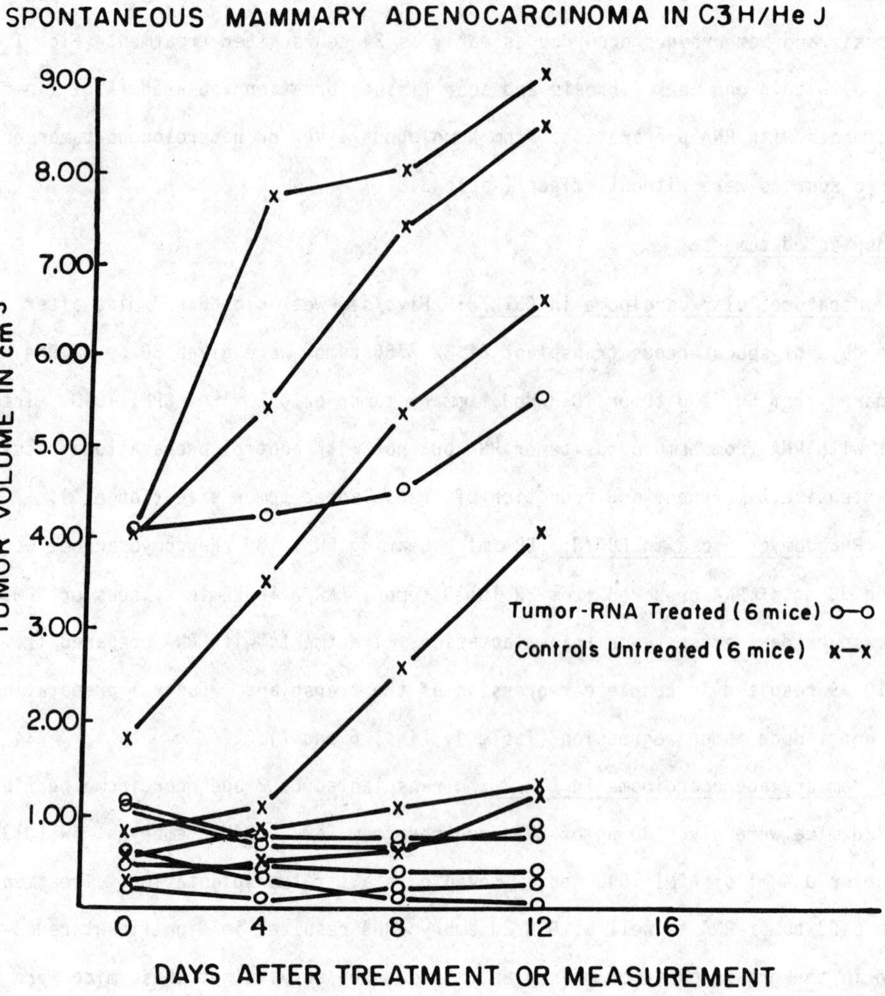

Fig. 5 Tumor growth in spontaneous mammary adenocarcinoma - bearing C3H/HeJ mice given single i.v. injection of 30 μg of RNA prepared from C3H/HeJ mammary tumor (o——o) or untreated (x——x).

TABLE 3

EFFECT OF RNA ON SPONTANEOUS C3H/HeJ MAMMARY ADENOCARCINOMAS

Source of RNA	No. mice with evidence of tumor suppression	Histological observations[2]
C3H/HeJ mammary	20/20[1,3]	Extensive necrosis, necrobiosis, hemorrhages and fibrosis
C3H/HeJ embryos	10/10[3]	Extensive necrosis, necrobiosis, hemorrhages and fibrosis
C3H/HeJ liver	0/10	Some hemorrhages
AKR/J leukemic tissue	1/10	Necrosis
A/J embryos	2/10	Necrosis
RPMI 1640 medium	0/10	No apparent effect
No treatment	0/10	No apparent effect

[1]Treatment consisted in each case of 10-15 μg of RNA in 0.2 ml RPMI 1640 medium.

[2]All tissue specimens were coded for histological evaluation.

[3]$P < 0.01$ when compared to control groups.

TABLE 4

EFFECT OF RNA ON TRANSPLANTED SOLID TUMORS

Tumor[1] and host strain	Source of RNA and treatment[2]	No. mice with evidence of tumor suppression[3]
BW 7756, C57L/J	BW 7756	10/10
	C3H/HeJ mammary tumor	0/10
	RPMI 1640	0/10
BW 10139, CBA/J	BW 10139	20/20
	AKR/J leukemic tissue	0/20
	RPMI 1640	0/20
CaD2, DBA/2J	CaD2	20/20
	DBA/2J embryo	20/20
	BW 10139	0/20
	RPMI 1640	0/20
6C3HED, C3H/HeJ	6C3HED	10/10
	C3H/HeJ embryo	10/10
	CaD2	1/10
	RPMI 1640	0/10
	C3H/HeJ mammary tumor[4]	0/10

[1]The tumor transplants were about 1.5 - 2 cm in diameter at the time of treatment.

[2]Each treatment consisted of a single caudal vein injection of 30 μg of RNA in 0.2 ml RPMI 1640.

[3]Reduction in tumor size or complete elimination of the transplant.

[4]In this group the tumor weight remained as high as in the controls but extensive tumor cell necrosis was observed in all mice.

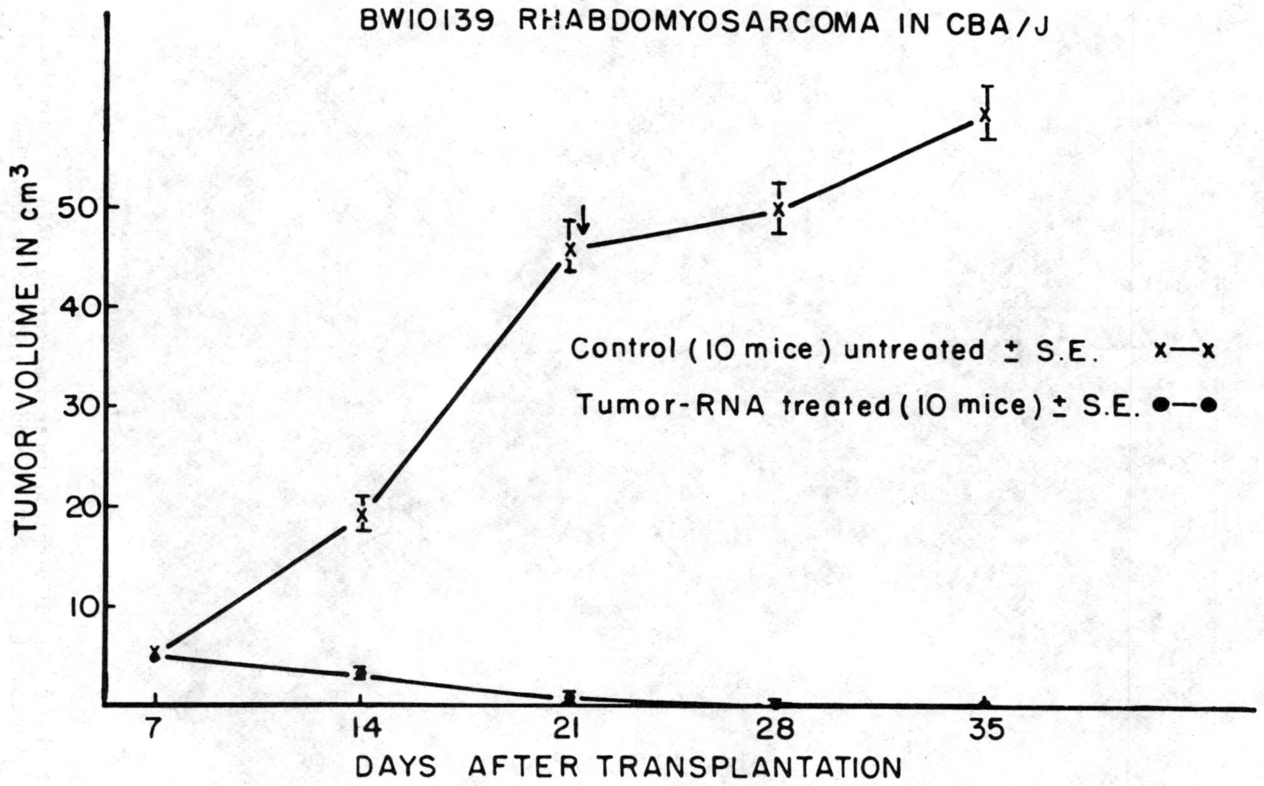

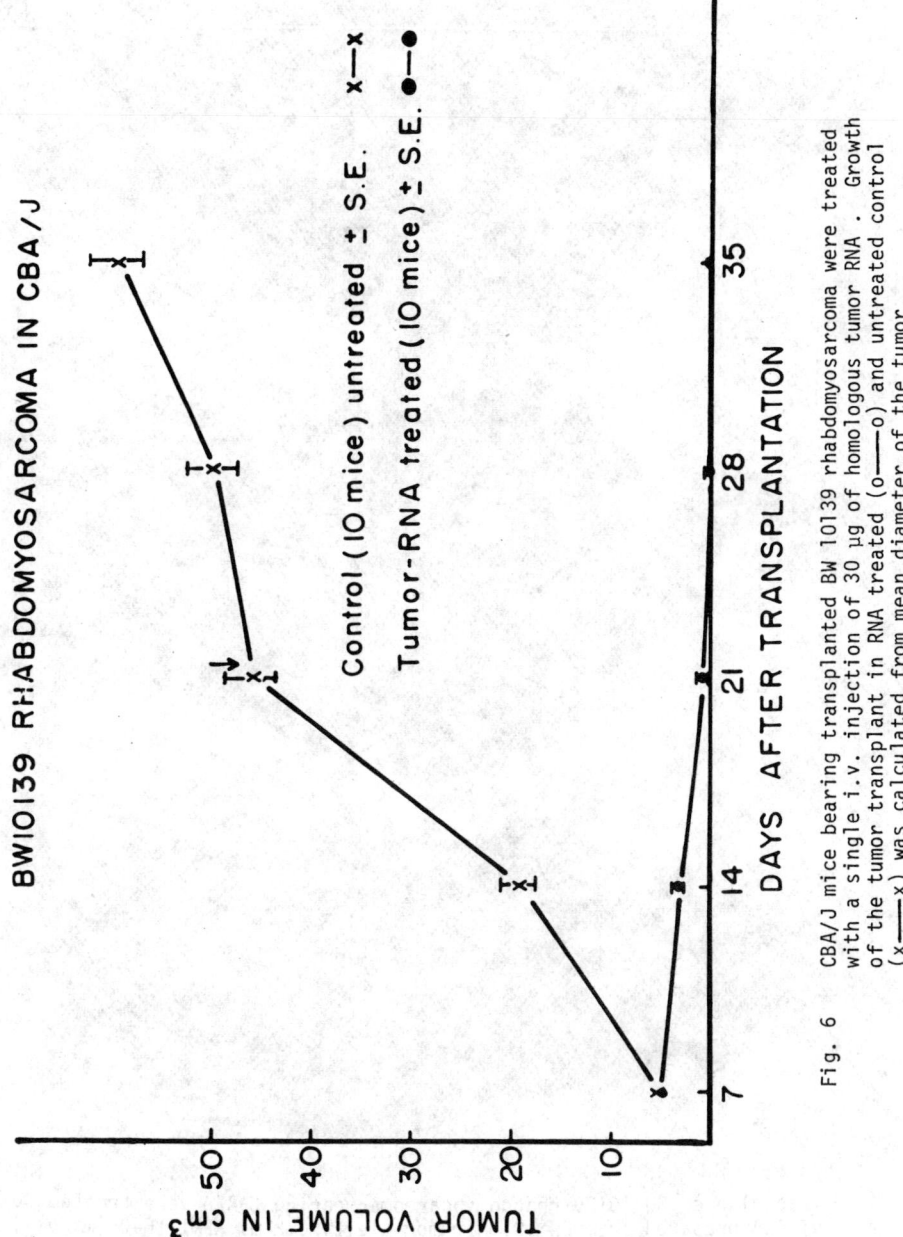

Fig. 6 CBA/J mice bearing transplanted BW 10139 rhabdomyosarcoma were treated with a single i.v. injection of 30 μg of homologous tumor RNA . Growth of the tumor transplant in RNA treated (o——o) and untreated control (x——x) was calculated from mean diameter of the tumor.

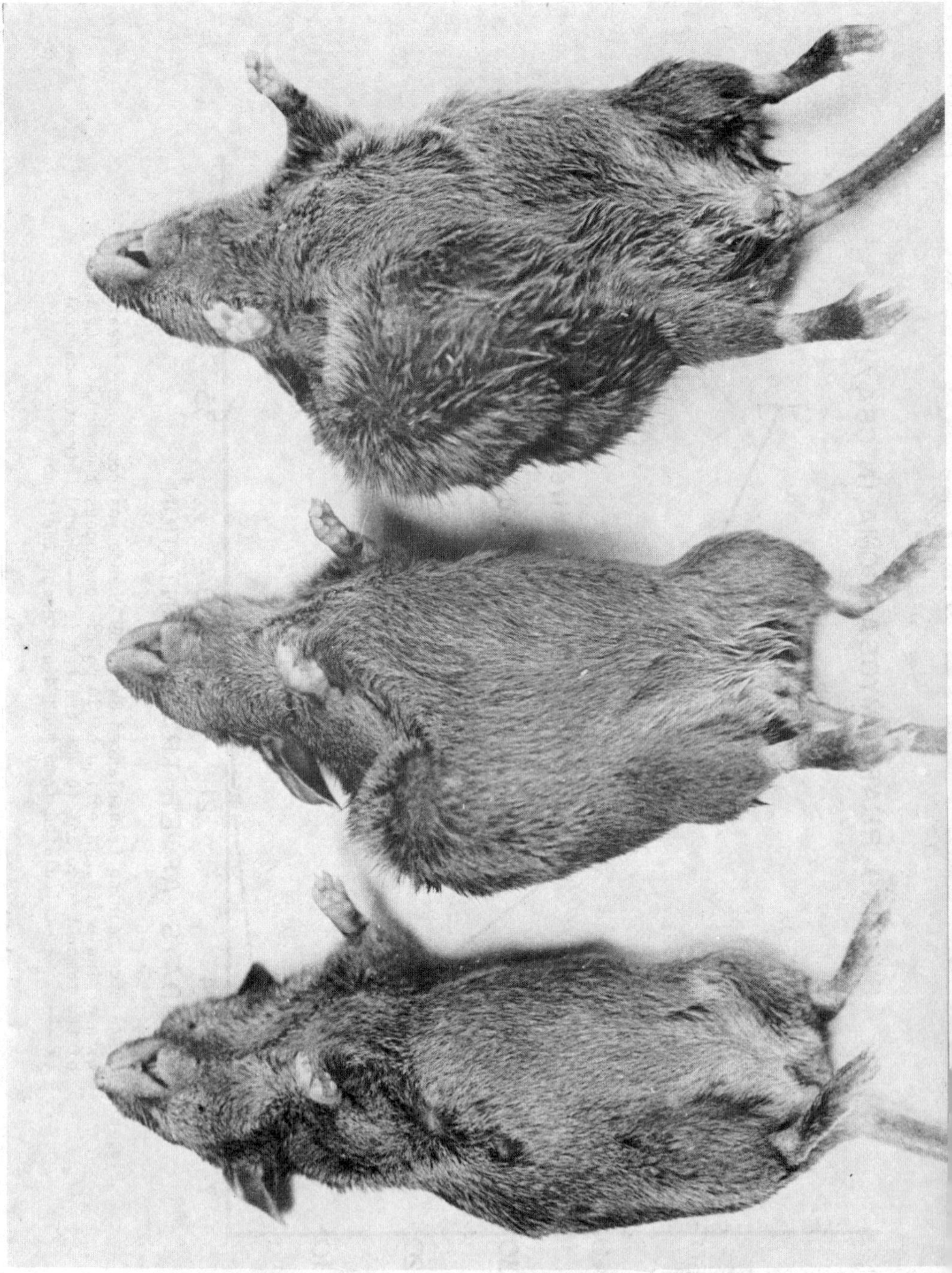

Fig. 7 Transplanted BW 10139 rhabdomyosarcoma bearing CBA/J were treated 30 μg of RNA prepared from BW 10139 tumor (left) 0.2 ml RPMI 1640 (middle) or 30 μg of RNA from AKR/J leukemic lymphoid organs (far right) on 7th day of transplantation. Treatment with homologous tumor RNA resulted in the complete regression of the transplant. Treatments with RPMI 1640 medium or leukemic RNA were ineffective. The picture was taken 22 days after transplantation.

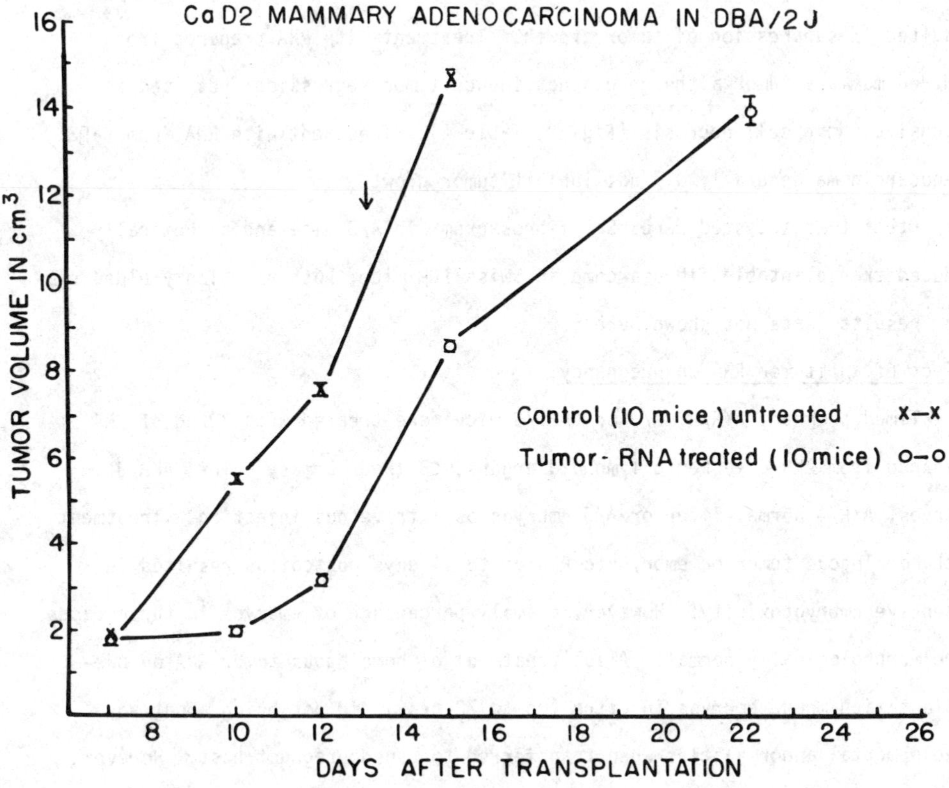

Fig. 8 Growth of CaD2 mammary adenocarcinoma in DBA/2J mice. Seven days post transplantation CaD2-bearing DBA/J mice were given i.v. injection of 30 μg of CaD2 RNA (o——o). Controls did not receive any treatment (x——x). Arrow indicates the day 80% of mice were dead.

and histopathological observations were made on tumor tissue and other visceral organs. Treatments with RNA prepared from 6C3HED as well as C3H/HeJ embryos resulted in suppression of tumor growth. Treatment with RNA prepared from C3H/HeJ mammary tumor although did not induce tumor regression, resulted in extensive tumor cell necrosis (Fig. 9, Table 4). Treatment with RNA from CaD2 adenocarcinoma generally did not inhibit tumor growth.

Other tumors tested were: Sal fibrosacroma in A/J mice and a chemically induced transplantable fibrosarcoma in Swiss/ICR mice, both of which yielded similar results (data not shown here).

Effect of antitumor RNA on pregnancy:

Timed pregnant AKR/J, A/J or CAFl/J mice were treated with 30 μg of RNA prepared from AKR/J leukemic lymphoid organs, C3H/HeJ mammary tumor, AKR/J embryos, AKR/J normal liver or A/J embryos as intravenous injection. Treatment with homologous tumor or embryonic RNA up to 14 days postcoitus resulted in extensive embryotoxicity. However, a small percentage of embryos in these cases were morphologically normal. Also, treatment of homologous tumor RNA on pre-implantation mouse embryos in vitro (up to 72 hrs.) did not bring about any developmental abnormalities when transferred to psuedopregnant host. However, treatment with homologous tumor or embryo RNA through either i.v. or intra-uterine route resulted in extensive degeneration of pre-implantation embryos (15). The exact cause of resorption of embryos is not known. Part of embryo-toxicity data is given in Table 5.

Immunogenicity of antitumor RNA: This was tested in ten week-old AKR/J and C3H/HeJ mice by the footpad swelling assay (delayed type hypersensitivity assay). Mice were given 15 μg of RNA in 30 μl PBS from AKR/J leukemic lyphoid organs or C3H/HeJ mammary tumor tissue by intradermal injection in left footpad, and the footpad thickness measured with a micrometer. The footpad thickness was again measured after 24 hours. Three days after sensitization the right hind footpad

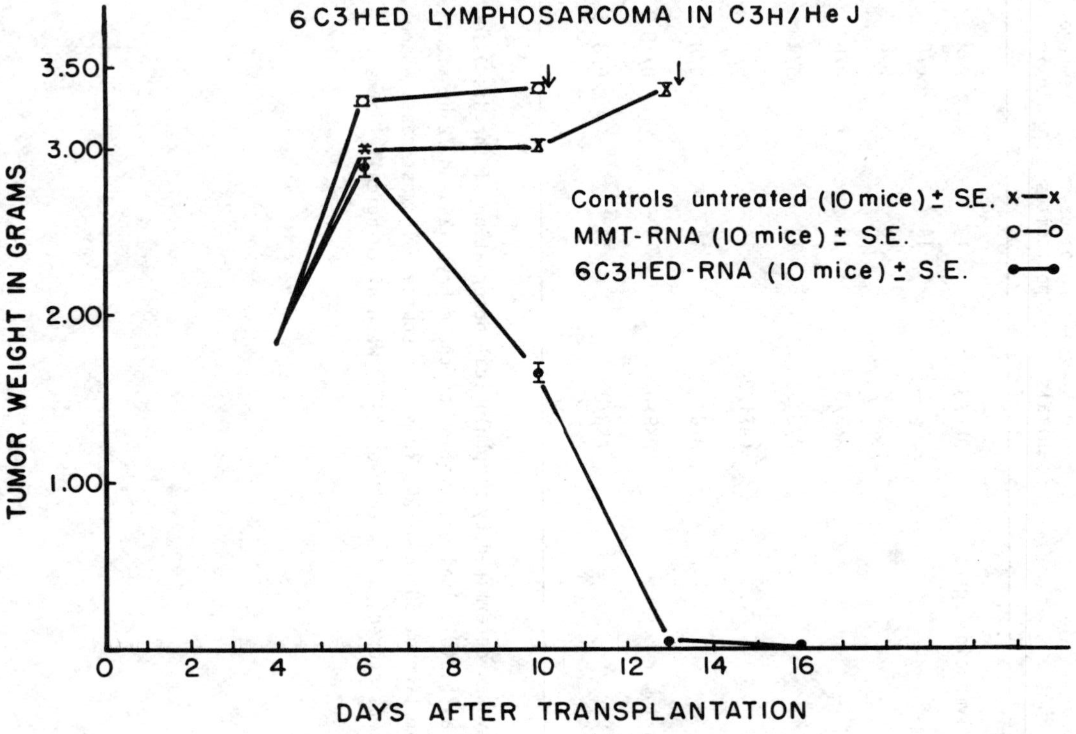

Fig. 9 Growth of 6C3HED lymphosarcoma in C3H/HeJ mice. 6C3HED transplanted C3H/HeJ mice were treated on 4th day of transplantation with 30 μg or RNA prepared from 6C3HED tumor (o——o) or spontaneous mammary adenocarcinoma of C3H/HeJ mice (□——□), or no treatment (x——x). Treated and control mice were sacrificed periodically to obtain wet tumor weight. Arrow indicates the day when all (100%) mice were dead.

TABLE 5

EFFECT OF ANTITUMOR RNA ON PREGNANCY

Source of RNA	Strain of test animals	No.of days postcoitus	Number of animals showing embryo toxicity[1]
AKR/J leukemic lymphoid organs	AKR/J	12	8/8
		14	6/8
		16	0/8
AKR/J embryos	AKR/J	12	8/8
A/J embryos	A/J	14	6/9
A/J embryos	CAF1/J	14	0/9
AKR/J embryos	A/J	14	0/8
C3H/HeJ mammary tumor	AKR/J	14	0/10
AKR/J normal liver	AKR/J	14	0/8

[1]Embryo toxicity was determined by morphological observation of the discussected uterus as well as histological observation of the embryos 24 and 72 hr after a single i.v. injection of 30 μg of RNA from each source. In most cases where embryotoxicity was scored, a few embryos (30%) were histologically normal.

thickness was determined for each mouse. The mice were then challenged with 15 μg of the same or different RNA injected into the right hind footpad. The final measurements of right hind footpad was taken 24 hours post-challenge. The specific swelling was calculated which was the difference between initial and final hind footpad thickness minus the difference between the initial and final left footpad thickness.

In C3H/HeJ mice, RNA isolated from C3H/HeJ mammary tumor induced significant DTH reaction while heterologous RNA from AKR/J leukemic tissue was not immunogenic in the footpad swelling assay. However, in AKR/J mice (due to extensive individual variation in response) there was no significant difference in footpad swelling between homologous and heterologous RNA (Table 6).

Effect of pre-immunization on leukemia incidence in AKR/J mice:

Fifty 4 week-old AKR/J mice were injected intravenously with 30 μg of RNA from AKR/J leukemic lymphoid organs. Control mice received the same does of RNA from C3H/HeJ mammary tumor tissue. These mice were observed for leukemia incidence. When moribund, the mice were autopsied and the viscera processed for histopathological observation.

The rate of leukemia incidence in AKR/J mice treated with RNA from AKR/J leukemic tissue was lower than that of AKR/J mice treated with RNA from C3H/HeJ mammary tumor (Fig. 10).

Mitogenicity of RNA: We employed a micromethod developed in our laboratory (16) for murine peripheral lymphocyte culture to determine the possible lymphocyte stimulatory potency of RNAs from different mouse strain origin. Arterial blood from different strains of mice were cultured in the presence and absence of homologous tumor RNA, homologous RNA from normal tissues, heterologous RNA, yeast tRNA or PHA. ^{3}H-TdR incorporation into the acid precipitable fraction of lyphocytes were monitored according to a standard previously described procedure (16). RNAs from AKR/J leukemic tissues and C3H/HeJ mammary tumor caused an

TABLE 6

IMMUNOGENICITY OF ANTI-TUMOR FACTOR

Source of factor	Strain of mice[1] tested	Mean specific footpad swelling[2]	SE	p[3]
AKR/J	C3H/HeJ	0.66	0.242	<0.05
C3H/HeJ	C3H/HeJ	1.74	0.409	
AKR/J	AKR/J	1.57	0.468	>0.05 <0.2
C3H/HeJ	AKR/J	0.49	0.246	

[1]Ten mice of each strain at 10-13 weeks of age were tested in each experiment. Two such experiments were performed using two batches of RNA preparations.

[2]Specific increase in footpad thickness (0.1 mm).

[3]Student's t-test was used.

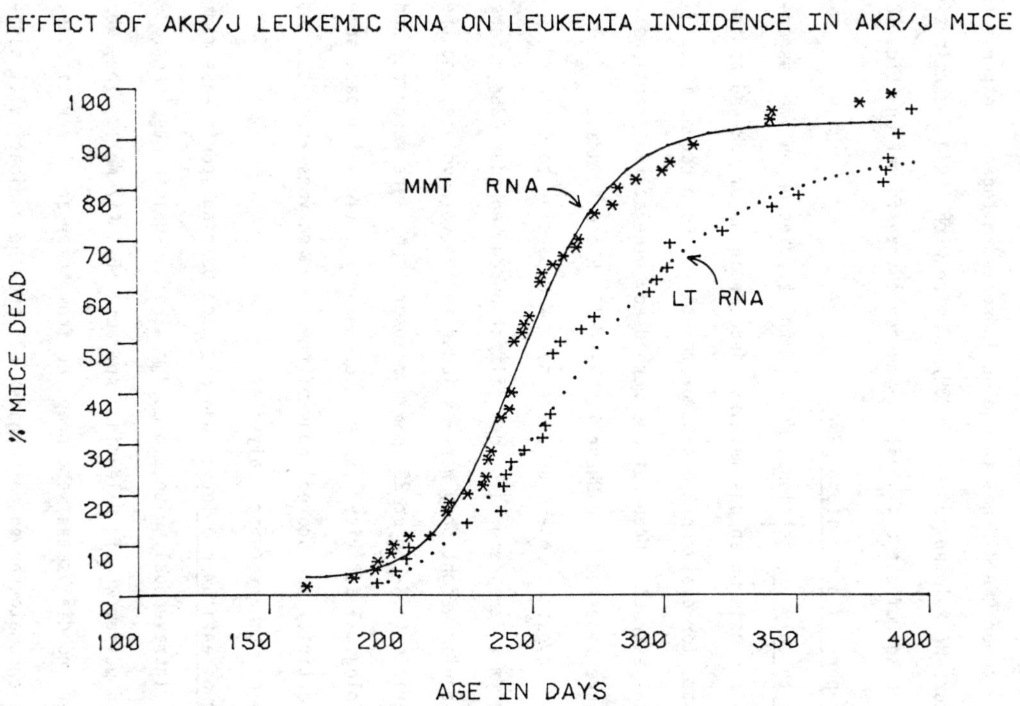

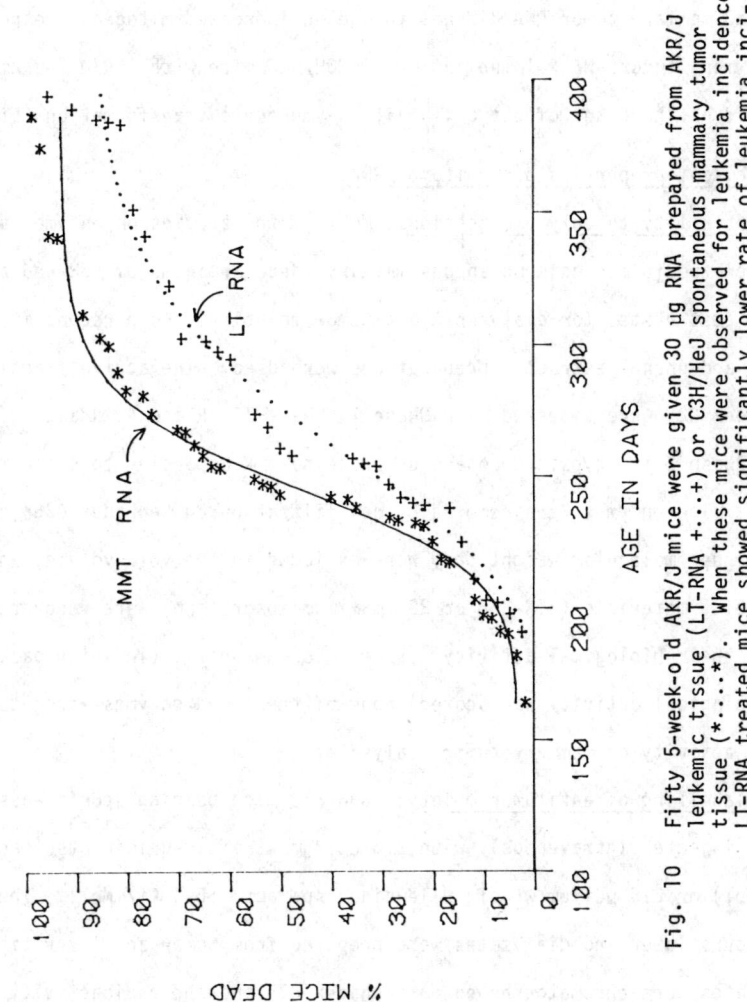

Fig. 10 Fifty 5-week-old AKR/J mice were given 30 μg RNA prepared from AKR/J leukemic tissue (LT-RNA +...+) or C3H/HeJ spontaneous mammary tumor tissue (*....*). When these mice were observed for leukemia incidence, LT-RNA treated mice showed significantly lower rate of leukemia incidence.

increased incorporation of ^{3}H-TdR into peripheral blood lymphocytes but only in respective homologous strains. No similar increase of labelled thymidine incorporation by AKR/J leukemic tissue RNA into peripheral blood lymphocytes occurred from related (RF/J) or unrelated (C58/J and C3H/HeJ) strains of mice. The increase in ^{3}H-TdR incorporation induced by homologous RNA was considerably lower than that of PHA (Fig. 11). Also, prior immunization of C3H/HeJ mice with C3H/HeJ mammary tumor RNA did not induce an increased mitogenic response to C3H/HeJ mammary tumor RNA. Immunization of C3H/HeJ mice with AKR/J leukemic tissue RNA also did not have any effect on in vitro lymphocyte transformation (Fig. 12).

Some biochemical properties of antitumor RNA:

Susceptibility to enzyme digestions: Most of the studies on enzyme susceptibility were performed on spontaneous mammary adenocarcinoma of C3H/HeJ as the source and test system for dialyzable antitumor moieties. Both reconstituted dialysates and phenol extracted preparations were used. The active fractions from Sephadex G25 were digested with DNase I, DNase II, RNase A, RNase B, RNase T_1, RNase III, trypsin, pronase or proteinase K according to standard procedures. The enzymes were removed by gel filtration on Sephadex G25 (because of their higher molecular weight they were excluded in the void volume) and the U.V. absorbing materials (254-268 or 280 maximum absorption) were recovered and tested for their biological activity. Except for RNase III (in which case some loss of biological activity was scored) none of the above enzymes abrogated the biological activity of mammary tumor dialysate.

Radiolabelling of antitumor moiety: C3H/H3J mice bearing spontaneous mammary tumor were injected intravenously with 2.5 μCi/gm wt of ^{3}H-uridine (sp. act. 26.2 Ci/mMol) or 2.5 μCi/gm wt of ^{3}H-leucine (sp. act. 16.4 Ci/mMol). They were killed 6 hours later and dialysates were prepared from tumor and liver tissues. The dialysates were chromatographed on Sephadex G25 and the radioactivity associated with each fraction was monitored. Biologically active fractions were

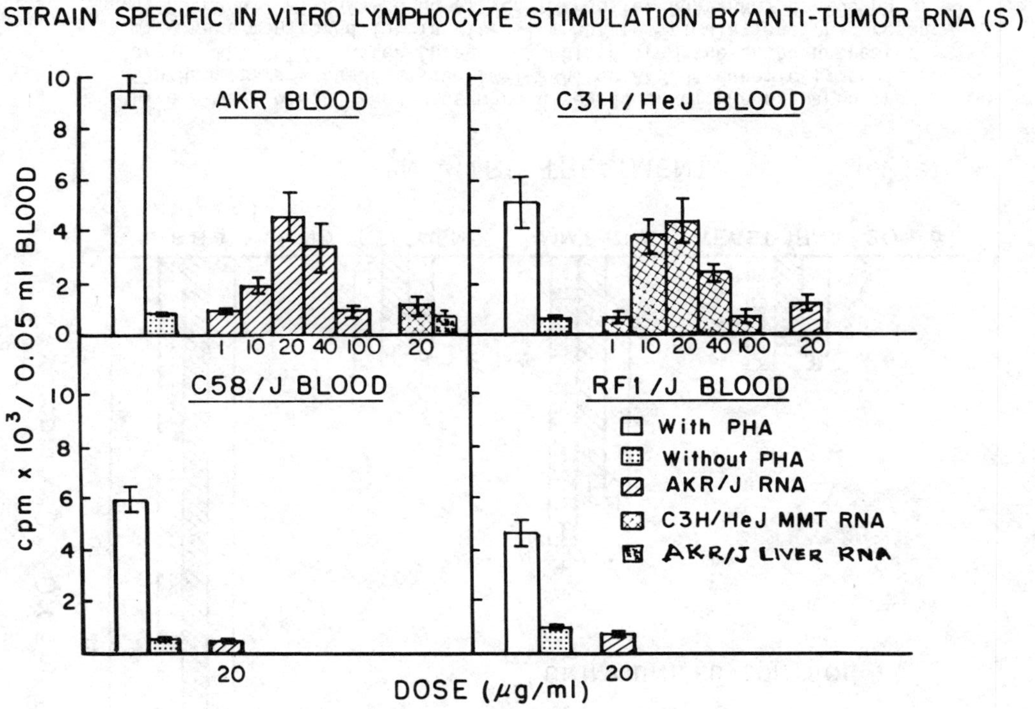

Fig. 11 Whole blood cultures from 8-10 week-old AKR/J, C3H/HeJ, C58/J and RF1/J were prepared as described in the text. In vitro lymphocyte stimulation was monitored in the presence of different concentrations of anti-tumor RNAs from AKR/J leukemic tissue, C3H/HeJ spontaneous mammary tumor tissues, normal liver tissues or PHA-M. A minimum of 40 cultures were done for each dose.

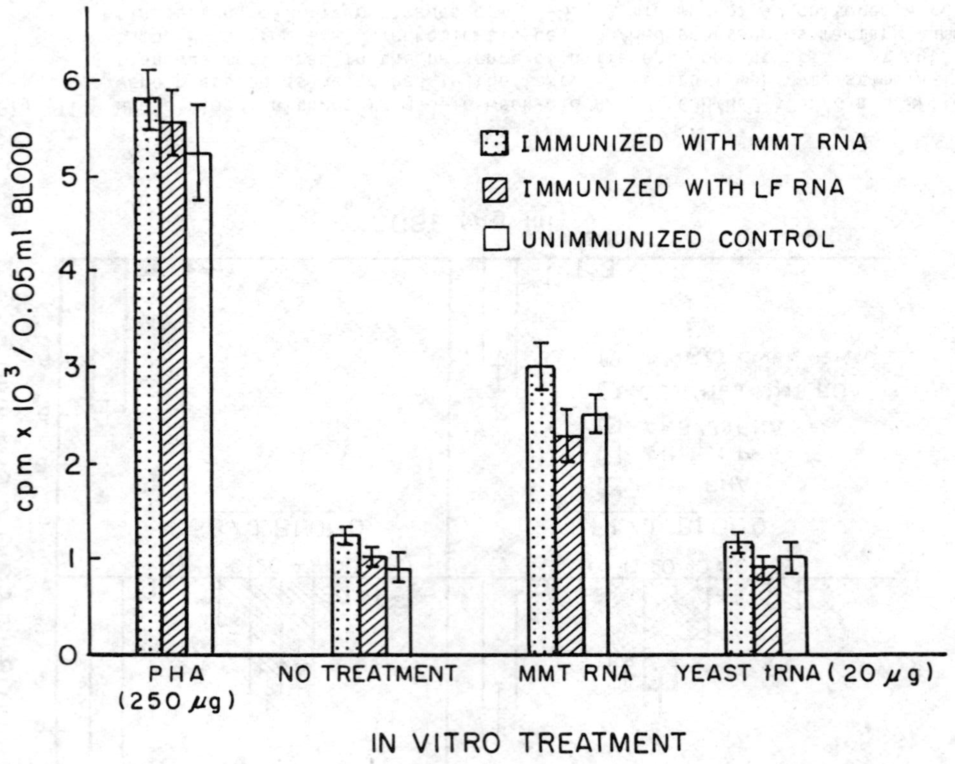

Fig. 12 Five 8-week-old C3H/HeJ mice were given 30 μg of RNA prepared from C3H/HeJ mammary tumor tissue (MMT-RNA) or AKR/J leukemic lymphoid organs (LT-RNA). After three weeks blood from these and untreated mice were monitored for in vitro lymphocyte stimulation, the presence of 20 μg of MMT-RNA, 20 μg of yeast transfer RNA(tRNA) or 250 μg of PHA-M. There was no apparent increase in labelled thymidine incorporation in the case of pre-immunized mice.

located and digested with RNase A and were re-chromatographed. ^{3}H-uridine was mainly incorporated into the major peak of fractions 10-16 whereas only little of ^{3}H-leucine was recovered (Fig. 13). RNase digestion eliminated the radioactivity from fractions 13-16 completely and to a small extent also from fractions 11 and 12 (Fig. 14).

Other properties: The biologically active fraction was orcinol reaction positive, formed single band in polyacrylamide gel electrophoresis, showed a density of 1.57 in KI density gradient, and showed increase in UV absorption with increase in temperature.

Discussion:

We report specific antitumor effects of RNAs isolated from both murine tumors and embryos. Although this effect seems to have high biological specificity, the fact that RNA from C3H/HeJ embryos was effective against both spontaneous mammary adenocarcinoma and transplantable 6C3HED lymphosarcoma of C3H/HeJ mice suggest that perhaps the RNA is a fetal gene product re-expressed in tumor cells (Tables 3 and 4). The mode of action of specific RNA is yet unknown, but there are several reports in literature indicating the role of RNA in cellular immune response (3-9). There is extensive evidence on the inhibition and stimulation of tumor transplants by prior injection of tumor cell fractions (17). Since the dialyzable antitumor RNA is immunogenic (Table 6) it seems possible that it acts via reticuloendothelial system.

Our data on enzyme susceptibility implicate the possibility that the active moiety could be a double stranded RNA. This conclusion is derived from the following observations: a) ^{3}H-uridine gets incorporated into the biologically active fraction and was resistant to TNase A or T_1 digestion, b) there was a considerable loss of biological activity following treatments with RNase III but not with RNase A, RNase B or T_1, c) phenol extracted, RNase digested and rechromatographed fractions were Orcinol reaction positive, had a buoyant density

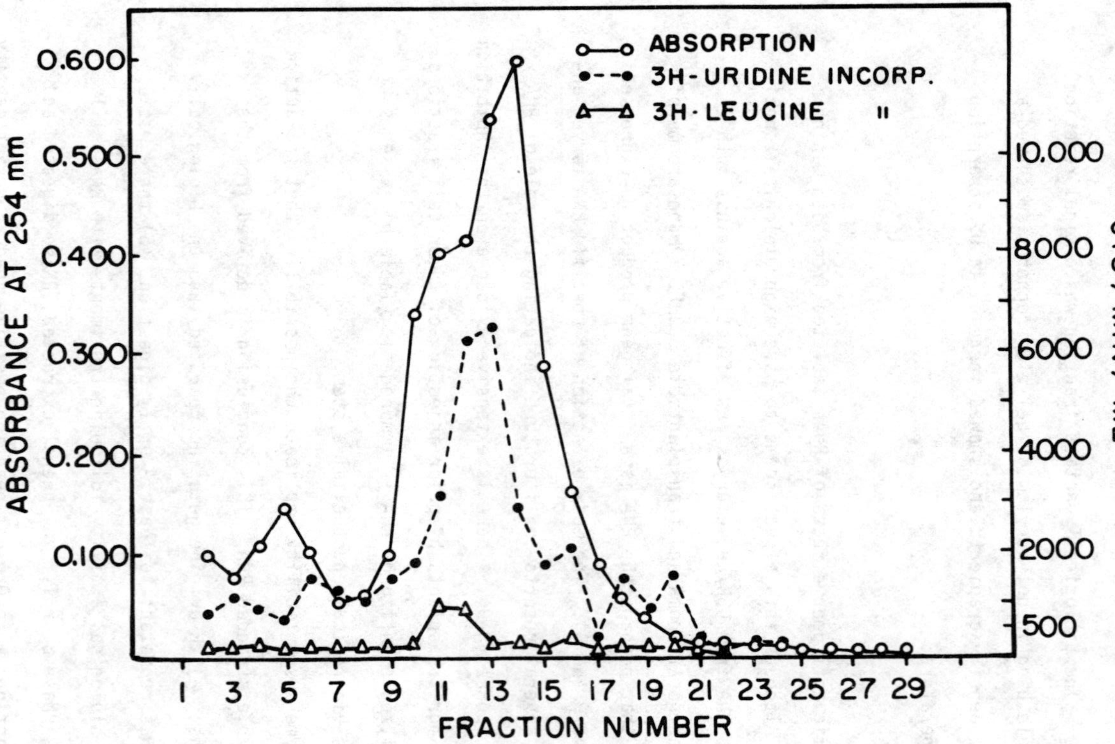

Fig. 13 Spontaneous mammary tumor bearing mice were injected i.v. with ^{3}H-uridine or ^{3}H-leucine as described in the test. The dialysates were chromatographed and monitored for UV absorption and radioactivity. Aliquots of each fraction were also tested for biological activity in spontaneous mammary tumor bearing mice. ^{3}H-uridine was incorporated into the major peak of fractions 10-16 whereas only little ^{3}H-leucine was incorporated.

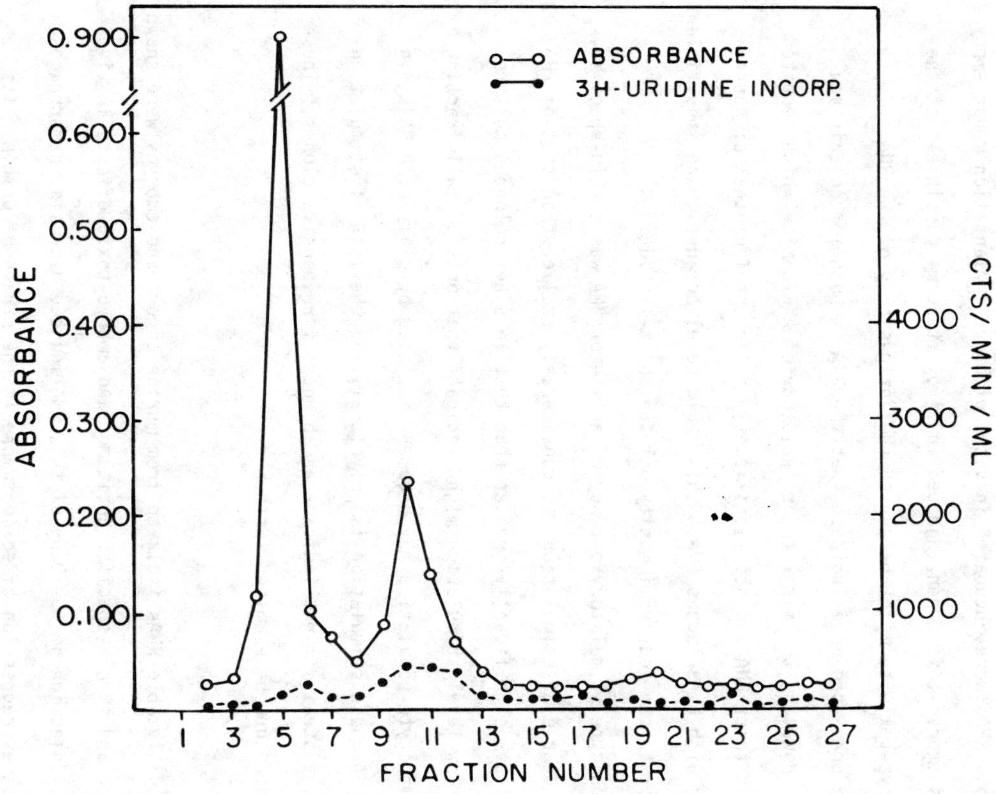

Fig. 14 Fractions 10-16 from ^{3}H-uridine prelabelled mammary tumor cell dialysates (see Fig. 13) were digested with RNase A for 6 hours and were rechromatographed on Sephadex G25. The first large peak represents RNase A and the second small peak represents the active moiety.

of 1.57 in KI gradient and showed increased UV absorption with increase in temperature. However, our data do not exclude the possibility that a very small peptide might be still attached to the RNA which might confer the observed biological specificity.

Embryo-toxicity of dialyzable RNA reported by us and others (13,15) although quite striking, the cause is greatly unknown. Since there was no embryo-toxicity when homologous RNAs were incubated in vitro with pre-implantation mouse embryos (while it was embryo-toxic when injected in vivo) it seems possible that the observed embryo-toxicity might be due to the antigenicity of the RNA.

The data presented here indicate that the RNA is both immunogenic and blastogenic. However, the final tumor incidence of AKR/J mice which received homologous antitumor RNA was not significantly different from that of control group treated with heterlogous RNA. This could be either due to an insufficient dose of RNA or to the induced immunity only be short-lasting.

The lymphocyte stimulatory potency of antitumor RNA was strain-specific. Although the degree of stimulation was considerably smaller than that of PHA but, nonetheless, significantly greater than that in a heterologous culture system. The low level stimulation might indicate that only a small fraction of the total lymphocyte population may become stimulated by dialyzable RNAs, in contrast to PHA whose stimulation is polyclonal, and the lack of stimulation in heterologous strains might indicate an absence of responsive clones. These conclusions need now be examined further.

Summary:

Certain dialyzable RNAs isolated from murine tumore and embryos were shown to have source and strain-specific antitumor and ambryo-toxic effects in vivo. A single i.v. injection of 15-30 μg of this RNA induced necrosis, hemorrhages and in some cases regression of solid tumors in the strain of origin. Its presence in both tumors and embryos suggests that it may result from re-expression of embryonic genes in tumor cells. Also, the RNA was shown to be immunogenic and blastogenic in its strain of origin. Preliminary studies indicate that the active moiety might be a double stranded RNA with approximate molecular weight of 6000. The exact mechanism by which these molecules bring about biological effects is still unknown.

ACKNOWLEDGEMENTS:

R. Pottathil wishes to thank Dr. D.J. Lang for his constant support and encouragement.

We also wish to thank Dr. H.J. Heiniger and Dr. H.W. Chen for their expertise and advice.

REFERENCES

1. Fishman, M. and Adler, F.L. 1963. Antibody formation initiated in vitro. II. Antibody synthesis in X-irradiated recipients of diffusion chambers containing nucleic acid derived from macrophages incubated with antigen. J. Exp. Med. 117:595-602.

2. Cohen, E.P. and Parks, J.J. 1964. Antibody production by non-immune spleen cells incubated with RNA from immunized mice. Science 157:1567-1569.

3. Mannick, J.A. and Egdhal, R.H. 1964. Transfer of heightened immunity to skin homografts by lymphoid RNA. J. Clin. Invest. 43:2166-2177.

4. Lawrence, H.S. 1955. The transfer in humans of delayed skin sensitivity to Streptococal M substance and to tuberculin with disrupted leukocytes. J. Clin. Invest. 34:219-230.

5. Spitler, L.E. 1979. Transfer factor in immunodeficiency diseases. Ann. N.Y. Acad. Sci. 332:228-235.

6. Pilch, Y.H., Ramming, K.P., and DeKernion, J.B. 1977. Preliminary studies of specific immunotherapy of cancer with immune RNA. Cancer 40:2747-2757.

7. Wang, B.S., Dickers, P.J., and Mannick, J.A. 1978. Kinetics of the transfer of tumor-specific cytotoxicity with immune RNA. Clin. Immunol. Immunopath. 9:218-228.

8. Ramming, K.P. and Pilch, Y.H. 1971. Transfer of tumor-specific immunity with RNA: inhibition of growth of murine tumor isografts. J. Nat. Cancer Inst. 46:735-750.

9. Friedman, H. 1979. RNA modulating specific and non-specific immune response. Ann. N.Y. Acad. Sci. 332:187-195.

10. Raveendran- P., Batra, B.K., and Menon, I.R. 1974. Specific in vivo stimulation of lymphocytes by tumor cell extract. Indian J. Med. Res. 62:1025-1031.

11. Menon, I.R., Raveendran, P., and Batra, B.K. 1974. Specific rejection and regression of solid tumors by tumor cell extract. Indian J. Med. Res. 62:1032-1038.

12. Batra, B.K., Raveendran, P. and Menon, I.R. 1975. Mechanism of tumor regression by tumor cell extract. Indian J. Med. Res. 63:31-35.

13. Pottathil, R. and Meier, H. 1977. Antitumor effects of RNA isolated from murine tumors and embryos. Cancer Res. 37:3280-3286.

14. Kloet, S.R. and Andrean, A.G. 1971. Buoyant density gradient centrifugation of RNA and DNA in alkali iodide solutions. Biochim. Biophys. Acta. 247:572-575.

15. Batra, B.K., Raveendran, P. and Maharajan, V. 1977. Response of mouse embryos to tumor cell dialysates. Indian J. Med. Res. 65:572-575.

16. Heiniger, H.J., Wolf, J.M., Chen, H.W., and Meier, H. 1973. A micromethod for lymphoblastic transformation of mouse lymphocytes from peripheral blood. Proc. Soc. Exp. Biol. Med. 143:6-11.

17. Snell, G.D., Kloudman, A.M., Taylor, A., and Douglass, P. 1946. Inhibition and stimulation of tumor homiotransplants by prior injection of lyophilized tumor tissue. J. Nat. Cancer Inst. 6:303-366.

FACTORS INVOLVED IN THE TRANSCRIPTION OF PURIFIED EUKARYOTIC GENES BY RNA POLYMERASES II AND III

R.G. Roeder, D. Engelke, D. Lee, D. Luse, T. Matsui, S. Ng, J. Segall, B. Shastry, P. Weil

Departments of Biological Chemistry and Genetics,
Division of Biology and Biomedical Sciences,
Washington University, St. Louis, MO.

We have previously established reconstituted cell-free transcription systems in which purified genes are accurately transcribed by purified class II and III RNA polymerases in the presence of crude cellular extracts or fractions thereof. The DNA templates for which accurate transcription initiation has been shown in these systems include those containing various adenovirus 2 (early and late) and mammalian β globin transcription units (class II) and those containing adenovirus 2 VA RNA and *Xenopus* tRNA and 5S RNA genes (class III). Various results indicate that the transcription of these genes by the corresponding class II and class III RNA polymerases is mediated by general transcription factors (in the extracts) which are neither tissue or species specific. To further understand the nature and mechanism of action of these factors we have begun to fractionate the extracts from *Xenopus* oocytes and from human KB cells. These studies indicate the presence of several distinct factors (fractions) required for the transcription (initiation and termination) of purified class III genes and several distinct factors required for the transcription initiation on class II genes. Those include components which are absolutely required for specific transcription as well as factors which appear to act by supressing random transcription and factors which act as general stimulatory factors. Thus far one factor, specific for 5S genes, has been purified (from *Xenopus* oocytes) to apparent homogeneity and shown to interact specifically with an intragenic 5S DNA control region. A further analysis of this factor, and the further purification and analysis of the other class II and class III transcription factors, will be presented.

RNAs IN *XENOPUS* OOCYTES

Michael Rosbash, Lynn Golden and Ulrich Schäfer

Department of Biology

and

Rosenstiel Basic Medical Sciences Research Center

Brandeis University

Waltham, Massachusetts 02254

U.S.A.

Amphibian oocytes have been an object of study for some time. Much of the interest in these germ cells comes from three features which makes them unusual by comparison to somatic cells. First, they undergo a prolonged period of growth in the absence of cell division, a general characteristic of panoistic oocytes. This period of growth results in a cell of enormous size, approximately 1 millimeter in diameter or a volume approximately 10^6 times that of a normal somatic cell. Second, this mature oocyte is a precursor to the fertilized egg and contributes much of the material which it accumulates during oogenesis to the zygote. There is therefore a great deal of interest among developmental biologists in these accumulated molecules. Third, the nuclei (germinal vesicles) of amphibian oocytes contain in their tetraploid genomes an unusual configuration called lampbrush chromosomes. These chromosomes have been scrutinized from the cytological point of view and are clearly in the process of intense and active RNA synthesis. The precise nature of the RNA transcripts synthesized by lampbrush chromosomes, however, has not been rigorously defined. Moreover, the relationship of the RNA transcribed from lampbrush chromosomes to the steady state "stored" mRNAs found in these same oocytes is obscure at best. This point will be discussed in more detail below.

Some clues as to the function and significance of the RNAs found in oocytes can be gleaned from a consideration of the amount and nature of these RNAs. As mentioned above, the mature amphibian oocyte is an extraordinarily large cell. Each mature oocyte contains approximately 4 micrograms of ribosomal RNA, or approximately 10^{12} ribosomes. On a per volume basis, therefore, the number of ribosomes is more or less normal (10^6 ribosomes/somatic cell volume equivalent). The quantity and nature of oocyte poly(A) containing RNA (mRNA) has also been investigated. Mature _Xenopus_ oocytes contain approximately 0.5% of their RNA as p(A) RNA[1]. Recent data from this laboratory on highly purified (banded) stage 6 mature oocytes suggest that this original measurement was somewhat high. Our present value is approximately 0.25-0.3% p(A) RNA. The complexity of this RNA has also been examined in some detail. Measurements with single copy DNA as well as RNA-cDNA kinetic measurements suggests that there exist approximately 2 x 10^4 different p(A) RNA molecules of approximately 10^6 copies each[2]. These values are also not very different from what is found in somatic cells when one normalizes for the enormous volume of an oocyte. In addition to these $\sim10^4$ low abundant mRNAs, there are also a number of moderately abundant mRNAs present at 10 to 100 times the intracellular concentration of the rare mRNAs. It is interesting that no genuine abundant mRNA component has been found in oocytes. In somatic cells, abundant mRNAs consist not only of terminally differentiated sequences such as hemoglobin or vitellogen mRNA, but also of mRNAs present in undifferentiated cells which code for proteins such as actin and tubulin. An abundant mRNA is usually present at thousands of copies per cell in normal somatic tissues and represents approximately 1% or more of the mRNA. No RNAs at these levels have been found in oocytes. At first glance, this is surprising since 2-Dimensional gels of _in vivo_ labeled oocyte proteins contain a relatively small number of major spots (including actin and tubulin) when exposed for a short time; these spots must be the major proteins synthesized by the oocyte polysomes[3]. We have confirmed these data by labeling with other amino acids and by labeling the oocytes for as short as five minutes. Neither of these variations effect the basic conclusion that there are a number of major proteins the mRNAs for which must be relatively abundant (greater or equal to $\sim$1% of the total polysomal mRNA each) on the oocyte polysomes. When total oocyte mRNA is purified and added as a template to an _in vitro_ protein synthesis system, an entirely different pattern of proteins is produced. The most outstanding feature of this pattern is the absence of protein spots which are even 1/10 as intense as the spots obtained by labeling the oocytes *in vivo*[4]. This relatively uniform pattern of spots is consistent with the data mentioned above, i.e., the absence of an abundant class of mRNAs by RNA-cDNA hybridization. The two different kinds of *in vitro* analyses are therefore in agreement. One way of reconciling the two *in vitro* analyses with the very different nature of the proteins produced *in vivo* is to suggest the presence of a strong degree of translational control in the oocyte which operates at a sequence specific level. More precisely, the data suggest that many (most) of the low abundant or rare mRNAs are undertranslated *in vivo* by comparison to other mRNAs, such as actin or tubulin. This is the most straightforward interpretation of the discrepancy between the template activity of phenol extracted mRNA and the protein synthesis in the cells from which these mRNAs have been extracted. While such a hypothesis is not mutually exclusive with published data which suggest that many, and perhaps most of the oocyte ribosomes are inactive and incapable of being activated *in vivo*[15], they do suggest that the protein synthetic machinery is selective, at either a quantitative or qualitative level, with respect to the mRNAs that are translated.

A second major difference between oocyte poly(A) RNA and somatic poly(A) RNA is its size. In an attempt to gain further insight into the nature of oocyte poly(A) RNA we have fractionated this RNA on sucrose gradients in order to examine a number of phenomena which might be related to size. The size of this RNA is large (2000-2500 nucleotides) and is approximately twice as large as poly(A) RNA isolated from somatic cells from either liver or tadpoles. Interestingly, all size classes of the oocyte poly(A) RNA are equally active in *in vitro* protein synthesis, suggesting that there is nothing unusual about even the large RNA from the point of view of its ability to act as a template in an *in vitro* protein synthesizing system.

How might one account for these data? To do so requires assembling a number of seemingly diverse observations from the literature. I will attempt to do so here because the data provide a hypothesis by which the large size of oocyte mRNA is understandable and suggest experiments (presently underway) which will test such a hypothesis. Perry's laboratory has shown that if one fractionates L cell mRNA by size, large mRNA is preferentially - and strongly preferentially - of the low abundant variety by comparison to small mRNA[5]. This is consistent with our observation that oocyte poly(A) RNA is large and missing an abundant component. Firtel's laboratory has isolated two unusual genomic clones from Dictyostelium discoidium which are complementary to mRNA[6,7]. These clones each contain a different small repeated segment which is found at the 5' end of their respective mRNAs. These mRNAs are therefore transcribed from "interspersed" DNA. This structure or organization within an mRNA is contrary to the prevailing dogma which indicates that mRNAs are either entirely single copy or entirely repeated. Both of the two repeat families have a large number of mRNAs to which they are complementary; the data indicate that each of these mRNAs has a repeated family member at the 5' end and a different single copy region which makes up the rest of the mRNA. Interestingly from our point of view, these "interspersed" sequences are all low abundant mRNAs and they are all considerably larger than average Dictyostelium mRNA. Finally, the Britten-Davidson laboratory has recently demonstrated the presence of repeated sequences in stored sea urchin oocyte p(A) RNAs[8]. Taking these data together the following rather simple and testable picture emerges. The rare L cell mRNAs are large because they contain repeated sequences at their 5' ends. These repeated sequences render these rare mRNAs larger than average mRNA, i.e., larger than other "normal" mRNAs from which these repeats are absent. The prediction is therefore that RNAs which we have characterized from Xenopus laevis oocytes and which are preferentially large and contain preferentially rare sequences will also contain a portion of repeated DNA, probably at the 5' end of the mRNA molecule. We are in the process of testing this prediction. While these experiments are not yet completed, results already obtained are consistent with the schema presented above. We have successfully screened a small collection of ovary and tadpole cDNA clones with cDNA. The cDNA used to screen these clones was of two types: cDNA synthesized in the usual manner with oligo dT as primer and cDNA synthesized with random calf thymus oligonucleotides as primer. The rationale in these experiments is as follows: We know that cDNA synthesized in the usual way (with oligo dT as the primer and poly(A) RNA from oocytes as template) is almost entirely (greater than 95%) complementary to unique DNA. If there exists a substantial population of mRNAs which contain repeated sequences far from the 3' end of the mRNA molecules, these repeated sequences might be poorly represented in "normal" cDNA, the synthesis of which underrepresents the 5' ends of the mRNAs as a template. By using oligonucleotides from calf thymus DNA as primer, a more representiative population of cDNA molecules should be synthesized and, in particular, the cDNA synthesized should not be restricted or preferentially biased towards the 3' ends of the mRNA molecules. Consequently, such calf thymus primed cDNA should contain an enriched fraction of interspersed repeated sequences as compared to oligo dT primed cDNA, if these interspersed mRNA molecules do indeed exist. To verify the existence of such molecules, and to identify cDNA-containing clones which contain such repeated sequences, we have screened cDNA clones with these two kinds of cDNA. Clones which fulfill the above expectation (i.e., hybridize well with random calf thymus-primed cDNA and poorly with oligo dT-primed cDNA) have been identified and isolated.

The fact that we find cDNA clones which hybridize well with one kind of cDNA and not with the other implies that these reactive recombinant DNA molecules contain repeated sequences which are present in mRNAs which are not entirely repeated. Otherwise, the oligo dT cDNA would work equally well. Consistent with this expectation are preliminary results in which the cDNA clones have been used individually as probes on Northern gels[9]. Such a procedure allows the identification of the size of individual poly(A) RNAs which will hybridize to each recombinant DNA probe. In contrast to all other cDNA clones which give a positive signal on Northern gels (discussed below), these specific cDNA clones react with large numbers of poly(A) RNA of diverse sizes and therefore give a "smear". This is precisely the result expected for recombinant DNA clones with the properties described above. These clones are under further examination.

This general technique -- the analysis on Northern gels of RNAs complementary to individual recombinant DNA molecules -- is yielding a large amount of information on the nature of mRNAs in amphibian oocytes. A second question which we have investigated in this way, albeit in a somewhat preliminary fashion, is the putative presence of globin sequences in oocytes. This laboratory has previously published a report in which analysis by RNA-cDNA hybridization suggested the presence of both adult and tadpole globin sequences in *Xenopus* oocytes[10]. In an attempt to confirm and extend this observation, we have utilized *Xenopus* globin cDNA clones, a generous gift from the laboratory of Jeff Williams in London[11], and the Northern gel procedure described above in an attempt to identify and locate globin RNA sequences in *Xenopus* oocytes. This we have failed to do. The precise nature of these experiments deserves some comment since this failure in actuality sets a lower limit on the number of globin transcripts which can be present in the oocyte. Very precise reconstruction experiments argue that this procedure can detect one picogram of β major globin mRNA in a background of 1 microgram of random poly(A) RNA. In other words, we can detect $\sim$one globin sequence in 10^6 non-globin sequences. Experiments with this level of sensitivity have failed to reveal any hybridization of the β major globin cDNA clone to oocyte poly(A) RNA. Experiments of somewhat less sensitivity (10^{-5}-10^{-6}) have also failed to reveal any hybridization of a second cloned β minor globin gene to oocyte RNA. This level of sensitivity must be put into some perspective. As mentioned above, classical hybridization experiments suggest that there are approximately 2×10^4 different poly(A) RNA sequences present in Xenopus oocytes. Since there exist as well some RNAs of greater abundance (approximately 50% of the total poly(A) RNA), these considerations suggest that individual low abundant poly(A) RNAs should be each present at between 10^{-4} and 10^{-5} of the total poly(A) RNA. Experiments described below with random cDNA clones are consistent with this expectation and suggest that individual low abundant RNAs are indeed present at this level. This level of 10^{-4}-10^{-5} is the approximate level at which we previously identified adult and tadpole globin sequences in oocytes.

This level is between 10 and 100 times greater than the lower limit which our reconstruction experiments have indicated we can detect. Consequently, while globin sequences could still be undetectably present at 10^{-6} of the poly(A) RNA population (this represents 10^4 mRNA molecules per cell) we feel that this is unlikely. In any case, our ability to detect globin sequences using cloned probes suggests that our previous hybridization results, at least for adult globin sequences, are incorrect. The only reasonable possibility which could account for this seeming inconsistency is that the globin sequences stored in oocytes are heterogeneous in size, like the cloned repeated sequences described above.

Were this the case, these sequences could be present at a considerable frequency on a per cell basis and be undetectable on Northern gels. This remaining possibility is currently under investigation. It is interesting that there are a number of other cDNA clones, synthesized from p(A) RNAs from somatic sources, which also fail to hybridize to oocyte p(A) RNA. It is additionally worthy to note that the use of cloned probes makes a dramatic difference in the nature of the result obtained. If cDNA, synthesized from globin mRNA, is used as the probe on Northern gels "apparent" globin mRNA sequences are detected in oocyte RNA at a level of approximately 10^{-5} of the total poly(A) RNA. Since the cloned β major globin probe is unable to detect such sequences at even a 10 fold greater level of sensitivity, we are forced to conclude that this "reactive" cDNA is either a contaminant present in the 9S globin mRNA purified from Xenopus reticulocytes, or alternatively indicates the presence of other minor globin RNA species present in Xenopus oocytes. In any case, our tentative conclusion at the moment is that β major globin mRNA is either heterogeneous in size or absent (by comparison to "average" low abundant poly(A) RNA) in Xenopus oocytes. The latter interpretation suggests that the RNAs present in oocytes are indeed under some form of differential gene expression.

All of the notions presented above, and in particular the significance of heterogeneous RNAs which react to cloned cDNA sequences, requires that individual cDNA clones hybridize reproducably to poly(A) RNAs of unique size. In general, this is the case. In order to characterize further the nature of these stored poly(A) RNAs and more precisely the nature of their accumulation during oogenesis, we have analyzed these RNAs as a function of oocyte stage. Why pursue such experiments other than to control for the interesting and unusual results described above? The answer is historical and lies in results obtained some years ago on the nature of the accumulation of oocyte poly(A) RNA during oogenesis[1].

At this time, we showed that the total amount of oocyte poly(A) RNA remained unchanged throughout the entirety of vitellogenesis. During this period of time, when lampbrush chromosomes are highly active and when oocyte RNA synthesis is actively taking place, the oocyte increases in diameter more than ten fold and therefore increases in volume more than a thousand fold. No increase in poly(A) containing RNA can be detected during this growth period. In contrast, a large increase in ribosomal RNA continues to take place as oogenesis proceeds.

The most likely interpretation of these results was that the amount of poly(A) RNA which is increasing in concentration is similar to the amount of poly(A) RNA which is decreasing as oogenesis proceeds. In other words, differential gene expression on lampbrush chromosomes is indeed taking place; different mRNAs are being transcribed at different times. As some mRNAs increase in concentration, other mRNAs decrease in concentration to account for the relative constancy of the total poly(A) RNA per cell during this process.

Nothing could be further from the truth. Within experimental error, RNAs complementary to 14 of 17 clones, individually examined on Northern gels, remain constant in concentration on a per cell basis during all of the vitellogenesis. The other three clones are exceptional in that "their RNAs" increase dramatically, approximately 500 fold, during vitellogenesis. Interestingly, all three of these cDNA clones have proved to be mitochondrial and are therefore complementary to mitochrondrial poly(A) RNAs. These are the exceptions which prove the rule and indicate that the lack of change in concentration of individual poly(A) RNAs throughout vitellogenesis is indeed correct.

These data also provide additional information, some of which is relevant to the issues raised above. While some of the cDNA clones hybridize to two or three bands on Northern gels (the significance of which is not clear) no clones, other than those described above, hybridize detectably to RNAs which are heterogeneous in size. This is also the case for the hybridization of these clones to liver and tadpole RNA. Not only does the hybridization to these somatic RNAs reveal distinct bands, but also the RNA bands are identical in size in oocyte and somatic RNA. This suggests that further processing of the oocyte p(A) RNA sequences in somatic cells either does not occur or does so on a small minority of mRNAs.

In contrast to the fixed concentrations at which these RNAs are maintained during oogenesis, their concentrations vary widely in somatic cells. Some of these poly(A) RNAs become much more abundant in either tadpole or liver RNA while others become much more rare. This heterogeneity in behavior suggests that the poly(A) RNAs which we have examined are indeed a "random" collection of sequences and not some unusual class selected on the basis of some specific and unknown property. In general, sequences which increase in concentration in somatic cells do so in both tadpole RNA and liver RNA; likewise, sequences which decrease in concentration in somatic cells also do so in both tadpole and liver RNA. These data, although preliminary, suggests that there might be some reality to the notion that the concentrations of mRNA sequences in oocytes is important.

In order to extend these observations to a larger number of sequences, we have screened several hundred ovary and tadpole cDNA clones by Grunstein-Hogness colony hybridization[12] with cDNA (the traditional oligo dT cDNA). These data confirm, in an albeit less precise way, the data obtained with Northern gels described above. When screened with cDNA made from stage 2 and stage 6 oocytes, almost all the clones hybridize equally well with cDNA made from stage 6 oocyte poly(A) RNA and from stage 2 poly(A) RNA. While these data do not give any information about the behavior of low abundant poly(A) RNAs (since cDNA clones complementary to such sequences do not hybridize with enough cDNA to give a detectable signal on Grunstein-Hogness colony hybridization -- this is precisely the reason for which we screened the clones on Northern gels) they are consistent with the data mentioned above. Also consistent with these data is the fact that only 6 exceptional clones hybridize much better with stage 6 than stage 2 cDNA. These are all mitochondrial cDNA clones and include the three clones mentioned above. All of these data bring to mind a paper that this laboratory published[13] in which we were unable to detect any turnover of oocyte poly(A) RNA synthesized during early oogenesis. These two different approaches complement each and suggest that most and perhaps all of the stored oocyte poly(A) RNA is indeed synthesized prior to vitellogenesis in small oocytes and then stored for the remainder of oogenesis. It is of additional interest to note that when very small (stage one) oocytes are analyzed for the presence of these individual poly(A) RNAs, it is possible to detect lower concentrations than are present in the larger vitellogenic oocytes. These are presumably the stages at which these poly(A) RNAs sequences are being actively accumulated. The ratio of any 2 RNA species remains constant during this time when the sequences are increasing in concentration. In other words, the ratio of the concentration of sequence A and sequence B is constant even in small oocytes. This suggests that during these early stages of oogenesis these sequences are coordinately controlled in that relative concentrations are maintained while absolute concentrations increase. As described above, this is in marked contrast to the behavior of these poly(A) RNAs subsequent to fertilization when their relative concentrations become random.

Where does all of this lead? Although these data underscore the importance of recombinant DNA technology to the understanding of molecular events in development, they only contribute to the confusion or the contradictions which surround lampbrush chromosomes. The most straightforward propositions which arise from this work is that lampbrush chromosomes may synthesize different sequences than those stored as stable poly(A) RNAs in oocytes, or that this synthesis does not contribute to this stable pool. These kinds of steady-state measurements cannot distinguish between these two possibilities. Two other approaches suggest that "normal" poly(A) RNA sequences are not the product of lampbrush chromosome transcription. First, the RNAs synthesized by amphibian oocyte germinal vesicles have been analyzed by a number of criteria and are unusual. In particular, there is a very high fraction of U residues in labeled *Triturus* nuclear RNA[14]. Second, the nascent RNAs found on lampbrush chromosomes do not hybridize *in situ* to cDNA clones. These transcripts should be sufficiently abundant to provide a good if not excellent substrate for the *in situ* hybridization of individual cloned cDNA sequences. Such experiments have been, up until the present time, a failure in our laboratory and in the laboratory of Joe Gall. While it is of course possible that such a failure has been due to technical difficulties associated with the *in situ* hybridization, it is also possible that the sequences present in cDNA clones -- sequences which react well with poly(A) RNAs on Northern gels -- are simply not present in the nascent transcripts on lampbrush chromosomes. This latter notion is consistent with the data we have presented at this symposium which suggest that the time of synthesis of these oocyte poly(A) RNAs precedes substantially the time of maximum loop extension of lampbrush chromosomes which takes place during and throughout vitellogenesis. If all of this is true, and we emphasize that these considerations are speculative, they underscore the mystery which surround lampbrush chromosomes. Why is there so much transcriptional activity associated with these unusual chromosomes and what is the function, if any, of the transcriptional products. The resolution of this issue will almost certainly have profound implications for the study of oogenesis and early development. In addition, it may have implications for laboratories which use stage 6 oocyte nuclei for the injection of cloned DNA. Clearly, transcription from cloned DNA will ultimately be viewed in the context of the transcription which takes place on the resident, endogenous chromosomes.

REFERENCES

1. Rosbash and Ford, J. Mol. Biol., 85, 87-101, (1974).
2. Perlman and Rosbash, Developmental Biology, 63, 197-212 (1978).
3. DeRobertis and Gurdon, Proc. Natl. Acad. Sci. USA, 74, 2470-2474 (1977).
4. Rosbash, manuscript in preparation.
5. Meyuhas and Perry, Cell, 16, 139-148 (1979).
6. Kimmel and Firtel, Cell, 16, 787-796 (1979).
7. Kindle and Firtel, Nuc. Acids Res., 6, 2403-2422 (1979).
8. Constantini et al., Cell, 15, 173-187 (1978).
9. Alwine et al., Methods in Enzymology, Volume 68, editor Ray Wu, 220-242 (1979).
10. Perlman et al., Proc. Natl. Acad. Sci. USA, 74, 3835-3839 (1977).
11. Hentschel et al., Develop. Biol, 72, 350-363 (1979).
12. Grunstein and Hogness, Proc. Natl. Acad. Sci. USA, 72, 3961-3965 (1975).
13. Ford et al., Develop. Biol., 57, 417-426 (1977).
14. Sommerville, Gene Activity in the Lampbrush Chromosomes of Amphibian Oocytes, pp. 79-156. in Internation Reivew of Biochemisty, Biochemistry of cell Differentiation II, 15, Ed. J. Paul. University Park Press, Baltimore, MD USA (1977).
15. Laskey et al., Cell, 11, 345-351 (1977).

SPECIFIC POLYPEPTIDE SYNTHESIS MEDIATED BY mRNAs IN CELL CULTURE

Shih Ying-Shien

Institute of Developmental Biology, Academia Sinica

ABSTRACT

Poly A^+ mRNAs were isolated from young calf liver and used for treatment to control and Ca^{2+}-treated embryonic kidney cell culture. These mRNAs were found to stimulate amino acid (AA) incorporation in cell culture and this stimulation was proportional to mRNA concentrations. The newly-synthesized polypeptides yielded positive reaction with purified bovine albumin antiserum. The immunoprecipitates were analyzed by SDS polyacrylamide gel electrophoresis. Both staining and autoradiogram of the gel revealed the presence of albumin. Counter immunoelectrophoresis against bovine albumin antiserum indicated that the newly-synthesized polypeptides gave a single precipitin line in the same location as that produced by bovine albumin control. The functional study of liver mRNAs was compared with that of lens poly A^+ mRNAs. The latter induced the synthesis of specific lens crystalline. Treatment of cell culture with actinomycin D prior to addition of mRNAs inhibited the total AA incorporation but not the mRNA induced specific polypeptide synthesis. (The mRNA samples used in our experiments had no albumin contamination and the incubation medium had no detectable activity of ribonuclease.) We have also examined the effect of histone and polylysine on macromolecular transport as well as on AA incorporation. Based on these observations, a model is proposed for macromolecular transport and the molecular mechanisms for mRNA induced polypeptide synthesis will be discussed.

INTRODUCTION

The function of mRNA is clearly defined in cell-free protein synthesis systems. Foreign mRNA mediated morphological genetic changes *in vivo* in neurospora (1,2), in microorganisms (3), and in eukaryotic systems such as goldfish (4,5), chicken embryos (6) and mice (7) have been reported. Recent developments in monolayer cell culture techniques have provided a useful system for the study of mRNA functions. Furthermore, monolayer-growing cells can be used to bridge the gaps between *in vitro* and *in vivo* systems. Most recently, a novel system was described for studying the control of protein synthesis in a monolayer culture of chicken embryo liver cells (8). Plasma proteins are synthesized at rates comparable to that *in vivo*.

In this paper, we report the effect of calf liver mRNA on albumin synthesis in serum-free media of chicken embryo kidney cell culture systems.

MATERIALS AND METHODS

Preparation of calf liver RNAs: All the glass materials were autoclaved. Calf liver polysomes were prepared by a modification of the method of Chen, Lavers and Spector (9). Usually, 100 g calf liver was homogenized in approximately 200 ml, containing 0.34 M sucrose and buffer A: 0.05 M-Tris, pH 7.6 0.025 M KCl, 0.008 M $MgCl_2$ and 0.008 M mercaptoethanol. After homogenization, an additional 200 ml of buffer A containing 0.34 M sucrose was added and the preparation was centrifuged at 17,000 g for 20 min at 4°C. The supernatant was then centrifuged at 100,000 g for 2 hours at 4°C. The polysomes were suspended in 0.1 M Tris-HCl (pH 9.0), 0.1 M NaCl, 1 mM EDTA, at a concentration of 20 A260 units/ml and then made 1% in sodium dodecyl sulfate (SDS). An equal volume of phenol: chloroform: isoamyl alcohol (50:50:1) was added, the mixture was shaken vigorously for 10 min at room temperature, and the phases were separated by centrifugation at 12,000 x g for 10 min. The aqueous phase was back extracted with phenol again, and made 0.3 M in LiCl. Crude polysomal RNA was precipitated by the addition of two volumes of ethanol and was allowed to stand at -20°C overnight.

Oligo(dT)-cellulose chromatography was carried out by modification of the procedure described by H. Aviv and P. Leder (10). The RNA pellet was washed twice with 80% ethanol and 100 A_{260} units of total RNA dissolved in 0.01 M Tris-HCl, pH 7.5, 0.5 M KCl. The solution was then applied to a 2 ml (about 0.5 g dry weight) of oligo(dT)-cellulose column previously equilibrated with the application buffer. The materials retained by the column were eluted in two steps with buffer of reduced ionic strength. The first elution buffer contained 0.01 M Tris-HCl, pH 7.5, 0.5 M KCl; the second, neutralized water. The materials eluted were immediately precipitated by the addition of 0.3 M LiCl and two volumnes of 100% ethanol.

Sucrose gradient centrifugation. The centrifugation of the total RNA was conducted in a linear 15-30% sucrose gradient in 0.05 M Tris-HCl, pH 7.4. Sedimentation was carried out at 38,000 rpm/min for 20 hrs at 4°C in a SW41 rotor. Standard RNA markers, E. coli 18S and 28S rDNA and Lens 10S RNA were run in parallel gradients. From this gradient, we collected two fractions of RNA: rRNA and 7-14S messenger RNA fraction, then applied it to a second centrifugation. The 9- 14S messenger RNA fraction region was collected for subsequent usage (11, 12).

Assay for in vitro protein synthesis: The cell-free protein synthesizing system was prepared from wheat germ lysate. Routine assay mixture contained in 50 µl: 28.9 mM HEPES, pH 7.6, 2.75 mM dithiothreitol, 64 mM KCl, 2 µg creatine kinase, 20 µM GPT, 8 mM creatine phosphate, 3.5 mM magnesium acetate, 20.7 µM (^{3}H) leucine at 1.50 µCi/mMole, 20 µg calf liver mRNA, 400 µg lysate protein. Lysate dilutent: 20 mM HEPES, 120 mM KCl, 5 mM magnesium acetate, and 1 mM dithiothreitol (13). Incubation was performed at 25°C for 60 min.

Preparation of cell culture: The preparation of kidney cell suspensions from 12-14 day-old chicken embryos, which were perfused through the heart. Kidneys were cut into small pieces and placed in a sterile dish containing 0.05% collagenase. (Collagenase buffer: NaCl 390mg, KCl 50 mg, $CaCl_2 \cdot 2H_2O$, 70 mg, HEPES 2,400 mg, 1 N NaOH 6.6 ml, 50 mg collagenase in 100 ml of double distilled water, pH 7.6.) The material was pipetted in and out of a Pasteur pipette for 10 minutes at room temperature. The cell suspension was centrifuged at 400-800 rpm for 1 min. The supernatant was removed and 10 ml of washing buffer (NaCl 830 mg, KCl 50 mg, $CaCl_2$ 18 mg, HEPES 240 mg, 1 M NaOH 0.55 ml in 100 ml of double distilled water, pH 7.4) was added. This was allowed to stand for 5 min, then the cells spun down and resuspended in washing buffer. After 5 min at room temperature, the cells were spun down again. Medium 199 was used to wash the cells and the cells were resuspended in medium supplemented with insulin 0.2 μg/ml. Each culture flask contained about 7×10^5 cells in 5 ml of medium, containing 0.2 μg/ml insulin.

Treatment of chicken embryonic kidney cells:

(1) Treatment of different fractions of RNA. The kidney cell cultures were treated with different fractions of RNA in the presence of radioactive amino acid (^{3}H)-leucine 2 μCi/ml (specific activity: 348 mCi/mM). The fractions of RNAs employed were A peak(A), B peak(B), messenger fraction from sucrose gradient (M), ribosomal fraction from sucrose gradient (R), and total RNA(T). Control experiments were the same except that RNAs were omitted. After 24 hr incubation with RNAs at 37°C with 5% CO_2, culture cells were harvested and the medium saved. The cells were centrifuged down and washed with 0.05 M Tris buffer pH 7.5 several times. 0.5 ml of 0.05 M Tris buffer pH 7.5 was added

to the cells before homogenization, then centrifuged at 3000 rpm 30 min. 10% TCA was added to the aliquot and precipitable radioactivity was measured by adding scintillation fluid and counting in a scintillation counter. Another portion of aliquots was run for counter immuno-electrophoresis and slab gel electrophoresis.

(2) Study of the kinetics of the amino acid incorporation. The culture flask contained about 7 x 10^5 cells in 5 ml of medium, containing 0.2 μg/ml insulin. Prior to the treatment, 3 O.D. liver mRNA was dissolved in 1 ml medium and the medium sucked out from the culture flask. Liver mRNA and 5 μCi ^{3}H-leucine was applied to the culture flask to treat culture cells at room temperature for one hour. 4 ml medium was then added to the flask and incubated at 37°C. After certain time intervals (0.6, 12, 18, 24, 30, 36, 42 hours), the cells were collected and washed with 0.05 M Tris buffer pH 7.5 several times and homogenized. 10% TCA was added to the aliquot and precipitable radioactivity measured.

(3) Study of the effect of amino acid uptake by different concentrations of calf liver and calf lens poly(A)-containing mRNA. We used different concentrations of calf liver (0.5, 1, 2 and 3 A260 units) and calf lens mRNA to treat the culture chicken kidney cells separately. Each culture flask contained about 7 x 10^5 cells. A given amount of mRNA and 5 μCi ^{3}H-leucine was added to the flasks. The treatment and counting methods were the same as described above.

(4) Study of the effect of histone and polylysine on mRNA-mediated amino acid incorporation. 25 μg of polylysine or histone was added separately into 120 μg of liver mRNA. The mixture was studied spectrophotometrically. The reaction mixture and 5 μCi ^{3}H-leucine

were used to treat culture cells. After 24 hr incubation, the TCA precipitation method was used to study the amino acid incorporation.

(5) <u>Study of the effect of various inhibitors on mRNA-mediated amino acid incorporation</u>. One culture flask contained about 7 x 10^5 cells in the TC 199 medium and cultured at 37°C. Actinomycin D (1 μg/ml) and streptoval (4 μg/ml) and 5 μCi ^{3}H-leucine were added separately to the culture cells. The treated cells were incubated at 37°C for 24 hrs.

(6) <u>Study of the effect of calcium treatment on the mRNA-induced ^{3}H-leucine incorporation</u>. The treated cells (about 6 x 10^7 cells/ml) were chilled in the ice box, washed twice with 25 mM Tris-HCl, pH 7.6 containing 0.01 M NaCl and 5 mM $MgCl_2$.

Preparation of Na-Ca-Mn solution:

NaAe	0.40 gm	
		adjust pH with HAc
$CaCl_2$	0.33 gm	

Then add 1.385 gm of $MnCl_2.4H_2O$, make up to 100 ml in 25 mM Tris HCl, pH 7.6, and filter sterilized. Cells were suspended in 10-20 ml of Na-Ca-Mn solution for 20 min in ice. The cells were centrifuged down and resuspended in 5 ml of Na-Ca-Mn solution, then centrifuged down and liver mRNA (3 O.D.) radioactive amino acid (5 μCi ^{3}H-leucine) added, together with 1 ml of culture medium kept at O°C for 5 min. They were then transferred to 37°C for 5 min and suspended in 5-10 ml of culture medium and incubated. The experimental culture cells were compared with that of the control.

Counter Immunoelectrophoresis: Counter immunoelectrophoresis was carried out by the method of Romas, et al (14). Glass slides, 75 x 50 mm, were first coated with dilute agarose solution and dried in the oven. 7 ml of 1 per cent agarose in Babital buffer (Barbituric acid 20.7 g, NaOH 3.8 g to 2 liters of double distilled water, pH 8.2) was poured on the coated slides and stored in moist chambers in the cold room before use. Parallel rows of cells were cut. Samples were placed in the left well (cathode) and antibody for bovine albumin in the right well (anode). The electrophoresis was run at 80 volts. The buffer temperature was maintained at 23°C throughout. After washing with saline (0.9% NaCl), the slides were dried at 60°C and stained by P-sulfobenzene-azo-o-sulfo-benzene azo-B-naphthol-3,6-disulfonic acid and washed with 5% acetic acid.

Vertical slab gels for SDS electrophoresis: The gel solutions were prepared on ice by adding all reagents, except the TEMED, and degassed under vacuum for 15 minutes. Then TEMED was added to initiate polymerization. Acrylamide solution was pumped into the sandwiched slate until it just touched the lower corner of the tilted PVC gel forming plates. A flow rate of 0.5 to 3.0 ml/min was used. We continued to add acrylamide while slowly pushing the elevated end of the gel up to the notch of the forming plate to allow for shrinkage of the gel during polymerization. A sample well forming comb was inserted between the glass plates and the unit placed on the leveling table. Using a pasteur pipet, the stacking gel was added at both ends of the well forming comb. The well forming combs were filled up to the acrylic notch, making it unnecessary to overlayer the stacking gel. After polymerization, 2 ml of stacking gel buffer was added to the notch around the comb, gently flexing, and the comb removed. The sample slots were rinsed twice with stacking gel buffer. The cams and the support sealing bars were removed from the unit.

Culture of the chicken kidney cells: 3 A260 liver or Lens mRNA and 5 µCi ^{3}H-leucine were added to the culture. After 24 hr incubation, cells were collected and washed with buffer (0.05 M Tris pH 7.5) several times. Then the cells were homogenized. Anti-chicken albumin antibody was then added and kept in a refrigerator overnight. The next day, the precipitate was spun down. The supernatant was saved and anticalf albumin antibody was added. It was again left in a refrigerator overnight, and the precipitate centrifuged. The precipitate was run on the SDS gel electrophoresis.

RESULTS

Culture of Chicken embryonic cells: Four-day old chicken embryonic kidney cells were cultured in TC 199 medium. Fig. 1 shows the cells beginning to spread at day 2 (1B). At day 7 (1D), the cells already formed a confluent sheet. Usually, the 7-day cell culture was used for the experiment.

Assay for the translational activity: In order to test the fidelity of RNA used in the experiment, a wheat germ cell-free protein synthesizing system was employed. Table 1 shows the results of this experiment. It is interesting to note that the A peak from oligo(dT)-cellulose exhibited stimulation. Both B peak and mRNA fractions collected from the 9 to 14S region of the sucrose gradient yielded a similar highest stimulation, comparable to that of the highly purified hemoglobin mRNA.

RNA-mediated amino acid incorporation: Table 2 shows the TCA precipitate counts in the intercellular media after 24 hr incubation with various fractions of calf liver mRNA. Again, both B peak and sucrose gradient-purified mRNA fractions gave the highest specific activity, while the activity from that of the total or Ribosome RNA was at the background level. The data

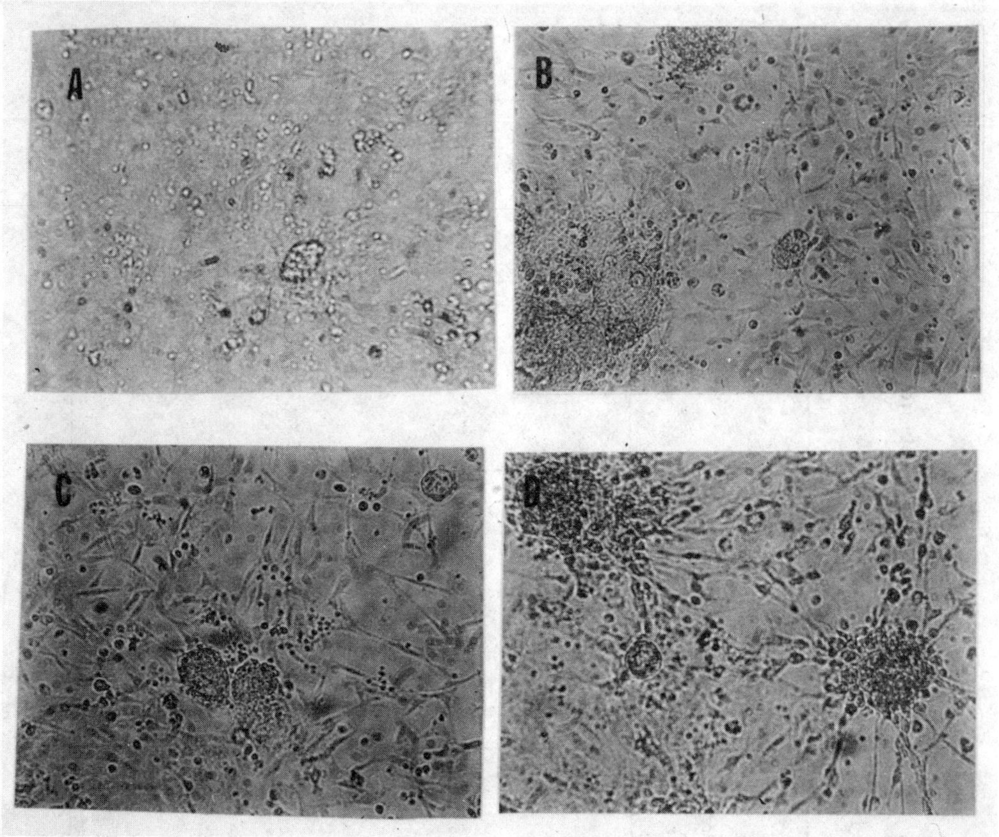

Fig. 1 The Chicken Embryonic Kidney Cell Culture
1B shows the cells beginning to spread at day 2,
1D at day 7 the cells already formed a confluent sheet.

Table 1 **Translational Assay of Various RNAs in a Wheat-germ Cell-free System**

RNAs	Incorporation (CPM*)	RNA Concentration (A 260 nm)	Stimulation per O.D. (A 260 nm)
Control	120.06	0	
Total RNA	402.83	0.4	8.4
A peak	655.0	0.4	13.625
B peak	999.57	0.25	33.28
m RNA fraction	901.33	0.5	15.0
Ribosome RNA	372.04	0.5	6.18
Hemoglobin mRNA	823.92	0.5	13.72

Table 2 RNA Mediated Amino Acid Incorporation in Chicken Embryonic Kidney cells Culture (A) TCA ppt Count in the Intercellular Media

RNAs	Concentration of RNA added (O.D./ ml)	Incorporation cpm (in media)	Stimulation (Fold)	Stimulation per O.D. (%)
Control	0	21.65		
Total RNA	10	83.25	3.8	38
A $peak_1$	10	76.33	4.7	47
A $peak_2$	10	127.56		
B peak	3	104.27	4.8	160
mRNA Fraction	5	99.31	4.6	92
Ribosome RNA	10	82.64	3.8	38

in Table 3 represent the amino acid incorporation induced by various RNA preparations in the intracellular extract. It is apparent that the B peak fraction exhibited the highest specific activity, almost twice that of the medium, indicating the site of RNA-mediated incorporation. The mRNA fraction from the sucrose gradient yielded higher specific activity than that of the A peak component, which gave an activity slightly higher than the background level of total and rRNA preparations.

In order to maximize the experimental condition, we carried out the study of amino acid incorporation kinetics. Fig. 2 shows that the incorporation is linear as a function of liver mRNA concentration. Using specific lens mRNA as a control, essentially the same result was obtained (Fig.3). Fig.4 shows the time kinetics of the stimulation, while 3 A260 calf liver mRNA was used in each case.

Characterization of RNA-induced polypeptides: The nature of the newly-synthesized polypeptides was characterized by SDS polyacrylamide gel electrophoresis and by counter immunoelectrophoretic techniques. Fig. 5 shows that the newly-synthesized polypeptides induced by poly (A)-containing mRNA fractions migrated to a position corresponding to that of calf albumin. The results from counter immunoelectrophroresis indicate that there is a complementary interaction between antigen and antibody. The first row in Fig. 6 represents the standard complementary interaction between calf albumin and anticalf albumin antibody, while the second row is the control. It is clear that there is definitely complementary interaction between the mRNA fraction from the sucrose gradient (third row, Fig. 6), the B peak fraction (fifth row, Fig. 6), and the anticalf albumin antibody. It is interesting to note that there is a precipitation line formation (fourth and sixth rows, Fig. 6)

Table 3 RNA Mediated Amino Acid Incorporation in Chicken Embryonic Kidney Cells Culture

(B) Incoperation of Amino Acid in the Intracellular Extract

RNAs	Concentration of RNA added (O.D./ml)	Incorporation cpm (Cells Extract)	Stimulation (Fold)	Percentage of Stimulation per O.D. (%)
Control	0	11.53		0
Total RNA	10	43.22	3.7	37
A $peak_1$	10	71.05	5.7	57
A $peak_2$	10	61.15		
B peak	3	93.87	8.1	270
mRNA Fraction	5	54.33	4.7	94
Ribosome RNA	10	38.80	3.36	33.6

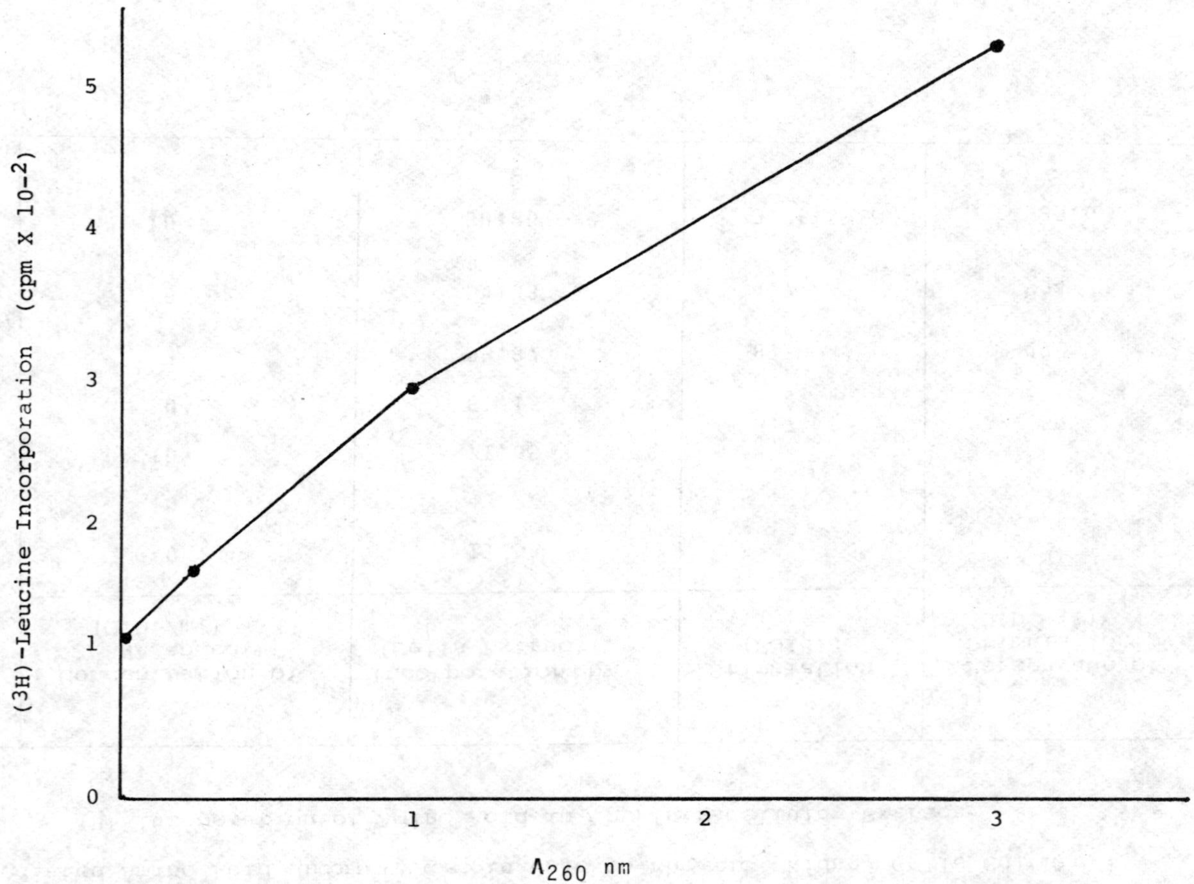

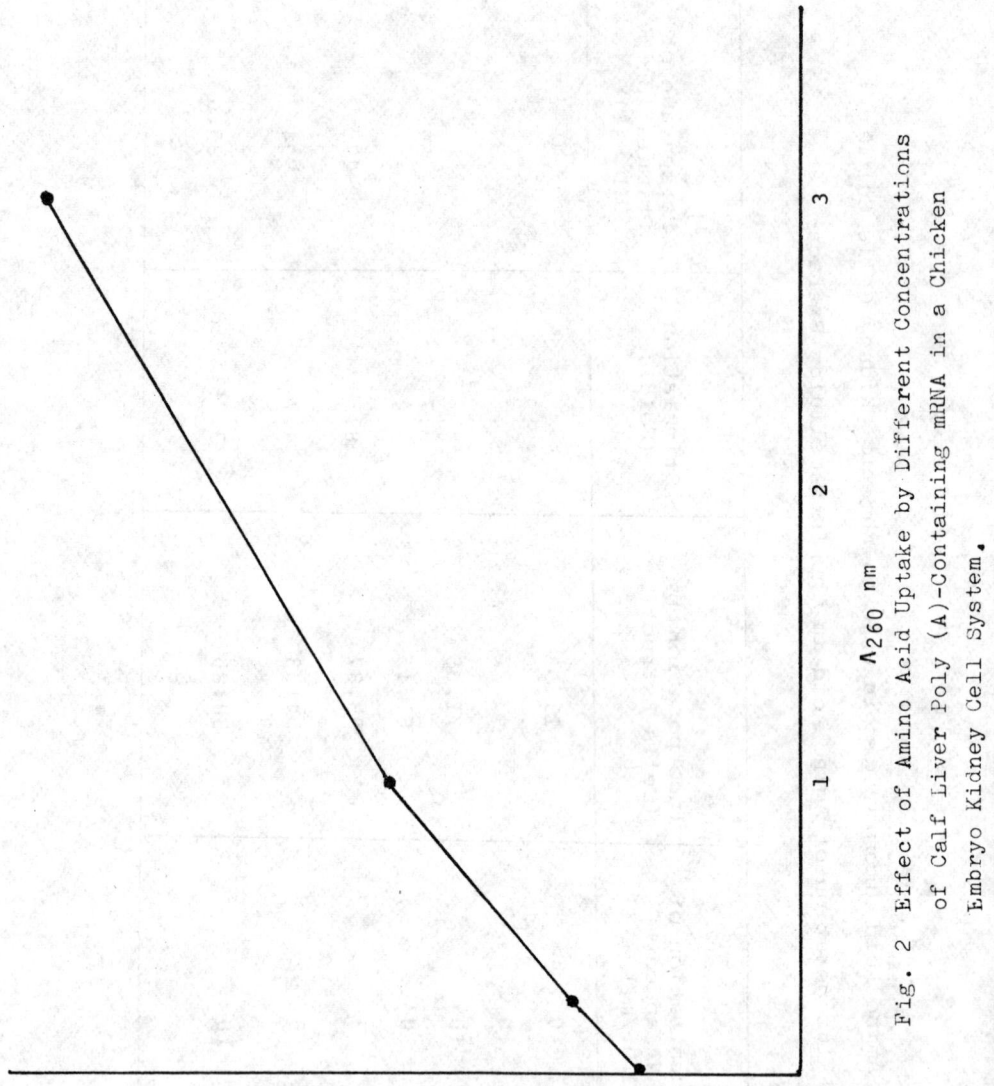

Fig. 2 Effect of Amino Acid Uptake by Different Concentrations of Calf Liver Poly (A)-Containing mRNA in a Chicken Embryo Kidney Cell System.

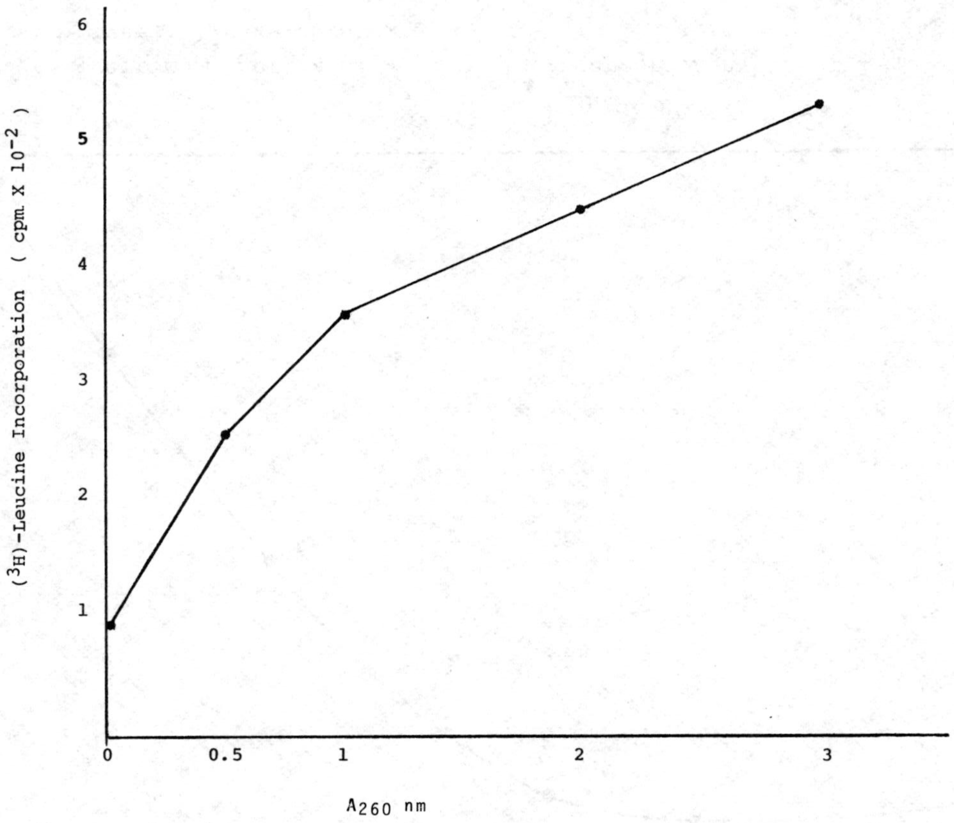

Fig. 3 Amino Acid Incorporation as a Function of Lens Poly (A)-Containing mRNA Concentration Incubated in the Chicken Embryo Kidney Cell System.

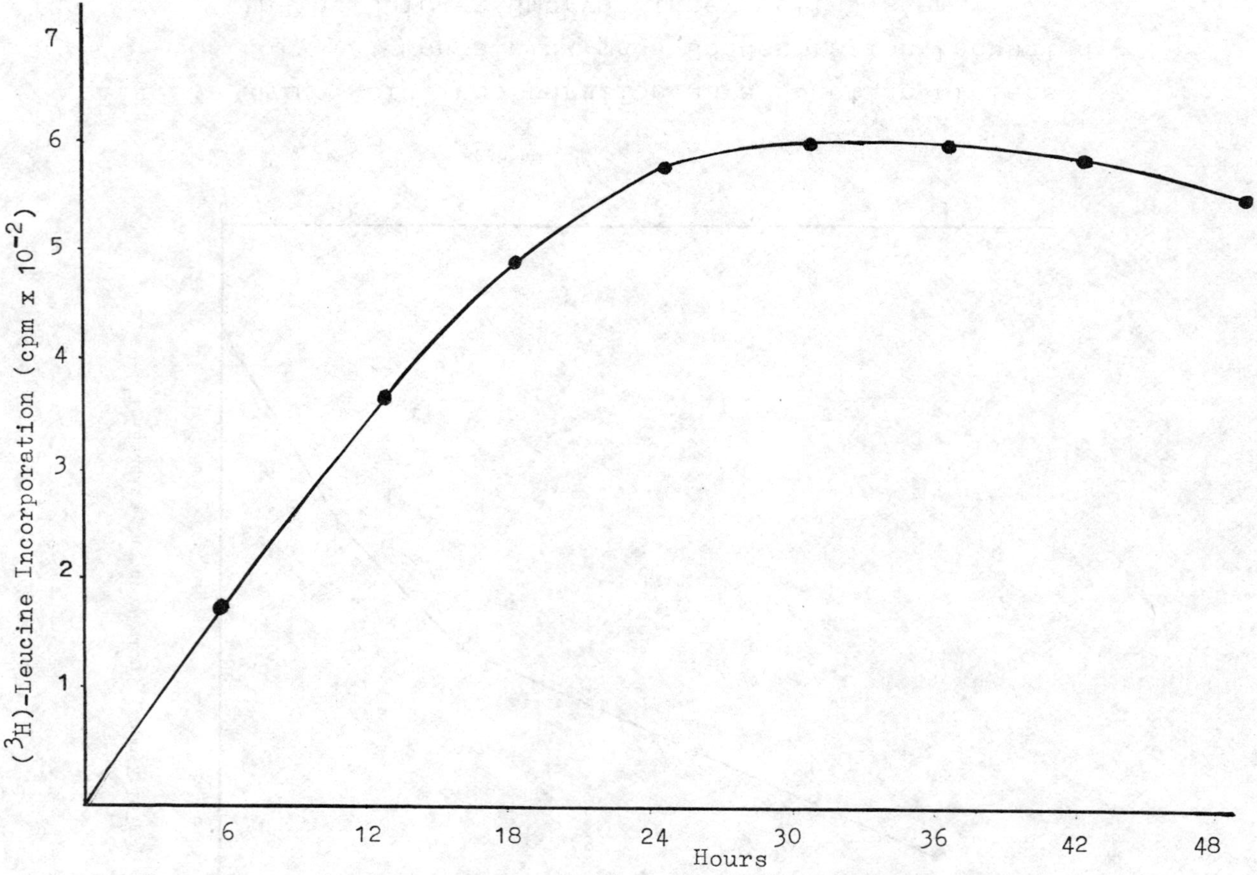

Fig. 4 Reaction Kinetics of Amino Acid Incorporation Effected by Addition of Liver Poly(A)-Containing mRNAs.

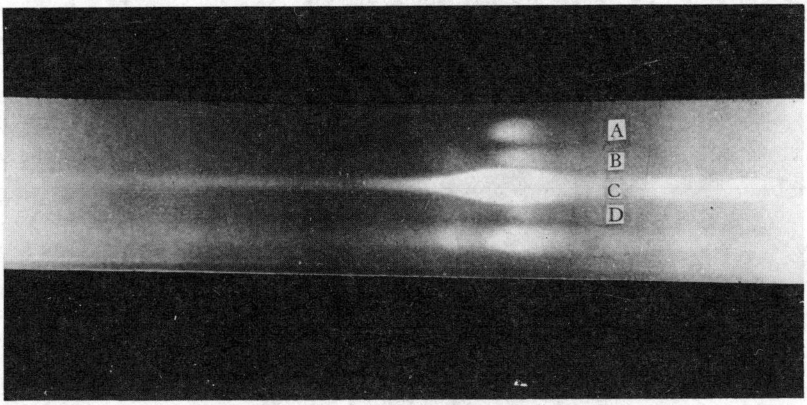

Fig. 5 Eletrophotogram of the Newly-Synthesized Polypeptides Induced by Poly(A)-Containing mRNA fractions

(A) (B) B peak fraction,

(C) Calf albumin (as a control),

(D) mRNA fraction from sucrose gradient.

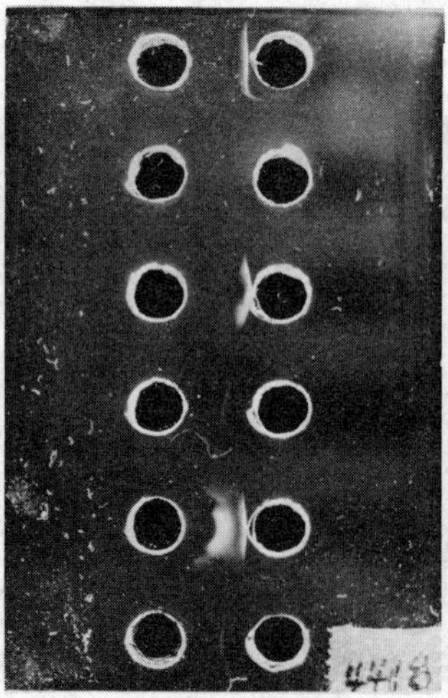

Fig. 6 Immunoeletrogram of Complementary Interaction between Samples and Anticalf Albumin Antibody Explanation See Text.

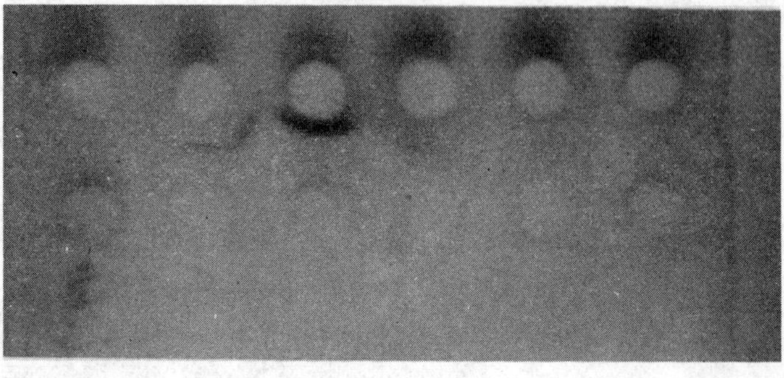

Fig. 7 Immunoelectrogram of Complementary Interaction between Samples and Antichicken Albumin Antibody

(A) The standard complementary interaction between chicken albumin and antichicken albumin antibody,

(B) Complementary interaction between B peak fraction and antichicken antibody.

between the A peak messenger fractions and the anticalf albumin antibody. When the antichicken albumin antibody was employed, a complementary precipitation line was also formed (Fig. 7).

Facilitated macromolecular transport by interaction with basic polypeptides: A recent report (16) suggests that polylysine can enhance the rate of RNA intercellular transport. Before testing it out in a cell culture, we determined the interaction *in vitro* spectrophotometrically. Since histone is a nuclear protein containing polylysine rich fractions, it is also included in our studies. Fig. 8 and Fig. 9 represent the interaction spectra of mRNA with histone and polylysine, respectively. In both cases, the intensive hypochromicity indicates strong interaction between the ligands and mRNAs.

The interaction mixtures of mRNAs and basic polypeptides were then incubated with the chicken kidney cell culture system. The results in Table 4 suggest that polylysine exhibited no effect on the RNA-mediated amino acid incorporation, whereas the interaction with histone suppressed about 60% of the activity.

Molecular mechanisms of RNA-induced polypeptide synthesis: Actinomycin D and streptoval were employed to test the possible mechanisms for the RNA-mediated peptide synthesis. Table 5 summarizes the results of such experiment. Actinomycin D at 1 μg/ml inhibited more than 90% amino acid incorporation. On the other hand, streptoval did not exert any significant effect on the incorporation system. The kinetics of RNA-induced incorporation after treatment by actinomycin D were also studied. Fig. 10 shows the incorporation inhibited in a linear fashion at mRNA concentration ranging from 0.5 to 4.0 A260/ml of embryonic chicken kidney cell culture. Fig. 11 shows that the uptake of amino acid is about ten times higher in the calcium-treated cell system than that of the untreated control system.

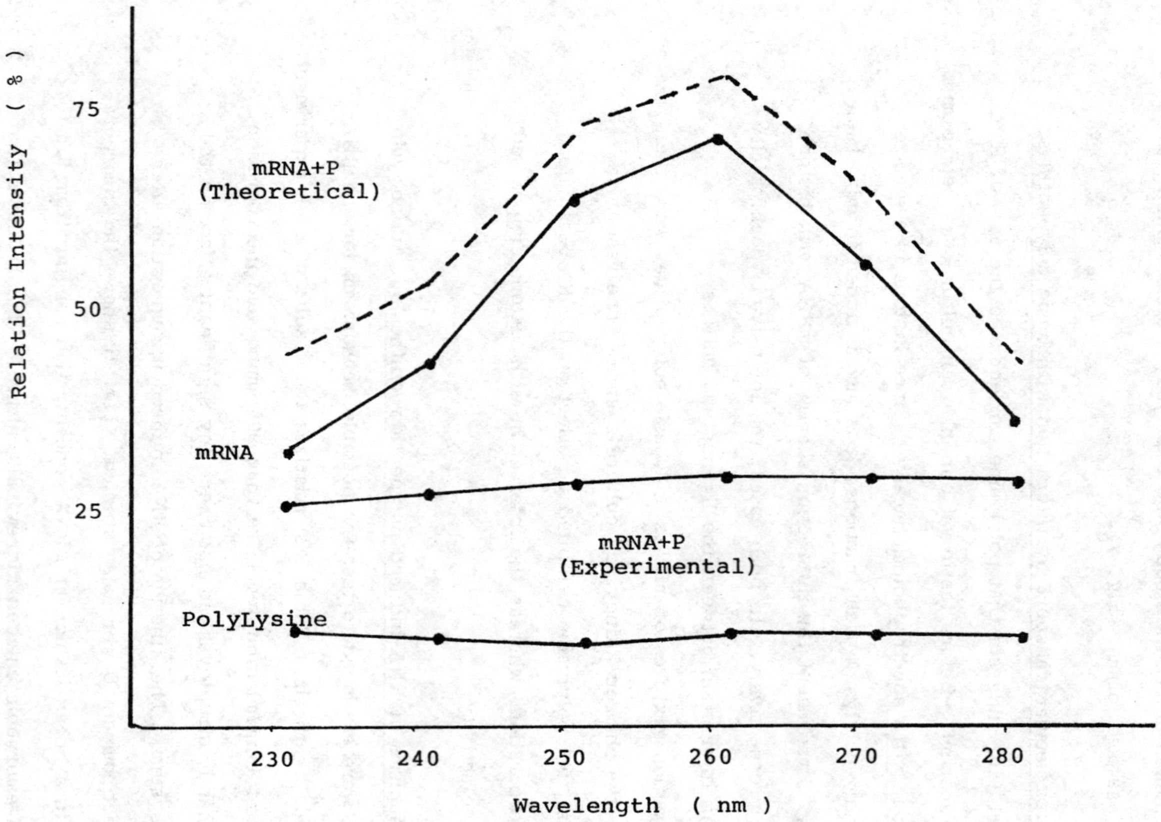

Fig. 8 Interaction Spectra of Poly-L-Lysine and Liver Poly (A)-Containing mRNA.

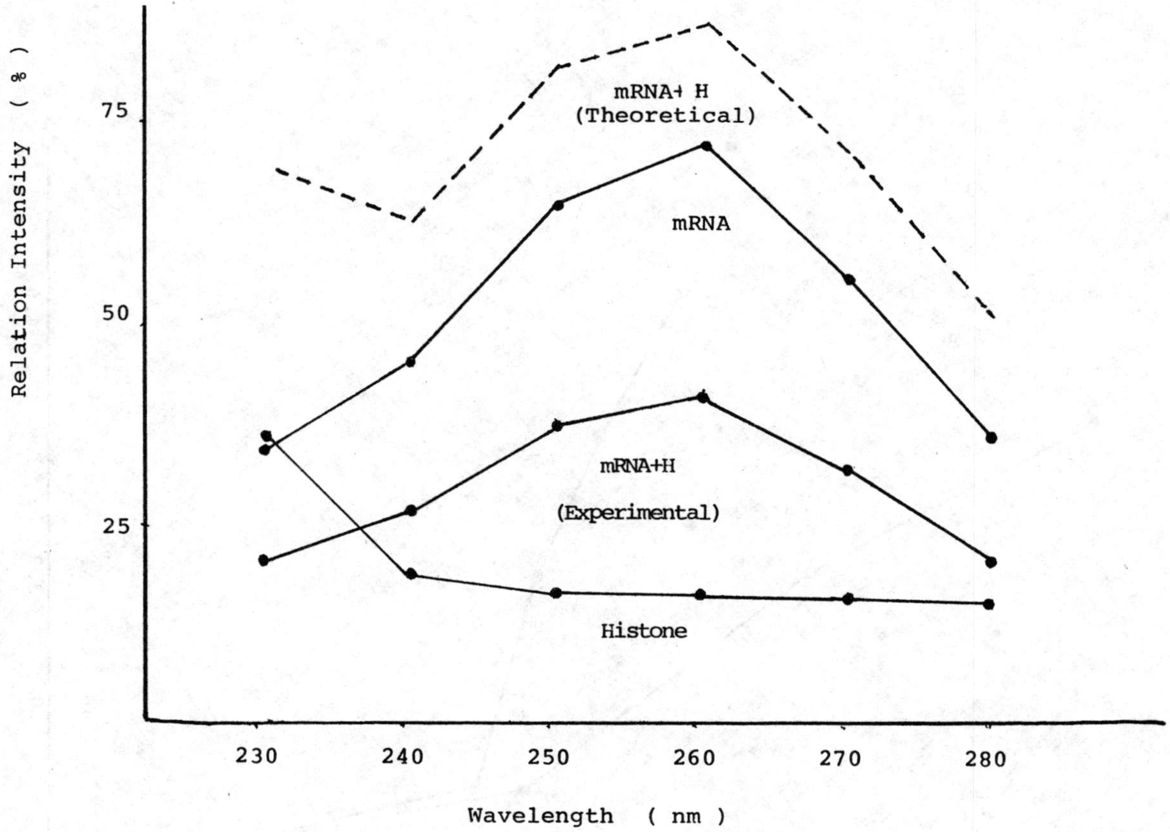

Fig. 9 Interaction Spectra of Histone and Liver Poly (A)-Containing mRNA.

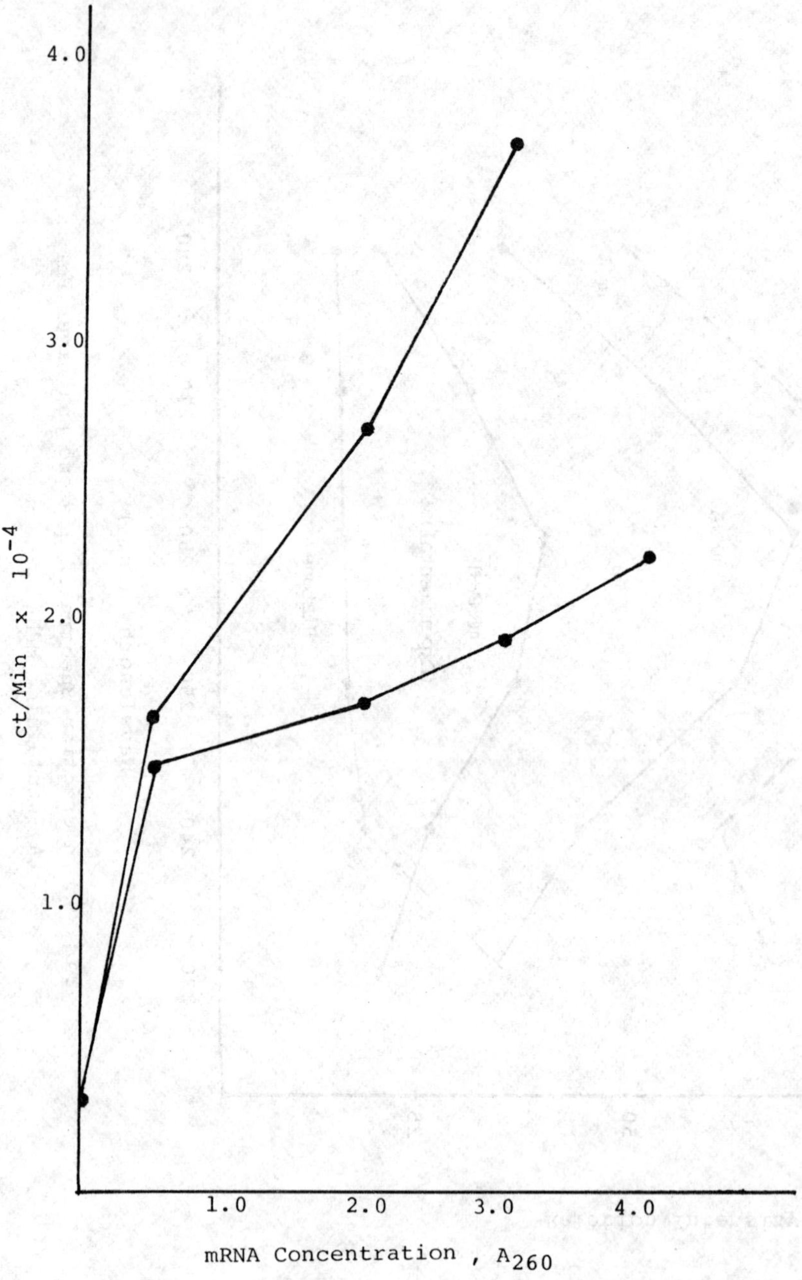

Fig. 10 Concentration Dependence of mRNA on (^{3}H)-Leucine Incorporation in the Presence and Absence of Actinomycin D.

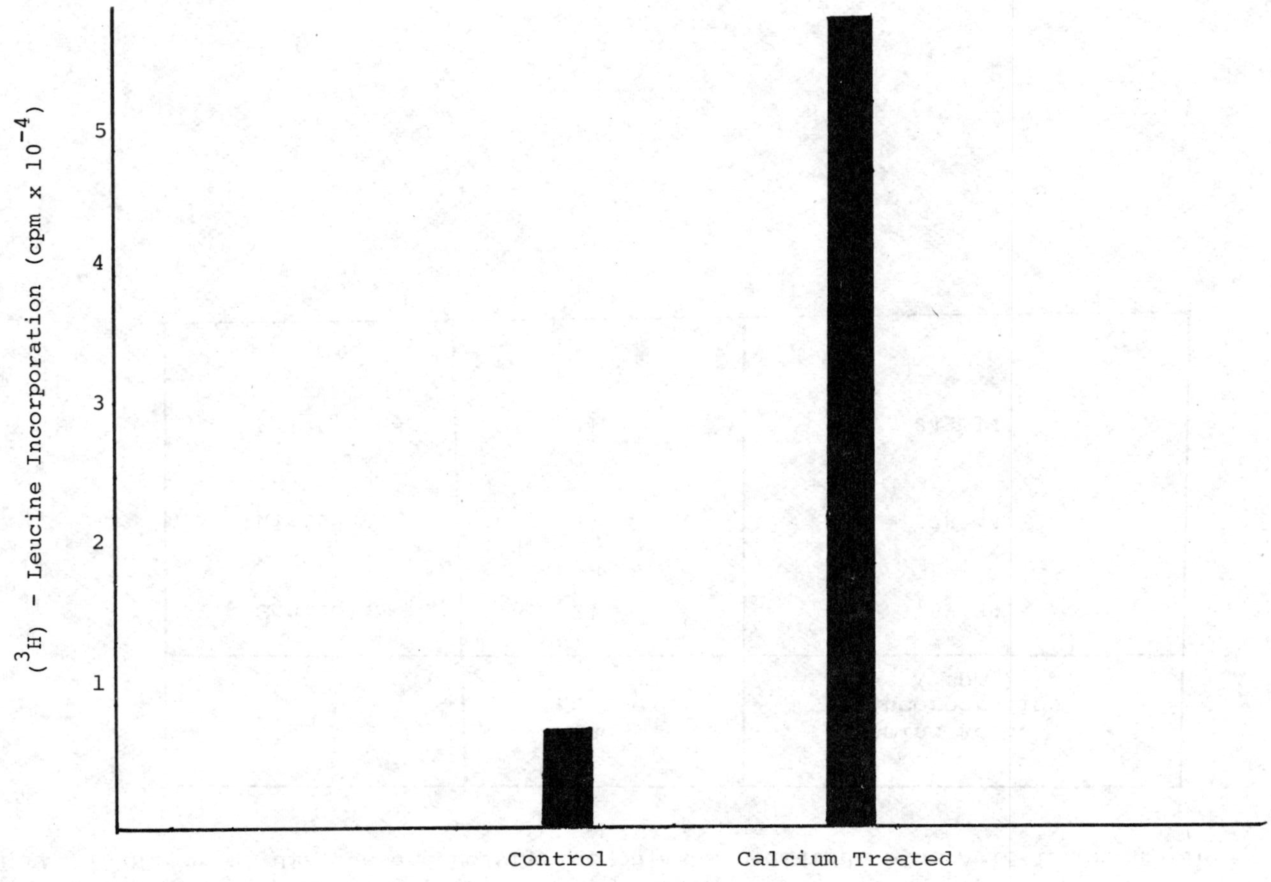

Fig. 11 Effect of Ca^{++}-Treatment on the mRNA-Induced (^{3}H)-Leucine Incorporation in Cell Culture.

Table 4 Effect of Histone and PolyLysine on mRNA-Mediated Amino Acid Incorporation

	Time (hours)	Amino Acid Incorporation (cpm)
Control	24	752.39
Histone	24	278.44
PolyLysine	24	843.14

Table 5 **Effect of Various Inhibitors on mRNA Mediated Amino Acid Incorporation***

	Incorporation (CPM)
Control	468.18
Actinomycin D 1 ug/ml	44.15
Streptoval 4 ug/ml	419.445

The present communication indicates that calf liver mRNA may contain poly(A) population. The fact that the B peak mRNA fraction possesses higher translation activity than that of mRNA fractions from sucrose gradient suggests the possible contamination of ribosomal RNA fragments.

It is interesting to note that the culture media contain newly-synthesized polypeptides. This data suggests the transport of the newly-synthesized materials from the cells to the culture media. This result is also consistent with the possibility that the newly-synthesized polypeptides are transport proteins in nature. However, the higher specific activity in the cellular extracts than that of the media indicates the intracellular nature of RNA induced incorporation.

The results of SDS polyacrylamide gel electrophoresis suggest that the newly-synthesized proteins are directed by added calf liver mRNA and lens mRNA. Data from immunoelectrophoresis show they contain albumin inducible products. We proposed to interpret the results partially as a consequence of gene activation. This concept is further substantiated by the experiments of actinomycin D inhibition and streptoval resistance. These data indicate the newly-synthesized mRNAs are involved in the production of polypeptides and the reverse transcriptase directed complementary DNA is likely not involved.

CONCLUSION

Our preliminary results indicate that mRNA incubated in a cell culture system can induce polypeptide synthesis linearally as a function of mRNA concentration. The newly-synthesized polypeptides have an electrophoretic mobility in SDS gel similar to that of the authentic albumin. Counter immunoelectrophoretic data suggest that the newly-synthesized polypeptides can react with antisera prepared against calf and chicken albumin, thus indicating both chicken and calf albumin are synthesized. Polylysine has no effect on the system, while the histone suppressed three quarters of the incorporation.

Finally, our overall data indicate that mRNA incubated in a cell culture can not only direct specific protein synthesis, but also can activate the host specific gene expression.

ACKNOWLEDEMENT

I wish to thank Dr. M. C. Niu of Temple University, and Dr. John H. Chen of New York University. This work was carried out in their laboratories while I received a visiting scholarship from the Academia Sinica.

REFERENCES

1. Mishra, N.C., Szabo, C.T. and Tatum, E.L. (1973), in "The Role of RNA in Reproduction and Development", eds. Niu, M.C. and Segal, S.J. (North-Holland Publ. Co., Amsterdam), pp. 259-268.

2. Mishra, N.C., M.C. Niu and Tatum, E.L. (1975), Proc. Natl. Acad. Sci. USA 72, 642-645.

3. Beijinski, M. and Plawecki, M. (1973), in "Role of RNA in Reproduction and Development", eds. Niu, M.C. and Segal, S. J. (North-Holland Publ. Co., Amsterdam), pp. 203-224.

4. Tung, T.C. and Niu, M.C. (1973), Sci. Sin. 16, 377-384.

5. Tung, T.C. and Niu, M.C. (1975), Sci. Sin. 18, 223-228.

6. M.A.Q. Siddiqui (1980), This proceedings.

7. Shih-Fang Yang and M.C. Niu (1977), Proc. Natl. Acad. Sci. USA, 74, No. 5, pp. 1894-1898.

8. Gerd Grieninger and S. Granick (1975), Proc. Natl. Acad. Sci. USA, 74, No. 12, pp. 5007-5011.

9. Chen, J. H., G. C. Lavers, and A. Spector (1974), Exp. Eye Res. 18, 189-199.

10. Aviv, Haim and Philip Leder (1972), Proc. Natl. Acad. Sci. USA, 69, pp. 1408-1412.

11. Evans, M.J. and Lingrel, J.B. (1969), Biochemistry 8, pp. 3000-3005.

12. Lee, S.Y., Mendecki, J. and Brawerman, G. (1971), Proc. Natl. Acad. Sci. USA, 68, pp. 1331-1355.

13. Chen, J.H. and A. Spector (1977), Proc. Natl. Acad. Sci. USA, 74, No. 12, pp. 5448-5452.

14. Romas, N.A., K. C. Hsu, P. Tonashefsky, M. Tannenbaum (July 1978), Urology, XII, No. 1.

15. Cleveland, D.W., S. G. Fisher, N.W. Kirschner, and U.K. Laemmli (1977), The Journal of Biological Chemistry, 252, No. 3, pp. 1102-1106.

16. Kolody, Gerald M. (1979), The Journal of Cell Biology, 83, No. 2, Part 2, RT 2324.

CONTROL OF CARDIAC MUSCLE GENE EXPRESSION IN EARLY CHICK EMBRYO

M.A.Q. Siddiqui[1], H-H. Arnold[2], A. K. Deshpande[3], H.J. Drabkin[4],

S.B. Jakowlew and S.K. Narula

Department of Biochemistry

Roche Institute of Molecular Biology, Nutley, N.J. 07110 U.S.A.

[1]To whom all correspondence should be addressed.

[2]Present address: Department of Toxicology, University of Hamburg, Hamburg,2000, Germany

[3]Present address: Population Council, Rockefeller University, New York, N.Y. 10021

[4]Present address: Department of Biology, Massachusetts Institute of Technology, Boston, Mass. 02139

ABSTRACT

Heart muscle development, which is an early event in chick embryonic development, provides a model system to investigate the mechanism(s) regulating the transition of precardiogenic embryonic cells to highly differentiated cardiac myocytes. We have recently reported that a low molecular-weight RNA from the embryonic chick heart (7S CEH-RNA) plays a role in control of heart muscle differentiation, since addition of this RNA induces a specific mode of change in the chick blastodermal cells similar to myogenic process.

Two approaches were undertaken to investigate the nature of molecular events controlling the induction of heart muscle differentiation both _in situ_ as well as the RNA-induced system _in vitro_. First, as a means of obtaining specific probes for monitoring the transcriptional activity for muscle specific genes during specific stages of development, the mRNAs for various muscle proteins and the inducer 7S CEH-RNA were isolated, purified, and characterized. DNA complementary to these RNAs were prepared and amplified by molecular cloning techniques. As a second approach, a detailed structural analysis of the DNA probes and of 7S CEH-RNA was undertaken. Furthermore, orientation of muscle specific genes with respect to the 7S gene within the physical map of the chick genome is being analyzed by restriction mapping of the DNAs and the subsequent hybridization with chick DNA restriction fragments.

Thus, the availability of specific DNA probes and the structural characterization of these molecules would facilitate the analysis of the mechanism underlying embryonic heart muscle differentiation and the assessment of the role 7S CEH-RNA plays during development.

INTRODUCTION

The understanding of molecular aspects of cell communication and differentiation in embryogenesis requires a comprehensive analysis of the mechanism(s) regulating the genomic activity of a specific population of cells as a function of development. For a particular developmental pathway, the mechanism of differential gene activity can be studied by identifying and isolating the genes that code for proteins involved. However, the complexity of eukaryotic genome precludes a direct isolation of structural genes and elucidation of the regulating elements. One way to circumvent this problem is to isolate the mRNAs coding for the developmentally regulated marker proteins and use the DNA probes generated by transcription of the mRNA to monitor the transcriptional activity during development. Several laboratories have studied regulation of gene expression in embryonic cells, but many of these studies have been concerned with cells which were partly differentiated, as such, the genome was previously active for transcription of specific proteins. Therefore, the question as to how the initial activation of a selected population of genes occurs remains unanswered.

According to the current concepts, changes in embryonic development, which are caused by differential gene activity, are controlled in some way by elements in cytoplasm of the fertilized eggs (1-3). These molecules play a key role in determining the early pattern of morphogenetic changes. Morphogenesis and cell differentiation in turn are the results of derepression of gene activity. The informational molecules must, therefore, be involved, in some way, in control of transcriptional activity of the genome. The existence of such gene controlling elements has been elegantly demonstrated in frog oocytes using nuclei transplantation technique (4). Models have been proposed in which proteins and/or RNA are implicated as regulatory elements (1-3, 5, 6). Although to our knowledge no clear evidence is yet available in support of the main theme of the models, there is ample documentation on manipulation of gene expression in general mediated by both protein and RNA. A particularly relevant example is that of a protein in Mexican axolotl synthesized during oogenesis which appears to be essential for activation of nuclear genes required during gastrulation (7). The effect of non-histone proteins on transcription of the stage specific chromatin is well documented (8, 9). The involvement of RNA has been postulated in control of several cellular functions, such as the specification of antibody production (10), interferon induction (11), in causing specific disease symptoms (12), and in others where the RNAs are implicated in regulation of cell development and differentiation (13-16; see refs. 17-20 for reviews). In spite of these intriguing examples, there is no clear definition of the physiological role of RNA and/or protein in regulatory processes. Embryonic system which permit identification and functional analysis of the potential regulatory macromolecules would be highly suitable for elucidation of the underlying mechanism.

In the post-gastrulation stage, stage 5 (21) of the chick embryo, the precardiogenic cells are known to be localized in bilateral regions adjacent to Hensen's node (Fig. 1) (22, 23). This apparently homogeneous and undifferentiated population of cells is already programmed to develop into

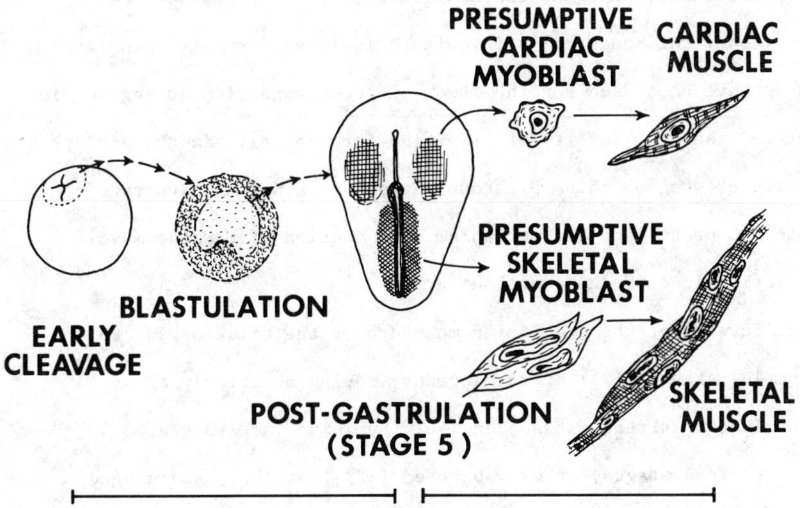

Fig 1. Chick embryonic muscle development. The presumptive cardiac muscle cells are localized on lateral sides (shaded oval areas) adjacent to Hensen's node in stage 5 chick embryo.

cardiac myoblasts, since the explants when grown in culture in a variety of media or as chorioallantoic transplants differentiate into well-defined cardiac muscle tissue. Although cells in culture isolated from differentiated muscle tissues have been used for the study of gene expression during muscle differentiation, the availability of primitive form of cells in the precardiogenic region of stage 5 chick blastoderm provides a unique advantage as a model system to probe into the mechanism of induction of cardiac muscle gene transcription in embryonic development.

It is believed that the precardiac mesoderm of the chick begins to invaginate during stage 4 (21) of development, moving anteriorly along the growing primitive streak. Based on [^{3}H] thymidine labeled grafts obtained from various stages, it was observed (22) that the precardiogenic region extends at this stage halfway between anterior and posterior ends of the streak. When tissue fragments isolated by transverse cuts 0.4-0.6 mm posterior to Hensen's node are grown separately _in vitro_, the posterior fragment, the postnodal piece (PNP), does not develop into any histologically identifiable structure (24-30). In contrast, the anterior portion (AP) containing Hensen's node develops readily into axial structures (Fig. 2). The PNP when combined with the AP however, develops into embryonic tissues, including a well-defined heart tubular structure, suggesting that the anterior portion carries the putative information needed for the precardiac cells located in the PNP to transform into cardiac muscle tissue. Later, it was shown that the PNP cells can also be induced to develop into cardiac

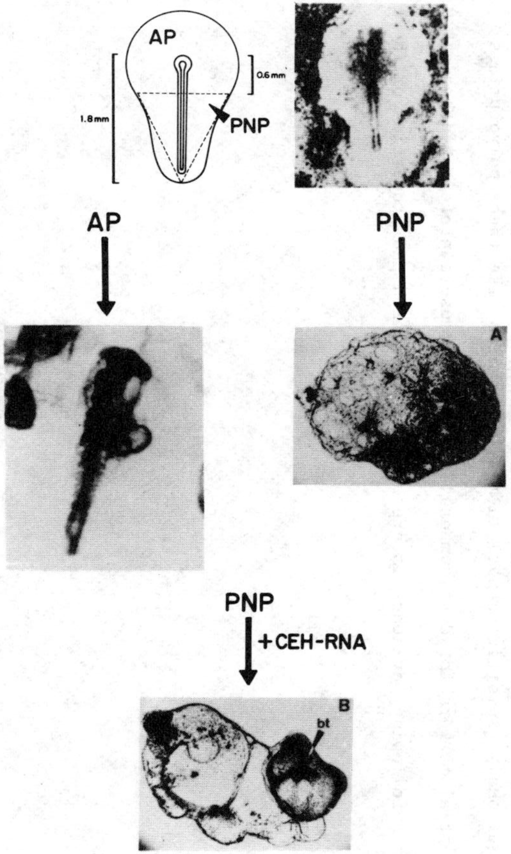

Fig 2. Stage 4 chick blastoderm characterized by a fully developed primitive streak (1.8 mm) is obtained after incubation of fertile chicken eggs for 20-22 hr at 38°C as described earlier (34). The explants, the post-nodal piece (PNP) and the anterior piece (AP) are obtained by transection of the blastoderm at 0.6 mm posterior to Hensen's node. AP when cultivated _in vitro_ (34) develops into embryonic structures including a well-defined heart tube, whereas the PNP under identical conditions fails to develop into any histologically identifiable structures (A). Addition of the inducer RNA obtained from the 16-day-old chick embryonic heart (see text and ref.34) to the PNP culture promotes the formation of the heart tube (bt) which starts spontaneous and rhythmic pulsations (B). The development appears to be similar to the embryonic cardiogenic process (see text and ref.34). Addition of the same RNA to the AP causes no apparent change in development.

muscle-like tissue when the explants are cultured in the presence of RNA isolated from differentiated chick embryonic heart (31-33). The difficulty in understanding the mechanism of this novel mode of differentiation was the lack of information on the nature and properties of the inducing agent and the biochemical criteria for establishing unequivocally the morphogenetic transition. We have reexamined the phenomenon critically, and have recently reported (34-37) that a distinct low-molecular-weight RNA (7S CEH-RNA) appears to play a crucial role in embryonic heart muscle formation, since addition of the RNA to PNP cells does indeed promote a specific mode of changes that is similar to embryonic myogenic process. Neither synthetic polynucleotides nor a variety of RNA isolated under identical conditions from several sources could replace the RNA from chick heart tissue.

In order to relate the process of embryonic heart cell differentiation to changes in specific gene activity in both the RNA-induced system *in vitro*, as well as *in situ*, our experimental approach is to monitor the syntheses of mRNAs for several muscle specific marker proteins in parallel with that of 7S CEH-RNA as a function of early embryonic differentiation (Fig. 3). Cardiac muscle differentiation is marked by the production of large quantities of contractile proteins, the myosin heavy and light chain polypeptides, tropomyosin, troponin, and actin subunits. While actin is synthesized ubiquitously, specific changes in myosin subunits make these polypeptides potential markers for developmentally regulated muscle protein synthesis. In the following, evidence is presented on isolation and properties of the

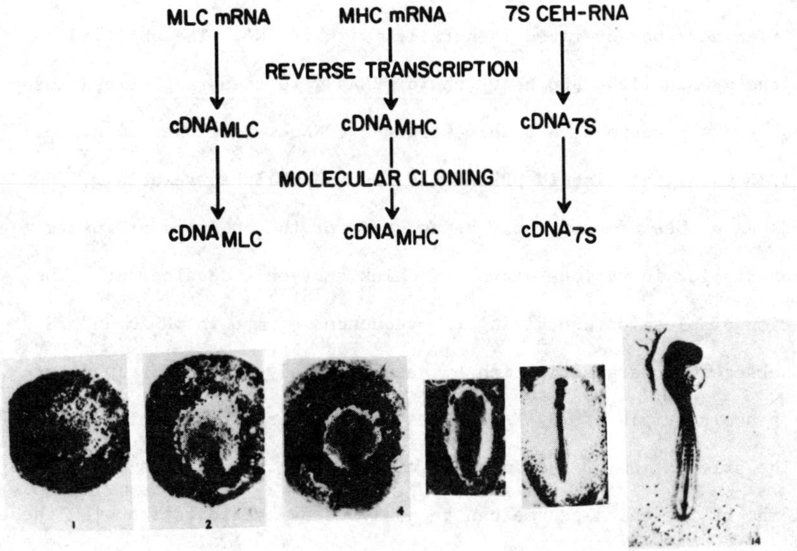

Fig 3. Scheme for quantitation of muscle specific RNAs synthesis during early stages of chick embryonic development (see text): MLC mRNA, myosin light-chain messenger RNA; HCM mRNA, myosin heavy-chain messenger RNA; cDNA, complementary DNA. The early stages of chick embryonic development are shown on bottom.

inducer RNA and its effects on pre-cardiac cells of PNP in culture. The cardiac muscle specific mRNAs, myosin light and heavy chain mRNAs were also purified and characterized in parallel with 7S RNA. The purified 7S RNA, and the myosin light and heavy chain mRNAs were then transcribed into cDNAs (38) and the respective double stranded cDNAs were cloned in _E. coli_ strain X1776 using the plasmid pBR322 (39, and unpublished results). The amplified DNA probes are currently being used for the analysis of muscle gene transcription in various stages of chick embryonic development. The construction of plasmids containing the sequences of myosin mRNAs and 7S RNA and the subsequent characterization by sequence analysis would help determine the genomic origin of 7S RNA and its relation to other muscle specific mRNAs. The orientation of myosin genes with respect to 7S genes within the physical map of the chick genome can be analyzed by hybridization with the chick DNA restriction fragments. These data would be highly useful in assessing the physiological role of 7S RNA in regulation of chick heart muscle induction.

RNA INDUCED DIFFERENTIATION AND ITS CHARACTERIZATION

Earlier workers had concluded (31-33) that the RNA capable of inducing specific changes in PNP might have mRNA-like properties. We reexamined the phenomenon and found that the transition is indeed dependent upon the addition of an RNA fraction isolated from embryonic heart tissue to culture medium supporting the PNP explants (34). The inducing molecule appears to be a low-molecular weight RNA, 7S RNA, with distinct properties. The inducer RNA does not appear to be an mRNA. 7S RNA has been purified and its structural characterization is currently under investigation. In order to understand fully the mechanism of induction of embryonic heart muscle gene transcription in chick embryo and the role the putative inducer RNA plays in the process, we asked the questions listed below:

(i) What is the nature of RNA-induced changes in the PNP cells?

(ii) Is the transition in PNP dependent upon a specific RNA molecule?

(iii) Is the RNA taken up by the PNP cells?

(iv) Do all PNP cells respond to inducer RNA?

(v) What are the functional and structural properties of inducer RNA?

(vi) Is the inducer RNA a physiological entity?

In the following, we shall attempt to answer these questions, at least in part, and summarize experiments currently in progress.

(i) What is the nature of RNA-induced changes in PNP?

In previous studies, identification of RNA-induced changes was based mainly on gross morphological criteria and on acquisition of spontaneous beating. In addition to these changes, we redefined the PNP differentiation by the following criteria:

(a) Appearance of myofibrils and glycogen granules: It is well established that appearance of myofibrils and glycogen particles are two identifying characteristics for differentiation of embryonic muscle cells *in vivo*, in cell suspensions, and *in vitro* (40, 41). Ultrastructural examination of RNA-treated and control PNPs showed normal cell components, e.g. endoplasmic reticulum, nuclei, mitochondria, Golgi complex etc., but the RNA-treated PNPs which exhibited spontaneous pulsations contained highly differentiated myocytes similar in ultrastructural complements to embryonic myocardial cells. Clearly evident (Fig. 4) were the orderly arranged myofibrils with thick and thin filaments and Z bands.

Because of its localization in cardiac muscle cells of young chick heart, glycogen has been used as a marker for heart muscle cell differentiation (41, 42). Histochemical examination of chick embryonic heart tissue and the beating PNPs revealed rich deposition of glycogen particles, whereas the control PNP were relatively free of glycogen (Fig. 5). The relatively slight localization of glycogen in the peripheral region of the control PNP is typical of endodermal cells.

(b) Synthesis of myosin and actin polypeptides: The major components of myofibrils are the proteins, myosin heavy and light chain, and actin, which are organized in interdigiting filaments. When the PNPs were labeled with [^{3}H] alanine, there was a 2.5-3.0 fold increase in label incorporated into myosin polypeptides. Fig 6 shows fractionation of partially purified

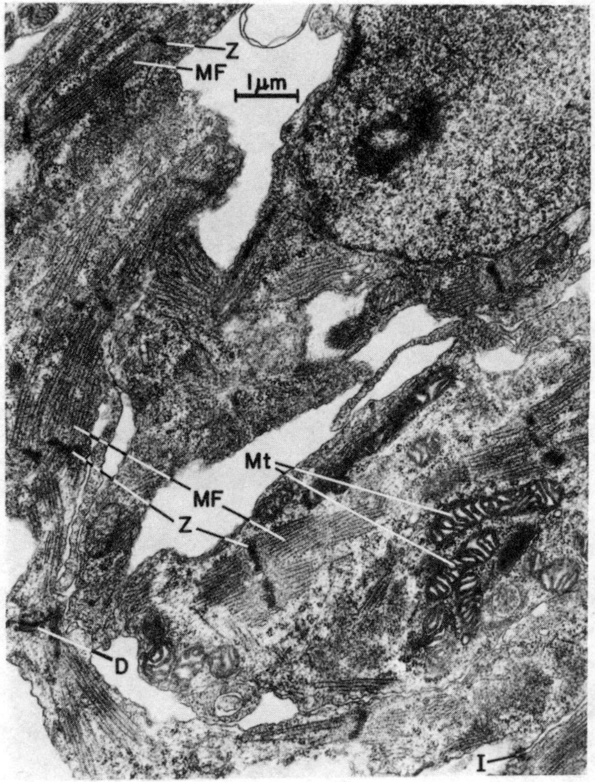

Fig 4

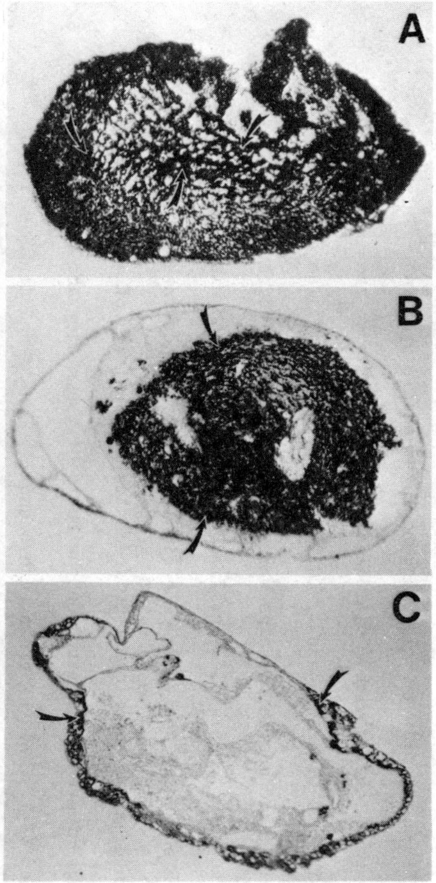

Fig 5

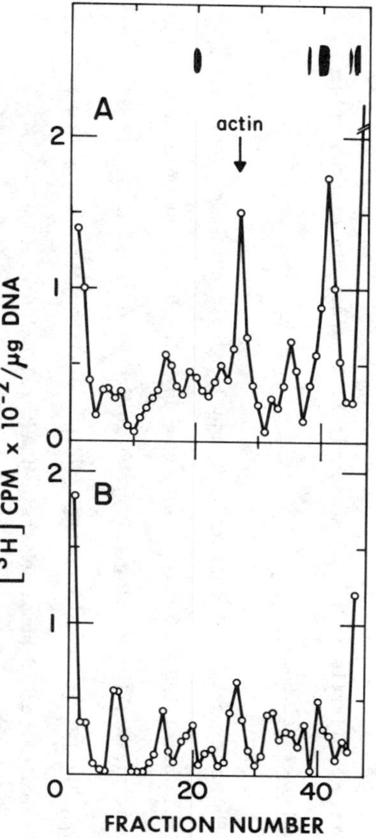

Fig 6

Fig 4. Electron micrographs of cells of PNP explants. The PNPs cultivated in the presence of CEH-RNA were examined for ultrastructural components. Myofibrils (MF) D; desmosomes; Z, Z bands; Mt, mitochondria; For details see Ref.34.

Fig 5. Photomicrographs of cultured PNP and 7-day-old chick embryonic heart tissue after PAS staining. (A) Seven-day-old chick embryonic heart ventricular tissue; (B) beating PNP grown in presence of bound II CEH-RNA (0.25 A_{260}/ml) for 96 hr; (C) control PNP grown without RNA for 96 hr. Arrows indicate the PAS-positive stained materal. 100 x. (See Ref.34)

Fig 6. SDS-Acrylamide gel electrophoresis of muscle proteins. The PNP proteins were labeled with [^{3}H]alanine, added directly to the incubation medium with and without bound II (0.25 A_{260}/ml) CEH-RNA. The explants were pooled, and muscle proteins were isolated and examined on 5.6% acrylamide gels. Nonlabeled purified myosin and actin obtained from 16-day-old chick embryonic hearts was run simultaneously on separate gels (see top stained gel for myosin; arrow indicates the position of actin). The counts per minute were plotted on the basis of the micrograms of DNA determined (34). (A) PNP grown with bound II CEH-RNA and differentiated into beating tissues; (B) PNP grown without CEH-RNA. Fraction No.1 represents the bottom of the gel.

muscle proteins on SDS/polyacrylamide gel. Although protein extracted from the control PNPs also appeared to contain myosin and actin-like polypeptides, the level of incorporation was significantly lower. It must be pointed out here that actin and myosin-like polypeptides are also found in non-muscle and premyogenic cells even though no myofibrillar structures are present (43).

(c) Acetylcholinesterase activity: The myogenic process involves, in addition to increased synthesis of muscle proteins, elaboration of specialized membrane components and the enzyme acetylcholinestrase. Cells of myogenic origin differentiated in culture in absence of neuronal elements produce acetylcholinestrase similar to muscle tissue <u>in situ</u> (44). When the PNP were cultured in presence of the inducer RNA and the beating PNPs were pooled and processed for acetylcholinestrase activity there was 3.4-fold increase in the rate of hydrolysis of acetylcholine compared to the non-beating PNP (Fig. 7), and a 6-fold increase over that of the unincubated PNPs. The increase was always found in the same batches of explants that also exhibited specific morphological and biochemical changes, whereas the PNPs grown in absence of inducer RNA or with carrier RNA distinctly lacked these properties. Histochemical examination for acetylcholinestrase provides a direct visualization of the locale of the enzyme during various stages of development. The explants when stained according to Karnovsky and Root (45) after incubation with the inducer RNA and the carrier RNA separately exhibited differences between the intensity and distribution of the enzyme positive granular material (Fig. 8). The unincubated PNP also showed a pattern similar to the incubated control PNPs.

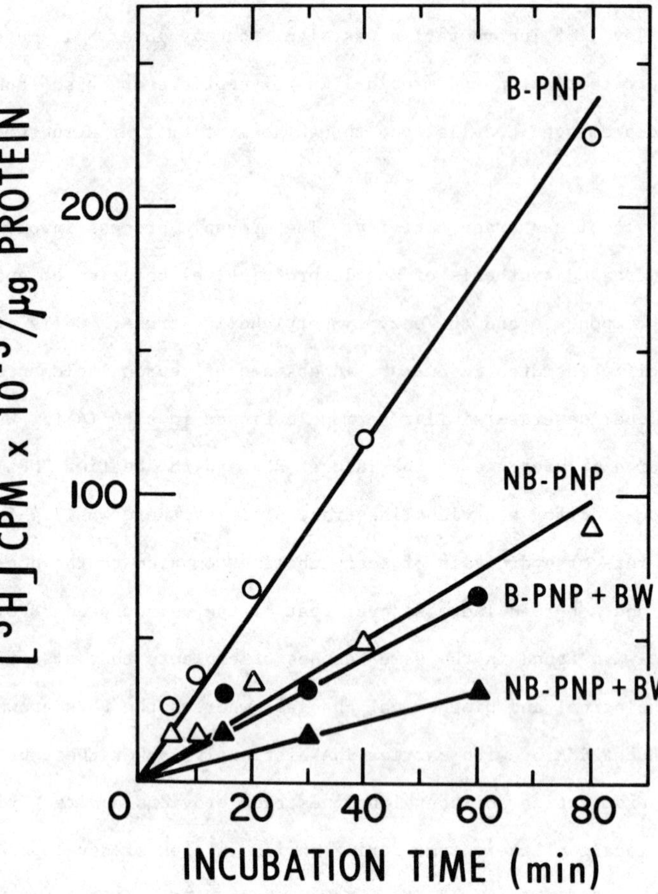

Fig 7. Acetylcholinesterase activity in 16-day-old chick embryonic heart. Kinetics of acetylcholinesterase hydrolysis (40 μg of homogenate protein was used per time point in the presence and absence of inhibitor, BW at 2.10^{-4} M) B PNP, beating PNP; NB PNP, non-beating.

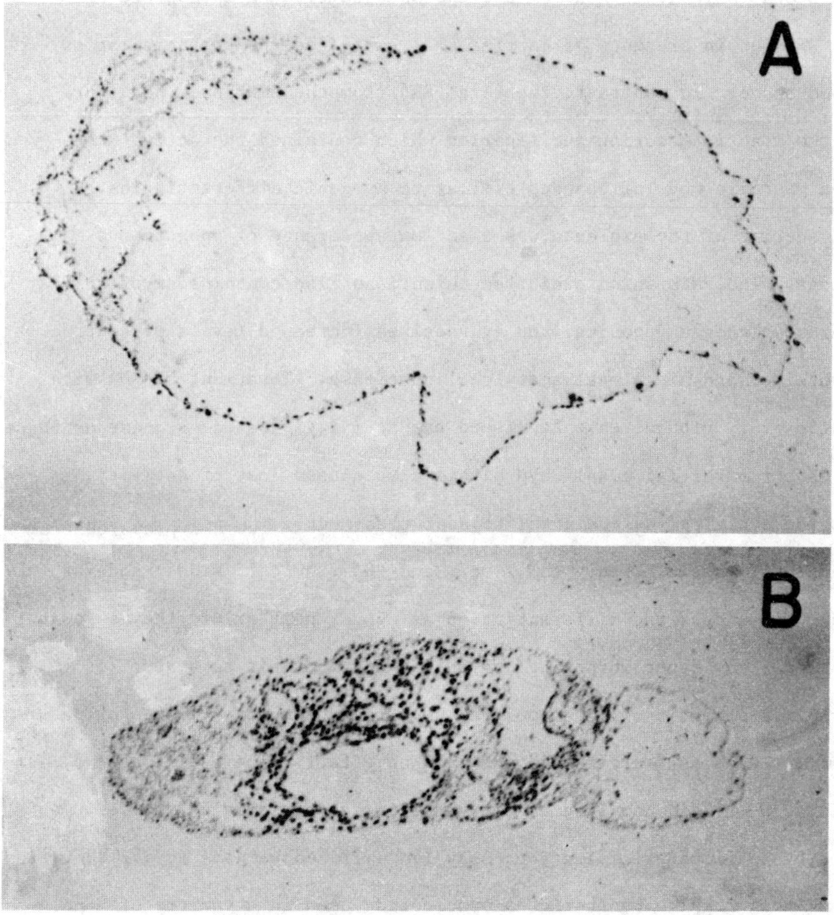

Fig 8. Transverse section of PNP explants after histochemical staining reaction. PNP explants cultivated in presence and absence of the inducer CEH-RNA were stained histochemically (36). The in toto stained tissues were then embedded in paraplasts and cut into 5 μM thick sections which were examined and photographed under light microscope. Arrow indicates the acetylcholinesterase positive stain deposits. A, control PNP section (no CEH-RNA); B, section from differentiated PNP (plus CEH-RNA).

(ii) Is the transition dependent upon a specific RNA molecule?

As shown in Table I, none of the explants cultivated in absence of inducer RNA (or in presence of carrier RNA) exhibited the characteristics described above. In contrast, the total RNA from the embryonic heart or any of the fractions during purification which contained the active RNA caused a specific mode of changes similar to myogenic differentiation. Initial scoring of the explants was based on appearance of spontaneous pulsations. The PNPs which exhibited pulsations also contained myofibrillar structures, glycogen deposits, and synthesized increased levels of contractile proteins, and high level acetylcholinesterase. Treatment of RNA with pancreatic or T_1 ribonuclease abolished the activity, whereas pronase or DNase had no effect. Partial alkali hydrolysis also caused loss of activity, whereas the activity was resistant to heat treatment. The RNA prepared from chick embryonic brain, kidney and liver, and RNA from calf kidney and rat liver did not induce the differentiation as above, neither did they support, in our hands, any other morphological change attributable to a specific mode of differentiation. Total poly(A) RNA from non-muscle tissues or purified mRNA such as δ-crystallin mRNA had no effect. The addition of δ-crystallin mRNA did not lead to δ-crystallin synthesis in PNP cells when examined by an immunoprecipitation assay (unpublished results in collaboration with J. Piatagorsky). Synthetic polynucleotide used in a variety of concentrations had no effect.

Thus, it appears that the induction of myogenic-like changes in PNPs are indeed dependent upon the presence of a specific RNA in the culture medium.

TABLE I

RNA-dependent differentiation in PNP

Additions	A_{260}[a]	N° of PNP Used	N° of PNP Differentiated[b]
Chick embryonic heart RNA:			
Total RNA	15.00	94[c]	60[c]
Unbound RNA	5.00	10	0
Unbound RNA	15.00	51	0
Bound I RNA	0.25	8	0
Bound I RNA	0.50	9	0
Bound II RNA	0.07	9	3
Bound II RNA	0.14	9	5
Bound II RNA	0.28	42[c]	30[c]
Bound II, fraction 1	0.01	19	16
Bound II, fraction 1	0.02	6	5
Bound II, fraction 1	0.05	9	6
Bound II, fraction 2	0.03	6	0
Bound II, fraction 2	0.07	6	0
Bound II, fraction 3	0.02	6	1
Bound II, fraction 3	0.08	8	3
Bound II, fraction 4	0.05	4	9
Bound II, fraction 1 'a'	0.005	20[c]	15[c]
Bound II, fraction 1 'a'	0.01	9	6
Bound, fraction 1'a' (heat treatment)	0.01	9	6
Bound II, fraction 1'a'(pan.RNase treated)	0.01	9	0
Bound II,fraction 1'a'(RNase treated)	0.01	8	0
Chick embryonic brain RNA:			
Unbound RNA	11.00	8	0
Bound RNA	0.25	8	0
Bound RNA	0.50	8	0
Rat liver RNA:			
Unbound RNA	10.00	8	0
Bound RNA	0.03	8	0
Bound RNA	0.60	8	0

TABLE I cont.

	A_{260}[a]	N° of PNP Used	N° of PNP Differentiated[b]
Calf kidney RNA:			
Unbound RNA	12.000	8	0
Bound RNA	0.25	8	0
Bound RNA	0.50	8	0
δ-crystallin mRNA	0.34	6	0
Rous sarcoma viral RNA	0.90	6	0
Rabbit macrophage RNA	3.00	6	0
Bombyx mori rRNA	15.00	6	0
Bombyx mori tRNA	12.00	16[c]	0
E. coli rRNA	12.00	9	0
E. coli tRNA	12.00	21[c]	0
Total chick embryonic RNA (DNase treated)	15.00	3	2
Total chick embryonic RNA (pronase treated)	15.00	3	2
poly(A)	0.005	9	0
poly(A)	0.01	9	0
poly(A)	0.05	6	0
poly(A)	0.10	6	0
poly(C)	0.09	3	0
poly(G)	0.09	3	0
poly(I)	0.10	3	0
Control (without RNA)	---	57[c]	0
Control (with carrier RNA)	15.00	90[c]	0

Culture technique, addition of RNA, incubation conditions, enzymic digestions and heat treatment are all described in detail elsewhere (34-36).

[a]The final concentration was adjusted to 15 A_{260} per ml by addition of carrier RNA. *E. coli* or *Bombyx mori* rRNA and tRNA, in a ratio of 10:1 was used as carrier RNA.

[b]Initial scoring of the differentiated explants was based on the beating property of the PNP cells. The beating and the non-beating explants were pooled separately and then divided randomly into batches which were then subjected to ultra-structural histochemical and biochemical examinations as described earlier (34-36).

[c]The number of explants represents a cumulative total of several experiments.

(iii) Do all PNP cells respond to Inducer RNA?

It is believed that during the early stages of development, i.e. stage 3+ to 4- (21), the precardiogenic cells lie in the bilateral regions of the epiblast adjacent and approximately halfway between the anterior and posterior ends of the growing primitive streak (23). Precardiac mesoderm begins to invaginate at stage 4 moving anteriorly along the growing primitive streak to form well defined bilateral regions at stage 5. These regions are centered at the level of Hensen's node, halfway between the node and the lateral edges of area pellucida. When tissue fragments, isolated by sequential cuts at intervals of 0.1 mm beginning with 0.4 mm and up to 1.2 mm posterior to Hensen's node, are grown in presence of inducer RNA, there was a gradient in potential to develop heart-like tissue (Fig. 9). The fragments obtained by cutting below 0.8 mm posterior to Hensen's node failed to develop into structures similar to fragments containing the anterior portions of PNP cells. This led us to speculate that heart-forming cells are located in area extending up to 0.8 mm behind the Hensen's node, which is in agreement with the heart fate mapping experiments (22) using [^{3}H] thymidine labeled grafts taken from various regions of the blastoderm.

(iv) Is the RNA taken up by the PNP cells?

Substantial evidence does exist on the uptake and transfer of RNA by eukaryotic cells (17). Mildly disaggregated cells were incubated with [^{125}I] labeled 7S RNA for 4 hr in medium-199 and the cells were washed free of RNA by repeated washings till no more radioactivity was released. RNA was extracted from the cells and examined by electrophoresis on polyacrylamide gel

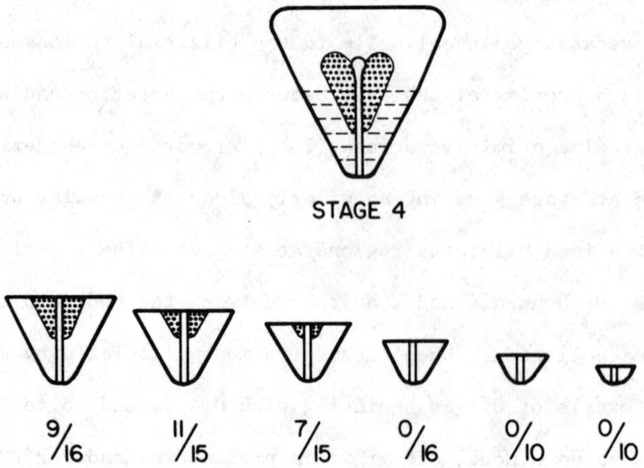

Fig 9. Sequential transection of chick stage 4 blastoderm was done as indicated by dotted lines. The resulting post-nodal explants (second row) were then grown in culture as usual (34). The number of beating explants out of the total explants used is shown in bottom. Shaded area represents an approximation of precardiac region in stage 4 blastoderm.

in formamide. The RNA isolated from the PNP migrated identically to the RNA added (unpublished results). A fingerprint analysis of the two RNA showed identical profiles. The autoradiographic examination of the sectioned PNP revealed a random distribution of grains in the cells. One can deduce from these experiments that 7S RNA might enter the PNP cells in an intact form, although one cannot exclude the possibility that the RNA visualized as grains had undergone modifications prior to or after its uptake by the cells. Further experiments are needed to determine the intracellular fate of 7S RNA and to ascertain whether the uptake occurred in macromolecular form or after specific modifications.

(v) <u>What are the structural and functional properties of inducer RNA?</u>

The partially purified 7S RNA migrates with a mobility of about 260 ± 20 nucleosides on 7.5% polyacrylamide gel containing 99% foramide (Fig. 10). Nucleoside composition analysis was done after post-labeling of the RNA digest with potassium [^{3}H] borohydride as described earlier (35) or by 5'-end labeling with [^{32}P]ATP (unpublished results). Purified $tRNA^{Val}$ was used as control. As shown in Table II, the RNA contains 53% adenosine residues, suggesting that about one-third of the molecule is comprised of poly(A) stretch(es). An independent measurement of poly(A) length by 3'-end labeling of the molecule with 5'-^{32}pCp and RNA ligase confirmed that the 3'-end terminal poly(A) is approximately 100 ± 20 residues long. The RNA was also digested with RNase-H in the presence of oligo(dT) and the product was examined on polyacrylamide-formamide gel (Fig. 11). The results were consistent with the removal from one end of approximately 100 ± 20 residues.

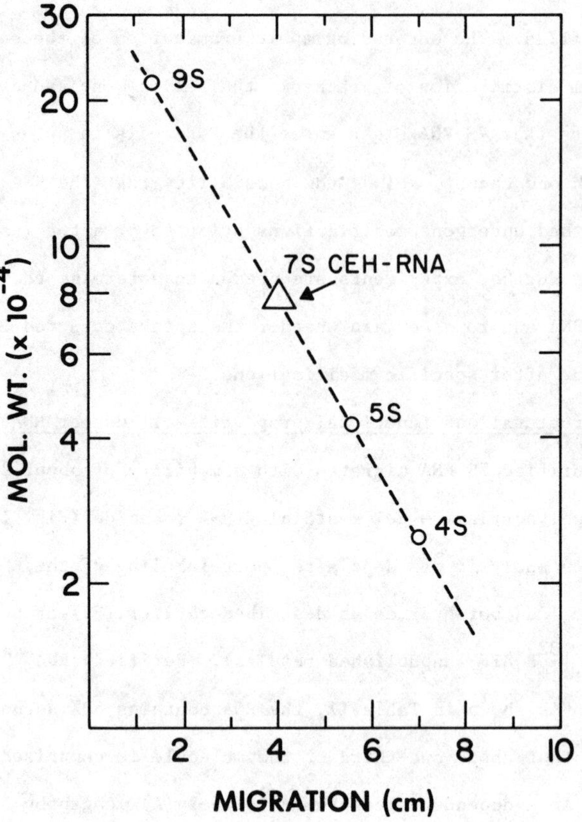

Fig 10. Relationship between molecular weight and electrophoretic mobility. Electrophoresis of 7S CHE-RNA was done on 7.5% polyacrylamide gel in formamide, in presence of marker 4S, 5S and 9S RNAs (35).

TABLE II

CEH-RNA Fraction 1 A	$tRNA_1^{Val}$ (Escherichia coli) Experimental	Actual
M%	residues/76	nucleosides
16.0	10.6	9 + 1 (s^4U)
53.6	14.4	14
15.2	24.3	23
15.1	22.8	23
0	0.84	1
0	0.99	1
0	0.84	1
0	0.99	1
0	0.76	1
0	0.99	1

See ref. 35 for details

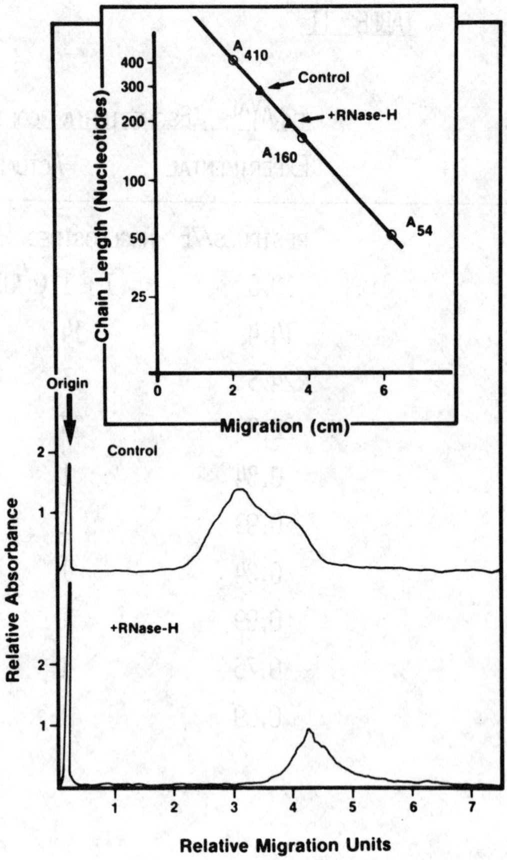

Fig 11. Polyacrylamide gel electrophoresis of 7S RNA before and after RNase-H digestion. Digestion of [^{125}I] labeled 7S RNA was carried according to Vournakis et al. (49) and the product was analyzed by 7% polyacrylamide-formamide gel electrophoresis (35). Poly(A) of known chain length were used as markers. Bottom, densitometer tracings of control and RNase-H treated 7S RNA.

Had the poly(A) residues been interspersed the resulting molecules would be two or more with a faster mobility. Fig. 12 represents the end group analysis of 5'- and 3'-end labeled 7S RNA. The 5'-end labeled RNA was subjected to P-1 or snake venom phosphodiesterase digestion and the products were separated by PEI thin-layer chromotography (46). Almost 95% of the label migrated as 5'-pA. When 7RNA was labeled at 3' terminus with ^{32}pCp and RNA ligase, digested with RNase T_2, and analyzed on PEI plates, 90% of the label was transferred to adenosine. Thus both the terminal residues appeared to be unmodified adenosine. The 5'-terminus must be uncapped, since tobacco acid pyrophosphatase treatment was not necessary for end-labeling.

5'-end labeled 7S RNA was used for the partial base sequence analysis. The labeled RNA was first separated on 7% polyacrylamide-urea gel (Fig. 13) and the 7S RNA band eluted from the gel was subjected to mobility shift analysis after a partial digestion with enzyme P-1 (46). The single tract (Fig. 14) suggests that the purity of the RNA is at least 80%. Further base sequence analysis was done by gel electrophoretic separation of the partial digests of the RNA with base specific nucleases (unpublished results). A typical gel profile is shown in Fig. 15. The combined results of sequence analysis for the first 40 5'-nucleosides is shown in Fig. 16. There are, however, ambiguities in reading the sequence from the sequence gel arising probably from two recurring problems. In several positions simultaneous cuts by both pancreatic and T_1 ribonuclease occur. If the RNA was impure, one would expect every position to have multiple base specific cuts. Clearly this is not the case. A second ambiguity comes from the fact that the RNA is very G-C rich near the 5'-terminus, as such, a considerable secondary

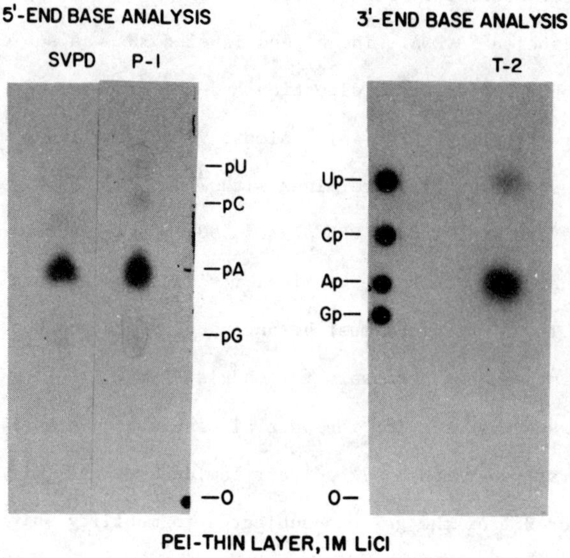

Fig 12. End group analysis of 7S RNA. 3'- and 5'-end group labeling was done with ^{32}pCp and γ-ATP^{32} respectively using RNA ligase and polynucleotide kinase (46, and unpublished results). The RNA was digested with enzymes as indicated, and the products were analyzed by separation on PEI thin-layer plate (46). Nucleoside 5'- and 3'-monophosphates were used as markers.

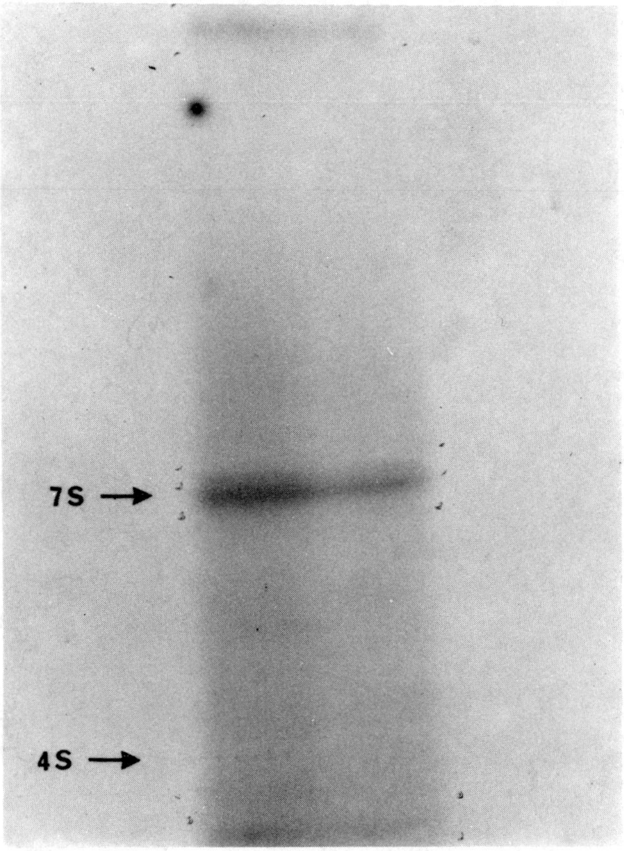

Fig 13. Polyacrylamide gel electrophoresis of purified 7S RNA. RNA was labeled with γ-ATP^{32} (46,50) and separated on 7% apolyacrylamide-urea gel (46) (unpublished results).

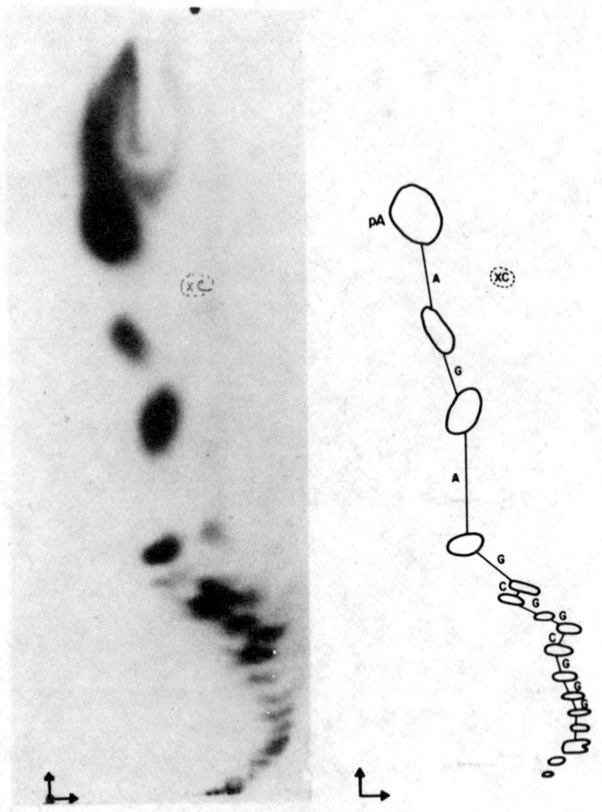

Fig 14. Mobility shift analysis of 7S RNA. The RNA was 5'-end labeled as described in text 46,50) and analyzed by two-dimensional separation after the partial digestion with enzyme P-1 (46).

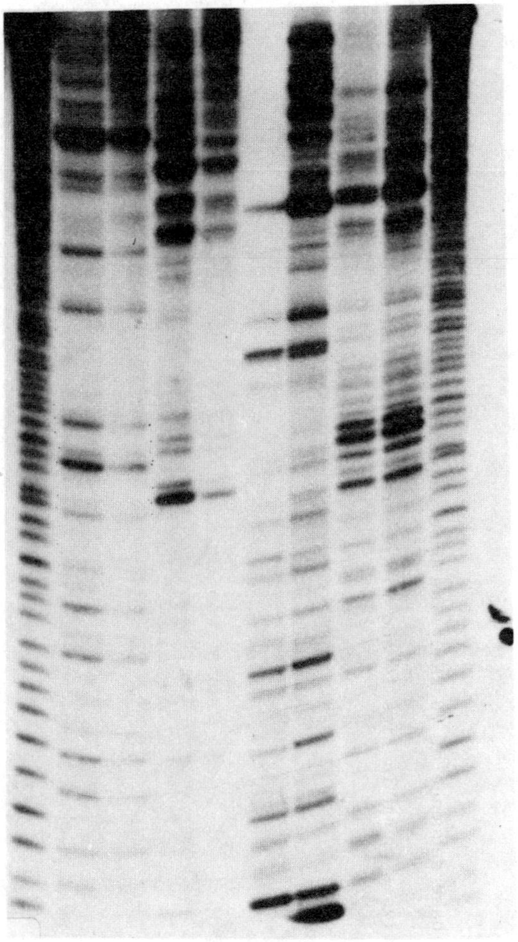

Fig 15. Polyacrylamide gel electrophoretic separation of the partial digests of 5'-end labeled 7S RNA. The RNA was labeled as described in text (and in ref 50) and digested with enzymes T_1, U_2, pancreatic RNase, and physarium RNase under conditions described by Maxam and Gilbert (50). The digests were analyzed by 10% polyacrylamide-urea gel electrophoresis (50). The lanes on both sides represent the partial digest with formamide of 5S RNA from _B. mori_. The next two lanes (starting from left) represent digestion with two concentrations of T_1, followed by U2, pancreatic RNase and Physarium RNase (see text for further details).

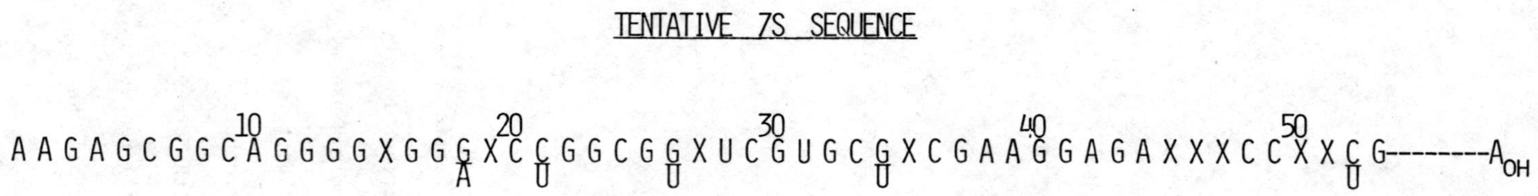

Fig 16. Partial 5' base sequence of 7S RNA. The base sequence was deduced from the analyses shown in Figs 14 and 15 (see ref. 54 for further details).

structure could occur and the aberrant bands may be due to "snap-back" of the denatured RNA digest fragments during electrophoresis. Further work on sequence determination is currently in progress.

Since the 7S CEH-RNA contains a 3'-terminal poly(A) tract, it is reasonable to assume that the RNA might have an mRNA-like activity. However, repeated attempts to translate the RNA in a variety of cell-free translation assays were not successful. Supplementing protein factors from the homologous chick heart tissue were of no avail. One of the objectives in undertaking the base sequencing analysis above was to ascertain whether the initiator signal, AUG, can be located in the 5'-region of the molecule. As shown in Fig. 16, there is no initiator codon present in the RNA up to at least 44 nucleotides. Furthermore, the alignment of the bases into any possible reading frames results in a large number of nonsense codons.

We have previously observed that 7S RNA inhibits both homologous and heterologous poly(A) RNA translations in cell-free systems (47). This suggested to us that the observed activity of 7S RNA in inducing a specific mode of differentiation in PNP might be related in some way to an effect in translation process <u>in vivo</u>. The RNA also shares some characteristics previously described for the low-molecular-weight RNA species isolated from chick muscle (14), <u>A. salina</u> embryos (16), rabbit reticulocytes (15), and Novikoff heptoma cells (48). These RNAs have been shown to participate as modulators of protein synthesis in cell-free translation assays and, therefore, are implicated to play a role in regulation of cell development and differentiation. Although 7S RNA is capable of inhibiting effectively the translation of both homologous and heterologous mRNA without any apparent selection of a specific mRNA or a group of mRNAs, the inhibition appears to be effected by the adenine-rich segment of 7S RNA, since the removal of non-poly(A) fragment of 7S RNA produced the same kinetics of inhibition as the intact molecule (47). These results would, therefore, suggest that a control at the level of translation is possibly not one of the mechanisms of action of 7S RNA in causing heart-like differentiation in the embryonic PNP cells.

(vi) Is the inducer RNA a physiological entity?

Although substantial evidence is available on the existence of distinct low-molecular-weight RNA species in a variety of organisms, the possibility that 7S RNA is a degradation product derived during RNA extraction cannot be excluded. Several lines of evidence, however, suggest that the RNA is not produced via a non-specific degradation of large-molecular-weight RNA species during extraction procedures. The relative amount of 7S RNA obtained at a particular stage is constant regardless of the procedure employed for RNA extraction. Preparation of RNA under conditions optimized for isolation of functionally intact mRNAs (unpublished results) yield 7S RNA in amounts comparable with those utilizing phenol extraction of total chick heart homogenate.

For an unambiguous conclusion on the origin of 7S RNA, however, further structural and functional characterization would be required. Yet it appears that 7S RNA is a physiologically relevant entity of RNA with a possible function in communication of regulatory processes during development. Our current experiments (see below) are designed to ascertain whether the RNA shares homology with one or more of the mRNAs in chick heart or whether it is transcribed as a separate class of molecule. A study on biosynthesis of 7S RNA in parallel with that of muscle specific mRNAs would help in assessing the functional role of 7S RNA in embryonic heart muscle development.

In the following section we shall describe the isolation, purification, and characterization of myosin light and heavy chain mRNAs and the preparation of the respective cDNA probes. The cloned DNA probes will then be used to monitor the transcriptional activity for the mRNAs in conjunction with that of 7S RNA as a function of embryonic development.

ISOLATION AND PURIFICATION OF RNAs

Myosin mRNAs: Myosin light chain mRNAs (MLC_1 and MLC_2 mRNA) were isolated from the MLC synthesizing polysomes of the 16-day old chick embryonic heart tissue by immunoadsorption of total polysomes to MLC specific antibodies prepared in our laboratory (38). The experimental procedure is outlined in Table III. The RNA from immunoadsorbed polysomes was fractionated by centrifugation on isokinetic sucrose gradients (Fig. 17a). As shown in Fig. 17b, the RNA fraction sedimenting between 9S and 14S contained MLC mRNA-like activity. The 9-14S fraction was further purified by a second sucrose gradient centrifugation and oligo (dT) cellulose chromatography. The purified fraction directed the translation of polypeptides that were indistinguishable from the authentic myosin light chains, MLC_1 and MLC_2 when examined by two-dimensional separation using isoelectrofocusing and polyacrylamide gel electrophoresis (Fig.18). The RNA migrated as two distinct bands on polyacrylamide-formamide and on agarose-urea gels (Fig. 19). The electrophoretic mobilities of these two RNA species correspond to 360,000 and 320,000 daltons accounting for approximately 1090 and 980 nucleosides per chain. The molecular weight of the translation products of these two RNA were 24,000 and 18,000 respectively, identical to those of MLC subunits of 16-day old chick embryonic heart. This would suggest that the mRNAs contain 370 and 430 nucleosides more than expected to code for these polypeptides respectively.

MLC mRNA can also be partially purified by repeated sucrose gradient centrifugations of total poly(A) RNA (see below). Upon translation of the

TABLE III

PROCEDURE OF MLC mRNA PURIFICATION

(i) Isolation of total polysomes in presence of Heparin

(ii) Incubation of polysomes in presence of MLC antibodies

(iii) Adsorption of antibody-polysome complex to an insoluble matrix of goat (anti-rabbit) IgG

(iv) Extensive washing of matrix to remove non-specifically bound material

(v) Elution of the bound polysomes with EDTA and recovery of RNA

(vi) Purification of MLC mRNA by successive sucrose gradient centrifugation and oligo(dT) cellulose chromatography

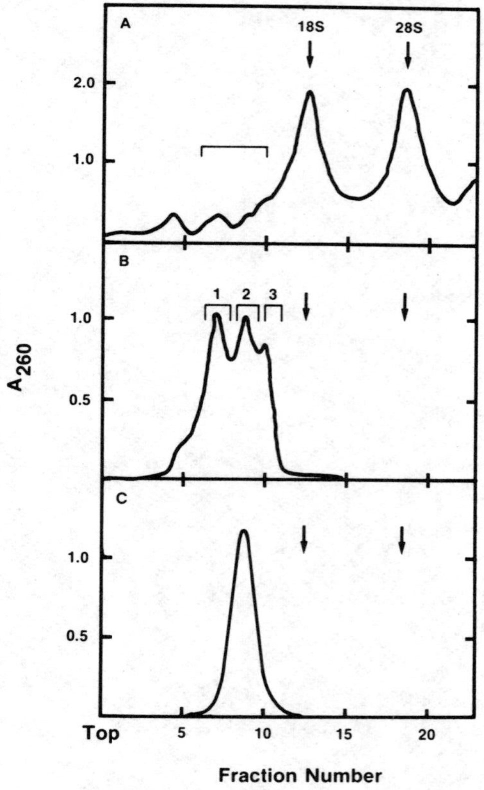

Fig 17A. Sedimentation profiles of RNA from MLC-synthesizing polysomes on isokinetic sucrose gradient. (A) Twenty A_{260} units of immunoprecipitated MLC-polysomal RNA was sedimented through a 15-35.9% isokinetic sucrose gradient (38). The gradient was fractionated into 0.6-mL aliquots and scanned by an ISCO density gradient fractionator Model 640. The RNA sedimenting between 9 and 14S, as indicated was pooled from several gradients, concentrated, and run again on a separate gradient under identical conditions (B). The material from peaks 1, 2, and 3 was recovered separately and subjected to translational assays using the rabbit reticulocyte lysate. Peak 2 RNA which displayed the template activity for MLC_1 and MLC_2 exclusively was centrifuged again on sucrose gradient as above after two successive passages through oligo(dT)-cellulose as described in Materials and Methods (c). The arrows indicate the positions of 18S and 28S rRNA.

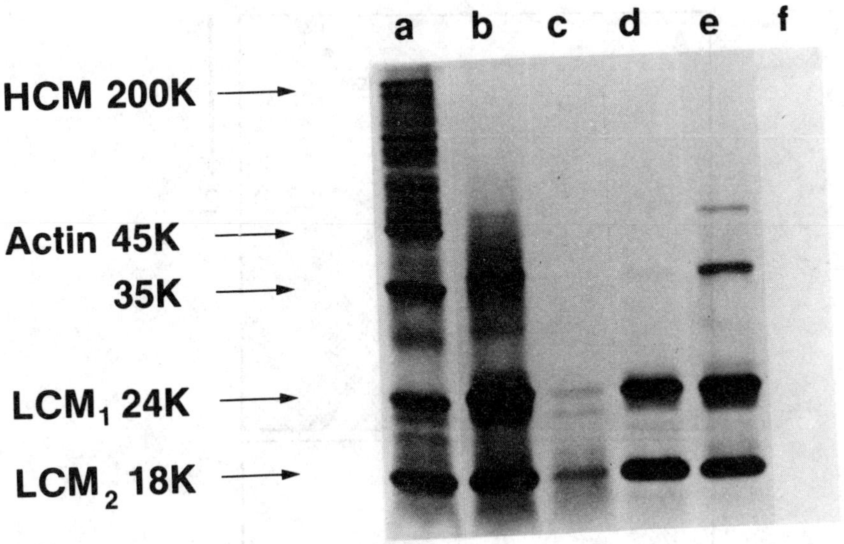

Fig 17B. $NaDodSO_4$-polyacrylamide gel electrophoresis of translation products of chick heart mRNA. mRNA-dependent translation was performed in micrococcal nuclease treated rabbit reticulocyte lysate. Five-microliter aliquots were taken at the end of incubation and applied directly to a 12% $NaDodSO_4$-polyacrylamide gel. Nonlabeled marker proteins were run simultaneously on the same gel. The positions of marker proteins were ascertained by staining the gel in Coomassie Blue. The gel was then processed for fluorography, exposed to X-ray film developed after 48 h. (a) Total polysomal RNA; (b) immunoabsorbed polysomal RNA; c, d, and e are peak 1, 2, and 3 RNA respectively: (f) endogenous nuclease-treated lysate (without RNA); HCM, heavy chain myosin; LCM_1 and LCM_2 light chain myosin subunits 1 and 2, respectively.

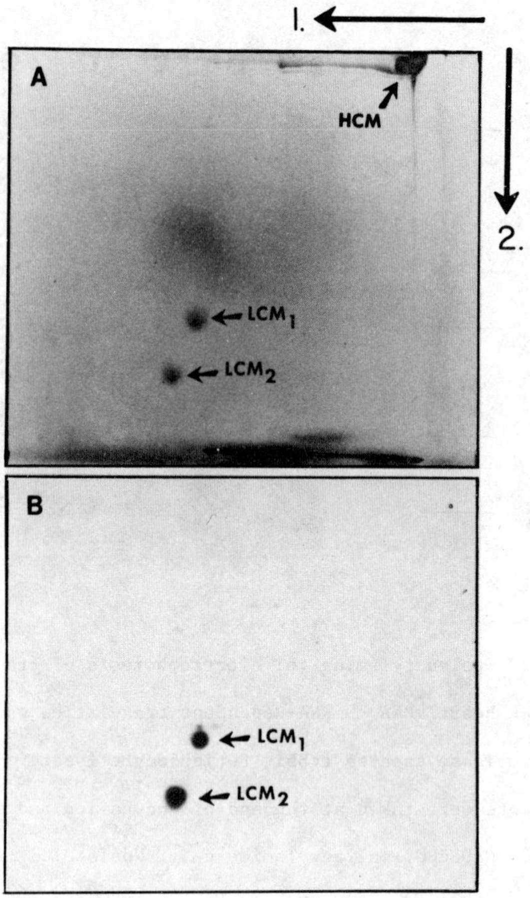

Fig 18. Two-dimensional separation of labeled translation products and purified myosin subunits. A 5-μL aliquot (≃40000 cpm) of the rabbit reticulocyte lysate translation mixture with the purified MLC mRNA and purified myosin preparation was subjected to the two-dimensional separation by isoelectrofocusing in the first dimension and polyacrylamide gel electrophoresis in the second dimension(38). The labeled products were identified by fluorography (51) and the marker MLC subunits by staining with Coomassie Blue. (A) Marker purified myosin subunits; (B) [^{35}S] methionine-labeled translation products of MLC mRNA. Migration on the first dimension was from right to left and on the second dimension from top to bottom. HCM, heavy-chain myosin; MLC_1 and MLC_2, light-chain myosin subunits 1 and 2.

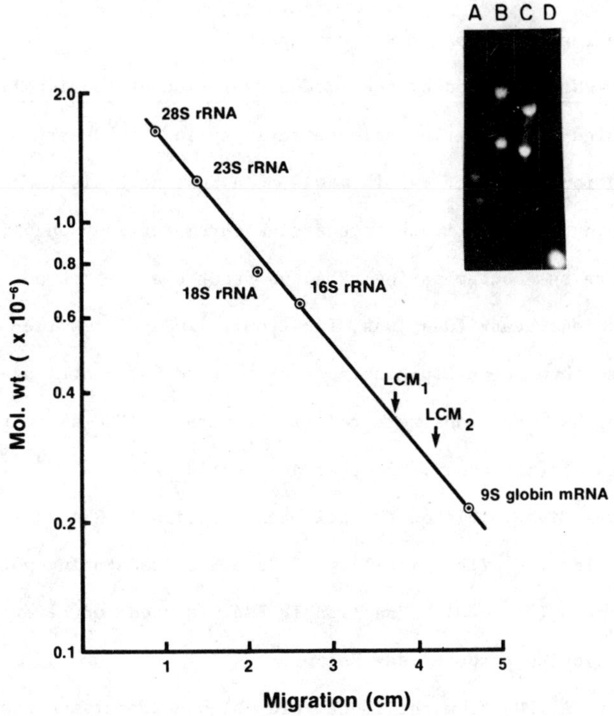

Fig 19. Molecular weight estimation of MLC mRNA by polyacrylamide-formamide gel containing 98% formamide along with the marker RNAs of known molecular weights. The RNAs were located on gel under UV light after ethidium bromide treatment or by staining (38). The inset shows the separation of RNAs on an agarose-urea gel as detailed in Materials and Methods. (A) MLC mRNA; (B) 18S and 28S rRNA; (C) 16S and 23S rRNA; (D) 4S RNA.

resulting fractions (Fig. 20), fraction 12 RNA contained the mRNA activity for MLC_2, whereas fraction 11 exhibited activity for both MLC_1 and MLC_2. For the purpose of cloning, fraction 12 RNA was used.

Although the MLC mRNAs obtained by the immunoadsorption of total polysomes were pure, the high level of ribonuclease activity in chick heart tissue precluded isolation of intact myosin mRNAs on a routine basis. When total poly (A) containing RNA purified by successive chromatography on oligo (dT) cellulose to remove the contaminating rRNA was fractionated by two successive sucrose gradient centrifugations, first on a linear formamide-containing gradient and then on an iso-kinetic sucrose gradient centrifugation in SDS, the slow moving RNA fractions were relatively pure for MHC activity (Fig.21). Based on translation assay, no other polypeptide but the MHC of about 200,000 dalton was discernible on the gel when fraction 18 RNA was used as template. The RNA migrated with a mobility of 28-30S on denaturing polyacrylamide or agarose gels (Fig. 22). Fraction 18 RNA was used for reverse transcription and for cloning purpose (see below).

<u>7S CEH-RNA</u>: Total poly (A) RNA from the 16-day old chick heart tissue was passed through oligo (dT) cellulose column under standard conditions using the two-step elution of bound RNA (35). The resulting three fractions, the unbound material, the intermediate fraction and the bound RNA (Fig. 23) were tested for the inducing activity using a wide range of concentrations. The PNP culture conditions and characterization of differentiation is described in ref. 34-36. Bound II RNA was the only fraction that possessed the capacity to cause a specific mode change in the PNP cells. Bound II RNA fraction was then

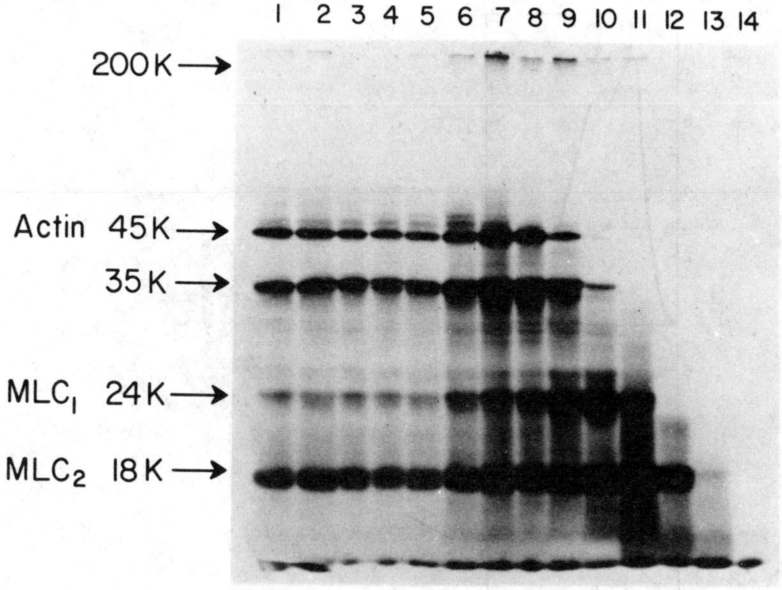

Fig 20. $NaDodSO_4$-polyacrylamide gel electrophoresis of translation products of poly(A) RNA fractions. Twenty A_{260} units of poly(A) RNA recovered from the sucrose gradient fractions sedimenting between the 16S rRNA marker and the top of gradient was run on an isokinetic $NaDodSO_4$-sucrose gradient. RNA in each fraction was ethanol precipitated and redissolved in 100 μL of H_2O. A 1-μL aliquot from each fraction was used for translation in the rabbit reticulocyte lysate system, and a 5-μL aliquot of the mixture was examined on 10% polyacrylamide gel containing $NaDodSO_4$ along with marker proteins (38). The gel was developed for fluorography (51).

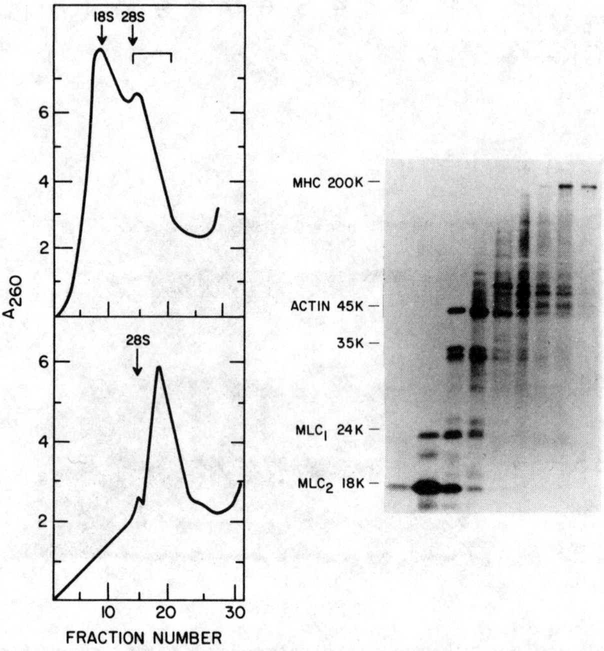

Fig 21. Isolation and purification MHC mRNA. Total poly(A) containing RNA was fractionated by successive sucrose gradient centrifugations (unpublished results, see text) and the fractions were analyzed by translation assay (38). The translation products were examined by polyacrylamide gel electrophoresis (38, and unpublished results).

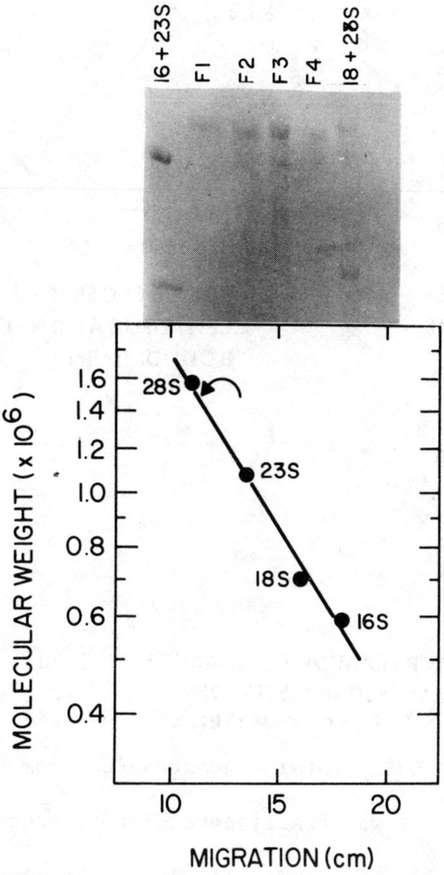

Fig 22. Agarose gel electrophoresis of purified MHC mRNA. MHC mRNA was isolated and purified as shown in Fig 21 and the fractions from the gradient were run on 1.0% agarose-urea gel along with marker rRNAs (unpublished results). F1, F2, F3, F4 represent fraction 18, 17, 16, 15 from the second gradient centrifugation.

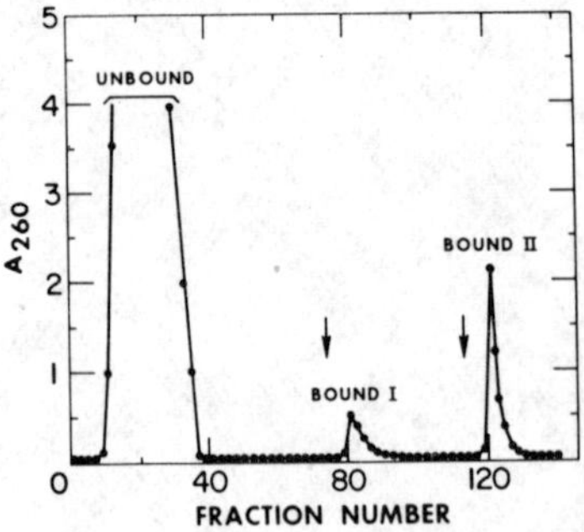

OLIGO (dT) CELLULOSE CHROMATOGRAPHY OF TOTAL CEH-RNA

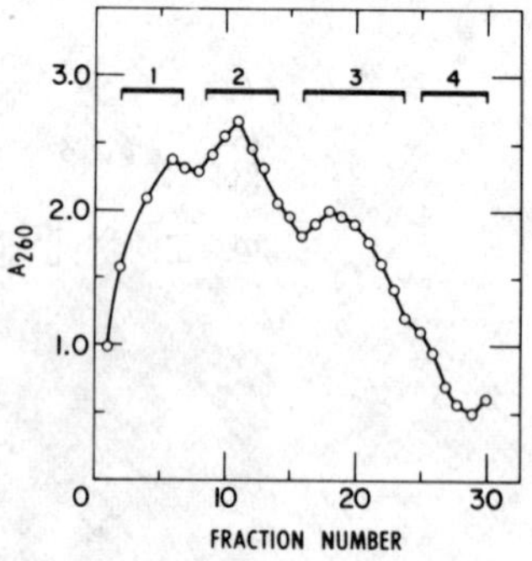

SDS-SUCROSE GRADIENT CENTRIFUGATION OF BOUND II RNA

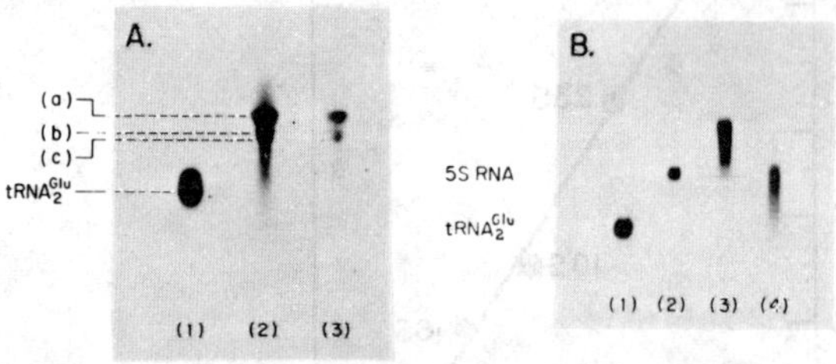

POLYACRYLAMIDE-FORMAMIDE GEL ELECTROPHORESIS OF BOUND II FRACTION 1 RNA

Fig 23. Purification of 7S RNA. Total RNA extracted from the 16-day-old chick embryonic heart (34, 35) was fractionated into the unbound, bound I and bound II RNA fractions by oligo (dT) cellulose column chromatography as described earlier. The bound II fraction was then centrifuged on 15-30% sucrose density gradient (35). The sedimentation was from left to right. The RNA from fraction 1 was pooled and purified further on a 7.0% polyacrylamide gel in 98% formamide (A) with tRNA$_2^{Glu}$ from E. coli as markers (A-1). The resulting three bands (a), (b), and (c) were recovered from the gel; band (a) was pooled separately from the bands (b) and (c), which were mixed together. Re-electrophoresis of the fractions, (a) and (b) + (c) shows minimal cross contamination (B). E. coli tRNA$_2^{Glu}$ (1) and 5S rRNA (2) were used as markers. B (3), Fraction (a); B (4), Fraction (b) + (c).

subjected to SDS/sucrose gradient centrifugation. Of the four fractions from the gradient as indicated, the top fraction (Fraction 1) exhibited the ability to induce differentiation. Fraction 1 was then run on 7.0% polyacrylamide containing 99% formamide. The RNA was resolved in at least three fractions. RNA from the top band was eluted from the gel separately, while the RNA from bands b and c were pooled. Re-electrophoresis of these fractions showed that the fractions were minimally cross-contaminated. The RNA from these fractions were eluted from the gel and passed through oligo (dT) cellulose for the second time in order to remove the gel pieces and other contaminants. The bound RNA thus recovered was tested on the PNP. Less than 10% RNA usually passed through the second oligo (dT) cellulose and could not be recovered in amounts to test its activity. The bound RNA from fraction '1a' was the only active fraction. Although for most purposes, RNA in fraction '1a' was used directly, the RNA was further purified on a 7.5% polyacrylamide gel containing formamide or on 5-7% polyacrylamide-urea gels for the purpose of structural analysis (see Fig. 13). Under similar electrophoretic conditions, the unlabeled fraction '1a" RNA resolved into two poorly separated components. The faster moving major band was 4.5 x more active than the slow moving component. The titration with various amounts of RNA added to PNP culture showed that some activity in the latter RNA fraction was apparently due to cross contamination (unpublished results). The fast moving active RNA fraction is referred to from here on as 7S CEH-RNA. A typical titration curve for partially purified 7S RNA for its activity on PNP is shown in Fig. 24.

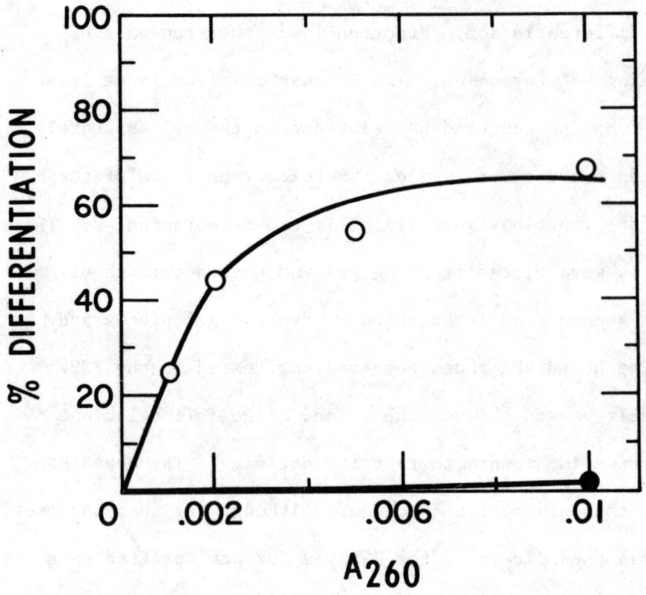

Fig 24. Effect of varying amount of 7S RNA fraction 1 'a' on PNP. The PNP explants were cultivated in the presence of varying amount of the inducer RNA and examined for differentiation after 96 hr of incubation (34). A minimum of 18 and a maximum of 38 explants were used for each point. o, untreated RNA; ●, RNA pretreated with insoluble pan RNase (see ref.22).

cDNA SYNTHESIS AND CLONING

cDNAs were synthesized using AMV reverse transcriptase and mRNAs or 7S RNA as templates. Under the conditions chosen, 0.3-0.4 μg of cDNAs for MLC mRNA and 7S RNA were made per μg of template RNA (38). A high proportion of cDNA for MLC mRNA were of the average length of 1050 nucleosides when Fraction 11 RNA was used as template (Fig. 25), although some batches of AMV transcriptase yielded cDNA of substantially smaller sizes. The S1 digestion of cDNA indicated that between 8-12% of the DNAs contained double-stranded structure. For further characterization of the cDNA, kinetics of reassociation of cDNA with the template mRNA was examined. As shown in Fig. 25, the reassociation took place within the range of 100 Rot values displaying a pseudo first order reaction rate. The Rot-1/2 observed was 6.3×10^{-3}. The Rot-1/2 observed for globin cDNA:mRNA hybridization was 2×10^{-3} in agreement with previously observed values. The complexity of MLC mRNA, based on the hybridization kinetics of globin and MLC back hybridizations, was 2225 nucleosides, which is in excellent agreement with the combined sizes of the two MLC mRNAs (MLC_1 = 1090 nucleosides and MLC_2 = 980 nucleosides) based on electrophoretic separation on denaturing gels. This suggested that MLC mRNA in Fraction 11 consists of two molecular species of high purity, as indicated by the translation assay, and the sequences of both mRNA are present in the cDNA transcripts. The potential of cDNA for MLC mRNA as a probe for quantitation of MLC mRNA sequences present in various RNA preparations was investigated by hybridization with the polysomal RNA and poly (A) RNA of chick heart muscle (Fig. 25). The

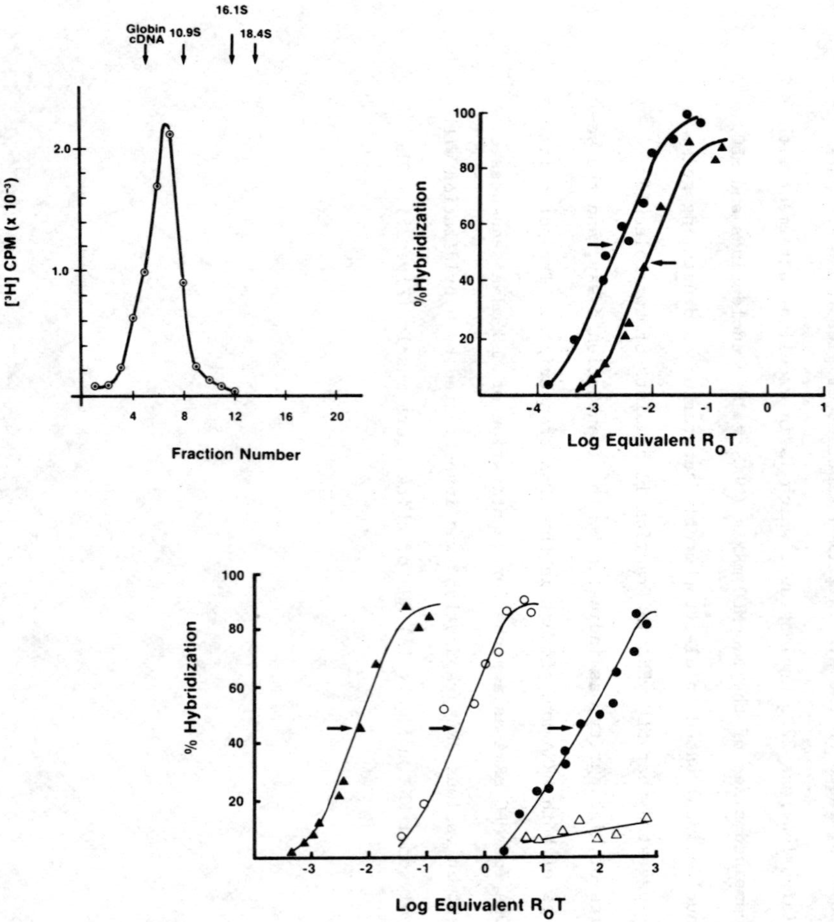

Fig 25. (Top left)

Alkaline sucrose density gradient centrifugation on MLC cDNA. [^{3}H]cDNA was prepared using AMV reverse transcriptase and purified MLC mRNA template (38). The DNA was centrifuged on a 10-30.9% alkaline sucrose isokinetic gradient for 21 h at 40000 rpm in a Beckman SW 41 rotor. The sample was preheated at 80°C for 30 min. Aliquots of 5 μL diluted with H_2O and neutralized with CCl_3COOH were assayed for radioactivity. ϕχ 174 DNA which sediments at 18.4S, 16.9S, and 10.9S under these conditions was used as a marker. The migration of β-globin cDNA run on a parallel gradient is shown by the arrow.

(Top right)

(38) Kinetics of hybridization of purified mRNA with the [^{3}H]cDNA. Purified MLC mRNA and β-globin mRNA were hybridized with their respective cDNA and the hybrids were assayed as described (38). (●) β-Globin cDNA: mRNA (▲) MLC cDNA:mRNA. (Bottom) Hybridization kinetics of [^{3}H] cDNA with purified MLC mRNA, poly(A) RNA, and total polysomal RNA of the chick heart. RNA from total polysomes, poly(A) RNA and the purified MLC mRNA were prepared as described (38). Rat liver RNA was a gift of Drs. Innis and Miller. Hybridization was carried to the indicated R_{ot} values by varying the RNA concentration and incubation times of the respective reactions. Arrows indicate the $R_{ot}1/2$ values. (▲) MLC mRNA; (o) poly(A) RNA; (●) total polysomal RNA; (Δ) rat liver RNA.

Rot-1/2 were 0.35 and 39 for the two RNAs respectively, suggesting that MLC mRNAs comprise 2.0% and 0.02% of poly (A) RNA and polysomal RNA respectively.

The cDNA synthesized using Fraction 12 RNA (Fig. 20) was enriched in large size transcripts after selection of gradient fractions which eliminated slower sedimenting small cDNA fragments. The cDNA synthesized had an average chain length of 950±50 nucleosides, almost the same size as the MLC_2 mRNA (980 nucleosides). A second size selection after the tailing reaction was essential, since the terminal transferase apparently contained endonucleolytic activity. The cDNA was readily converted to double-stranded cDNA by using AMV reverse transcriptase and the digestion with S1 nuclease under chosen conditions rendered ds-cDNA without hairpin loops. The ds-cDNA was then cloned in *E. coli* strain X1776 using the Hind III cleavage site in the plasmid pBR322. Of the 12 transformants that contained large size DNA inserts, clone pML10 insert was 950±50 base pairs in length. On the basis of size estimates of MLC mRNA (38), there are 340 extra nucleosides in the mRNA than required to code for MLC_2 polypeptide. On the assumption that most of these non-coding nucleosides are located in the 3' terminus, more than half of the coding region of MLC_2 mRNA must still be represented in the clone DNA insert. The clone pML10 was identified by hybridization with a highly pure cDNA probe for MLC_2 nRNA. The presence of MLC_2 mRNA sequence in pML10 DNA was further documented by hybrid-arrested translation assay (Fig. 26). Clone pML10 was first digested with enzyme EcoRI and the product after heat denaturation was hybridized in formamide with total poly(A) RNA from the chick embryonic heart and then translated in rabbit reticulocyte lysate. EcoRI-treated control plasmid pBR322

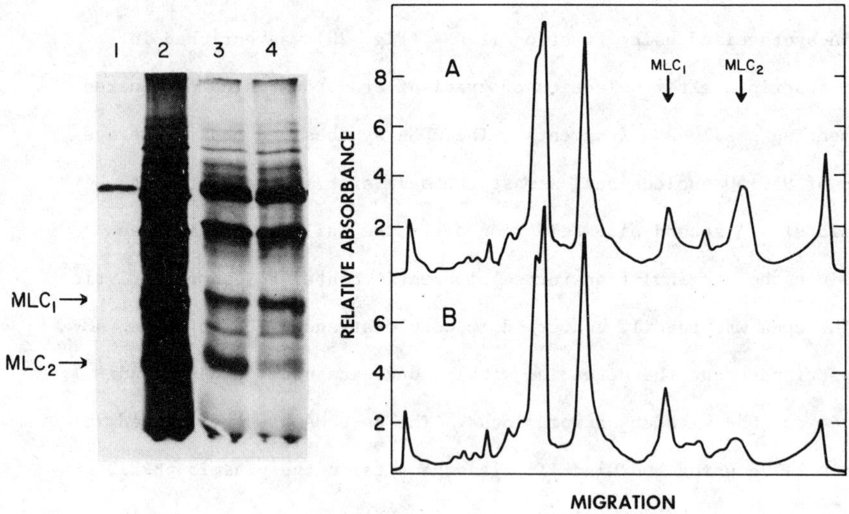

Fig 26. Cell-free translation products of control and hybrid-arrested poly(A) RNA. Clone pML10 DNA and plasmid pBR322 DNA after digestion with EcoRI were hybridized with total poly(A) RNA of the 16-day-old chick embryonic heart as described (39). The material after recovery by several ethanol precipitations was subjected to the cell-free translation assay in the rabbit reticulocyte lysate. The products were analyzed by $NaDodSO_4$-polyacrylamide gel electrophoresis and examined by fluorography. Lane 1, micrococcal nuclease treated lysate; lane 2, translation products of total poly(A) RNA; lane 3, translation products of total poly(A) RNA after incubation with pBR322 DNA; lane 4, translation products of total poly(A) RNA after hybridization with clone pML10 DNA. Panels A and B represent the densitometric tracings of lanes 3 and 4 respectively.

DNA was also hybridized with another batch of the same RNA under identical conditions. As shown in the Fig. 26, the translation activity of clone pML10 DNA containing the mRNA fraction was identical to mRNA to which the control DNA was added with the exception of one polypeptide that migrated identically with MLC_2. The synthesis of MLC_2 was reduced by more than 75%. The residual counts in the MLC_2 band might be due to incomplete hybridization. The neutralization of a selective polypeptide is nevertheless a clear demonstration for the qualitative presence of MLC_2 specific DNA sequence in the clone pML10 DNA.

The synthesis of myosin heavy chain (MHC)cDNA was relatively inefficient. Two to six percent of total RNA input was transcribed into cDNA. Variations in ionic conditions, NTP concentration and temperature did not significantly improve the cDNA yield. Fig. 27 shows the fractionation of MHC cDNA on an alkaline sucrose gradient and by electrophoresis on denaturing gel. The largest transcripts appeared to be approximately 3,500 nucleosides in chain length. The kinetics of reassociation between the cDNA specific for Fraction 18 RNA (Fig. 21) and purified 28S rRNA of the chick suggested that a substantial proportion of the cDNA transcripts represented 28S rRNA. Lambda DNA containing chick rRNA sequences (a kind gift from Ann Skalka) also hybridized efficiently with Fraction 18 cDNA on a Southern gel (unpublished results). Apparently, repeated chromatography of poly(A) RNA on oligo(dT) cellulose did not totally remove the contaminating rRNA and the poor template activity of MHC mRNA contributed to an increased synthesis of cDNA for rRNA. The ds-cDNA made with Fraction 18 RNA as template was used for cloning and the resulting transformants are being screened with probes containing pure rRNA sequences and MHC mRNA.

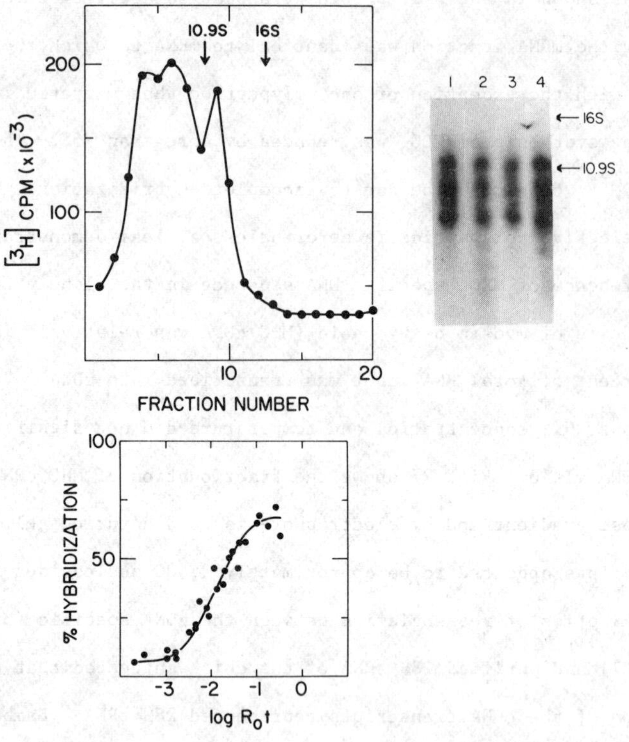

Fig 27. MHC cDNA synthesis and its characterization. MHC cDNA was prepared using partially purified MHC mRNA (see Fig 21) and AMV reverse transcriptase and analyzed on an alkaline sucrose gradient using $\phi\chi174$ DNA as markers. [^{32}P] labeled cDNA for MHC prepared simultaneously was analyzed on 1% agarose gel (top right). Lanes 1, 2 indicate incubations done at 37°C in presence and absence of 40 mM KCl respectively and 3, 4 were the incubation at 46°C with and without KCL. Back hybridization of cDNA with MHC mRNA (fraction 18) (see Fig 21) (unpublished results) is shown on bottom.

7S CEH-RNA contains a 3'-terminal poly(A) tail of approximately 100 residues. As such, the RNA serves as an efficient template for transcription using AMV reverse transcriptase. Fig. 28 shows that the majority of 7S transcripts sediments in alkaline sucrose with a mobility of about 300 nucleosides. The cDNA was rendered double-stranded and cloned in *E. coli* X1776 using the Pst I site in plasmid pBR322. The resulting clones are currently being characterized as potential probes for screening of 7S sequence.

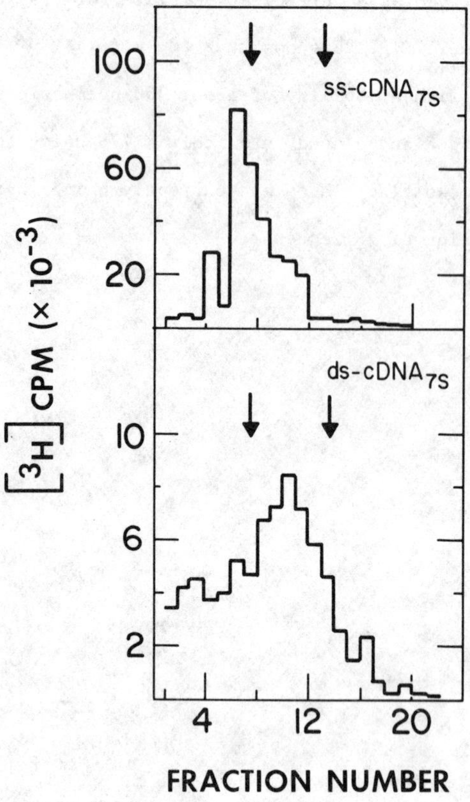

Fig 28. Single-stranded and double-stranded cDNA synthesis for 7S RNA. ss-cDNA and ds-cDNA for 7S RNA were prepared using AMV reverse transcriptase (see text) and analyzed by alkaline sucrose gradient centrifugation using marker DNA sedimenting with a mobility of approximately 300 and 600 nucleosides (left and right arrows respectively) (unpublished results).

SUMMARY

To summarize briefly, we have demonstrated unequivocally the existence of a distinct low-molecular-weight RNA species that is capable of causing a specific mode of differentiation in the early embryonic cells of the chick blastoderm. We have also shown that the differentiation is similar to the embryonic myogenic process and is dependent upon the 7S RNA since addition of a variety of other RNA with similar structural features is not effective.

7S RNA is an authentic component of chick embryonic heart tissue. It is apparently absent in other chick tissue examined. We have, however, observed that a low-molecular-weight RNA fraction from the chick embryonic leg muscle can also cause morphogenetic changes in PNP similar to 7S RNA. One apparent difference was the lack of rhythmic pulsations, although occasional twitchings were recorded. Since we do not at present have on hand a sensitive assay to discriminate between the leg muscle development and the cardiac muscle development, we assume that the transition was like the leg muscle. This correlation for the presence of inducer RNAs in cardiac muscle and leg skeletal muscle tissues and the presence of the respective myogenic competence in PNP cell population is more than coincidence, since premyogenic cells for skeletal muscle are known to be localized in the postnodal region of the early chick blastoderm.

The functional significance of 7S RNA remains unknown. Although the RNA contains a 3'-poly(A) tail, it fails to translate in a variety of cell-free translation systems. The absence of the initiator codon, at least in the 5' one-fourth of the molecule, also suggests that 7S RNA may not be an mRNA.

The relationship between the 7S RNA induced biological transition in the PNP cells and the inhibition of *in vitro* translation due to the RNA is not clear. Since the product of RNase digestion of 7S RNA results in a loss of its inducing activity and the fact that poly(A) alone, at several concentrations, cannot mimic the effects of 7S RNA, it would be reasonable to assume that a control at the level of translation based on the observations on *in vitro* translation assays can be eliminated as one of the possible mechanisms of action of the RNA on PNP. However, a mechanism envisaging a selective discrimination of specific mRNA translation under distinct physiological conditions, which are not necessarily duplicated by the rabbit reticulocyte lysate assay, cannot be excluded.

7S RNA nevertheless is a potential candidate for a regulator of muscle gene transcription during embryogenesis. The understanding of the actual role 7S RNA plays in the process would require a comprehensive analysis of the structural and functional properties of the RNA.

REFERENCES

1. Davidson, E.H. (1968) "Gene Activity in Early Development" Acad. Press., N.Y.

2. Raff, R.A. (1977) Bioscience 27, 394-401.

3. Davidson, E.H. and Britten, R.J. (1971) J. Theoret. Biol. 32, 123-130.

4. DeRobertis, E.M. and Gurdon, J.B. (1977) Proc. Nat. Acad. Sci. U.S.A. 74, 2470-2474.

5. Dickson, E. and Robertson, H.D. (1976) Cancer Res. 36, 3387-3393.

6. Davidson, E.H. and Britten, R.J. (1979) Science 204, 1052-1059.

7. Brothers, A.J. (1976) Nature 260, 112-115.

8. Thompson, J.A., Stein, J.L., Kleinsmith, J.L. and Stein, G.S. (1976) Science 194, 428-431.

9. Stein, G.S., Stein, J.L., Perk, W.D., Detkel, S., Lichtler, A.C., Shephard, E.A., Jansing, R.L. and Phillips, I.R. (1979) Arch. Biol. Med. Exper. 12, 439-455.

10. Pilch, V.H., Fritz, D., Waldman, S.R., and Kern, D.H. (1975) Current Topics in Microbiology and Immunology 72, 157-190.

11. Colby, C., Stollar, B.D. and Simon, M.I. (1971) Nature New Biol. 229, 172-174.

12. Diener, T.D., Smith, D.R. and O'Brien, M.J. (1972) Virology 48, 844-846.

13. Slavkin, H. and Crossant, R. in "The Role of RNA in Reproduction and Development" (Niu and Segal, Eds.) pp.247-258 (1973), North-Holland, Publishers.

14. Bester, A.J., Kennedy, D.S. and Heywood, S.M. (1975) Proc. Nat. Acad. Sci. U.S.A. 72, 1523-1527.

15. Bogdavsky, D., Herman, W. and Schapira, E. (1973) Biochem. Biophys. Res. Commun. 54, 25-32.

16. Lee-Huang, S., Sierra, J.M., Navanjo, R., Filipowicz, W. and Ochoa, S. (1977) Arch. Biochem. Biophys. 180, 276-287.

17. Bhargava, P. and Shanmugan, G. in Prog. Nucleic Acid Res. and Mol. Biol. (Davidson, J.N. and Colon, W.E., eds.) vol. 11, pp.103-192 (1971) Acad. Press, N.Y.

18. Davidson, E.H. and Britten, J.R. (1973) Quant. Rev. Biol. 48, 565-612.

19. Siddiqui, M.A.Q., Deshpande, A.K., Arnold, H-H., Jakowlew, S.B. and Crawford, P.A. (1977) Brookhaven Symp. in Biol. 29, 62-81.

20. Siddiqui, M.A.Q., in "Genes and Protein Structure" (Weissbach,H., Siddiqui, M.A.Q. and Krauskopf, M., eds.) Acad. Press (In Press).

21. Hamburger, V. and Hamilton, H.L. (1951) J. Morphol. 88, 49-92

22. Rosenquist, G.C. (1970) Develop. Biol. 22, 461-465.

23. DeHaan, R.L., Dunning, J.O., Hirakow, R., McDonald, T.F., Rash, J., Sachs, H., Wildes, J., Wolf, K.M. (1970) in Annual Rept. of the Director, Carnegie Ins. Washington Yearb. 70, 66-84.

24. Hunt, T.E. (1931) J. Exp. Zoology 59, 395-427.

25. Rudnick, D. (1938) Anat. Rec. 70, 351-368.

26. Rawles, M.E. (1943) Physiol. Zool. 16, 22-44.

27. Spratt, N.T. Jr. (1952) J. Exp. Zool. 120, 109-130.

28. Mulherkar, L. (1958) J. Embryol. Exp. Morphol. 6, 1-14.

29. Butros, J. (1967) J. Embryol. Exp. Morphol. 17, 119-130.

30. Chauhan, S.D.S. and Rao, K.V. (1970) J. Embryol. Exp. Morphol. 23, 71-78.

31. Butros, J. (1965) J. Embryol. Exp. Morphol. 13, 119-128.

32. Sanyal, S. and Niu, M.C. (1966) Proc. Nat. Acad. Sci. U.S.A. 55, 743-750.

33. Niu, M.C. and Deshpande, A.K. (1973) J. Embryol. Exp. Morphol. 29, 455-501.

34. Deshpande, A. K. and Siddiqui, M.A.Q. (1977) Develop. Biol. 58, 230-247.

35. Deshpande, A.K., Jakowlew, S.B., Arnold, H-H., Crawford, P.A. and Siddiqui, M.A.Q. (1977) J. Biol. Chem. 18, 6521-6527.

36. Deshpande, A.K. and Siddiqui, M.A.Q. (1978) Differentiation 10, 133-137.

37. Siddiqui, M.A.Q., Arnold, H-H., Deshpande, A.K., Jakowlew, S.B. and Crawford, P.A. (1979) Arch. Biol. Med. Exper. 12, 331-348.

38. Arnold, H-H. and Siddiqui, M.A.Q. (1979) Biochemistry 18, 647-654.

39. Arnold, H-H. and Siddiqui, M.A.O. (1979) Biochemistry 18, 5641-5647.

40. Manasek, F. J. (1968) J. Morphol. 125, 329-365.

41. Pollinger, I.S. (1972) Exp. Cell. Res. 76, 243-252.

42. Manasek, F. J. (1969) J. Embryol. Exp. Morphol. 21, 265-275.

43. Orkin, R.W., Pollard, T.D. and Hay, E.D. (1973) Develop. Biol. 35, 388-394.

44. Hauschka, S.D. (1968) in "The Stability of the Differentiated State" H. Urspring, ed.) vol. 1 pp. 32-57, Springer-Verlag, Berlin.

45. Karnovsky, M.J. and Roots, L.J. (1964) Histochem. Cytochem. 12, 219-221.

46. Silberklang, M., Gillum, A.M. and Rajbhandary, U.L. in "Methods of Enzymology" (Moldave & Grossman, eds.) vol. LIX, pp.58-109 (1979) Acad. Press. N.Y.

47. Arnold, H-H., Innis, M.A., Siddiqui, M.A.Q. (1978) Biochemistry 17, 2050-2054.

48. Rao, N.J., Blackstone, M. and Busch, H. (1977) Biochemistry 16, 2756-2762.

49. Vournakis, J.H., Efstrahadis, A. and Kajotos, F.C. (1975) Proc. Nat. Acad. Sci. U.S.A. 72, 2959-2963.

50. Maxam, A. M. and Gilbert, W. in "Methods of Enzymology" (Moldave & Grossman, eds.) Acad. Press, N.Y. (In Press).

51. Bonner, W.M. and Laskey, R.A. (1974) Europ. J. Biochem. 46, 83-88.

THE POSSIBLE MODE OF TRANSMISSION FOR "INDUCTIVE RNA" DURING EPITHELIAL-MESENCHYMAL INTERACTIONS

H.C. Slavkin and M. Zeichner-David, Laboratory For Developmental Biology, Graduate Program in Craniofacial Biology, and Department of Biochemistry, School of Dentistry, University of Southern California, Los Angeles, California 90007 U.S.A.

The phenomena of determination during embryogenesis remains one of the most fascinating and as yet poorly understood issues in developmental biology. How cells become differentiated remains an elusive and as yet somewhat obscure scientific problem. One promising postulate is that cells become differentiated from discrete and localized "inductive" events which translate into cellular differentiation. Induction has the strict requirement of cell-to-cell interactions. Further, it is assumed that one cell type will "induce" a responsive cell type to acquire a pattern of gene expression which is both original to a discrete interaction and the result of a specific interaction. Our observations suggest that direct cell-to-cell contact between embryonic epithelium and mesenchyme mediate the transduction of spatial and positional information. Experiments of epithelial-mesenchymal interactions clearly indicate that mesenchymal cells become competent to induce adjacent epithelia to differentiate into specific phenotypes. Using heterochronic and heterologous tissue recombinations, such as mouse dental mesenchyme and chick pharyngeal arch epithelium, mesenchymal specificity results in the induction of tooth forms and the expression of the ameloblast phenotype. During these unorthodox heterologous tissue recombinations <u>in vitro</u> direct cell-to-cell contacts are formed prior to the expression of the specific epithelial phenotype. Indirect evidence suggests that a low molecular weight "regulatory RNA" may mediate the transduction of information from mesenchyme to the responding epithelia. The RNA may be associated with discrete matrix vesicles formed from the mesenchymal cells. Available information argues against the possibility of diffusible molecules transmitted through the extracellular matrix interposed between heterotypic tissue interactants. Using the embryonic tooth organ our laboratory suggests that mesenchyme-derived inductive signals are transmitted by direct cell-to-cell contact.

I. INTRODUCTION

The phenomena of determination during embryogenesis remains one of the most fascinating and as yet poorly understood issues in developmental biology. During the past three decades significant attention and progress has been generated regarding the problem of cellular differentiation in multicellular organisms and the mechanisms controlling gene expression in differentiated cells. However, little progress has been made regarding how eucaryotic cells become differentiated. What process or set of processes are involved in selective determination of a particular phenotype? Although the problem of determination in eucaryotic cellular systems was enthusiastically pursued by Spemann (1) Mangold (2) and Weiss (3,4), our collective understanding of how cells become differentiated remains an elusive and as yet a somewhat obscure scientific problem.

Several mechanisms have been suggested which attempt to explain the available data and which have been used rather successfully to foster renewed scientific interest in the problem of how cells become differentiated. These mechanisms include the suggestion of (1) chemical gradients generating morphogenetic fields within the embryo, and (2) discrete and localized "inductive" events which translate into cellular differentiation. These suggested mechanisms differ both in their requirement for cellular interactions and in the nature of the cellular interactions involved (5-8).

Induction has a strict requirement for cell-to-cell interactions. Further, it is assumed that one cell type will "induce" a responsive cell type to acquire a pattern of differential gene expression and a related phenotype which is both original to a discrete interaction and the result of a specific interaction. Induction <u>is not</u> the amplification of already predetermined gene expression; <u>it is not</u> merely a quantitative enhancement of gene expression. Induction must invoke qualitative changes within the responding cell type which are non-reversible and characteristic of a unique cell phenotype.

An interesting concept has recently emerged which attempts to integrate the suggested mechanisms of morphogenetic gradients within the embryo and embryonic induction. In this context, Wolpert has stated that "epithelial-mesenchymal interaction is the transfer of positional information" (9). During embryogenesis clones of dissimilar cells express different and changing quanta of positional information. When a responsive cell responds to an inductive event, according to Wolpert, "positional information is the interpretation of epigenetic signals" from the immediate microenvironment of the responding cell (9). Despite these challenging and fascinating explanations for the phenomena of determination during embryogenesis, a number of painfully pertinent issues remain as yet unresolved. What are the molecular "epigenetic signals"? What is the mode of transmission for these inductive signals? What mechanism provides for the transduction of positional or inductive information during development ? What might the role be for RNA during epithelial-mesenchymal interactions?

II. RNA AS AN EPIGENETIC SIGNAL DURING DEVELOPMENT

The expression of cellular genetic information is a profound process in developmental biology. During embryogenesis, including gastrulation, neurulation and subsequent organogenesis, cells become determined as a function of spatial and temporal information. Spatial and temporal positional information and the subsequent genesis of pattern and form both constitute a significant linkage between genetics, molecular biology and biochemistry, and morphology. Within this broad and complex problem area is the critical postulate that RNA and/or polynucleotides mediate embryonic induction by representing the transfer of a determinant morphogen between heterotypic cells. A mechanism which has received considerable attention is the possibility that an informational macromolecule is transferred from an inducing cell to a responding cell. RNA has been particularly suggested as a candidate in this regard (7, 10-19).

III. SUGGESTED REGULATORY RNA MOLECULES

A number of RNA molecules have been detected in vivo and in vitro associated with the extracellular microenvironment during cell-to-cell interactions (20-21). Relatively small molecular weight RNAs, some of which have been found to be methylated, have been found within the extracellular matrix during embryonic epithelial-mesenchymal interactions (22-23), as well as in the media of cells in culture (24-26). Evidence has been found which indicates that RNA is associated with outer cell surfaces (26). Despite this information, very little data has been obtained which unequivocally demonstrates (1) the direct transfer of RNA between homotypic and/or heterotypic cells in vivo or in vitro, (2) the isolation and characterization of this "informational RNA," (3) the direct demonstration of this discrete RNA upon differential gene expression during development, (4) the direct demonstration of "causality" between the transfer of this informational RNA and spatial and temporal positional values for the responding cell, and (5) direct evidence indicating that this process actually occurs in vivo during a normal developmental process.

Despite these qualifications, a number of suggested "regulatory RNA molecules" are pertinent to our discussion. Double-stranded RNA have been isolated from eucaryotic cells (27-29). These RNAs are low molecular weight (4-14S), appear to originate in the nucleus and hybridize to cellular DNA. They appear to be transcribed from reiterated DNA sequences. The functional significance of these RNAs is as yet unknown.

A number of low molecular weight RNAs have been demonstrated (4-9S) and suggested to have a regulatory function in eucaryotic cell differentiation (30-32). Whereas, many low molecular weight RNAs have been found within the nucleus and/or the cytoplasm of cells, and have been shown to be present in amounts equivalent to rRNA, their function(s) remains unknown.

Another approach has been to search for single-stranded RNAs as possible regulator RNAs. A number of such RNAs have been isolated and suggested to function in differential gene activity within eucaryotic cells. The concept of regulatory RNA has recently been supported through the elegant studies which demonstrate a low molecular weight RNA (7S) from embryonic chick heart capable of inducing a tissue-specific set of changes within progenitor cardiac cells *in vitro* (11, 14, 15). This RNA seems to be heart-specific and is not found in other embryonic chick tissues. The RNA sediments at 7S, contains detectable poly (A) sequence and is not translatable in a cell-free system (14-15). In addition, two low molecular weight "regulatory" RNAs have recently been described which are operant during *Artemia salina* embryogenesis (33). One of these RNAs is a "translational inhibitor," whereas the other RNA is a "translational activator." During embryonic chick myogenesis a translational control RNA (tcRNA) has been identified which might regulate the initiation and translation of mRNAs (34).

IV. TRANSFER OF RNA BETWEEN CELLS IN "CONTACT"

Examples of cell-to-cell communication exist throughout metazoan developing systems. The nature of this communication, however, is known to involve a number of different and unique modes of transmission and reception including: (1) direct cell-to-cell contacts, (2) short-range cell-to-cell interactions mediated through an interposed extracellular matrix, (3) and long-range intercellular communication mediated by humoral factors such as those associated with polypeptide and/or steroid signals (e.g. the effects of pancreatic insulin on lens cell elongation) (see reviews 6, 8, 35-39)(Figure 1).

Several recent reviews are available which provide a critical assessment of the transfer of nucleotides and RNAs between cells in contact (20, 21, 40, 41). The assumed "information passing between heterotypic cells in vivo and in vitro is further postulated to be highly specific. The specificity might be mediated by the highly specific base sequences and conformational properties of ribonucleic acids or polynucleotides. Several lines of investigation have been very provocative. One approach has been to examine the role of gap junctions in cell-to-cell communication. Available evidence suggests that gap junctions represent ionic coupling between cells achieving metabolic cooperation and a mechanism by which molecules and/or ions can be exchanged during significant phases of development (41). Gap junctions appear to mediate the flow of ions between similar and dissimilar cells, the intercellular passage of fluorscein molecules, and the autoradiographic demonstration of the transfer of radio-labeled molecules from one cell to another including nucleotides, polynucleotides and low molecular weight RNAs (20, 24, 41-43). Gap junctions appear in most developing eucaryotic cell systems coincident with major changes in cell differentiation; they are observed within and amongst homotypic cell types as well as between heterotypic cell types. Available information suggests that gap junctions or equivalent forms of intercellular communication might mediate the transfer of RNAs between cells providing both spatial and temporal specificity.

Metabolic cooperation is a form of cell-to-cell communication in which one cell may directly transfer a specific molecule to an immediately adjacent cell. The formation of these direct cell-to-cell contacts, such as gap junctions, occur in vivo and in vitro (43). Metabolic cooperation has been classically defined as the phenomena in which the phenotype of a mutant cell is corrected

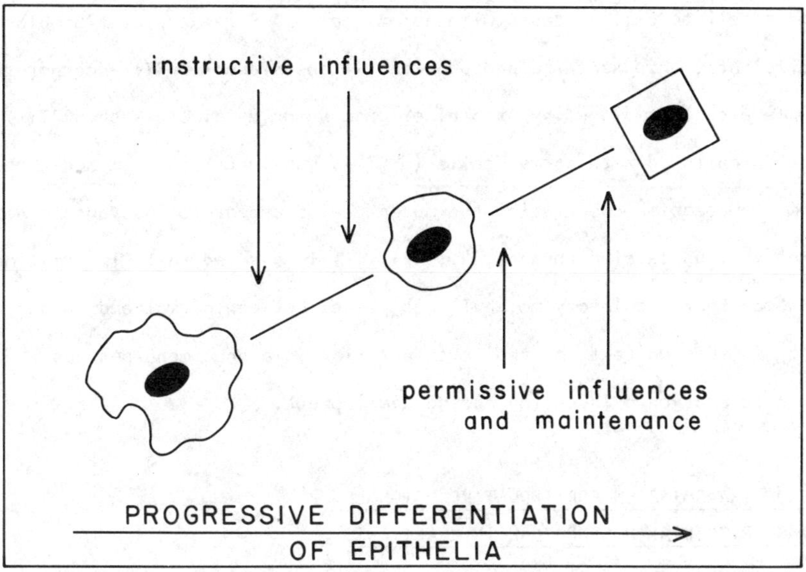

Fig. 1 Diagrammatic representation of the possible modes for transmission of secondary embryonic induction during epithelial-mesenchymal interactions (6, 37).

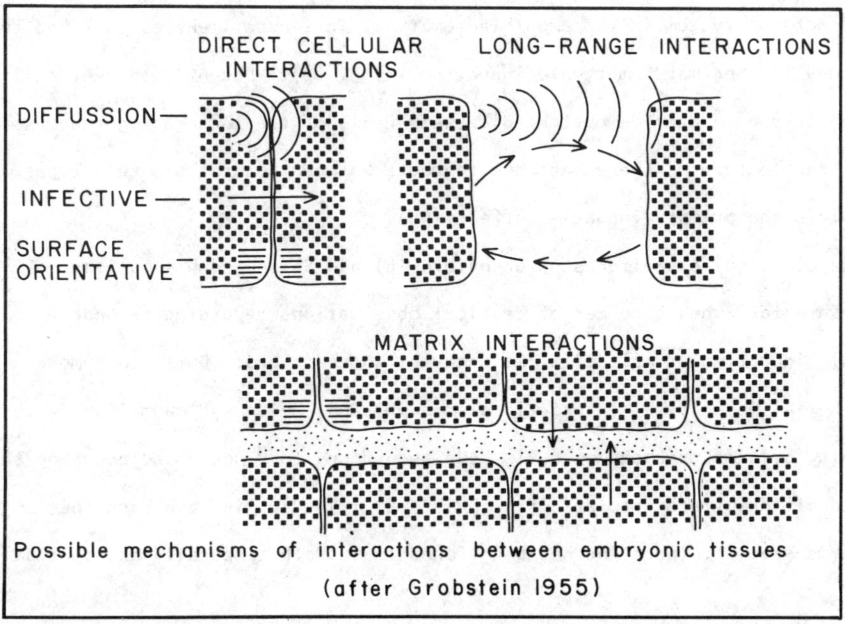

Fig. 2 The significance of instructive cell-to-cell interactions during epithelial-mesenchymal interactions (e.g. skin, thyroid, salivary gland, mammary gland, thymus, tooth organ). See extensive discussions by Professor Saxen and his colleagues (37).

by intimate cell-to-cell contacts with normal cells. For example, fibroblast cells deficient in hypoxanthine phosphoribosyltransferase (HPRT) incorporate very low levels of tritiated hypoxanthine, guanine or uridine. When HPRT(-) mutant cells are co-cultured with normal cells (HPRT+), gap junctions form and through some form of metabolic cooperation the mutant cells appear to incorporate normal amounts of various labeled purines (41, 42). It is assumed that "informational RNA" or some other regulatory molecule was passed between normal and mutant cells (20, 21). In this context, metabolic cooperation is a phenomena perhaps comparable to embryonic induction during development.

V. MODE OF TRANSMISSION FOR INDUCTIVE "SIGNALS" DURING EPITHELIAL-MESENCHYMAL INTERACTIONS

The classical experiment of Spemann and Mangold (44) demonstrated that the invaginating chorda-mesoderm during gastrulation is responsible for the induction of the overlaying presumptive medullary plate (neuroectoderm) to become the central nervous system in the amphibian embryo. This experiment established the conditions for "primary embryonic induction" (1-3, 44). Recently, direct cell-to-cell contacts have been demonstrated between cells of the inducing chorda-mesoderm and adjacent ectoderm; these intimate contacts between heterotypic cell types may mediate the primary inductive effect (45).

The classical experiments of Grobstein (6) and Saxen and his colleagues (37) have established a number of critical observations regarding secondary embryonic induction or epithelial-mesenchymal interactions. During a number of epithelial-mesenchymal interactions in amphibian, avian and mammalian developing systems mesenchymal cells have been found to "induce" adjacent epithelial cells to differentiate into specific phenotypes (37-40). In each case the mesenchyme-specific induction has been found to be both temporally and spatially-specific (3, 6, 9, 35-40). In examples using heterotypic tissue recombinations, for example embryonic rabbit mammary gland mesenchyme and embryonic chick dorsal skin epithelium, the mesenchyme "induces" the epithelium to express a phenotype which is complementary to that of the inductive mesenchyme; mammalian mammary mesenchyme induces avian skin epithelium to differentiate into branching tubular structures containing acinar cells allegedly synthesizing casein (46).

What is the mode of transmission for inductive signals during epithelial-mesenchymal interactions? Is the "inductive signal" a diffusible substance or is direct cell-to-cell contact required? A number of transfilter experiments indicate that the inductive signal is a relatively large macromolecule, that transmission of the signal requires nascent RNA and protein synthesis, the transmission is energy-dependent, and that short-range cell-to-cell interactions or direct cell-to-cell contacts are likely modes for transmission (37). Dialysis tubing of 3,000 or 10,000 molecular weight exclusion inhibits transfilter epithelial-mesenchymal interactions (37). Nuclepore filters having pore sizes of 0.2 to 0.8 um diameter support positive inductive interactions (37). All available information currently argues for direct cell-to-cell contact mediated by particles or cellular processes which are 0.1 um in diameter or larger (6, 9, 19, 35-40). It is also appropriate to emphasize that the responding epithelial cells must show DNA replication in order to respond to the mesenchymal inductive signal (37) (Figures 1, 2, 3).

VI. EPITHELIAL-MESENCHYMAL INTERACTIONS DURING TOOTH MORPHOGENESIS: A MODEL SYSTEM

A. Epithelial and Mesenchymal-Specific Phenotypes

The embryonic mammalian tooth organ is formed by a series of reciprocal epithelial-mesenchymal interactions which result in the acquisition of highly

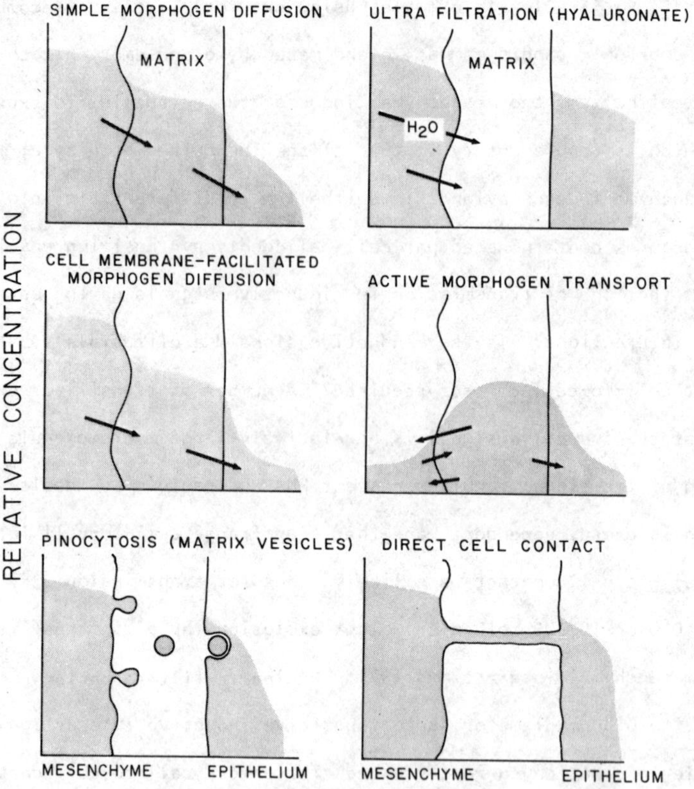

Fig. 3 The possible mode of transmission for purine bases, nucleosides, nucleotides, polynucleotides and "inductive RNA" during epithelial-mesenchymal interactions. Diffusion is a physical-chemical process by which molecules move across cell membranes in direct proport on to the electrochemical gradient(s) established across the membrane. Transfer is determined by Fick's law. Simple diffusion occurs down a concentration gradient and is not energy-dependent. Ultra Filtration can result from an influx of hyaluronic acid resulting in a change or shift in hydrostatic pressure. Cell membrane-facilitated morphogen diffusion suggests that molecules are transferred via a cell membrane component such as a cell process or matrix vesicle. Active morphogen transport implies that energy is required to maintain transmission activity. During pinocytosis mesenchymal cell processes form matrix vesicles which transport inductive RNAs to the adjacent epithelia; matrix vesicle contents are then internal ized by the epithelia. Direct cell-to-cell contact suggests that instructive molecules can be transmitted from mesenchyme cell process to adjacent epithelial cells via gap-like junctions.

specific phenotypes as well as unique patterns of cells integrated into tooth form (39). In the course of early embryogenesis, cranial neural crest cells migrate into the forming craniofacial complex and differentiate into dental papilla mesenchyme as well as other crest cell-derivatives. The progenitor tooth mesenchyme differentiate in juxtaposition to an ectodermally-derived epithelium that subsequently becomes the enamel organ epithelium. These two tissue types interact throughout the early stages of tooth morphogenesis. The interface between these heterotypic tissues provides a spatial and temporal orientation.

Prior to overt terminal differentiation, the dental papilla mesenchyme and adjacent enamel organ epithelia represent dividing and expanding populations of cells. Those cells aligned in juxtaposition to the interface between epithelium and mesenchyme express a set of unique phenotypes. The mesenchymal cells at the interface with epithelium produce type I collagen, cell surface fibronectin and hyaluronate-rich glycosaminoglycans (47). The adjacent epithelial cells produce type IV collagen, laminin and glycosaminoglycans associated with the basal lamina (47-48). The basal lamina is an epithelial-derived cell surface specialization deposited at the interface between the inner enamel epithelia and the adjacent cranial neural crest-derived mesenchyme (49)(Figures 4-6).

During subsequent epithelial-mesenchymal interactions, the mesenchymal cells at the interface align themselves into a sheet or "epithelia" of mesenchyme to form the progenitor odontoblast cell layer. These cells differentiate into non-dividing and highly polarized odontoblast cells which synthesize and secrete extracellular matrix macromolecules including types I and III collagen, dentine phosphoprotein and proteoglycans (49). Mesenchymal cell differentiation precede that of the adjacent epithelium by approximately 48 hours (49). The adjacent

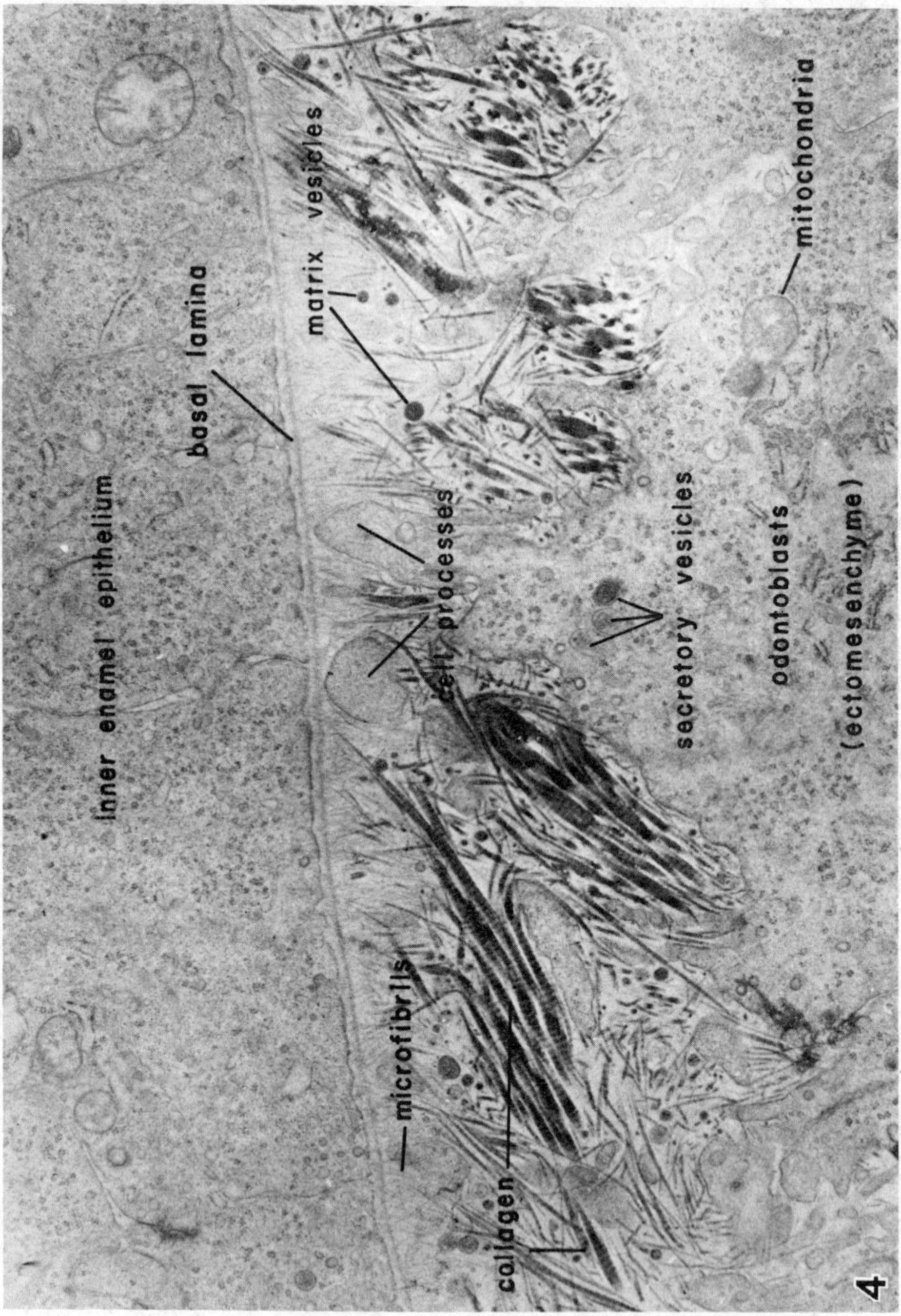

Fig. 4 Early phases of epithelial-mesenchymal interactions during tooth formation (see Text for discussion). A number of positive cellular interactants are indicated (magnification x 5000).

(Figure reduced to $\frac{7.5}{10}$ during printing.)

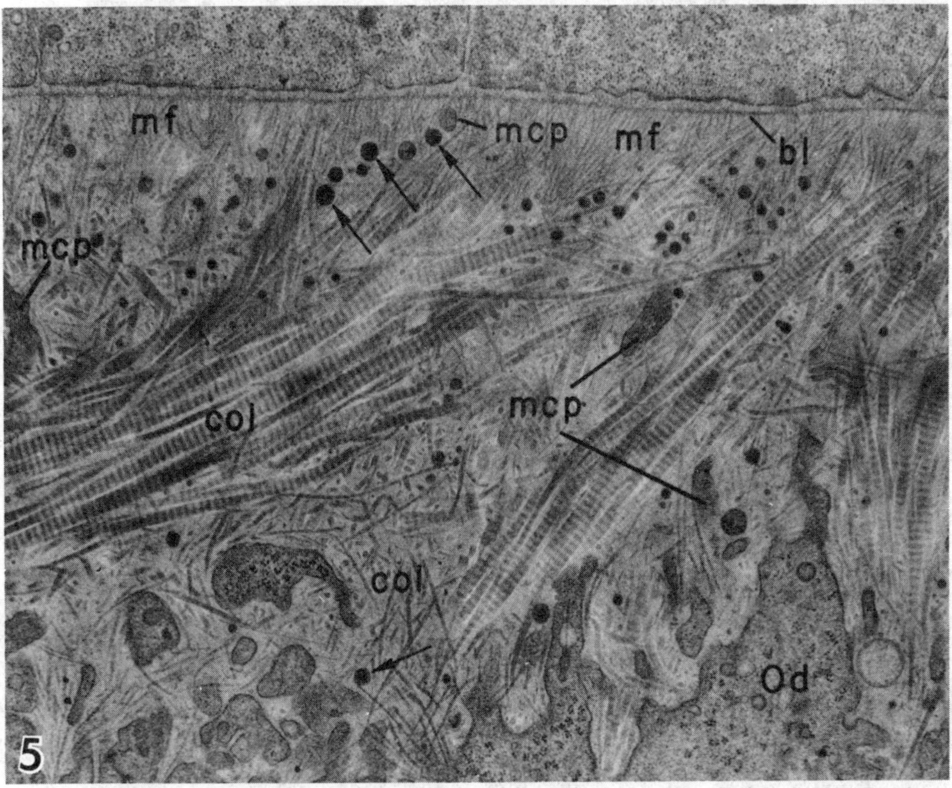

Fig. 5 During later phases of epithelial-mesenchymal interactions, a number of matrix vesicles (arrows) accumulate along the undersurface of the basal lamina (bl) with associated microfibrils (mf). Note the mesenchymal cell processes (mcp) and type I collagen (col) associated with these progenitor odontoblast cells (od) (magnification x16,000).

(Figure reduced to $\frac{8}{10}$ during printing.)

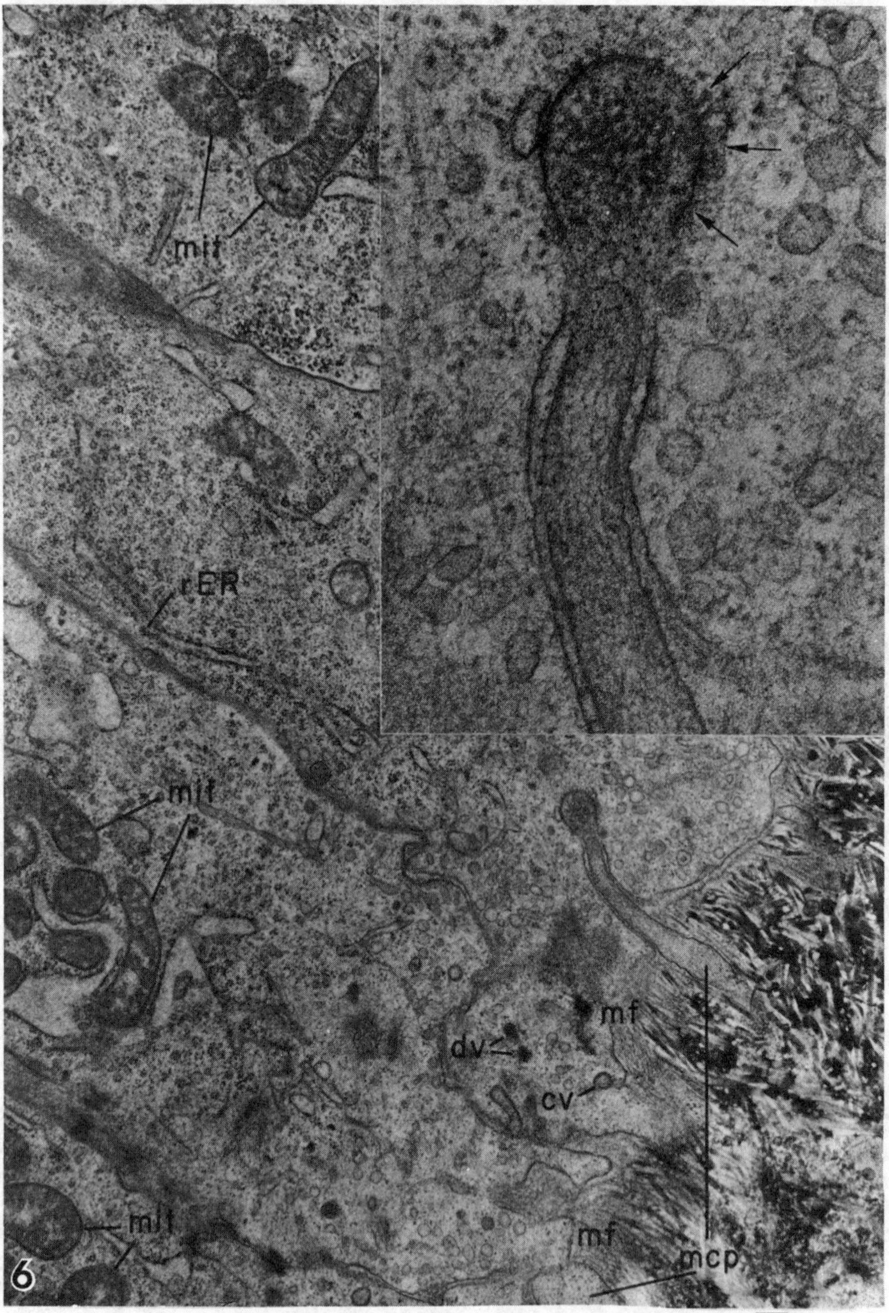

Fig. 6 Prior to epithelial differentiation into non-dividing, elongated ameloblasts, mesenchymal cell processes (mcp) form direct cell-to-cell contacts (insert) with adjacent epithelium. Mitochondria (mit), rough endoplasmic reticulum (rER), dense vesicles (dv), coated vesicles (cv), microfibrils (mf) (magnification x 19,500). Note the arrows within the Insert indicating electron-dense granules (magnification x100,000).

(Figure reduced to $\frac{7}{10}$ during printing.)

inner enamel epithelia differentiate into non-dividing and highly polarized ameloblasts which synthesize and secrete extracellular matrix macromolecules of amelogenins (enamel proteins) and glycosaminoglycans (50-52). Both the dentine and the enamel organic matrices subsequently mineralize.

B. Mesenchymal Specificity in Heterologous Tissue Recombinations

One very interesting question regarding the nature of epithelial-mesenchymal interactions during tooth morphogenesis is the issue of determination. When during development are the mesenchymal and epithelial phenotypes determined? When during development is the mesenchyme inductive? Using inbred mouse strains it is known that the dental papilla mesenchyme is highly inductive from Theiler stage 23 through stage 25 (e.g. the so-called "cap stage" of tooth development). For example, Theiler stage 25 dental papilla mesenchyme obtained from a molar tooth organ by enzymatic dissociation of the tooth analagen has been shown to induce embryonic mouse footpad epidermis to differentiate into secretory ameloblasts (53).

Embryonic induction has been shown to experimentally mediate epithelial-mesenchymal tissue recombinations when the sources of each tissue type represents diverse phylogenetic origin. A number of fascinating heterochronic (diverse chronology of development) and heterotypic (diverse tissue sources such as chick/mouse chimeras) have demonstrated the conservation of mesenchyme-specific embryonic induction (1, 46, 54-55). Recently, Kollar and Fisher demonstrated that dental papilla mesenchyme from Theiler stage 25 mouse embryos (circa days 16-18 gestation) induced 5-day-old chick embryonic pharyngeal arch epithelium to differentiate into secretory ameloblasts (56). These investigators have suggested that the loss of teeth in Aves did not result from a loss of genetic information coding for enamel polypeptides in the avian epithelium, but rather the epithelial-mesenchymal interaction in chick embryos is unable to express odontogenesis.

Is there reason to believe that the unique sequences for enamel proteins are present in Avian chromatin? No teeth are present in lower chordates or fossil ostracoderms! In all jawed vertebrates teeth are universally present except where secondarily lost as in turtles and modern birds (60). Teeth have been identified within the fossils of certain Jurassic and Cretaceous birds. However, most experts agree that modern birds are <u>toothless</u> and do not contain dental vestiges. Therefore, during evolution the reptilian genes for enamel proteins may have been delegated during the genesis of modern birds, or modern birds do contain unique sequences for enamel proteins but do not express amelogenesis due to an alteration in epithelial-mesenchymal interactions required for tooth formation. The recent findings by Kollar and Fisher (56) would indicate that modern birds (5-day-old chick embryonic pharyngeal arch epithelium) do contain the structural genes for enamel matrix proteins. However, if present, these highly conserved sequences are usually repressed during embryogenesis. Yet another interpretation is that during these chick/mouse heterologous tissue recombinations, dental papilla mesenchyme from the mouse transferred "informational RNA" which specifically induced amelogenesis within a responding avian ectodermal derivative. In either case, each postulate can be tested using modern methodology currently available in molecular biology.

C. <u>Direct Cell-To-Cell Contact During Epithelial-Mesenchymal Interactions</u>

Available evidence suggests that epithelial-mesenchymal interactions during tooth development require direct cell-to-cell contact between heterotypic tissues for some duration (circa 24-48 hours) in order to accomodate the instructive phases of differentiation and subsequent morphogenesis (6, 35, 37, 39, 49, 57). Direct cell-to-cell contacts have been demonstrated during tooth development <u>in vivo</u> and <u>in vitro</u> in positions of orientation prior to ameloblast differentiation (37, 39, 49, 57, 58). In addition to mesenchymal cell processes positioned within epithelial clefts following the degradation of the epithelial-derived basal lamina (58), discrete mesenchyme-derived matrix vesicles (circa 0.1 um in diameter) are also deposited along the undersurface of the preameloblast cells and in contact with the progenitor ameloblast cells (19, 39, 40, 49, 58, 59). Using ultracytochemical methods (40) and biochemical analysis we have determined that some of these matrix vesicles contain methylated, low molecular weight RNAs (19, 22-23).

D. Transmission For "Inductive RNA" During Epithelial-Mesenchymal Interactions.

During embyronic tooth morphogenesis RNA has been detected within the extracellular matrix interposed between epithelium and adjacent mesenchyme (19, 22, 23, 40). Approximately 4-7 ug low molecular weight, methylated RNA per 24 mg (dry weight) extracellular matrix was obtained (23). More recently, the development of techniques to physically isolate matrix vesicles from the extracellular matrix enabled the subsequent identification of RNA within these vesicles (19). The RNA within the matrix vesicle is resistant to RNAse A and T_1 digestion.

Organ culture experiments suggested that extracellular matrix RNAs were critically dependent upon nascent transcription. Preincubation of physically isolated tooth organs with dactinomycin, followed by pulse-chase additions of a number of different labeled RNA precursors, prevented silver grain density from being detected by light microscopic autoradiography localized over the extracellular matrix (22, 23). Descriptive studies using tritiated methyl-methionine, cytosine, uridine, adenosine or guanosine produced qualitatively similar results (19).

More recently, we have attempted to demonstrate the incorporation of tritiated uridine into molecules localized within mesenchymal cell processes and matrix vesicles in discrete positions relative to those regions related to embryonic induction during tooth morphogenesis (see Figures 4-6). The in vivo incorporation of tritiated uridine by progenitor odontoblast cells (mesenchyme) and into matrix vesicles derived from these cells, has been analyzed by high-resolution electron microscopic autoradiography. Special attention has been focused on the unique extracellular matrix organelle; the matrix vesicle derived from the mesenchymal cell processes (Figures 4-5). Experiments were designed to determine whether, during the phases of epithelial differentiation prior to cessation of DNA replication, matrix vesicles associated with the mesenchyme incorporate (5,6-^{3}H) uridine. Moreover, in studies ranging from 5 minutes to 24 hours, following the intraperitoneal injection of tritiated uridine to newborn mice, we assessed the relative distribution of silver grains relative to the epithelial-mesenchymal interface. Initial observations showed that silver grains were immediately localized over mesenchymal cell nuclei as well as mesenchymal cell processes and matrix vesicles adjacent to the extracellular matrix interface (Figures 7-8). Within 15 minutes following the administration of the tritiated uridine, silver grains were localized over matrix vesicles and mesenchymal cell processes immediately adjacent to inner enamel epithelial cells (Figure 9).

Caution must be exercised when considering these observations. The matrix vesicles are 0.1 um in diameter and the silver grains per se are 0.1 um in diameter. The resolution of this technique is limited and it is very difficult to localize the specific origin of the radioactivity as detected by autoradiography of such small structures. In addition, it is not known that the tritiated uridine is incorporated into RNA as visualized in these preparations. However, analysis of the number of significantly labeled matrix vesicles and mesenchyme

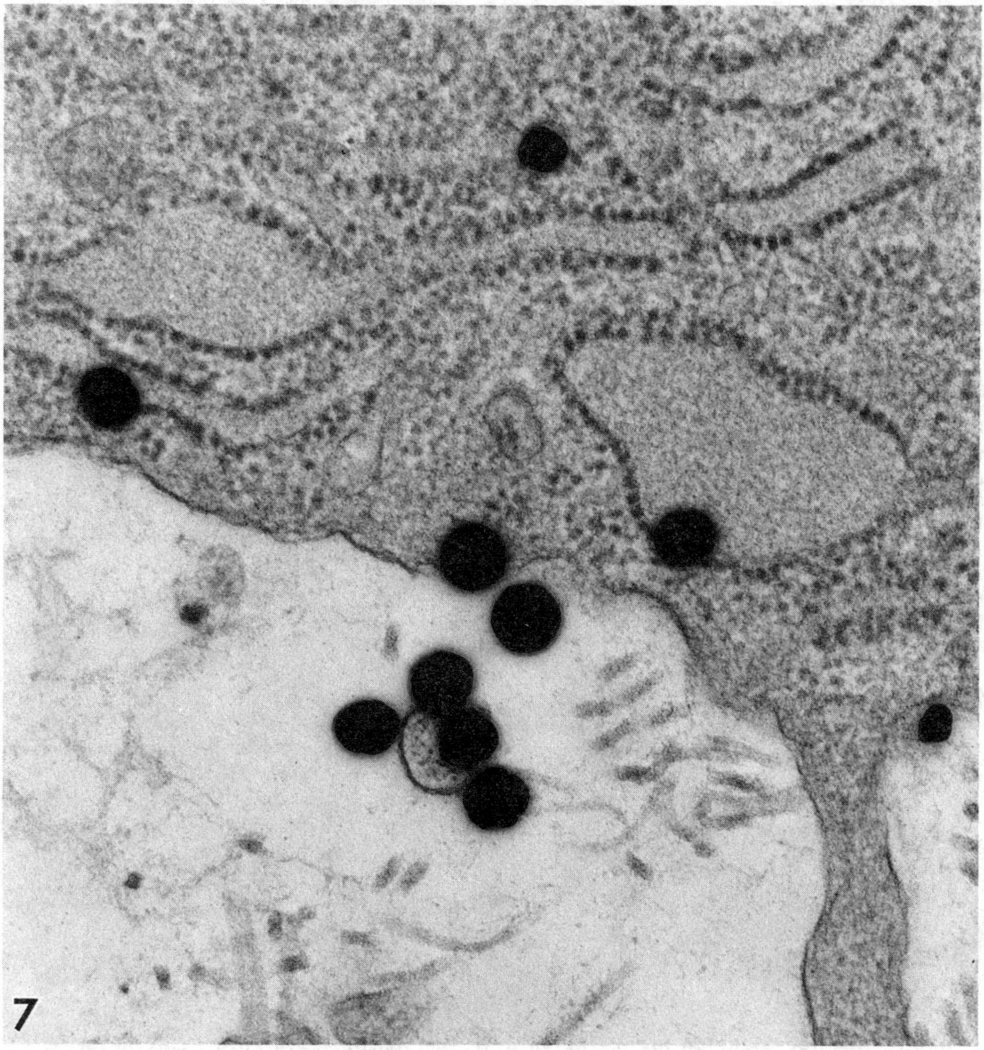

Fig. 7 Matrix vesicles form from preodontoblast cells following short-term pulse of (^3H) uridine injected into newborn mouse pups in vivo. The silver grain localization over a unit area containing matrix vesicles was significantly higher than comparable extracellular matrix areas devoid of vesicles. This electron photomicrograph of high-resolution autoradiography represents 5 minutes following the intraperitoneal administration of 25uCi (5,6-^3H) uridine (specific activity 55 Ci/mmde, New England Nuclear, Boston) (magnification x 90,000).

(Figure reduced to $\frac{8}{10}$ during printing.)

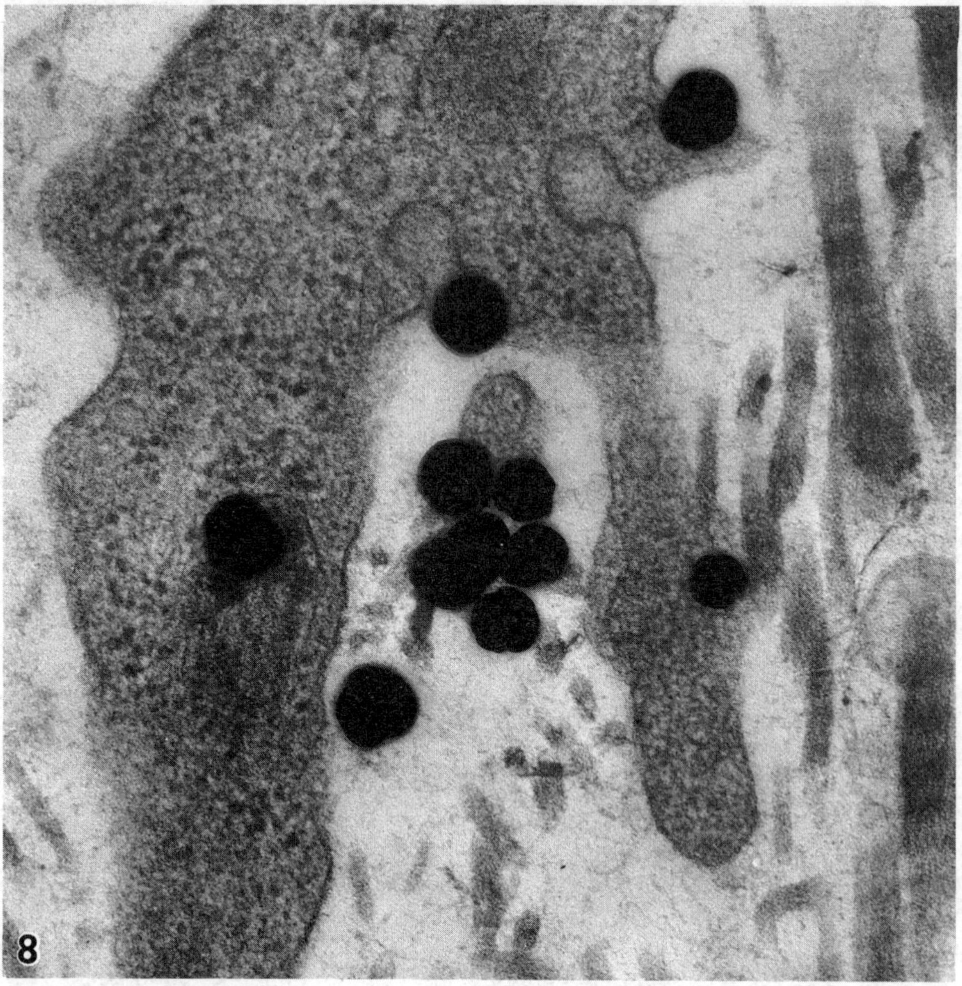

Fig. 8 The localization and relative distribution of silver grains associated with mesenchymal cell processes and matrix vesicles 15 minutes following the administration of (^{3}H) uridine (magnification x120,000).
(Figure reduced to $\frac{7}{10}$ during printing.)

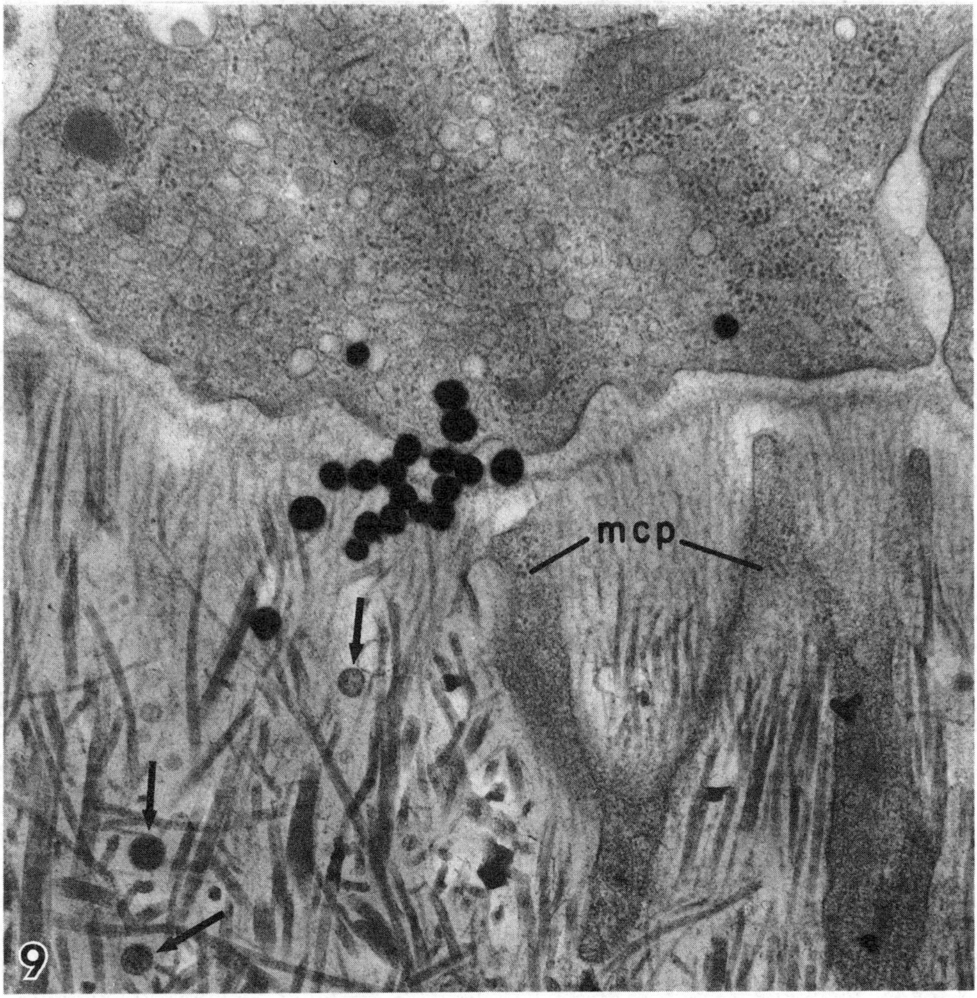

Fig. 9 Silver grains were often localized in discrete regions along the interface between mesenchyme and epithelium. Pulse/chase studies, ranging from 5 minutes to 24 hours in vivo, suggest that silver grains appear to be transmitted from mesenchyme cell processes (mcp) and matrix vesicles (arrows) to the adjacent epithelial cell surfaces (magnification x44,000). (Figure reduced to $\frac{8}{10}$ during printing.)

cell processes observed at various times after (^{3}H) uridine injection indicates that incorporation takes place very rapidly and within critical regions of cell-to-cell interactions. In regions of overt cytodifferentiation of odontoblast and ameloblast cells, silver grains were not observed within the extracellular matrix.

The progression of epithelial-mesenchymal interactions, including those associated with tooth morphogenesis, is arrested by puromycin, cycloheximide and dactinomycin (37, 39). During epithelial-mesenchymal interactions associated with tooth formation, electron microscopic cytochemistry revealed matrix vesicles which contain ribonucleoprotein particles (40). In the chordamesoderm-neuroectoderm interface in amphibia embryos electron microscopy has revealed extracellular matrix ribonucleoproteins (61). This observation was confirmed by Tarin (62). A number of investigators have identified extracellular "organelles" containing ribonucleoproteins in locations associated with secondary embryonic induction (19, 23, 25, 40, 49, 59, 61-62). RNA may be transmitted from the inducing tissue to the responding tissue type during direct cell-to-cell interactions (Figure 10).

V. PROSPECTUS

Recent studies of epithelial-mesenchymal interactions indicate that direct cell-to-cell contact is a requirement for organogenesis. In a number of developing organ systems it has been shown that mesenchyme "induces" epithelium to differentiat into a cell type complementary to that of the mesenchyme. Heterochronic and heterotypic tissue recombinations across vertebrate species indicate that mesenchyme specificity is highly conserved. For example, embryonic rabbit mammary gland mesenchyme has been shown to induce avian epithelium to differentiate into mammary gland-like structures (46). More recently, embryonic mouse dental papilla

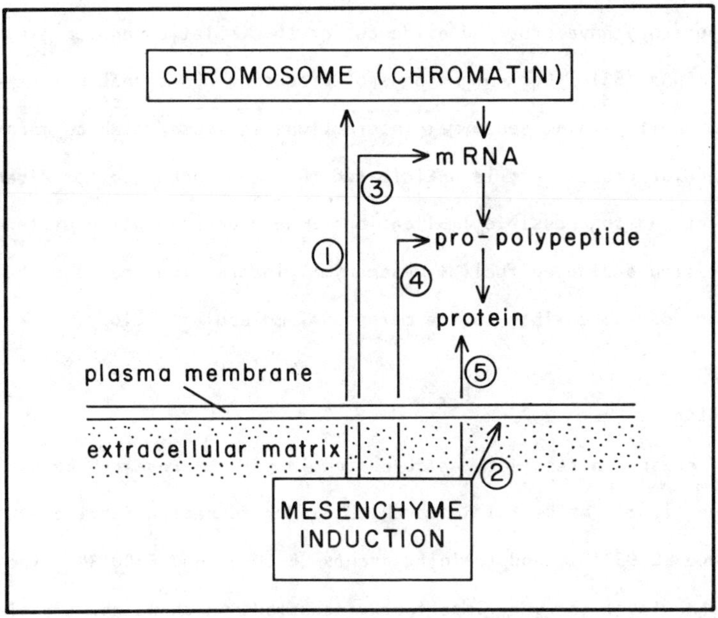

Fig. 10 Mesenchymal induction of epithelial cell differentiation may have effects at a number of of different sites within the responding epithelia including: (1) transcription, (2) cell surface, (3) the processing of hnRNAs, (4) translation and (5) post-translational modifications. On the basis of recent heterologous tissue recombinations within and amongst vertebrate species, we now assume that mesenchymal induction represents the transmission of an "inductive RNA" which regulates (somehow) de novo transcriptional events. Suggested systems to test this assumption include rabbit mammary gland mesenchyme induction of avian epithelium (using casein mRNA-directed cDNA probes), and mouse dental mesenchyme induction of avian epithelium (using enamel mRNA-directed cDNA probes).

mesenchyme has been shown to induce avian epithelium to differentiate into enamel matrix producing cells - secretory ameloblasts (56). Recent investigations in our laboratory have provided evidence for the isolation and partial characterization of enamel mRNAs (63). Current methods of producing cDNAs enable more rigorous analysis of epithelial-mesenchymal interactions in mouse/chick chimeras using molecular probes. It is anticipated that the mechanism for direct cell-to-cell contact and the possible implications toward understanding differential gene expression during epithelial-mesenchymal interactions may soon be less obscure and more accessible to the methods of molecular biology.

ACKNOWLEDGEMENTS

This research assessment was supported in part by research grants from the National Institute for Dental Research, U.S. Public Health Service (DE-02848, DE-03569 and DE-03513), and training grants DE-0094 and DE-00134. The authors wish to acknowledge the very important discussions which we had the pleasure of sharing with Drs. George Martin, Edward J. Kollar and Jean-Victor Ruch related to selected topics in this review. The authors wish to acknowledge our appreciation to Bonnie Sudderth for preparation of the manuscript and to Pablo Bringas, Jr. for the illustrations and electron photomicrographs.

REFERENCES

1. H. Spemann, Embryonic development and induction (Yale University Press, New Haven, (Conn.), 1938).

2. O. Mangold, Acta Genet. Med. Gemellol. 10:1 (1961).

3. P. Weiss, Int. Rev. Cytol. 7:391 (1958).

4. P. Weiss, Yale J. Biol. Med. 19:235 (1947).

5. L.G. Barth and L.J. Barth, Develop. Biol. 28:18 (1972).

6. C. Grobstein, Nat. Cancer Inst. Monogr. 26:279 (1967).

7. M.C. Niu, In: Cellular Mechanisms in Differentiation and Growth (D.Rudnick, ed.), pp. 155-171, Princeton University Press, New Haven (Conn.) 1956.

8. L. Wolpert, Curr. Top. Dev. Biol. 6:183 (1971).

9. L. Wolpert, In: Recent Research Trends in Prenatal Craniofacial Malformations (R. Pratt, ed.), North Holland/American Elsevier, New York (in press).

10. J. Brachet, Arch. Biol. 53:207 (1942).

11. A.K. Desphande, L.C. Niu and M.C. Niu, In: The Role of RNA in Reproduction and Development (M.C. Niu and S.J. Segal, eds.) pp. 229-246, North Holland/American Elsevier, New York 1973.

12. N.C. Mishra, M.C. Niu and E.L. Tatum, Proc. Nat. Acad. Sci. (USA) 72:642 (1975).

13. G. Czihak and S. Hőrstadius, Develop. Biol. 22:15 (1970).

14. H.H. Arnold, M.A. Innis and M.A.Q.Siddiqui, Biochem. 17:2050 (1978).

15. A.K. Deshpande and M.A.Q. Siddiqui, Differentiation 10:133 (1978).

16. H. Amos, Biochem. Biophys. Res. Comm. 5:1 (1961).

17. P. Bhargava and G. Shanmugan, In: Progress in Nucleic Acid Research and Molecular Biology (J.N. Davidson and W.E. Cohn, eds.) pp. 103-192, Academic Press, New York (1971).

18. N. Stebbing, Cell Biol. Internat. Reports 3:485 (1979).

19. H.C. Slavkin and R. Croissant, In: The Role of RNA in Reproduction and Development, (M.C. Niu and S.J. Segal, eds.) pp. 247-258, North Holland/American Elsevier, New York (1973).

20. G.M. Kolodny, In: Cell Communication (R.P. Cox, ed.) pp. 97-111, John Wiley & Sons, New York (1974).

21. G.M. Kolodny, In: Extracellular Matrix Influences on Gene Expression (H.C. Slavkin and R.C. Greulich, eds.) pp. 75-84, Academic Press, New York (1975).

22. H.C. Slavkin, P. Bringas, and L.A. Bavetta, J. Cell. Physiol. 73:179 (1969).

23. H.C. Slavkin, P. Flores, P. Bringas and L.A. Bavetta, Develop. Biol. 23:276 (1970).

24. G.M. Kolodny, L.A. Culp and L.J. Rosenthal, Exp. Cell Res. 73:65 (1972).

25. J.W. Beierle, Science 161:798 (1968).

26. M. Bennett, E. Mayhew and L. Weiss, J. Cell. Physiol. 74:183 (1969).

27. L. Montagneir, C.R. Academ. Sci. (Ser. D) 267:1417 (1968).

28. P.H. Duesberg and C. Colby, Proc. Nat. Acad. Sci. (USA) 64:396 (1969).

29. R. Stern and R. Friedman, Nature (Lond.) 226:612 (1970).

30. T.S. Ro-Choi, and H. Busch, In: The Molecular Biology of Cancer (H. Busch, ed.) pp. 241-276, Academic Press, New York (1974).

31. S. Frederiksen, T. Pederson, P. Hellung-Larson and J. Engberg, Biophys. Acta 340:64 (1974).

32. M. Zeichner and D. Breitkreutz, Arch. Biochem. Biophys. 188:410 (1978).

33. S. Lee-Huang, J. Sierra, R. Naranjo, W. Filipowicz and S. Ochoa, Arch. Biochem. Biophys. 180:276 (1977).

34. S. Heywood and D. Kennedy, Biochem. 15:3314 (1976).

35. N.K. Wessells, Tissue Interactions and Development, Benjamin/Cummings, Menlo Park (California) 1977.

36. A.A. Moscona and R.E. Hausman, In: Cell and Tissue Interactions, (J.W. Lash and M.M. Burger, eds.) pp. 173-186, Raven Press, New York (1977).

37. L. Saxen, M. Karkinen-Jaaskelainen, E. Lehtonen, S. Nordling and J. Wartiovaara, In: The Cell Surface in Animal Embryogenesis and Development (G. Poste and G.L. Nicolson, eds.) pp. 331-408, North-Holland/American Elsevier, New York (1976).

38. D. McMahon and C. West, In: The Cell Surface in Animal Embryogenesis and Development (G. Poste and G.L. Nicolson, eds.), pp. 449-494, North-Holland/American Elsevier, New York (1976).

39. H.C. Slavkin, J. Biol. Buccale 6:189 (1979).

40. H.C. Slavkin, In: Developmental Aspects of Oral Biology (H.C.Slavkin and L.A. Bavetta, eds.) pp. 165-201, Academic Press, New York (1972).

41. J.D. Pitts, In: International Cell Biology 1976-1977, (B.R. Brinkley and K.R. Porter, eds.) pp. 43-49, Rockefeller Press, New York (1977).

42. J.H. Subak-Sharpe, R.R. Burke and J.D. Pitts, J. Cell Sci. 4:353 (1969).

43. N.B. Gilula, O.R. Reeves and A. Steinbach, Nature (Lond) 235:262 (1972).

44. H Spemann and H. Mangold, Arch. Entwicklungsmeek. Org. 100:599 (1924).

45. H. Grunz and J. Staubach, Differentiation 14:59 (1979).

46. A.Y. Propper and L. Gomot, Experentia 29:1543 (1973).

47. I. Thesleff, H.J. Barrach, J.M. Foidart, A. Vaheri, R.M. Pratt and G.R. Martin, Develop, Biol. (in press).

48. M. Osman, H. Lesot and J.V. Ruch, Develop. Biol., (in press).

49. H.C. Slavkin, Oral Sci. Rev. 4:1 (1974).

50. H. Graver, R. Herold, T. Chung, P. Christner, C. Pappas and J. Rosenbloom, Develop. Biol. 63:390 (1978).

51. H.C. Slavkin, B. Weliky, W. Stellar, M.D. Slavkin, M. Zeichner-Ganz, P. Bringas, H. Hyatt-Fischer, M. Shimizu and M. Fukae, J. Biol. Buccale 6:309 (1979).

52. H. Guenther, R.D. Croissant, S.E. Shonfeld and H.C. Slavkin, Biochem. J. 163:591 (1977).

53. E.J. Kollar and G.R. Baird, J. Embryol. exp. Morph. 24:159 (1970).

54. C.H. Waddington, J. Exp. Biol. 11:224 (1934).

55. B. Garber, E.J. Kollar and A.A. Moscona, J. Exp. Zool. 168:455 (1968).

56. E.J. Kollar and C. Fisher, Science 207:993 (1980).

57. I. Thesleff, E. Lehtonen and L. Saxen, Differentiation 10:71 (1978).

58. H.C. Slavkin and P. Bringas, Develop. Biol. 50:428 (1976).

59. H.C. Slavkin, R.D. Croissant, P. Bringas, P. Matosian, P. Wilson, W. Mino and H. Guenther, Fed. Proc. 35:127 (1976).

60. A.S. Romer and T.S. Parsons, The Vertebrate Body. pp. 241-245, W.B. Saunders Co., Philadelphia (1978).

61. R.O. Kelley, J. Exp. Zool. 172:153 (1969).

62. D. Tarin, J. Anat. 111:1 (1972).

63. M. Zeichner-David, B.G. Weliky and H.C. Slavkin, Biochem. J. 185:489 (1980).

THE CHARACTERISTICS OF CRUCIAN LENS MESSENGER RNA

Song Dexiu and John H. Chen*

Institute of Developmental Biology, Academia Sinica, and

*Departemnt of Biochemistry, New York University, Dental Center

Abstract Crucian fish lens mRNA was isolated by oligo(dT)-cellulose chromatography and their characteristics studied. Results show that fish mRNA exists as a heterologous population containing both poly(A)-plus and poly(A)-minus molecular species, similar to that of the calf lens. The fish lens mRNA was shown to direct the synthesis of polypeptide chains of crystallin in a wheat germ cell-free system. Translation products of the fish lens 14S mRNA showed three discrete peaks which have the same electrophoretic mobilities as those of the fish authentic crystallin in an alkaline urea gel electrophoretic system. These three bands are comparable with the A_2, B_2 and B_3 chains of fish lens crystallin. This result indicates that the 14S mRNA may be a polycistronic messenger. These unique structural features of the lens mRNA may be involved in the development and differentiation of the lens tissue.

INTRODUCTION

The lens is a highly specialized tissue. It is composed of a few crystallins. In 1964, Spector et al (1) showed that approximately 50% of the total protein synthesis was directed toward the production of α-crystallin polypeptides in an calf lens organ culture system. They also showed that α-crystallin is composed of 4 subunits, A_1, A_2 and B_1, B_2 (2). Chen et al have isolated calf lens 10S and 14S mRNA by Zonal centrifugation (3), and showed that calf lens α-crystallin 10S and 14S mRNAs coded for the B_2 chain and A_2 chain, respectively. The 10S mRNA is theoretically comparable to the size of the B_2 chain, but the 14S mRNAs contains approximately twice the required number of nucleotides necessary for A_2 chain synthesis. Therefore, they suggested that it may contain an additional cistron (4).

Crucian fish is a form of lower vertebrate. It is interested to see whether it contains the same class of mRNAs and possesses similar characteristics. We have isolated the crucian lens mRNA by oligo (dT)-cellulose chromatography and have characterized the translation products. These studies indicated that crucian lens mRNAs can also exist as a heterologous population and polycistronic messenger.

MATERIALS AND METHODS

Preparation of Lens Polysomal RNA

Lens polysomal RNA were prepared by a modification of the method of Chen et al (5). All handling was performed at 0-4°. Crucian lenses were homogenized in 1 volume of ice cold medium A, containing 0.05M Tris-HCl, pH 7.6, 0.025M KCl, 0.008M $MgCl_2$, 0.008M mercaptoethanol, 0.34M sucrose. Nuclei, mitochondria, and cell debris were removed by centrifugation at 17,000 g for 20 min at 4°C. Deoxycholate and Triton X-100 were added to the supernatant to make a final concentration of 1% and 0.1%, respectively. The solution was then centrifuged at 100,000 g for 90 min at 4°C. The polysomal pellet was dissolved in 0.05M Tris-HCl, pH 9.0 and then extracted with phenol: chloroform (V/V, 1:1), containing 0.1% 8-hydroxyquinoline and 0.5% SDS. The top aqueous fluid was deproteinized once more with phenol: chloroform. The aqueous phase was adjusted to 0.3M LiCl and ethanol precipitated at -20°C. The precipitate was washed with 70% ethanol and dissolved in the equilibrating buffer system used for affinity chromatography.

Oligo (dT)-Cellulose Chromatography

All operations were done at room temperature. Glassware and reagents (except for oligo(dT)-cellulose) were autoclaved. Oligo(dT)-cellulose columns were washed exhaustively with distilled water followed by 0.5M NaCl 0.01M Tris-HCl, pH 7.5 high salt equilibrating buffer to remove unbound affinity ligand and other foreign materials. 30-50A_{260} units of lens polysomal RNA dissolved in the application buffer was applied to a 3 x 1 cm column and the unbound RNA (peak A) was eluted by high salt buffer. After all unbound A_{260} material was removed, the bound RNA (peak B) was eluted with neutralized water. RNA from the pooled fractions was adjusted to 0.3M LiCl, precipitated and washed with ethanol.

Sucrose Gradient Centrifugation

The peak A and peak B RNA fractions were separately layered on a 15-30% linear sucrose gradient containing 0.05M Tris-HCl pH 7.4. Sedimentation was carried out at 38,000 rev/min for 20 hr at 2°C in a SW41 rotor. Fractions were collected, and their A_{260} was determined. Standard RNA markers, rat liver 30S, 18S and 5S RNA (6), were run in parallel gradients.

Cell-Free Protein Synthesizing System and Assays

Preparation of Wheat-Germ Extracts. Wheat germ was prepared by the methods of Marcu and Dudock (7), and Roberts and Paterson (8). All manipulations were done at 4°C. Wheat germ (2g) was ground in a chilled mortar with an equal weight of sand for 1 min, and 4 ml of a solution containing 20 mM HEPES (pH 7.6), 100 mM KCl, 1 mM magnesium acetate, 2 mM $CaCl_2$, and 6 mM 2-mercaptoethanol. The homogenate was centrifuged at 30,000g for 10 min at 0-2°C, and the supernatant was collected avoiding both the surface layer of fat and the pellet. The S-30 fraction was made up 3.5 mM magnesium acetate and preincubated for 15 min at 30°C with 1 mM ATP, 20 μM GTP; 2 mM dithiothreitol, 8 mM creatine phosphate, and 40 μg of creatine phosphokinase per ml of homogenate. The S-30 fraction was passed through a column (21 x 1 cm) of Sephadex G-25, equilibrated with 20 mM HEPES (pH 7.6), 120 mM KCl, 5 mM magnesium acetate, and 6 mM 2-mercaptoethanol at a flow rate of 1.5 ml/min. The peak of the turbid fractions was pooled and dispensed into polypropylene tubes in liquid nitrogen, and stored in -80°C freezer.

Assay for in vitro Protein Synthesis. Lens mRNA was assayed in 50 μl reaction mixture (9) containing 20 mM HEPES (pH 7.4), 1 mM DTT, 1 mM ATP, 20 μM GTP, 8 mM creatine phosphate, 200 μg creatine kinase (products of the Tongfung Reagents Factory, Shanghai, China) 2.5 mM MgAe, 100 mM KCl, 25 μM (each) of 19 non-radioactive amino acids, 0.25 μCi (311 μCi/μmole) ^{14}C-leucine, 6 μg of lens mRNA, 20 μl of wheat germ lysate, and 60 μM of of spermine. Incubation was at 29°C for 90 min, and 1 ml of ice-cold 5% Cl_3COOH was added. The mixture was cooled at 0°C for 5 min, and then the precipitate was collected over a 0.45 μm pore-size Millipore filter, washed three times with 3 ml (each) of 5% Cl_3 COOH, dried, and counted in a liquid scintillation counter. All determinations were done in duplicates.

Characterization of Cell-Free Products

The volumes of the assay described above were scaled up to 500 μl and incubated for 90 min. The reaction mixture was made 70 mM with EDTA (pH 7.4) and 20 μg/ml with pancreatic ribonuclease A. This was incubated at 37°C for 15 min and centrifuged at 3000 rpm for 50 min. The supernatant was precipitated with 10% TCA, and then washed twice with 5% TCA, thereafter with ethanol, ethanol-ether (1:1 V/V), and ether. The products were analyzed on polyacrylamide gels.

Alkaline urea gel electrophoresis. Electrophoresis on urea gels was performed (10) in 8.5 x 0.5 cm cylindrical glass tubes at 3 mA/gel for 5 hours. The gels contain: 10% acrylamide, 0.2% Bis 0.028% TEMED, 0.15% ammonium persulfate, 0.2M Tris 1.5 M glycine (pH 9.8), 1.8% 2-mercapto-ethanol and 7 M urea. Gels and buffers were prepared before use. Pre-electrophoresis was performed at 5 mA/gel for 1 hour. Sample was solubilized in 7 M urea, 8 mM 20mercaptoethanol, 5% sucrose. After electrophoresis, the gels were fixed in 50% TCA overnight, stained for standard fish α-crystallin markers of A_1, A_2, A_3 and B_1, B_2, B_3 with amino black for 30 min. The gels were destained in 2% acetic acid. Gels of the radioactive products obtained by response to added 14S mRNA were sliced into 1-2 mM sections and solubilized in 0.2 ml 60% perchloric acid and 0.4 ml of 30% hydrogen peroxide and kept at 60°C for 4-5 hours. Then 10 ml of scintillation liquid (600 ml toluene, 300 ml Triton X-100, 100 ml H_2O .4 g PPO, 0.05 g POPOP) were added. Radioactivity was determined by a Beckman LS 9000 liquid scintillation counter.

RESULTS

Oligo (dT)-cellulose Chromatography of Crucian Lens Polysomal RNA

When approximately 100 A_{260} units of lens polysomal RNA was applied to the oligo (dT)-cellulose column, the profile shown in Fig. 1 was obtained. The unbound RNA, peak A, was eluted from the column with high salt buffer. Bound RNA, peak B, was eluted with neutralized water, amounting to approximately 3-7% of the A_{260} material added to the column.

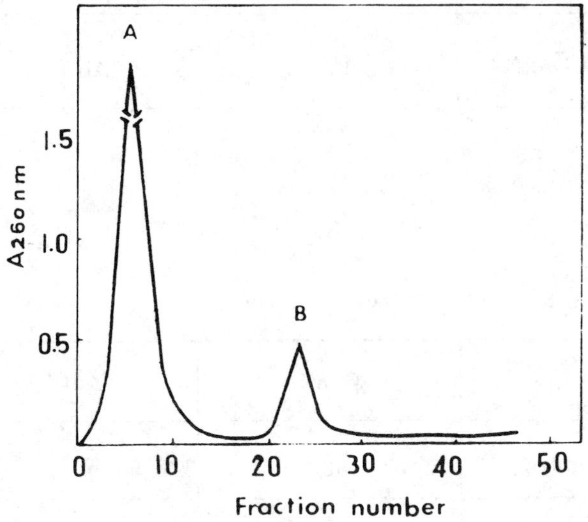

Fig. 1. Chromatography on Oligo (dT)-cellulose of lens polysomal RNA. Peak A, unbound RNA, was eluted from the column using high ionic strength buffer. Peak B, bound RNA eluted with neutralized water.

Table 1 shows the approximately equal distribution of poly $(A)^+$ mRNA and poly $(A)^-$ mRNA in both fish and calf lens tissues. The biological activities of these mRNA populations were determined in a wheat germ system and their respective products characterized by urea-polyacrylamide gel electrophoresis. Table 2 shows the biological activities between the sucrose gradient purified poly $(A)^+$ and poly $(A)^-$ mRNAs. It is interested to note that approximately the same magnitude of stimulation was obtained.

Table. 1

Heterologous nature of lens crystallin mRNA

mRNA Fraction \ Species	Crucian	Calf
poly$(A)^+$mRNA	51	52
poly$(A)^-$mRNA	49	48

Table. 2

Incorporation of ^{14}C-Leucine (cpm)

mRNA Fraction \ Species	Fish	Calf
poly$(A)^+$ mRNA	1.0×10^3	0.5×10^3
poly$(A)^-$ mRNA	0.9×10^3	

Sucrose Gradient Centrifugation

Centrifugation of the B peak RNA fraction in a linear 15-30% sucrose gradient showed five peaks, 30S, 18S, 14S, 10S and 4S RNAs (Fig. 2). It is clear from the sedimentation profile that the 10S and 14S mRNAs are discrete molecular species.

The 10S and 14S fractions were pooled and further fractionated through another sucrose gradient. Most of the purified mRNAs sedimented as discrete 10S and 14S species with no apparent contamination by larger or smaller components.

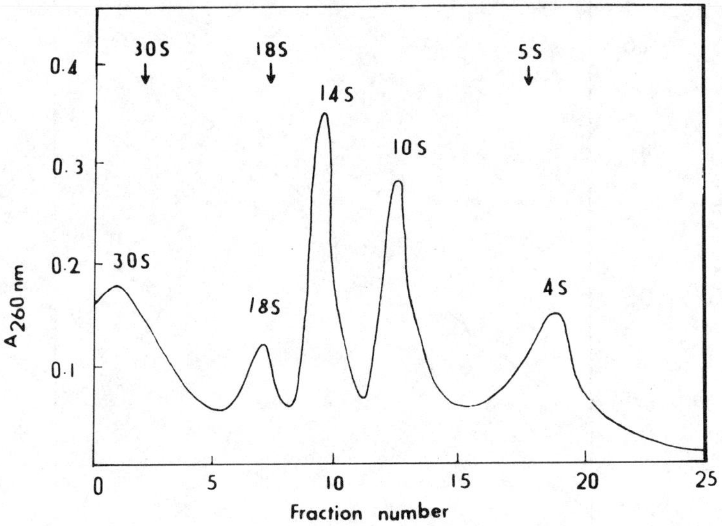

Fig. 2. Centrifugation of the peak B RNA fraction was carried out on a 15-30% linear sucrose gradient at 38,000 rev/min for 20 hr at 2°C in a SW 41 rotor. Arrows indicate the sedimentation peak positions of marker RNAs, ribosomal 30S, 18S and 5S, run in a parallel gradient.

Assay for in vitro Protein Synthesis

Figs. 3 and 4 show the kinetics of *in vitro* protein synthesis obtained with the 14S mRNA. The incorporation of ^{14}C-leucine into TCA precipitable material was approximately proportional to the mRNA added up to 6.0 μg/50 μl of incubation solution (Fig. 3).

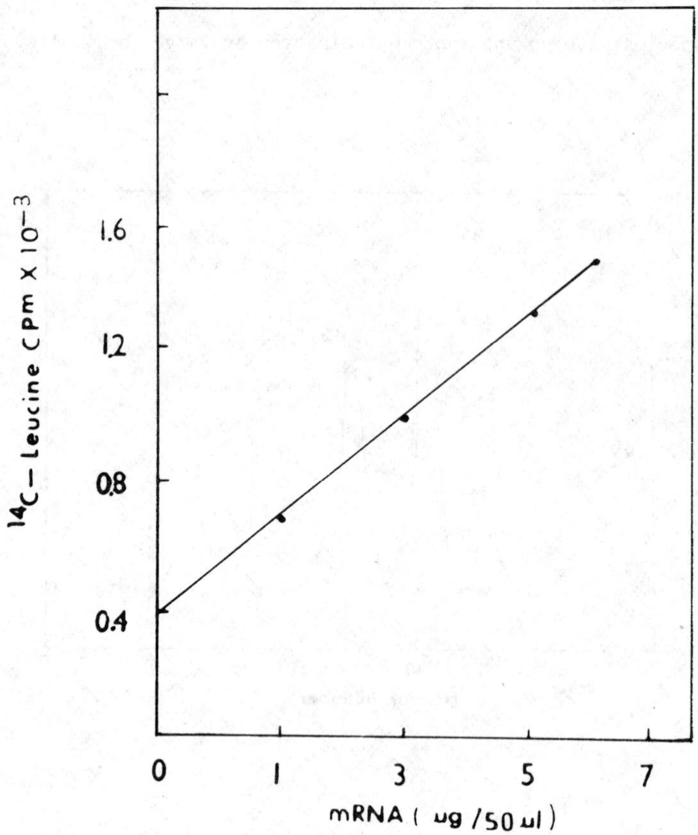

Fig. 3. Effect of Oligo (dT)-cellulose-purified RNA on protein synthesis. Incorporation of (^{14}C)-leucine into protein was measured in 50μl reaction mixtures after 90 min incubation. The incorporation of (^{14}C)-leucine into TCA precipitable material was approximately proportional to the mRNA added up to 6.0 μg/50 μl of incubation solution.

Under these conditions, approximately a four- to five- fold stimulation of ^{14}C-leucine incorporation was obtained. At mRNA concentration of 7 μg/50 μl, an inhibitory effect was observed. All mRNA concentrations tested gave increased incorporation for incubation periods up to 90 min at 29°C (Fig. 4).

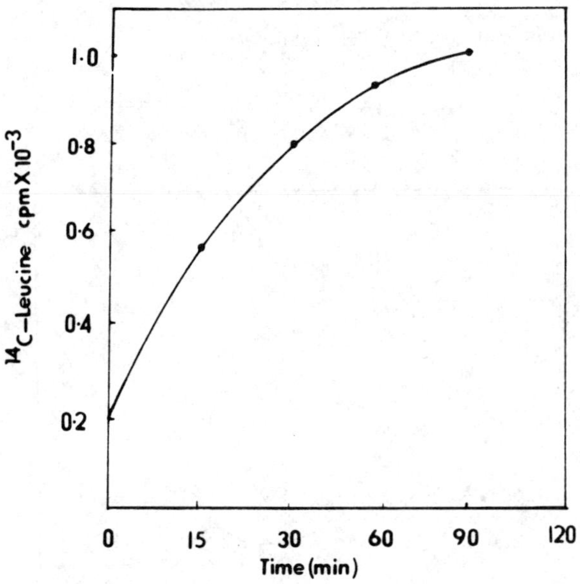

Fig. 4. Time course of ^{14}C-Leucine incorporation by lens mRNA. Several mRNA concentrations tested all gave increased incorporation for incubation periods up to to 90 min at 29°C.

Fig. 5. shows the translational activity of poly(A) RNA after sucrose gradient fractionation. The mojority of transtionation activity resided in the 14S and 10S RNA fraction.

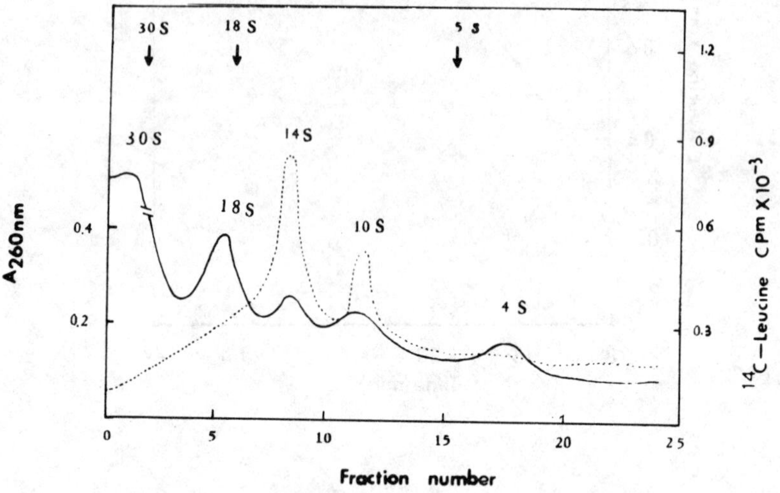

Fig. 5. Protien Synthesis of peak A RNA after sucrose gradient centrifugation. (———); (^{14}C)-Leucine incorporated in reponse to a 50 µl aliquot of each fraction in the wheat germ cell-free system, (········); After peak A RNA was analyzed by sucrose gradient centrifu- -gation, it contains five absorbing peaks, 30, 18, 14,10,4s fractions.

Characterization of in Vitro Translation Products

The ^{14}C-leucine labeled translation products were characterized by polyacry -lamide gel electrophoresis in 7M urea, pH 8.9.Crucian lens crystallins isolated by procedures of Spector et al (11), were characterized in a parallel gel. A clear separation of the six polypeptide chains of crystallin, A_1, A_2, A_3 and B_1, B_2, B_3 was obtained (Fig. 6A). The alkaline urea electrophoretic pattern obtained with translation products of lens B peak mRNA contained four peak. They have the same electrophoretic mobility as those of the A_2, A_3 and B_2, B_3 chains (Fig. 6B).

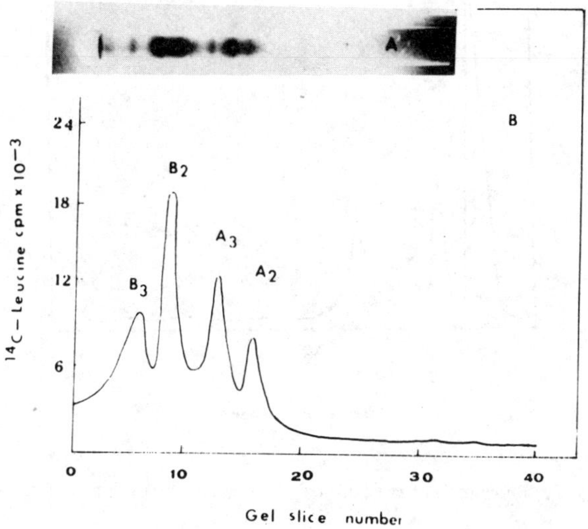

Fig. 6 (A) Electrophoretic separation of the polypeptide chain of crucian lens α-crystallin on polyacrylamide gel containing 6M urea at pH8.9. Six polypeptide chains, A_1, A_2, A_3, and B_1, B_2, B_3, appeared. (B) Tracing of the radioactive products obtained by reponse to added Oligo (dT)-Cellulose-purified lens B peak mRNA.

Fig. 7 shows characterization of cell-free translation products by 14S mRNA. It showed 3 peaks, A_2, B_2, and B_3; they have the same electrophoretic mobility as those of the A_2, B_2 and B_3 chains of crucian lens crystallin. This figure shows 14S mRNA may be a polycistronic component. With the 10S mRNA as template, the polypeptide produced migrated with the authentic A_3 chain of crucian Crystallin (Fig. 8)。

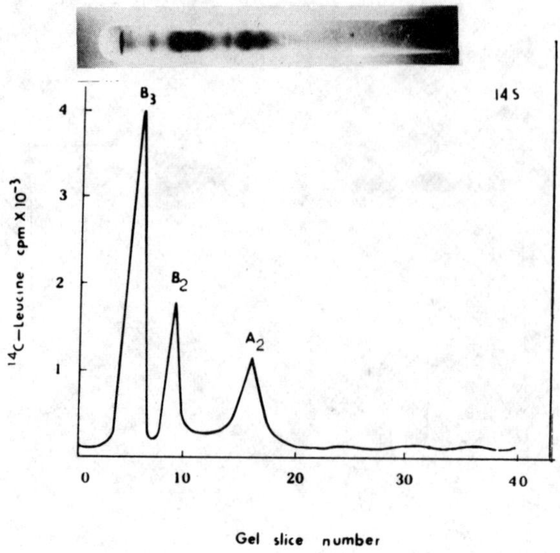

Fig. 7. Characterization of 14S mRNA cell-free translation products by urea/ polyacrylamide gel electrophoresis. Top panel represents the stained gel while the profiles represent radioactivity tracing.

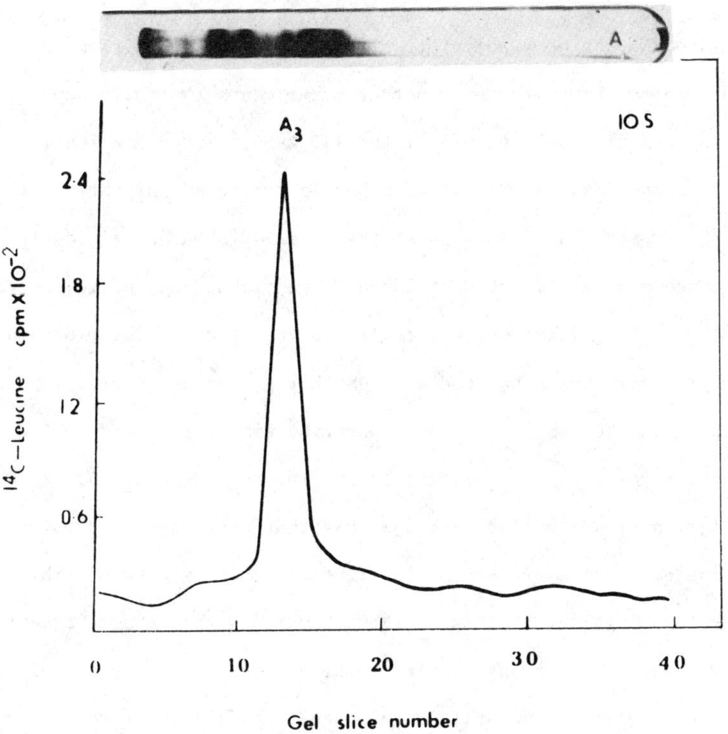

Fig. 8. Characterization of cell-free translation products directed by 10S mRNA by urea/ polyacrylamide gel electrophoresis.

DISCUSSION

In the past several years there have been a virtual revolution in the study of eukaryotic mRNAs. Among the notable recent achievements are the positive identification of mRNA precursors in HnRNA(12), the enuceration of the DNA sequences from which mRNA is Transcribed(13, 14), and the finding that most mRNAs in eukaryotes contain Poly(A) (15, 16) at the 3'end and methylated bases(17). These Discoveries aside from providing a powerful tool for mRNA isolation, have generated a large body of research into the properties and metabolism of mRNA. Poly(A) was added post-transcriptionally by a terminal addition enzyme, and becomes shorter as the mRNA ages, but it is not known whether this loss of oligo A goes to completion. The finding that crucian lens 10S and 14S mRNAs contain both Poly $(A)^+$ and Poly$(A)^-$ molecular species raises the question of its biogenesis. Above all, the results reported in this communication suggest that crucian crystallin 14S mRNA can be a polycistronic messenger. In addition to the observation reported here, there are other experiments that also support the concept of the polycistronic nature of the 14S mRNA. Two-dimensional fingerprints of the RNase T_1 digest of ribosome protected initiation sites of the calf alpha-crystallin mRNAs suggest that the 14S mRNA has a closely related sequence complexity but contains different abundance of initiation sequence(18). This difference may reflect a variation in the ribosome-protected region for the second codon. Another pertinent observation is that both calf lens 10S and 14S mRNAs transcribe similarly sized cDNA with reverse transcriptase(19). This can be interpreted as being the result of a possible termination signal on the 14S mRNA that prevents the complete reading of the template by reverse transcriptase. It is also possible that they are two initiation sites on the 14S mRNA so that the reverse transcriptase may read both cistrons simultaneously. The probablity of multiple initiation sites(20) or structural or sequence stops(21) for reverse transcriptase has been suggested for RNA genome that direct the synthesis of smaller cDNA transcripts. The relevance of polycistronic mRNA to the expression of information in the lens tissue is not clear at this point. It may be involved in the regulation of the development and differentiation of the lens tissue.

REFERENCES

1. Spector, A. and Kinoshita, J. H.(1964) Invest. Ophthal. 3, 517

2. Spector, A. and Travis, D. (1966) J. Biol. Chem. 241, 1290

3. Chen, J. H., Lavers, G. and Spector, A. (1973) Biochem. Biophys. Res. Comm.52, 767

4. Chen, J. H. and Spector, A. (1977) Proc. Nat'l Acad. Sci. 74,5488

5. Chen, J. H. Lavers, G. and Spector, A. (1974) Exp't. Eye Res. 18, 189

6 Kurland,C. G. (1960) J. Mol. Biol. 2, 285

7. Marcu, K. and Dudock, B. (1974) Nucleic Acids Research, 1, 1385

8. Roberts, D. E. and Peterson, B. M. (1973) Proc. Nat'l Acad. Sci. 70, 2330

9. Darnbough, F. and Ford, J. (1976) Development Biology, 50, 285

10. Schoemakers, J. G., Matze,M. R., van Poppel, M. and Bloemendal. (1969) Int'l J. Protein Research 1, 19

11. Spector, A. Li, L. K., Augusteyn, R. C., Schneider, A. and Freund, T. (1971) Biochem. J. 124, 337

12. Diez, J. and Brawerman, G. (1974) Proc. Nat'l Acad. Sci. 71, 4091

13. Britten, R. J. and Kohne, D. E. (1968) Science 161, 529

14. Greenberg, J. R. and Perry, R. F. (1971) J. Cell Biol. 50, 774

15. Adesnik, M. M., Salditt, W. Thpmas, M. and Darnell, J. E. (1972) J. Mol. Biol. 71, 21

16. Gr-enberg, J. R. and Perry, R. P. (1972) J. Mol. Biol. 72, 91

17. Shatkin, A. J. (1974) Proc. Nat'l Acad. Sci. 71, 3204

18. Chen, J. H. and Spector, A. (1977) Biochemistry 16, 499

19. Chen, J. H. Lavers, G. and Spector, A. (1975) Biochem. Biophys. Res. Commun. 52, 767

20. Taylor, J. M.,Gartin, D. E., Levinson, W. E. and Bishop,J. H., and Goodman, H. M. (1975) Biochemistry 13, 3159

21. Haseltine, W. A., Kleid, D. G., Panet, A., Rothenberg, E. and Baltimore, D. (1976) 106, 109

MEMBRANE DIFFERENTIATION OF CARDIAC MYOBLASTS INDUCED *IN VITRO* BY AN RNA-ENRICHED FRACTION FROM ADULT HEART

N. Sperelakis, M. J. McLean, J. -F. Renaud, and M.C. Niu

(Department of Physiology University of Virginia School of Medicine Charlottssville, VA 22908, U. S. A.)

and

(Department of Biology Temple University Philadelphia, PA. 19122, U. S. A.)

Abstract Young (2-3-day-old) embryonic chick myoblasts (myocardial cells) ordinarily do not continue to differentiate in culture, i. e., their development is arrested *in vitro*. The cells in organ-cultured intact hearts and in cell cultures of trypsin-dispersed cells (spherical reaggregate cultures) exhibit spontaneous slowly-rising (about 15 V/sec) action potentials which are insensitive to tetrodotoxin (TTX). That is, they do not normally gain TTX-sensitive fast Na^+ channels *in vitro*. However, incubation with purified adult heart extracts enriched in RNA (AHE-RNA) for 6-14 days caused the appearance of fast Na^+ channels *de novo* in such young myocardial cells. These AHE-RNA-treated cells, when stimulated, fired fast-rising (e. g., 100 V/sec) TTX-sensitive action potentials from large stable resting potentials. The induction of fast Na^+ channels was blocked by cycloheximide , suggesting dependence on protein synthesis, Adult chicken liver extract was not effective, and pre-treatment of the AHE-RNA with ribounclease (protease-free) destroyed the inductive activity. Poly-adenylate (poly-A) was not effective. Thus, the membrane properties of cardiac myoblasts can be made to further differentiate *in vitro* by addition of exogenous heart extracts enriched in RNA.

In 20-hr-old chick embryonic blastoderm, the portion

anterior to Hensen's node contains the cells which are destined to give rise to the heart, whereas the cells in the posterior portion (the post-nodal piece, PNP) normally do not. When the PNP was isolated and exposed to AHE-RNA , a spontaneously contracting tubular structure was formed *in vitro* within 1 week. Typical cardiac action potentials, both spontaneous and evoked, were recorded by intracellular microelectrodes from cells within these tubules. Actinomycin D or cycloheximide prevented the inducing effect of the extract. Induction of contracting tubules and excitability did not occur in PNPs untreated or treated with liver extract. These results suggest that cells not originally destined to become myocardial cells can be induced to gain many of the properties of heart cells by an RNA-enriched fraction from adult heart, and this induction is dependent upon protein synthesis.

In conclusion, it appears that RNA-enriched extracts of adults hearts can alter the membrane properties of embryoinc cardiac myoblasts and of some cell type of non-cardiac origin.

INTRODUCTION

The electrophysiological properties of embryonic chick cardiac myoblasts change progressively in the course of normal development (39 , 46 , 48). Therefore, the electrophysiological parameters, characteristic of different stages of heart development *in situ*, can serve as criteria for ascertaining the stage of differentiation of myocardial cells.

The anterior half of the flat 16-20 hr old chick embryo blastoderm contains bilateral "precardiac" areas (mesoderm) whose cells are destined to form the heart (32). If fragments of the precardiac areas (anterolateral blastoderm) are dissected and placed into organ culture for several days, a tubular heart develops (within a vesicle) that beats spontaneously, and from which electrical activity can be recorded (14); but further differentiation does not proceed *in vitro* (14). In normal development, twin tubular primordia are formed bilaterally from the precardiac mesoderm and fuse to form a single tubular heart (30). The tubular heart begins contracting spontaneously at 30-40 hr (9-19 somitestage), and begins to propel blood (30). The blood pressure is very low (1-2 mm Hg) at this stage, and the velocity of propagation of the peristaltic contraction wave is very slow (approx. 1 cm/sec in 3-day hearts)(30). Morphological changes occur so that chambers appear in the heart on about day 5. Circulation to the chorio-allantoic membrane is established on day 5, so that metabolism of the embryo becomes aerobic at that time. The nerves also arrive at the heart on about day 5 (30), but they do not become functional with respect to neurotransmitter release until considerably later (11 , 27). The heart rate of the chick embryo increases from about 50 beats/min on day 1.5 to the maximal value of about 220 beats/min on day 8.

The myocardial cells in young (2-3 days *in ovo*) hearts possess slowly rising (10-30 V/sec) action potentials preceded by pacemaker potentials (Fig. 1A). Hyperpolarizing current pulses do not greatly increase the rate of rise of the action potential, and excitability is not lost until the membrane is depolarized to less than -25 mV. The upstroke is generated by Na^+ influx through slow Na^+ channels which are insensitive to tetrodotoxin (TTX), a specific blocker of fast Na^+ channels (Fig. 1B). The young (2-3-day-old) cells have either no or only a few fast Na^+ channels. A substantial number of kinetically-fast Na^+ channels which are sensitive to TTX are present in the membrane on about day 4 or 5, and the maximal rate of rise of the action potential ($+\dot{V}_{max}$) is about 50-70 V/sec at this time. The fast Na^+ channels continue to increase in density progressively thereafter, until about day 18, when the adult maximal rate of rise of about 150 V/sec is achieved (Fig. 1E).

During an intermediate stage of development (from day 5 to day 8), fast Na^+ channels coexist with the large complement of slow channels. TTX reduces $+\dot{V}_{max}$ to about the value observed in 2-3-day hearts, i.e., 10-20 V/sec, but the action potentials and accompanying contractions persist (Fig. 1 C-D). After day 8, the action potentials are completely abolished by TTX (despite increased stimulus intensity)(Fig. 1F), and depolarization to less than -50 mV now abolishes excitability. This indicates that the action potential-generating channels consist predominantly of fast Na^+ channels, most of the slow Na^+ channels having been lost (functionally) so that insufficient numbers remain to support regenerative excitation. Addition of some positive inotropic agents increases the number of available slow channels (Ca^{++}-Na^+) in the membrane, and leads to regain of excitability in cells whose fast Na^+ channels have been inactivated (33 , 36 , 44).

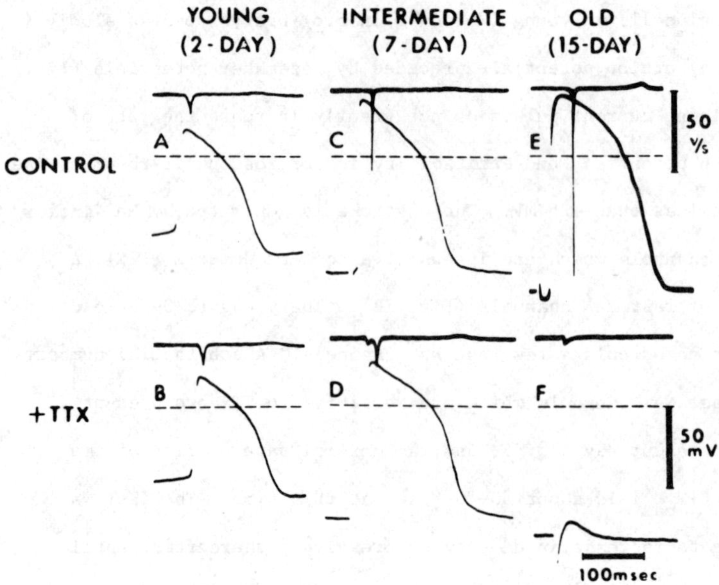

Fig. 1 . Development of sensitivity to TTX of intact embryonic chick hearts with increasing embryonic age. A-B: Intracellular microelectrode recordings from a 2-day-old heart before (A) and 20 min after (B) the addition of TTX (20 μg/ml). C-D: Recordings from a 7-day-old heart before (C) and 2 min after (D) the addition of TTX (2 μg/ml)(note depression of the rate of rise in D). E-F: Recordings from a 15-day-old heart prior to (E) and 2 min after (F) the addition of TTX (1 μg/ml). The action potentials were abolished and excitability was not restored by strong field stimulation in F. The upper traces give dV/dt; this trace has been shifted relative to the V-t trace to prevent obscuring dV/dt. The horizontal broken line in each panel represents zero potential. dV/dt calibration (in E) and voltage and time calibrations (in F) pertain to all panels. From Sperelakis and Shigenobu, 1972.

The resting potential also increases gradually during development, from about -40 mV on day 2 to about -80 mV by day 12. This change can be accounted for by an increase in K^+ permeability (P_K) during development (46, 48). As shown by resting potential versus log $[K]_o$ curves for hearts of different ages (Fig. 2), the low resting potentials in young hearts are caused by a high P_{Na}/P_K ratio of about 0.2, rather than to an internal $[K]_i$ greatly lower than that of old hearts. The $[K]_i$ level is about 130 mM in 3-day-old hearts compared to 150 mM in 15-day hearts; hence, the K^+ equilibrium potential (E_K) is about -91 mV (at a $[K]_o$ of 4 mM) in 3-day hearts. The P_{Na}/P_K ratio decreases to about 0.1 by day 5, and is between 0.05 and 0.01 by day 15. The increase in resting potential parallels the change in the P_{Na}/P_K ratio. The input resistance of the cells is high in young hearts (about 13 MΩ), and it declines to a final value of about 4.5 MΩ during development, and from this it was concluded that a low P_K was primarily responsible for the high P_{Na}/P_K ratio in young hearts. Carmeliet et al. (3, 4) have also shown from ^{42}K flux measurements that P_K is several fold lower in 6-day hearts than in 19-day hearts.

The presence of pacemaker potentials is determined by P_K in cells which have a low P_{Cl}, as myocardial cells do. The low P_K in young heart cells makes for automaticity, as well as accounts for the low resting potentials. The incidence of pacemaker potentials in cells in young hearts is very high, and this incidence decreases during development, roughly in parallel with the increase in P_K (44, 48). Automaticity is absent in the ventricular cells of the older hearts.

The (Na,K)-ATPase specific activity is low in the young chick hearts and increases progressively during development, reaching the final adult

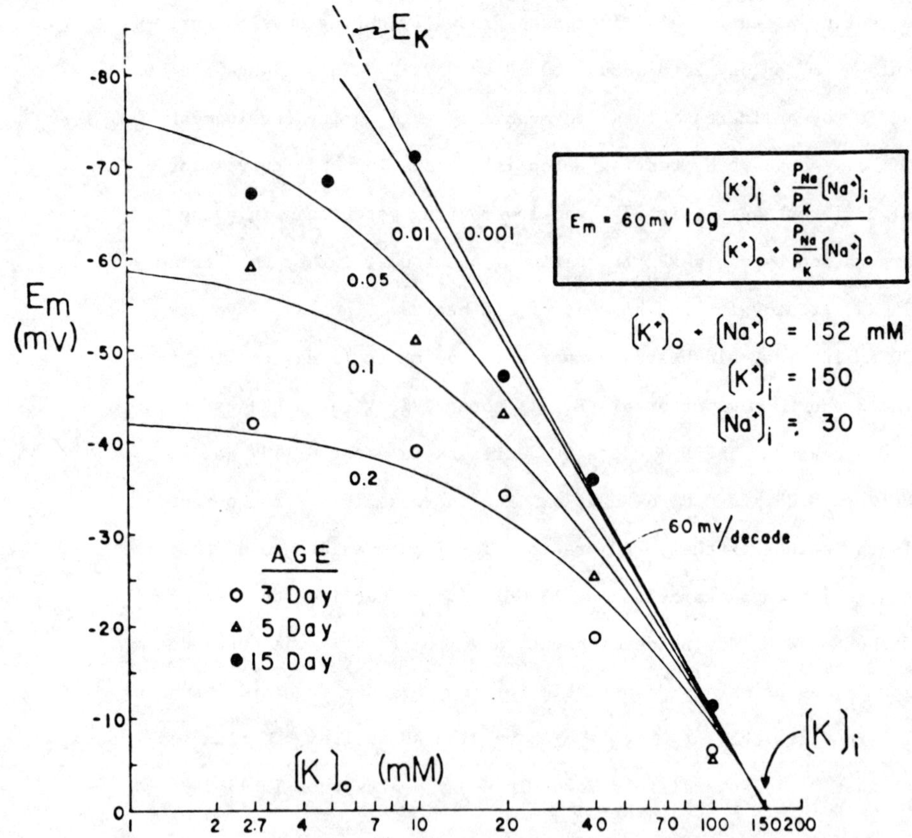

Fig. 2 . Resting potential E_m plotted as a function of the external K^+ concentration ($[K]_o$) on a logarithmic scale for data obtained from embryonic chick hearts at days 3 (o), 5 (Δ) and 15 (o). $[K]_o$ was elevated by substitution of K^+ for equimolar amounts of Na^+. Continuous lines give theoretical calculations from the constant-field equation (inset) for P_{Na}/P_K ratios of 0.001, 0.01, 0.05, 0.1, and 0.2. Calculations were made assuming $[K]_i$ and $[Na]_i$ values shown. For a P_{Na}/P_K ratio of 0.001, the curve is linear over the entire range with a slope of 60 mV/decade, i.e., it closely follows the K^+ equilibrium potential (E_K) calculated from the Nernst equation. The data for the 3-day heart follow the curve for a P_{Na}/P_K ratio of 0.2, those for 5-day heart follow the theoretical curve for 0.1, and those for the 15-day heart follow the curves for 0.01 to 0.05. The estimated intracellular K^+ activities ($[K]_i$) obtained by extrapolation to zero potential are nearly the same for all ages. From Sperelakis and Shigenobu, 1972.

value by about day 18 (40). Although the Na-K pump capability is thus low in young hearts, it is sufficient to maintain a relatively high $[K]_i$. The Na-K pump is aided in the task by the low P_K: that is, in young hearts, the pump capability is low but the ion leak is correspondingly low.

The cyclic AMP level is very high in young hearts, and it decreases during development, first rapidly and then more slowly, reaching the final adult level by about day 16 (18 , 29). Since elevation of cyclic AMP leads to an increase in the number of available slow Ca^{++}-Na^{+} channels (33 , 36), the decrease in cyclic AMP level during development could be related to the concomitant decrease in density of available slow channels (in the absence of positive inotropic agents). We have proposed (33 , 36 , 45) that a protein constituent of the slow channel must be phosphorylated in order for the channel to be available for voltage activation, and that this is one mechanism of action of those positive inotropic agents which elevate cyclic AMP (activation of cyclic AMP-dependent protein kinase).

Electron microscopy of young chick hearts shows that there are only few and short myofibrils (Fig. 3A). The sarcomeres are not complete, and the myofibrils run in all directions. There is an abundance of ribosomes and rough endoplasmic reticulum, and large pools of glycogen are found in the cells. As development progresses, the number of myofibrils increases and they become aligned, as illustrated in Figure 3B. By day 18, the ultrastructure of the myocardial cells is quite similar to that of cells in adult hearts.

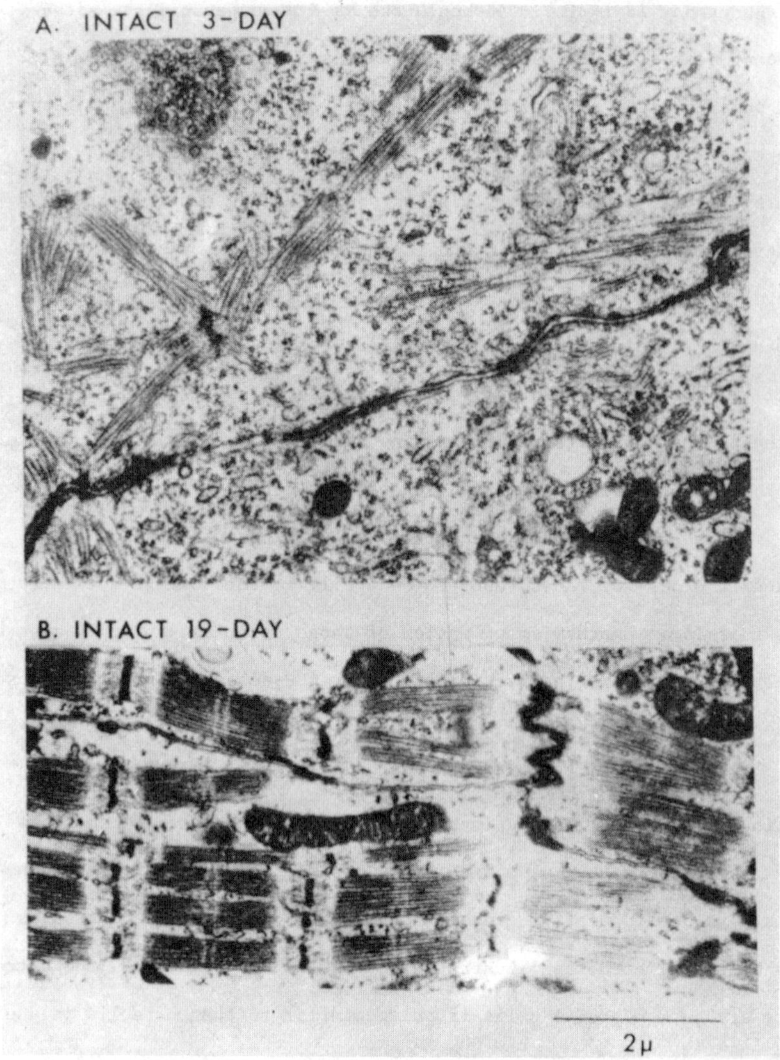

Fig. 3 . Cell ultrastructure of young (3 days *in ovo*) and old (19 days *in ovo*) intact embryonic chick heart *in situ*. A: 3-day ventricular cells demonstrating paucity and nonalignment of myofibrils; ribosomes are abundant in the cytoplasm; the contiguous cells are held in close apposition by desmosomes. B: 19-day ventricular cells with abundant and aligned myofibrils. A convoluted intercalated disk appears between contiguous cells. Calibration given in B also applies to A.

METHODS

I. Organ cultured hearts.

Spontaneously contracting heart tubes were removed from 60 h (Hamburger & Hamilton, stage 17) chick embryos. Each young heart (ventricular segment only) was placed in a Falcon plastic culture tube with 1-2 ml of Rinaldini culture medium (10% chick embryo extract, 20% horse serum, and 70% Tyrode's solution, without antibiotics). The tubes were incubated in a roller drum apparatus at 37°C for 4-14 days. The medium was not changed during the entire culture period. Other details of the procedure have been described (37).

II. Cultured heart cells.

Spherical reaggregate cultures were prepared by the methods previously described (13,16,23,41, 43). Tubular hearts (60 h) were dissociated into single cells by stirring in Ca^{++} and Mg^{++}-free Tyrode's solution containing 0.05% (w/v) trypsin and 100% glucose. At 20 min intervals, the cloudy cell-containing supernatant was decanted into chilled culture medium (10% horse serum, 40% Puck's N-16 nutrient solution, 50% Hank's balanced salt solution, and penicillin/streptomycin (50 U/ml)). After washing, the cells were plated at a density of about 2.5 x 10^5/ml on cellophane. Within several days of incubation at 37°C, the cells reaggregated to form small spheres (0.1-0.5 mm diameter). Most aggregates untreated with RNA contracted spontaneously. The aggregates were cultured for a total period of 10 days, at which time they were studied electrophysiologically. The medium was not changed during the culture period. The morphology of the reaggregate cultures compared to monolayer cultures is illustrated in Fig. 4 for heart cells dissociated from 19-day-old embryonic hearts.

III. Embryo blastoderm.

Post-nodal pieces (PNPs) were dissected from the embryonic chick blastoderm at the primitive streak stage (Hamburger and Hamilton, stage 4), as

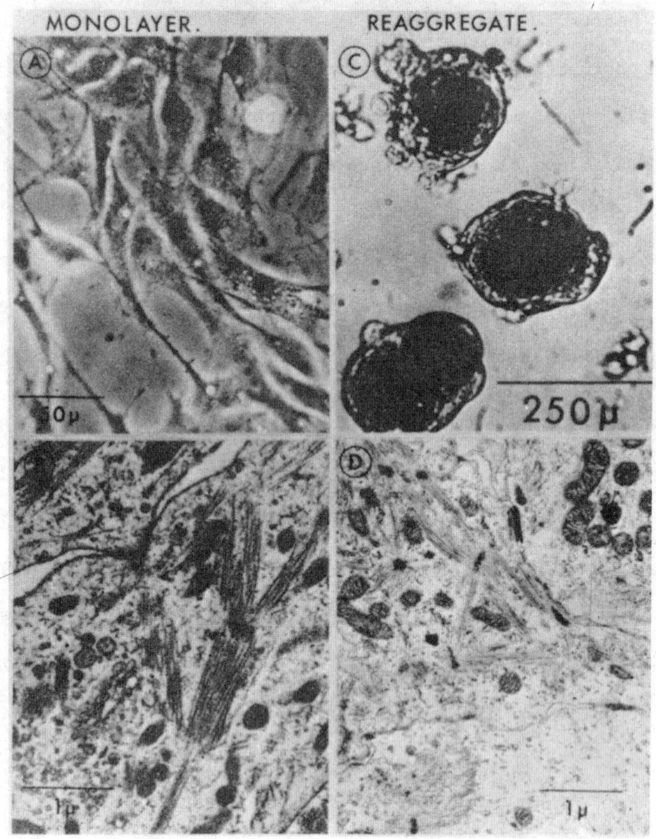

Fig. 4 . Micrographs of 19-day-old embryonic chick ventricular cells cultured as monolayers (A-B) and in spherical reaggregates (C-D). A: Low-power phase contrast micrograph showing cell arrangement of the monolayer cultures. B: Electron micrograph showing paucity and nonalignment of myofibrils in the monolayer cells. C: Low-power light micrograph showing several cultured spherical reaggregates. D: Electron micrograph of several cells in the reaggregates showing sparse, incomplete, and nonaligned myofibrils.

described by Niu and Deshpande (24)(Fig.12, top left diagram). The blastoderm was cut transversely at a level 0.6 mm caudal to Hensen's node (20). Few cardiogenic cells are present in the post-nodal region. The PNPs were transferred to a support made from vitelline membrane mounted on a glass ring and placed in a watch glass. Nutrient medium was prepared by homogenizing a whole egg in Ringer's solution, and then centrifuging the suspension to remove the yolk platelets (2000 rpm for 30 min at $4^{o}C$). The supernatant was mixed with an equal volume of Pannett-Compton solution, and 1 ml was placed in the watch glass outside the ring.

IV. RNA-enriched extracts.

The RNA-enriched extracts prepared from adult chicken hearts (AHE-RNA) were kindly provided to us by Dr. M.C. Niu of Temple University. The procedures were described in detail elsewhere (9 , 24). In brief, the hearts were homogenized in a Waring blender in buffer solution (50 mM Tris-HCL, 5 mM EDTA, 75 mM NaCl, and 0.5% sodium dodecyl sulfate (SDS)) at pH 8.3. The homogenate was extracted with buffer-saturated phenol/chloroform/isoamyl alcohol (0.495:0.495:0.01). After centrifugation, the aqueous phase was siphoned off, and one-half volume of buffer was added to the phenol and interphase layers. The material was then mixed well and centrifuged again. The aqueous phase was combined with that of the first extraction, and again extracted with fresh chloroform/phenol/isoamyl alcohol. After centrifugation, two volumes of 95% ethanol were added to the aqueous phase to precipitate the nucleic acids overnight. The nucleic acids were pelleted by centrifugation, dissolved in saline (pH 7.4), and reprecipitated with ethanol several times to remove the SDS completely. DNAase was used to destroy contaminant DNA; the Dische reaction revealed residual contamination by DNA of less than 1%. Protein analysis by the Lowry method indicated protein contamination was less than 1%. UV spectrophotometry gave minimal and maximal absorption, respectively, at 230 and 250 nm, characteristic of nucleic acids. Polyadenylate-containing mRNA was partially purified by the technique of Schutz et al. (34) employing cellulose column chromatography. The RNA-containing fractions isolated by this method have been shown to stimulate protein synthesis in a cell-free system. Adult chicken liver extract (RNA-enriched) was prepared similarly for use in control experiments.

The AHE-RNA (0.1 ml from a stock solution of 80 OD/ml, about 25 OD/mg RNA) was added into the media of the organ-cultured hearts and the reaggregate cultures immediately after placement in the culture vessels. For some control experiments, AHE-RNA was preincubated with protease-free pancreatic ribonuclease (RNAase (Sigma): 100 μg/ml for 2 h at 37°C) in order to destroy the RNA. In some experiments, cycloheximide (Sigma; 1 μg/ml) was added to the culture medium along with the RNA.

For the PNP experiments, aliquots of RNA-containing fractions (0.2 ml from a stock solution of 80 OD/ml (about 25 OD/mg RNA)) were added to the medium. The PNPs were exposed to the RNA-extract for only 24-36 h, after which time the medium was changed to standard medium without RNA. The medium was changed daily thereafter. The PNPs rounded up to form small vesicles within 24 h, and, if induction occurred, spontaneously contracting (heart-like) structures appeared within the vesicles within 3-5 days (Fig.12, upper right photograph). Contractions were never seen before 6 days of incubation. The PNPs were incubated for 7-12 days prior to electrophysiological experimentation.

Oligo-dT bound RNA. For special experiments, the RNA extract was further purified by binding to oligodeoxy-thymidylate-cellulose (oligo-dT). These experiments employed only the bound II fraction of poly-A-containing RNA prepared from 16-day-old embryonic chicken hearts, as described by Deshpande &

Siddiqui (8). This fraction had maximal adsorption at 260 nm and had an A260/A280 ratio of about 2. Bound II RNA from several different batches (proven effective in inducing differentiation in post-nodal pieces) was generously supplied by Doctors Siddiqui and Deshpande. One batch of bound II RNA prepared for testing at high concentrations (8 OD/ml) was found to be effective in the organ-culture system.

In these experiments, 2-3-day-old chick hearts (ventricles only) were preincubated in Tyrode solution (components autoclaved prior to mixing to destroy ribonuclease) containing the RNA for 4 h. The preincubated tissues were then placed on a vitelline membrane supported by a metal grid in a Falcon plastic organ-culture dish, and the RNA-containing solution was placed in the depression beneath the grid. Egg nutrient medium (25% whole egg homogenate, 75% Tyrode; see (24)) was added until the fluid level was in contact with the metal grid. Fluid and nutrients were able to reach the tissue by diffusion through the vitelline membrane. The tissues were incubated in a moist-air incubator gassed with 95% O_2 - 5% CO_2 for 2 weeks prior to electrophysiological experiments.

V. Electrophysiological recording.

Standard intracellular recording techniques were used (39). Glass capillary microelectrodes were filled with 3 M KCl, and had resistances of about 30-60 MΩ. The microelectrode preamplifiers used were model VF-1 (WP Instruments), and the signals were displayed on a Tektronix 565 oscilloscope. A DC bridge circuit (39) was used to allow simultaneous potential recording and passage of polarizing current pulses. Action potential rates of rise were determined from photographs taken at high sweep speed and by use of a passive RC differentiating network.

The organ-cultured hearts or spherical reaggregate cultures were transferred to a heated bath filled with fresh culture medium. The pH was held constant at 7.4 throughout the experiments by blowing humidified gas (95% O_2 - 5% CO_2) over the chamber. Contractions were visually observed through a dissecting microscope. Drugs were added directly to the bath from concentrated stock solutions. The degree of TTX (obtained from CalBiochem) sensitivity (concentrations of 0.1 - 4 μg/ml) was classified as completely sensitive (action potential abolished), partially sensitive (action potential persisted with reduced rate of rise of about 10 V/sec), and insensitive (action potential rate of rise and overshoot virtually unaffected).

For microelectrode penetrations, the PNPs were placed in a chamber containing Ringer's solution (pH 7.4 at 37°C), and the vesicles were cut open to expose the contracting tubes. Where tubes were not visible, penetrations were made at random into the vesicle itself to demonstrate absence of excitability in non-induced tissue.

Results

I. Organ-Cultured Young Chick Embryonic Hearts Arrested in Development

Organ culture of young embryonic hearts that are arrested in development *in vitro* allows one to assay for factors that affect differentiation during normal development *in situ*. When 3-day-old chick embryonic hearts, which have not yet become innervated, were placed intact into organ culture for 10-14 days, these muscle cells did not continue to differentiate electrically or morphologically (28,42,47). The impaled cells had low resting potentials and slow-rising TTX-insensitive action potentials (Fig. 5A-D). that is, their properties were identical to those of fresh noncultured 3-day-old hearts. They did not gain fast Na^+ channels. These hearts retained their slow Na^+ channels, as evidenced by the fact that Mn^{++} (1 mM) did not block the action potentials, but D600 (verapamil analog) did (Fig. 5C-D). Thus, the cells were arrested in the stage of differentiation attained at the time of explantation. The same was true of young hearts grafted on to the chorio-allantoic membrane of a host chick and thereby blood perfused (28)(Fig. 5E-H). Therefore, something in the *in situ* environment of the heart in the developing embryo must control the appearance of the fast Na^+ channels. Although the innervation reaches the heart on about day 5, it is unlikely that a trophic influence of the neurons triggers the next step in differentiation, because some cells contain a small complement of fast Na^+ channels on days 3 and 4, i.e., before innervation reaches the heart.

When organ-cultured young hearts were incubated with the RNA-enriched extracts obtained from adult chicken hearts, fast Na^+ channels appeared *de novo*, the resting potentials increased, automaticity ceased, and the action potentials were rapidly rising and completely sensitive to TTX (19 , 21)(Fig. 6). The results of the RNA experiments on organ cultured hearts are summarized in Table 1B.

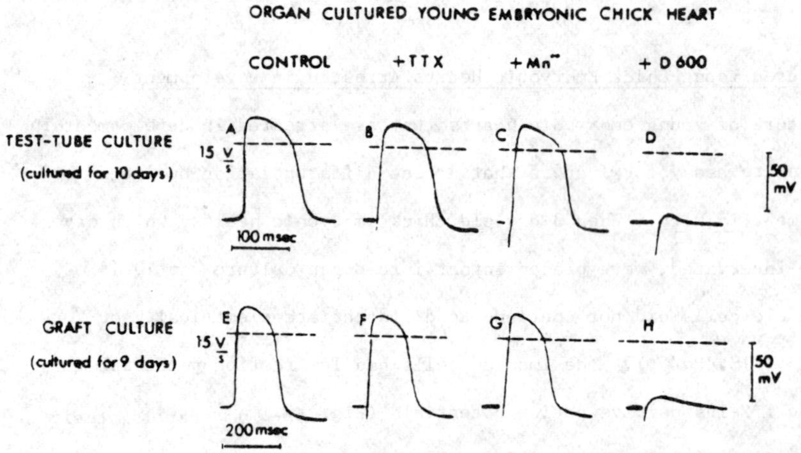

Fig. 5 . Lack of gain of fast Na^+ channels in organ-cultured young (3-day-old) embryonic chick hearts: A-D: Tissue in test-tube culture for 10 days. E-H: Tissue in graft culture for 9 days. As shown in A and E, the control action potentials had slow rates of rise of about 15 V/sec. Addition of tetrodotoxin (TTX; 2 μg/ml) had little or no effect on the rate of rise of the action potential (B, F). Addition of Mn^{++} (1 mM), a known blocker of Ca^{++} current, did not affect the action potentials (C, G). However, D600 (10^{-6} M), an analog of verapamil and a known blocker of slow channels, did abolish the action potentials (D, H). Different time scale in E-H compared to A-D. Taken from Renaud and Sperelakis, 1976.

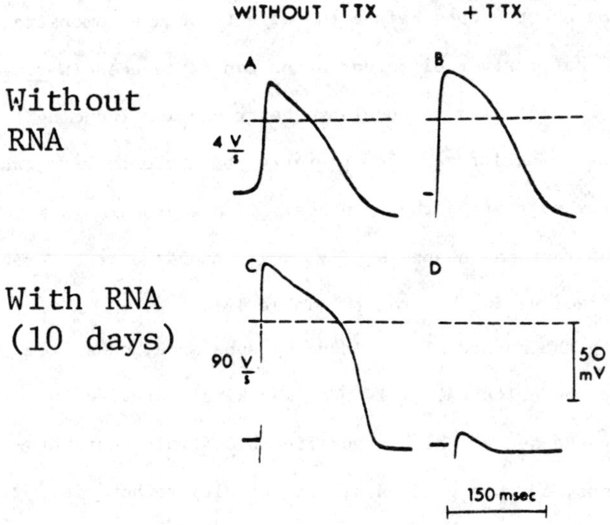

Fig. 6 . Electrophysiological properties of young cardiac myoblasts (ventricular) of 2.5-day-old intact hearts placed in organ culture for 10 days in the absence (A-B) and presence (C-D) of RNA obtained from adult chicken hearts (8 OD/culture tube). A: Control recording (resembling that of the fresh non-cultured young heart) from a cell in an organ-cultured heart not treated with heart RNA. B: Action potential persisting in the same cell 15 min after addition of TTX (1.0 μg/ml); stimulation was required to elicit the action potential due to suppression of automaticity by TTX. C: Intracellular recording from a cell in a heart treated with heart RNA showing high resting potential and rapidly-rising action potential (+$\dot{V}_{max}$ of 90 V/sec) (resembling properties of the fresh non-cultured old embryonic heart). D: TTX (0.1 μg/ml) abolished all excitability despite intense stimulation. Calibrations in D pertain throughout. The horizontal broken lines give zero potential. Electric field stimulation applied in B-D.

There was a lag period of about 6 days before the effects became demonstrable (Fig. 7). These findings indicate that further differentiation can be induced in vitro. Pretreatment of the RNA extract with ribonuclease destroyed its inducing effect (Fig. 8)(Table 1B). The induction was also prevented by cycloheximide, an inhibitor of protein synthesis, thus suggesting that synthesis of new protein is required for the appearance of the fast Na^+ channels (Fig. 9). An RNA extract prepared from chicken liver was not effective in this regard (Table 1B).

To further elucidate the nature of the inducing substance, mRNA from adult heart was added to the cultures. This mRNA was kindly provided by Drs. Siddiqui and Deshpande and was highly purified by affinity chromatography on oligo-dT columns, sucrose gradients, and gel electrophoresis. Partial differentiation occurred in organ-cultured young hearts incubated with this material, namely $+\dot{V}_{max}$ increased and the action potentials were abolished by TTX. These effects are not due to nonspecific binding of polyanions, as indicated by failure of poly-adenylate (Poly-A) to induce differentiation.

In summary, young embryonic chick hearts in organ culture tend to retain the electrical properties that they originally possessed at the time of placement into culture. However, the addition of an RNA-enriched extract, obtained from adult chicken hearts, to the culture medium allowed further differentiation to proceed in vitro.

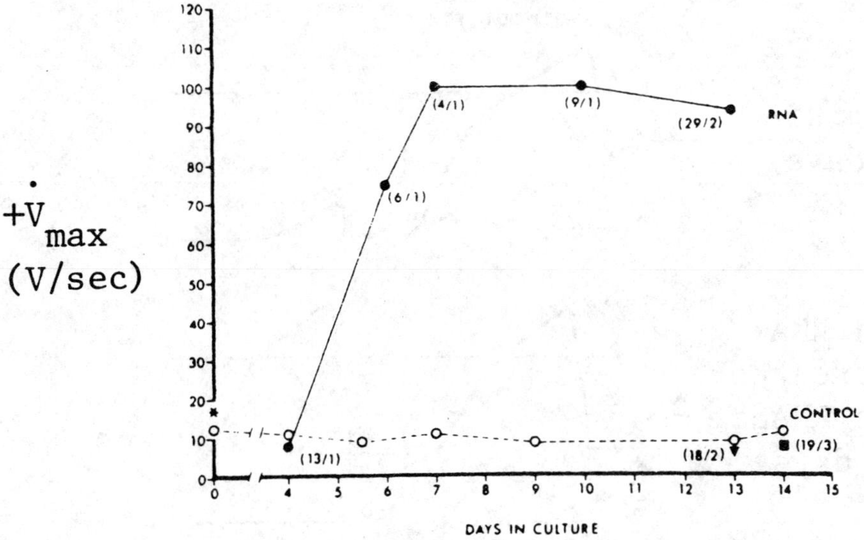

Fig. 7 . Changes in maximal rate of rise of the action potentials ($+\dot{V}_{max}$) of organ-cultured young embryonic hearts, incubated with and without exogenous adult heart RNA extract, as a function of time *in vitro*. Open circles: Data taken from control (untreated) hearts. Filled circles: Data obtained from RNA-treated hearts. Filled triangle: Data for hearts treated with RNA pre-incubated with ribonuclease (RNAase). Filled square: Data from hearts incubated with both RNA and cycloheximide. Numbers in parentheses give the number of penetrations for the RNA-treated hearts. Asterisks at zero days in culture are values taken from Sperelakis & Shigenobu, 1972. Action potential amplitude (not plotted) also increased substantially in the hearts treated with RNA-extract.

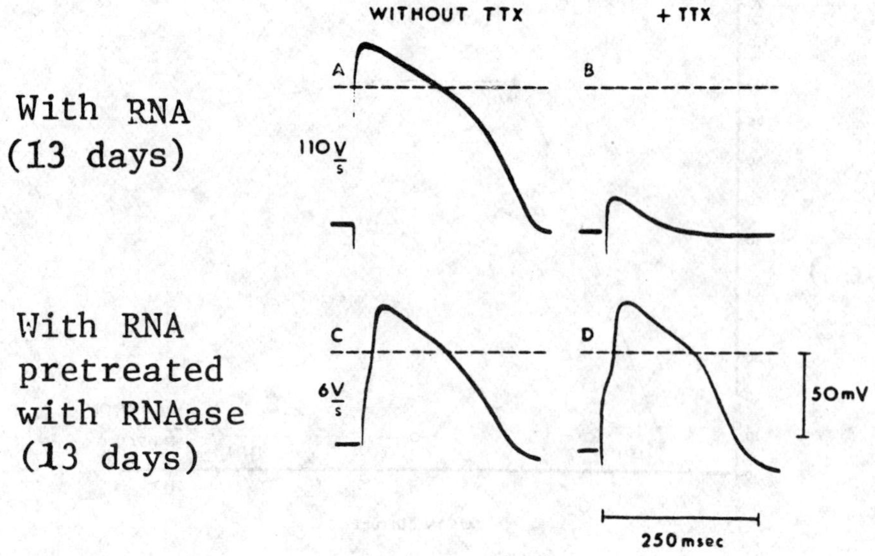

Fig. 8. Prevention of the heart RNA-induced alterations in membrane properties by pretreatment of the RNA extract with ribonuclease. A-B: Recordings from an organ-cultured 2.5-day-old heart made at 13 days after addition of heart RNA to the culture medium. A: Control action potential ($+\dot{V}_{max}$ of 110 V/sec). B: Abolition of excitability by TTX (0.1 μg/ml). C-D: Recordings from an organ-cultured 2.5-day-old heart made at 13 days after addition of heart RNA pretreated with ribonuclease (RNAase). C: Slowly-rising (6 V/sec) action potential followed by a hyperpolarizing afterpotential. D: Action potential unaltered after prolonged exposure to TTX (4 μg/ml). Electrical stimulation applied in all panels.

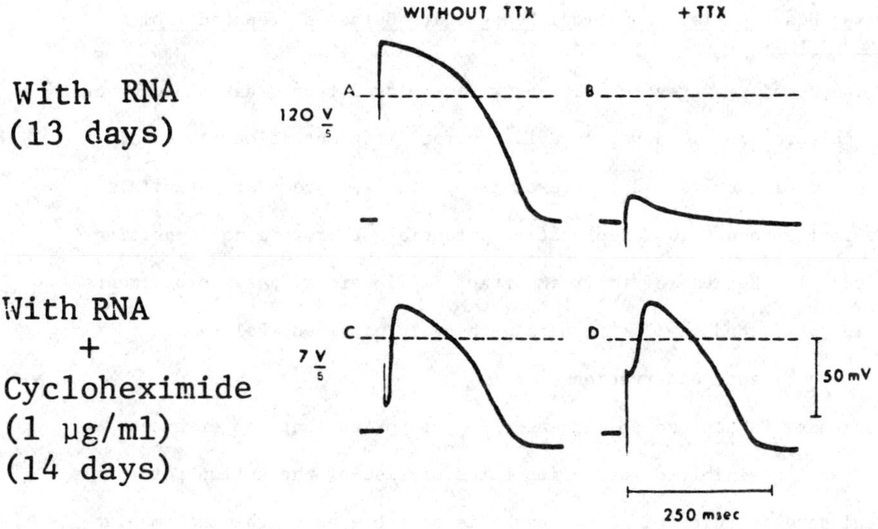

Fig. 9 . Prevention of the effect of the heart RNA extract by cycloheximide, indicating dependence on protein synthesis. A-B: Recordings from an organ-cultured 2.5-day-old heart 13 days after addition of heart RNA to the culture medium. A: Rapidly-rising (120 V/sec) control action potential fired from a high resting potential. B: TTX (0.1 μg/ml) abolished excitability despite intense field stimulation. C-D: Recordings from a young heart (2.5-day-old) cultured for 14 days with both cycloheximide (1 μg/ml) and RNA. C: Slowly-rising (7 V/sec) action potential (followed by a hyperpolarizing afterpotential). D: Action potentials persisted in the same cell 15 min after addition of TTX (4 μg/ml). Field stimulation applied throughout.

II. Cultured Heart Cells: Spherical Reaggregate Cultures Prepared from Young Hearts.

When spherical reaggregate cultures were prepared from 3-day-old embryonic hearts and cultured for 10 days, no evidence for differentiation was obtained (19, 21, 22, 43). The impaled cells had low resting potentials, pacemaker potentials, and slowly-rising (about 10 V/sec) action potentials that were not sensitive to TTX (Fig. 10). Because of their important implications, these experiments have been repeated, and similar results were again obtained (Pelleg and Sperelakis, unpublished observations).

In contrast, DeHaan and co-workers (5) reported that reaggregates of 4-day-old cells exhibited increasing rates of rise of the action potentials over several days in culture. Since day 4 is on the edge of the intermediate period, it is possible that genes coding for the production of protein for fast Na^+ channels had been activated prior to culturing. Alternative possible explanation for the discrepancy in the findings between these two laboratories include the presence of chick embryo extract in the culture medium used by DeHaan et al. (see ref. 5).

As in the case of organ-cultured young hearts, when the spherical reaggregate cultures prepared from 3-day-old embryonic hearts were incubated with RNA-enriched extracts obtained from adult chicken hearts, the cells proceeded to differentiate after a lag period of several days. There were large stable resting potentials and rapidly-rising TTX-sensitive action potentials (19, 21)(Fig. 11, C-D). Cycloheximide (1 μg/ml) added to the culture medium prevented the inductive effect of the heart RNA extract. Liver RNA extract failed to trigger differentiation of the heart cells. The results of the RNA-extract experiments on cultured heart cells are summarized in Table 1, part C.

These findings are consistent with the view that differentiation of immature cardiac myoblastic cells ordinarily does not proceed spontaneously *in vitro*, but that under appropriate conditions, the cells can be induced to do so.

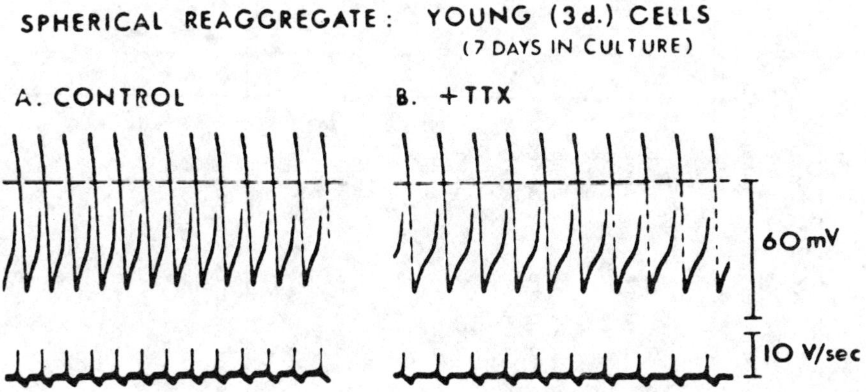

Fig. 10. Spherical reaggregate culture prepared from young (3-day-old) embryonic chick hearts illustrating the lack of electrical differentiation of the cells in vitro. A: Control intracellular microelectrode recording before tetrodotoxin (TTX) addition showing the low resting potential with spontaneous slowly-rising action potentials, each preceeded by a pacemaker potential. B: In presence of TTX (1 μg/ml, 10 min) showing no effect on the overshoot or rate of rise of the action potentials (TTX often depresses automaticity to varying degrees). Modified from McLean et al., 1976.

Table 1. Comparison of electrophysiological properties of chick embryonic myocardial cells developing *in situ* with those of cells cultured *in vitro* in the presence and absence of an RNA-enriched extract from adult chicken heart.

	N	Resting potential (mV)	$+\dot{V}_{max}$ (V/sec)	TTX sensitivity	Incidence of pacemaker potentials
A. Fresh non-cultured intact hearts					
(a) Young (day 2)		-40	< 20	None	High
(b) Old (day 16)		-80	> 100	Complete	Absent
B. Organ cultured young (2-3-day-old) hearts					
(a) Control	24/4	-60 ± 1	6 ± 1	None	High
(b) Heart RNA (6-14 days)	48/6	-76 ± 1	90 ± 4	Complete	Absent
(c) Heart RNA plus cycloheximide	19/3	-54 ± 2	6 ± 1	None	High
(d) Heart RNA + RNAase	18/2	-57 ± 2	6 ± 1	None	Absent
(e) Liver RNA	7/1	-55 ± 2	7 ± 1	None	High
(f) Poly-adenylate	12/3	-56 ± 2	6 ± 1	None	Absent
C. Cultured heart cell reaggregates					
(a) Control	4/3	-56 ± 2	< 10	None	High
(b) Heart RNA	6/4	-70 ± 2	> 100	Complete	Absent
(c) Heart RNA plus cycloheximide	3/3	-50 ± 2	< 10	None	High
(d) Liver RNA	3/1	-40 ± 2	< 10	None	High

Figures in column labelled N give number of cells penetrated over number of hearts or reaggregates. Resting potential and $+\dot{V}_{max}$ data expressed as mean ± 1 S.E.

Data from intact hearts were taken from Sperelakis & Shigenobu (46).

TTX sensitivity refers to effect on $+\dot{V}_{max}$.

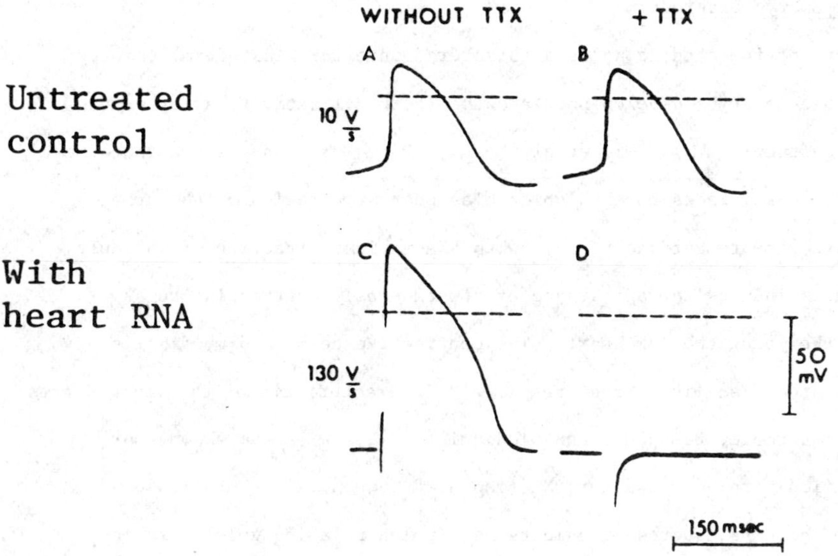

Fig. 11. Induction of changes in membrane electrical properties by heart RNA extract in cultured spherical reaggregates of cells trypsin-dispersed from young (2.5-day-old) embryonic chick hearts. A-B: Control reaggregate incubated without RNA extract. A: Spontaneously-firing slowly-rising (10 V/sec) action potential preceded by a pacemaker potential; low resting potential and maximum diastolic potential ($-E_{max}$). B: Action potential in the same cell unaltered after prolonged exposure to TTX (4 µg/ml). C-D: Records from a reaggregate cultured with heart RNA extract. C: Rapidly-rising (130 V/sec) action potential fired from a high stable resting potential. D: TTX (0.1 µg/ml) rapidly abolished all excitability in the same cell, and contractions ceased concomitantly. Stimulation applied in C-D.

III. Chick Embryo Blastoderm.

Explants of the cardiac anlagen (bilateral anterior blastoderm) removed from 20-h chick embryos develop spontaneous electrical activity after several days in culture (14). Niu et al. (6 , 24 , 25) have shown that treatment of post-nodal pieces of 19-h chick blastoderms, a region which normally does not give rise to cardiac tissue, with RNA-enriched fractions from adult chicken hearts induces the appearance of spontaneously contracting tubes resembling the young tubular heart. The contractile cells had myofibrils and intercalated disc-like structures (24), characteristic of the young hearts (42). Furthermore, Deshpande and Siddiqui (8 , 9) have shown that highly purified cardiac messenger RNA (isolated sequentially on oligodeoxythymidylate columns, sucrose gradients, and polyacrilamide gels) also induces the spontaneously contracting cardiac-like tissue in isolated post-nodal blastoderm. We therefore decided to determine whether the contracting structures induced in the RNA-treated post-nodal pieces exhibit typical cardiac action potentials.

Treatment of PNPs with heart RNA-enriched extract led to the induction of contracting structures (tubules) in 9 of the 14 PNPs tested (Table 2) (20). Electrical recordings were made from all nine induced tubules after 7-10 days of incubation. By day 12, most of the vesicles collapsed leaving dense cell masses that appeared necrotic.

Recordings from the contracting tubules revealed spontaneously firing action potentials in all cells impaled. Examples from four different PNPs are shown in Figure 12 . Penetrations were difficult to make, probably due to small cell size. The resting potentials ranged between -35 and -80 mV, with a mean of -38 ± 3 mV (M ± 1 SE)(Table 2). The maximum rates of rise of the action potentials were usually slow, ranging between 2 and 100 V/s,

Table 2

Summary of electrophysiological data obtained from post-nodal pieces of chick embryo blastoderm treated with RNA-enriched extracts from adult heart.

Addition	Number of PNPs treated	Number of PNPs with contracting tubules	Resting potential (mV)	Action potentials
None	10	0	< -15	absent
Heart RNA extract	14	9	-38 ± 3(43)	present (rate of rise 8 ± 2 V/sec)
Heart RNA extract + cycloheximide	8	0	< -15	absent
Heart RNA extract + actinomycin D	4	0	< -15	absent
Liver RNA extract	22	0	< -15	absent

The spontaneously contracting tubules were formed within about 1 week.

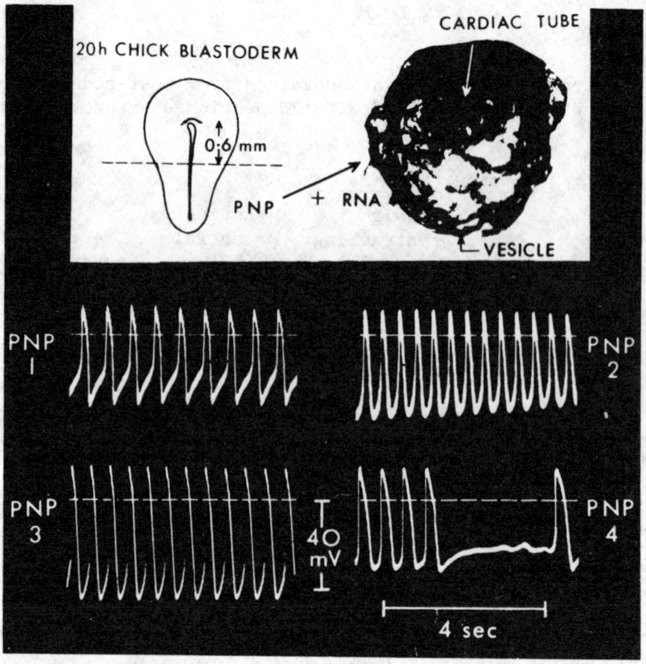

Fig. 12. Electrophysiological recordings from contracting tubules induced by an RNA-containing fraction from adult chicken hearts in post-nodal pieces (PNPs) of embryonic chick blastoderm. Upper left diagram: Diagram of the early blastoderm (20 h) showing the level at which the area pellucida was transected to obtain the PNPs. Upper right photograph: Photomicrograph of a vesicle formed from a PNP treated with the RNA-enriched fraction for 7 days. The vesicle contained a spontaneously contracting structure (cardiac tube). Lower records: Examples of intracellular electrical recordings obtained from four different preparations treated with RNA-enriched extracts. These cardiac-like action potentials were obtained from the spontaneously contracting tubules. Time and voltage calibrations apply to all tracings.

with a mean of 8 ± 2 V/s. Pacemaker potentials were observed in many of the impaled cells (e.g., see Fig. 10, PNP 1), and the duration (90% repolarization value) of the action potentials was about 100-150 ms. The action potentials fired in synchrony with the contractions of the vesicles (observed through a dissecting microscope). The frequency of the spontaneous contractions ranged between 20 and 150 beats/min, and was often irregular. In one PNP, in which spontaneous activity was not present, the cells fired action potentials and contracted in response to field stimulation. In general, the properties of the cells resembled those of the young tubular hearts developing normally *in situ*.

Microelectrode penetrations into the five PNPs which did not show inductions (in response to the extract) revealed no resting potentials greater than -15 mV, and no action potentials could be obtained even under intense stimulation.

All ten control PNPs, which received no RNA-extract, failed to develop contracting structures (Table 2). Action potentials were absent, and the resting potentials were less than -15 mV in cells impaled in the PNP walls.

PNPs treated with adult chicken liver RNA-extract survived well (without collapse of the vesicle) for up to 14 days in culture. However, these PNPs did not develop contracting regions, and no evidence of excitability was obtained in any of the cells penetrated (Table 2).

In order to determine whether the effect of the heart RNA-extract was dependent on protein synthesis, cycloheximide (1 μg/ml) was added to the medium of eight PNPs at the same time as the heart RNA-extract. All eight PNPs thus treated failed to show the development of contracting regions, and no excitable properties were detected in the penetrated cells (Table 2). Similarly, four PNPs treated with actinomycin D (1 μg/ml) in addition to the heart RNA-extract failed to develop contracting regions, and none of the penetrated cells showed signs of excitability.

Discussion

Organ-Cultured Hearts and Cultured Heart Cells

The present experiments showed that incubation *in vitro* with RNA-enriched extracts from adult chicken hearts (AHE-RNA) induced fast Na^+ channels both in organ-cultured young (2.5-day-old) hearts and in spherical reaggregates of trypsin-dispersed young cells. The young cardiac myoblasts exhibited only slowly-rising action potentials, indicating that they possessed little or no fast Na^+ channels at the time of culturing. The RNA extract caused a shift to TTX-sensitive action potentials with fast upstroke velocities (> 80 V/sec). Since changes similar to those induced by AHE-RNA occur during normal development *in situ*, the exogenous substance brought about further differentiation of the young embyronic cells whose development had been arrested by placement *in vitro*. This resulted in the achievement of adult-like electrophysiological properites. Without AHE-RNA addition, fast Na^+ channels did not increase in number: instead, the untreated cells retained TTX-insensitive slow channels.

Loss of pacemaker activity and increased action potential amplitude served as additional evidence of the induced differentiation. Other adult-like properties observed in the cells exposed to the RNA extract included the presence of functional beta-adrenergic receptors (shown by the ability of TTX-blocked hearts to generate slow electrical responses when stimulated in the presence of isoproterenol) and brief chronaxies, indicative of high excitability.

The present findings show that reaggregated cardiomyoblasts from 2-3-day-old embryonic chick hearts, untreated with AHE-RNA, exhibited arrested development *in vitro*. DeHaan et al. (5) also found that reaggregates of 2.5-day-old embryonic chick heart cells continued to beat in the presence of TTX, but they reported that reaggregates of 4-day-old cells achieved faster rising velocities (i.e., gain fast Na^+ channels) and TTX sensitivity *in vitro*, although automaticity persisted. Since Sperelakis & Shigenobu (32,40) have shown that TTX sensitivity first appears on about days 4 or 5 of development *in situ*, marking the beginning of the intermediate period, the cells used by DeHaan et al. may have already begun the synthesis of fast Na^+ channel proteins. (Since the rate of development *in ovo* is highly dependent on temperature (30), small differences in incubator temperature could account for large differences in actual state of development.) Thus, the difference in differentiative state achieved prior to culturing could account for the discrepancy observed in culture between 2.5-day and 4-day cells.

Although it is possible that the observed changes in membrane electrical properties reported here were caused by some non-nucleic acid contaminant, e.g., RNA-associated protein (12, 49, 50, 51), the evidence suggests that the active principal is an RNA: the inductive capacity was destroyed by pretreatment with protease-free ribonuclease. The failure of adult liver RNA extracts to produce the differentiative alterations suggests a specific dependence on material of cardiac origin. The poly-A experiment suggests that non-specific interaction by a poly-anion, e.g. at the cell surface, is unlikely to have triggered the observed changes.

The AHE-RNA extract run through oligo-dT columns resulted in partial induction only. Although TTX sensitivity appeared, the highest rates of rise achieved were only 30-50 V/sec, and automaticity remained. One explanation for this partial effect is that some of the active material was destroyed during preparation. The AHE-RNA/oligo-dT extract used was potent in inducing differentiation in post-nodal pieces. Although cAMP has also been shown to induce cardiac-like differentiation in post-nodal pieces (7), cAMP does not induce differentiation in organ-cultured young hearts (38).

In the present experiments, there was a lag period before the inductive effect of the AHE-RNA became manifest. The first evident changes were observed on day 6, and the peak effect occurred several days after that. This indicates that continued presence of the exogenous substance is not required; rather, the substance may trigger intracellular changes in the first few hours, which then become expressed gradually over a prolonged time course.

The inductive mechanism of the active principal remains to be elucidated. Since exogenous genetic material has been shown to be taken up by cells (2 , 26), one possibility is that uptake of the active principal by the cells could lead to the synthesis of new membrane proteins which are constituents of the fast Na^+ channels. The prevention of the effect of AHE-RNA by cycloheximide suggests that the induced changes in membrane properties are mediated by protein synthesis. McDonald et al. (17) have also reported the involvement of protein synthesis in regulation of membrane properties. Benzer & Raftery (1) have shown that the TTX-binding component of nerve, presumably the fast Na^+ channel, is a protein. Thus, the arrested development in organ culture is probably due to the absence of an extrinsic factor (normally present in situ) which exercises control over synthesis of protein components of the fast Na^+ channels.

One mechanism that may be excluded for the induction of fast Na^+ channels by the RNA extract is a change in cyclic AMP level. For example, addition of cyclic AMP to organ-cultured 2.5-day embryonic chick hearts failed to increase $+\dot{V}_{max}$ or otherwise induce differentiation (Shigenobu & Sperelakis (38) and Renaud, McLean & Sperelakis, unpublished observations). Furthermore, addition of cyclic AMP, dibutyryl cyclic AMP or N^6-monobutyryl cyclic AMP (100 μg/ml, final concentration daily) failed to induce appearance of fast-rising TTX-sensitive action potentials in monolayer-cultured (16-day-old) embryonic chick heart cells which had partly de-differentiated <u>in vitro</u> (McLean and Sperelakis, unpublished observations).

Chick Embryo Blastoderm

The results demonstrate that isolated post-nodal pieces of chick blastoderm, tissue that normally does not give rise to heart tissue, will develop spontaneously contracting tubes, when treated with RNA-enriched extracts of adult chicken heart. These contracting regions exhibited electrical excitability and cardiac-like action potentials. Thus, it may be inferred that the induced structure is indeed heart-like on the basis of spontaneous contractions, presence of myofibrils (6 , 24 , 25), and presence of cardiac-like action potentials. The resting potentials and action potentials recorded from the contracting structures were similar to those recorded from young embryonic chick tubular hearts on embryonic days 2 or 3 (46). The resting potentials were only about -40 mV, and the action potentials had slow maximal rates of rise of about 10 V/s. The observed pacemaker potentials in most cells penetrated were also similar to what is found in young intact hearts (46), as were the action potential shape and duration (of about 130 ms). The brief survival time (of about 12 days) of the PNPs in culture may have limited the extent of the membrane electrical differentiation observed.

The results suggest that myocardial membrane properties may be acquired *in vitro* by normally inexcitable cells, since few cardiogenic cells are present in the post-nodal region. The cells of the post-nodal region normally give rise to non-heart cells, although some myosin-synthesizing cells (10) and some cells which contribute to normal cardiogenesis (31 , 32) are present. However, unless treated with exogenous adult heart RNA-enriched extract, the PNPs did not develop contracting regions from which action potentials could be recorded. Thus, one possibility is that some non-excitable cells were induced to become heart cells.

The observed inductions must have been triggered by some substance in the added material. Deshpande and Siddiqui (8) have recently achieved cardiac induction in the PNPs with mRNA (isolated by affinity chromatography on oligo-dT columns) concentrations of about 0.01 OD/ml, indicating 100-1000 times purification over the material used here. Thus, RNA may be the active principle in the extract used in the present experiments. The blockade of the effect of the extract by cycloheximide and by actinomycin D suggests that protein synthesis is required for the effect. It is possible that the exogenous RNA molecules are taken up by the cells and perhaps lead to *de novo* synthesis of new mRNA in the target cell nuclei; subsequent translation of these mRNA molecules may lead to synthesis of membrane proteins involved in the electrogenesis of the action potentials. The observed lag period of about 6 days would be consistent with such an interpretation. However, since addition of cyclic AMP also gives rise to contracting structures in isolated PNPs (7), it appears that more than one substance can produce the induction.

SUMMARY

Young (2-3 day-old) embryonic chick myoblasts (myocardial cells) do not continue to differentiate in culture. This is true of organ-cultured intact hearts and of cell cultures of trypsin-dispersed cells (spherical reaggregate cultures). The cells in both types of cultures exhibit spontaneous slowly-rising (about 15 V/sec) action potentials which are insensitive to tetrodotoxin (TTX). That is, they do not normally gain TTX-sensitive fast Na^+ channels *in vitro*.

Incubation with purified adult heart extracts enriched in RNA (AHE-RNA) for 6-14 days caused the appearance of fast Na^+ channels *de novo* in such young myocardial cells whose development had been arrested *in vitro*. These AHE-RNA -treated cells, when stimulated, fired fast-rising (100 V/sec) TTX-sensitive action potentials from large stable resting potentials. The induction of fast Na^+ channels was blocked by cycloheximide, suggesting dependence upon protein synthesis. Adult chicken liver extract was not effective, and pre-treatment of the AHE-RNA with protease-free ribonuclease destroyed the inductive activity, consistent with identification of the active principal as RNA. Poly-adenylate (poly-A), a poly-anionic oligonucleotide, was not effective, suggesting that the induction was not due to non-specific interaction with the cell surface. Thus, the membrane properties of cardiac myoblasts can be made to further differentiate *in vitro* by addition of exogenous heart extracts enriched in RNA.

In 19-20-h-old chick embryonic blastoderm, the portion anterior to Hensen's node contains the cells which are destined to give rise to the heart, whereas the cells in the portion of the blastoderm posterior to Hensen's node (the post-nodal piece, PNP) normally do not. As found by Niu and associates, when the PNP was isolated (by cutting 0.6 mm posterior to Hensen's node) and exposed to the RNA-enriched extract obtained from adult chicken heart (AHE-RNA), a spontaneously contracting tubular structure was formed *in vitro* within 1 week.

Typical cardiac action potentials, both spontaneous and evoked, were recorded by intracellular microelectrodes from cells within these contracting tubules. Actinomycin D or cycloheximide prevented the inducing effect of the extract. Induction of contracting tubules and excitability did not occur in untreated controls of in PNPs treated with RNA-containing extracts from adult chick liver. These results suggest that cells not originally destined to become myocardial cells can be induced to gain many of the properties of heart cells by an RNA-enriched fraction from adult heart, and that this induction is dependent upon protein synthesis.

In conclusion, it appears that RNA-enriched extracts of adult hearts can alter the membrane properties of embryonic cardiac myoblasts and of some cell type of non-cardiac origin.

Acknowledgments

The authors wish to thank Dr. M.C. Niu for the generous supply of the RNA-enriched fractions and for his enthusiastic support of this study. Drs. Siddiqui and Deshpande also generously supplied RNA fractions.

The electrophysiological experiments were carried out in the laboratory of Dr. N. Sperelakis by Dr. M. J. McLean and Dr. J.-F. Renaud. At the time, McLean was a Predoctoral Trainee supported by an NIH training grant (HL-05815), and Renaud was a Postdoctoral Fellow supported by the D.G.R.S.T. of France. The research was supported by a grant from the National Institutes of Health (HL-18711).

REFERENCES

1. Benzer, T.I. and M.A. Raftery. Solubilization and partial characterization of the tetrodotoxin binding component from nerve axons. Biochim. Biophys. Res. Commun. 51: 939-944, 1973.

2. Bhargava, R.M. and G. Shanmugam. Uptake of nonviral nucleic acids by mammalian cells. Prog. Nucl. Acid Res. and Molec. Biol. 11: 103-192, 1971.

3. Carmeliet, E.E., C.R. Horres, M. Lieberman and J.S. Vereecke. Developmental aspects of potassium flux and permeability of the embyronic chick heart. J. Physiol. (London) 254: 673-692, 1976.

4. Carmeliet, E.E., C.R. Horres, M. Lieberman and J.S. Vereecke. Potassium permeability in the embryonic chick heart: change with age, external K and valinomycin. In: Developmental and Physiological Correlates of Cardiac Muscle (eds., M. Lieberman & T. Sano), Raven Press, New York, 1976, pp. 103-116.

5. DeHaan, R.L., T.F. McDonald and H.G. Sachs. Development of tetrodotoxin sensitivity of embryonic chick heart cells in vitro. In: Developmental and Physiological Correlates of Cardiac Muscle (eds., M. Lieberman & T. Sano), Raven Press, New York, 1976, pp. 155-168.

6. Deshpande, A.K., L.C. Niu and M.C. Niu. Requirement for informational molecules in heart formation. In: The Role of RNA in Reproduction and Development (eds., M.C. Niu & J.S. Segal), North Holland, Amsterdam, 1973, pp. 229-246.

7. Deshpande, A.K. and M.A.Q. Siddiqui. Differentiation induced by cyclic AMP in undifferentiated cells of early chick embryo in vitro. Nature (Lond.) 263: 588-591, 1976.

8. Deshpande, A.K. and M.A.Q. Siddiqui. A reexamination of heart muscle differentiation in the postnodal piece of chick blastoderm mediated by exogenous RNA. Develop. Biol. 58: 230-247, 1977.

9. Deshpande, A.K. and M.A.Q. Siddiqui. Acetylcholinesterase differentiation during myogenesis in early chick embryonic cells caused by an inducer RNA. Differentiation 10: 133-137, 1978.

10. Ebert, J.D. Analysis of the synthesis and distribution of the contractile protein, myosin, in the development of the heart. Proc. Nat. Acad. Sci. (Washington) 39: 333-344, 1953.

11. Enemar, A., B. Falck and R. Hakanson. Observations on the appearance of norepinephrine in the sympathetic nervous system of the chick embryo. Devel. Biol. 11: 268-283, 1965.

12. Jacobson, A.G. and J.T. Duncan. Heart induction in salamanders. J. Exp. Zool. 167: 79-103, 1968.

13. Jongsma, H.J., M. Masson-Pevet and J. de Bruyne. Synchronization of the beating frequency of cultured rat heart cells. In: Developmental and Physiological Correlates of Cardiac Muscle (eds., M. Lieberman & T. Sano), Raven Press, New York, 1976, pp. 185-196.

14. LeDouarin, G., G. Obrecht and E. Coraboeuf. Déterminations régionales dans l'aire cardiaque présomptive mises en evidence chez l'embryon de poulet par la methode microélectrophysiologique. J. Embryol. Exp. Morph. 15: 153-157, 1966.

15. Lieberman, M., T. Sawanobori, N. Shigeto and E.A. Johnson. Physiologic implications of heart muscle in tissue culture. In: Developmental and Physiological Correlates of Cardiac Muscle (eds., M. Lieberman & T. Sano), Raven Press, New York, 1976, pp. 139-154.

16. McDonald, T.F. and H.G. Sachs. Electrical activity in embryonic heart cell aggregates: Developmental aspects. Pflügers Arch. 354: 151-164, 1975.

17. McDonald, T.F., H.G. Sachs and R.L. DeHaan. Tetrodotoxin desensitization in aggregates of embryonic chick heart cells. J. Gen. Physiol. 62: 286-302, 1973.

18. McLean, M.J., R.A. Lapsley, K. Shigenobu, F. Murad and N. Sperelakis. High cyclic AMP levels in young embryonic chick hearts. Devel. Biol. 42: 196-201, 1975.

19. McLean, M.J., J.-F. Renaud, M.C. Niu and N. Sperelakis. Membrane differentiation of cardiac myoblasts induced in vitro by an RNA-enriched fraction from adult heart. Exp. Cell Res. 110: 1-14, 1977.

20. McLean, M.J., J.-F. Renaud and N. Sperelakis. Cardiac-like action potentials recorded from spontaneously-contracting structures induced in post-nodal pieces of chick blastoderm exposed to an RNA-enriched fraction from adult heart. Differentiation 11: 13-17, 1978.

21. McLean, M.J., J.-F. Renaud, N. Sperelakis and M.C. Niu. Messenger RNA induction of fast sodium ion channels in cultured cardiac myoblasts. Science 191: 297-299, 1976.

22. McLean, M.J. and N. Sperelakis. Rapid loss of sensitivity to tetrodotoxin by chick ventricular myocardial cells after separation from the heart. Exp. Cell Res. 86: 351-364, 1974.

23. McLean, M.J. and N. Sperelakis. Retention of fully differentiated electrophysiological properties of chick embryonic heart cells in culture. Develop. Biol. 50: 134-142, 1976.

24. Niu, M.C. and A.K. Deshpande. The development of tubular heart in RNA-treated post-nodal pieces of chick blastoderm. J. Embryol. Exp. Morph. 29: 485-501, 1973.

25. Niu, M.C., A.K. Deshpande and L.C. Niu. Poly (A)-enriched RNA as activator in embryonic differentiation. Proc. Soc. Exp. Biol. (N.Y.) 147: 318-322, 1974.

26. Niu, M.C., L.C. Niu and S.F. Yang. In vivo uptake of RNA and its function in the castrate uterus. In: The Role of RNA in Reproduction and Development (eds., M.C. Niu & S.J. Segal), North Holland Publishing Co., Amsterdam, 1973, pp. 90-109.

27. Pappano, A.J. Development of autonomic neuroeffector transmission in the chick embryo heart. In: Developmental and Physiological Correlates of Cardiac Muscle (eds., M. Lieberman & T. Sano), Raven Press, New York, 1976, pp. 235-248.

28. Renaud, J.-F. and N. Sperelakis. Electrophysiological properties of chick embryonic hearts grafted and organ-cultured in vitro. J. Molec. Cell. Cardiol. 8: 889-900, 1976.

29. Renaud, J.-F., N. Sperelakis and G. LeDouarin. Increase of cyclic AMP levels induced by isoproterenol in cultured and non-cultured chick embryonic hearts. J. Molec. Cell. Cardiol. 10: 281-286, 1978.

30. Romanoff, A. The Avian Embryo: Structure and Functional Development. Macmillan, New York, 1960.

31. Rosenquist, G.C. Localization and movement of cardiogenic cells in the chick embryo: heart-forming portion of the primitive streak. Develop. Biol. 22: 461-475, 1970.

32. Rosenquist, G.C. and R.L. DeHaan. Migration of precardiac cells in the chick embryo: A radioautographic study. Contr. Embryol. Carneg. Instn. 38: 111-121, 1966.

33. Schneider, J.A. and N. Sperelakis. Slow Ca^{++} and Na^{+} responses induced by isoproterenol and methylxanthines in isolated perfused guinea pig hearts exposed to elevated K^{+}. J. Molec. Cell. Cardiol. 7: 249-273, 1975.

34. Schutz, G., M. Beato and P. Feigelson. Isolation of eukaryotic messenger RNA on cellulose and its translation in vitro. Biochem. Biophys. Res. Commun. 49: 680-689, 1972.

35. Shigenobu, K. and N. Sperelakis. Development of sensitivity to tetrodotoxin of chick embryonic hearts with age. J. Molec. Cell. Cardiol. 3: 271-286, 1971.

36. Shigenobu, K. and N. Sperelakis. Ca^{++} current channels induced by catecholamines in chick embryonic hearts whose fast Na^{+} channels are blocked by tetrodotoxin. Circ. Res. 31: 932-952, 1972.

37. Shigenobu, K. and N. Sperelakis. Failure of development of fast Na^{+} channels during organ culture of young embryonic chick hearts. Develop. Biol. 39: 326-330, 1974.

38. Shigenobu, K. and N. Sperelakis. Prolongation of the action potential plateau of embryonic chick hearts organ cultured in the presence of cyclic AMP. Jap. J. Pharmacol. 25: 481-484, 1975.

39. Sperelakis, N. Electrical properties of embryonic heart cells. In: Electrical Phenomena in the Heart (ed., W.C. DeMello), Academic Press, New York, 1972, pp. 1-61.

40. Sperelakis, N. (Na^+,K^+)-ATPase activity of embryonic chick heart and skeletal muscles as a function of age. Biochim. Biophys. Acta 266: 230-237, 1972.

41. Sperelakis, N. Cultured heart cell reaggregate model for studying cardiac toxicology. Environmental Health Perspectives 26: 243-267, 1978.

42. Sperelakis, N., M. Forbes, K. Shigenobu and S. Coburn. Organ-cultured chick embryonic heart cells of various ages. Part II, Ultrastructure. J. Molec. Cell. Cardiol. 6: 473-483, 1974.

43. Sperelakis, N. and M.J. McLean. Electrical properties of cultured heart cells. In: Recent Advances in Studies on Cardiac Structure and Metabolism v. 12 (eds., T. Kobayashi, Y. Ito & G. Rona), University Park Press, Baltimore, 1978, pp. 645-666.

44. Sperelakis, N. and M.J. McLean. The electrical properties of embryonic chick cardiac cells. In: Fetal and Newborn Cardiovascular Physiology. Vol. 1, The Heart and Blood Flow (eds., L.D. Longo & D.D. Reneau), Garland Press, New York, pp. 191-236, 1978.

45. Sperelakis, N. and J.A. Schneider. A metabolic control mechanism for calcium ion influx that may protect the ventricular myocardial cell. Am. J. Cardiol. 37: 1079-1085, 1976.

46. Sperelakis, N. and K. Shigenobu. Change in membrane properties of chick embryonic hearts during development. J. Gen. Physiol. 60: 430-453, 1972.

47. Sperelakis, N. and K. Shigenobu. Organ-cultured chick embryonic heart cells of various ages. Part I, Electrophysiology. J. Molec. Cell. Cardiol. 6: 449-471, 1974.

48. Sperelakis, N., K. Shigenobu and M.J. McLean. Membrane cation channels--changes in developing hearts, in cell culture, and in organ culture. In: Developmental and Physiological Correlates of Cardiac Muscle (eds., M. Lieberman & T. Sano), Raven Press, New York, 1976, pp. 209-234.

49. Spirin, A.S. The second Sir Hans Krebs Lecture. Informosomes. Europ. J. Biochem. 10: 20-35, 1969.

50. Tiedemann, H. Biochemical aspects of primary induction and determination. In: Biochemistry of Animal Development (ed., R. Weber), Academic Press, New York, 1967, pp. 3-55.

51. Yamada, T. Embryonic Induction. In: The Chemical Basis of Development (ed., W.D. McElroy & B. Glass), The Johns Hopkins Press, Baltimore, 1958, pp. 217-238.

THE STRUCTURES OF THE α-FETOPROTEIN AND ALBUMIN GENES IN THE MOUSE

Shirley M. Tilghman, Michael B. Gorin, Dimitris Kioussis,

Fern Eiferman, Paul van de Rijn and Robert S. Ingram

Institute for Cancer Research

The Fox Chase Cancer Center

Philadelphia, Pa. 19111

Abstract

The structures of the α-fetoprotein (AFP) and albumin genes in the mouse have been analyzed by restriction endonuclease mapping and electron microscopy. Both genes are organized similarly into 15 coding segments interrupted by 14 intervening sequences. The sizes of the corresponding coding segments in each gene are identical, lending support to the hypothesis that the two genes were derived from a common ancestral gene. However, no homology between coding segments was observed. Both the sizes and the nucleotide sequence of flanking and intervening sequences have diverged significantly as well. An examination of the sizes of the coding segments in each gene revealed a thrice repeated domain, consisting of 4 coding segments. These correspond to the 3 domains which Brown (Fed. Proc. 35, 2141-2144) has identified in both human and bovine albumins, based on the regular spacing of cysteine-cysteine disulfide bridges. A closely analogous structure for murine AFP can be drawn, using the amino acid sequence derived from the sequencing of 3 murine AFP cDNA clones. From a comparison of the murine AFP with either human or bovine albumin, a 50% conservation of amino acid sequence was observed, extending throughout the protein, but excluding the NH_2 terminus.

We propose that the AFP and albumin genes arose as the result of a duplication of a common ancestral gene. In addition, the presence of 3 closely related domains in these genes suggests that the ancestral gene itself arose as the consequence of a triplication of a more primordial domain.

Introduction

The major protein component in the serum of the developing fetus is α-fetoprotein (AFP), which is synthesized by the embryonic liver and yolk sac (1-3). It is replaced after birth by serum albumin, whose synthesis is initiated in liver before birth at the time when the synthesis of AFP is beginning to decline (4,5). A close evolutionary relationship between these two proteins has been suggested by several investigators, based on this reciprocal cross-over in their synthesis. In addition, the two proteins are believed to perform similar physiological roles, that is, to control the osmotic pressure of the intravascular fluid and to bind a variety of metabolites (6,7). They are structurally very similar: in size, amino acid composition and immunocross-reactivity (8,9). The very limited amount of amino acid sequence data available on AFP has prevented a thorough assessment of the degree of homology between AFP and albumin, and the current results are contradictory. A comparison of the first 25 amino acids at the amino termini of murine AFP and albumin revealed no significant homology (10). On the other hand, when 59 amino acids derived from equivalent internal peptides of human AFP and albumin were compared, a 50% agreement in amino acid sequence was observed (11).

We have used two different criteria to test the hypothesis that the AFP and albumin genes arose from a common ancestor. The first is based on a direct comparison of the structures of the genes themselves. Should albumin and AFP be duplicated genes, one might expect that the basic organization of their coding sequences be maintained even though significant divergence in the nucleotide sequence may have occurred. For example, the α globin gene, which diverged from the β globin gene at least 300 million years ago, has retained no nucleotide sequence homology with the β globin gene. However, the organization of the two genes is identical, in that they are composed of 3 identically sized coding segments which are interrupted by two intervening sequences occurring at the same positions (12). Thus, a careful comparison of the organization of the AFP and albumin genes within a species should prove useful in assessing the likelihood that they share a common ancestry.

The second approach has been to determine the complete amino acid sequence of murine AFP, by sequencing three chimeric cDNA plasmids that contain over 90% of the AFP mRNA sequence (13). In this way, a direct comparison of the primary sequences of both proteins can be made.

Experimental Procedures

Mouse genomic libraries, prepared from random partial *Hae*III-*Alu*I digests of Balb/c embryonic DNA or partial *Hae*III digests of the plasmacytomas MOPC41 and S107 DNAs and the outer *Eco*R1 fragments of the vector Charon 4A (14,15) were generously provided by Dr. J. G. Seidman. The libraries were screened by the procedure of Benton and Davis (16) using DNA fragments derived from restriction endonuclease digests of the chimeric AFP or albumin cDNA plasmids (13, and unpublished results). Selected phage were purified and grown in LB media using the *E. coli* strain DP50*sup*F as the host.

Restriction endonuclease mapping of phage DNA was performed using conditions recommended by New England Biolabs except that albumin was omitted. Restriction fragments were analyzed by agarose and acrylamide gel electrophoresis and stained with 25 μg/ml ethidium bromide. The restricted DNA was transferred from agarose gels to nitrocellulose filters by the method of Southern (17). Pretreatment, hybridization, and washing of the filters were performed as described by Wahl et al. (18).

To form RNA-DNA duplexes, either phage DNA or plasmid DNA (20 μg/ml), linearized with a suitable restriction endonuclease, was heated with 20 μg/ml mRNA in 80% formamide, 0.15M NaCl, 0.1M Tricine, pH 8.0, 10mM EDTA at 80°C for 5 min, followed by incubation at 52°C for 2 to 3 hr.

Heteroduplexes between the subcloned EcoR1 genomic fragments in pBR322 were formed between linearized DNAs at a 1:1 molar ratio after denaturation in 0.3N NaOH and neutralization for 1 hr at room temperature in 0.1M Tris HCl pH 7.5, 10 mM EDTA at a DNA concentration of 5-10 μg/ml. All hybrids were spread for electron microscopy as described previously (19) and were viewed at 16,000X magnification.

DNA sequencing was performed on pAFP1 and pAFP2 (13) by the method of Maxam and Gilbert (20).

Results

In order to isolate complete copies of the AFP and albumin genes, we screened several mouse genomic libraries derived from Balb/c DNA, using cloned cDNA copies of the AFP and albumin mRNA sequences (13,25, and unpublished results). Four recombinant phage, together comprising over 35 kb of mouse genomic DNA, hybridized specifically to AFP cDNA probes. These were designated λAFP7, 8, 13 and 14, and were shown by restriction endonuclease mapping to overlap each other in the manner shown in Fig. 1A. Similarly, two recombinant phage containing segments of the albumin gene, termed λalb 5 and 6, were cloned, and are represented in Fig. 1B.

To determine the 5' to 3' orientation of these genomic clones, and to identify the location of coding regions within them, a combination of restriction endonuclease mapping and Southern hybridization procedures were used, as is illustrated in Fig. 2, for λAFP7. When λAFP7 DNA was cleaved with EcoR1, 6 internal fragments were generated, labelled EcoA-F in Fig. 2. That the band at 4.75 kb constituted a doublet was demonstrated by cleavage of the EcoR1 digest with Xho I, which does not cleave the outside λCh4A arms, and cleaves only the

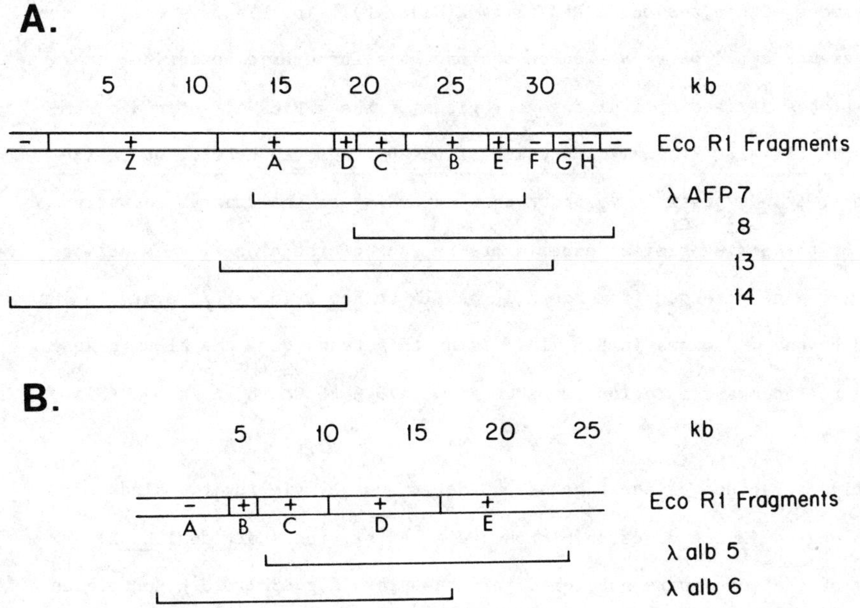

Figure 1: The Mapping of AFP and Albumin Genomic Clones

The relationship between recombinant phage containing the AFP (A) or albumin genes (B) and their designation is illustrated at the bottom of each figure. The individual EcoR1 fragments referred to in the text are labelled. The "+" signs identify those fragments which hybridized to AFP or albumin cDNA probes, and which constitute the extent of the coding region in each gene.

Eco B fragment within the mouse DNA (Fig. 2, lane b). In this way, each of the EcoRl fragments could be separated on agarose gels for hybridization to specific probes derived from different regions of the AFP mRNA, shown diagrammatically in Fig. 2. By analyzing which of the 6 fragments hybridized to the cDNA probes, it was possible to order them according to the diagram shown in Fig. 1. By performing similar experiments on each of the clones, we concluded that the AFP gene extended from Eco Z in λAFP14 to Eco E in λAFP7, 8 and 13, as indicated by the "+" signs in Fig. 1A. Using this technique, the albumin gene was shown to encompass a region from Eco B in λalb 6 to Eco E in λalb 5 (Fig. 1B).

The fine structure of these genes was determined by viewing two kinds of hybrid molecules in the electron microscope. First, the individual EcoRl fragments of the genes were subcloned into the plasmid pBR322. Heteroduplexes were then formed between these plasmids and the appropriate cDNA clones. Each plasmid was first linearized with Sal I, which does not cleave within the inserted DNAs, but does cut pBR322 DNA at a site 650 bp from the EcoRl site. In this way, the orientation of the heteroduplex structures, visualized in the electron microscope, could be immediately determined.

In Fig. 3, an example of a heteroduplex between AFP Eco A and pAFPl is shown, along with its 5' to 3' orientation. pAFPl is a cDNA clone which contains a 960 bp insert in the HindIII site of pBR322 that includes most of the 5' AFP mRNA sequence (13). Four loops of single-stranded DNA, 2 outside "substitution" loops and 2 internal "deletion" loops are connected by 3 separate homologous regions, representing three coding sequences in Eco A. The presence of single-stranded AFPl DNA at both ends of the heteroduplex shows that the three coding segments in Eco A fall within the AFP mRNA sequence, and that the gene must extend on either side of Eco A.

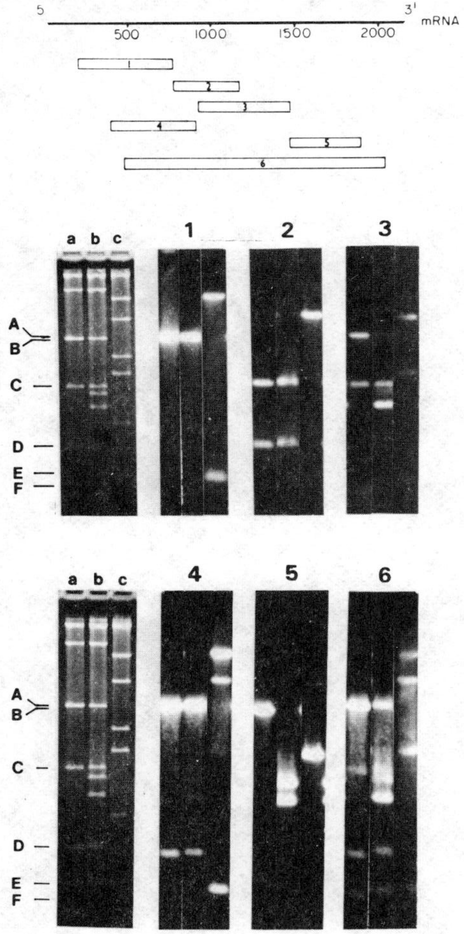

Figure 2   Orientation of EcoRl Fragments within λAFP7

Specific restriction fragments, identified in the numbered boxes of the upper diagram, were prepared from pAFP1 and pAFP2 (probes 1-5) or pAFP3, (probe 6). The regions of AFP mRNA from which they were derived are indicated. In the gels below, 0.5 μg of aliquots of λAFP7 DNA were digested with EcoRl (lane a), EcoRl/Xho I (lane b), or Sac I (lane c) and electrophoresed on replica 1% agarose gels. Ethidium bromide staining of the gels are shown on the left. Sizes of restriction endonuclease fragments were determined by comparison to fragments of a HindIII digestion of λcI857 and a HaeIII digestion of Col EI. The EcoRl fragments, lettered A-F, are 4.75, 4.75, 2.75, 1.48, 1.1 and 0.95 kb in size. In panels 1 through 6 are hybridizations of the digests of λAFP7 DNA with probes 1 through 6, respectively.

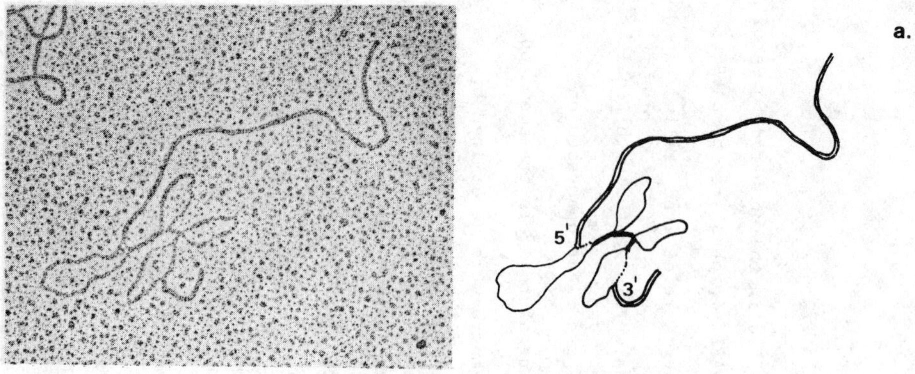

Figure 3. Electron Microsopy of Heteroduplexes Between pAFP1 and AFP Eco A

The plasmids were linearized by cleavage with Sal I. The designation of 5' and 3' ends of the hybrid molecules correspond to the 5' and 3' ends of the AFP mRNA. In the drawings beside the micrographs, the double lines represent the 3.71 kb and 0.62 kb homoduplexes of pBR322 DNA. Single narrow lines indicate single-stranded genomic DNA and dotted lines the single stranded cDNA segments in pAFP1. Heavy lines represent the regions of homology between the AFP cDNA and the AFP genomic DNA.

When suitable cDNA clones were not available, hybrids between mRNA and individual EcoRl genomic fragments were used to visualize the structure of the genes, such as that shown in Fig. 4, between AFP Eco B and AFP mRNA. Five coding segments could be identified, which were interrupted by four single-stranded DNA loops, indicating the presence of intervening sequences. The RNA tails at the 5' and 3' ends of the hybrid also confirmed that the mRNA coding sequence spanned either side of Eco B.

From a series of such experiments, a total of 15 coding segments, separated by 14 intervening sequences were identified in both the albumin AFP genes. The sizes of these gene segments are compiled in Table 1.

Discussion

The observation that AFP and serum albumin were similar in both their structural and biological properties led to the hypothesis that the two genes which encoded them arose from a common ancestor. Our experiments were designed to test this hypothesis by directly comparing the organization of the genes themselves. From the data shown in Table 1 we show that the organization of coding segments in the two genes is indeed identical and that they most probably had a common ancestor.

Both genes are comprised of 15 coding segments, each of which has a close parallel, in terms of size, in the other gene. In contrast, there is much greater variation in the sizes of the 14 corresponding intervening sequences. Yet while there is obviously considerable evolutionary constraint placed upon the size of each coding segment, the nucleotide sequences within them have diverged to the point where there is no detectable cross-hybridization between the two mRNAs.

The most striking observation made by comparing the sizes of the gene segments in Table 1 was the apparent regularity in the pattern of sizes of

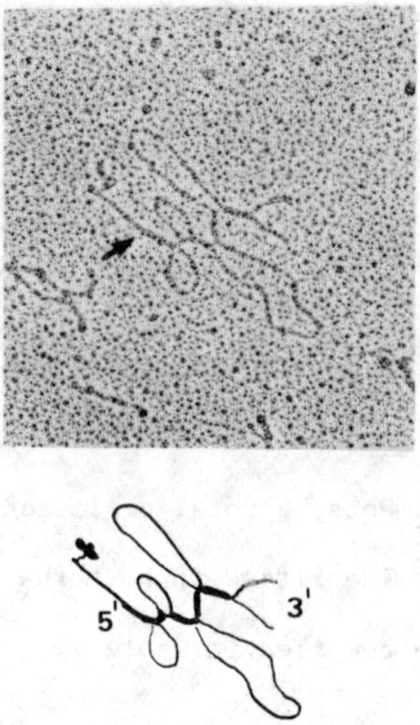

Figure 4 Electron Microscopy of Hybrids Formed Between AFP mRNA and AFP Eco B

The genomic insert of pAFP Eco B was excised with EcoRl and hybridized to 18S yolk sac RNA as described in the text. The 5' and 3' ends of the mRNA are noted on the drawing. The arrow indicates the approximate end in the homology between the RNA and DNA, which is immediately adjacent to the 5' end of the DNA fragment.

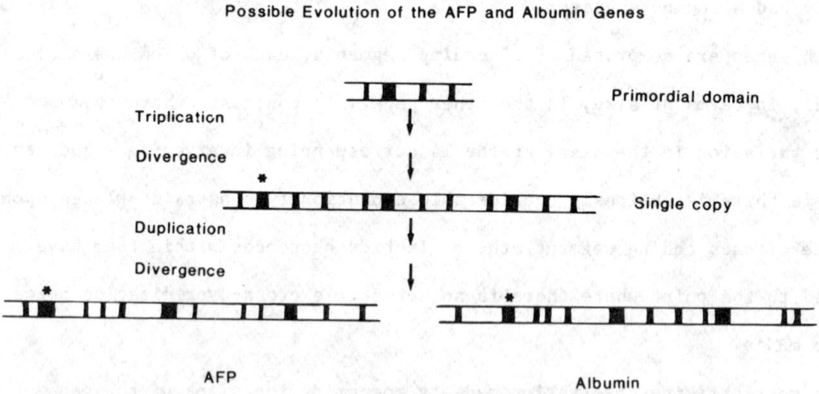

Figure 5 A Model for the Evolution of the AFP and Albumin Genes

The diagram illustrates the expansion of the putative 4 coding segments of the primordial domain by a series of unequal crossing-over events to generate the putative ancestral gene. This gene underwent a duplication to generate the AFP and albumin genes. Only the 12 internal coding segments of each gene are presented in the final diagram.

Table 1

The Sizes of Coding and Intervening Sequences in the AFP and Albumin Genes

I	Coding		II	Intervening	
	AFP	Albumin		AFP	Albumin
1	114 ± 32	110 ± 29	1-2	890 ± 166	1138 ± 61
2	53 ± 14	75 ± 27	2-3	882 ± 165	582 ± 69
3	148 ± 37	121 ± 48	3-4	3324 ± 470	2021 ± 102
4	218 ± 52	240 ± 49	4-5	1045 ± 126	308 ± 51
5	144 ± 26	170 ± 70	5-6	954 ± 198	1343 ± 83
6	104 ± 22	118 ± 33	6-7	1260 ± 107	1077 ± 120
7	133 ± 29	137 ± 35	7-8	1610 ± 120	826 ± 72
8	230 ± 23	218 ± 65	8-9	1830 ± 160	1045 ± 147
9	154 ± 67	136 ± 30	9-10	755 ± 96	1380 ± 93
10	125 ± 35	125 ± 28	10-11	635 ± 62	1200 ± 118
11	135 ± 45	153 ± 31	11-12	1603 ± 143	558 ± 86
12	280 ± 70	222 ± 42	12-13	1072 ± 122	1277 ± 129
13	175 ± 50	122 ± 36	13-14	548 ± 102	1016 ± 77
14	69 ± 27	62 ± 26	14-15	230 ± 61	804 ± 96
15	149 ± 33	112 ± 31			

The data are presented as base pairs ± SD and were derived from measurements of 15-20 molecules such as those shown in Figs. 2 and 3. Taken from Gorin and Tilghman (25) and unpublished results.

coding segments in both genes. A triplicated repeat, composed of four coding segments of sizes ~140, 220, 140 and 100 bp in length, was clearly evident, as indicated by the brackets in Table 1. This is particularly significant in light of the proposal by Brown (22), based on the regularity of cysteine residues in bovine and human albumin, that the protein is composed of three equal domains, each of which provides a different functional binding site.

If the domain structure in albumin is important to the function of the protein, one would predict that it would be present in AFP. Consequently, to make this comparison, we sequenced over 90% of the coding sequence of mouse AFP, by applying the Maxam and Gilbert procedure to pAFP1, 2 and 3 (Gorin et al., manuscript in preparation). When we re-drew the amino acid sequence of mouse AFP in a manner similar to that which Brown (22) used to illustrate the domains of human and bovine albumins, we found that AFP could also be folded on the basis of regularly spaced doublet disulfide bridges into a triplicated 2-dimensional array. Thus we suggest that the triplicated domains seen in the two genes results in a similar arrangement in the proteins. It also raises the possibility that the ancestral gene itself arose as the result of a triplication in a single functional domain which would have been composed of 4 coding segments alone (Fig. 5).

This mode of evolution, where intervening sequences provide the flexibility necessary for the amplification, by unequal crossing over, of functional domains to assemble new genes has been proposed by Gilbert (23) and Darnell (24). An alternative proposal by Gilbert (23), that coding segments can also be shuffled as functional building blocks within the genome in order to assemble new genes, does not, superficially, appear to apply in this case.

The determination of the amino acid sequence of AFP also provided us with an opportunity to assess its degree of homology to albumin. A 50% conservation in amino acid sequence between murine AFP and either human or bovine albumin extends from amino acid 25 to the carboxy termini of the proteins, and lends further support to the proposal that these are closely related proteins. Interestingly, we agree with Ruoslahti and Terry (11) that the first 25 amino acid residues in AFP bear no significant resemblance to those in albumin. As it is likely that these residues are encoded by the second coding segment in each gene, which falls outside the borders of the internal triplicated domains, it raises the possibility that the two coding segments which make up the 5' terminus of each gene were not included in the duplication that led to the establishment of the two genes, but were acquired separately. This seems somewhat unlikely in light of the fact that they are the same size in both genes. An alternative model, that there is little selective pressure on this region of the protein is possible, although the amino termini of several albumins are as closely related as the rest of the protein, suggesting that they have not been able to drift preferentially. It may be that this region of each protein is involved in specific functions which are not common to both.

REFERENCES

1. Abelev, G. I. (1971) Adv. Cancer Res. 14, 295-358.
2. Abelev, G. I. (1974) Transplant. Rev. 20, 3-37.
3. Ruoslahti, E. and Seppala, M. (1979) Adv. Cancer Res. 29, 276-346.
4. Tamaoki, T., Muira, K., Lin, T. and Banks, P. (1976) in Oncodevelopmental Gene Expression (Fishman, W. H. and Sell, S., eds.) Academic Press, New York, p. 115-122.
5. Sala-Trepat, J. M., Dever, J., Sargent, T. D., Thomas, K., Sell, S. and Bonner, J. (1979) Biochemistry 18, 2167-2178.
6. Peters, T., Jr. (1977) Clin. Chem. (Winston-Salem, N.C.) 23, 5-12.
7. Peters, T., Jr. (1975) in The Plasma Proteins, ed. Putnam, F. W. (Academic Press, New York) 2nd Ed., Vol. 1, pp. 133-181.

8. Watabe, H. (1974) Int. J. Cancer 13, 377-388.

9. Ruoslahti, E. and Engvall, E. (1976) Proc. Natl. Acad. Sci. USA 73, 4641-4644.

10. Peters, E. H., Nishi, S. and Tamaoki, T. (1978) Biochem. Biophys. Res. Comm. 83, 75-82.

11. Ruoslahti, E. and Terry, W. D. (1976) Nature 260, 804-805.

12. Leder, A., Miller, H., Hamer, D., Seidman, J. G., Norman, B., Sullivan, M. and Leder, P. (1978) Proc. Natl. Acad. Sci. USA 75, 6187-6191.

13. Tilghman, S. M., Kioussis, D., Gorin, M. B., Garcia Ruiz, J. P. and Ingram, R. S. (1979) J. Biol. Chem. 254, 7393-7399.

14. Blattner, F. R., Williams, B. G., Blechl, A. E., Denniston-Thompson, K., Faber, H. E., Furlong, L.-A., Grunwald, D. J., Kiefer, D. O., Moore, D. D., Schumm, J. W., Sheldon, E. L. and Smithies, O. (1977) Science 196, 161-169.

15. Maniatis, T., Hardison, R. C., Lacy, E., Lauer, J., O'Connell, C., Quon, D., Sim, G. K. and Efstratiadis, E. (1978) Cell 15, 687-701.

16. Benton, W. D. and Davis, R. W. (1977) Science 196, 180-182.

17. Southern, E. M. (1975) J. Mol. Biol. 98, 503-517.

18. Wahl, G. M., Stern, M. and Stark, G. R. (1979) Proc. Natl. Acad. Sci. USA 76, 3683-3687.

19. Tiemeier, D. C., Tilghman, S. M. and Leder, P. (1977) Gene 2, 173-191.

20. Maxam, A. and Gilbert, W. (1977) Proc. Nat. Acad. Sci. USA 74, 560-564.

21. Kioussis, D., Hamilton, R., Hanson, R. W., Tilghman, S. M. and Taylor, J. M. (1979) Proc. Natl. Acad. Sci. USA 76, 4370-4374.

22. Brown, J. R. (1976) Fed. Proc. 35, 2141-2144.

23. Gilbert, W. (1978) Nature 271, 501.

24. Darnell, J. E., Jr. (1978) Science 202, 1257-1260.

25. Gorin, M. B. and Tilghman, S. M. (1980) Proc. Natl. Acad. Sci. USA 77, 1351-1357 .

HORMONAL CONTROL OF THE RELEASE OF STORED mRNA FOR ORNITHINE DECARBOXYLASE

Claude A. Villee

Department of Biological Chemistry, and Laboratory of Human Reproduction and Reproductive Biology, Harvard Medical School, Boston, Massachusetts.

Abstract Ornithine decarboxylase (ODC) is an unusual enzyme; both it and its mRNA have exceptionally short half-lives, about 15 minutes. The target tissues of most, if not all, hormones respond to that specific hormone with an early and marked increase in ODC activity. While investigating the induction of adenocarcinomas of the hamster kidney by long-term estrogen administration we found that hamster kidney cells in culture show a marked increase in ODC activity when diethylstilbestrol (DES, 10^{-9}M) is added to the medium. This rise in ODC activity appears to represent *de novo* synthesis of the enzyme, for it is inhibited by the addition of cycloheximide (100-500 ng/ml) to the medium. Actinomycin D (500 ng/ml) added to the medium does not inhibit the increase in ODC activity but actually increases it. This level of actinomycin D does inhibit the *de novo* synthesis of RNA, for it inhibits the incorporation of ^{3}H-uridine into RNA. ODC activity in the kidney cells and in a cell line derived from mouse thigh muscle, 1929, is increased by added dibutyryl cAMP and by the phosphodiesterase inhibitor, methyl isobutyl xanthine. These increases in activity are also inhibited by cycloheximide. The second enzyme in the biosynthetic pathway for polyamines, S-adenosyl methionine decarboxylase (SAMDC), is also increased in activity by DES and by actinomycin D; these increases are inhibited by cycloheximide. The mRNA for ODC and for SAMDC may be present in the cell in some bound, inactive state and released by the hormonal stimulus or by actinomycin D and undergo translation to produce the active enzyme. This phenomenon may parallel the masked messenger RNA present in the oocyte which is freed upon fertilization and translated to produce the proteins synthesized early in development. Such a mechanism could account for the rapid increase in ODC activity following hormonal stimulation and might be involved in the unusually short half life of ODC.

INTRODUCTION

Questions regarding the biological roles of the polyamines spermine and spermidine and ornithine decarboxylase, the first enzyme in their biosynthetic pathway, have occupied the attention of biologists and biochemists for several decades. Ornithine decarboxylase (ODC; L-ornithine carboxylyase; E.C. 4.1.1.17) catalyzes the conversion of ornithine to putrescine, a precursor in the synthesis of spermine and spermidine. Ornithine decarboxylase is an unusual enzyme; both it and its mRNA have exceptionally short half-lives, about 15 minutes. It has been postulated that the positively-charged polyamines may have some role in stabilizing the negatively-charged nucleic acids within the cell (1). It seems significant that the concentrations of polyamines are elevated in embryonic tissues, cancers and other actively growing tissues (2).

The ubiquitous distribution of polyamines, their responses to hormonal stimuli, and their possible involvement in both the transcriptional and translational events of cellular growth processes have been reported repeatedly (3). The finding that an increased activity of ornithine decarboxylase is a feature of the early response of many target tissues to their respective hormones is especially intriguing. Thus the ornithine decarboxylase activity of the prostate is stimulated by testosterone (4). The ornithine decarboxylase activity of the immature rat uterus is stimulated by estradiol (5,6) as is the ornithine decarboxylase activity of the chick oviduct (7). The ornithine decarboxylase activity of the liver is stimulated by growth hormone (8) and the ornithine decarboxylase of the adrenal is stimulated by ACTH (9). Ovarian ornithine decarboxylase activity rises sharply during the preovulatory surge of LH (10) and the rise of activity can be prevented by injecting anti-LH before proestrus. Ovarian ornithine decarboxylase increases sharply one hour after the administration of LH. This appears to be due to the *de novo* synthesis of the enzyme, for the rise is prevented by actinomycin D or cycloheximide (11). The ornithine decarboxylase activity of the immature rat testis is stimulated by LH, FSH or cyclic AMP (12,13). Increased ornithine decarboxylase activity in response to hormonal stimulation has also been demonstrated in fibroblasts (14), granulosa cells (15) and hepatocytes (16) in culture. The activity of the enzyme is increased in essentially all rapidly growing cells such as tumor cells. Ornithine decarboxylase activity is increased in the kidneys of hamsters that have been implanted with estrogen pellets for the induction of renal adenocarcinomas; it is twice normal after one month and four times normal after four months. The ODC activity of the kidney tumor after nine months of estrogen implants is five-fold greater than the normal kidney. It seemed of interest to determine whether estrogens added *in vitro* to kidney cells in culture would induce a comparable increased activity of ornithine decarboxylase.

EXPERIMENTAL PROCEDURES

Cell cultures were prepared from normal hamster kidneys or from an established baby hamster kidney (BHK-21) cell line. The target cells of estrogen in the kidney were shown to be the epithelial cells of the proximal convoluted tubules (17); hence, the fibroblasts of the primary cell culture were selectively eliminated by adding D-valine to the medium while the cells formed a subconfluent monolayer. The cells were seeded in 100 x 20 mm Falcon plastic culture dishes at a density of 2 to 3 x 10^6 cells per dish in 10 ml of Dulbecco's modified Eagle's

Medium containing 10% fetal calf serum and antibiotics (streptomycin 100 μg/ml, penicillin 100 μg and fungizone 0.25 μg/ml of medium). The cells were grown under 95% air, 5% CO_2 in a moisture saturated atmosphere until they formed a subconfluent monolayer.

At the beginning of each experiment the medium was removed and replaced with serum-free Dulbecco's medium with or without diethylstilbestrol or the compound to be tested. The cells were then incubated at 37°C, usually for three hours. At the end of the incubation, the medium was removed by suction and 5 ml of TED buffer (25 mM Tris pH 7.4, 0.1 mM EDTA and 1 mM dithiothreitol) was added to each dish and the cells were scraped from the dish with a rubber policeman. The cells from four to six dishes were combined and centrifuged at 200 x g at 4° for 10 minutes to sediment the cells. The cells were washed again with TED buffer, centrifuged, and then homogenized in 1-2 ml of TED buffer. The homogenate was centrifuged at 29,000 x g at 4°C for 10 minutes to sediment nuclei and cell debris. The supernatant fraction was analyzed for protein by the dye-binding method of Bradford (18) and for ornithine decarboxylase activity by the method of Jänne and Williams-Ashman (19) using 1 ^{14}C ornithine. The supernatant fractions were incubated with labeled ornithine for one hour at 37° in a tube containing hyamine hydroxide in a center well and closed with a thin rubber cap. After an hour 0.5 ml of 1N sulfuric acid was injected through the rubber cap into the incubation mixture to drive any remaining labeled carbon dioxide over into the hyamine hydroxide. The hyamine hydroxide in the center well was transferred directly to scintillation vials containing omnifluor formulated scintillation fluid and counted. Each assay was carried out in triplicate using different amounts of the supernatant fraction to assure that the assay was linear.

S-adenosyl methionine decarboxylase activity was measured by the method of Feldman, Levy and Russell (20). BHK-21 cells were incubated three hours with or without DES and were recovered, washed, homogenized and centrifuged as in the measurement of ornithine decarboxylase. The supernatant fraction was incubated with ^{14}C-carboxy labeled S-adenosyl L-methionine in stoppered test tubes fitted with a center well containing hyamine hydroxide to absorb the labeled carbon dioxide released in the reaction.

RESULTS

The ornithine decarboxylase activity of primary cell cultures of kidney from both control and DES-implanted hamsters was very low and no response to DES added to the medium was observed. Cultures of BHK-21 cells were incubated with and without DES for periods ranging from 1 hour to 6 days. The ornithine decarboxylase activity begins to increase within 60 minutes of the addition of diethylstilbestrol and reaches its peak at three hours (Figure 1). The ODC activity had returned nearly to normal by 24 hours despite the continued presence of DES in the medium. The three hour incubation period proved to be optimal and was used in all subsequent experiments. Another series of experiments demonstrated that a concentration of 10^{-9} M DES was optimal and that greater concentrations such as 10^{-7} M produced a lesser stimulation of ornithine decarboxylase activity than 10^{-9} M (Figure 2).

The stimulation of ornithine decarboxylase by added DES was inhibited in a dose-related fashion by cycloheximide added to the incubation medium at

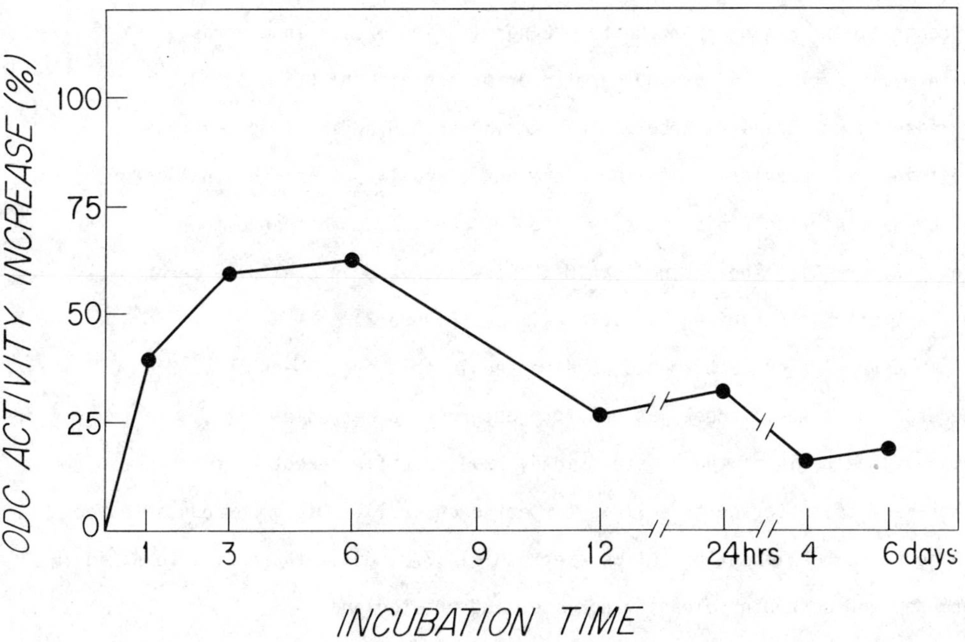

Figure 1. Ornithine decarboxylase activity in BHK-21 cells in culture as a function of the time after the addition of diethylstilbestrol, 10^{-9} M, to serum-free Dulbecco's modified Eagle's medium.

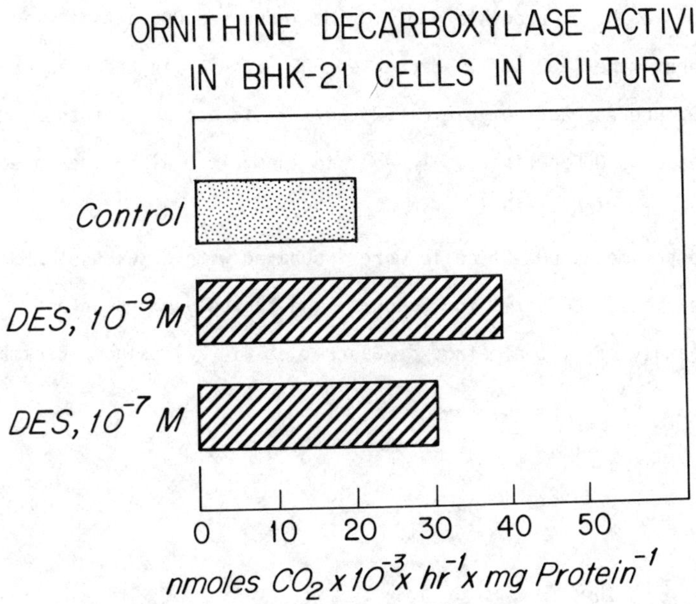

Figure 2. Ornithine decarboxylase activity in BHK-21 cells in culture three hours after the addition of diethylstilbestrol. Each value is the mean of nine observations.

concentrations ranging from 100 to 500 ng/ml (Figure 3). In contrast, actinomycin D, 100, 250 or 500 ng/ml did not inhibit the DES induced increase in ornithine decarboxylase. Actinomycin D added alone increased ornithine decarboxylase activity in the BHK-21 cells but the combination of DES and actinomycin D did not stimulate ODC activity more than either one alone (Figure 4). The actinomycin D-induced increase in ornithine decarboxylase activity is inhibited by cycloheximide (Figure 5).

Several steroid hormones were tested in this system and at 10^{-9} M the order of effectiveness was DES progesterone, > estradiol 17β, > testosterone. The cells of the hamster kidney have specific receptors for progesterone and testosterone as well as for estrogens (21). Other compounds tested and found to be without effect were estradiol 17α, dexamethasone, luteinizing hormone, gonadotropin releasing hormone, and prostaglandin $F_{2\alpha}$.

A cell line (L929) derived from mouse thigh muscle was similarly exposed in culture for three hours to DES, progesterone, actinomycin D or cycloheximide. The results paralleled those for the BHK-21 cell line (Figure 6). Diethylstilbestrol 10^{-9}M, or progesterone, 10^{-9}M, increased ODC activity in the cells. Cycloheximide (500 ng/ml) decreased ODC activity in the presence or absence of DES, whereas actinomycin D (500 ng/ml) did not inhibit the DES-induced increase in ODC activity. Indeed, actinomycin D alone increased ODC activity three- to eight-fold.

In other experiments BHK-21 cells were incubated with or without DES 10^{-9}M, or cycloheximide, 500 ng/ml for 3 hours, and the polyamine content of the cells was analyzed by the method of Gehrke, et al (22) using a Beckman

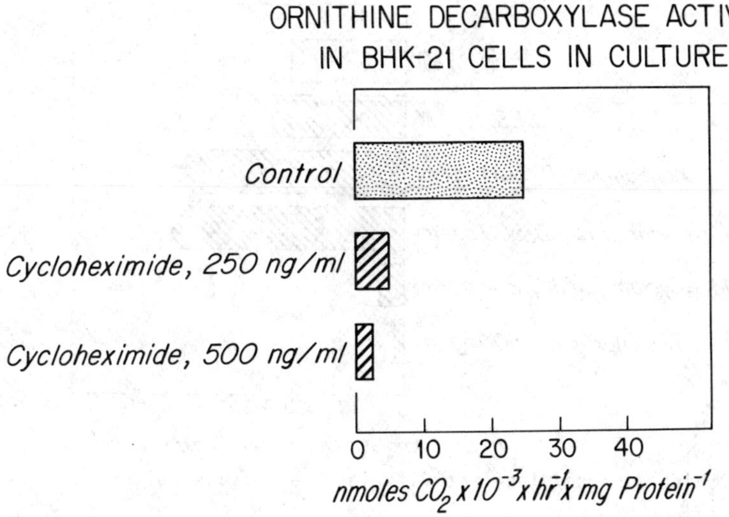

Figure 3. The effect of added cycloheximide on the activity of ornithine decarboxylase in BHK-21 cells in culture. Cells were grown in Dulbecco's modified Eagle's medium containing 10 per cent fetal calf serum and antibiotics until they formed a subconfluent monolayer. The medium was removed and replaced with serum-free modified Eagle's medium with or without cycloheximide and the cells were incubated for three hours. Each value is the mean of eight observations.

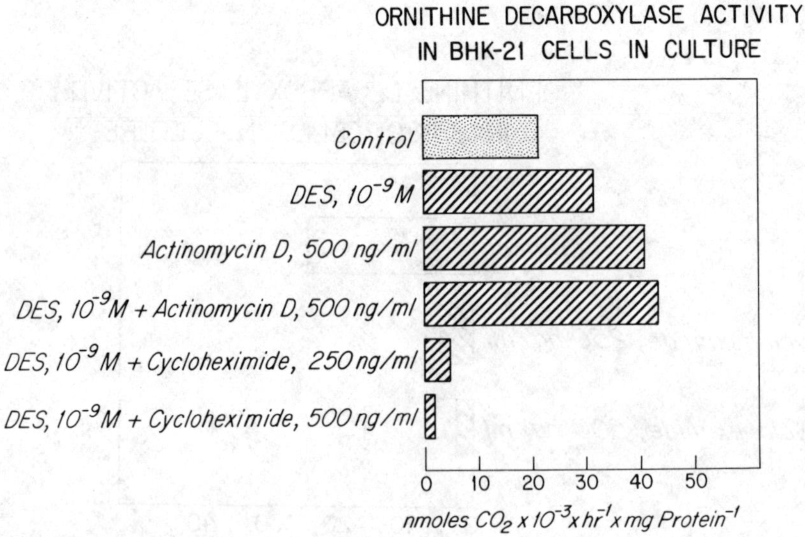

Figure 4. Ornithine decarboxylase activity in BHK-21 cells cultured in serum-free modified Eagle's medium three hours after the addition of diethylstilbestrol, actinomycin D, or cycloheximide. Each value is the mean of six to eight observations.

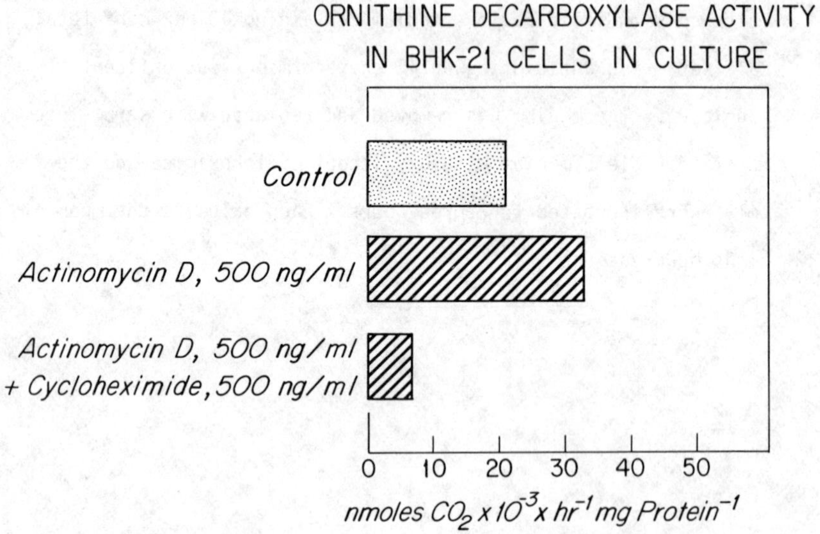

Figure 5. Inhibition of the actinomycin D stimulates increase in ornithine decarboxylase activity by added cycloheximide. The BHK-21 cells were harvested after being exposed for three hours in culture to the compounds. The incubation medium was serum-free modified Eagle's medium. Each value is the mean of nine observations.

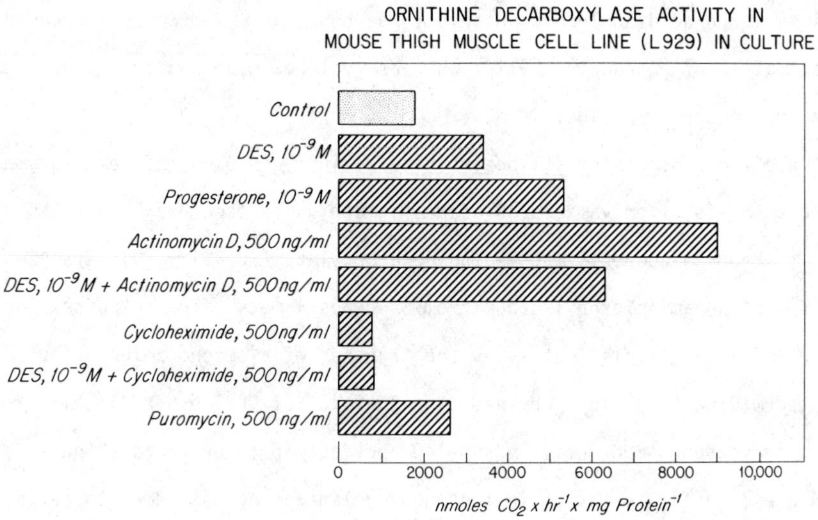

Figure 6. Effects of diethylstilbestrol, progesterone, actinomycin D and cycloheximide on ornithine decarboxylase activities on the mouse thigh muscle cell line, L929, in culture. The cells were grown in modified Eagle's medium containing 10 per cent fetal calf serum and antibiotics until a subconfluent monolayer was formed. The medium was removed and replaced with serum-free modified Eagle's medium, with or without the compounds to be tested, and the cells were incubated for three hours. Each value is the mean of six to ten observations.

Model 121 MB amino acid analyzer. Within three hours 10^{-9} M DES had increased the concentration of putrescine by 10% whereas cycloheximide had decreased it to 45% of the control value.

The effects of diethylstilbestrol and actinomycin D on the next enzyme in the biosynthesis of polyamines, S-adenosyl methionine decarboxylase (SAMDC) were also tested. S-adenosyl methionine decarboxylase (EC 4.1.1.50) catalyzes the transfer of an amino propyl group from S-adenosyl methionine to putrescine to form spermidine. It also catalyzes the transfer of a second amino propyl group to spermidine to yield spermine. The addition of DES to BHK-21 cells in culture increased SAMDC activity some 40 per cent in three hours (Figure 7). This increase in activity was inhibited when cycloheximide was added to cells in culture. In contrast, the addition of actinomycin D or puromycin to the cells in culture did not inhibit the DES induced increase in SAMDC activity, but increased it even more. In other experiments it was clear that cycloheximide inhibits the stimulation of ODC activity induced by actinomycin D as well as that induced by diethylstilbestrol. The specific inhibitor of RNA polymerase II, α-amanitin, when added to cultures of BHK-21 cells in amounts of 2.5 to 500 ng/ml, inhibited ODC activity and inhibited the increase in ODC activity produced by the addition of DES or actinomycin D (Figure 8).

The ornithine decarboxylase activity of a number of tissues (13,30) is increased by the addition of cyclic AMP or by the addition of methylated xanthines such as methyl isobutyl xanthine (MIX). These inhibit phosphodiesterases which convert cyclic AMP to adenosine monophosphate and thus increase the concentration of cyclic AMP within the cell. The ODC activity of the BHK-21 cells was increased in a dose-related fashion by methyl isobutyl

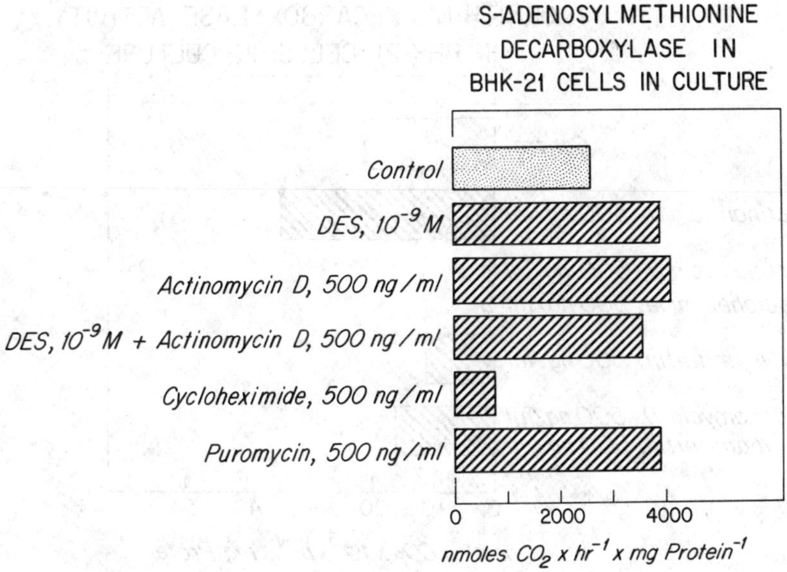

Figure 7. Effects of diethylstilbestrol, actinomycin, cycloheximide and puromycin on the S-adenosyl methionine decarboxylase activity of BHK-21 cells in culture. The cells were grown in modified Eagle's medium containing 10 per cent fetal calf serum and antibiotics until they formed a subconfluent monolayer. The medium was removed and replaced with serum-free modified Eagle's medium, with or without the compounds to be tested, and the cells were incubated for three hours. Each value is the mean of six to ten observations.

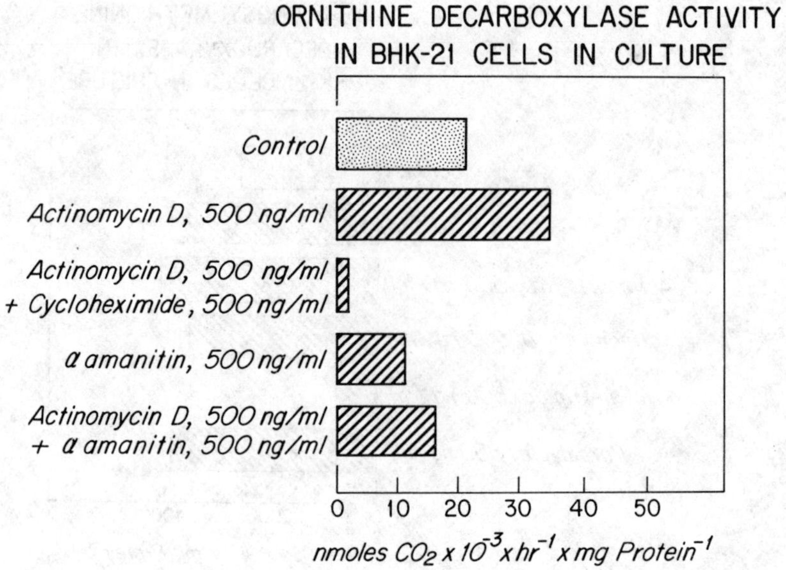

Figure 8. Effects of α amatin on the activity of ornithine decarboxylase in BHK-21 cells in culture. The BHK-21 cells were grown in Dulbecco's modified Eagle's medium containing 10 per cent fetal calf serum and antibiotics until they reached a subconfluent monolayer. The medium was removed and replaced with serum-free modified Eagle's medium with or without the components to be tested. The cells were incubated for three hours, harvested and ornithine decarboxylase activity was measured by the method of Janne and Williams-Ashman (19). Each value is the mean of eight to twelve observations.

xanthine (Figure 9) and by dibutyryl cyclic AMP. The increased ODC activity resulting from the added cyclic AMP or the added phosphodiesterase inhibitor was not decreased by added actinomycin D but was greatly decreased by added cycloheximide. Thus these experiments provide further evidence that the increased ODC activity involves protein synthesis but may not require the *de novo* synthesis of RNA at that time.

DISCUSSION

Both steroidal and synthetic estrogens, as well as progesterone and testosterone, increased the ornithine decarboxylase activity in cultures of baby hamster kidney cells. This increase was evident within one hour and reached its peak at three to six hours, after which the activity declined. The activation of ornithine decarboxylase is inhibited in a dose-related fashion by cycloheximide which suggests that the increased enzyme activity represents *de novo* synthesis of the enzyme rather than the activation of a preformed enzyme precursor.

The finding that actinomycin D did not inhibit the DES-induced increase in ornithine decarboxylase activity of hamster kidney cells in culture whereas cycloheximide did inhibit it was unexpected. The sample of actinomycin D used was shown to be effective in inhibiting RNA synthesis in rat decidual tissue. A sample of actinomycin D from another source was also shown not to inhibit the increased ODC activity in response to DES. This level of actinomycin D does inhibit the *de novo* synthesis of RNA in these cells, for it inhibits the incorporation of ^{3}H-uridine into RNA by BHK cells in culture to less than

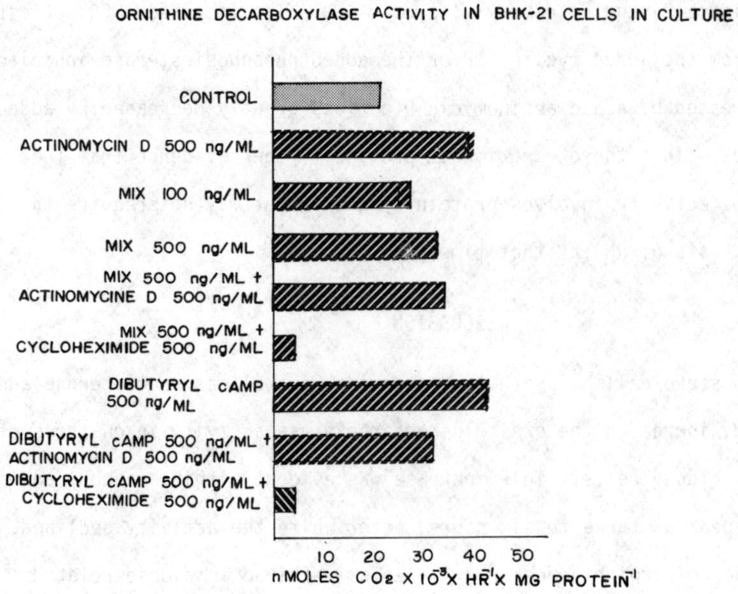

Figure 9. Effects of dibutyryl cyclic AMP and methyl isobutyl xanthine on the activity of ornithine decarboxylase in BHK-21 cells in culture. The BHK-21 cells were grown in Dulbecco's modified Eagle's medium containing 10 per cent fetal calf serum and antibiotices until they reached a subconfluent monolayer. The medium was removed and replaced with serum-free modified Eagle's medium with or without the components to be tested. The cells were incubated for three hours, harvested and ornithine decarboxylase activity was measured by the method of Janne and Williams-Ashman (19). Each value is the mean of six to ten observations.

30% of the value in control cells. Thus the DES-induced increase in ODC activity appears to involve protein synthesis, for it is inhibited by cycloheximide; however, it does not appear to involve RNA synthesis at that moment for it is not inhibited by actinomycin D. These findings suggest that the mRNA for ODC may be synthesized earlier and be retained in the cell in some inactive form until it is released by the stimulus provided by the estrogen or by actinomycin D. It then undergoes translation to produce the active enzyme. This hypothesis parallels the masked maternal messenger RNA hypothesis suggested for the mRNAs involved in coding for protein synthesis in early embryonic life in the sea urchin and other organisms (23,24). These are produced in the oocyte before fertilization and retained in an inactive form until fertilization occurs. They are then released and translated to produce the proteins synthesized early in development. The unusually rapid response of ODC activity to the estrogenic stimulus and the remarkably short half-life of the enzyme would be consistent with this theory

In a comparable system Goldstein, et al (25) found that infection of a primary mouse kidney cell culture with polyoma virus resulted in a biphasic increase in the activities of ornithine decarboxylase and S-adenosyl-methionine decarboxylase. The addition of actinomycin D, 10 μg/ml, increased ODC activity at two hours.

Ornithine decarboxylase has a number of unusual properties such as the very brief (10-20 min) half-life of the enzyme (26) and of its messenger

RNA (27) which suggests that it may play some unique role in growth processes. Manen and Russell (28) suggested that ornithine decarboxylase may be the labile protein with a short half-life that modulates the activity of RNA polymerase I. They could purify ornithine decarboxylase by using an RNA polymerase I affinity chromatography column and then showed that adding the purified ODC increased the initial rate of the RNA polymerase I reaction by increasing the rate of initiation of synthesis of RNA chains. If ODC were the initiation factor for RNA polymerase I and were present in the cell tightly bound to it, this might provide an explanation for our observation that actinomycin D produces an increase in activity of ODC in the kidney cells in culture. The actinomycin D could be visualized as interacting with the RNA polymerase I-ODC complex (which does not have ODC enzyme activity) and releasing the ODC from the polymerase so that it can exert its catalytic activity. However, since this mechanism does not involve protein synthesis it should not be inhibited by cycloheximide.

This unusual effect of actinomycin D may be restricted either to the type of cell in the BHK-21, L929 and mouse kidney cell cultures or to cells growing in culture. The observation (25) that actinomycin increased the ornithine decarboxylase activity of a primary mouse kidney cell culture lends support to the hypothesis that this response may require that the cells be growing in culture. With cells *in vivo* the response of testicular ornithine decarboxylase to FSH or LH (20) and the response of ovarian ornithine decarboxylase to LH (11) is inhibited by actinomycin D as one might expect. Our experiments showed that actinomycin increases ornithine decarboxylase activity but at the same time inhibits RNA synthesis as measured by the incorporation of uridine into RNA. The simplest hypothesis that explains this finding is that the mRNAs for ornithine decarboxylase and for S-adenosyl methionine decarboxylase are preformed and present in some bound inactive state in the cells in culture. The addition of diethylstilbestrol or of actinomycin D appears to free or activate those mRNAs so that they are translated to form the active enzyme. The inhibition of the incorporation of uridine into RNA by added actinomycin may reflect primarily an inhibition of synthesis of ribosomal RNA rather than mRNA, but an inhibition of rRNA synthesis would not account for the increased enzyme activity produced by the addition of actinomycin D. The increase in ornithine decarboxylase activity seems to involve some relatively nonspecific mechanism since it can be induced by diethylstilbestrol, by cyclic AMP and by inhibitors of phosphodiesterase, by actinomycin D or by puromycin. It does involve protein synthesis since it is inhibited by cycloheximide.

ACKNOWLEDGEMENT

I want to express my appreciation to my colleagues, Dr. Y.C. Lin and Miss Janet Loring, who participated in carrying out these experiments. These investigations were supported in part by a grant from the National Cancer Institute, CA24,615.

REFERENCES

1. Cohen, S.S. Introduction to the Polyamines, Prentice-Hall, Englewood Cliffs, N.J. (1971).
2. Russell, D.L. Polyamines in Normal and Neoplastic Growth, Raven Press, N.Y. (1973).
3. Williams-Ashman, H.G., A. Corti and A.R. Sheth. In Normal and Abnormal Growth of the Prostate (M. Gobel, ed.), pp. 222-239, C.C. Thomas, Springfield, Ill. (1975).
4. Pegg, A.E., D.H. Lockwood and H.G. Williams-Ashman. Biochem. J. 117:17 (1970).
5. Kaye, A.M., I. Icekson and H.R. Lindner. Biochem. Biophys. Acta 252:150 (1971).
6. Villee, C.A. and J.M. Loring. Adv. Enzyme Regulation 13:137 (1975).
7. Cohen, S., B.W. O'Malley and M. Stasny. Science 170:336 (1970).
8. Janne, J. and A. Raina. Biochem. Biophys. Acta 174:769 (1969).
9. Richman, R., C. Dobbins, S. Voina, L. Underwood, D. Mahaffee, H.J. Gitelman, J. VanWyk and R.L. Ney. J. Clin. Invest. 52:2007 (1973).
10. Kobayashi, Y., J. Kupelian and D.V. Maudsley. Science 172:379 (1971).
11. Kaye, A.M., I. Icekson, S.A. Lamprecht, R. Gruss, A. Tsafriri and H.R. Lindner. Biochemistry 12:3072 (1973).
12. Reddy, P.R.K. and C.A. Villee. Biochem. Biophys. Res. Comm. 65:1350 (1975).
13. Villee, C.A. and J.M. Loring. In Structure and Function of the Gonadotropins (K.W. McKerns, ed.), pp. 215-313, Plenum Publishing Co., New York (1978).

14. Haselbacher, G.F. and R.F. Humbel. Cell Physiol. 88:239 (1976).

15. Osterman, J. and J.M. Hammon. Endocrinol. 101:1335 (1977).

16. Klingensmith, M.R., A.F. Freifeld, A.E. Pegg and L.S. Jefferson. Endocrinol. 106:125 (1980).

17. Pantic, V., J. Li and C. Villee. Cytobiologie 9:89 (1974).

18. Bradford, M. Anal. Biochem. 72:248 (1976).

19. Janne, J. and H.G. Williams-Ashman. J. Biol. Chem. 246:1725 (1971).

20. Feldman, M.J., C.C. Levy and D.H. Russell. Biochemistry 11:671 (1972).

21. Li, S.A., J.J. Li and C.A. Villee. Ann. N.Y. Acad. Sci. 286:369 (1977).

22. Gehrke, C.W., K.C. Kuo and R.L. Ellis. J. Chromatography 143:345 (1977).

23. Gross, P.R. and G.N. Cousineau. Exper. Cell Res. 33:368 (1964).

24. Spirin, A.S. Curr. Topics Devel. Biol. 1:1 (1966).

25. Goldstein, D.A., O. Heby and L.J. Marton. Proc. Nat. Acad. Sci. 73:4022 (1976).

26. Russell, D.H. and S.H. Snyder. Mol. Pharmacol. 5:253 (1969).

27. Villee, C.A. and J.M. Loring. In The Role of RNA in Reproduction and Development (M.C. Niu and S.J. Segal, eds.), pp. 167-182, American Elsevier, N.Y., (1973).

28. Manen, C.A. and D.H. Russell. Science 195:505 (1977).

29. Villee, C.A. and J.M. Loring. In Reproductive Processes and Contraception (K. McKerns, ed.), Plenum Publishing Co., New York (1980).

30. Manen, C.A. and D.H. Russell. Life Sciences 17:1769 (1975).

INDUCTION OF A LYMPHOBLASTOID CELL LINE WITH IMMUNE RNA

D. Viza

Laboratorire d'Immunobiologie, Faculté de Médecine, 15, rue de l'Ecole de Médecine, 75006 Paris, France.

Y. H. Pilch

Surgical Oncology Service, Department of Surgery, University of California, San Diego, School of Medicine, 225 W. Dickinson Street, San Diego, CA 92103, U. S. A.

G. Pizza

Ospedale M. Malpighi, Via P. Palagi 9, Bologna, Italy.

Abstract Immune RNA extracted from the lymphoid organs of immunized sheep (Is-RNA) was used for the induction of the LDV/7 human lymphoblastoid cell line. RNA extracted from the induced cell line (Ic-RNA) carries the same immunological information as the inducing sheep RNA extract. The Ic-RNA is capable of converting "naive" human lymphocytes to cytotoxic effector cells to the same extent as the Is-RNA preparation used for the induction of the LDV/7 cell line. The Is-RNA information is incorporated into LDV/7 cells in a stable fashion and it can be recovered from the lymphoblastoid cells several weeks after induction. It is therefore suggested that Is-RNA information can be replicated by a xenogeneic cell line

---

Dedicated to the memory of Professor G. Bilski-Pasquier, in appreciation of the stimulus and generous help he has provided to the first author.

during its replication.

The incubation of the cell line with Is-RNA induces the appearance of antigen receptors (AR) on the surface of the LDV/7 cells, which are specific for the antigens used to immunize the Is-RNA donor sheep. The specificity of the AR was shown by several methods: immunofluorescence, immunoperoxidase, immune cytoadherance.

The specificity of some of the gamma-globulins spontaneously produced by the LDV/7 cell line was changed following incubation with Is-RNA and specific IgG for antigens specified by the Is-RNA were found. Thus, it seems that Is-RNA information can be expressed by human lymphoblastoid cells.

The <u>in vitro</u> induction system presented here could be used for the study of the incorporation of information carried by xenogeneic immune RNA and it should provide insight into the mechanisms controlling the ingress and processing of immunological information carried by I-RNA.

The idea of using RNA to transfer information to living cells and organisms is not novel. Several observations suggested that syngeneic or xenogeneic RNA can be incorporated into living cells and induce the synthesis of new proteins and M.C. Niu was one of the first to investigate this phenommenon (5,10, 22-28, 36).

Fishman's experiments, reported in the early 1960's (7,8), introduced the concept of Immune RNA (I-RNA). It was shown that RNA extracts from rat macrophages which had been incubated, in vitro, with T2 phage could transfer anti-T2 sensitivity to unsensitized rat lymphocytes. Furthermore, I-RNA obtained from immune macrophages was capable of transferring to spleen cells the information for synthesizing anti-T2 immunoglobulins.(1)

These observations established that it is possible to transfer immunological information from immunocompetent or "informed" cells to naive cells, <u>in vitro</u>, or to normal, non-immune animals, <u>in vivo</u> by means of RNA extracts obtained from lymphoid tissues (spleen, lymph nodes, peritoneal exudate cells) of animals that have been previously sensitized to a particular antigen. (For a review on I-RNA see

Ref. 27, also Ref. 6 and 9).

Since the first pioneering work of the 1960's, numerous experiments confirming or invalidating the original observation have been performed. However, as is usually the case in science, new controversial observations will be followed, unless they are outrightly rejected, by those who will appraise them, copy them, and confirm them until they become fashionable. The second wave of criticism leading to rejection usually follows a few years later, unless sheer oblivion shrouds the discovery, for objective criticism is far from being the essential quality of scientists. Rather, passionate reactions govern our professional relations underlined by fierce competition for prestige, grants and publications.

The story of RNA as a vehicule for transferring biological information in general and immune information in particular has not been an exception to this typical scenario. After the fascination created by the initial observations the criticisms mounted. The main objection concerning the I-RNA has been that RNA is not responsible for the transfer, but rather it acts as a "superimmunogen", the antigen being wrapped up by the nucleic acid, this presentation antigen/nucleic acid being far more potent immunogen than the antigen alone. As is usually the case in science, and elsewhere, truth can be found in contradicting theses. So although the "superimmunogen" complex is a reality, it is well accepted today by the unbiased scientists that I-RNA can transfer immunological information without the aid of an antigenic moiety.

The experiments we are reporting here provide evidence that I-RNA information can be incorporated into an established lymphoblastoid cell line in a stable fashion, and that this results in a "reprogramming" of the expression of the immunological information of the cells. I-RNA may induce new specific membrane receptors and change the specificity of the gamma-globulins secreted by the cell. It should be emphasized that the I-RNA information incorporated into the cell line can be retrieved after several cell divisions by re-extracting RNA.

REPLICATION OF I-RNA INFORMATION BY THE LDV/7 LYMPHOBLASTOID CELL LINE

A few years ago, we demonstrated that immunological information carried by transfer factor (TFd) (a dialysable extract obtained from immune lymphocytes) could be used to induce the lymphoblastoid cell line LDV/7 (41,42). A transfer factor of the same immunological specificity could be retrieved from the induced cells after several weeks of continuous growth. It was thus possible to replicate this immunological active informational molecule in tissue culture.

The induction procedure has been described previously (41). Briefly, transfer factor is obtained according to the method of Lawrence by disrupting immune leucocytes and dialysing the disrupted cells against distilled water (for a review article on transfer factor see Ref. 20). The informational activity is recovered in the dialysate, which is subsequently freeze-dried and aliquoted in RPMI 1640 medium. The lymphoblastoid cell line LDV/7 was established in our Paris laboratory from the peripheral blood lymphocytes of a healthy 65-year-old male blood donor. TFd obtained from 5×10^7 peripheral blood lymphocytes is usually incubated with 5×10^7 LDV/7 cells. The cells are grown in RPMI 1640 medium supplemented with 10% of foetal calf serum until a total number of $10^9 - 10^{10}$ cells is obtained. LDV/7 cells are kept at a concentration of $3 - 5 \times 10^5$ cells per ml. Since they divide every 30 hours, fresh medium is added every other day, so that the cell number does not exceed 5×10^5 cells per ml.

The dialysates obtained from the induced lymphoblastoid cells (TFdl) carry the same immunological information as the inducing TFd, and have been used extensively in our clinical department for the treatment of patients suffering from transitional-cell carcinoma of the bladder, hypernephroma, and chronic active hepatitis (32-35). All patients were monitored for several immunological parameters, e. g. E rosette formation, PHA and CON-A stimulation and leucocyte migration inhibition factor secretion in presence of the corresponding tumour antigens. The activity of the TFdL produced *in vitro* has been convincingly shown both by in vitro tests and clinical efficiency.

These observations prompted us to investigate whether I-RNA could be replicated in a similar fashion by the LDV/7 cell line. Transfer factor and I-RNA share some similar functional properties. They both can convert "naive" lymphocytes to specific immunologic activity, *in vitro* and *in vivo*, and thus induce the appearance of new clones. It is therefore plausible to assume that the "informed" lymphocytes continue to carry the I-RNA information after several cell divisions. Lymphoblastoid cell lines are a good candidate for the replication of I-RNA information during their own replication, providing of course they are capable of incorporating it in the first instance.

The induction of LDV/7 cells with I-RNA has already been described (43). I-RNA was obtained from the spleens and lymph nodes of sheep immunized with human melanoma or colon carcinoma cells or with Keyhole Limpet Haemocyanine (KLH). It was extracted in San Diego, code-labelled and shipped to Paris in dry ice. All experiments were carried out in double blind fashion. 5×10^7 LDV/7 cells were incubated with 1 mg of I-RNA in 1 ml of RPMI 1640 medium without serum for 1 to 2 hours. The cells were subsequently grown at 5×10^5 cells per ml in RPMI 1640 medium containing 10% of foetal calf serum for several weeks until the desired number of cells was obtained (usually 2×10^9). At the end of the culture period, the cells were pelleted and kept frozen at -20°C until they were shipped to San Diego for RNA extraction.

Approximately 1.5 mg of RNA is obtained from 10^9 cells. The final RNA preparations obtained from induced or uninduced LDV/7 cells contained 7.6—8.9 mg/ml of RNA contaminated by 0.5 - 1.6% DNA and 0.8 - 2.92% protein. All the RNA preparations were tested on sucrose-density-gradients and gave a nondegraded profile (30). (The sucrose gradient profile of I-RNA allows one to estimate the degradation which may have occurred during extraction and which may affect biological activity. A degraded profile shows an increase in the low molecular weight 4S peak.)

The biological activity of the I-RNA produced in tissue culture was compared to the biological activity of the inducing Is-RNA preparation in an *in vitro* microcytotoxicity test. This assay measures the conversion of allogeneic naive peripheral blood lymphocytes obtained from healthy blood donors to cytotoxic effector cells after incubation with a specific I-RNA (12, 15, 39).

Cultured human melanoma and colon carcinoma cells were used as target cells in these experiments. They were labelled with ^{125}IUdR and subsequently plated at 2×10^3 cells per 0.1 ml in Falcon Microtest II plates. Three hours later, I-RNA incubated or control, unincubated "naive" lymphocytes are added to the various wells in 0.2 ml of medium at a lymphocyte/target cell ratio of 200:1 and incubated for 48 hours at 37°C. At the end of this period, the wells are washed and the number of viable tumour cells adhering to wells is evaluated by measuring the radioactivity of each well in a gamma-counter. Six replicate wells are set for each lymphocyte preparation. The mean and standard errors are calculated for each group of wells and the cytotoxicity index (CI) is determined by the following formula:

$$CI = \frac{(\text{cpm no RNA}) - (\text{cpm RNA})}{(\text{cpm no RNA})}$$

where cpm no RNA is the cpm remaining from tumour cells exposed to lymphocytes not incubated with I-RNA, and cpm RNA is the cpm remaining from tumour cells exposed to lymphocytes incubated with I-RNA.

The results obtained for the various groups were statistically compared using the Student's t-test for unpaired data. In eight experiments, it was shown that I-RNA obtained from LDV/7 cells after induction with Is-RNA was capable of inducing "naive" allogeneic lymphocytes to become significant ($P < 0.001$) cytotoxic for human tumour target cells (melanoma or colon carcinoma).

Figs. 1 and 2 show the results from four experiments. It is worth emphasising that an experiment is considered positive if the CI obtained with lymphocytes incubated with post-induction Immune RNA (Ic-RNA) is not different from the CI obtained with lymphocytes incubated with the sheep Immune RNA (Is-RNA) used for inducing the LDV/7 cells, providing, obviously, that CIs obtained after incubating lymphocytes with Is-RNA and Ic-RNA are significantly higher than the CIs obtained with unincubated lymphocytes or with lymphocytes incubated with control RNAs obtained either from unimmunized animals or from uninduced cells.

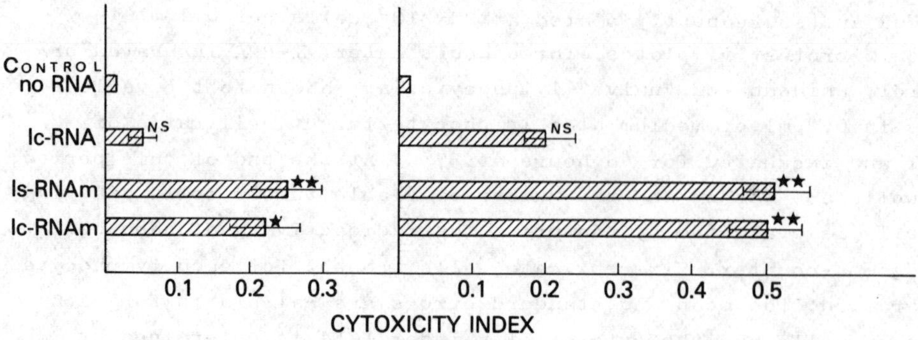

CONTROL
no RNA
Ic-RNA
Is-RNAm
Ic-RNAm
NS
NS
0.1
0.2
0.3
0.1
0.2
0.3
0.4
0.5
CYTOXICITY INDEX

Fig 1

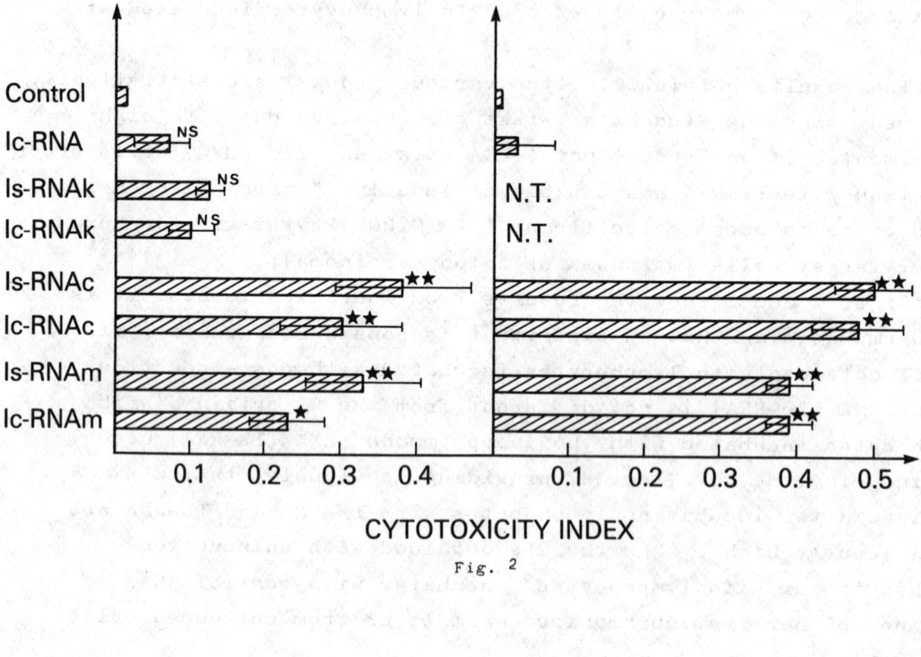

Control
Ic-RNA
Is-RNAk
Ic-RNAk
Is-RNAc
Ic-RNAc
Is-RNAm
Ic-RNAm
NS
NS
NS
N.T.
N.T.
0.1
0.2
0.3
0.4
0.1
0.2
0.3
0.4
0.5
CYTOTOXICITY INDEX

Fig. 2

Figs. 1 and 2

Four experiments in which the cytotoxicity for ED-H melanoma target cells of aliquots of allogeneic human lymphocytes from the peripheral blood of a healthy donor following incubation with various Is-RNAs and their corresponding Ic-RNAs was determined . 5 x 10^7 lymphocytes were incubated with Is-RNA extracted from sheep immunized with KLH (Is-RNAk), human colon carcinoma cells (Is-RNAc) human melanoma cells (Is-RNAm) and with the corresponding Ic-RNAs produced after induction of aliquots of LDV/7 cells. Ic-RNA from uninduced LDV/7 cells was used as a control. (NT = not tested). The term "allogeneic lymphocytes" is used in reference to the donor of the tumour cells used for the immunization. The melanoma target cells used for the cytotoxicty tests were also allogeneic with respect to the melanoma tumour donor used for the immunization of the sheep. The inductions of LDV/7 cells with Is-RNA were performed in Paris. The Ic-RNA extractions and cytotoxicity assays were performed in San Diego. All samples were shipped code-labelled and frozen in dry ice.

* : $P < 0.01$ ** : $P < 0.005$ NS = not significant

These results suggest that xenogeneic Is-RNA can transfer immunological information to human lymphoblastoid cells, and that this information can be replicated during cell division in culture. Furthermore, this information can be retrieved at the end of a culture period which may last several weeks.

It is worth pointing our that the antigen targets responsible for the cytotoxic activity induced by the various I-RNA preparations are not solely tumour associated antigens but also HL-A specificities. However, what is important in the context of these experiments is the similarity of specific activity between the Is-RNA and the corresponding Ic-RNA obtained from the LDV/7 cells.

The mechanisms by which the Is-RNA replicates in the cultured cells are not obvious. Experiments described below ruled out the possibility that the Is-RNA remained in messenger-form within the lymphoblastoid cells, a situation which may have presented some analogies with the one described in amphibian eggs where xenogeneic mRNA continues translation in the host-cell (18).

Kern et al (16) and Pilch and Nayak (unpublished) have shown that I-RNA from the lymphoid organs of guinea pigs immunized with mouse tumour cells induces the production of anti-tumour immunoglobulins when it is incubated with normal "naive" mouse spleen cells, in vitro. The immunoglobulins detected are only of mouse origin and no guinea pig idiotypes has ever been detected.

These observations prompted us to investigate the expression of specific antigen receptor sites on the surface of LDV/7 cells induced with Is-RNA as well as the production of specific immunoglobulins by LDV/7 cells following incubation with Is-RNA.

INDUCTION OF SPECIFIC MEMBRANE ANTIGEN RECEPTORS

It is well established that xenogeneic I-RNA is capable of informing lymphocytes in vitro to recognize antigens to which they have not previously been exposed (e.g. [9,28,29,34]). Since the LDV/7 cells appeared to incorporate Is-RNA information, it seemed important to search for the expression of new specific antigen receptors (AR) on their surface, following incubation with I-RNA. Several techniques were used to study this phenomenon and some of the preliminary results obtained have already been reported elsewhere (42).

Immune cytoadherence

Sheep red blood cells (SRBC) were mixed with an equal volume of 0.005% tannic acid solution and incubated for 10 minutes at 37°C. After washing in phosphate buffered saline (PBS), the cells were suspended in PBS at a 1:20 ratio (volume per volume). 1 ml of this suspension was mixed with 5 ml of PBS containing 1 mg/ml of KLH and incubated for 10 minutes at 37°C, in order to fix the antigen on the SRBC membranes. After washing, the antigen-coated SRBC were resuspended in 5 ml of PBS. Washed LDV/7 cells were mixed with antigen-coated SRBC and incubated for two hours at 37°C. Samples were taken at the end of the incubation period, placed between glass slides and cover slips, and counted under a microscope. LDV/7 cells to which more than 3 SRBC were attached were considered to be "rosetting" cells.

LDV/7 cells, induced with Is-RNAk (from the lymphoid organs of sheep immunized with KLH), showed a significantly higher number of rosetting cells, when they were incubated with SRBC coated with KLH, than did LDV/7 cells induced with I-RNA obtained from non-immunized animals, or with I-RNA from sheep immunized with an antigen different than the one used for coating the SRBC, e.g. complete Freund's adjuvant (CFA) (Is-RNAf). Some results are shown in the scattergram presented in Figure 3.

Immunofluorescence

KLH labelled with tetramethylrhodamineisothiocyante, isomer G, (RITC) was used to localize the KLH membrane receptors by direct immunofluorescence. 200 μg/ml of RITC labelled KLH was incubated with 6 x 10^5 LDV/7 cells for 30 minutes at 37°C. LDV/7 cells are incubated under the same conditions with unlabelled KLH. The latter was revealed by indirect immunofluorescence using a rabbit anti KLH antiserum conjugated to fluorescein isothiocyante, isomer I (FITC). A goat anti-rabbit FITC-labelled antiserum was used in the "sandwich" indirect technique to reveal unlabelled rabbit anti-KLH antibodies fixed on the surface of LDV/7 cells incubated with KLH.

LDV/7 cells induced with Is-RNA and incubated with RITC,

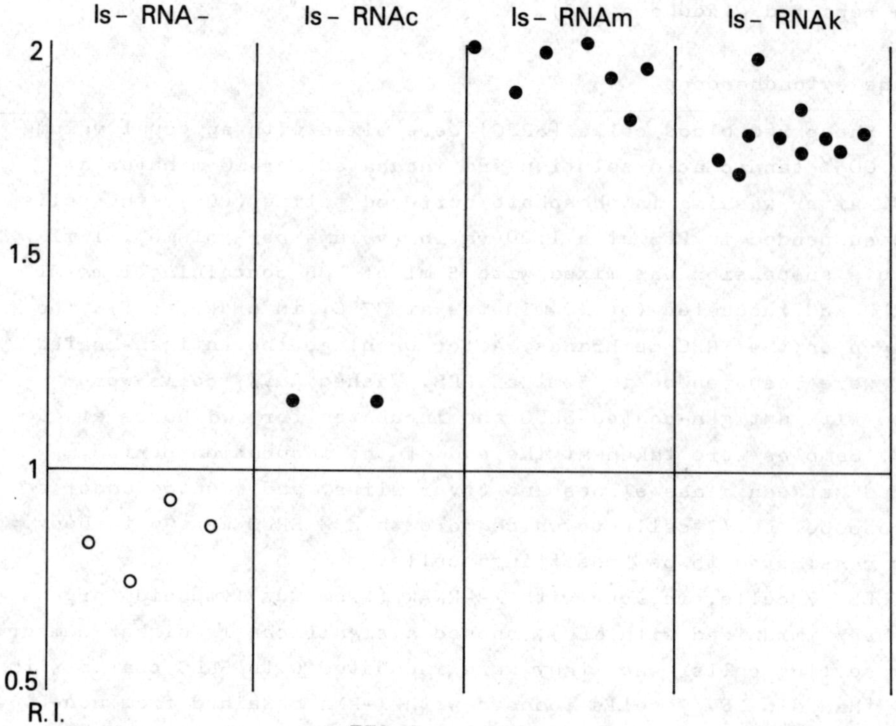

FIG 3

Fig. 3 Scattergram of Rosetting Indices (R.I.) obtained in several experiments with LDV/7 cells induced by Is-RNA. Four different RNA preparations were used for the inductions: Control RNA prepared from the lymphoid organs of an unimmunized sheep (Is-RNA-), RNA from a sheep immunized with KLH (Is-RNAk), RNA from a sheep immunized with human malignant melanoma tissues, KLH and Freund's complete adjuvant (CFA) (Is-RNAm), and RNA from a sheep immunized with human colon tumour cells, KLH and CFA (Is-RNAc).

The R. I. is calculated as follows:

$$\text{R.I.} = \frac{\text{Number of rosettes with the induced cells}}{\text{Number of rosettes with the non-induced cells}}$$

Both induced and uninduced cells were incubated with KLH coated SRBC. Note that Is-RNAc gives much lower R. I. (almost identical to controls) than does Is-RNAm or Is-RNAk. Variations in the induction efficiency of the different I-RNA preparations and/or in the effectiveness of various induction techniques using the same I-RNA preparation are not unusual in this system.

labelled KLH showed fluorescent staining in the direct test. Similarly, in the indirect test, unlabelled KLH was revealed by using either fluorescent rabbit anti-KLH antiserum, or by applying unlabelled rabbit anti-KLH antiserum followed by fluorescent goat anti-rabbit antiserum. (Tabel 1)

Cells induced with Is-RNAf from the lymphoid organs of a sheep immunized with CFA, or Is-RNA from an unimmunized sheep, did not show AR for KLH on their surface and remained unstained. Specific immune blocking (I.B.) confirmed the specificity of the staining (Table 1). Approximately 25 - 30% of the cells from the induced cultures fixed KLH on their surface as detected in immunofluorescence.

Peroxidase-diaminobenzidine (PO-DAB) (11)

10^6 LDV/7 cells, incubated in 1 ml of a solution of KLH (200 μg/ml), after thorough washing, were fixed with Karnovsky fixative (14) for 5 minutes and afterwards exposed to a rabbit anti-KLH antiserum. After further thorough washing, the cells were exposed to a goat anti-rabbit antiserum conjugated with horse radish peroxidase. The preparation was exposed to a saturated solution of 3-3' diaminobenzidine-base free (DAB) for 15 minutes at 20°C, and then fixed in Karnovsky fixative for 30 minutes. The pellet was then dehydrated and embedded in Epon 812 (21). Thin sections were obtained using a Reichert microtome and examined unstained, or lead-contrasted in Phillips 300 electron microscope.

LDV/7 cells induced with Is-RNAk and incubated with KLH and stained with PO-DAB showed, on electron microscopy, a thick dark line on the outer surface of their membranes, corresponding to the peroxidase reaction. Figure 4 shows a peroxidase positive cell next to a negative cell from the same culture. Approximately 30% of LDV/7 cells were found to be peroxidase positive following treatment with Is-RNAk.

INDUCTION OF SPECIFIC IMMUNOGLOBULIN FORMATION

Following induction with Is-RNAk, the LDV/7 cells were grown and supernatants from both induced and uninduced cultures were collected and, after 1000 fold concentration, tested directly

	Non-Induced Cells	Is-RNA Induced Cells
Direct staining with RITC labelled KLH	–	+
I.B. : pre-incubation with KLH followed by RITC labelled KLH	–	–
Indirect staining: incubation with KLH followed by FITC labelled rabbit anti-KLH antiserum	–	+ +
I.B. : pre-incubation with KLH, then unlabelled rabbit anti-KLH antiserum, followed by FITC labelled rabbit anti-KLH antiserum	–	–
Indirect staining: incubation with KLH, followed by rabbit anti-KLH antiserum and then followed by an FITC labelled goat anti-rabbit γ-globulin antiserum	–	+ + +

Table I

Immunofluorescence testing of AR for KLH on the membrane of uninduced LDV/7 cells and of LDV/7 cells induced with I-RNAk. The specificity of the reaction is demonstrated by immune blocking (I. B.).

The degree of fluorescent staining is expressed as the percentage of fluorescing cells as follows:

+ > 25%

++ > 40%

+++ > 50%

FITC Fluorescein isothiocyanate Isomer I

RITC Tetramethylrhodamine isothiocyanate Isomer G. 820(F)

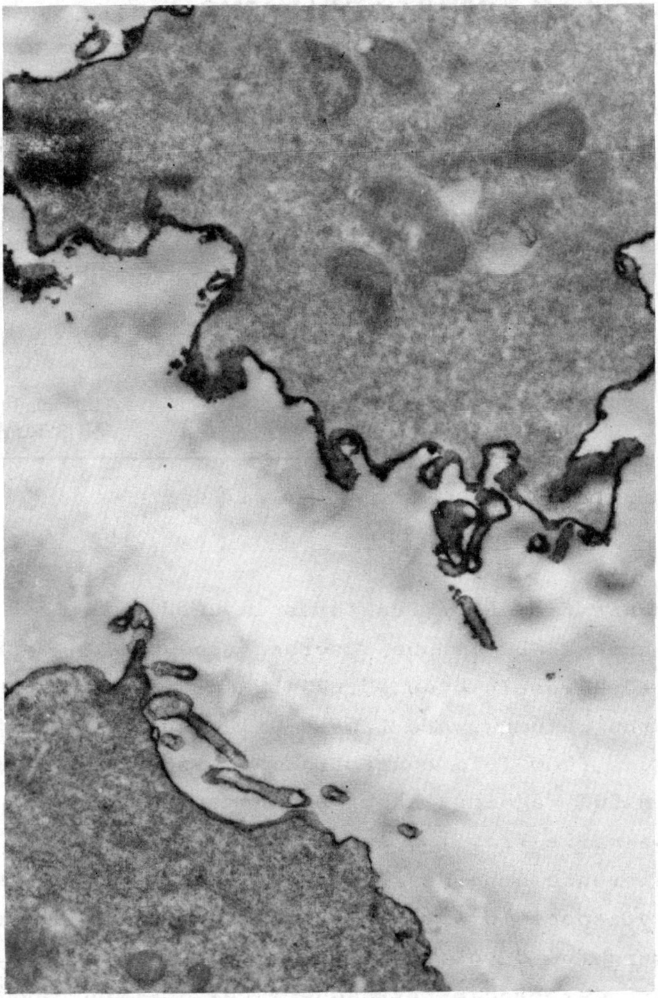

Fig. 4

Top: Immunoperoxidase reactivity on the surface of LDV/7 cells induced with Is-RNAk showing KLH fixation on the membrane. Approximately 30% of the cells display peroxidase reactivity as is shown on the upper cell.
Bottom: Non-reactive cell is shown at the bottom of the figure.

KLH AFFINITY COLUMNS

ADSORPTION OF SUPERNATANT FROM LDV/7 INDUCED WITH:	TESTED IN COE AGAINST: KLH	ANTI-IGG	ANTI-IGM	ROSETTING INDEX	ng OF γ-GLOBULINS 10^{-1} ml OF SUPERNATANT
Is-RNAk	+	+	+	1.40	6.25
Is-RNAkm	+	+	+	1.99	2.50
Is-RNAf	−	−	−	0.83	0
UNINDUCED LDV/7	−	−	−	1.00	0
RABBIT ANTI-KLH SERUM	+	NT	NT	1.39	3.90 mg/ml OF ANTISERUM

Table II

Supernatants from LDV/7 cultures induced with Is-RNA, were passed through KLH immune adsorbent columns made from CN activated sepharose 4 B on which KLH was attached. The eluates from the columns were freeze-dried and reconstituted in PBS so that a 1000 fold concentration was obtained. They were tested in COE against KLH, anti-human IgG and anti-human IgM. The cells from the same cultures were tested in immune cytoadherence and the R. I. showed a correlation with the presence of specific antibodies in the culture supernatants, i. e. R. I. were found positive in the cultures were specific antibodies were detected by cytoadherence. The approximate amount of specific antibody in the culture media is shown in the right column. The lymphocytes and the serum from a rabbit immunized with KLH were used as positive controls. Is-RNAf was obtained from a sheep immunized with CFA alone, whereas Is-RNAm was obtained from a sheep immunized with CFA, KLH and human melanoma tissue.

in cross-over electrophoresis (COE) against KLH. Furthermore, supernatants from induced and uninduced cells were filtered through KLH immunoadsorbent columns made from CN activated sepharose 4B (2,37) on which were attached either KLH or rabbit anti-human IgG gamma-globulins. Batches of carbodiimide beads (17) on which KLH was attached were also used for the removal of specific anti-KLH gamma globulin from the LDV/7 supernatants. Eluates from the columns and the carbodiimide beads, were concentrated 1000 fold and tested in COE and radial diffusion against KLH and against rabbit anti-human IgG antibodies.

The results obtained with antibodies, concentrated by affinity chromatography, are summarized in Table II. They clearly suggest that xenogeneic I-RNA can induce the production of specific IgG antibodies in human lymphoblastoid cells in culture.

The eluates were subsequently tested in radial diffusion and COE against goat anti-sheep IgG in an attempt to detect the presence of sheep IgG molecules (i.e. idiotypes of the same species as that of the I-RNA donor). In all cases, no trace of sheep immunoglobulin was detected.

DISCUSSION

The controversy concerning the mode of action of I-RNA and the putative role that traces of antigen may play in the transfer of immunity is not over. However, the work of Paque et al, extensively discussed elsewhere (27), has provided convincing evidence that minute amounts of antigen associated with I-RNA are not responsible for the immunological activity. The observations reported here show that xenogeneic I-RNA can transfer immunological information to cells from a lymphoblastoid cell line in a stable manner and that this information is replicated within the induced cells. These data tend to exclude the presence of antigenic material in the final Ic-RNA extract. Indeed the dilution factor in the tissue culture replication system is such that there is no possibility of the presence of a significant quantity of antigen in the Ic-RNA. Furthermore, incubation of the LDV/7 cells with various antigens - e.g. KLH or melanoma soluble antigen (44) - does not result in the production of either specific AR or specific IgG in the tissue culture medium. Thus there is no possibility of direct sensitization of the LDV/7 cells by the presence of an antigen in the tissue culture medium.

Several hypotheses could be proposed in order to explain the I-RNA replication observations. If one accepts that I-RNA is a messenger RNA (3), it is possible to imagine that this mRNA can be incorporated into lymphoblastoid cells and replicated in synchrony with the replication of host cells, while it translates for its specific product, e.g. immunoglobulin (27). This proposition would be consistent with other observations that mRNA can replicate and translate in a xenogeneic cell, namely in amphibian eggs (18). This proposition seems unlikely however, since the antigen specific gamma-globulins detected in the tissue culture supernatants of the induced cells were all of human origin, and no trace of sheep gamma-globulin idiotypes was detected.

It is worth mentioning here that the supernatants were all collected several days after induction. It is, therefore, possible that early sheep-gamma-globulin synthesis which might occur in this system, comparable to the one described by Niu (38), would have subsided by the time the supernatants were collected. This could account for the apparent discrepancy of our data with those of Adler et al (1), who reported the presence of I-RNA donor cell allotypes after I-RNA induction.

An RNA-dependent RNA-polymerase could be responsible for the maintenance of Is-RNA within the LDV/7 cells. Since this is rather unlikely, one may speculate on the presence of a reverse transcriptase which transfers the Is-RNA information into the DNA of LDV/7 cells. This could account for the stability of the I-RNA information within the cell line. Both these hypotheses have been discussed by Jachertz (13).

However, we rather favor the hypothesis that a derepression mechanism might be involved during induction of LDV/7 cells by Is-RNA. That mRNA may have such a derepression activity at the DNA level has been previously proposed by Niu (24,25). Other investigators have suggested that low molecular weight nuclear RNA may have a similar derepression activity (19). The observations of Brown & Weiss (4) showing that a heterokaryon obtained by the fusion of a rat-hepatoma cell with a mouse-lymphocyte produces mouse-liver enzymes may also be explained by the same derepression mechanism discussed here, i. e. rat mRNA would be derepressing the corresponding liver enzyme genes in the mouse nuclei. Similar derepression mechanisms are involved when a mouse uterus produces mouse albumin after incubation with bovine mRNA (38). The production of bovine albumin in this system subsides a few hours after the induction, whereas the mouse albumin synthesis, which starts later than the bovine albumin synthesis, continues for a much longer period. Several other observations showing the inductive properties of xenogeneic and/or allogeneic RNA extracts in various embryonic and/or tumour systems could be explained by specific derepression mechanisms (5,10,22).

The implications that these observations may have in future developments are of some importance. We live in the era of biological technology and the name of the present game is celled genetic engineering. It consists in the transfer of eucaryotic information to procaryotes aiming at its replication and translation in order to collect the final product. Our data suggest that it should be possible to reprogramme the expression of the genome of mammalian cells maintained in tissue culture. Human lymphoid cells in continuous culture could be the first example of such reprogramming. They could thus be induced to produce for example allogeneic immunoglobulins of desired specificity, a product which could find immediate use as diagnostic and/or therapeutic tool.

Acknowledgements: The work in Paris was supported by NATO (Grant No. 1242), the Ligue Nationale Française contre le Cancer, the ARBE and it was carried out despite the refusal of support by agencies who have decided a priori that this type of work was impossible.

REFERENCES

(1) Adler, V. L., Fishman, M., Dray, S. Antibody formation initiated in vitro III. Antibody formation and allotypic specificity directed by ribonucleic acid from peritoneal exudate cells. J. Immunol. 97, 554, 1966.

(2) Axen, R., Porath, J., Ernback, S. Chemical coupling of peptides and proteins to polysaccharides by means of cynogen halides. Nature 214, 1302, 1967.

(3) Bilello, P., Fishman, M., Koch, G. Evidence that immune RNA is messenger RNA. Cell. Immunol. 23, 309, 1976.

(4) Brown, J. E., Weiss, M. C. Activation of production of mouse liver enzymes in rat hepatoma-mouse lymphoid cell hybrids. Cell, 6, 481, 1975.

(5) Deshpande, A. K., Siddiqui, M. A. Q. Acetylcholinesterase differentiation during myogenesis in early chick embryonic cells caused by an inducer RNA. Differentiation, 10, 133, 1978.

(6) Immune RNA in Neoplasia, Fink, M. A. (ed.) London: Academic Press 1976.

(7) Fishman, M. Antibody formation in vitro. J. Expt. Med. 114, 837, 1961.

(8) Fishman, M., Adler, F. L. Antibody formation initiated in vitro. J. Exp. Med. 117, 595, 1963.

(9) RNA in the immune response. Friedman, H. (ed.) Vol. 207, Ann. N. Y. Acad. Sci. 1973.

(10) Galand, P., Ledoux, L. Uptake of Exogenous Ribonucleic Acid by Ascites Tumour Cells. 43, 391, 1966.

(11) Graham, R. C. and Karnovsky, M. J.: The early stages of absorption of injected horseradish peroxidase in the proximal tubules of mouse kidney: Ultrastructural cytochemistry by a new technique. J. Histochem. Cytochem., 14, 291, 1966.

(12) Han, T., Takita, H., Marabella, P. C., Mittelman, A. In vitro transfer of tumour specific immunity with human immune RNA. Clin. Exp. Immunol. 19, 219, 1975.

(13) Jachertz, D. Flow of information and gene activation during antibody synthesis. In: RNA in the Tumor Response. Friedman, H. (ed.) Vol. 207, 122. Ann. N. Y. Acad. Sci. 1973.

(14) Karnovsky, M. J.: A formaldehyde glutaraldehyde fixative of high osmolarity for use in electron microscopy. J. Cell. Biol. 27, 137A, 1965.

(15) Kern, D. H., Fritze, D., Drogemuller, C. R., Pilch, Y. H. Mediation of cytotoxic immune responses against human tumor-associated antigens by xenogeneic immune RNA. J. Natl. Cancer Inst. 57, 97, 1976.

(16) Kern, D. H., Pilch, Y. H. Antitumor antibody synthesis in vitro induced in mouse spleen cells by xenogeneic immune RNA. J. Natl. Cancer Inst. 60, 599, 1978.

(17) Khorana, H. G. The use of dicyclohexylcarbodiimide in the synthesis of peptides. Chem. Ind. London. 1087-1088, 1955.

(18) Knowland, J. S., Gurdon, J. B., Laskey, R. A. Injection of messenger RNA into living cells and its application to the study of gene action in Xenopus laevis. In: The Role of RNA in Reproduction and Development. Niu, M. C., Segal, S. J. (eds.) p. 110, Amsterdam, North Holland, 1972.

(19) Krause, M. O., Ringuette, M. Low molecular weight nuclear RNA from SV40-transformed W138 cells; effect on transcription of W138 chromatin in vitro. Biochem. Biophys. Res. Comm. 76, 796, 1977.

(20) Lawrence, H. S. Transfer Factor in Cellular Immunity. The Harvey Lectures. Series 68, 239, 1974.

(21) Luft, J. H.: Improvements in epoxy resin embedding methods. J. Biophys. Biochem. Cytol., 9, 409, 1961.

(22) McLean, M. J., Renaud, J. F., Sperelakis, N. Cardiac-like action potentials recorded from spontaneously-contracting structures induced in post-nodal pieces of chick blastoderm exposed to an RNA enriched fraction from adult heart. Differentiation, 11, 13, 1978.

(23) Niu, M, C. Effects of Ribonucleic Acid on Mouse Ascites Cells. Science, 131, 1321, 1960.

(24) Niu, M. C., Cordova, C. C., Niu, L. C., Radbill, C. L. RNA induced Biosynthesis of Specific Enzymes. Proc. Nat. Acad. Sci. 48, 1964, 1962.

(25) Niu, M. C. Mode of action of the exogenous ribonucleic acid in cell function. Nat. Canc. Inst. Monog. 13, 167, 1964.

(26) Niu, M. C., Niu, L. C., Guha, A. The entrance of exogenous RNA into the mouse ascites cell. Proc. Soc. Exp. Biol. and

Med. 128 550, 1968.

(27) Paque, R. E. Methods for isolation and assessment of RNA mediating cellular sensitivity to tumor and nontumor antigens. Methods in Canc. Res. 15,279, 1978.

(28) Pilch, Y. H., Ramming, K. P. Transfer of tumor immunity with ribonucleic acid. Cancer, 26, 630, 1970.

(29) Pilch, Y. H., Ramming, K. P., Deckers, P. J. Induction of anti-cancer immunity with RNA. N. Y. Acad. Sci. 207, 409, 1973.

(30) Pilch, Y. H., Ramming, K. P., Deckers, P. J. Studies in mediation of tumor immunity with 'Immune' RNA. In: Methods in Cancer Research. H. Busch (ed.) Vol. IX, p. 195. New York Academic Press 1973.

(31) Pilch, Y. H., Fritze, D., De Kernion, J. B., Ramming, K. P., Kern, D. K. Immunotherapy of cancer with immune RNA in animal models and cancer patients. N. Y. Acad. Sci. 277, 592, 1976.

(32) Pizza, G., Viza, D., Boucheix, Cl. and Corrado, F. 'In Vitro' production of a transfer factor specific for transitional cell carcinoma of the bladder. Br. J. Cancer 33, 606, 1976.

(33) Pizza, G., Viza, D., Boucheix, Cl. and Corrado, F. Studies with 'in vitro' produced transfer factor. In: Transfer Factor, M. S. Ascher, A. A. Gottlieb and L. H. Kirkpatrick (eds.) Academic Press pp. 173-184, 1976.

(34) Pizza, G., Viza, D., Boucheix, Cl. and Corrado, F. Effect of 'in vitro' produced transfer factor on the immune responses of cancer patients. Eur. J. Cancer 13, 917, 1977.

(35) Pizza, G., Viza, D., Roda, A., Aldini, R., Roda, E., Barbara, L. Transfer Factor for the treatment of chronic active hepatitis. New Eng. J. Medicine, 23, 1332, 1979.

(36) Sanyal, S., Niu, M. C. Effects of RNA on the developmental potentiality of the posterior primitive streak of the chick blastoderm. Proc. Nat. Acad. Sci. 55, 743, 1966.

(37) Sheenan, J. C., Hess, G. P. A new method for forming peptide bonds. J. Amer. Chem. Soc. 77, 1067, 1955.

(38) Shih-Fang Yang, Niu, M. C. Albumin synthesis in mouse uterus in response to liver mRNA. Proc. Nat. Acad. Sci. 74, 1894, 1979.

(39) Veltman, L. L., Kern, D. H., Pilch, Y. H. Immune cytolysis of human tumor cells mediated by xenogeneic 'immune' RNA. Cell. Immunol. 13, 367, 1974.

(40) Viza, D., Goust J-M., Moulias, R. and Muller-Berat, Nicole. 'In Vitro' production of transfer factor. Scandinavian J. Immunol., 3, 892, 1974.

(41) Viza, D., Goust, J-M., Moulias, R., Trejdosiewicz, L. K., Collard, Avril and Muller-Berat, Nicole. 'In Vitro' production of transfer factor by lymphoblastoid cell lines. Transplant. Proc., VII, n°1, supplement 1, 329, 1975.

(42) Viza, D., Boucheix, Cl., Phillips, J., Cesarini, J-P., phillips, T. M., Lewis, M. G. and Pilch, Y. Replication de l'ARN-Immun par des cellules lymphoblastoides humaines. C. R. Acad. Sc. Paris, 284, 2565, 1977.

(43) Viza, D., Boucheix, Cl., Kern, D. H., Pilch, Y. Replication of immune information from xenogeneic RNA by human lymphoblastoid cells in culture. Differentiation, 11, 181, 1978.

(44) Viza, D. and Phillips J. Identification of an antigen associated with malignant melanoma. Int. J. Cancer, 16, 312 317, 1975.

IMMUNOTHERAPY OF CANCER WITH IMMUNE RNA

Bosco Shang Wang, Glenn D. Steele Jr., Jerome P. Richie,
Peter J. Deckers and John A. Mannick

Department of Surgery, Peter Bent Brigham Hospital,
Harvard Medical School, Boston, Massachusetts 02115
and Department of Surgery, Boston University School
of Medicine, Boston, Massachusetts 02118.

Abstract

Transfer of immune responsiveness with I-RNA has been demonstrated in a variety of experimental systems. Although immunoprophylaxis with I-RNA is successful in some animal tumor systems, therapeutic attempts with I-RNA to control an already established tumor are disappointing. In order to evalute the effect of I-RNA therapy in cancer, we will present data from our previous animal works and also from a Phase I clinical trial in this communication.

We first attempted to augment hosts immune defenses with I-RNA against a primary syngeneic sarcoma in C3H/He mice. Results showed that mice treated with I-RNA had a decresed incidence of early tumor development. However, this effect appeared to be transient and all treated animals eventually died. This delay of tumor development with I-RNA was sensitive to the treatment with RNAse and also tumor-specific. We subsequently attempted to prevent a metastases from developing by I-RNA therapy after the primary tumor was surgically removed. Using a metastatic B16 melanoma in C57BL/6 mice, we significantly improved the survival rate of mice whose primary B16 were previously excised with I-RNA. This therapeutic effect was again RNAse sensitive and tumor specific. We finally undertook a Phase I clinical trial and 6 patients with far advanced metastatic renal cell carcinoma were treated with I-RNA. One patient had a complete remission by 6 months after therapy and remained without evidence of disease until 18 months. Two patients showed more than a 50% decrease in the size of their pulmonary metastases for 8 to 10 months. Two additional patients had a temporary stabilization in the growth of their metastases after therapy. One patient with brain metastases showed no discernible change in his tumor growth. Results from our study justify a randomized and prospective clinical trail with I-RNA.

During the past several years, our laboratory has been engaged in investigating the transfer of tumor-specific immunity with immune RNA (I-RNA) in an attempt to define the potential clinical applicability of I-RNA therapy to the treatment of human malignancy. In this communication, we will review some of our previous work in which I-RNA was used for treating animal tumors. Based upon the knowledge accumulated from animal studies, we have proposed and completed a Phase I clinical trial with I-RNA.

Mannick and Egdahl first reported from this laboratory that rabbit skin allografts were rejected in a second set fashion when test animals were previously treated with I-RNA from immunized donor rabbits (1). This transfer of transplantation immunity with I-RNA was found to be sensitive to treatment with RNase and also showed antigenic specificity. In 1971, Ramming and Pilch adopted this model to the transfer of tumor specific immunity with I-RNA (2). These authors immunized guinea pigs with a murine tumor and I-RNA was prepared from the lymphoid tissues of these guinea pigs. They incubated normal mouse lymphocytes with the I-RNA in vitro and then injected these cells into normal mice to whom a lethal dose of the tumor used for sensitizing guinea pigs was simultaneously given. A significant inhibition of the growth of tumor isografts was achieved in these treated animals. This sort of immunoprophylaxis experiment with I-RNA has been repeatedly shown to be successful. However, this experimental model is not entirely comparable to the clinical situation in which immunotherapy might be employed.

We therefore performed a series of experiments in an attempt to augment hosts defenses with I-RNA against an already established tumor isograft[3,4]. A BP-8 murine sarcoma, chemically induced by 3,4-benz(a)pyrene in C3H/HeN mice, was used. I-RNA was extracted from the lymphoid tissues of guinea pigs that had been immunized with BP-8 tumor cells. A group of normal C3H/HeN mice were injected with 10^3 BP-8 tumor cells in their right hind limbs. Each mouse was injected I.P. with 10^7 syngeneic normal spleen cells that had been incubated with I-RNA in vitro. Injections were given at days 5, 7, 9, 11 and 13 after initial tumor cell inoculation. The incidence of tumor take was recorded serially. Control mice included those receiving no treatment, untreated normal syngeneic spleen cells, spleen cells incubated with BP-8 I-RNA pretreated with RNase, spleen cells incubated with non-specific RNA's extracted from guinea pigs immunized with complete Freund's adjuvant (CFA) without tumor cells (CFA-RNA) or with an antigenically distinct BP-9 tumor (BP-9 I-RNA).

As summarized in Table 1, mice treated with BP-8 I-RNA had a decreased incidence of early tumor development which was initially statistically significant. However, the therapeutic effect of BP-8 I-RNA appeared to be transient since the tumor incidence among treated mice eventually equalled that of the control groups and all treated animals eventually died. In a second experiment (Table 2) BP-8 I-RNA was again shown to inhibit tumor development in mice although it was again a short-lived effect. It is of interest to note that pre-treatment with RNase abolished the effect of I-RNA, suggesting the necessity of an intact RNA molecule in this system. Furthermore, I-RNA extracted from guinea pigs immunized with CFA or a BP-9 tumor which was antigenically different from BP-8 tumor did not delay the appearance of BP-8 tumor isografts, indicating the specificity of the tumor immunity transferred by I-RNA.

Based upon these observations and those reported by others [5], we concluded that temporary interference with tumor growth has the maximum effect likely to be achieved by I-RNA therapy in recipients with progressively growing tumors. This finding in conjunction with the fact that most forms of cancer immunotherapy are ineffective when tumor burden is sizeable, forced us to alter our experimental approach. We proposed to attempt to prevent a metastases from developing by administering I-RNA therapy after the primary tumor was surgically removed.

Table 1

IMMUNOTHERAPY OF THE PRIMARY MURINE TUMORS WITH IMMUNE RNA

(Modified by permission of Ann. N.Y. Acad. Sci.)

Experimental Group	N[a]	Tumor Incidence (%) Days after Tumor Isograft 15	20	23	26	28	30	34
No treatment	20	25	30	40	60	70	80	80
Normal spleen cells[b]	17	29	48	53	82	82	82	82
Spleen cells + BP-8 I-RNA[c]	16	0	0	19	38	57	75	75
Spleen cells + CFA-RNA[d]	13	31	31	46	46	69	69	77

[a] Total number of mice tested in each group.

[b] Mice were injected with syngeneic spleen cells from normal mice.

[c] Mice were injected with normal spleen cells incubated with BP-8 I-RNA.

[d] Mice were injected with normal spleen cells incubated with CFA-RNA.

Table 2

IMMUNOTHERAPY OF THE PRIMARY MURINE TUMORS WITH IMMUNE RNA

(Modified by permission of Ann. N.Y. Acad. Sci.)

		Tumor Incidence (%)					
		Days after tumor isograft					
Experimental group	N[a]	14	16	19	23	26	29
No treatment	23	26	35	61	78	83	87
BP-8 I-RNA	22	0	23	32	59	59	61
BP-8 I-RNA+RNAase[b]	10	40	40	40	60	70	70
BP-9 I-RNA[c]	21	29	48	52	67	71	71

[a] Total number of mice tested in each group.

[b] BP-8 I-RNA was previously incubated with RNAase prior to testing.

[c] I-RNA was extracted from BP-9 tumor-sensitized guinea pigs.

In order to test this idea, the B16 melanoma in the C57BL/6 mouse strain was chosen. B16 tumor is immunogenic and also causes a high percentage of deaths from pulmonary metastases after the primary isograft is removed. Our experimental protocol is illustrated in Figure 1. Guinea pigs were immunized with B16 tumor emulsified in CFA and I-RNA was extracted from their spleens and lymph nodes two weeks later. A group of C57BL/6 mice were injected with 2×10^3 B16 tumor cells into their hind limbs. The limbs together with the B16 isografts were amputated when the tumors became palpable (approximately 2½ weeks after the initial B16 inoculation). At days 2, 4, 8 and 10 after tumor excision, each mouse received 75×10^6 syngeneic normal spleen cells that had been incubated with I-RNA or with control RNA's. The survival rate was recorded until 100 days after the excision of the primary B16 isografts. Dead animals were autopsied in order to prove the presence of pulmonary metastases.

As summarized in Table 3, control mice that had received untreated syngeneic normal spleen cells began to die within 20 days after the primary B16 isografts were excised and 92% of these mice were dead by 45 days after tumor resection. However, the survival rate was significantly improved when the mice were injected with B16 I-RNA treated spleen cells with only 48% deaths 100 days after tumor removal. The biologic activity of I-RNA was destroyed by pre-treatment with RNase, suggesting that an intact I-RNA molecule was required in order to achieve a therapeutic effect. CFA-RNA extracted from guinea pigs sensitized with CFA without B16 tumor had no effect, nor did 3LL I-RNA extracted from guinea pigs immunized with an antigenically distinct C57BL tumor, Lewis lung carcinoma (3LL). Furthermore, B16 I-RNA active in preventing animals from dying from B16 metastases did not improve the survival of mice that had had 3LL isografts excised, suggesting again the specificity of the tumor immunity transferred by I-RNA.

No animal manifested evidence of toxicity after administration of I-RNA treated syngeneic spleen cells. At autopsy, pulmonary tumor nodules were evident in all dead mice. However, no mouse surviving 100 days after excision of the primary tumor isograft had gross or microscopic evidence of pulmonary metastases. Although we did not

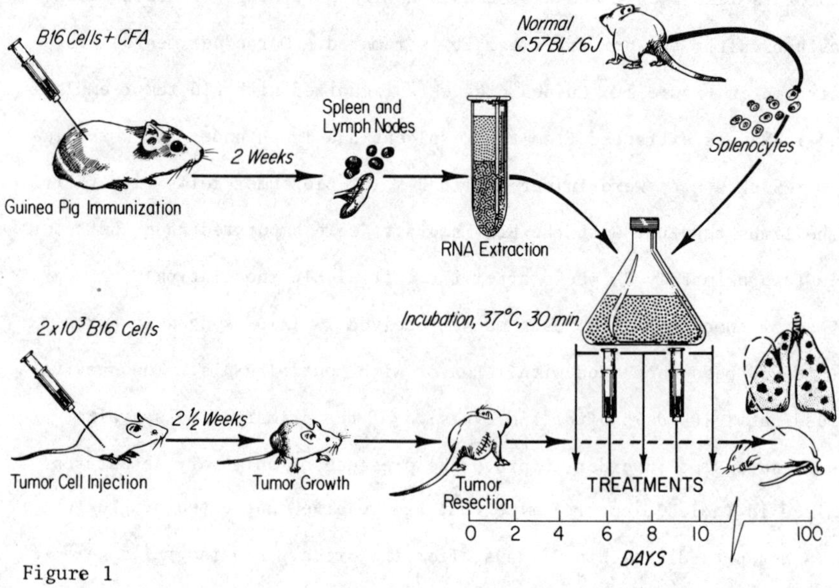

Figure 1

Immunotherapy of B16 melanoma with I-RNA. Guinea pigs were immunized with B16 melanoma and CFA. Two weeks later, their lymphoid tissues were removed and I-RNA was extracted from these lymphoid tissues. A group of C57BL/6 mice were inoculated with 2×10^3 B16 cells into their hind limbs. The limbs were amputated when B16 isografts became palpable (approximately 2½ weeks later). Each mouse then received 5 treatments with normal syngeneic spleen cells that had been incubated with I-RNA. Control animals received untreated spleen cells or spleen cells incubated with various control RNA's. The survival rate of each group was recorded until 100 days after the excision of the primary isografts. All dead animals were autopsied to prove the presence of pulmonary metastases (by permission of Ann. N.Y. Acad. Sci.).

Table 3

IMMUNOTHERAPY OF METASTATIC B16 MELANOMA WITH IMMUNE RNA

(Modified by permission of Ann. N.Y. Acad. Sci.)

		Animal Deaths (%)						
		Days after the excision of the primary B16 isografts						
Experimental group	N[a]	20	25	30	35	40	45	100
Normal spleen cells[b]	24	4	8	21	46	75	92	92
B16 I-RNA[c]	21	0	0	5	19	38	43	48
B16 I-RNA + RNAase[d]	22	0	9	23	46	82	91	91
CFA-RNA[e]	7	0	0	29	57	57	71	86
3LL I-RNA[f]	13	0	8	31	46	92	92	100
B16 I-RNA (against 3LL tumor)[g]	10	60	60	60	90	90	90	90

[a] Total number of mice tested in each group.

[b] Mice were injected with syngeneic spleen cells from normal mice.

[c] Mice were injected with normal spleen cells incubated with B-16 I-RNA.

[d] B-16 I-RNA was treated with RNAase prior to incubating with normal spleen cells.

[e] Mice were injected with normal spleen cells incubated with CFA-RNA.

[f] I-RNA was obtained from guinea pigs immunized with 3LL tumor.

[g] Normal pleen cells incubated with B-16 I-RNA were injected into mice that had had 3LL isografts amputated.

not achieve 100% prevention of death from B16 metastases, 40% improvement in survival was consistently seen in treated mice. Moreover, the long lasting effect (more than 100 days) of I-RNA therapy was encouraging with regard to its future clinical application. The B16 tumor model, in fact, closely resembled the clinical situation in which gross tumor is removed from the patient at surgery and in which I-RNA might be employed in an attempt to prevent distant metastases from developing in the future. Based upon the repeated success of I-RNA therapy in the B16 melanoma experiments [6,7], we have undertaken a Phase I clinical trial [8].

Six patients with far advanced metastatic renal cell carcinoma were chosen to be treated with I-RNA in this trial. I-RNA was extracted from guinea pigs immunized with tumors obtained from the patients at surgery. Two to three weeks after removal of the gross tumor, each patient had a Scribner arteriovenous shunt placed in his or her left forearm. Peripheral blood leukocytes were harvested by a Haemonetics-M30 leukophoresis machine. Lymphocytes were further purified by Ficoll-Hypaque gradient centrifugation and then incubated with I-RNA _in vitro_. After 60 minutes incubation, the lymphocytes were washed twice with normal saline and reinfused into the venous side of the arteriovenous shunt. Each patient received five treatments on an every-other-day schedule. No toxicity was noted in any of the patients during or after I-RNA therapy. Three patients tolerated repeated leukophoreses and reinfusion of autologous leukocytes following I-RNA treatment without continuous hospitalization.

The clinical course of each patient is summarized in Table 4. Patient #1 showed 50% regression of multiple pulmonary metastases 3 months after I-RNA therapy. She had a complete remission by six months and remained without evidence of disease until 18 months after therapy. Patients #2 and #3 showed more than a 50% decrease in the size of their tumor metastases for 8 and 10 months, respectively, after therapy. An additional two patients had a temporary stabilization in the growth of their renal cell carcinoma pulmonary metastases after therapy. Patient #6

Table 4

PATIENTS INCLUDED IN THE PHASE I I-RNA TRIAL
(By permission of Cancer)

Patient #	Diagnosis	Clinical Course			
		Before I-RNA	After I-RNA	Duration of Response	Present Status
1	Renal cell carcinoma	Lung mets ↑ [a]	Lung mets gone	18 months	Alive and well
2	Renal cell carcinoma	Lung mets ↑	Lung mets →		
		Bone mets ↑	Bone mets ↓		
		Scalene node ↑	Scalene node ↓	10 months	Dead at 12 months
3	Renal cell carcinoma	Lung mets ↑	Lung mets →		
		Liver mets ↑	Liver mets ↓		
		Right atrium			
		Tumor thrombus		8 months	Alive and well
4	Renal cell carcinoma	Lung mets ↑	Lung mets →	4 months	Dead at 10 months
5	Renal cell carcinoma	Lung mets ↑	Lung mets →	3 months	Dead at 6 months
6[b]	Renal cell carcinoma	Lung mets ↑			
		Brain mets ↑	Brain mets ↑	--	Dead at 1 month

[a] ↑ = progression, ↓ = regression, → = stabilization

[b] = This patient has been treated with I-RNA therapy only once.

with brain metastases from his renal cell carcinoma was treated with I-RNA therapy only once and showed no discernible change in tumor growth after therapy. Furthermore, we have recently demonstrated that serial samples of peripheral blood lymphocytes withdrawn from each patient at each treatment showed a progressive increase in cytolytic activity against a renal cell carcinoma line in vitro during the course of I-RNA therapy [9].

Since this Phase I trial was purposely non-randomized, no direct relationship between I-RNA therapy and the subsequent clinical course in any of these patients can be established. Nevertheless, our results show no toxicity nor enhancement of tumor growth by I-RNA therapy. Despite the fact that our patients all had far advanced tumors, their clinical courses after I-RNA therapy appeared promising when compared to the natural history of patients with metastatic renal cell cancers. We believe that our studies together with those of others [10,11], justify testing I-RNA immunotherapy in a randomized and prospective human trial.

As a means of immunotherapy in cancer patients, I-RNA has a number of attractions. It permits crossing the species barrier and thus does not require a human subject to be immunized as an I-RNA donor. Since I-RNA is incubated with autologous leukocytes prior to administration to patients, no foreign cells are required and thus potential graft-vs-host reactions resulting from histoincompatibility should not occur. Xenogeneic I-RNA is not given to patients directly since I-RNA-treated autologous leukocytes are washed prior to infusion into patients. Finally, I-RNA therapy appears to have no toxic effect on patients during and after therapy. Therefore, I-RNA seems to have advantages over a number of current immunotherapeutic modalities for human malignancy.

Acknowledgements

We thank P. Stuart, S. Onikul, E. Heacock, K. Collins, I. Saporoschetz, and M. Fallon for their technical assistance. We are also grateful to S. Lovett for her secretarial help.

References

1. Mannick, J.A. and Egdahl, R.H.: Transfer of heightened immunity to skin homografts by lymphoid RNA. J. Clin. Invest. 43:2166, 1964.

2. Ramming, K.P. and Pilch, Y.H.: Transfer of tumor specific immunity with RNA: Inhibition of growth of murine tumor isografts. J. Natl. Cancer Inst. 46:735, 1971.

3. Deckers, P.J., Wang, B.S., and Mannick, J.A.: Immunotherapy of murine tumors with immune RNA. Ann. N.Y. Acad. Sci. 277:575, 1976.

4. Wang, B.S. and Deckers, P.J.: The augmentation of concomitant tumor immunity with RNA. J. Surg. Res. 20:183, 1976.

5. Alexander, P., Delorme, E.J., Hamilton, L.D.G., and Hill, J.G.: Effect of nucleic acid from immune lymphocytes on rat sarcoma. Nature 213:569, 1967.

6. Wang, B.S., Onikul, S.R. and Mannick, J.A.: Prevention of death from metastases by immune RNA therapy. Science 202:59, 1978.

7. Wang, B.S., Steele, G., Onikul, S.R. and Mannick, J.A.: Immunotherapy with RNA in cancer. Ann. N.Y. Acad. Sci. 332:207, 1979.

8. Steele, G., Wang, B.S., Richie, J., Ervin, T., Yankee, R., and Mannick, J.A.: Results of xenogeneic immune RNA therapy in patients with metastatic renal cell carcinoma. Cancer (in press).

9. Steele, G., Wang, B.S., Richie, J., Ervin, T., Yankee, R., Fallon, M. and Mannick, J.A.: In vivo effects and parallel in vitro lymphocyte-mediated tumor cytolysis after Phase I xenogeneic I-RNA treatment of patients with widespread melanoma or metastatic renal cell carcinoma. Cancer Res. (in press).

10. Ramming, K.P., and deKernion, J.B.: Immune RNA therapy for renal cell carcinoma: Survival and immunologic monitoring. Ann. Surg. 186:459, 1977.

11. Pilch, Y.H., Ramming, K.P. and deKernion, J.B.: Preliminary studies of specific immunotherapy of cancer with immune RNA. Cancer 40:2747, 1977.

CHANGES IN SYNTHETIC RATES AND AMOUNTS OF EARLY EMBRYONIC HISTONE mRNA IN THE SEA URCHIN *

Eric S. Weinberg, Marvin B. Hendricks[1] and Kari Hemminki[2]

Department of Biology
University of Pennylvania
Philadelphia, PA 19104, U.S.A.

Abstract There is a shift in the species of histone proteins and mRNAs which are made during the development of the sea urchin embryo. It has been shown previously that the mRNAs synthesized during cleavage stages are different in size and sequence from those made in the late blastula and gastrula. Here we quantitate the relative amounts of the early histone mRNAs and rates of their synthesis. The amount of early-type histone mRNA increases 12-fold during the first 9 hr. of development (morula) and then decreases 3-fold in the subsequent 6 hr. (to pre-hatched blastula). After this time, little or no early mRMA is synthesized and the mRNA continues to decay to an undetectable level in the gastrula. The rate of synthesis of the early histone mRNAs per embryo is 0.024, 0.081, and 0.03 pg/min. for the 6, 9, and 12 hr. embryo respectively; the maximal rate per embryo, thus, being at the morula stage. On a per cell basis, however, the 6 hr. (16 cell) embryo is most active with a synthetic rate on a per gene basis of 1.9×10^{-6} pg/min., or 1-2 molecules of each of the 5 histone mRNAs being synthesized each minute on each histone gene repeat.

1 Current address: Hutchinson Cancer Research Center, 1124 Columbia Street, Seattle, Washington 98104.
2 Current address: Institute of Occupational Health, Haartmaninkatu 1, SF-00290 Helsinki 29, Finland.

* Supported by grant GM 27322 from the National Institutes of Health, U.S.A.

Introduction

During development of the sea urchin embryo, histones and histone mRNAs are synthesized extensively. In the first 24 hours after fertilization, the embryo develops from a single cell into a 300 - 400 cell blastula. It is therefore not surprising that chromatin proteins are major translational products of the embryo. In fact, histone mRNAs are the most prominant transcriptional products and histone proteins, the most prominant translational products produced in the embryo at this time (Kedes & Gross, 1969; Kedes et al., 1969).

The synthesis of histones continues in the post-blastula embryo (Ruderman & Gross, 1974; Cohen et al., 1975), even though the rate of cell division decreases markedly. In the second day of development, although the cell number only doubles, there is still a need for new histones as an additional 300 - 400 nuclei are formed. The H1, H2A and H2B histones made during this latter period are shown to be distinctly different proteins (Ruderman et al., 1974; Newrock et al., 1978) and all five classes of late embryonic histones are made on new mRNAs which are different in sequence from the early embryonic set (Kunkel & Weinberg, 1978; Grunstein, 1978; Childs et al., 1979). The mRNAs of early and late embryos thus derive from different gene sets and their appearance in the embryo is highly regulated.

Here we shall show that the synthesis of the early histones in the embryos of the sea urchin, *Strongylocentrotus purpuratus* is accompanied by transcription of new early embryonic histone mRNA and that the decrease in synthesis of these species during late blastula is due to decay of their mRNAs in the absence of their synthesis. We shall not address the question of regulation of synthesis of the late embryonic histone set here. The decrease in synthesis of early histones is due not merely to a rapid replacement of one mRNA class by another, but also involves a modulation of transciption of the early histone gene set.

Materials and Methods

Preparation of RNA

Eggs were collected from *S. purpuratus* (Pacific Biomarine Supply), fertilized, and cultured at 17° until the specified stage. Polysomal RNA was prepared from polysomal pellets and the RNA extracted as previously described (Weinberg & Overton, 1978). Total cytoplasmic RNAs were prepared essentially as described by Ruderman & Pardue (1977) and modified in our laboratory (Hieter et al., 1979). Total cellular RNA was prepared after incubation of embryos with 60μCi/ml of [^{32}P]-orthophosphate (275 Ci/mg; carrier free, ICN) in phosphate-free artificial sea water or with [2-^{3}H-] adenosine (26-29 Ci/mM, Amersham-Searle) for the designated time. Embryos labeled with [^{32}P] were washed in Ca^{+2}, Mg^{+2} -free sea water and lysed in 7-volumes of ice-cold diethylpyruvate carbonate (DEP)-treated 5.7M guanidine -HCl, 0.1M potassium acetate (pH 5.0). RNA was prepared essentially as described by Strohman et al., (1977). The [^{3}H]-adenosine-labeled RNA samples were prepared by the method of Holmes and Bonner (1973).

Cell-free translation

Wheat germ extracts were prepared by the method of Marcu and Dudock (1974). Cell-free synthesis was carried out in 50μl reactions which contained a final concentration of 20mM Hepes (pH 7.4), 2mM dithiolthreitol, 1mM ATP, 40μM GTP, 9mM creatine phosphokinase (Sigma), 60mM KCl, 2.5mM magnesium acetate, 30μM of each amino acid (with the exception of leucine), and 55μM spermine. The reactions also contained 15μl of the S-30 fraction and 5μCi -[^{3}H]-leucine (120 Ci/mM, Amersham-Searle). The reactions were incubated at 26° for 90 minutes, 3μl of β-mercapthoethanol added, and samples then stored at -70°C.

Electrophoresis of translation products

Triton X-100-acetic acid-urea gels were prepared as previously described

(Alfageme et al., 1974, Cohen et al., 1975) and modified for use with slab gels (Newrock et al., 1978). The gels were scavenged, prior to electrophoresis of protein, with 0.5M cysteamine, 5% acetic acid, 5% sucrose, and then with a solution of protamine sulfate in 5% acetic acid, 5% sucrose. Samples were subject to electrophoresis for 14-17 hours at 200 volts and the gels were processed for fluorography.

Electrophoresis of RNA and Northern transfer

The RNA was resolved into histone mRNAs by electrophoresis in acrylamide or agarose. 7.5% polyacrylamide gels were run in 7M urea similarly to the method described by Donis-Keller et al., (1977). 1.8% agarose gels were used for the Northern transfer procedure (Alwine et al., 1977). RNA samples were treated with glyoxal in DMSO as described by McMaster and Carmichael (1977). The RNA was transfered from the gel to DBM paper using a modification of the Alwine procedure (Wahl et al., 1979). Glyoxal was removed from the gel by soaking in 200ml of 50mM NaOH for 45 minutes and the gel neutralized by twice soaking in 200mM potassium phosphate (pH 6.5) for 10 minutes. ABM paper (Enzo Biochemicals) was treated with $NaNO_2$ and applied to the gel for the transfer.

Preparation of plasmid DNA and filter hybridizations

Hybridization was performed with a DNA recombinant plasmid containing a full length histone repeat inserted into pBR313 (Overton & Weinberg, 1978). DNA on 13mm Millipore filters was hybridized to labeled RNA in 3xSSC, 0.1% SDS in Denhardt's buffer (1966) at 65% for 18 hours. After hybridization, the filters were washed batchwise with two changes of 3xSSC, 0.1% SDS at 55% plus a 30 minute wash in 2xSSC containing RNAase (Sigma) and an additional wash in 2xSSC, both at room temperature. The filters were dried at 80° prior to scintillation counting.

For the standardization of hybridization efficiency, each hybridization reaction contained a known amount of [^{32}P-]-labeled 9S polysomal RNA prepared from 9 hour embryos. The amount of histone RNA in this preparation was estimated on polyacrylamide gels and in hybridization reactions wtih gel purified histone mRNA preparations. Such determinations indicated that the [^{32}P-]-9S preparation was 25% pure in histone mRNA. The extent of hybridization of the [^{32}P]-RNA in the presence of [^{3}H]-adenosine-labeled RNA, allowed a measurement of the efficiency of hybridization, and thus a determination of the percentage of [^{3}H]-RNA which is histone mRNA.

Determination of ATP and RNA specific radioactivity

The specific radioactivity was determined principally as described by Brandhorst and Humphreys (1971). The ATP samples stored in 70% ethanol were cleared by low speed centrifugation. An equal volume of H_2O was added, and the samples were treated with activated charcoal. The nucleotides eluted from the charcoal were spotted on polyethylene imine thin-layer plates, which were developed in 0.9 M LiCl containing 0.1 M sodium borate. The ATP spots were localized according to the migration of standard ATP samples. Only the lower half of the ATP spot was used because dATP migrated only slightly ahead of ATP. ATP samples were eluted in 0.1 ml of 0.6 M NaCl, and were used for the determination of ATP concentration by the luciferin-luciferase assay (Emerson and Huphreys 1971). The assayed samples were dissolved in Triton X-toluol scintillation fluid for the determination of radioactivity in ATP.

The specific radioactivity of RNA was determined using deproteinized and DNAase-treated RNA samples, which were resuspended in 3xSSC, 0.3% SDS (1xSSC: 0.15 M M NaCl, 0.015 sodium citrate, pH 7.0). An aliquot of the samples was transferred into 5 ml of ice cold H_2O containing 0.5 mg of bovine serum albumin. Perchloric acid was added to make the solution 0.7N, and after 10 minutes in ice, the solution was centrifuged at 10,000 g for 10 minutes. The pellet was washed once with 5 ml of ice-cold 0.7 N perchloric acid, neutralized and then hydrolysed in 0.3 N NaOH at 37° for 90 minutes. The sample was cooled on ice, and 0.14 ml of concentrated (70%) perchloric acid was added. After 10 minutes in ice, the sample was centrifuged at 10,000 g for 10 minutes. Absorbance ($A_{260\ nm}$) and radioactivity of the supernatant were determined to give the specific radioactivity of RNA.

Results

Timing of the shift in histone RNA populations

Two methods were used to show at which times the early and late mRNAs are present in the embryo. The *in vitro* translation of embryonic RNA in a wheat germ system and hybridization of the RNA using the Northern procedure (Alwine et al., 1979) gave essentially the same result. In the experiment illustrated in Figure 1, total cytoplasmic and polysomal RNAs of different developmental stages are used as templates in a wheat germ system. Shifts in the subtypes of H2A, H2B and H1 histones which are made *in vitro* are clearly seen to occur during this time period. The early embryonic (α) mRNAs of H2A, H2B, and H1 persist in the embryo through 23 hours of development but are absent in the 38 hour gastrula. Late embryonic forms (β,γ,δ) are barely seen at 9 hours but are prominant only at 15 hours. The band in the 9 hour sample in the position of H2A δ is a distinct protein, H2Aα′, which co-migrates in this gel with H2Aδ (Newrock et al., 1978). The change in distribution of the early and late mRNAs in each sample is similar for polysomal and total cytoplasmic RNAs used as templates. The shift in early to late forms is a gradual one but is virtually complete by the gastrula stage.

Early and late histone mRNAs can be distinguished by their electrophoretic mobilities in gels (Grunstein, 1978; Hieter, 1979; Childs et al., 1979). Figure 2 shows hybridization of RNAs transfered from agarose gels after electrophoresis using the Northern procedure. The RNA species were visualized using [^{32}P]-labelled plasmid DNA (pCO2) containing a complete repeat unit of the early histone genes (Overton & Weinberg, 1978). The late and early H1 species can be distinguished on the gel, as are the H4 species. The other three histone classes are difficult to resolve in the agarose gel. Hybridization of the pCO2 probe to late embryonic RNAs is expected to be faint, even with large

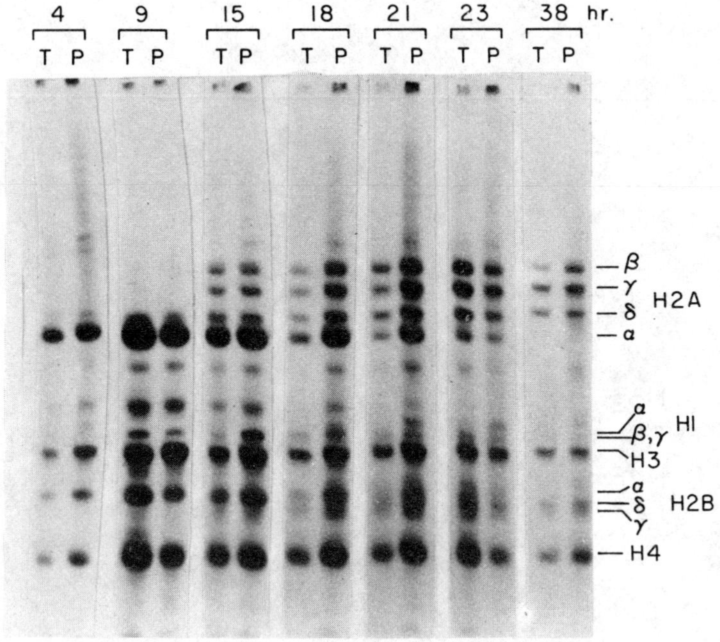

Figure 1. Translation Products Directed by Cytoplasmic and Polysomal RNA from Various Embryonic Stages of S. purpuratus.

Electrophoresis was carried out in a Triton-acid-urea gel containing 8.5M urea. In vitro translation products directed by polysomal RNA are denoted by "P" and products directed by cytoplasmic RNA are denoted by "T". The egg RNA sample is derived from a cytoplasmic RNA preparation. Translation products were made in the wheat germ system as described in Materials and Methods.

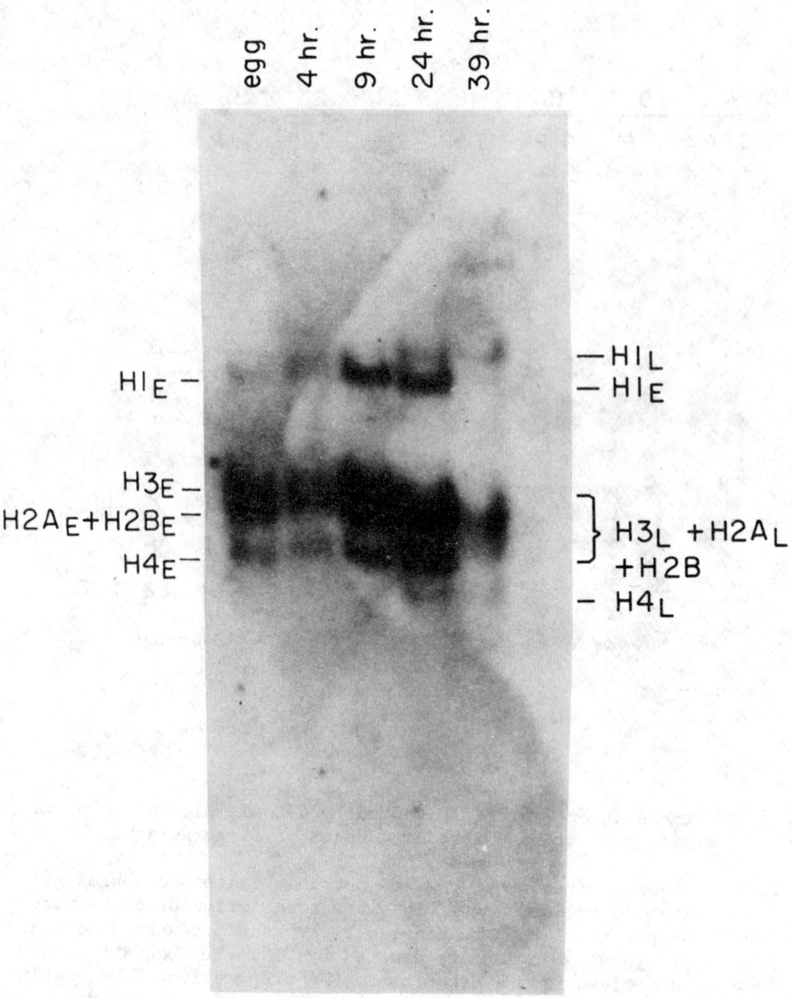

Figure 2. Northern Transfer of Histone mRNAs at Various Embryonic Stages of S. purpuratus.

Cytoplasmic and polysomal RNAs from embryos were glyoxalated, electrophoresed in a vertical 1.8% agarose gel and transferred to DBM paper. The gel was loaded with 22μg of 6-16S egg cytoplasmic RNA, 21μg of 9 hour whole cytoplasmic RNA, 21μg of 6-16S 24 hour polysomal RNA and 26μg of 6-16S 39 hour polysomal RNA. The RNAs, bound to DBM paper, were hybridized for 3.5 days at 33° with $4x10^{6}$ cpm/ml of ^{32}P-labeled 6.3Kb fragment ($4x10^{7}$ cpm/μg) derived from a Hind III digest of pCO2 histone DNA.

amounts of RNA, since there is extensive mismatching of the sequences (Kunkel & Weinberg, 1978; Hieter et al., 1979; Childs et al., 1979); nevertheless, late H1 and H4 are detected by the procedure. The early H1 and H4 mRNA are seen to have disappeared from the cell by the gastrula stage. The relative amounts of hybridization to the various mRNAs in this particular experiment cannot be taken to show quantitative changes in the amount of early mRNA since the amount of RNA originally added for each lane was uncontrolled. Ratios of early and late mRNA at any one time point also cannot be determined from this experiment since the relative efficiency of hybridization of pCO2 DNA to the early and late mRNAs is not known. The experiment does, however, show clearly the absence of early mRNA in the late embryo. The disappearance first demonstrated by the translational experiments cannot, therefore, merely be due to an inhibition of translational potential of early RNA.

Timing in the shift of RNA synthesis

The shift in synthesis of mRNA types is more abrupt than the change in mRNA populations. Figure 3 illustrates that between 14-17 hours of development there is a dramatic cessation of early mRNA synthesis. Early H4 mRNA is virtually unlabeled at 17-20 hours and is completely undetectable in later stages, although it is highly labeled in the 12-14 hour sample. Late mRNAs can be seen to be labeled as early as 9 hours of development so there is an overlap of at least 6 hours in the synthetic periods. The early histone genes which are reiterated several hundred - fold are apparently transcriptionally inactive in the late embryo.

Changes in amounts of early histone mRNA

Relative amounts of early histone RNA at different developmental stages were determined by both translation and hybridization assays. Gels such as those shown in Figure 1 were scanned densitometrically and the amount of

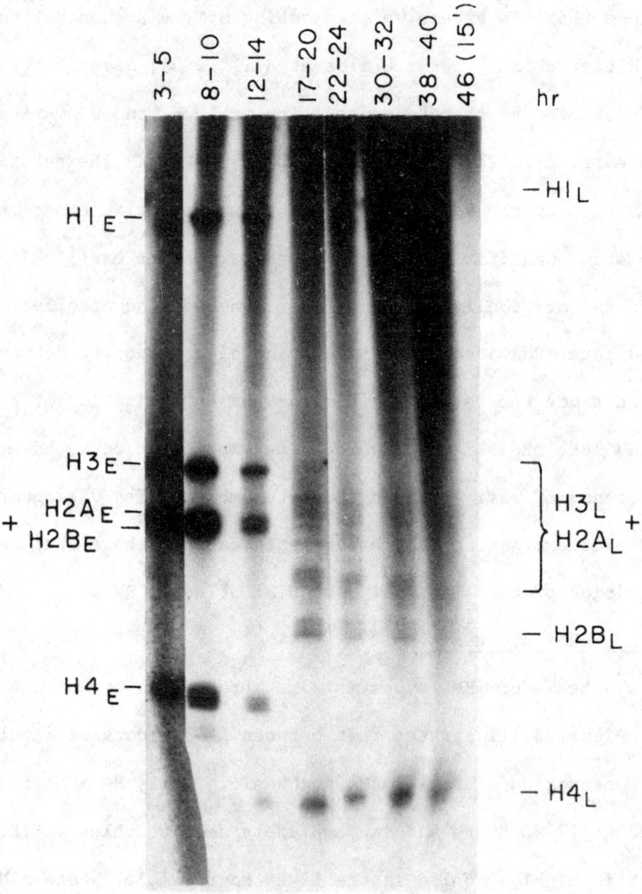

Figure 3. Gel Electrophoretic Patterns of Histone mRNAs Synthesized During Various Stages of Embryonic Development of S. purpuratus.

Embryos were labeled with [^{32}P] orthophosphate during the indicated time intervals following fertilization and whole cell RNA was isolated as described in Materials and Methods. Electrophoresis was performed on a 7M urea - 7.5% polyacrylamide gel and the gel was then dried and autoradiographed. Approximately equal amounts of RNA (mass units) were loaded in each lane up to the 17-20 hour sample. A composite of different X-ray exposures was used to show all the histone mRNA species.

histone translation product measured. The overall RNA content of the embryo does not change during development (Schmidt et al., 1948; Tocco et al., 1963) so that if the amount of total cytoplasmic RNA is measured, a comparison of specific translation activity can be calculated for the various samples.

Figure 4 presents the relative translation of histone RNAs as normalized by the amount of RNA used in the translation. Four stages are shown: egg, 4 hours, 9 hours, and 15 hours. During the first 4 hours the amount of histone mRNA per embryo increases by a factor of 1.5 over the stored histone mRNA. During the ensuing 5 hours, however, the amount per embryo increases 11.9-fold (although the amount per cell remains about the same). During the next 6 hours the amount per embryo decreases 3.2-fold, a decrease per cell of over 10-fold per cell.

The amount of mRNA was also quantitated by Northern hybridization. Known amounts of embryonic RNAs from 4 hour, 9 hour and 15 hour embryos were electrophoresed on a gel. The hybridization of pCO2 DNA was shown to be directly proporational to the amount of 9 hour RNA run as a dilution series. The results are virtually identical with the determinations made by specific translation and are also shown in Figure 4. These increases and decreases of early mRNA levels occur during a period of low late mRNA synthesis (when late histone protein synthesis in the embryo is nil). First there is a period of slow accumulation of early mRNA, followed by a period of 11-fold increase of the RNAs. Finally, there is a decrease of over 3-fold. Changes if these magnitudes are found for each individual histone mRNA as well, indicating a coordinate regulation of the amount of early histone mRNA in the embryo (data to appear elsewhere). The increase in histone mRNA during the first 9 hours of development was previously estimated to be 4-fold (Skoulchi & Gross, 1973) using less sensitive hybridization competition assays.

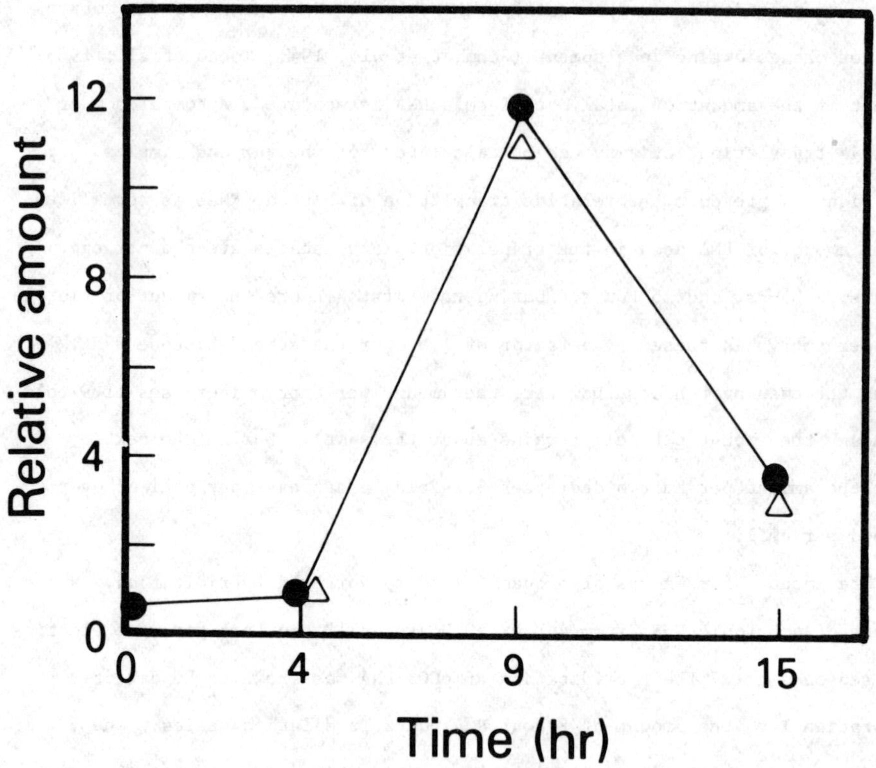

Figure 4. Relative Concentrations of Histone mRNA at Various Embryonic Stages

Two types of data are plotted here: measurements of relative histone mRNA concentrations by hybridization (triangles) and translation (dots). Relative hybridization values were calculated from densitometric tracings of the egg, 4 hour, 9 hour, and 15 hour lanes of a Northern gel similar to that of Figure 3. However, known amounts of each RNA were added in each case and the data was normalized for these amounts, The gel also contained dilutions of the 9 hour sample to demonstrate a linearity of the densitometric response at intensities observed. Relative translation data was obtained by densitometric tracings of gels similar to that shown in Figure 1. The densitometric areas of early histone products were summed and normalized for the amount of RNA used in the translation. All data are for total cytoplasmic DNA.

Rates of early mRNA synthesis

The probable explanation of changes in amount of histone mRNA is that the synthetic rate of these RNAs is high prior to 9 hours, and that after this period the bulk of the mRNA decays while only relatively small amounts are made. Superficially, the results of Figure 3 favor such a model. However, different quantities of mRNA were put on each lane in that experiment, and therefore, only the distribution of label between early and late mRNA is significant in comparing stages. The experiment was not controlled for possible differences in uptake of the precursor or changes in specific activity of the percursor pool. In order to accurately determine the rate of synthesis of histone mRNAs at different stages, we measured the incorporation of ^{3}H-adenosine into the ATP pool as well as into the histone mRNAs.

The determination of rate of histone RNA synthesis consisted of two separate measurements. First the rate of total RNA synthesis was ascertained by following the incorporation of radioactivity into both RNA and intracellular ATP during the first hour after the addition of [^{3}H]-adenosine to the culture, as described in Materials and Methods. The second procedure was to find out the percentage of radioactivity in the RNA which represented histone transcripts. This was done by the hybridization procedures described in Materials and Methods.

The results of the first set of experiments are shown in Figure 5a. The accumulation of label over the one hour period is not linear and is composed of two components as expected from previous results (Brandhorst & Humphreys, 1971, 1972). Synthesis of both rapidly labeled but unstable nuclear RNA and stable cytoplasmic RNA contribute to the accummulation kinetics. The accummulation curves presented here are interesting, however, in that the kinetics of accumulation for the three developmental stages are different.

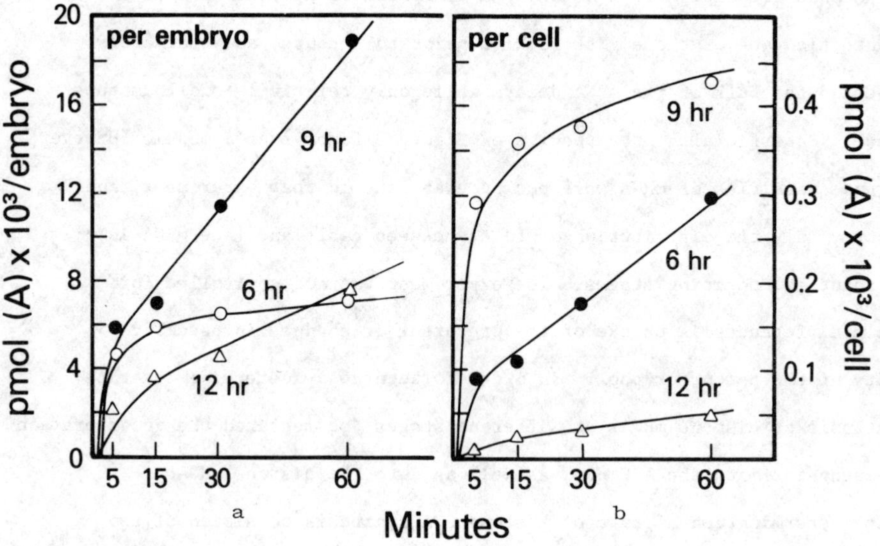

Figure 5. Synthesis of Total RNA at Different Developmental Stages

Embryos were labeled with ^{3}H-adenosine and RNA prepared as described in Materials and Methods. At each time point, the specific incorporation into intracellular ATP and RNA was determined. The amounts of RNA synthesized were calculated by determining the specific activity of the ATP.

Left panel: Incorporation per embryo. The values were obtained by measuring the A_{260} of each sample and converting cpm (thus pg)/A_{260} into pg/embryo by using the value of 3.1 ng RNA per embryo (Whitely, 1949). (5a)

Right panel: Incorporation per cell. Values shown in the left panel were normalized by dividing by the number of cells present in embryos at each stage: 16 cells (6 hour), 64 cells (9 hour), 160 cells (120 hour)(Hinegardner, 1967). (5b)

Absolute amounts of synthesis are expressed as pmol of adenosine incorporated per embryo. The cpm per sample are converted to pmol (A) by knowing the specific activity of ATP in the embryo at that time. Normalizaiton on a per embryo basis is done by measuring the A_{260} of the sample, using the conversion factor of 3.1 ng RNA per embryo (Whitely, 1949) and assuming that adenosine is present as 25% of the nucleosides. The initial synthetic rate on a per embryo basis is higher for the 6 and 9 hour embryos than the 12 hour embryos. The accummulation of new RNA of the more stable component is highest for the 9 hour embryo and lowest for 6 hour embryo. When the data is expressed on a per cell basis, however (Figure 5b), a different picture emerges. The initial rate of RNA synthesis is highest in the 6 hour embryo and decreases as the embryo divides. The nuclei are most active in transcription in early cleavage. The accummulation of more stable mRNA, however, on a per cell basis, appears to be fairly constant from 6 to 9 hours, as the slopes of the accummulation curves from 15 minutes to 60 minutes are about the same. At 12 hours the rate is about 1/2 of that at 9 hours. Further investigation of this phenomenon is necessary before more definite conclusions are drawn about differences in synthetic rates and stabilities at different stages.

The accummulation of [^{3}H]-adenosine into RNA at 15 minutes is taken to be a measure of initial rate of RNA synthesis. The rate is probably even higher since the most rapidly turning over RNA has a halflife of less than 15 minutes (Brandhorst & Humphrey, 1972). Measurements of labeled histone RNA in these same samples were used to construct the graph shown in Figure 6a. The rate of histone RNA synthesis is obviously highest in the 9 hour embryos and about 3 times that of the 6 hour and 12 hour embryos. On a per cell basis, the rate decreases with age of the embryo, with a dramatic fall in the rate per cell between 9 and 12 hours. These results are summarized

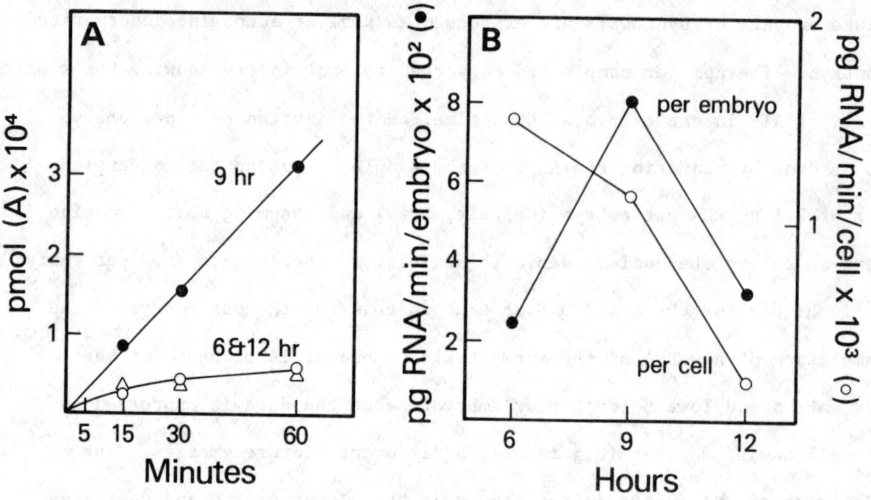

Figure 6. Amounts and Rates of Histone RNA Synthesized at Different Developmental Stages.

Left panel: Amounts of histone synthesized at various times after addition of ^{3}H adenosine. The samples used in Figure 5 were hybridized to histone DNA and the amount of histone RNA per embryo calculated as described in Materials and Methods. (6a)

Right panel: Values at 15 minutes from the left panel are used for rate determination. Values are presented on a per embryo basis or per cell basis (using cell numbers indicated in the legend to Figure 5). (6b)

in Figure 6b. The rate of synthesis of the early histone RNA changes during development and is roughly proportional to the amount of early histone RNA (compare Figures 4 and 6b). The slight difference in rate of synthesis per embryo for the 6 and 12 hour embryos shown in Figure 6b is due to the calculation being based on the 15 minute incorporation. The 30 and 60 minute calculations would show the rate of the 6 hour embryo to be slightly greater than the 12 hour embryo. The changes in rate of synthesis per cell shown in figure 6b, however, are highly significant.

Discussion

Decrease in level of early histone mRNA before time of switch

We have demonstrated that an important component in the switch of histone synthesis of the embryo is a decrease in early embryonic histone mRNA on a per embryo and per cell basis. The decrease occurs somewhat after 9 hours; by 15 hours the RNA per embryo has decreased three-fold, and by 38 hours it has disappeared, as determined by *in vitro* translation of total cytoplasmic RNA and hybridization by the Northern procedure. The change starts well in advance of any noticeable decrease in early histone protein synthesis and is accompanied by the synthesis of small amounts of late histone mRNAs. After 15 hours there is little early RNA synthesized, as shown in Figure 3. The halflife of early mRNAs after this time is calculated from quantitative translation experiments to be about 5 hours (Hendricks & Weinberg, in preparation) which is what has been found for the bulk of RNA in the embryo (Nemer et al., 1975; Galau et al., 1977). The turnover of the histone mRNA may be somewhat faster in the early blastula since a three fold decrease occurs in 6 hours (Figure 4), even in the presence of synthesis.

Modulation of rate of early histone mRNA synthesis

There are several models which might explain a varying rate of histone RNA synthesis: 1) Sets of genes of different frequencies may exist with each set having a characteristic promoter. In this case when high levels of transcript are needed, one bank of genes is activated; when a lower level is required, a less reiterated (or even unique) gene set responds to a new signal. 2) A reiterated gene set may be modulated by activating different numbers of the members of the gene set. 3) All members of the reiterated gene set may be transcribed, but the rates of initiation or elongation may vary. The data presented here would rule out the first possibility, since the early histone transcriptional rate is found to vary greatly, per cell. There is no evidence to distinguish the second and third possibilities, however. In calculating rates of synthesis per gene, we will assume all genes to be active at one time. Such determinations would thus give a lower limit of the rate: fewer genes would have to be read faster to achieve the same amount of RNA synthesized.

<u>Synthetic rates</u>

The maximal synthetic rate per cell was found at 6 hours. The 16 cell embryo is making histone RNA at 1.51×10^{-3} pg per minute per cell (See Figure 6b). Since there are 400 gene repeats per haploid genome, the rate of synthesis per gene is 1.89×10^{-6} pg per minute. This represents 3400 nucleotides per minute per gene repeat. About 2500 nucleotides are transcribed from each gene repeat unit (Childs et al., 1979); so every minute, 1.36 gene repeats are transcribed. Since the genes appear to be transcribed as 5 individual mRNAs, between 1 and 2 of each mRNA is produced per gene per minute in the 6 hour embryo. This is well below the theoretical maximal transcriptional rate estimated by Kafatos (1972) on the basis of packing limits of RNA polymerases. It is also considerably lower than the transcripitonal rate calculated for ovalbumin RNA synthesis in the stimulated chick oviduct (Palmiter, 1975). At the theoretical maximal rate of transcription each sea urchin embryonic cell would still have to have about 40-80 active histone genes to permit the high rates of histone RNA synthesis we observe. Yet, 800 gene copies in the diploid cell are well above this figure. Why are there so many histone genes even though the transcriptional rates could be obtained with far fewer copies if the assumptions of Kafatos (1972) are correct? Perhaps, the 6 hour embryo is not the time of greatest histone transcription per cell, but some other point in the life cycle of the urchin is even more demanding of rapid histone synthesis. More likely is that elongation rates for the sea urchin transcripts at 17° are much slower than at the 37° condition of warm-blooded animals. A 10-fold decrease in elongation and/or initiation rate of transcription would thus require the hundreds of gene copies in the cell to provide adequate transcription. Another possibility is that for either evolutionary or functional reasons, only a particular proportion of histone gene copies is used at one time in the cell.

Amount of histone RNA made per embryo

If we assume a fairly constant rate of transcription per cell over the period from 4 to 9 hours of development (not strictly true, see Figure 4), the total amount of histone RNA made during that time can be calculated. The embryo increases from 4 to about 64 cells during that period, a doubling approximately every 90 minutes. Although cell divisions are not truly synchoronous during this period, an estimate can be made of the cell numbers of embryos at 90 minute intervals starting at 4 hours: mainly 8 cells from 4-5 1/2 hours, 16 cells from 5 1/2-7 hours, 32 from 7-8 1/2, and 64 from 8 1/2-10 hours. Using the value of 1.5×10^{-3} pg per minute per cell, a total of 10.2 pg is calculated to be synthesized from 4 to 9 hours. This value assumes no turnover of histone mRNA during the period, but a half-life of 3-5 hours does not decrease the estimate by more than 20-30%. Since the amount of histone RNA increases 12-fold from 4-9 hours (Figure 4) and 10.2 pg is assumed to be synthesized during that time, the 4 hour embryo would have about 0.93 pg per embryo. The egg has about 2/3 of this value (Figure 4) or about 0.62 pg of stored histone RNA. Thus, about 0.1 pg of each histone mRNA would be stored in the egg. If these values are correct, the 9 hour embryo would have 0.36% of its total RNA as histone mRNA, while the unfertilized egg would contain only 0.02% stored histone mRNA. If the turnover of histone RNA is faster than estimated, the values would be somewhat lower. A slightly lower value would also result if the 25% decrease in rate of synthesis per cell from 6-9 hours is taken into consideration, instead of assuming a constant rate over the complete 5 hour period. Nevertheless, the value of 11.13 pg (10.2 + 0.93) for the 9 hour embryo seems a reasonable estimate.

The early histone gene system in the sea urchin thus represents an example of a gene which is highly active at one devleopmental stage, somewhat less active at another stage, and inactive in a third stage. Further exploration of the state of these genes in chromatin at these stages is being pursued to gain some idea of the difference between "active" and "inactive" genes.

References

Alfageme, C.R., Zweidler, A., Mahowald, A. and Cohen, L.H. (1974). J. Biol. Chem. 249 3729.

Alwine, J.C., Kemp, D.J. and Stark, G.R. (1977). Proc. Natl. Acad. Sci. USA 74 5350.

Brandhorst, B.P. and Humphreys, T. (1971). Biochemistry 10 877.

Brandhorst, B.P. and Humphreys, T. (1972). J. Cell Biol. 53 474.

Cohen, L., Newrock, K.M. and Zweidler, A. (1975). Science 190 994.

Childs, G., Maxson, R. and Kedes, L.H. (1979). Develop. Biol. 73 153.

Donis-Keller, H., Maxam, A., and Gilbert, W. (1977). Nucleic Acids Res. 4 2527.

Denhardt, D.T. (1966). Biochem. Biophys Res. Commun. 23 641.

Emerson, C.P. and Humphreys, T. (1971). Anal. Biochem. 40 254.

Galau, G.A. Lipson, E.D., Britten, R.J. and Davidson, E.H. (1977). Cell 10 415.

Grunstein, M. (1978). Proc. Natl. Acad. Sci., USA 75 4135.

Hieter, P.A., Hendricks, M.B., Hemminki, K. and Weinberg, E.S. (1979). Biochemistry 13 2707.

Hinegardner, R.T. (1967). In Methods in Developmental Biology. Fred H. Wilt and Norman K. Wessells, eds. (Thomos Y. Crowell Co., New York) pp. 139-155.

Holmes, D. and Bonner, J. (1973). Biochemistry 12 2330.

Kafatos, F.C. (1972). Curr. Topics Devel. Biol. 7 125.

Kedes, L.H. and Gross, P.R. (1969). Nature 223 1335.

Kedes, L.H., Gross, P.R., Cognetti, G. and Hunter, A.L. (1969). J. Molec. Biol. 45 337.

Kunkel, N.S. and Weinberg, E.S. (1978). Cell 14 313.

Marcu, K. and Dudock, B. (1974). Nucleic Acids Res. 1 1385.

McMaster, G.K. and Carmichael, G.G. (1977). Proc. Natl. Acad. Sci. USA 74 4835.

Nemer, M., Dubroff, L.M. and Graham, M. (1975). Cell 6 171.

Newrock, K.M., Cohen, L.H., Hendricks, M.B., Donnelly, R.J. and Weinberg, E.S. (1978). Cell 14 327.

Overton, G.C. and Weinberg, E.S. (1978). Cell 14 247.

Palmiter, R.D. (1975). Cell 4 247.

Ruderman, J. and Gross, P.R. (1974). Develop. Biol. 36 286.

Ruderman, J. and Pardue, M. (1977). Develop. Biol. 60 48.

Ruderman, J.V., Baglioni, C. and Gross, P.R. (1974). Nature 247 36.

Schmidt, G. Hecht, L. and Thanhauser, S.J. (1948). J. Gen. Physiol. 31 203.

Skoultchi, A. and Gross, P.R. (1973). Proc. Nat. Acad. Sci. USA 70 2840.

Strohman, R.C., Moss, P.S., Micou-Eastwood, J. and Spector, D. (1977). Cell 10 265.

Wahl, G.M., Stern, M. and Stark, G.R. (1970). Proc. Natl. Acad. Sci. USA 76 3683.

Weinberg, E.S. and Overton, G.C. (1977). In Methods in Cell Biology 19 (New York: Academic Press), pp 273-286.

STRUCTURE, EXPRESSION AND HORMONAL RESPONSIVENESS OF THE AUTHENTIC AND PSEUDO-OVALBUMIN GENES

by

Savio L.C. Woo, Ming-Jer Tsai and Bert W. O'Malley

from

Department of Cell Biology and
Howard Hughes Medical Institute
Baylor College of Medicine
Houston, Texas 77030

SUMMARY

We have isolated chicken DNA clones containing the complete chromosomal ovalbumin gene from a chicken gene library. Using DNA from these clones as hybridization probes, we have isolated recombinant phages from the same chicken gene library which contain two genes designated X and Y. These two genes are linked to the ovalbumin gene in the order 5'-X-Y-ovalbumin-3'. Both genes share limited sequence homology with the authentic ovalbumin gene and all 3 genes are comprised of 8 structural gene segments separated by 7 intervening sequences. Northern hybridization studies using the X and Y gene probes indicated the presence of putative precusor molecules in stimulated oviduct RNA preparations, which differ in size from those observed for ovalbumin. R_0t analysis has demonstrated that similar to the ovalbumin gene, the level of X and Y gene transcripts is differentially increased by the steroid hormone estrogen. The extent of hormonal responsiveness of the 3 closely related genes is in the relative order normalized to ovalbumin of OV: Y:X≃100:10:1. Pulse-labeling studies of these three closely linked genes suggest that in estrogen-stimulated oviduct, the markedly different steady-state levels of the X, Y and ovalbumin gene transcripts reflect their differential transcription rates.

INTRODUCTION

Expression of the ovalbumin gene in the chicken oviduct is regulated by steroid hormones (O'Malley and Means, 1974; Cox Haines and Emtage, 1974; McKnight, Pennequin and Schimke, 1975; Harris et al., 1975; Monahan, Harris and O'Malley, 1976 ; Palmiter et al., 1976; Hynes et al., 1977).

The chromosomal ovalbumin gene is comprised of 8 structural gene segments separated by 7 intervening sequences (Woo et al., 1978:

Dugaiczyk et al., 1978; Breathnach et al., 1978; Gannon et al., 1979). The structural and intervening regions of the gene are expressed in their entireties and the expression of both regions is dependent on estrogen (Woo et al., 1978; Roop et al., 1978). Using the S1 nuclease mapping method of Berk and Sharp (1977), we have determined that the 5'-and 3'-ends of ovalbumin precursor RNA are coincident with the capping and polyadenylation sites on the gene (Roop et al., 1979). Northern hybridization and pulse-chase experiments have also supported a model of mRNA maturation in which the entire ovalbumin gene is transcribed as a large precursor RNA molecule which is subsequently processed by excision of intervening sequence transcripts and proper ligation of structural sequence transcripts (Roop et al., 1979).

Related genes of unknown function but bearing sequence homology with the ovalbumin gene had recently been found to exist in the chick genome (Royal et al., 1979; Woo et al., 1979). These two genes, designated X and Y, are transcriptionally active in the chick oviduct. The ovalbumin gene and the X and Y "pseudogenes" are closely linked at the same chromosomal locus with the linkage order 5'-X-Y-ovalbumin-3'. It was thus of immediate interest to determine whether the expression of the X, Y and ovalbumin genes is coordinantly regulated by steroid hormones.

MATERIALS AND METHODS

Screening of the chicken gene library

All cloning and propagation procedures were carried out in a certified P3 facility in accordance with NIH guidelines for recombinant DNA research.

The phage titer of the chicken gene library was determined by serial dilution and approximately 10,000 phages were spread in soft agar onto each of 200 square agar plates. Screening of the recombinant phage plagues was carried out by a modification of the procedure of Benton and Davis (1977) as described by Woo, 1979.

The hybridization probes were prepared by the nick translation procedure of Maniatis et al. (1979) as modified in our laboratory (Lai et al., 1979).

DNA transfer and Southern hybridization analysis

Following restriction endonuclease digestion, DNA fragments were separated by agarose slab-gel electrophoresis and transferred to nitrocellulost filters by the procedure of Southern (1975). After transfer, the filters were dry baked for 2 hours at (68°) followed by incubation (68°C, 6 hours) in a 6XSSC solution containing 0.04% ficoll, 0.04% polyvinyl pyrrolidone, and 0.04% BSA (Denhardt, 1966). Hybridization with a [^{32}P]labeled probe was carried out in the same solution containing 0.5% SDS, 1 mM EDTA, for 12 hours at 68°C. Washing of the filters was carried out at 68°C with 3 x 2 hours washed in 1 x SSC, 0.5% SDS, and 1-hour wash in 1 x SSC without SDS. The filters were then exposed to X-ray film using Dupont cronex intensifying screens at -20°C.

Preparation of oviducts

Oviducts were obtained from White Leghorn chicks. For stimulated oviducts, chicks were implanted weekly with a 20 mg of diethylstilbestrol (DES) pellet subcutaneously (Sigma Chemical Co.) which provided continuous release of DES for 8-9 days. For withdrawn oviducts, chicks were subcutaneously injected daily with 2.5 mg of diethylstibestrol for 14 days and then withdrawn from all hormone for 14 days. For experiments involving acute stimulation with estrogen, chicks received an injection of 2.5 mg of DES on the 14th day of withdrawal and then 24 h later. Oviducts were collected at the indicated time intervals.

Electrophoresis of RNA and transfer to DBM paper for hybridization

Poly A-containing RNA was prepared from stimulated and withdrawn chick oviduct by extraction of total nucleic acids from tissues with phenol followed by oligo dT-cellulose column chromatography as described previously (Woo *et al*., 1975). The procedure for agarose gel electrophoresis in the presence of methylmercury hydroxide (Bailey and Dividson, 1976) and transfer of RNA to DBM and hybridization to [^{32}P]DNA probes were carried out by a modified procedure of Alwine *et al*., 1977 as described by Roop *et al*. (1978).

RNA excess hybridization analysis

The DNA probes were labeled with [^{3}H] as described by Roop *et al*. (1978). RNA-DNA hybridization reactions were performed in Kontes reaction vials containing 600-700 cpm [^{3}H]DNA probe (4×10^{6} cpm/μg) and 1-150 μg RNA in a 50 μl solution for various lengths of time. Following incubation, the samples were treated with S1 nuclease and the trichloroacetic acid prescipitable, S1-nuclease resistant, radioactive counts determined and plotted as previously described (Tsai *et al*., 1978; Roop *et al*., 1978).

Determination of transcription rates in oviduct tissue suspensions

Preparation of oviduct tissue suspensions and extraction of RNA was carried out as previously described (Swaneck *et al*., 1979). Incorporation of [^{3}H]uridine and [^{3}H]cytidine was carried out at 41°C in F-12 medium supplemented with 10^{-8} M estradiol, 1 μg/ml insulin and 10^{-8} M dexamethasone. Preparation of plasmid DNA-bearing filters was carried out as described by Tsai *et al*. (1978). Hybridization reactions were carried out in 0.6 M NaCl, 0.01 M HEPES (pH 7.0) and 2 mM EDTA at 68°C for 18 hours. Washing of the filters prior to determination of hybridized radioactive counts was carried out as described by Swaneck *et al*., (1979) and preparation of cRNA from recombinant plasmid templates was carried out as described by Roop et al. (1978).

Electronmicroscopic analysis of DNA-RNA hybrid molecules

DNA-RNA hybrids were formed by incubation of 10 μg/ml DNA with 1-200 μg/ml RNA in 70% formamide, 10 mM EDTA, 0.1 M NaCl, 0.1 M Tris-HCL, pH 7.6. The mixture was sealed in capillary tubes, incubated at 80°C for 5 minutes followed by incubation at 55°C for 3 hours. The incubated samples were immediately prepared for electronmicroscopic analysis as described previously (Dugaiczyk et al., 1979) and viewed with a Joel 100C electronmicroscope.

RESULTS AND DISCUSSIONS

Isolation of the ovalbumin and X, Y pseudogenes from a chicken gene library

Molecular structure of the chromosomal ovalbumin gene had previously been determined by restriction mapping and electronmicroscopic analyses of the individually cloned chicken DNA fragments containing overlapping regions of the gene (Figure 1). In order to obtain a clone containing the entire ovalbumin gene, several million phage plaques from a chicken gene library were screened for the ovalbumin gene. One of the positive clones, designated CL64, appeared to contain the entire natural ovalbumin gene. Upon EcoRI digestion, CL64 DNA yielded an 8.0-and a 5.5-kb DNA fragment in addition to the 2.4-, 1.8-, and 0.5-kb fragments.HindIII and Pst I digestion of this DNA generated the expected 3.2-and 4.5-kb bands, indicating the presence of additional chicken DNA sequences to the left of the 2.4-and 0.5-kb EcoRI DNA fragments of this ovalbumin gene clone.

In addition to CL64, other ovalbumin gene-containing recombinant phages were also obtained and characterized by restriction endonuclease mapping. One of these recombinants, designated CL26, contained additional sequences flanking the 5'-end of the ovalbumin gene. Some of the restriction

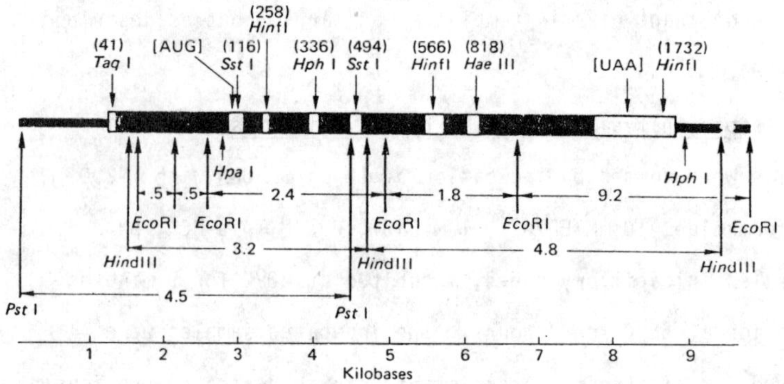

Figure 1: Physical map of the entire natural ovalbumin gene displaying some of the key restriction cleavage sites, the locations of the initiation and termination codons, and regions of interspersed structural and intervening sequences. This map is constructed from detailed analyses of various gene fragments, cloned in our laboratory: 2.4 Kb EcoRI DNA, 1.8 Kb EcoRI DNA 9.2 Kb EcoRI DNA, 3.2 Kb HindIII DNA, 4.5 Kb PstI DNA. Thin solid line, flanking gene sequence; thick solid line, intervening gene sequence; thick open line, structural gene sequence.

sites of these clones DNAs are presented in Figure 2. As depicted in the figure, the chicken DNA insert in CL26 extended the range of DNA approximately 11 Kb beyond the 5' terminus of the ovalbumin gene. A Bam HI-generated 2.1 Kb fragment from CL26 DNA was approximately 9Kb from the 5'-terminus of the ovalbumin gene and utilized as a probe to isolate clones from the same chick gene library containing DNA sequences further 5' from the ovalbumin gene. Several positive phages were identified and their DNAs characterized by restriction endonuclease mapping. The map of one of these clones, designated CL78, is presented in Figure 2. The inserted DNA in this clone extended the length of DNA about 25 Kb beyond the 5'-terminus of the ovalbumin gene, a sufficient length to contain all of the Y-gene and a portion of the X-gene sequences as reported by Royal *et al*. (1979).

In order to obtain the entire X gene sequence, the 5'-terminal 2.5 Kb *Eco* RI fragment of CL78 DNA was employed as a probe for a third round of library screening. One of the recombinant phages, CL36, contains 16 Kb of chick DNA sequences further 5' from the ovalbumin gene and includes the entire X gene (Fig. 2).

Molecular structure of the ovalbumin and Y, X pseudogenes

Demonstration of the presence of the ovalbumin, Y and X mRNA coding regions in CL64, CL78 and CL36 DNAs was accomplished by electronomicroscopic mapping after hybridization with total poly A-containing RNA extracted from hormonally stimulated chick oviducts. Figure 3 shows representative molecules of the ovalbumin Y and X gene structures. A total of seven single-stranded loop regions are evident in all 3 molecules as indicated on the corresponding line drawings. Such structures are routinely observed for the Y and X gene hybrids, indicating that the structural sequences of the Y and X genes, like the ovalbumin gene, are organized into eight

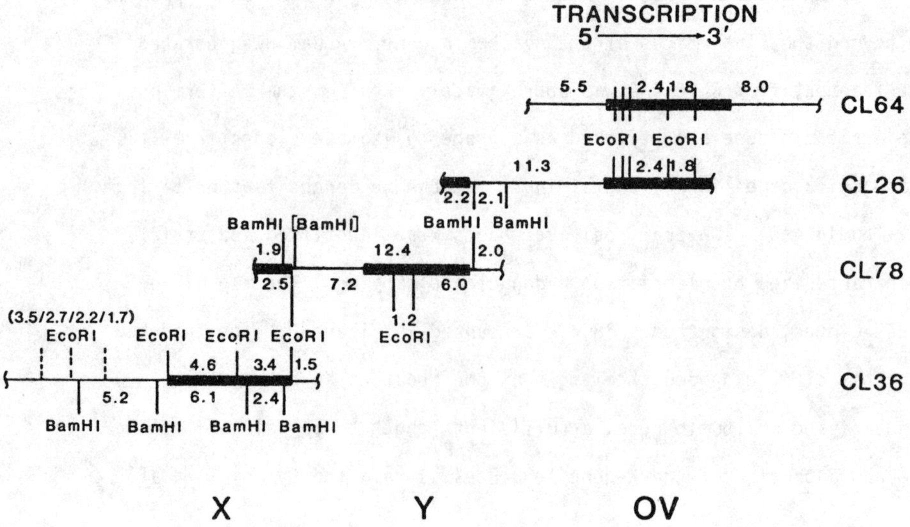

Figure 2: Restrictions maps of four recombinant phages isolated from a chicken gene library containing overlapping DNA regions of the ovalbumin gene locus in the chicken genome. The linear DNA maps of individual phages were constructed on the basis of restriction endonucleases digests with Bam HI, EcoRI and both enzymes together. The positions of these sites and the distance in kilobases are presented in the figure. The synthetic EcoRI sites used in construction of the recombinant phages are indicated by (ʃ), and the positions of the authentic and pseudogenes in individual phages are shown by (—). It should be noted that the [BamHI] site in CL78 DNA is not present in some of the other clones and is polymorphic. Such genotypic polymorphism had previously been observed in the chromosomal ovalbumin gene (Weinstock et al. 1978; Lai et al. 1979) and in the globin gene system (Jeffreys, 1979; Kan and Dozy, 1978). Since the exact positions of the EcoRI sites harboring the 5' terminal fragments in CL36 DNA have not been mapped, they are indicated by dashed lines.

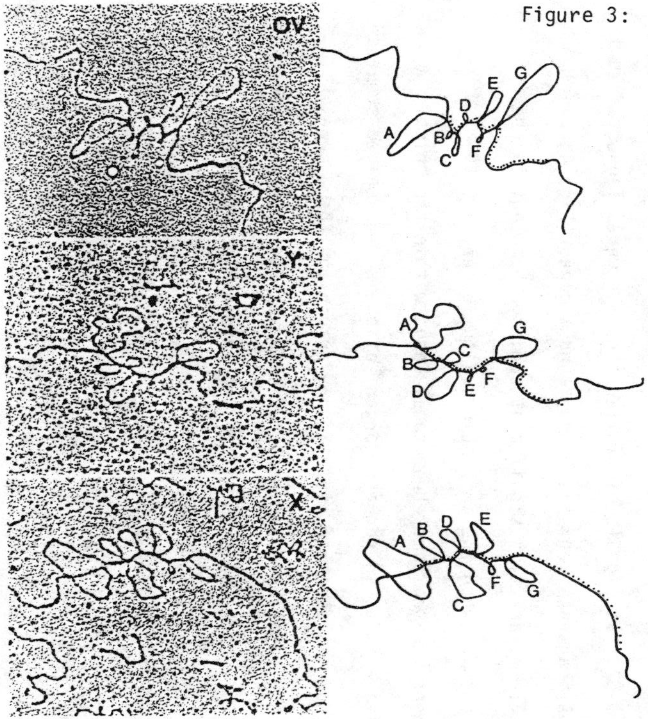

Figure 3: Electron micrographs and corresponding line drawings of hybrid molecules formed between polyA-containing RNA from stimulated chick oviduct and denatured CL64 DNA (upper panel), CL78 DNA (middle panel) and CL36 DNA (lower panel). (———), single-stranded DNA; (········) oviduct RNA. Hybridization was carried out under conditions that favored the formation of RNA/DNA hybrids but not renatured DNA molecules. Homologous regions between DNA and RNA molecules would form hybrids and intervening regions within DNA molecules not represented in the RNA molecules would be displaced to form single-stranded DNA loops. A total of seven loops designated A, B, C, D, E, F, and G are apparent within the ovalbumin and X genes as shown in the upper and lower panels. In the Y gene there are two small intervening loops (E and F) in addition to the five apparent loops (A, B, C, D, and G) as shown in the middle panel. The two small loops are present at the same locations in 10 individual molecules.

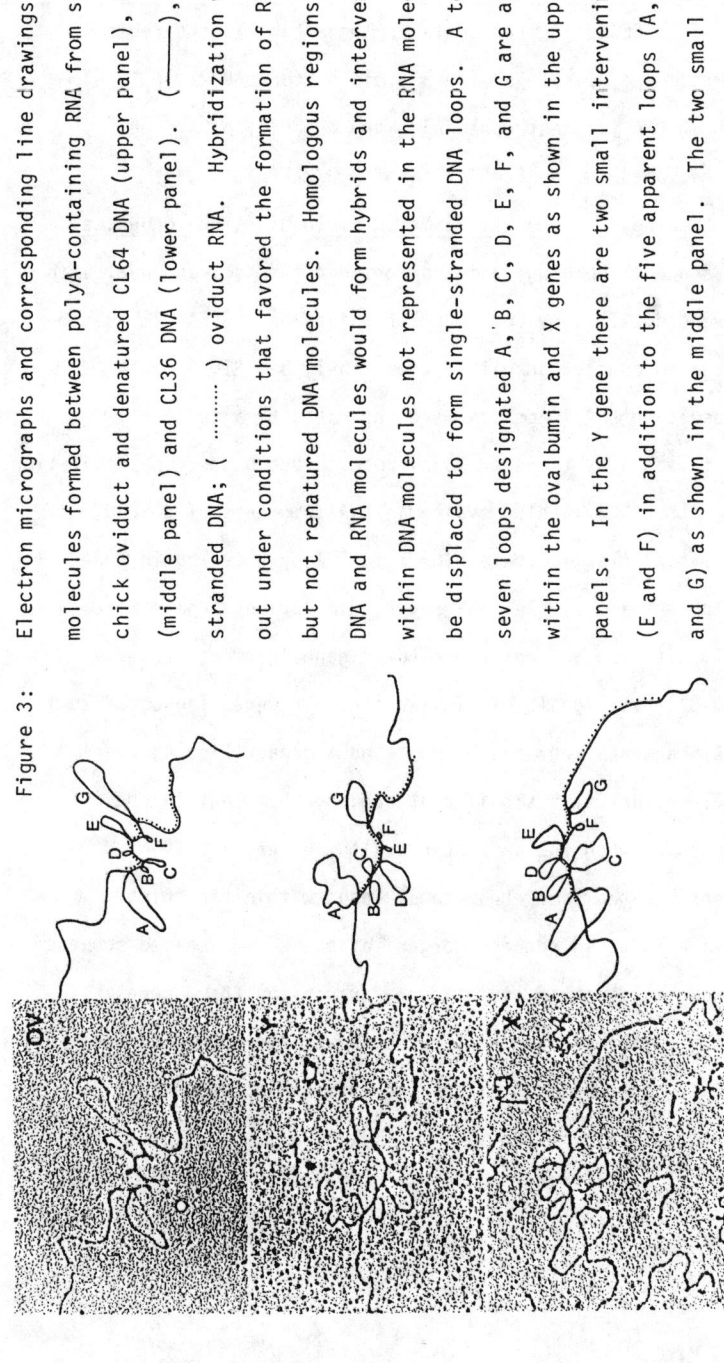

segments separated by seven intervening sequences.

To facilitate propagation, a 12 Kb DNA fragment containing the entire ovalbumin gene was obtained from CL64 DNA by partial Hind III digestion and cloned in the Hind III site of pBR322 (pOV12). A Bam HI-generated 12.4 Kb fragment from CL78 DNA that contains the entire Y gene region was sub-cloned in the Bam HI site of the plasmid vector pBR322 (pY12.4). Similarly, a 8.0 Kb DNA fragment containing the majority of X gene sequences was sub-cloned in the EcoRI site of pBR322 (pX8.0).

Regions of homology between the X, Y and ovalbumin gene sequences

Regions of sequence homology existing between the X, Y and ovalbumin gene sequences were detected by Southern hybridization. CL64, 78 and 36 DNAs were digested to completion with either EcoRI, Bam HI, or both enzymes. The resulting DNA fragments were separated by agarose gel electrophoresis followed by transfer to nitrocellulose paper. Hybridization was carried out with [^{32}P] labeled ovalbumin DNA fragments of pOV230, which is a recombinant plasmid containing a full length ovalbumin cDNA insert (McReynolds, et al., 1977). Strong hybridization signals were obtained not only with the authentic ovalbumin gene fragments (Fig. 4, Panel B, lanes 1-3), but signals of similar strength were also obtained with various DNA fragments containing the Y and X genes (Fig. 4, Panel B, lanes 4-9). There are thus significant sequence homology in the structural segments between the authentic ovalbumin gene and the X and Y pseudogenes in the same chromosomal locus within the chick genome. The similarities in sequence organization and the shared sequence homology would suggest a common evolutionary origin for the 3 genes. Moreover, when the experiment was repeated using as the hybridization probe labelled pOV12 DNA which contains all of the intervening sequences in

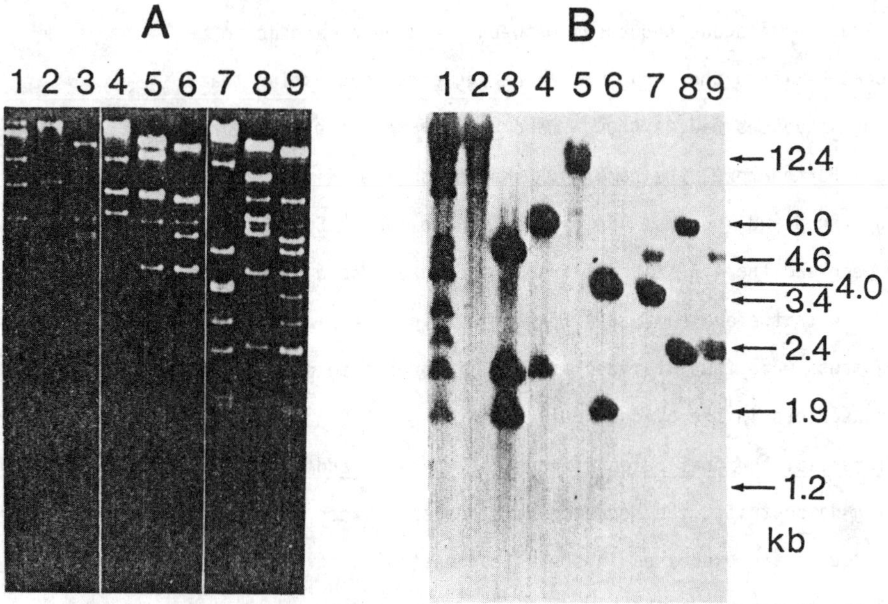

Figure 4: Sequence homology between the authentic and pseudo-ovalbumin genes by Southern hybridization. Panel A, ethidium bromide stain of an agarose gel containing CL64 (lanes 1-3), CL78 (lanes 4-6) and CL36 DNA (lanes 7-9) after digestion with EcoRI (lanes 1, 4, 7), BamHI (lanes 2, 5, 8), or both enzymes (lanes 3, 6, 9). Panel B, radioautogram of a Southern blot of the gel shown in Panel A using as the hybridization probe [^{32}P] labeled pOV230 DNA, which is a full length ovalbumin cDNA clone (McReynolds et al. 1977).

addition to the structure ovalbumin gene sequences, the hybridization signals obtained from the Y and X DNA fragments were relatively much weaker. Thus, homologous sequences between the 3 genes appear to be mainly located within the structural gene segments, suggesting that intervening sequences had diverged more rapidly than the structural sequences. Furthermore, the homology is also confined within certain regions on the X and Y genes. Specifically, the 1.2 Kb EcoRI fragment of the Y gene and the 4.6 Kb EcoRI fragment of the X gene hybridized only weakly with the ovalbumin DNA fragments (Figure 4, lanes 4 and 7). These fragments were thus utilized as specific probes to detect Y and X gene transcripts in the oviduct cell.

Differential hormonal responsiveness of the X, Y and ovalbumin genes

Having demonstrated the sequence relatedness between these genes and that all 3 genes are expressed in the estrogen-stimulated oviduct, it was important to determine the hormonal responsiveness of these linked and closely related genes. The relative concentrations of transcripts from each of the genes in total nuclear RNA preparations were quantitated using estrogen-stimulated and withdrawn chick oviducts. Presented in Figure 5 are results obtained from hybridization studies by R_0t analysis using $[^3H]$ labeled ovalbumin, $Y_{1.2}$, and $X_{4.6}$ probes. Relative transcript abundance in the RNA preparation was calculated on the basis of observed $R_0t_{1/2}$ values. Consistent with previous hybridization studies (Roop et al., 1978), a dramatic shift in $R_0t_{1/2}$ values were obtained with the $[^3H]$ovalbumin probe between stimulated and withdrawn RNA preparations (upper panel). The half R_0t transition values of these curves are indicative of an estrogen-mediated increase in the concentration of nuclear ovalbumin

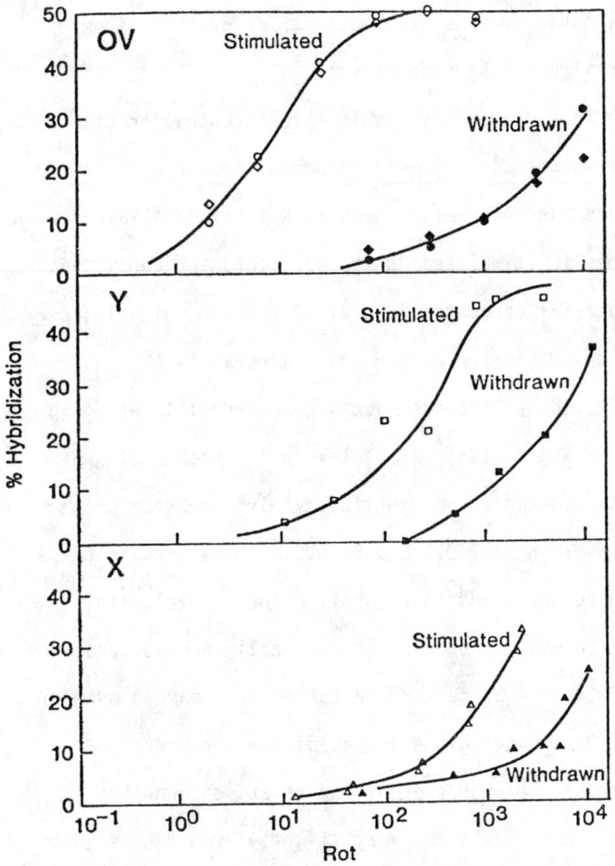

Figure 5: Steady-state levels of X, Y and ovalbumin gene transcripts in estrogen stimulated (open symbols) and withdrawn (closed symbols) oviduct nuclear RNA preparations. [^{3}H]-probes used in the upper panel were the EcoRI-generated 1.8 Kb (◇ — ◇) and 2.4 Kb (O - O) fragments of the ovalbumin gene (Dugaiczyk et al. 1978); the 1.2 Kb EcoRI fragment of the Y gene (□ – □) in the middle panel; and the 4.6 Kb EcoRI fragment of X gene (△ — △) in the lower panel.

gene transcripts in withdrawn versus stimulated chick oviducts from <1 to 240 molecules per cell nucleus, respectively.

A similar, though less dramatic shift in the concentration of Y and X gene transcripts within the same stimulated and withdrawn oviduct nuclear RNA preparations was observed. The half R_0t values obtained with the $Y_{1.2}$ and $X_{4.6}$ probes in stimulated RNA preparations are consistent with Y and X gene transcript concentrations of 12 and 2-3 molecules per oviduct cell nucleus, respectively (middle and lower panel).

Since the hybridization probes are mostly composed of intervening sequences, analysis of X and Y steady state levels in cytoplasmic RNA was not feasible. Thus, the corresponding intervening sequence probes for the authentic ovalbumin gene (pOV1.8 and pOV2.4) were used in these studies for comparison purposes, and our quantitation of ovalbumin, Y and X gene transcripts in oviduct nuclear RNA is really a measure of the intracellular concentration of precursor RNA rather than mature mRNA of the respective genes. Depending on the extent of the structural gene regions present in each of the probes used, the relative concentration of the final mature mRNA levels may vary slightly from the estimated value. The extent of such variations, however, should be minimal since similar results were obtained using the entire ovalbumin and Y gene DNAs as probes for hybridization. Thus, these results support the concept that there is a graded response in gene transcript accumulation to steroid hormone induction such that ovalbumin >Y >X.

Transcription rates of the X, Y and ovalbumin gene

As demonstrated above, the steady-state nuclear concentrations of X, Y and ovalbumin gene transcripts are quite different. Such observed differences could be due to a differential rate of turnover and/or a differential rate of gene transcription. In order to determine the relative transcription rate of these three genes, hormonally stimulated oviduct tissue was suspended in a defined medium and allowed to incorporate [^{3}H] cytidine and [^{3}H]uridine for a brief period.

The concentration of RNA transcribed from structural sequences in the authentic ovalbumin gene was determined by hybridizing the [^{3}H]RNA to filters containing pOV230 DNA. Approximately 0.1% of the RNA synthesized in these nuclei corresponds to mRNA sequences (Table 1).

The *in vitro* synthesis of RNA corresponding to the intervening sequences within the ovalbumin gene was determined by hybridizing labeled RNA synthesized in nuclei to filters containing pOV12 DNA (Table 1). An average of 0.058% of the total synthesized RNA binds to pOV12 filters, and competition with $mRNA_{ov}$ suggests that 95% of the counts bound (0.055% of the total counts) correspond to the intervening sequences in the 12 Kb fragment. These values are surprisingly similar to that obtained for the structural sequences (0.1%), if we consider that the accumulation of sequence mass from the intervening and structural regions in total RNA differs substantially by a factor of more than 100-under conditions of chronic estrogen stimulation (40,000 versus 300 moleucles per tubular gland cell; Harris *et al*., 1975). The observation that the amount of *in vitro* synthesized RNA corresponding to structural gene sequences was slightly greater than that for intervening sequences is consistent with the conclusion that some processing of RNA transcribed from the intervening sequences may be occurring in nuclei.

Having demonstrated the specificity of the *in vitro* transcription rate assay, the [^{3}H]labeled nuclear RNA was hybridized to filters containing either pOV12, pY12.4 or pX8.0 DNA. As a means of assessing

TABLE 1

In Vitro Transcription of Ovalbumin Gene Sequences in Chick Oviduct Nuclei

Filter	Competitor	^{3}H-RNA Hybridized[1] (cpm)	^{32}P-cRNA Recovery (%)	Hybridizable Gene Sequences (cpm)	Total RNA (%)	Average[2]
pOV230	-	1009	16.4	4499	0.110	0.10 ± 0.009
	$mRNA_{ov}$	271	-	-	-	-
pOV12	-	715	21.7	2295	0.056	0.058 ± 0.004
	$mRNA_{ov}$	712	20.7	2391	0.058	0.055 ± 0.003
	$mRNA_{ov}$ + RNA_{12}	217	-	-	-	-

1 Input [^{3}H]RNA = 4.1 x 10^{6} cpm.

2 Average and standard error of the mean determined from results of three independent experiments.

the percent recovery of potentially hybridizable material bound to the filter, [^{32}P]cRNA transcribed _in vitro_ from pOV12, pY12.4 and pX8.0 DNA was added to the hybridization reactions as an internal standard. The routine recovery of [^{32}P] internal standard also allowed a means of assessing the specificity of the hybridization reaction. As indicated in Table 2, [^{32}P]cRNA transcribed from each of the recombinant plasmid templates hybridized well with DNA filters containing homologous DNA. When excess unlabeled cRNA from the ovalbumin and Y gene-bearing plasmids was added as a competitor in the hybridization reaction, only the homologous competitor cRNA was capable of significantly reducing the hybridized counts. There was, however, substantive hybridization between X gene transcript and ovalbumin DNA due to sequence homology.

The hybridization values obtained with the [^{3}H]pulse labeled nuclear RNA preparations are summarized in Table 2. It is clear that significantly more label has been incorporated into ovalbumin transcripts than into Y and X gene transcripts. That these values obtained for the Y gene represent gene-specific hybridization is indicated by the lack of competition to the Y DNA filter when excess heterologous cRNA is added to the hybridization reaction. The apparent level of incorporations into X gene transcripts must be corrected for cross hybridization with the ovalbumin gene, and the final level after competition with ovalbumin cRNA should be a more accurate estimate. Given the gene-specific [^{3}H]counts hybridized, the total input of counts in the reaction and the percent recovery of potentially hybridizable counts obtained from internal standards, the percentage of newly synthesized nuclear RNA represented by specific gene transcript products was estimated. Under the conditions employed, the calculated values were 0.058%, 0.007% and 0.002% for ovalbumin, Y and X gene transcripts,

TABLE 2

Hybridization of Pulse Labeled [^{3}H]RNA Synthesized in Oviduct Tissue Slices to DNA Filters

Filter[1]	cRNA Competitor[2]	CPM Hybridized After Pulse[3]	Relative Recovery of ^{32}P Internal Standards	% of Total Sequence[4]
pOV12	---	4418	1.00	0.058
	pOV12	480	0.10	
	pY12.4	4550	0.83	
pY12.4	---	524	1.00	0.007
	pOV12	302	0.83	
	pY12.4	118	0.10	
pX8.0	---	292	1.00	0.004
	pOV12	174	0.63	0.002
	pY12.4	244	0.88	
	pX8.0	54	0.09	

1 7.5 x 10^6 cpm was used in every hybridization.

2 15 μg of competitor RNAs transcribed from various plasmid DNAs were used.

3 [^{32}P]RNAs transcribed from various plasmid DNAs were used as internal standards to monitor the recovery of hybridizable sequence. The recovery in this experiment was around 50%.

4 Each experiment has been carried out at least twice. The range of variation for a series of independent samples is ± 20%.

respectively. These values are consistent with the results obtained in the R_0t analysis studies described above, and suggest that the different steady-state levels of X, Y and ovalbumin gene transcripts reflect mostly differential rates of transcription of the 3 closely related genes.

Although the lebeling time was short (7.5 min), it was still possible that differential levels of ovalbumin, Y and X gene transcripts might be derived from differential rate of turnover. We have thus carried out the pulse-chase experiment by addition of excess unlabeled uridine and cytidine (5 mM each) to determine whether the difference can be due to RNA turnover. As shown in Table 3, the relative concentration of ovalbumin, Y and X gene transcripts in the chase samples is strictly proportional to that of the pulse labeled RNA transcripts. The overall decrease of readioactivity in the RNA transcripts of all three genes is most probably due to splicing of the precursor RNAs and simultaneous degradation of intervening sequence transcripts. These results would indicate that the ovalbumin gene and its pseudogenes are transcribed at different rates in the hormone stimulated oviduct tissue,and this difference is reflected in their respective concentration levels at steady state.

Although we do not know as yet whether the X and Y genes fulfill a functional role in the developing or mature oviduct tissue, the fact that the member genes of a closely linked gene cluster, existing in a presumably similar environment, are expressed as such different levels in response to steroid hormones is of great interest. While the precise cellular mechanism by which accumulation of pseudogene transcripts occurs remains to be determined, our studies designed to measure their rates of transcription in hormonally stimulated oviduct cells suggest that the steady-state levels of X and Y gene transcripts, like ovalbumin, are

TABLE 3

Hybridization of [^{3}H]RNA Synthesized in Oviduct Tissue Slices to DNA Filters after Chase

Filter	cRNA Competitor	CPM Hybridized After Chase[1]	Relative Recovery of ^{32}P Internal Standards
pOV12	---	926	1.00
	pOV12	110	0.07
	pY12.4	1026	0.76
pY12.4	---	114	1.00
	pOV12	90	0.76
	pY12.4	10	0.13
pX8.0	---	176	1.00
	pOV12	94	0.51
	pY12.4	112	0.84
	pX8.0	20	0.07

[1] Chase was for 30 minutes after addition of 5mM uridine and cytidine.

regulated to a large degree at the level of transcription. This being the case, one may wonder why the transcription rate of the X pseudogene is so poor compared to the authentic ovalbumin gene if in fact it has arisen by a genetic duplication event. Thus, it would be of immediate interest to compare the nucleotide sequence organization at putative regulatory sites located at the regions of initiation of transcription at each of these differentially expressed genes.

REFERENCES

Alwine, J. C., Kemp, D. J. and Stark, G. R. (1977). Proc. Natl. Acad. Sci. USA 74, 5350-5354.

Bailey, J. M. and Davidson, N. (1976). Anal. Biochem. 70, 75-85.

Benton, W. D. and Davis, R. W. (1977). Science 196, 180-182.

Berk, A. J. and Sharp, P. A. (1977). Cell 12, 721-732.

Breathnach, R., Benoist, C., O'Hare, Gannon, F. and Chambon, P. (1978). Proc. Natl. Acad. Sci. USA 75, 4853-4857.

Cox, R. F., Haines, M. E. and Emtage, S. (1974). Eur. J. Biochem. 49, 225-236.

Dodgson, J. B., Strommer, J. and Engel, J. D. (1979). Cell 17, 879-887.

Dugaiczyk, A., Woo, S. L. C., Lai, E. C. Mace, M. L., McReynolds, L. A. and O'Malley, B. W. (1979). Proc. Natl. Acad. Sci. USA 76, 2253-2257.

Gannon, F., O'Hare, K., Perrin, F., LePennec, J. P., Benoist, C., Cochet, M., Breathnach, R., Royal, A., Garapin, A., Cami, B. and Chambon, P. (1979). Nature 278, 428-439.

Harris, S. E., Rosen, J. M., Means, A. R. and O'Malley, B. W. (1975). Biochemistry 14, 2072-2081.

Hynes, N. E., Groner, B., Sippel, A. E., Nguyen-Huu, M. C. and Schutz, G. (1977). Cell 11, 923-932.

Kan, Y. W. and Dozy, A. M. (1978). Proc. Natl. Acad, Sci. USA 75, 5631-5635.

Lai, E. C., Woo, S. L. C., Dugaiczyk, A. D., Catterall, J. F. and O'Malley, B. W. (1978). Proc. Natl. Acad. Sci. USA 75, 2205-2209.

Lai, E. C., Woo, S. L. C., Bordelon-Riser, M. E., Fraser, T. H. and O'Malley, B.W. (1980). Proc. Natl. Acad. Sci USA 77, 224-248.

McKnight, G. S., Pennequin, P. and Schimke, R. T. (1975). J. Biol. Chem. 250, 8105-8110.

Monahan, J. J., Harris, S. E. and O'Malley, B. W. (1976). J. Biol. Chem, 251, 3738-3748.

O'Malley, B. W. and Means, A. R. (1974). Science 182, 610-620.

Palmiter, R. D., Moore, P. B., Mulvihill, E. R. and Emtage, S. (1976). Cell 8, 557-572.

Roop, D. R., Nordstrom, J. L., Tsai, S. Y., Tsai, M. J. and O'Malley, B. W. (1978). Cell 15, 671-685.

Roop, D. R., Tsai, M. J. and O'Malley, B. W. (1980). Cell 10, 63-68.

Royal, A., Garapin, A., Cambi, B., Perrin, F., Mandel, J. L., LeMeur, M., Bregegegre, F., Gannon, F., LePennec, J. P., Chambon, P. and Kourilsky, P. (1979). Nature 279, 125-132.

Southern, E. M. (1975). J. Mol. Biol. 98, 503-517.

Swaneck, G. E., Nordstrom, J. L., Kreuzaler, F., Tsai, J. J. and O'Malley, B. W. (1079). Proc. Natl. Acad. Sci. USA 76, 1049-1053.

Tsai, S. Y., Roop, D. R., Tsai, M. J., Stein, J. P., Means, A. R. and O'Malley, B. W. (1978). Biochemistry 17, 5773-5779.

Weinstock, R., Sweet, R., Weiss, M., Cedar, H. and Axel, R. (1978). Proc. Natl. Acad. Sci. USA 75, 1299-1303.

Woo, S. L. C., Rosen, J. M., Liarakos, C. D., Choi, Y. C., Busch, H., Means, A. R., Robberson, D. L. and O'Malley, B. W. (1975). J. Biol. Chem. 250, 7027-7039.

Woo, S. L. C., Dugaiczyk, A., Tsai, J. J., Lai, E. C., Catterall, J. F. and O'Malley, B. W. (1978). Proc. Natl. Acad. Sci. USA 75, 3688-3692.

Woo, S. L. C., Tsai, S. Y., Tsai, M. J., Lai, E. C., Mace, M. L. and O'Malley, B. W. (1979). Biochem. Biophys. Res. Commun. 89, 997-1005.

Woo, S. L. C. (1979), In: Methods in Enzymology, Vol. LXVIII (ed. Ray Wu), pp. 389-395.

THE PLASMID DNA REPLICATION CYCLE OF STREPTOMYCES GRISEUS

Xue Yugu, Zhuang Zenghui, Zhu Yingfang, Xu Yi and Dong Kening

Institute of Microbiology
Academia Sinica
Beijing, China

Abstract A series of electron micrographs showing various stages of replication of plasmid DNA of *Streptomyces griseus* have been obtained. Based upon an analysis of these electron micrographs, the plasmid DNA replication cycle of *S. griseus* is believed to involve at least three steps:

1. Primary initiation step.
2. Replication step.
3. Termination step.

From the average contour length of the plasmid DNA molecule measuring 1.3 μm and a generation time, 120 minutes including the replication time, about 15.5 minutes, it is calculated that the plasmid genome comprises 3.8×10^3 nucleotides pairs and the rate of chain growth of the replica would be four nucleotides per second. As one complete rotation is required for every ten nucleotides added to the growing replica chain, the replicating plasmid DNA would rotate at about 24 rpm to perform each replication.

INTRODUCTION

Our previous studies have shown that there are several different patterns of plasmid DNA molecules of *S. griseus* No.45 using electron microscopic (EM) observations. The molecular weight of the plasmid was calculated from the average plasmid DNA contour length. The thermal denaturation temperature (Tm) value and G-C percentage of the plasmid DNA were also measured(1,2). This paper is concerned with EM observations of the plasmid DNA replication cycle of *S. griseus* which will be of value for further studies of the mechanism of plasmid replication and the relationship of plasmid replication of S. griseus to the expression of genes involved in streptomycin synthesis.

MATERIAL AND METHODS

1. Strains: Streptomyces griseus No. 45.
2. Media: As previously described[2].
3. Cultural conditions:

S. griseus was transfered to seed growth medium in Erlenmeyer flasks, incubated at 28°C on a rotatory shaker, 200-220 rpm, for 24 hrs. Five percent of the seed solution was inoculated or a piece of agar culture from a slant was cut and directly to the growth flask and incubated at 28°C for about 40-48 hrs.

4. Method for isolation of plasmid DNA and sampling for electron microscope observation: As previously described[2].

RESULTS AND DISCUSSION

The plasmid DNA used was isolated from the growing population of a 40 to 48-hour culture of S. griseus No.45. Its generation time is about 120 minutes at 28°C[3]. The generation time comprises the time for accumulation of an "initiator" complement of constant size per origin, the time for a round of plasmid DNA replication and the time between the end of a round of replication and following cell division. While we observed and calculated the different replication forms from 163 plasmid DNA configurations under EM, we found that there were 104 in a primary initiation stage, 21 in a replication stage and 38 in the termination stage. The percentage of replication stage is estimated to be 21/163 = 12.88%. The time necessary of this stage is calculated as 120 x 12.88%, that is 15.5 minutes or 930 seconds.

A set of electron micrographs representing various stages of the plasmid DNA replication cycle of S. griseus have been obtained. (Fig 1) Based upon an analysis of these electron micrographs, the plasmid DNA replication cycle of S. griseus is believed to involve at least three steps.

(1) Primary initiation step.

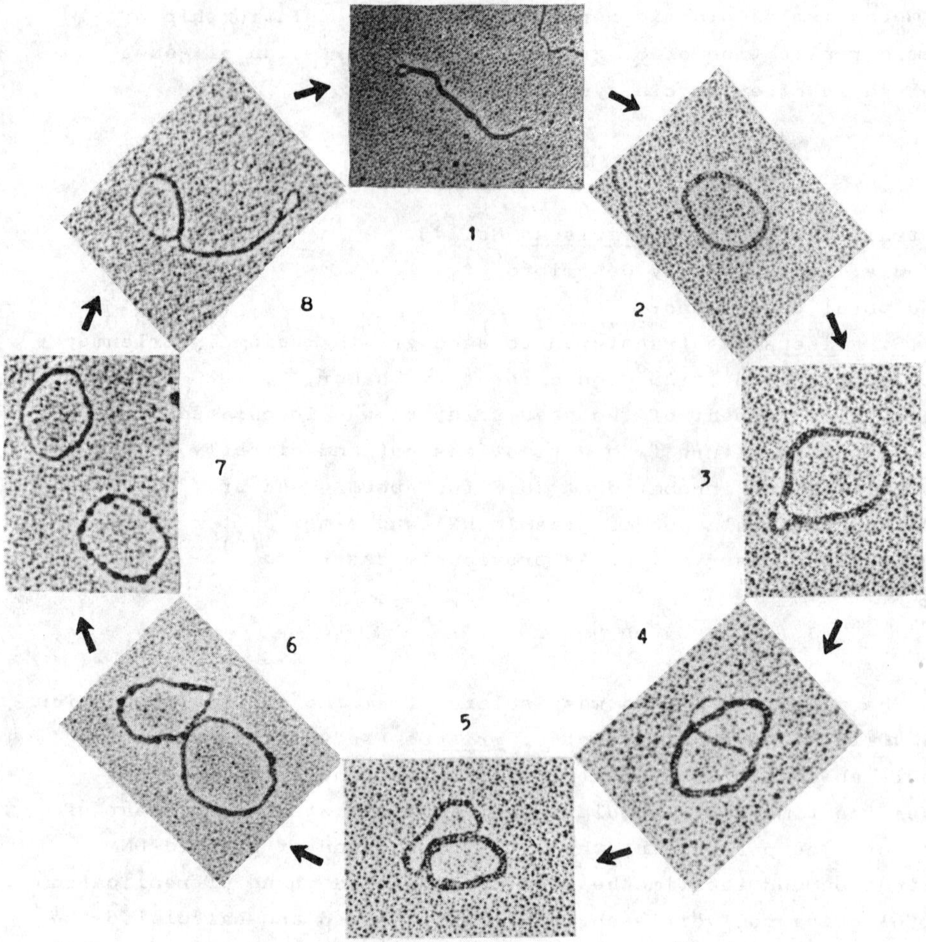

Fig 1. Electron micrographs showing plasmid replication cycle of <u>Streptomyces</u> <u>griseus</u> x6000

(Figure reduced to $\frac{7.5}{10}$ during printing)

A supercoiled duplex monomer, (Fig. I (1)) is released to open circular monomers (Fig. I (2)). Once a nick is induced into a single strand, this permits separation of the two parental strands into an open circular monomer[5] (Fig. I (2)).

(2). Replication step.

The double strands begin to separate at the position opposite to the nick of the complementary strand, develop a D-loop and form an initial point P_1 of replication, (Fig. I (3)). The synthesis of complementary replica strand appears to proceed unidirectionally from a single origin P_1 along both strands of the template direct to P_2 of the replication Y-fork. The distance between P_1 and P_2 is increased to nearly one half on DNA contour length and then decreased until two mature covalently closed duplex DNA monomers are nearly completed (Fig. I (3)-(6)).

(3). Termination step.

Separation of two open circular DNA monomers (Fig. I (7)) from the replication origin to form partial supercoiled duplex DNA monomers by partial rotation of the relica (Fig. I (8)), and continue the rotation to form supercoiled duplex DNA monomers (Fig. I (1)).

From these EM observations of plasmid DNA of S. griseus, it can be concluded that the plasmid DNA of S. griseus is a closed circular double strand structure.

Since the DNA double helix contains ten nucleotide base pairs per 34 Å or $3.4 \times 10^{-3}\ \mu m$, and knowing the average contour length of the smaller plasmid DNA of S. griseus to be 1.3 μm, we can now calculate that the plasmid DNA of S. griseus comprises

$$1.3 \times 10/3.4 \times 10^{-3} = 3.8 \times 10^{3} \text{ b.p.}$$

In the case of stringent replication of plasmid DNA of S. griseus, one cycle of replication can be achieved in a period about 15.5 minutes or 930 seconds at 28°C. Then the rate of chain growth of the replica nucleotide chains would have to be at least $1.3 \times 10/3.4 \times 10^{-3} \times 930 = 4$ b.p/second. In other words, this represents an increase of four nucleotides per second. It is thus considered that, while in replicating in a semiconservative manner, the parental double helix must rotate to allow separation of its complementary nucleotide strands. As one complete rotation is required for every ten nucleotides added to the growing replica chain, then in the case of one replicating Y-fork, the replicating plasmid DNA would rotate a full round in at least 4 x 60/10 = 24 rpm.

Whether the replication model of *S. griseus* plasmid DNA and the characteristics of plasmid DNA such as the nucleotide base pairs, the rate of chain growth, rotation of replica nucleotide chain and its replication time reported above is of general significance in *Streptomyces* and their relations to the expression of genes in streptomycin synthesis await further study.

ACKNOWLEDGMENTS

We should like to thank comrades He Nengbo, Chen Jianzhang, Lin Caichan, Xue Hanhuang and Su Shucheng of Institute of Microbiology of Province Guangdong, Guangzhou, for their invaluable help in preparing most of these electron micrographs.

References

1. Xue Yugu et al: *Acta Microbiol Sinica* 18 (3): 195-201, 1978.
2. Xue Yugu et al: *Acta Microbiol Sinica* 18 (4): 287-292, 1978.
3. Zhuang Zenghui et al: *Acta Genetica Sinica* 7 (4), 1980.
4. Helmstetter, C. S. Cooper et al: C.S.H. Symposia Quantit. Biol. Vol 33, p 809. 1968.
5. Roth, T. F. and D. R. Helinski, P.N.A.S. 58: 650, 1967.

EGG mRNA ENCODED PROTEIN SYNTHESIS: IMMUNOCHEMICAL DETECTION OF ALBUMIN AND HEMOGLOBIN*

Yu Jiankang M.C. Niu**

(Institute of Developmental Biology, Academia Sinica)

Abstract: Poly A-attached RNAs were isolated from carp eggs and tested in the cell-free wheat germ translational system. The reaction products were analyzed with methods involving immunoprecipitation, polyacrylamide gel electrophoresis and peptide mapping analysis. The results revealed that carp egg mRNAs contain subfractions encoding the synthesis of albumin and hemoglobin.

INTRODUCTION

It has long been known that stored or preformed mRNAs are present in unfertilized eggs of amphibians and sea urchins. These mRNAs are stored in the form of ribonucleoprotein particles. After fertilization, they are activated and used as templates for protein synthesis. The kind of mRNAs present in eggs is a subject of current interest. So far as the authors are aware, those we know of are related to cell division, for instance, histone mRNA (see Gross in this volume) and tubulin mRNA.(1) Back in 1973, we isolated poly A-attached RNAs from ovarian and mature eggs of crucian and carp. When they were injected into fertilized goldfish eggs, the developed goldfish had their caudal fin changed from the goldfish's veil-shape to the carp's single.(2) In order to see if egg mRNA had also mediated the alteration of other organs, liver of the induced single-tailed goldfish was first analyzed, using lactate dehydrogenase (LDH) as a genetic marker. Among the cathodally migrated bands of liver specific LDH-C_4, a new intermediate band appeared in liver of the goldfish injected with egg mRNAs as well as carp liver mRNAs.[3] The latter finding leads to the proposal that carp egg mRNAs contain a component responsible for the formation of liver. In order to further support this hypothesis, egg mRNAs were tested in the cell-free wheat germ translational system and the reaction products analyzed. Here we report the data demonstrating that carp egg mRNAs contain subfractions capable of encoding the synthesis of albumin and hemoglobin.

EXPERIMENTAL METHODS AND RESULTS

(1) Isolation and purification of carp egg mRNAs: Polysomal RNAs of carp eggs were isolated and applied to an oligo(dT)-cellulose column as desbribed previously.[4] There were three peaks: A, B, C (Fig. 1), and peak C was the partially purified mRNAs of carp eggs. Poly A-attached mRNAs amount to about 1-2% of the total polysomal RNAs.

(2) In vitro translation of carp egg mRNAs: A crude extract was prepared from wheat germ of the Peking Agricultural University strain No. 139 and stored in a freezer (Revco) -80°C.[4] The incorporation of amino acids into acid insoluble materials (proteins) was assayed in the cell-free wheat germ translational system according to methods of Maious Dudock.[5] Total volume of reaction mixture was 50 µl. It was incubated for 30 min at 30°C. Radioactivity of the labelled product was counted with a Beckman LS 9000 liquid scintillation counter. Fig. 2 shows the time course of the carp egg poly A-attached RNA encoded protein synthesis. It

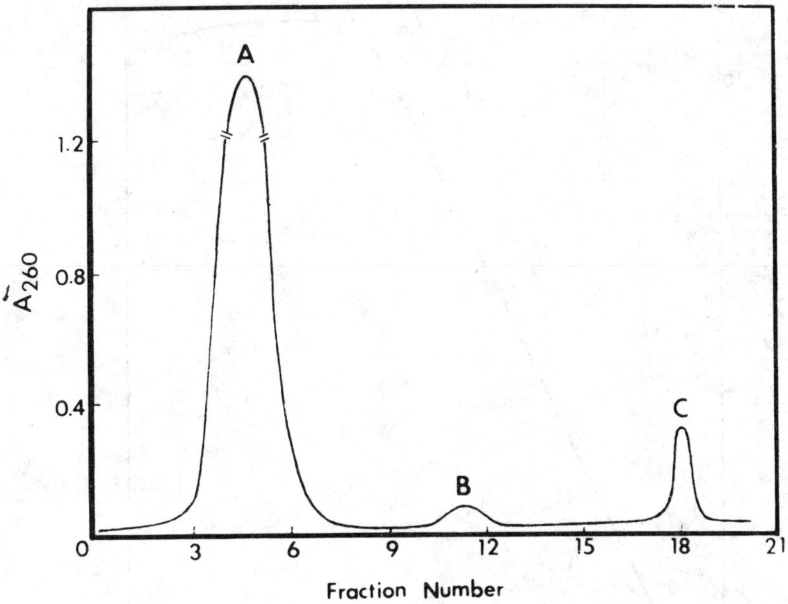

Fig. 1 Oligo(dT)-cellulose chromatography of carp egg polysomal RNA. Peak A: material that is not retained by the column; Peak B: material eluted with buffer A (0.2M KCl, 0.01M Tris-HCl pH 7.5), Peak C: material eluted with buffer B (0.01M Tris-HCl pH 7.5).

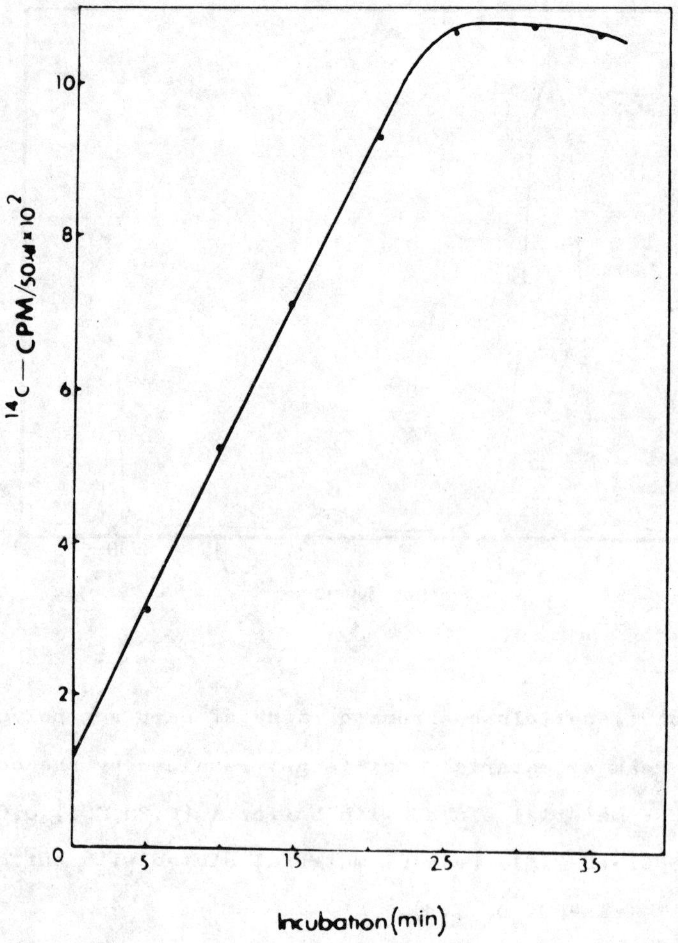

Fig. 2 Time course of carp egg mRNA encoded incorporation of ^{14}C-leucine into protein.

can be seen that under our experimental conditions 30 min of incubation was optimal. The rate of the RNA programmed protein synthesis varied in accordance with the concentration of egg mRNA (Fig. 3). Under our experimental system, 5-6µg per 50 µl of the incubation mixture was optimal.

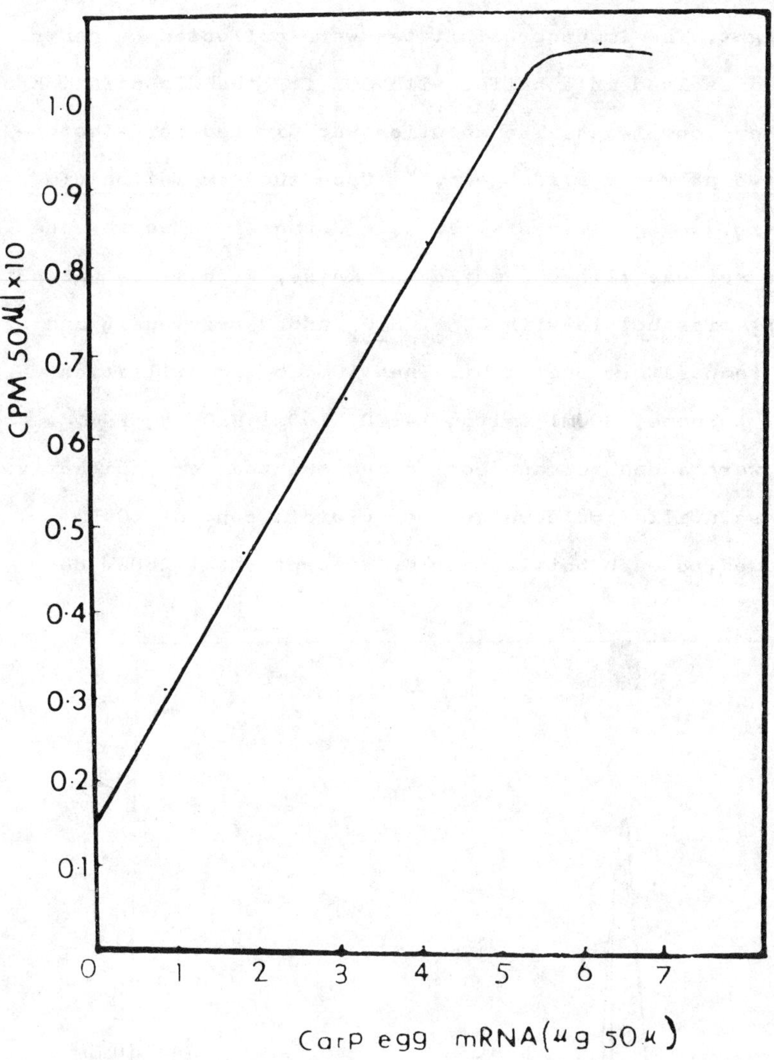

Fig. 3 The effect of mRNA concentration on egg mRNA encoded incorporation of ^{14}C-leucine into protein.

(3) Polyacrylamide gel electrophoresis: For this experiment, the volume of the reaction mixture was increased to 500 µl and five of them were pooled and centrifuged. To the supernatant we added either authentic carp albumin and rabbit antiserum against carp albumin or authentic hemoglobin and rabbit antiserum against carp hemoglobin. The mixture was incubated at 37°C for 1 hour and kept

at 4°C overnight. The immunoprecipitates were collected by centrifugation and dissolved in a buffer with SDS for the dissociation of antigen-antibody complexes. The solution was divided for electrophoresis on two polyacrylamide gels.(6) Upon the completion of electrophoresis, one gel was stained with Coomassie blue for one hour. Another gel was sliced 2 mm in thickness. Each slice was put into a scintilation bottle with 0.5ml H_2O_2 added previously and kept at room temperature overnight. Then 10ml of scintillation liquid (600ml toluene, 300ml Triton X-100, 100ml H_2O, 4g PPO, 0.05g POPOP) were added to each bottle and counted for radioactivity by Beckman's scintillation counter. For protein content, 0.5ml 1N HCl was added to each bottle, and UV absorption at 280nm de-

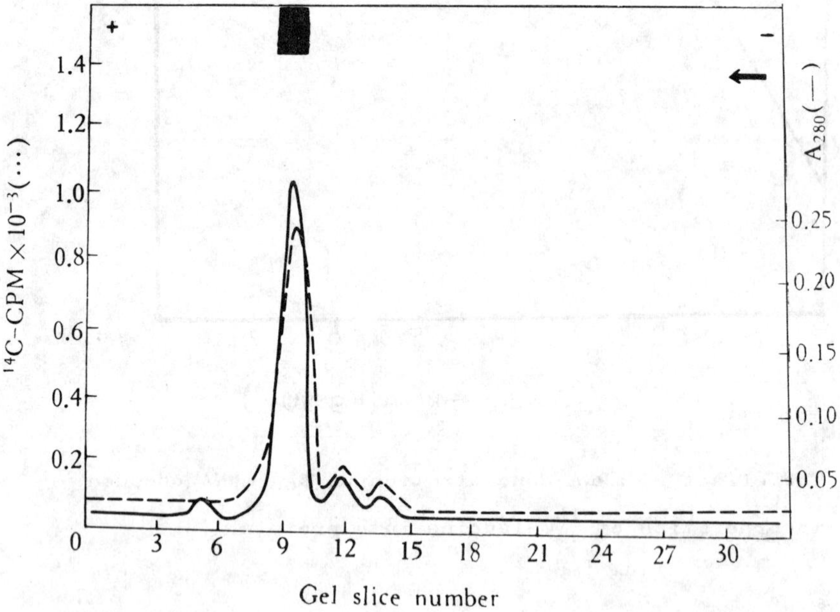

Fig. 4 Polyacrylamide gel electrophoresis of anti-albumin serum immunoprecipitable material from reaction products. A_{280} • —— • , CPM o ---- o, and albumin is located in top band.

termined with a Gilford spectrophotometer. The electrophoretic patterns of anti-albumin and anti-hemoglobin sera immunoprecipitable materials from the reaction products are shown respectively by figures 4 and 5. In both cases, the newly synthesized protein

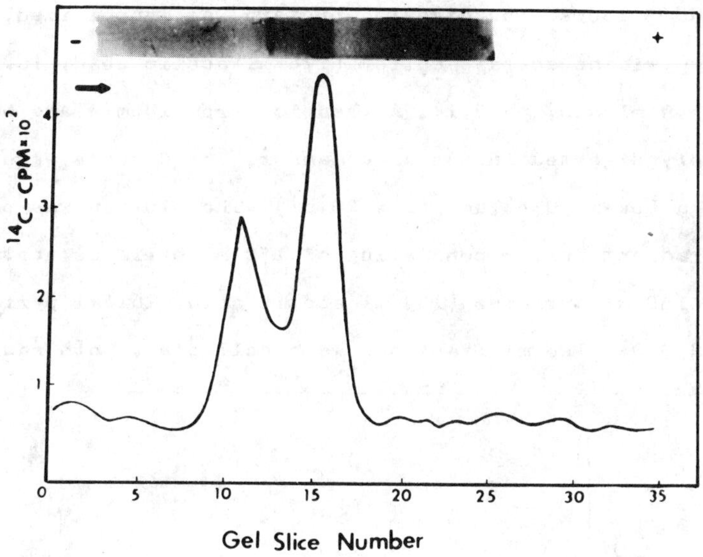

Fig. 5 Sodium dodecyl sulfate/polyacrylamide gel electrophoresis of anti-globin serum immunoprecipitable material from reaction products in the cell-free wheat germ translational system. The arrow shows the direction of migration. CPM o —— o, and hemoglobin is located in top band.

(^{14}C-labelled) and the authentic albumin or hemoglobin migrated at the same speed. This means that the newly synthesized protein is immunologically and electrophoretically indistinguishable from the authentic albumin or hemoglobin.

(4) <u>Tryptic peptide analysis</u>: Tryptic peptide analysis was carried

out by a modified procedure of Aviv and Leder [7] and Bloemendal.[8] The immunoprecipitates were dissolved in 1 ml of 0.1M NH_4HCO_3 (pH 8.0). After adding trypsin (40μg/ml), the mixture was incubated at 37°C for 16 hours. Additional trypsin (40μg/ml) was added and the incubation continued 5 hours. The digested protein was lyophilized, and dissolved in pyridine acetate buffer (278 ml acetic acid, 16.1 ml pyridine, 705.9 ml H_2O, pH 3.1). Authentic carp albumin and hemoglobin were separately digested in the same manner. The digests were chromatographed on a Dowex 50 column (1 x 20 cm). The elution was carried out with a gradient buffer consisting of 150 ml pyridine acetate (pH 3.1) and 150 ml pyridine (139 ml acetic acid, 161 ml pyridine, 700 ml H_2O pH 5.0). Two ml fractions were collected. Both radioactivity

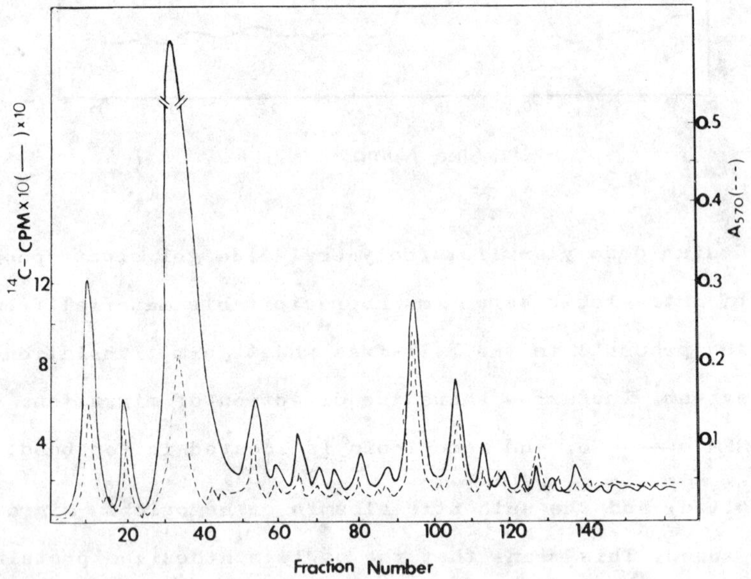

Fig. 6 Ion exchange chromatography of peptides produced by tryptic digestion of anti-carp-albumin serum precipitated ^{14}C-leucine labelled reaction product (. ——— .) and carp albumin (A_{570} o ---- o).

and absorbance at wave length 570 nm were determined. Fig. 6 shows the distribution of peptides (various sizes) obtained from trypsin digests of the newly synthesized and authentic albumin. The correspondance of the peptide maps between these two albumins suggests their similarity in chemical composition. The same analysis was applied to the newly synthesized hemoglobin. The results are shown in figure 7. Except for one point, the two peptide maps are also comparable.

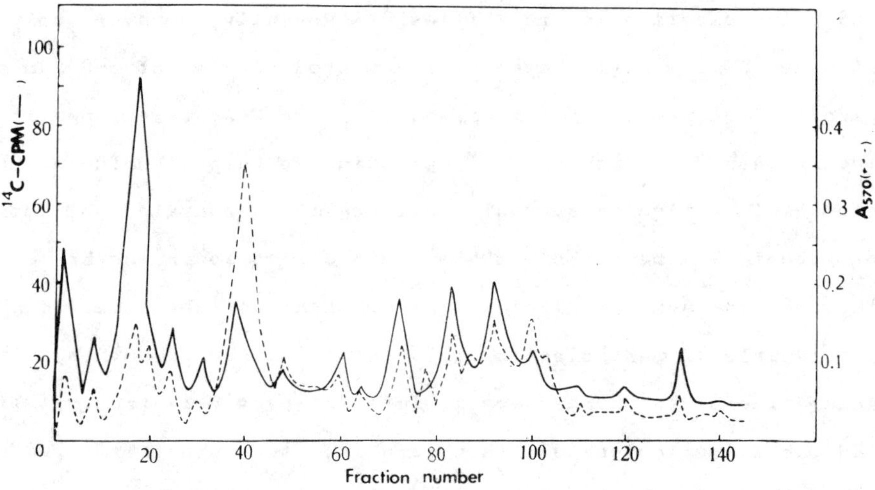

Fig. 7 Ion exchange chromatography of peptides produced by tryptic digestion of anti-carp-hemoglobin serum precipitated ^{14}C-leucine labelled reaction product (. ——— .) and carp hemoglobin (A_{570} o ---- o).

DISCUSSION

It was T.A. Morgan who emphasized cytoplasmic effects on the nucleus of developing egg.[9] Since then many investigators have studied different aspects of these effects. The methods used include nuclear transplantation, cell fusion and micro-injection. While studies on nuclear transplantation and cell fusion show respectively the effect on the nucleus of developing eggs and in cell culture, micro-injection reveals cytoplasm mediated repair of genetic defects in axolotl[10] and fruit flies.[11] These authors provide evidence to show that the active agent in cytoplasm is not species specific, for instance, some unspecific protein in axolotl[12] and polynucleotide in fruit flies.[13] Recently, however, poly A attached RNAs were isolated from advanced oocytes of crucian and carp. In the preceeding experiments, these RNAs were found to encode the synthesis of such specific proteins as carp albumin and hemoglobin. This finding suggests that egg mRNAs contain subfractions of albumin and hemoglobin-mRNAs. In order to learn whether or not the albumin- and hemoglobin-mRNA components of the stored egg mRNAs are capable of participating in organ or tissue formation, liver- and reticulocytes-mRNAs were isolated respectively from rat liver and rabbit reticulocytes. They were injected into fertilized goldfish eggs. Liver of the rat-liver mRNA injected goldfish possessed two cathodally migrated bands, one of which was greatly affected by the injected mRNA. The red blood cells of the rabbit reticulocyte mRNA injected goldfish were found to possess also rabbit hemoglobin, and LDH bands of the rabbit erythrocytes.[14] Based upon these findings that (1) egg mRNAs encodes the synthesis of carp albumin and hemoglobin and (2) rat liver and rabbit reticulocytes RNAs inside goldfish eggs participate respectively in the development of liver and erythrocytes, the authors conclude that carp egg mRNAs contain subfractions of liver- and erythrocyte- forming mRNAs.

* The authors wish to thank Mr. Shen Shoukun and Mr. Wu Naihu for their excellent technical assistance.

** Honorary Research Professor, Institute of Zoology, Academia Sinica and Professor of Biology of Temple University, U.S.A.

REFERENCES

(1) Raff, R.A., Colot, H.V., Selvig, S.E. & Gross, P.R., Nature, 221 (1972), 214.

(2) Tung, T.C. & Niu, M.C., Scientia Sinica, 16 (1973), 377.

(3) Niu, M.C. & Tung, T.C., Ibid, 10 (1977), 803.

(4) Niu, M.C., Yu Jiankang & Song Dexiu, ibid, 23 (1980), 510.

(5) Maious, K. & Ducock, Nucleic Acids Research, 1 (1974), 1385.

(6) Yang, S.F. & Niu, M.C., Proc. Nat. Acad. Sci., 74 (1977), 1894.

(7) Aviv & Leder, ibid, 69 (1972), 1408.

(8) Auton, Berns, J.M. & Bloemendal, H., Methods in Enzymology, 30 (1974), 675.

(9) Morgan, T.H., 1934 Embryology and Genetics, New York: Columbia Univ. Press.

(10) Briggs, R. & Justus, J. T. J. Exp. Zool. 167 (1968) 105.

(11) Garen, A. & Gehring, W., ibid, 69 (1972), 2982.

(12) Briggs, R., J. Exp. Zool., 181 (1972), 271.

(13) Okada, Kleimman & Schneiderman, Develop. Biology, 37 (1974 b), 55; 39 (1974 c), 286.

(14) Niu, M.C., Niu, L.C. & Xue, K.H., American Zoologist, 19, no. 3 (1979), 978.

PURIFICATION AND CHARACTERIZATION OF PROTAMINE mRNA FROM HYBRID CARP TESTIS

Zhao Hui-Zhi and Lin Wan-Lu

Institute of Biophysics, Academia Sinica, Peking, China

SUMMARY

The poly $(A)^+$ protamine mRNA was isolated and purified from postmitochondrial supernatant of hybrid carp testis. Addition of this mRNA to a wheat germ cell-free system significantly stimulated both (^{14}C) arginine and (^{14}C)isoleucine incorporation into acid-insoluble material. Polyacrylamide-urea gel (pH 2.0) electrophoresis showed that these labeled amino acids were incorporated in vitro into synthesized polypeptide in response to this mRNA identical with authentic protamine. This mRNA migrated as a single band in polyacrylamide-urea gel (pH 8.3) and the size of this mRNA as estimation by its mobility in compare with markers of tRNA (4S) and 5S RNA was close to 6S.

INTRODUCTION

Protamine mRNAs are among the smallest mRNAs known with about 300-350 nucleotides (1). They have all of the structural features and functional capabilities of general eukaryotic mRNA but also possess a unique feature. Thermal denaturation and T_1 nuclease resistance studies (2.3.4.) indicate that they possess considerable secondary structure. Protamine synthesized during the late stages of spermatogenesis using protamine mRNA as template can displace histones and play an uniqe role in the packing and derepression of sperm DNA (5.6.7.). It is evident that this mRNA must play a significant role in the regulation of protein synthesis and cellular differentiation. Thus protamine mRNAs are good material for studying the relationship between structure and function of eukaryotic mRNA, protein synthesis and cellular differentiation, as well as the control of the expression of the genes.

We used a simple column chromatography procedure (8) to isolate and purify milligram quantities of poly(A)$^+$ protamine mRNA from hybrid carp testis. Thus, many advantages can be provided for studying the structure and function of this class of mRNA. In addition we have developed an improved method for low pH polyacylamide-7M urea slab gel electrophoresis to characterize translation product of poly(A)$^+$ protamine mRNA in wheat germ cell-free system. In this paper we describe the procedure of purification and characterization of poly(A)$^+$ protamine mRNA from hybrid carp testis in detail.

EXPERIMENTAL PROCEDURES

In all cases solutions were prepared using distilled water. All solutions and glasswares were sterilized before use.

Preparation of Postmitochondrial Supernatant

Hybrid carp (red carp×mirror carp, mature in one year) testis at the protamine stage was collected in March, 1979 at Nanning, Guanxi, China, and stored at -80°C. The tissue was homogenized in 2 vol. of buffer containing 75mM NaCl, 5mM Tris-HCl, pH7.6, and 25mM EDTA (disodium salt) in a Waring Blendor at full speed for 3 min at 4°C. The homogenate was centrifuged at 7,000 rpm to remove nuclei and debris. The postnuclear supernatant after filtration through cheesecloth was centrifuged further for 30 min at 18,000 rpm to remove the postmitochondrials. The postmitochondrial supernatant was used as a source for protamine mRNA preparation. All operation was carried out at 0-4°C as described by Gedamu et al. (8).

DEAE-Cellulose Chromatography

The procedure was performed at room temperature (about 20°C) (8). The postmitochondrial supernatant was adjusted to 0.25M NaCl, 0.5% SDS, 10mM Tris-HCl, pH7.6 and applied to a DEAE-cellulose column (2.5cm×25cm) equilibrated in the application buffer without SDS. The column was then washed with the application buffer to remove most of the cellular basic protein, which did not bind to DEAE-cellulose. The low-molecular-weight RNA fraction bound to DEAE-cellulose was eluted with 0.5M NaCl, 0.5% SDS, 10mM Tris-HCl, pH7.6, as a single peak. Then the high-molecular-weight RNA was eluted with the same buffer containing 1M NaCl. The low-molecular-weight RNA fractions were pooled for further purification.

Oligo(dT)-Cellulose Chromatography

The low-molecular-weight RNA fractions eluted from the DEAE-cellulose column with the buffer containing 0.5M NaCl was applied directly to on the oligo(dT)-cellulose column (1g, Type 7, P-L, Biochemicals, Inc.) equilibrated in 0.5M NaCl, 10mM Tris-HCl, pH7.6, at room temperature (8.9.). After the unbound poly(A)$^-$RNA passed through the column, the column was washed with 0.5M NaCl, 10mM Tris-HCl, pH7.6, followed by the same buffer containing 0.1M NaCl at 4^oC. Then the poly(A)$^+$ RNA was eluted with water. The poly(A)$^+$ RNA was made 0.24M in ammonium acetate (10) precipitated with 2 Vol. of ethanol, dried, dissolved in water and stored at- 10^oC or -20^oC.

Polyacrylamide Gel Electrophoresis of Poly(A)$^+$ protamine mRNA

The method of Brownlee and Cartwright (11) was modified for analysis of poly(A)$^+$ protamine mRNA on polyacrylamide slab gel (pH 8.3). Polyacrylamide slab gels (32cm x 15cm x 0.15cm) were prepared in glass plates sealed in plastic bag. Gel solution contained 7% acrylamide, 0.35% N, N' - methylene bisacrylamide, 7M urea, 90mM Tris-borate, pH8.3 and 0.1mM EDTA. The reservoir buffer was the same as the gel buffer. The gel was pre-electrophoresed for 2 hr at 500 V prior to loading samples, and then electrophoresed for 4 hr at 700 V under completely denaturing condition. The gels were stained with Stains-All (12) for detecting RNA zones.

Assay of Translation Activity of Poly(A)$^+$ Protamine mRNA in Vitro

The poly(A)$^+$ protamine mRNA activity was assayed in a wheat germ cell-free system as described previously (13. 14). A wheat germ cell-free system (S-26) was prepared as described (14). The poly(A)$^+$ protamine mRNA activity was assayed in a 50 μl of reaction mixture that contained 20 μl of wheat germ S-26, 25mM Tris-HCl, pH7.6, 70mM KCl, 2.5mM magnesium acetate, 2mM DTT, 1mM ATP, 0.1mM GTP, pyruvatekinase (0.2mg/ml), phosphoenolpyruvate (2.0mg/ml), 40 μM (each) of 18 nonradioactive amino acids except arginine and isoleucine, (^{14}C) arginine (312mCi/mmol) and (^{14} C)isoleucine (312mCi/mmol). The amount of the poly(A)$^+$

protamine mRNA used was listed in Table1.The reaction mixtures were incubated at 30°C for 60 min. At the end of incubation, an aliquote of 15 μl reactoin mixture was taken and saved for characterization of the translation products after added carrier protamine by acid gel electrophoresis (pH 2.0). The rest of the reaction mixture was treated by addition of TCA to measure the reactivity.

Characterization of Translation Product of the Poly(A)$^+$ Protamine mRNA in Vitro

A low pH polyacrylamide slab gel electrophoresis was developed for characterization of translation product of the protamine mRNA in wheat germ cell-free system(14). Acid polyacrylamide slab gels (19cm×10cm×0.15cm) were prepared with 7% acrylamide, 0.35% N, N'-methylene bisacrylamide, 7M urea, 94mM citric acid, 12mM Tris-H_3po_4, pH 2.0. The reservoir buffer was the same as the gel buffer. Following pre-electrophoresis this gel for 1.5 hr at 150 V, an aliquote of 15 μl reaction mixture taken from the cell-free protein synthesis reaction mixture added carrier protamine was loaded directly on the acid slab gel. At the same time, a sample of 15 μl cell-free protein synthesis reaction mixture without the poly(A)$^+$ protamine mRNA was loaded directly on the adjacent slot of the gel to serve as control. Electrophoresis was carried out for 2 hr at 340 V at room temperature. The sheets of gels were stained in Coomassie brilliant blue solution (15) for 30 min and washed with water to remove excess dyes on the surface of gels. Then the rapidly stained gels were cut into 2 mm slices. Each slice was solubilized in 300 μl of hydrogen peroxide (30% W/V) and 150 μl perchloric acid (60% W/V) by incubation at 60°C for 5 hr and counted in 5 ml of Bray scintillation fluid.

RESULTS

We have used the method described by Gedamu et al. (8) to isolate and purify milligram quantities of poly(A)$^+$ protamine mRNA from 1 kg of hybrid carp testis. Our experimental operation is outlined in Scheme 1.

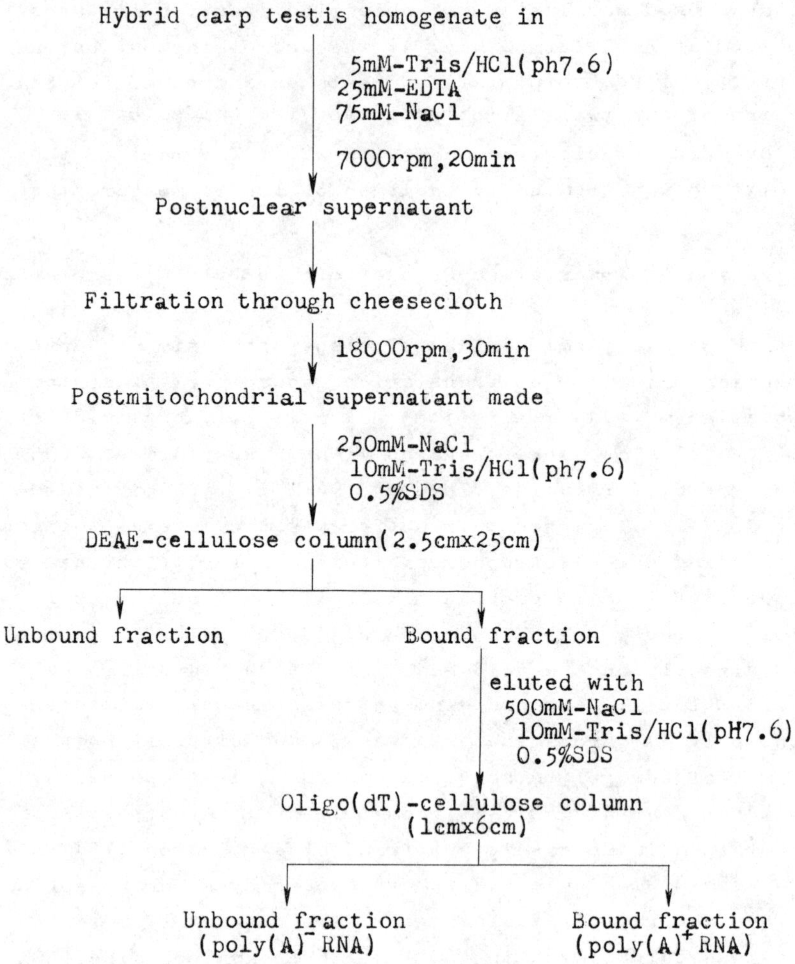

Scheme 1. Scheme for the isolation of protamine mRNA from hybrid carp testis

Using the above procedure, postmitochondrial supernatant was chromatographed on the DEAE-cellulose column to isolate the low-molecular-weight RNA.Fig. 1 shows a typical absorption profile of low-molecular-weigh RNA fraction eluted from the DEAE-cellulose

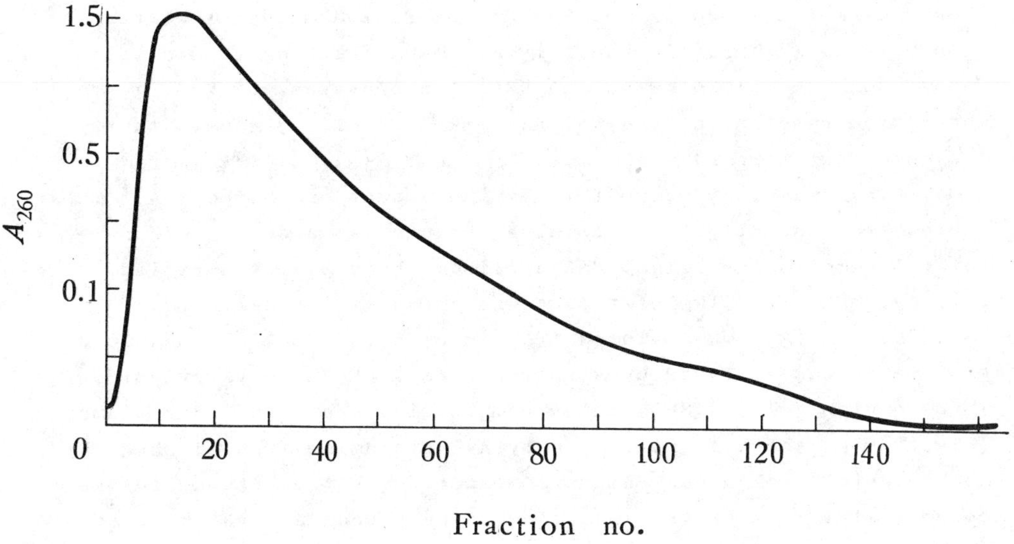

Figure 1. Chromatography of hybrid carp testis postmitochondrial supernatant on a DEAE-cellulose column.
Hybrid carp testis postmitochondrial supernatant in 0.25M NaCl, 0.5% SDS, 10mM Tris-HCl, pH7.6, was applied to a DEAE-cellulose column (2.5cm x 25 cm) equilibrated in the application buffer without SDS. After washing out unbound fraction with the application buffer, the RNA containing mRNA and acid protein were eluted with 0.5M NaCl, 0.5% SDS, 10 mM Tris-HCl, pH7.6. About 3,500 A_{260} units of low-molecular-weight RNA/kg of tissue was obtained.

column with 0.5M NaCl, 0.5% SDS, 10mM Tris-HCl, pH7.6. The low-molecular-weight RNA fraction was further chromatographed on the oligo (dT)-cellulose column. Only one peak of the poly(A)$^+$ mRNA was eluted with water, after the column was subsequently washed with 0.5M NaCl, 0.1M NaCl to remove all of unbond material. As shown in Fig. 2 there was no 260 nm absorption material eluted with 0.1M NaCl. Threfore the poly(A)$^+$ mRNA fraction seemed to be free from high-molecular-weight RNAs and proteins. U. V.-absorption spectrum of the poly(A)$^+$ mRNA in Fig. 3 showed that the characteristic ratios, A_{260}/A_{230}, A_{260}/A_{280} and A_{280}/A_{260} were 2.04, 2.17 and 0.46, respectively. The amount of the poly(A)$^+$ mRNA eluted from the oligo(dT)-cellulose column was about 1% of the total low-molecular-weight RNA applied. The yield of purified poly(A)$^+$ mRNA from 1kg of tissue was approximatly 1-1.9 mg.

The poly(A)$^+$ mRNA eluted from an oligo(dT)-cellulose column with water was analyzed by electrophoresis on 7% polyacrylamide-7M urea slab gel (pH 8.3) under completely denaturing conditions (Fig. 4). As shown in Fig. 4, poly(A)$^+$ protamine mRNA prepared by the oligo(dT)-cellulose column in tandem with the DEAE-cellulose column (slot 1) migrated as a single sharp band and the size of this mRNA as estimated by its mobility as compared with markers of tRNA(4S) and 5S RNA was close to 6S. Other closely spaced minor bands were not found near the major band.

Addition of the poly(A)$^+$ protamine mRNA to the wheat germ S-26 cell-free protein synthesis system stronly stimulated both (^{14}C)arginine and (^{14}C)isoleucine incorporation into acid-insoluble material. Table 1 shows the results that the poly(A)$^+$ protamine mRNA translation activity was assayed in the wheat germ S-26 system. As shown in table 1, the stimulation of both (^{14}C)arginine and (^{14}C)isoleucine incorporation by the poly(A)$^+$ protamine mRNA was found to be approximatly 8-fold.

The polypeptide products synthesized in the wheat germ translational system in response to the poly(A)$^+$ protamine mRNA were analysed by polyacrylamide-7M urea slab gel electrophoresis (pH 2.0). In the presence of this mRNA a major radioactive band was found to co-migrated with carrier protamine as shown in the stained gel (Fig. 5a). However, two other faint radioactive band near the upper part of the gel far from the major band were also observed. These two faint radioactive bands just co-migrated with

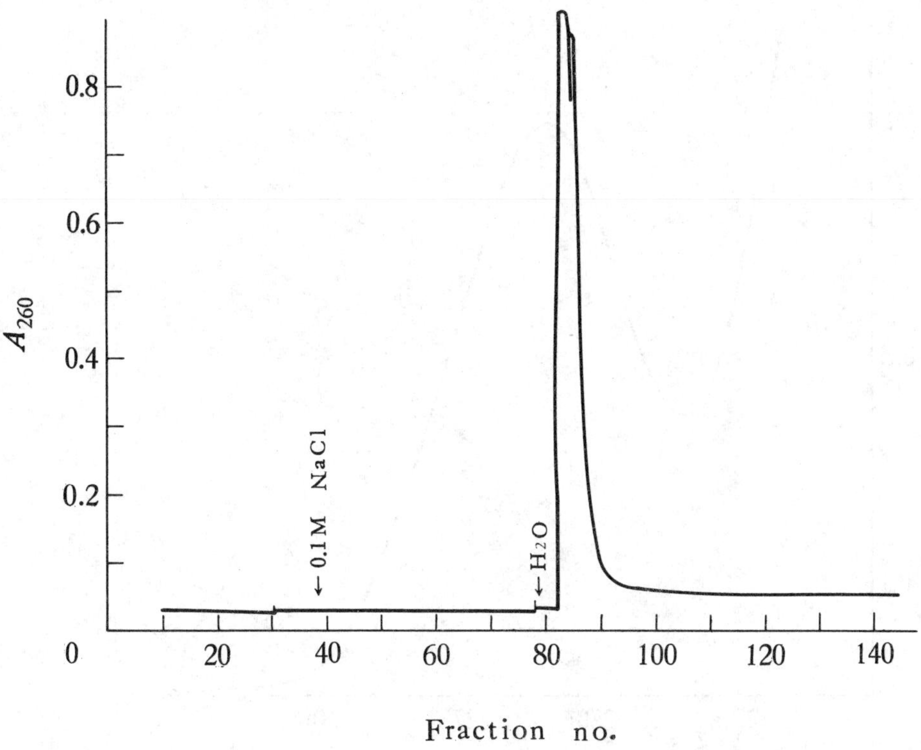

Figure 2. Chromatography of low-molecular-weight RNA species from the DEAE-cellulose column on an oligo(dT)-cellulose column. The low-molecular-weight RNA fraction from the DEAE-cellulose column with 0.5M NaCl, 0.5% SDS, 10mM Tris-HCl, pH7.6, (Fig. 1) was applied directly to an oligo(dT)-cellulose column at room temperature (8.9.). The column was washed with 0.5M NaCl, 10mM Tris-HCl, pH7.6, at 4°C, to remove unbound fraction and SDS, followed by 0.1M NaCl, 10mM Tris-HCl, pH7.6. The poly(A)$^+$ mRNA was finally eluted with water.

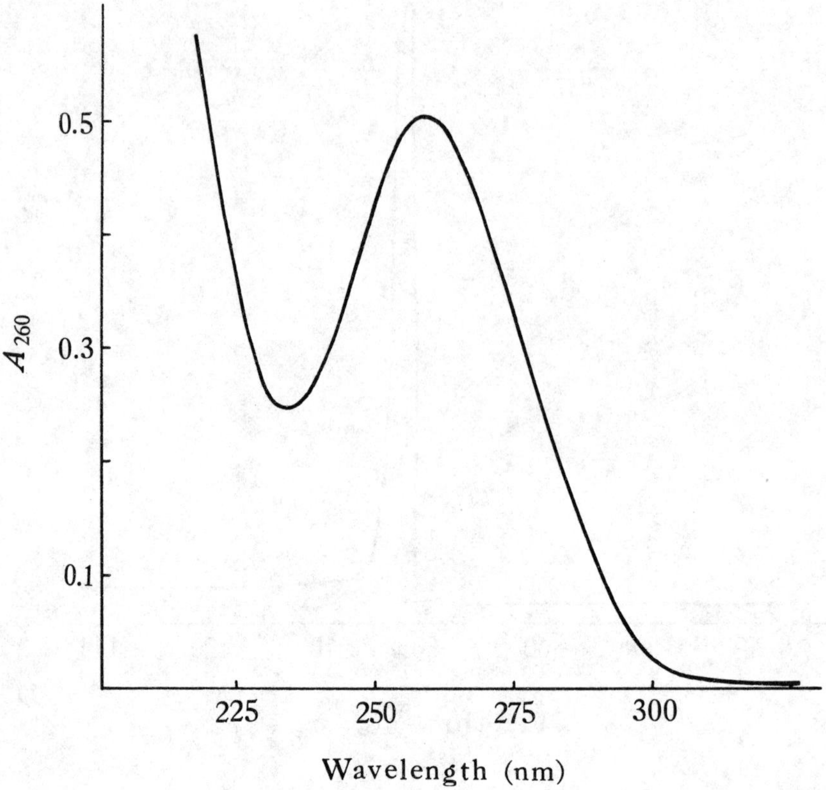

Figure 3. Spectrum of the poly(A)$^{+}$ mRNA eluted from the oligo (dT)-cellulose column with water.
The spectrum was determined in the HITACHI 556 double beam spectrophotometer.

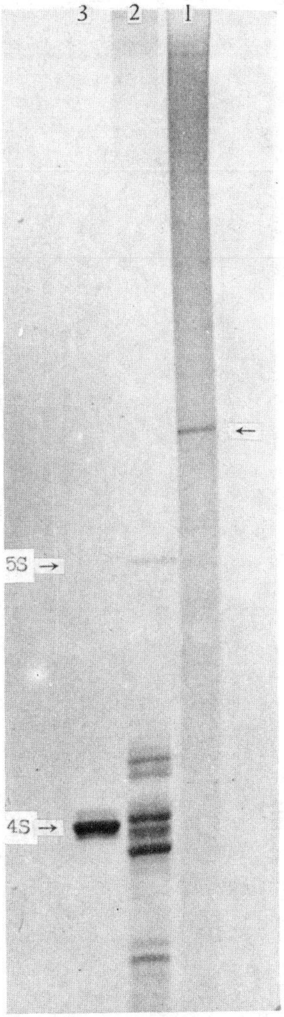

Figure 4. Analysis of the poly(A)$^{+}$ mRNA on a 7% polyacrylamide-7M urea slab gel.
The poly(A)$^{+}$ protamine mRNA was analysed on 7% polyacrylamide-7M urea slab gel for electrophoresis (pH8.3) (11). Electrophoresis of this mRNA samples was performed as described in experimental procedures section under completely denaturing condition. Slot 1: 30 μg of poly(A)$^{+}$ protamine mRNA. Slot 2: 5S RNA from rat liver. Slot 3: 4S tRNA^{ala} from yeast. Positions of the poly(A)$^{+}$ protamine mRNA, 5S and 4S RNA were indicated by arrows.

Table 1. Activities of poly(A)$^+$ protamine mRNA fraction in a wheat germ cell-free system

Sample	Amount of RNA added (μg)	(^{14}C)Arginine and (^{14}C)Isoleucine incorporated (C.P.m./20 μl)	Specific radioactivity (C.P.m/μg)
First oligo(dT)-cellulose bound RNA fraction	0.04	8,685	190,125
Second oligo(dT)-cellulose bound RNA fraction	0.09	19,887	220,966

The poly(A)$^+$ protamine mRNA eluted from the oligo(dT)-cellulose column in tandem with the DEAE-cellulose column were assayed in the wheat germ cell-free system as described in experimental procedures section. Both (^{14}C) arginine and (^{14}C) isoleucine incorporated into hot trichloroacetic acid-precipitable polypeptides were determined in a 20 μl sample. Radioactivity incorporated into the endogenous protein (1,080 C.P. m) was subtracted.

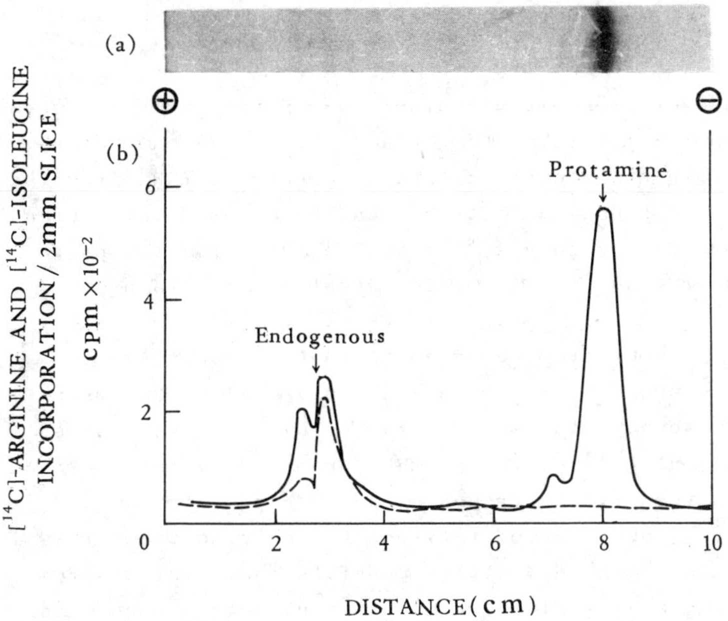

Figure 5. Analysis of polypeptides synthesized in the wheat germ cell-free system in the presence of the poly(A)$^+$ protamine mRNA by polyacrylamide-7M urea gel electrophoresis.
Reaction mixtures (100 μl) containing 40 μl of wheat germ S-26 were incubated with the poly(A)$^+$ protamine mRNA under the conditions described in experimental procedures section. 20 μl of the reaction mixture was taken for measurement of trichloroacetic acid-precipitable radioactivity. 15 μl of the above reaction mixture added carrier protamine (10 μg/ μl) was loaded directly to 7% polyacrylamide-7M urea slab gel (pH 2.0). Analysis was performed as described in experimental procedures section.———, product obtained in the presence of the poly(A)$^+$ protamine mRNA. _____, product due to endogenous mRNA.

two faint bands that were only present in the control. Thus they should be endogenous protein synthesized in response to the endogenous mRNA (Fig. 5b).

DISCUSSION

In order to overcome some disadvantages of the methods of phenol extraction and polyribosomes,Gedamu et al. (8) have developed an improved method for purification of protamine mRNA from the postmitochondrial supernatant of trout testis. We used this method to isolate and purify the poly(A)$^+$ protamine mRNA from the postmitochondrial supernatant of hybrid carp testis and obtained similar results.

It is also apparent from our results that the poly(A)$^+$ protamine mRNA are present in the low-molecular-weight RNA fraction of the postmitochondrial supernatant of hybrid carp testis. However, when the poly(A)$^+$ protamine mRNA prepared by this procedure was analysed by electrophoresis in 7% polyacrylamide-7M urea gel (pH 8.3) under completely denaturing condition, they migrate as a single sharp band. It is possible that this is the character of this class of fish. Perhaps the closely spaced minor band near the major band were too faint so that they could not be seen in the pattern of gel electrophoresis. Thus it may provide the favorable conditions for obtaining higly purified poly(A)$^+$ protamine mRNA.

It has already been shown that arginine content in protamine is very high, about 60-80% (17). However, we found that the incorporation of (^{14}C)arginine into polypeptide products synthesized in response to the protamine mRNA was more less whichever cell-free translation systems are used (18). The reason is still not clear at present. Therefore when the translation activity of the poly(A)$^+$ protamine mRNA was assayed in the wheat germ cell-free system, we used both (^{14}C)arginine and (^{14}C)isoleucine. These results have been compared with the results using only (^{14}C)arginine and it has been shown that radioactivity incorporated into trichloroacetic acid-precipitable material in response to the poly(A)$^+$ protamine mRNA has increased by about 4-fold. The poly(A)$^+$ protamine mRNA samples stored at -11°C or -20°C for two months possessed the same translation activity. It appears that the poly(A)$^+$ protamine mRNA is more stable.

In order to analyse the polypeptide products synthesized in

the wheat germ cell-free system, We have developed an improved low pH polyacrylamide-7M urea slab gel electrophoresis. Under this condition, the cell-free protein synthesis samples could be loaded directly on acid slab gels for electrophoresis without any treatment because labeled free amino acids did not co-migrate with any protein and satisfactory fractionation could be obtained. At the same time various basic protein samples were separated to test the versatility and reproducibility of the system. The results have shown that various basic proteins can be fractionated successfully besides the cell-free synthetic products.

For obtaining highly purified poly(A)$^{+}$ protamine mRNA, We also used both phenol extraction and polyribosome method (9. 18.) except the column chromatography (8). It will be seen from our results that the column chromatography (8) would apear to be suitable for purification of low-molecular-weight size classes of mRNA. We think that this method has some advantages: (1) it is a faster and more efficient method; (2) the yield and purify of mRNA are very high; (3) large quantities of tissue can be processed and milligram quantities of highly purified protamine mRNA can be obtained. Specifically, the carp testes are obtained very easy. Many advantages can be provided for purifying enough amount of purified poly(A)$^{+}$ protamine mRNA. These are neccessary for work involving extensive structural and functional studies of mRNA.

ACKNOWLEDGEMENTS

We thank Mr. C. C. Cheng for support; Mr. C. P. Wai for help during the early stage of the work; Mr. K. Y. Wang., Mr. Y. H. Lu and Ms. R. Z. Kou for help in preparing the figures; Dr. J. H. Chen and Prof. C. L. Tson for their direction; Prof. Y. Liu and Dr. S. H. Hsiao for carefully reading the manuscript.

References

1. Iatrou, K., Gedamu, L. and Dixon, G. H. (1979) Can. J. Biochem. 57. 945-956.
2. Dixon, G. H., Davies, P. L., Ferrier, L. N., Gedamu, L. and Iatrou, K. (1977) in The Proceedings of the International Symposium on Molecular Biology of the Mammalian Genetic Apparatus, Vol. 1, 355-379 (editor P. O. P. T'so), Elsevier, North Holland.
3. Davies, P. L., Dixon, G. H., Ferrier, L. N., Gedamu, L. and Iatrou, K. (1976) Prog. in N. A. Res. and Mol. Biol. 19, 135-155.
4. Davies, P. L. and Dixon, G. H. (1979) Nucleic Acids Res. 7, 2323-2345.
5. Dixon, G. H. (1972) Karolinska Symposia on Research Methods in reproductive Endocrinology, 5th Symposium, 130-154.
6. Ling, V. and Dixon, G. H. (1970) J. Biol. Chem. 245, 3035-3042.
7. Felix, K. (1960) Advanc. Protein chem. 15, 1-56.
8. Gedamu, L., Iatrou, K. and Dixon, G. H. (1978) Biochem. J. 171, 589-599.
9. Aviv, H. and Leder, P. (1972) Proc. Natl. Acad. Sci. U. S. A. 69, 1408-1412.
10. Osterburg, H. H., Allen, J. K. and Finch, C. E. (1975) Biochem. J. 147, 367-368.
11. Brownlee, G. G. and Cartwright, E. M. (1977) J. Mol. Biol. 114, 93-117.
12. Dahlberg, A. E., Dingman, C. W. and Peacock, A. C. (1969) J. Mol. Biol. 41, 139-147.
13. The Virus Replication Study group, Institute of Microbiology, Academia Sinica, Peking (1976) Acta Biochimica et Biophysica Sinica 8, 179-186.
14. Roberts, B. E. and Paterson, B. M. (1973) Proc. Natl. Acad. Sci. U. S. A. 70, 2330-2334.
15. Lichtner, R. and Wolf, H. U. (1979) Biochem. J. 181, 759-761.
16. Fairbanks, G., Sleck, T. L. and Wallach, D. F. (1971) Biochemistry 10, 2606-2617.
17. Ando, T., Yamasaki, M. and Suzuki, K. (1973) Mol. Biol. Biochem. Biophys. 12.
18. Gedamu, L. and Dixon, G. H. (1976a) J. Biol. Chem. 251, 1446-1454.

ON THE KINETICS OF NUCLEIC ACID CHANGES AND THE FORMATION OF LONG-LIVED mRNA DURING EMBRYOGENESIS OF RICE

Zhu Zhi-ping   Wang Mei-li*   Shen Rui-juan
Tang Si-hua

(Shanghai Institute of Plant Physiology, Academia Sinica)

The plant embryo is the developing rudiments of young individual, and it is also the prelude of the whole life of the plant. During embryogeny, accompanied with the morphological changes, the futher developmental programme are progressing, according to the prescribed developmental "blue print" stored in cell nuclei of embryo, Therefore the pattern of nucleic acid changes during embryogenesis is of much interest to us.

I. Kinetics of nucleic acid during embryogenesis of rice

By means of microbiochemical analysis(1), we determined the kinetics of total nucleic acid, DNA, RNA and protein content changes during embryogenesis of rice, over the period 6th to 30th days from anthesis(2).

During the embryogenesis of rice, the first, second, third and fouth leafprimordia appeared in 6, 7, 9 and 13 days respectively after anthesis and the embryonic differentiation was almost completed within 15 days after anthesis(3).

In this period the total nucleic acid (Fig. 1), DNA and RNA contant (Fig. 2) per embryo increased quickly, especially from 6 to 9 days, concomitant with embryonic differentiation(2). The degeneration of integument and other surrouding embryonic tissues took place on 6th to 7th days after anthesis(3) might be related to the active synthesis of both DNA and RNA in the early

---

* Hangchow University, Dept. of Biology.

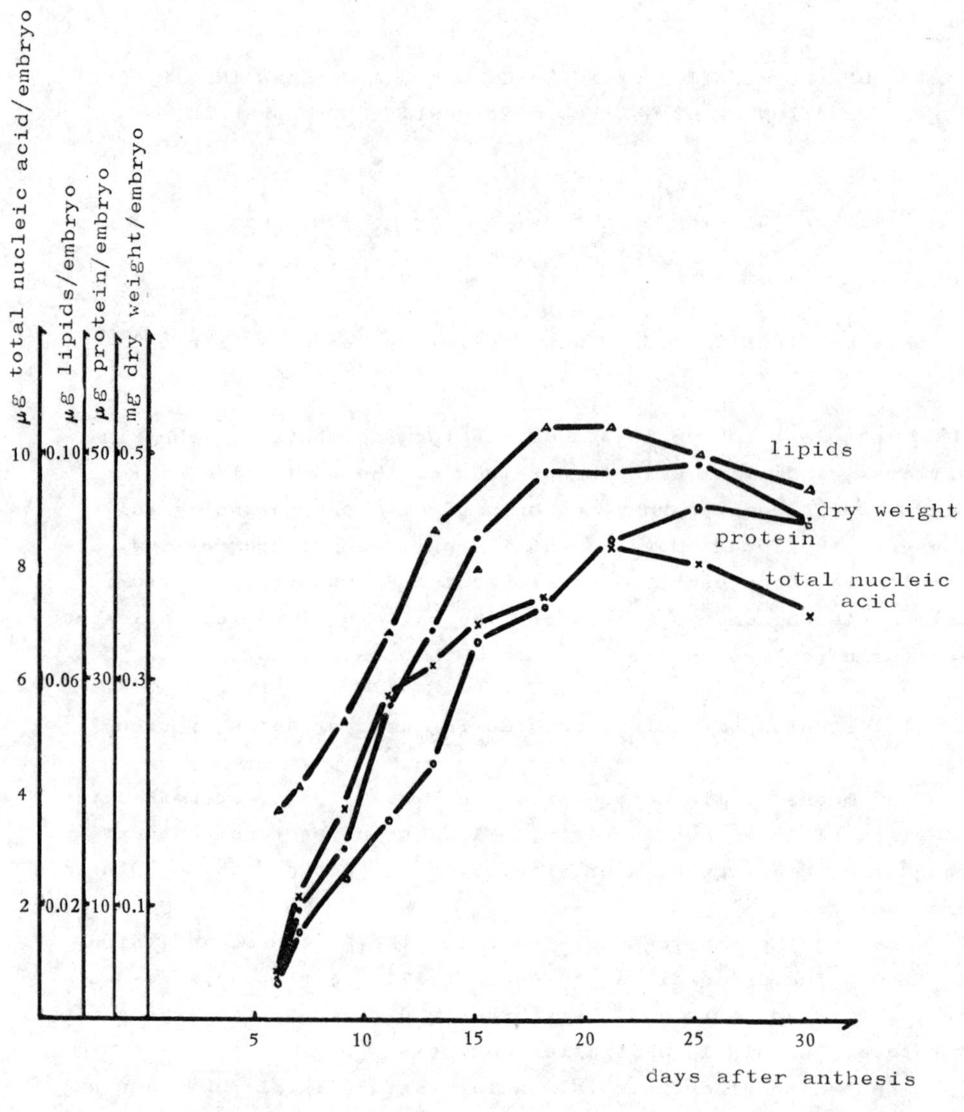

Fig. 1 The kinetic changes of per embryo protein, total nucleic acid, dry weight and lipids content during embryogenesis of rice.

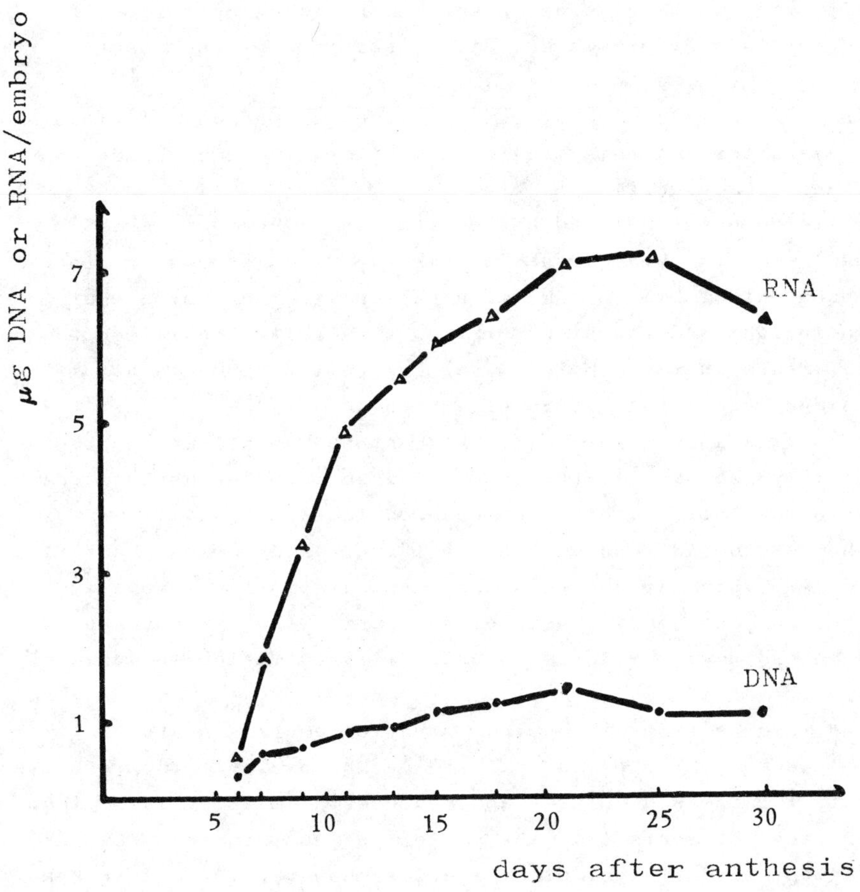

Fig. 2 The kinetic changes of per embryo DNA, RNA content during embryogenesis of rice.

stage of embryogenesis. At this time, the molecules of nucleic acid or its hydrolysis products might be transfered to embryo directly or as intermediates in the synthesis of nucleic acid and accelerated the growth and differentiation of the young embryo.

The rate of increase of nucleic acids slowed down from 18 to 30 days after anthesis. In the same time, the embryo was dehydrated and became getting mature.

While both the DNA and RNA content on dry weight basis declined[2] during embryogenesis (Fig. 3). The decrease of relative content of DNA and RNA might be due to the increase of cell inclusions and the structural material. It also reflected the difference in ratio between various cell components at diferent stages of diffentiation of embryo.

As to the amount per cell, DNA content was almost constant during embryogenesis[2]. (Fig. 4), because the DNA was located mainly in the cell nuclei which changed little in quantity.

Whereas the RNA content per cell increased during the early stage of embryogenesis and was then kept at similar level[2] (Fig. 4). The active synthesis of RNA during the early stage of embryogenesis may be related to cell differentiation and organogenesis.

The results from biochemical study of embryogenesis of wheat[4] were much like that of rice[2] and barley[5], except its RNA content increased quicker and reached the maximun later than that of rice. It seems that the pattern of nucleic acid changes is different in different plant species. However the active synthesis of RNA during early stage of embryogenesis and the cessation of this synthesis long before maturation are common to both rice and wheat embryos and are therefore a matter of great interest.

II. On the formation of long-lived mRNA during embryogenesis of rice

An important feature of development during maturation of the embryo is the gradual loss of water from the cells. The mechanism by which the embryo keeps a relatively low level of metabolism in the quiescent state and yet remains the capability to be activated in a short time after imbibition, is one of the profound

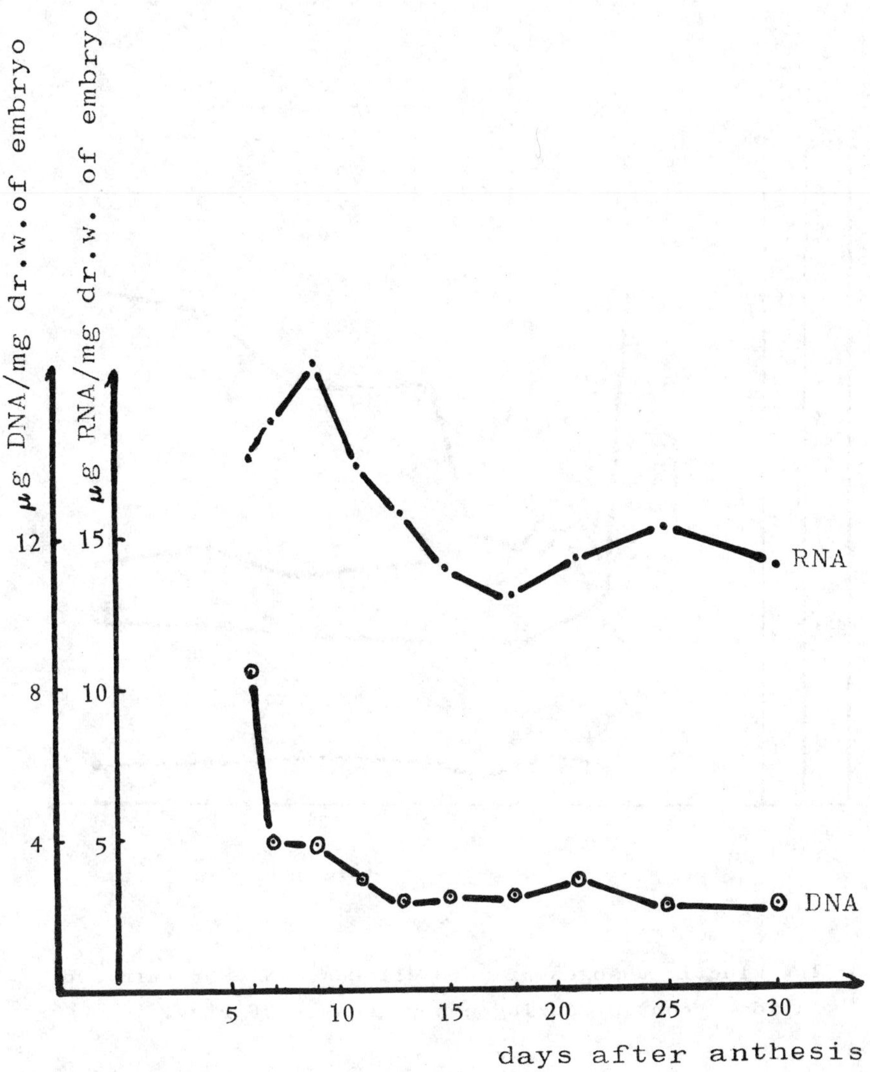

Fig. 3 The kinetic changes of DNA, RNA content on the basis of dry weight of embryo during embryogenesis of rice.

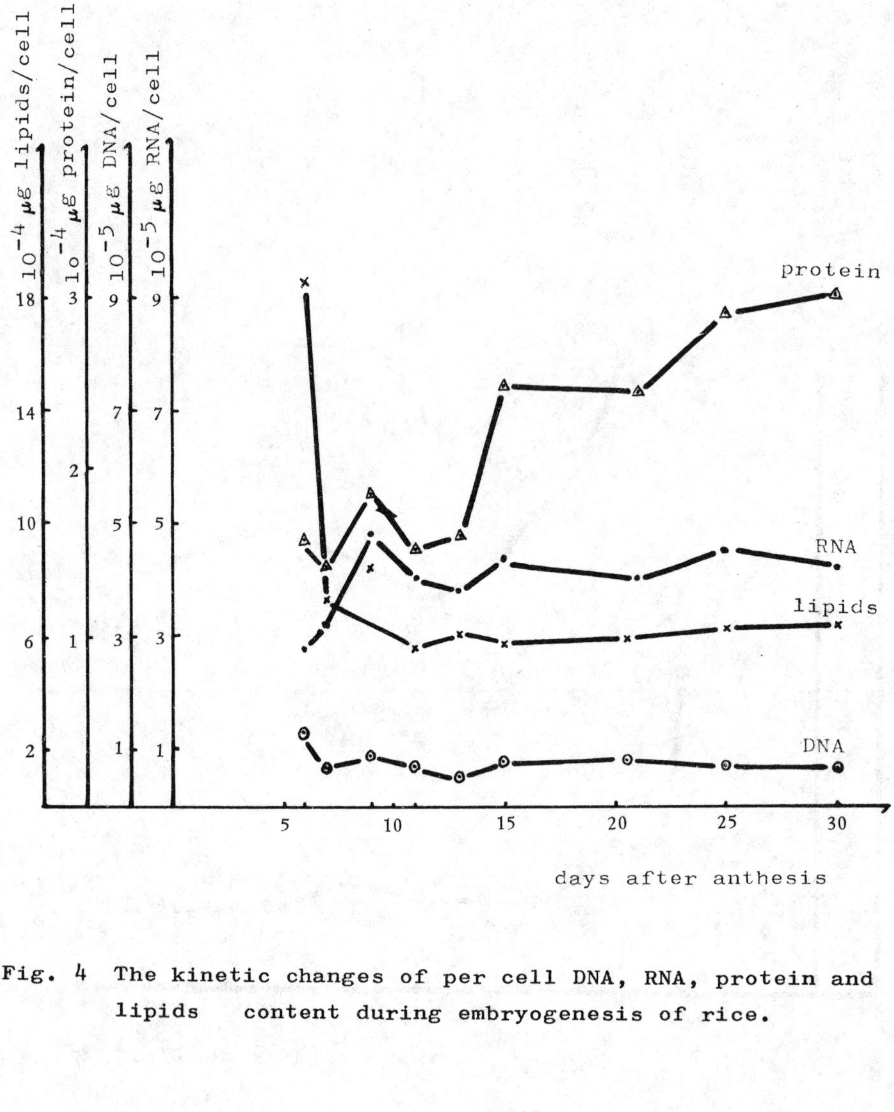

Fig. 4 The kinetic changes of per cell DNA, RNA, protein and lipids content during embryogenesis of rice.

mysteries in the biochemistry of germination. Therefore the RNA storage during maturation of the embryo is quite remarkable.

In order to study this problem, we approached by inhibitor experiments.

The experiments on growth response to inhibitors are of great interest. When treated with actinomycin D (200 μg/ml), all the seeds germinated into rootless seedlings, because of strong inhibition of root elongation, while the protrusion of radicles and coleoptiles were not sensitive to the inhibitor treatment. In same time, the same concentration of cycloheximide completely inhibited the germination of rice (Fig.5). It seems that the synthesis of, at least, a part of enzyme or protein required for germination may be not translated from newly synthesised mRNA.

Accordingly, peroxidase, one of the enzymes active in germination, was determined after treatment with different inhibitors. During germination, the peroxidase activity of untreated embryos appeared on the first day. Later, its activity per embryo increased steadily up to the first five days of germination. With actinomycin D (200μg/ml) treated embryos, the emzyme activity was promoted in similar manner, except that during the 4th and 5th days of germination which was influenced only slightly (Fig. 6)[6].

Since actinomycin D was a transcription inhibitor, the information for synthesis peroxidase in the early stage of germination in the presence of it must be translated from a mRNA formed before the actinomycin D treatment and stored in mature embryo, which was therefore a long-lived one.

In the same time, cycloheximide in similar concentration completely inhibited the activity of peroxidase during germination. It also indicated that the peroxidase present at the time of germination was a newly synthesised and not a preexisting one(Fig.6)

The similar results were also reported in cotton seeds[7-8] and other plant seeds[9].

Futhermore, the investigation by polyacrylamide electrophoresis showed that peroxidase isozyme of rice embryo germinated for two days had only one band. Then, the number of bands increased to six or more three days after germination. When rice seeds germinated for two days were treated with cycloheximide (50μg/ml) and determined on the third day, its peroxidase isozyme had only a single band, as on the second day. But under treatment of ac-

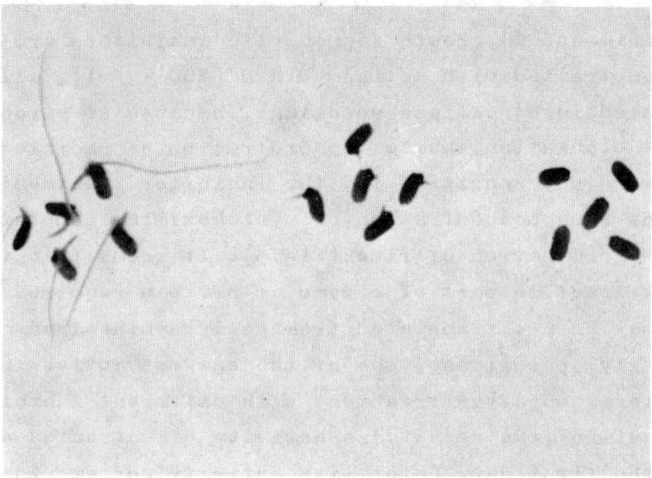

1 2 3

Fig. 5 The effect of actinomycin D (200 μg/ml) and cycloheximide (200 μg/ml) on the germination of rice seeds (after three days germination).

1. CK
2. actinomycin D treatment
3. cycloheximide treatment

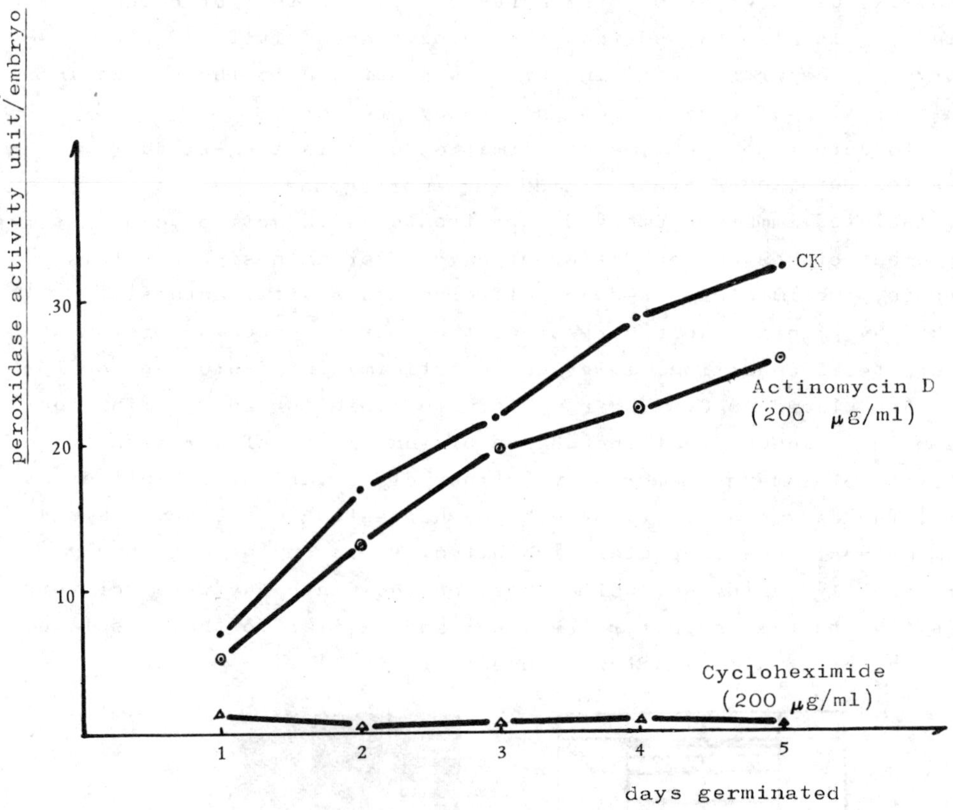

Fig. 6 The effect of actinomycin D (200 μg/ml and cycloheximide (200 μg/ml) on the peroxidase activity of rice embryo during germination.

tinomycin D (200 μg/ml), the number of peroxidase isozyme in embryos two days after germination increased gradually to more then six, as shown by electrophoresis, just as that of check (Fig. 7). It also showed that the peroxidase activity of rice embryo in the presence of inhibitor was not due to the change in configuration of enzyme protein already present.

To determine the time of formation of this long-lived mRNA for peroxidase synthesis during embryogenesis, the isolated immature embryos were treated with actinomycin D at different stages and on different days after anthesis. For this purpose, the immature seeds on different days after anthesis marked by painting were harvested, the embryos were then removed under steril condition, immersed in actinomycin D solution for 15 min, and placed on 0.8% agar plate with inhibitor for germination. The results showed that the degree of inhibition of peroxidase activity of immature embryo by actinomycin D varied with different time. The enzyme activity of embryos harvested on the seventh day from anthesis was completely inhibited. Those on the eighth day were greatly inhibited, while those on the ninth day were not very sensitive to the inhibitor treatment and similar to those isolated on 17th, or 25th days after anthesis (Table 1)[6].

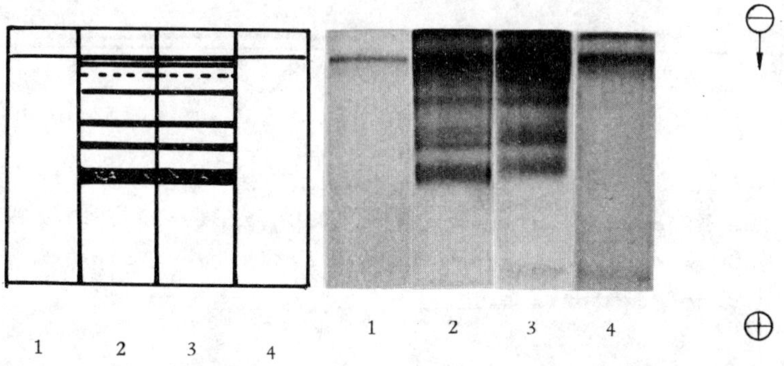

Fig. 7 The effect of actinomycin D and cycloheximide on the zymogram of peroxidase isozyme of germinated rice embryo.

1. CK two days germinated
2. CK three days germinated
3. three days germinated, but treated with actinomycin D (200 μg/ml) on second day
4. three days germinated, but treated with cycloheximide (50 μg/ml) on second day of germination

Table 1 The effect of actinomycin D on the peroxidase activity of pregerminated isolated immature rice embryo. (peroxidase activity unit/embryo)

days after anthesis	days for pregermination	CK -Act.D	treatment +Act.D (200μg/ml)
7	3	1.60	0
7	5	2.30	0
8	3	7.28	1.44
8	5	6.56	2.16
9	3	11.28	8.24
17	3	27.84	17.60
25	3	29.60	21.28

According to the above results, we postulated that the peroxidase synthesis during germination was mediated by a stored long-lived mRNA, which was transcribed in the immature embryo at the stage of eight to nine days after anthesis, i. e., the stage of 0.09-0.17 mg dry weight and the formation of second to third leaf primordia and vascular trace, but was translated until imbibition. The long-lived mRNA formed in the early stage of embryogenesis may play an important role in the regulation of germination and the adaptation to the environment.

References

1) Madison, J. T,, Thompson, J. F., Menuster, A-M. E Deoxyribonucleic aicd, ribonucleic acid, protein and uncombined amino acid content of legume seeds during embryogeny. Ann. Bot. 40 (168): 745-756, 1976.

2) Zhu Zhi-ping, Shen Rui-juan, Tang Si-hua Studies on developmental biology of embryogenesis in higher plants II. The Biochemical changes during embryogeny in Oryza sativa L. Acta Phytophysiologia Sinica 6 (2): 141-148, 1980.

3) Tang Si-hua, Shen Rui-juan, Cao Mei-sheng Studies on developmental biology of embryogenesis in higher plants I. Growth patterns of embryogenesis in rice and wheat Acta Phytophysiologia Sinica 6 (1): 57-63, 1980.

4) Zhu Zhi-ping, Shen Rui-juan, Tang Si-hua Studies on developmental biology of embryogenesis in higher plants III. Kinetic changes of nucleic acids and protein during embryogenesis of wheat. Acta Botanica Sinica 22 (2) 1980, (in press)

5) Duffus, C. M. and Rosie R. Biochemical changes during embryogeny in Hordeum distichum. Phytochemistry 14 (2): 319-323, 1975.

6) Zhu Zhi-ping, Wang Mei-li, Shen Rui-juan, Tang Si-hua The effect of actinomycin D and cycloheximide on the formation of peroxidase during embryogenesis and germination of rice (Oryza sativa L.) A preliminary report Kexue tongbao 1980. (in press)

7) Ihle J. N. and Dure III L. S. Synthesis of a protesase in germinating cotton cotyledons catalyzed by mRNA synthesized during embryogenesis. B. B. R. C. 36 (5); 705-710, 1960.

8) Ihle J. N. and Dure III. L. S. The developmental biochemistry of cotton seed embryogenesis and germination III. Regulation of the biosynthesis of enzymes utilized in germination J. B. C. 247 (16) 5048-5055 1972

9) Payne. P. I. The long-lived messenger ribonucleic acid of flowering — plant seeds. Biol. Rev. 51 329-363 1976

Author Index